Lecture Notes in Computer Science 3686

Commenced Publication in 1973
Founding and Former Series Editors:
Gerhard Goos, Juris Hartmanis, and Jan van Leeuwen

T0189934

Sameer Singh   Maneesha Singh
Chid Apte   Petra Perner (Eds.)

Pattern Recognition and Data Mining

Third International Conference on
Advances in Pattern Recognition, ICAPR 2005
Bath, UK, August 22-25, 2005
Proceedings, Part I

 Springer

Volume Editors

Sameer Singh
Maneesha Singh
Loughborough University
Research School of Informatics
Loughborough LE11 3TU, UK
E-mail:{s.singh/m.singh}@lboro.ac.uk

Chid Apte
IBM Corporation
1133 Westchester Avenue, White Plains, New York 10604, USA
E-mail: apte@us.ibm.com

Petra Perner
Institute of Computer Vision and Applied Computer Sciences, IBaI
Körnerstr 10, 04107 Leipzig, Germany
E-mail: ibaiperner@aol.com

Library of Congress Control Number: Applied for

CR Subject Classification (1998): I.5, I.4, H.2.8, I.2.6-7, I.3.5, I.7.5, F.2.2, K.5

ISSN   0302-9743
ISBN-10   3-540-28757-4 Springer Berlin Heidelberg New York
ISBN-13   978-3-540-28757-5 Springer Berlin Heidelberg New York

Springer is a part of Springer Science+Business Media

springeronline.com

© Springer-Verlag Berlin Heidelberg 2005
Printed in Germany

Typesetting: Camera-ready by author, data conversion by Scientific Publishing Services, Chennai, India
Printed on acid-free paper   SPIN: 11551188   06/3142   5 4 3 2 1 0

Preface

This LNCS volume contains the papers presented at the Third International Conference on Advances in Pattern Recognition (ICAPR2005) organised in August, 2005 in the beautiful city of Bath, UK. The conference was first organised in November 1998 in Plymouth, UK and subsequently in March 2001 in Rio, Brazil. The conference encouraged papers that made significant theoretical and application based contributions in pattern recognition. The emphasis was on open exchange of ideas and shared learning. The papers submitted to ICAPR2005 were thoroughly reviewed by up to three referees per paper and less than 40% of the submitted papers were accepted. The papers have been finally published as two volumes of LNCS and these are organised under the themes of Pattern Recognition and Data Mining (which included papers from the tracks of Pattern Recognition Methods, Knowledge and Learning, and Data Mining), and Pattern Recognition and Image Analysis (which included papers from the Applications track). From the conference technical programme point of view, the first volume contains papers on pattern recognition, data mining, signal processing and OCR/ document analysis. The second volume contains papers from the Workshop on Pattern Recognition for Crime Prevention, Security and Surveillance, Biometrics, Image processing and Medical Imaging.

ICAPR2005 was run in parallel with the International Workshop on Pattern Recognition for Crime Prevention, Security and Surveillance that was organised on the 22nd of August, 2005. This workshop brought together a number of excellent papers that focussed on how pattern recognition techniques can be used to develop systems that help with crime prevention and detection. On the same day, a number of tutorials were also organised. Each tutorial focussed on a specific research area and gave an exhaustive overview of the scientific tools and state-of-the-art research in that area. The tutorials organised included the topics of Computational Face Recognition (given by Dr. Babback Moghaddam, MERL, USA), 2-D and 3-D Level Set Applications for Medical Imagery (given by Dr. Jasjit Suri, Biomedical Technologies, USA; Dr. Gilson Antonio Giraldi, National Laboratory of Computer Science, Brazil; Prof. Sameer Singh, Loughborough University, UK; and Prof. Swamy Laxminarayan, Idaho State University, USA.), Geometric Graphs for Instance-Based Learning (given by Prof. Godfried Toussaint, McGill University, Canada) and Dissimilarity Representations in Pattern Recognition (given by Prof. Bob Duin, and Elzbieta Pekalska, Delft University of Technology, The Netherlands).

The conference also had three plenary speeches that were much appreciated by the audience. On the first day of the conference, Prof. David Hogg from the University of Leeds, UK gave an excellent speech on learning from objects and activities. On the second day of the conference Prof. Ingemar Cox from University College London, UK gave the second plenary speech. On the final day of the conference Prof. John Oommen from Carlton University, Canada gave a plenary speech on the general problem of syntactic pattern recognition and string processing.

ICAPR was a fully reviewed and well-run conference. We would like to thank a number of people for their contribution to the review process, especially the Program Chairs, Tutorial Chair Dr. Majid Mirmehdi and Workshops Chair Prof. Marco Gori. The members of the program committee did an excellent job with reviewing most of the papers. Some papers were also reviewed by academics who were not in the committee and we thank them for their efforts. We would also like to thank the local arrangements committee and University of Bath Conference Office for their efforts in ensuring that the conference ran smoothly. In particular, our thanks are due to Dr. Maneesha Singh, Organising Chair and Mr. Harish Bhaskar, Organising Manager who both worked tirelessly. The conference was supported by British Computer Society and a number of local companies within the UK. We would like to thank Springer-Verlag in extending their support to publish the proceedings as LNCS volumes. Finally, we thank all the delegates who attended the conference and made it a success.

August 2005   Sameer Singh
                                                                 Maneesha Singh
                                                                      Chid Apte
                                                                    Petra Perner

Organization

Executive Committee

Conference Chair: Sameer Singh
(Loughborough University, U.K)

Program Chairs: Chid Apte
(IBM, New York, USA)

Petra Perner
(University of Leipzig, Germany)

Organizing Chair: Sameer Singh
(Loughborough University, U.K)

Maneesha Singh
(Loughborough University, U.K)

Organizing Manager: Harish Bhaskar
(Loughborough University, U.K)

Tutorials and Demonstrations: Majid Mirmehdi
(University of Bristol, UK)

Workshops: Marco Gori
(University of Siena, Italy)

Program Committee

Edward J. Delp
Purdue University,USA

Mohamed Cheriet
University of Quebec,Canada

Horst Haussecker
Intel, USA

Nozha Boujemaa
INRIA, France

Christophe Garcia
France Telecom R&D, France

J. Ross Beveridge
Colorado State University, USA

Roger Boyle
University of Leeds, UK

Xiang "Sean" Zhou
Siemens Corporate Research Inc., USA

Adnan Amin
University of New South Wales, Australia

Kobus Barnard
University of Arizona at Tucson, USA

Hans Burkhardt
University of Freiburg, Germany

Witold Pedrycz
University of Alberta, Canada

Patrick Bouthemy
IRISA, France

Xiaoyi Jiang
University of Munster, Germany

XiaoHui Liu
Brunel University, UK

David Maltoni
University of Bologna, Italy

Sudeep Sarkar
University of South Florida, USA

Mayer Aladjem
Ben-Gurion University, Israel

Jan Flusser
Academy of Sciences of the Czech Republic, Czech Republic

Vladimir Pavlovic
Rutgers University, USA

Jean-Michel Jolion
INSA, France

Ingemar Cox
University College London, UK

Michal Haindl
Academy of Sciences of the Czech Republic, Czech Republic

Luigi Cordella
University of Napoli, Italy

Ales Leonardis
University of Ljubljana, Slovenia

Ata Kaban
University of Birmingham, UK

Mike Fairhurst
University of Kent, UK

Sven Loncaric
University of Zagreb, Croatia

Boaz Lerner
Ben-Gurion University, Israel

Mohamed Kamel
University of Waterloo, Canada

Peter Tino
University of Birmingham, UK

Richard Everson
University of Exeter, UK

Hiromichi Fujisawa
Central Research Laboratory, Hitachi, Japan

Ian Nabney
Aston University, UK

Wojtek Krzanowski
University of Exeter, UK

Andrew Martin
University College London, UK

Steve Oliver
University of Manchester, UK

Edoardo Ardizzone
University of Palermo, Italy

David Hoyle
University of Exeter, UK

David Parry-Smith
Purely Proteins, UK

Malcolm Strens
Qinetiq, UK

Gerhard Rigoll
Munich University of Technology, UK

John McCall
Robert Gordon University, UK

Mark Last
Ben Gurion University, Israel

Rachel Martin
Shimadzu-Biotech, UK

Theo Gevers
University of Amsterdam, Netherlands

Herv Bourlard
Swiss Federal Institute of Technology

Mads Nielsen
University of Kobenhavn, Denmark

Mario Figueiredo
Inst. for Telecommunication, Portugal

Mohamed Kamel
University of Waterloo, Canada

Matthew Turk
University of California, USA

Jonathan Hull
Ricoh Innovations Inc., USA

Nicu Sebe
University of Amsterdam, Netherlands

Paulo Lisboa
Liverpool John Moores University, UK

Ana Fred
Inst. of Telecommunication, Portugal

Steve Maybank
Birkbeck College, UK

Mario Vento
University of Salerno, Italy

Andrew Webb
Qinetiq, UK

Fabio Roli
University of Cagliari, Italy

John McCall
The Robert Gordon University, UK

B.S.Manjunath
University of California, USA

Heinrich Niemann
Universitaet Erlangen-Nuernberg, Germany

Table of Contents – Part I

Pattern Recognition and Data Mining

Signal Processing

OCR/Document Analysis

Enhancing Trie-Based Syntactic Pattern Recognition Using AI Heuristic Search Strategies

Ghada Badr[1] and B. John Oommen[2]

[1] Carleton University, Ph.D student,
School of Computer Science, Carleton University, Ottawa, Canada : K1S 5B6
badrghada@hotmail.com
[2] Carleton University, *Fellow of the IEEE*,
School of Computer Science, Carleton University, Ottawa, Canada : K1S 5B6
oommen@scs.carleton.ca

Abstract. This paper[1] [5] deals with the problem of estimating, using enhanced AI techniques, a transmitted string X^* by processing the corresponding string Y, which is a noisy version of X^*. We assume that Y contains substitution, insertion and deletion errors, and that X^* is an element of a finite (possibly large) dictionary, H. The best estimate X^+ of X^* is defined as that element of H which minimizes the Generalized Levenshtein Distance $D(X, Y)$ between X and Y, for all $X \in H$. In this paper, we show how we can evaluate $D(X, Y)$ for every $X \in H$ simultaneously, when the edit distances are general and the maximum number of errors is not given *a priori*, and when H is stored as a *trie*. We first introduce a new scheme, Clustered Beam Search (CBS), a heuristic-based search approach that enhances the well known Beam Search (BS) techniques [33] contained in Artificial Intelligence (AI). It builds on BS with respect to the pruning time. The new technique is compared with the Depth First Search (DFS) trie-based technique [36] (with respect to time and accuracy) using large and small dictionaries. The results demonstrate a marked improvement up to (75%) with respect to the total number of operations needed on three benchmark dictionaries, while yielding an accuracy comparable to the optimal. Experiments are also done to show the benefits of the CBS over the BS when the search is done on the trie. The results also demonstrate a marked improvement (more than 91%) for large dictionaries.

1 Introduction

1.1 Problem Statement

We consider the traditional problem involved in the syntactic Pattern Recognition (PR) of strings, namely that of recognizing garbled words (sequences). Let

---

[1] Patent applications have been filed to protection the intellectual property and the results contained in this paper.

S. Singh et al. (Eds.): ICAPR 2005, LNCS 3686, pp. 1–17, 2005.

Y be a misspelled (noisy) string obtained from an unknown word X^*, which is an element of a finite (possibly, large) dictionary H, where Y is assumed to contain Substitution, Insertion and Deletion (SID) errors. Various algorithms have been proposed to obtain an appropriate estimate X^+ of X^*, by processing the information contained in Y, and the literature contains hundreds (if not thousands) of associated papers. We include a *brief* review here. In what follows, we assume that the dictionary is stored as a trie.

1.2   Contribution of the Paper

Most techniques proposed to prune the search in the trie have applied the so-called "cutoff" strategy to decrease the search. The "cutoff" is based on the assumption that the maximum permitted error is known *a priori*, and is more useful when the inter-symbol costs are of form 0/1. In this paper, we show how we can optimize non-sequential PR computations by incorporating heuristic search schemes used in AI into the approximate string matching problem. First, we present a new technique enhancing the Beam Search (BS), which we call the Clustered Beam Search (CBS), and which can be applied to any tree searching problem[2]. We then apply the new scheme to the approximate string matching when the dictionary is stored as a trie. The trie is implemented as a Linked List of Prefixes (LLP) as shown in [27]. The latter permits *level-by-level* traversal of the trie (as opposed to traversal along the "branches"). The newly-proposed scheme can be used for Generalized Levenshtein distances and also when the maximum number of errors is not given *a priori*. It has been rigorously tested on three benchmarks dictionaries by recognizing noisy strings generated using the model discussed in [28], and the results have been compared with the acclaimed standard [32], the Depth-First-Search (DFS) trie-based technique [36]. The new scheme yields a marked improvement (of up to 75%) with respect to the number of operations needed, and at the same time maintains almost the same accuracy. The improvement in the number of operations increases with the size of the dictionary. The CBS heuristic is also compared with the performance of the original BS heuristic when applied to the trie structure, and the experiments again show a surprising improvement of more than 91%. Furthermore, by marginally sacrificing a small accuracy in the general error model, or by permitting an error model that increases the errors as the length of the word increases (as explained presently), an improvement of more than 95% in the number of operations can be obtained.

2   The State-of-the-Art

The literature contains hundreds of papers which deal with the Syntactic PR of strings/sequences. Excellent recent surveys about the field can be found in [12], [25].

---

[2] The new scheme can also be applied to a general graph structure, but we apply it to the trie due to the dominance of the latter in our application domain, i.e., approximate string matching.

2.1   Dictionary-Based Approaches

Most of the time-efficient methods currently available require that the maximum number of errors be known *a priori*, and these schemes are optimized for the case when the edit distance costs are of a form 0/1. In [14], Du and Chang proposed an approach to design a very fast algorithm for approximate string matching that divided the dictionary into partitions according to the lengths of the words. They limited their discussion to cases where the error distance between the given string and its nearest neighbors in the dictionary was "small".

Bunke [10] proposed the construction of a finite state automaton for computing the edit distance for *every* string in the dictionary. These automata are combined into one "global" automaton that represents the dictionary, later used to calculate the nearest neighbor for the noisy string when compared against the active dictionary. This algorithm requires time which is linear in the length of the noisy string. However, the number of states of the automaton grows exponentially. Oflazer [26] also considered another method that could easily deal with very large lexicons. To achieve this, he used the notion of a **cut-off** edit distance: this measures the minimum edit distance between an initial substring of the incorrect input string, and the (possibly partial) candidate correct string. The cutoff-edit distance required *a priori* knowledge of the maximum number of errors found in Y and that the inter-symbol distances are of a form 0/1, or when general distances are used, a maximum error value.

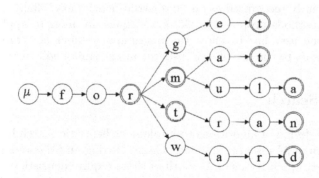

Fig. 1. An example of a dictionary stored as a trie with the words {for, form, fort, fortran, formula, format, forward, forget}

Baeza-Yates and Navarro [6] proposed two speed-up techniques for on-line approximate searching in large indexed textual databases when the search is done on the vocabulary of the text. The efficiency of this method depends on the number of allowable error value.

The literature[3] also reports some methods that have proposed a filtering step so as to decrease the number of words in the dictionary that need to be

---

[3] More details about the state-of-art can be found in [4] and [5].

considered for calculations. One such method is "the similarity" keys method [35] that offers a way to select a list of possible correct candidates in the first step. This correction procedure, proposed in [35], can be argued to be a variant of Oflazer's approach [26]. The time required depends merely on the permitted number of edit operations involved in the distance computations.

A host of optimizing strategies have also been reported in the literature for methods which model the language probabilistically using N-grams, and for Viterbi-type algorithms [9], [17], [37], [3]. These methods do not explicitly use a "finite-dictionary" (trie or any other) model, and so we believe that it is not necessary to survey them here. The same is also true for methods that apply to error correct parsing [2], [31] and grammatical inference [24], where the dictionary is represented by the language generated by a grammar whose production probabilities are learnt in the "training" phase of the algorithm.

2.2   Dictionaries Represented as Tries

The *trie* is a data structure that can be used to store a dictionary when the dictionary[4] is represented as a set of words. Words are searched as a character by character basis. Figure 1 shows an example of a trie for a simple dictionary of words {for, form, fort, forget, format, formula, fortran, forward}.

With regard to traversal, the trie can be considered as a graph and can be searched using any of the possible search strategies applicable to AI problems. The literature includes two possible strategies that have been applied to tries, namely the Breadth First Search strategy [18], [27] and the Depth First Search strategy [36], currently recognized as an "industrial benchmark" [32].

Although the methods proposed in [36] are elegant, in order to apply these cutoff principles the user has to know the maximum number of errors, K, *a priori*, and also resort to the use of $0/1$ costs for inter-symbol edit distances.

3   Heuristic Search

Heuristics and the design of algorithms to implement heuristic search have long been a core concern of AI research [22], [30]. Game playing and theorem proving are two of the oldest applications in AI: both of these require heuristics to prune spaces of possible solutions. It is not feasible to examine every inference that can be made in a domain of mathematics, or to investigate every possible move that can be made on a chessboard, and thus a heuristic search is often the only practical answer. It is useful to think of heuristic algorithms as consisting of two parts: the heuristic measure and an algorithm that uses it to search the state space[5].

---

[4] In terms of notation, A is a finite alphabet, H is a finite (possibly large) dictionary, and μ is the null string, distinct from λ, the null symbol. The left derivative of order one of any string $Z = z_1 z_2 \ldots z_k$ is the string $Z_p = z_1 z_2 \ldots z_{k-1}$. The left derivative of order two of Z is the left derivative of order one of Z_p, and so on.

[5] When we refer to heuristic search we imply those methods that are used for solving the problems possessing ambiguities or which are inherently time consuming. In both

3.1   Beam Search (BS)

The simplest way to implement a heuristic search is through a *Hill-Climbing* (HC) procedure [22], [30]. HC strategies expand the current state in the search space and evaluate its children. The best child is selected for further expansion and neither its siblings nor its parent are retained. The search halts when it reaches a state that is better than any of its children. The algorithm cannot recover from failures because it keeps no history. A major drawback to HC strategies is their tendency to become stuck at local maxima/minima.

To overcome the problems of HC, and to provide better pruning than BFS, researchers have proposed other heuristics such as the *Beam Search*[6] (BS) [33]. In BS, we retain q states, rather than a single state as in HC, and these are stored in a single pool. We then evaluate them using the objective function. At each iteration, all the successors of all the q states are generated, and if one is the goal, we stop. Otherwise, we select the best q successors from the complete list and repeat the process. This BS avoids the combinatorial explosion of the Breadth-First search by expanding only the q most promising nodes at each level, where a heuristic is used to predict which nodes are likely to be closest to the goal and to pick the best q successors.

One potential problem of the BS is that the q states chosen tend to quickly lack diversity. The major advantage of the BS, however, is that it increases both the space and time efficiency dramatically, and the literature includes many applications in which the BS pruning heuristic has been used. These applications are include: handwriting recognition [15], [20], [23], Optical Character Recognition (OCR) [16], word recognition [19], speech recognition [8], [21], information retrieval [38] and error-correcting Viterbi parsing [2].

3.2   Proposed Clustered Beam Search (CBS)

We propose a new heuristic search strategy that can be considered as an enhanced scheme for the BS. The search will be done level-by-level for the tree structure[7], where by "level" we mean the nodes at the same depth from the start state. At each step we maintain $|A|$ small priority queues, where A is the set of possible clusters that we can follow in moving from one node to its children[8]. The lists maintain only the best first q nodes corresponding to each cluster, and in this way the search space is pruned to have only $q|A|$ nodes in each level.

CBS is like BS in that it only considers some nodes in the search and discards the others from further calculations. Unlike BS, it does not compare all the nodes

---

these cases, we seek a heuristic to efficiently prune the space and lead to a good (albeit suboptimal) solution.

[6] When we speak about Beam Search, we are referring to *local* beam search, a combination of an AI-based local search and the traditional beam search [33] methodologies.

[7] For the case where the graph is a general structure (and not a tree), we need to maintain the traditional *Open* and *Close* lists that are used for Best First Search [22], [30].

[8] For example, A can be the set of letters in the alphabet in the case of the trie.

in the same level in a single priority queue, but rather compares only the nodes in the same cluster. The advantage of such a strategy is that the load of maintaining the priority queue is divided into $|A|$ priority queues. In this way, the number of minimization operations needed will dramatically decrease for the same number of nodes taken from each level.

As we increase q, the accuracy increases and the pruning ability decreases. When the evaluation function is informative, we can use small values for q. As q increases, the cost associated with maintaining the order of the lists may overcome the advantage of pruning. The possibility of including a larger number of nodes per level increases with the new CBS scheme when compared to the BS leading to increased accuracy.

3.3   The Proposed CBS Algorithm and Its Complexity

The pseudo code for the proposed algorithm is shown in Algorithm 1. The saving of the CBS over the BS appears in step 8, in which we only maintain a small queue.

If the number of nodes after pruning in any level is $O(q|A|)$, the number of children considered for pruning in the next level is $O(q|A|^2)$, which is then divided among the $|A|$ clusters. For each cluster, we need to maintain a priority queue containing q nodes. The search cost will be $O(ch(q|A|^2)\log(q))$, where c is the cost of calculating the heuristic function per node, and h is the maximum height of the tree (maximum number of levels). If the $|A|$ queues can be maintained in parallel, then the cost will be reduced to $O(ch(q|A|)\log(q))$. This is in contrast with the BS where the search cost will be $O(ch(q|A|^2)\log(q|A|))$, because the corresponding BS maintains only one queue of length $O(q|A|)$. The benefits of CBS increases as $|A|$ increases while the performance of the BS will decrease.

4   The CBS for Approximate String Matching

For approximate string matching the problem encountered involves both ambiguities and the excessive time required as the dictionary is large. Observe that there is no exact solution for the noisy string that one is searching for, and at the same time the process is time consuming because one has to search the entire space to find it. We thus seek a heuristic to determine the nearest neighbor to the noisy string, and one which can also be used to prune the space. The ambiguity of the problem can be resolved by several methods as explained previously. One of the methods is the Depth-First-Search trie-based heuristic that uses the dynamic equations and edit distance calculations described in the next section.

We attempt to describe a heuristic search for the approximate string matching problem to also prune the search space when the inter-symbol distances are general, and the maximum number of errors cannot be known *a priori*. We will first present the heuristic measure, the data structures used to facilitate the calculations, and finally, the algorithm as applied to approximate string matching.

Algorithm 1. CBS

Input:
a. The tree to be searched, T, with clusters, A.
b. The matrix $Q[|A| \times q]$ to maintain the priority queues.
c. The beam width, q, considered for each cluster.
d. The goal we need to search for.
Output: Success or Failure.
Method:
1: Form $|A|$ single-element queues, $Q[1, \ldots, |A|]$, where each queue $Q[i]$ contains the child in cluster i of the root node.
2: **while** Q is not empty and the goal is not found **do**
3:    Determine if any of the elements in Q is the goal.
4:    **if** goal found **then**
5:       **return** Success.
6:    **else**
7:       **for** each Q[i], where $1 \leq i \leq |A|$ **do**
8:          Sort the nodes originally in $Q[i]$ in terms of the heuristic, keeping only the first best q nodes.
9:          Replace each of the q nodes by its successors, and add each successor to the end of the corresponding $Q[c]$ according to its cluster c.
10:    **end for**
11:   **end if**
12: **end while**
13: **return** Failure.
14: **End Algorithm CBS**

4.1   The Heuristic Measure

In string-processing applications, the traditional distance metrics quantify $D(X, Y)$ as the minimum cost of transforming one string X into Y. This distance is intricately related to the costs associated with the individual edit operations, the SID operations. These inter-symbol distances can be of a form 0/1, parametric, or entirely symbol dependent. If they are symbol dependent [18], [34], they are usually assigned in terms of the confusion probabilities:

$$d(x, y) = - \ln[Pr(x \to y) \div Pr(x \to x)]$$
$$d(x, \lambda) = - \ln[Pr(x \text{ is deleted}) \div Pr(x \to x)]$$
$$d(\lambda, x) = Kd(x, \lambda), \tag{1}$$

where K is an empirically determined constant.

We have discovered that there are two possible heuristic functions that can be used to measure the similarity between strings. Both those heuristic functions can also be used as a measure to prune the search space. These measures are:

Algorithm 2. CBS-LLP-based Scheme

Input:
a. The dictionary, H, represented as a Linked Lists of Prefixes, LLP, with alphabet, A.
b. The matrix QM to maintain the priority queues.
c. The width of the beam considered for each alphabet q.
d. The noisy word, Y, for which we need to find the nearest neighbor.
Output: X^+, the nearest neighbor string in H to Y.
Method:
 1: Start from the first level in LLP, which contains the root of the trie.
 2: Initialize *minnode* to null, which stores the node representing the string that is nearest neighbor to Y so far.
 3: Initialize *childq* to 1, which is the number of nodes to be considered in the current level.
 4: **while** *childq* $\neq 0$ **do**
 5:     Initialize QM to be empty.
 6:     **for** for each node n in the *childq* nodes of the current level **do**
 7:         Get the character c represented by node n.
 8:         Calculate the edit distances $D(X, Y)$, for the string represented by node n and using the column information stored in the parent of n.
 9:         Add n to $QM[c]$ if it is one of the best q nodes already in the list according to the distance value. If the distance value of n is equal to one of the q nodes, one of the solutions we use is to extend the $QM[c]$ to include n.
10:         **if** n is an accept node, i.e., a word in the dictionary **then**
11:             Compare $D(X, Y)$ with the minimum found so far and if it has lower edit distance value, store n in *minnode*.
12:         **end if**
13:     **end for**
14:     Move all the children of the best nodes in QM to the beginning of the next level in LLP, if any, and store their number in *childq*.
15:     Increment current level to the next level.
16: **end while**
17: **return** the string X^+, corresponding to the path from *minnode* to the root.
18: **End Algorithm CBS-LLP-based Scheme**

- Heuristic function F_1: The *Edit distance* $D(X, Y)$
  F_1 can be computed using the dynamic programming rule:

$$D(x_1 \ldots x_N, \ y_1 \ldots y_M) = \min [\ \{D(x_1 \ldots x_{N-1}, \ y_1 \ldots y_{M-1}) + d(x_N, y_M)\},$$
$$\{D(x_1 \ldots x_N, \ y_1 \ldots y_{M-1}) + d(\lambda, y_M)\},$$
$$\{D(x_1 \ldots x_{N-1}, \ y_1 \ldots y_M) + d(x_N, \lambda)\}], \quad (2)$$

  where $X = x_1 x_2 \ldots x_N$ and $Y = yx_1 y_2 \ldots y_M$.
  Recognition using distance criteria is obtained by essentially evaluating the string in the dictionary which is "closest" to the noisy one as per the metric under consideration.
  These dynamic equation are exactly the ones that are used for the DFS-trie-based technique [36], where the actual inter-symbol costs are of a form 0/1,

and the transposition evaluation is added to the dynamic equation. Both the trie and the matrix are needed in the calculations. When the DFS is used in the calculations, the reader will observe that only a single column will have to be calculated at any given time, and this only depends on the previously calculated column already stored in the matrix (thus preserving the previous calculations).

– Heuristic function F_2: The *Pseudo-distance* $D_1(X, Y)$

The second heuristic function, F_2, is the pseudo-distance proposed by Kashyap *et al.* [18]. For each character in Y, the pseudo-distance is calculated for the whole trie (level-by-level), and for each character of Y that is processed, we consider two additional levels of the trie.

4.2   Characteristics of the Heuristic Functions

The first heuristic function, F_1, seems to be very effective for a CBS, as will be seen presently. The problem with using F_2 as a pruning measure is that it needs an additional parameter that has to be tuned, in addition to the parameter q. This additional parameter is the length of the prefix of Y that has to be processed after which we can start applying the pruning strategy - to avoid early removal of entire portions of the trie. Indeed, if we used the pseudo-distance as a measure for pruning, we believe that it will not yield the same accuracy as when the measure $D(X, Y)$ is used, except if excessive tuning is permitted, thus rendering it impractical.

4.3   Data Structures Used

There are two main data structures used to facilitate the computations:

– *The Linked List of Prefixes (LLP)*: To calculate the best estimate X^+, we need to divide the dictionary into sets of prefixes. Each set $H^{(p)}$ is the set of all the prefixes of H of length equal to p, where $1 \leq p \leq N_m$, and N_m is the length of longest word in H. More precisely, we want to process the trie level-by-level. The trie divides the prefixes and the dictionary in the way that we want, and further represents the FSM. The problem in the trie structure is that it can be implemented in different ways, and it is not easily traversed level-by-level. We need a data structure that facilitates the trie traversal, and that also leads to a unique representation which can always be used to effectively compute the edit distances or pseudo-distances for the prefixes. We called this data structure the Linked Lists of Prefixes (LLP).

The LLP can be built from the trie by implementing it as an ensemble of linked lists, and where all the lists at the same level are "coalesced" together to make one list. Simultaneously, the LLP also permits us to keep the same parent and children information. The LLP consists of a linked list of levels, *where each level is a level in the corresponding trie*. Each level, in turn, consists of a linked list of all prefixes that have the same length p. The levels are ordered in an increasing order of the length of the prefixes, exactly as in the case of the trie levels. Figure 2 shows the corresponding LLP for the trie of Figure 1. The links for children and parent are omitted in the figure for simplification.

Within each entry of the LLP, we keep a link to the column information that is needed for the calculations required for the next children. In this way, we need only to maintain this column information for the first *childq* nodes of each level. This column link is set to NULL if the node is already pruned. The storage requirement for the LLP is the same as the trie, in addition to the links between children in the same level, and between the different levels themselves and the links to the column information.

- *The Queues Matrix (QM)*: This matrix structure is used during the pruning done for each level in the LLP. A newly initialized QM matrix is needed for each level. The matrix, QM (of dimension $|A| \times q$), can be used to maintain the $|A|$ priority queues and keep pointers to the best q nodes in each cluster. Each entry in the matrix keeps a pointer to a node in the LLP, and all the pointers in the matrix will be to nodes in the same level. The space required for this matrix is $O(q|A|)$.

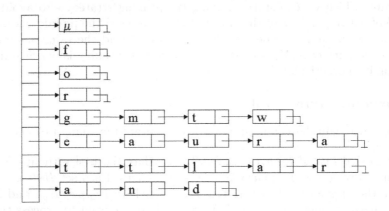

Fig. 2. The corresponding LLP for the trie with words {for, form, fort, fortran, formula, format, forward, forget}

The two data structures are used simultaneously, in a conjunctive manner, to achieve the pruning that is needed. This will be illustrated in more detail in the next section.

4.4 Applying the CBS

The pseudo code, as shown in Algorithm 2, illustrates how the CBS can be applied to approximate string matching, namely the CBS-LLP-based scheme. It also shows how the proposed data structures, presented in the previous section, can be used to facilitate the calculations. The LLP helps to maintain the list

Table 1. Statistics of the data sets used in the experiments

	Eng	Dict	Webster
Size of dictionary	8KB	225KB	944KB
number of words in dictionary	964	24,539	90,141
min word length	4	4	4
max word length	15	22	21

of the best nodes and to achieve the pruning expediently. Moving the nodes in the same lists will not affect the trie order at all, but helps us to effectively maintain information about which nodes are to be processed and which are to be discarded. The QM also helps us to maintain the queues and to retain the required column information.

5   Experimental Results

To investigate the power of our new method with respect to computation we conducted various experiments on three benchmark dictionaries. The results obtained were (in our opinion) remarkable with respect to the gain in the number of computations needed to get the best estimate X^+. By computations we mean the number of addition and minimization operations needed. The CBS-LLP-based scheme was compared with the acclaimed DFS-trie-based work for approximate matching [36] when the maximum number of errors was not known *a priori*.

Three benchmark data sets were used in our experiments. Each data set was divided into two parts: a *dictionary* and the corresponding *noisy file*. The dictionary was the words or sequences that had to be stored in the trie. The noisy files consisted of the strings which were searched for in the corresponding dictionary. The three dictionaries we used were as follows:

- Eng[9]: This dictionary consisted of 946 words obtained as a subset of the most common English words [13] augmented with words used in computer literature. The average length of a word was approximately 8.3 characters.
- $Dict$[10]: This is a dictionary file used in the experiments done by Bentley and Sedgewick in [7].
- *Webster's Unabridged* Dictionary: This dictionary was used by Clement *et. al.* [1], [11] to study the performance of different trie implementations.

The statistics of these data sets are shown in Table 1. The alphabet is assumed to be the 26 lower case letters of the English letters. For all dictionaries we removed words of length smaller than or equal to 4.

Three sets of corresponding noisy files were created using the technique described in [29], and in each case, the files were created for three specific error

---

[9] This file is available at www.scs.carleton.ca/~oommen/papers/WordWldn.txt

[10] This file can be downloaded from www.cs.princeton.edu/~rs/strings/dictwords

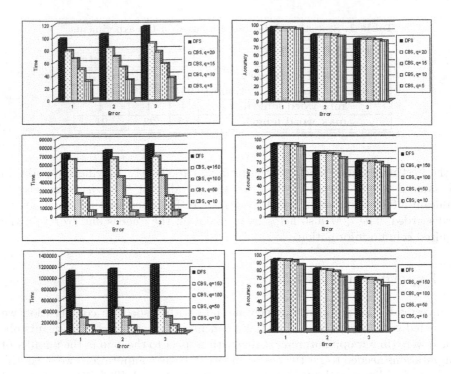

Fig. 3. The results for comparing the **CBS-LLP**-based method with the **DFS-trie**-based method for (top) the **Eng** dictionary, (middle) the **Dict** dictionary, and (bottom) the **Webster** dictionary. The time is represented by total number of operations in millions.

characteristics, where the latter means the number of errors per word. The three error values tested were for 1, 2 and 3, referred to by the three sets SA, SB, and SC respectively.

Each of the three sets, SA, SB and SC were generated using the noise generator model described in [28]. We assumed that the number of insertions was geometrically distributed with parameter $\beta = 0.7$. The conditional probability of inserting any character $a \in A$ given that an insertion occurred was assigned the value 1/26, and the probability of deletion was 1/20. The table of probabilities for substitution (typically called the confusion matrix) was based on the proximity of character keys on the standard QWERTY keyboard and is given in [28][11].

The two algorithms, the DFS-trie-based and our algorithm, CBS-LLP-based, were tested with the three sets of noisy words for each of the three dictionaries. We report the results obtained in terms of the number of computations (additions and minimizations) and the accuracy for the three sets. The calculations were done on a Pentium V processor, 3.2 GHZ. Figure 3 shows a graphical representation of the results. The figures compares both time and accuracy. The numbers are shown in

---

[11] It can be downloaded from www.scs.carleton.ca/~oommen/papers/QWERTY.doc

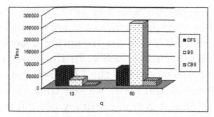 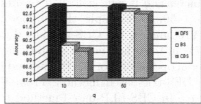

Fig. 4. The results for comparing the **CBS-LLP**-based method with the **BS-LLP**-based method for the **Dict** dictionary when applied to set **SA**. The time is represented by total number of operations in millions.

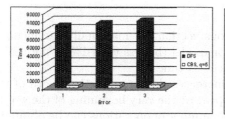 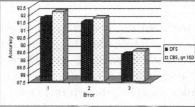

Fig. 5. The results for comparing the **CBS-LLP**-based method with the **DFS-trie**-based method for the **Dict** dictionary when the optimized **error model** is used and $q = 5$. The time is represented by total number of operations in millions.

millions. The results show the significant benefit of the CBS-based method with respect to the number of computations, while maintaining excellent accuracy. For example consider the *Webster* dictionary, for the SA set, and $q = 100$: the number of operations for DFS-trie-based is 1,099,279, and for the CBS-LLP-based method is 271,188 representing a savings of 75.3%, and a loss of accuracy of only 0.5%. For the *Dict* dictionary, for the SA set, and $q = 100$, the number of operations for the DFS-trie-based is 72,115, and for the CBS-LLP-based method is 44,254 which represents a savings of 36.6%, and a loss of accuracy of only 0.2%. When $q = 50$, the number of operations for the CBS-LLP-based method is 21,366, representing a savings of 70.4%, with a loss of accuracy of only 0.5%. There is always a trade-off between time and accuracy but the loss of accuracy here is negligible compared to the "phenomenal" savings in time.

To show the benefits of the CBS over the BS, we show the results when applying the BS for the *dict* dictionary in Figure 5, when the approximately equivalent width (number of nodes taken per level) is considered for the BS. The width is considered approximately equal when we approximately equate the number of addition operations. The figure shows only the result when applied to set SA, as the results for the other sets are representative of the other sets too. From the figures the reader will observe the tremendous gain in the minimization operations needed when the same ordering technique is used for arranging all the priority queues. The results are shown only for $q = 10$ and $q = 50$, because if we increase q it will yield bad results for the BS which is much worse than when $q = 50$. For example, the number of operations for BS-LLP-based method is 256,970, and for the CBS-trie-based method is 21,366, representing a savings of 91.7% in total number of operations with accuracy 92.3%. In this case, the number of operations for the BS method is much more than the 72,115 operations of the DFS-trie-based method. From Figure 3 (middle), we see that we can increase q in CBS-LLP-based method to 100 and get an accuracy of 92.6 with savings of 36.6% in total number of operations with respect to DFS-trie-based method. This is not feasible by the applying the BS method.

6   Optimizing Computations When Changing the Error Model

A we see from the results in the previous section, by marginally sacrificing a small accuracy value for the general error model (by less than 1%) a noticeable improvement can be obtained with respect to time.

By permitting an error model that increases the errors as the length of the word increases (i.e., the errors do not appear at the very beginning of the word), an improvement of more than 95% in the number of operations can be obtained, which is, in our opinion, absolutely amazing. This is because if errors are less likely to appear at the very beginning of the word, the quality of pruning, with respect to accuracy, will be more efficient at the upper levels of the tree. Thus we can utilize a small value for q, the width of the beam, and as a result achieve more pruning. All our claims have been verified experimentally as shown in Figure 5. The results are shown for $q = 5$, (which is a very small width) demonstrating very high accuracy. For example, for the set SA, the number of operations for DFS-LLP-based method is 74,167, and for the CBS-trie-based method is just 3,374, representing a savings of 95.5%. This has obviously great benefits if the noisy words received are not noisy at the beginning, in which case we still need to apply approximate string matching techniques. Even here we would like to make use of the approximately exact part at the very beginning of the word which is variant from one word to another, where we cannot use partition of the noisy word to apply exact match.

7   Conclusion and Future Work

In this paper we have presented a feasible solution for the problem of estimating a transmitted string X^* by processing the corresponding string Y (a noisy version of X^*), an element of a finite (possibly large) dictionary H, when the whole dictionary is considered simultaneously.

First, we proposed a new AI-based search called the Clustered Beam Search (CBS) that can be considered an enhancement of the Beam Search (BS) used in AI. The new scheme can be used to search a graph more efficiently with respect to the number of operations needed. The CBS was applied to dictionary-based approximate string matching, where the dictionary is stored using a trie.

Secondly, we proposed a new representation of the trie, namely the Linked List of Prefixes (LLP), to facilitate the computations. The new implementation strategy helps prune the search space more efficiently and dramatically decreases the number of operations needed to get the nearest neighbor to a noisy string Y.

Thirdly, the CBS-LLP-based approximate string matching has been compared with the acclaimed Depth First trie-based technique proposed by Shang *et. al.* [36] using big and small dictionaries. The results demonstrates a significant improvement with respect to the number of operations needed (up to 75%) while keeping the accuracy comparable to the optimum. It has also been compared with the BS-LLP-based method to show the benefits of CBS over BS. The results shows improvements of more than 90% when q equals 50.

Finally the new scheme, CBS-LLP-based, has also been tested using a new error model, where the error predominantly appears at the end of the string. The results demonstrates a great improvement of more than 95%, while keeping the accuracy the same.

As a future work, we would like to apply the proposed new search scheme, CBS, to approximate string matching when a probabilistic model, discussed in [29], is used in the computations instead of the edit distance model used here. We anticipate that we can get more accurate results and at the same time maintain the same pruning capability.

References

1. A. Acharya, H. Zhu, and K. Shen. Adaptive algorithms for cache-efficient trie search. *ACM and SIAM Workshop on Algorithm Engineering and Experimentation*, January 1999.
2. J. C. Amengual and E. Vidal. Efficient error-correcting viterbi parsing. *IEEE Transactions on Communications*, 20(10):1109–1116, October 1998.
3. J. C. Amengual and E. Vidal. The viterbi algorithm. *IEEE Transactions on Pattern Analysis and Machine Intelligence*, 20(10):268–278, October 1998.
4. G. Badr and B. J. Oommen. Enhancing trie-based syntactic pattern recognition using ai heuristic search strategies. Unabridged version of the present paper.
5. G. Badr and B. J. Oommen. Search-enhanced trie-based syntactic pattern recognition of sequences. 2005. Patent.
6. R. Baeza-Yates and G. Navarro. Fast approximate string matching in a dictionary. *in Proceedings of the 5th South American Symposium on String Processing and Information Retrieval (SPIRE'98), IEEE CS Press*, pages 14–22, 1998.
7. J. Bentley and R. Sedgewick. Fast algorithms for sorting and searching strings. *Eighth Annual ACM-SIAM Symposium on Discrete Algorithms New Orleans*, January 1997.
8. E. Bocchieri. A study of the beam-search algorithm for large vocabulary continuous speech recognition and methods for improved efficiency. *In Proc. Eurospeech*, 3:1521– 1524, 1993.
9. A. Bouloutas, G. W. Hart, and M. Schwartz. Two extensions of the viterbi algorithm. *IEEE Transactions on Information Theory*, 37(2):430–436, March 1991.
10. H. Bunke. Fast approximate matching of words against a dictionary. *Computing*, 55(1):75–89, 1995.

11. J. Clement, P. Flajolet, and B. Vallee. The analysis of hybrid trie structures. *Proc. Annual A CM-SIAM Symp. on Discrete Algorithms, San Francisco, California*, pages 531–539, 1998.
12. R. Cole, L. Gottieb, and M. Lewenstein. Dictionary matching and indexing with errors and don't cares. *Proceedings of the thirty-sixth annual ACM symposium on Theory of computing, Chicago, IL, USA*, pages 91–100, June 2004.
13. G. Dewey. *Relative Frequency of English Speech Sounds*. Harvard Univ. Press, 1923.
14. M. Du and S. Chang. An approach to designing very fast approximate string matching algorithms. *IEEE Transactions on Knowledge and Data Engineering*, 6(4):620–633, 1994.
15. J. T. Favata. Offline general handwritten word recognition using an approximate beam matching algorithm. *IEEE Transactions on pattern Analysis and Machine Intelligence*, 23(9):1009–1021, September 2001.
16. Z. Feng and Q. Huo. Confidence guided progressive search and fast match techniques for high performance chinese/english ocr. *Proceedings of the 16th International Conference on Pattern Recognition*, 3:89–92, August 2002.
17. G. D. Forney. The viterbi algorithm. *Proceedings of the IEEE*, 61(3):268–278, March 1973.
18. R. L. Kashyap and B. J. Oommen. An effective algorithm for string correction using generalized edit distances -i. description of the algorithm and its optimality. *Inf. Sci.*, 23(2):123–142, 1981.
19. P. Laface, C. Vair, and L. Fissore. A fast segmental viterbi algorithm for large vocabulary recognition. *in Proceeding ICASSP-95*, 1:560–563, May 1995.
20. C. Liu, M. Koga, and H. Fujisawa. Lexicon-driven segmentation and recognition of handwritten character strings for japanese address reading. *IEEE Transactions on pattern Analysis and Machine Intelligence*, 24(11):1425–1437, November 2002.
21. F. Liu, M. Afify, H. Jiang, and O. Siohan. A new verification-based fast-match approach to large vocabulary continuous speech recognition. *Proceedings of European Conference on Speech Communication and Technology*, pages 1425–1437, September 2001.
22. G. F. Luger and W. A. Stubblefield. *Artificial Intelligence Structure and Strategies for Complex Problem Solving*. Addison-Wesley, 1998.
23. S. Manke, M. Finke, and A. Waibel. A fast technique for large vocabulary online handwriting recognition. *International Workshop on Frontiers in Handwriting Recognition*, September 1996.
24. L. Miclet. Grammatical inference. *Syntactic and Structural Pattern Recognition and Applications*, pages 237–290, 1990.
25. G. Navarro. A guided tour to approximate string matching. *ACM Computing Surveys*, 33(1):31–88, March 2001.
26. K. Oflazer. Error-tolerant finite state recognition with applications to morphological analysis and spelling correction. *Computational Linguistics*, 22(1):73–89, March 1996.
27. B. J. Oommen and G. Badr. Dictionary-based syntactic pattern recognition using tries. *Procdings of the Joint IARR International Workshops SSPR 2004 and SPR 2004*, pages 251–259, August 2004.
28. B. J. Oommen and R. L. Kashyap. A formal theory for optimal and information theoretic syntactic pattern recognition. *Pattern Recognition*, 31:1159–1177, 1998.
29. B. J. Oommen and R. K. S. Loke. Syntactic pattern recognition involving traditional and generalized transposition errors: Attaining the information theoretic bound. Submitted for Pubication.

30. J. Pearl. *Heuristics : intelligent search strategies for computer problem solving.* Addison-Wesley, 1984.
31. J. C. Perez-Cortes, J. C. Amengual, J. Arlandis, and R. Llobet. Stochastic error correcting parsing for ocr post-processing. *International Conference on Pattern Recognition ICPR-2000*, 2000.
32. K. M. Risvik. Search system and method for retrieval of data, and the use thereof in a search engine. *United States Patent*, April 2002.
33. S. Russell and P. Norvig. *Artificial Intelligence: A Modern Approach.* Prentice Hall, 2003.
34. D. Sankoff and J. B. Kruskal. *Time Warps, String Edits and Macromolecules: The Theory and practice of Sequence Comparison.* Addison-Wesley, 1983.
35. K. Schulz and S. Mihov. Fast string correction with levenshtein-automata. *International Journal of Document Analysis and Recognition*, 5(1):67–85, 2002.
36. H. Shang and T. Merrettal. Tries for approximate string matching. *IEEE Transactions on Knowledge and Data Engineering*, 8(4):540–547, August 1996.
37. A. J. Viterbi. Error bounds for convolutional codes and an asymptotically optimum decoding algorithm. *IEEE Transactions on Information Theory*, 13:260–269, 1967.
38. J. G. Wolff. A scaleable technique for best-match retrieval of sequential information using metrics-guided search. *Journal of Information Science*, 20(1):16–28, 1994.

Mathematical Features for Recognizing Preference in Sub-saharan African Traditional Rhythm Timelines

Godfried Toussaint*

School of Computer Science, McGill University,
Montréal, Québec, Canada
godfried@cs.mcgill.ca

Abstract. The heart of an African rhythm is the *timeline*, a beat that cyclically repeats thoughout a piece, and is often performed with an iron bell that all performers can hear. Such rhythms can be represented as sequences of points on a circular lattice, where the position of the points indicates the time in the cycle at which the instrument is struck. Whereas in theory there are thousands of possible choices for such timeline patterns, in practice only a few of these are ever used. This brings up the question of how these few patterns were selected over all the others, and of those selected, why some are preferred (have more widespread use) than others. Simha Arom discovered that the rhythms used in the traditional music of the Aka Pygmies of Central Africa possess what he calls the *rhythmic oddity property*. A rhythm has the rhythmic oddity property if it does not contain two onsets that partition the cycle into two half-cycles. Here a broader spectrum of rhythms from West, Central and South Africa are analysed. A mathematical property of rhythms is proposed, dubbed *"Off-Beatness"*, that is based on group theory, and it is argued that it is superior to the rhythmic oddity property as a measure of preference among Sub-Saharan African rhythm timelines. The *"Off-Beatness"* measure may also serve as a mathematical definition of *syncopation*, a feature for music recognition in general, and it is argued that it is superior to the mathematical syncopation measure proposed by Michael Keith.

1 Introduction

It is useful for the mathematical analysis of cyclic rhythms to represent them as ordered sets of points on a circle. The points represent the onsets of notes in time. For example, the rhythm consisting of five onsets with corresponding time intervals of (3 2 3 2 2) units is illustrated in Figure 1 (a), where the onsets, starting at position "zero", are joined together with straight line segments to form a convex polygon. This rhythm is used in the traditional music of the *Aka* Pygmies of Central Africa [1], [2], [3]. The rhythm contains two intervals

---
* Research supported by NSERC.

S. Singh et al. (Eds.): ICAPR 2005, LNCS 3686, pp. 18–27, 2005.
© Springer-Verlag Berlin Heidelberg 2005

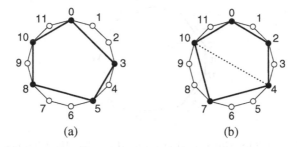

Fig. 1. (a) A rhythm used by the *Aka* Pygmies of Central Africa, (b) The *Seguiriya* rhythm used in the Flamenco music of Spain

of length three and three intervals of length two. The interval lengths consisting of one, two, and three units are strongly favoured in the rhythm timelines of much of Sub-Saharan traditional music [7], [8], [9]. Given only this constraint, there are usually different ways of arranging the intervals while maintaining their cardinality fixed (cyclic permutations). For example, in the preceeding rhythm of Figure 1 (a) the two intervals of length three may be placed side by side as in Figure 1 (b). The resulting rhythm, with time intervals (interval vector) given by (2 2 3 2 2), is the *Seguiriya* metric pattern used in the Flamenco music of Andalucia in Southern Spain [4].

The question then arizes as to how the *Aka* Pygmies evolved to prefer the first of these two rhythms over the second. Simha Arom [1] discovered that the *Aka* Pygmies use rhythms that have the *rhythmic-oddity* property [2], [3]. A rhythm consisting of an *even* number of time units, has the rhythmic-oddity property if no two onsets partition the cycle (entire time span) into two sub-intervals of equal length. Such a partition will be called an *equal bi-partition*. Note that the rhythm of the *Aka* Pygmies in Figure 1 (a) has the rhythmic-oddity property, whereas the *Seguiriya* rhythm of Figure 1 (b) does not.

A note is in order to clarify what we mean when we write that one rhythm is "preferred" over another. This preference has not been established scientifically in a laboratory by experiments conducted by music perception psychologists. Rather, it is inferred from the existence and ubiquity of the rhythms observed in the field by ethnomusicologists [1]. It is assumed for example that rhythms that are played, are preferred over those that are not.

Before proceeding it is useful to define some terminology. A *necklace* token is a circular chain of n elements of m different colors. Two n-element necklace tokens are considered to be one and the same necklace if one token can be rotated so that there is a one-to-one correspondence between the colors of their elements. However, a necklace token may not be turned over to find such a correspondence [6]. A *Lyndon word* is a non-periodic necklace [6].

The two rhythms depicted in Figure 1 with interval vectors (3 2 3 2 2) and (2 2 3 3 2) can be viewed as necklaces, where the intervals of the rhythms correspond to the elements of the necklaces and the two different lengths (2 and

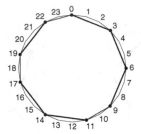

Fig. 2. The *Mokongo* rhythm used by the *Aka* Pygmies

3 units) of the intervals correspond to the two colors of the elements. In other words, both rhythms are distinct 5-element necklaces of two different colors with the additional constraint that the color of the 3-unit intervals is used exactly twice. This additional constraint is called *fixed density*. It is an interesting computational problem to generate efficiently all necklaces and Lyndon words under these constraints [10]. It follows from Pólya's Theorem [6], and the reader may easily verify, that the two rhythms of Figure 1 are the only necklaces possible composed of five elements of two colors with exactly two elements of one color. Both rhythms are also Lyndon words (non-periodic).

The *Aka* Pygmies also play two other rhythm timelines that use intervals of length two and three units in a time span of 24 units [1], [2], [3]. The first is the pattern (3 2 2 2 2 3 2 2 2 2 2) containing two intervals of length three and nine intervals of length two. There are five distinct necklaces of eleven elements of two colors with exactly two elements of one color. Using interval vector notation the five necklaces are: (3 3 2 2 2 2 2 2 2 2 2), (3 2 3 2 2 2 2 2 2 2 2), (3 2 2 3 2 2 2 2 2 2 2), (3 2 2 2 3 2 2 2 2 2 2), (3 2 2 2 2 3 2 2 2 2 2). Only the last one of these, selected by the *Aka* Pygmies, has the rhythmic oddity property, and none of the other four patterns are used by the *Aka* Pygmies.

The second 24-unit pattern used by the *Aka* Pygmies is (3 3 3 2 3 3 2 3 2) which contains six intervals of length three and three intervals of length two. This rhythm depicted in Figure 2 is called the *Mokongo*, and also has the rhythmic oddity property.

One might hypothesise that the rhythmic-oddity property is a good measure of preference used as a selection criterion (perhaps unconciously) for the adoption of rhythms by the peoples of Central Africa in general, and the *Aka* Pygmies in particular. One might even hope that this property could predict rhythm preference in other parts of Africa. However, for these purposes the rhythmic-oddity property has some limitations.

Consider for example the ten fundamental West (and South) African bell timelines composed of seven onsets in a time span of twelve units, with five intervals of length two and two intervals of length one (see [12] for more details). The ten rhythms and their interval vectors are as follows: *Soli* = (2 2 2 2 1 2 1), *Tambú* = (2 2 2 1 2 2 1), Bembé = (2 2 1 2 2 2 1), Bembé-2 = (1 2 2 1 2 2 2), *Yoruba* = (2 2 1 2 2 1 2), *Tonada* = (2 1 2 1 2 2 2), *Asaadua* = (2 2 2 1 2 1 2),

Sorsonet = (1 1 2 2 2 2 2), *Bemba* = (2 1 2 2 2 1 2), *Ashanti* = (2 1 2 2 1 2 2). All ten rhythms are obtained by suitable rotations of one of three necklaces as pictured in Figure 3. None of these ten rhythms, nor any of the other eleven that are not used but belong to the same three generating necklaces, has the rhythmic-oddity property. Furthermore, among the group of ten that are used, some are more widespread than others, but the rhythmic oddity property does not offer an explanation for this preference.

Turning to the *Mokongo* rhythm of the *Aka* Pygmies, there are ten distinct necklaces of nine elements of two colors with exactly six elements of one color. Using interval vector notation the ten necklaces are:

$$(3\ 3\ 3\ 3\ 3\ 3\ 2\ 2\ 2),\ (3\ 3\ 3\ 3\ 3\ 2\ 3\ 2\ 2),\ (3\ 3\ 3\ 3\ 3\ 2\ 2\ 3\ 2),\ (3\ 3\ 3\ 3\ 2\ 3\ 3\ 2\ 2),$$
$$(3\ 3\ 3\ 3\ 2\ 3\ 2\ 3\ 2),\ (3\ 3\ 3\ 3\ 2\ 2\ 3\ 3\ 2),\ (3\ 3\ 3\ 2\ 3\ 3\ 3\ 2\ 2),\ (3\ 3\ 3\ 2\ 3\ 3\ 2\ 3\ 2),$$
$$(3\ 3\ 3\ 2\ 3\ 2\ 3\ 3\ 2),\ (3\ 3\ 2\ 3\ 3\ 2\ 3\ 3\ 2).$$

The last of these is periodic and therefore not of interest in this context. Of the nine distinct Lyndon words remaining, the last three have the rhythmic oddity property. Therefore the rhythmic-oddity property fails to explain how the *Mokongo* pattern is favored over the other two.

Here a broader spectrum of rhythm timelines from West, Central, and South Africa are considered. A mathematical property of rhythms is proposed, dubbed "*off-beatness*", based on group generators, and it is argued that it is superior to the rhythmic-oddity property as a measure of preference in Sub-Saharan African rhythm. First we take a small detour to describe a generalization of the rhythmic oddity property that widens its applicability.

2 A Generalization of the Rhythmic Oddity Property

Simha Arom [1] defines the rhythmic oddity property in a strictly binary mode, i.e., a rhythm either has or does not have the rhythmic oddity property. This concept may be generalized by defining a multi-valued variable that measures the amount of rhythmic oddity a rhythm possesses. This variable (*rhythmic oddity*) is defined as the *number* of equal bi-partitions that a rhythm admits. The fewer equal bi-partitions a rhythm admits, the more rhythmic oddity it possesses. As in [1], this property makes sense only for time spans of even length. Figure 3 shows the three necklace patterns that generate the ten West and South African bell timelines, with all the bi-partitions (in dotted lines) contained in each. The necklaces are labelled with the name of the most representative of its rhythms. We see that the rhythms belonging to the *Sorsonet* necklace contain three equal bi-partitions, the rhythms belonging to the *Tonada* necklace contain two equal bi-partitions and the rhythms belonging to the *Bembé* necklace contain one such bi-partition.

The *Sorsonet* necklace is the least preferred of the three, yielding only two rhythms used in traditional music [12], the *Sorsonet* rhythm (1 1 2 2 2 2 2) used in West Africa and the Persian *kitaab al-adwaar* rhythm given by (2 2 2 2 1 1 2). The *Tonada* necklace is encountered more frequently, yielding two West African

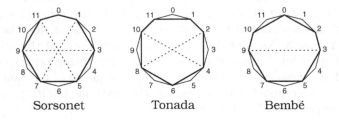

Fig. 3. The three necklaces comprising the ten 12/8 time bell patterns

rhythms, the *Tonada* = (2 1 2 1 2 2 2) and the *Asaadua* = (2 2 2 1 2 1 2), and one Persian rhythm, the *Al-ramal* = (2 2 2 2 1 2 1). The *Bembé* necklace is clearly preferred over the other two necklaces. All seven rhythms obtained by starting the cycle at every one of its seven onsets are heavily used [8], [9]. The six listed in the preceeding section are used predominantly in West African, South African, and Afro-American traditional music [7], [12]. The seventh rhythm, given by (1 2 2 2 1 2 2), is a *Bondo* rhythm used in Central Africa [1].

It is evident then that among this family of rhythms there is a marked preference for those that admit as few as possible equal bi-partitions, and thus a high degree of rhythmic oddity. Nevertheless, although this measure performs better than the rhythmic oddity property, it still has limitations. Among the seven rhythms determined by the *Bembé* necklace some are more popular than others. In fact, one of these, the *Bembé* = (2 2 1 2 2 2 1) is by far the most preferred of the seven, and is considered to be the African *signature* bell-pattern. The master drummer Desmond K. Tai has called it the Standard Pattern [5]. Afro-Cuban music has escorted it across the planet, and it is used frequently on the ride cymbal in jazz. Since all seven rhythms belonging to this necklace have exactly one equal bi-partition, even the multi-valued rhythmic-oddity measure does not discriminate among them, and thus does not favor the *Bembé* rhythm over the other six.

3   The *Off-Beatness* Measure

Consider first the rhythms defined in a 12-unit time span. A twelve-unit interval may be evenly divided (with no remainders) by *four* numbers greater than one and less than twelve. These are the numbers six, four, three and two. Dividing the twelve unit circle by these numbers yields a bi-angle, triangle, square, and hexagon, respectively, as depicted in Figure 4. African music usually incorporates a drum or other percussion instrument on which at least one or portions of these patterns are played. Sometimes the music is accompanied by hand-clapping rhythms that use some or all of these patterns. For example, the *Neporo* funeral piece of Northwestern Ghana uses the triangle, square, and hexagon clapping patterns [13] shown in Figure 4. In any case the rhythm has a steady fundamental beat which we may associate with position "zero" in the cycle. In polyrhythmic

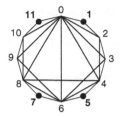

Fig. 4. The four positions not obtainable when dividing 12 by 6, 4, 3 or 2

music these four sub-patterns form the possible underlying pulses (sub-beats). Two of the patterns (bi-angle and square) are binary pulses and two (triangle and hexagon) ternary pulses. Therefore notes played in other positions are off-beat in a strong polyrhythmic sense. There are four positions not used by these four pulse patterns. They are positions 1, 5, 7, and 11. Onsets at these positions will be called *off-beat* onsets. A rhythm that contains at least one off-beat onset will be said to contain the *off-beatness* property. A measure of the *off-beatness* of a rhythm is therefore the number of off-beat onsets that it contains.

These off-beat positions (1, 5, 7, and 11) also have a group-theoretic interpretation. The twelve positions of the possible notes in the cycle form the cyclic group of order 12 denoted by C_{12}. The four off-beat position values correspond to the sizes of the intervals that have the property that if one traverses the cycle starting at, say "zero" in a clockwise direction in jumps equal to the size of one of these intervals, then one eventually returns to the starting point after having visited all twelve positions. Conversely, if the lengths of the jumps are taken from the complementary set (2, 3, 4, 6, 8, 9, 10) then the start point will be reached without having visited all twelve positions in the cycle. For this reason the elements (1, 5, 7, and 11) are called the *generators* of the group C_{12}.

Returning to the ten West-African bell patterns in 12/8 time discussed in the preceeding, recall that the *Bembé* rhythm is the most frequently used of these patterns. Among these ten rhythms, the highest value of off-beatness is three and only the *Bembé* realizes this value.

Since every cyclic group C_n has a set of generators, the off-beatness measure described in the preceeding generalizes to rhythms defined on n-unit time spans for other values of n. Although the measure works best for even values of n, it also has some applicability for odd n. On the other hand, if n is some prime number p then every number from 1 to $p-1$ is relatively prime to p. In such a case the measure is useless since all the onset positions from 1 to $p-1$ would be off-beatness notes under the present definition of off-beatness. For rhythms that have a 24-unit time span the eight off-beatness onsets are determined by the generators of C_{24}, namely, (1, 5, 7, 11, 13, 17, 19, 23).

The ten necklaces of nine elements of two colors with density six, of which the *Mokongo* rhythm of the *Aka* Pygmies is a member, are listed in Figure 5 along with the number of equal bi-partitions admitted by each. Also included (in the last column) are the off-beatness values of the rhythms in their canonical rotational positions. Of the four rhythms that have the rhythmic oddity property (zero bi-partitions) there are three Lyndon words of interest (shown shaded). Rhythms 7 and 9 have off-beatness values of two, whereas rhythm 8, the *Mokongo* has a value of three. Hence the off-beatness measure offers an explanation for selecting the *Mokongo* rhythm over the other two rhythms, both of which have the rhythmic oddity property.

No.	Necklace	Bi-Partitions	Off-Beatness
1	3 3 3 3 3 3 2 2 2	3	0
2	3 3 3 3 3 2 3 2 2	2	1
3	3 3 3 3 3 2 2 3 2	2	2
4	3 3 3 3 2 3 3 2 2	1	1
5	3 3 3 3 2 3 2 3 2	1	2
6	3 3 3 3 2 2 3 3 2	1	1
7	3 3 3 2 3 3 3 2 2	0	2
8	3 3 3 2 3 3 2 3 2	0	3
9	3 3 3 2 3 2 3 3 2	0	2
10	3 3 2 3 3 2 3 3 2	0	2

Fig. 5. The ten necklaces of nine elements of two colors with density six

The number of bi-partitions a rhythm possesses is invariant to cyclic rotations of the rhythm. In other words this property has the same value for a necklace as it does for all the rhythms it generates. By contrast, the off-beatness measure clearly does not have this property. However, there are several ways to obtain an off-beatness measure that *is* invariant under rotations. One natural method is to simply add the off-beatness values over all its onsets used as starting points. This yields an off-beatness value averaged over all starting points. If this is done for the Lyndon words with zero bi-partitions in Figure 5, the *Mokongo* necklace is still the winner with a score of 27. Necklace No. 9 comes second with a score of 26 and Necklace No. 7 is third with a score of 24.

Another (*worst-case*) rotation-invariant measure of off-beatness, counts the number of onsets that yield the minimum off-beatness value when these onsets are used as starting points in the rhythm. The smaller such a number is, the greater is the guarantee that any starting point of the necklace will yield a rhythm with high off-beatness value. If this number is calculated for necklaces 7, 8, and 9 in Figure 5, the *Mokongo* necklace is again the winner. The *Mokongo* necklace (no. 8 in the figure) has *one* starting onset that yields the minimum off-beatness value of two. Necklace No. 9 comes second with *two* starting onsets, and necklace No. 7 is third with *four* such onsets.

4   *Off-Beatness* as a Measure of Syncopation

The off-beatness measure may also serve as a precisely defined mathematical measure of syncopation. According to Michael Keith [6], "although syncopation in music is relatively easy to perceive, it is more than a little difficult to define precisely." Indeed, most definitions have in common that syncopation involves accenting an onset that is not normally accented, which begs for a definition of *normal*. Keith [6] proposed a mathematical measure of syncopation based on three types of events (notes) he calls *hesitation, anticipation,* and *syncopation,* where syncopation is the combination of hesitation and anticipation. To these events he gives the weights 1, 2, and 3, respectively. An anticipation occurs if the start of the note is "off the beat", whereas a hesitation occurs if the end of the note is "off the beat". What remains to be defined precisely is the notion of "off the beat". Unfortunately Keith defines "off the beat" only for meters with time spans equal to a power of 2 and a partition consisting of all 2's. His definition is as follows:

Let δ be the duration of the event (note) as a multiple of $1/2^d$ beats, and let S be the start time of the event (with time positions numbered starting with 0), expressed in the same units. Furthermore, let $D = \delta$ rounded down to the nearest power of 2. Then the *start* of the event is defined to be "off the beat" if S is *not* a multiple of D, and the *end* of the event is defined to be "off the beat" if $(S + \delta)$ is *not* a multiple of D. The syncopation value for the $i - th$ event (note) in the rhythmic pattern, denoted by s_i is defined as: $s_i = $ (2 if $start_i$ is off the beat) + (1 if end_i is off the beat) Finally, the measure of syncopation of a rhythmic pattern is the sum of the syncopation values s_i summed over all i.

It is interesting to compare the syncopation measure of Keith with the off-beatness measure proposed here. Since Keith's definition holds only for time spans of units equal to powers of 2, we cannot compare it to the off-beatness measure using the rhythms of the *Aka* Pygmies discussed in the preceeding, since they have time spans of 12 and 24 units. However, we may perform the comparison using the 4/4 time clave patterns studied in [11] which have time spans of 16 units. The six fundamental timelines are:

$$[\text{x . . . x . x . . . x . x . . .}] - \textit{Shiko}$$
$$[\text{x . . x . . x . . . x . x . . .}] - \textit{Son}$$
$$[\text{x . . x . . . x . . x . x . . .}] - \textit{Rumba}$$
$$[\text{x . . x . . x . . . x x . . . .}] - \textit{Soukous}$$
$$[\text{x . . x . . x . . . x . . x . .}] - \textit{Bossa-Nova}$$
$$[\text{x . . x . . x . . . x . . . x .}] - \textit{Gahu}$$

It is also interesting to compare these Afro-American timelines with one of the most popular ostinatos used in classical music given by:

$$[\text{x . x . x . . . x . . . x . . .}] - \textit{classical ostinato}$$

The values of Syncopation and off-beatness for these seven rhythmic patterns are shown in Figure 6. The syncopation measure makes global sense on these seven rhythms but is rather coarse and yields several questionable judgements. In contrast, the off-beatness measure shows remarkable agreement with human perception of syncopation. There is no doubt that the classical ostinato and *Shiko* patterns are less syncopated than the *Son, Rumba,* and *Soukous* patterns. There is also no doubt that the *Bossa-Nova* is more syncopated than all of these. Both measures support these conclusions. However, the *Rumba* is clearly

	Syncopation	Off-beatness
Classical	0	0
Shiko	1	0
Son	2	1
Rumba	2	2
Soukous	3	2
Bossa-Nova	3	2
Gahu	3	1

Fig. 6. A comparison of the syncopation and off-beatness measures

more syncopated than the *Son*, something the off-beatness measure bears out but the syncopation measure does not. It is difficult to decide which of *Soukous* or *Rumba* is the more syncopated of the two. The off-beatness measure reflects this difficulty whereas the syncopation measure judges the *Soukous* as being more syncopated. The *Bossa-Nova* feels more syncopated than the *Gahu*, something the off-beatness measure also bears out but the syncopation measure does not. There appears to be only one point in favor of the syncopation measure. The *Shiko* is more syncopated than the *classical ostinato*. Here the syncopation measure agrees more with human perception than does the off-beatness measure.

5   Concluding Remarks

The analysis of Sub-Saharan African rhythm timelines suggests that the off-beatness measure is a good mathematical predictor of the frequency (and thus preference) with which they are used in traditional music. It should be a useful feature for music information retrieval systems, as well as other applications.

The data (4/4 time *clave* patterns) analysed in the preceeding section suggest that the off-beatness measure also provides more agreement with human perception of syncopation than does the syncopation measure of Keith [6]. Therefore it should be a useful feature for recognizing syncopated music. It would be interesting to generalize Keith's measure to hold for general rhythmic patterns and to compare such a generalization with the off-beatness measure proposed here. Measuring off-beatness for rhythms with time spans consisting of n units, where n is a prime number such as 7, 11, 13, etc., is a challenging open problem.

References

1. Simha Arom. *African Polyphony and Polyrhythm*. Cambridge University Press, Cambridge, England, 1991.
2. Marc Chemillier. Ethnomusicology, ethnomathematics. The logic underlying orally transmitted artistic practices. In G. Assayag, H. G. Feichtinger, and J. F. Rodrigues, editors, *Mathematics and Music*, pages 161–183. Springer-Verlag, 2002.
3. Marc Chemillier and Charlotte Truchet. Computation of words satisfying the "rhythmic oddity property" (after Simha Arom's works). *Information Processing Letters*, 86:255–261, 2003.

4. José Manuel Gamboa. *Cante por Cante: Discolibro Didactico de Flamenco*. New Atlantis Music, Alia Discos, Madrid, 2002.
5. A. M. Jones. *Studies in African Music*. Oxford University Press, Amen House, London, 1959.
6. Michael Keith. *From Polychords to Pólya: Adventures in Musical Combinatorics*. Vinculum Press, Princeton, 1991.
7. Jeff Pressing. Cognitive isomorphisms between pitch and rhythm in world musics: West Africa, the Balkans and Western tonality. *Studies in Music*, 17:38–61, 1983.
8. Jay Rahn. Asymmetrical ostinatos in sub-saharan music: time, pitch, and cycles reconsidered. *In Theory Only*, 9(7):23–37, 1987.
9. Jay Rahn. Turning the analysis around: African-derived rhythms and Europe-derived music theory. *Black Music Research Journal*, 16(1):71–89, 1996.
10. Frank Ruskey and Joe Sawada. An efficient algorithm for generating necklaces with fixed density. *SIAM Journal of Computing*, 29(2):671–684, 1999.
11. Godfried T. Toussaint. A mathematical analysis of African, Brazilian, and Cuban *clave* rhythms. In *Proceedings of BRIDGES: Mathematical Connections in Art, Music and Science*, pages 157–168, Towson University, Towson, MD, July 27-29 2002.
12. Godfried T. Toussaint. Classification and phylogenetic analysis of African ternary rhythm timelines. In *Proceedings of BRIDGES: Mathematical Connections in Art, Music and Science*, pages 25–36, Granada, Spain, July 23-27 2003.
13. Trevor Wiggins. Techniques of variation and concepts of musical understanding in Northern Ghana. *British Journal of Ethnomusicology*, 7:117–142, 1998.

Empirical Bounds on Error Differences
When Using Naive Bayes

Zoë Hoare

School of Informatics, University of Wales Bangor,
Bangor, Gwynedd, LL57 1UT
mape01@bangor.ac.uk

Abstract. Here we revisit the Naïve Bayes Classifier (NB). A problem from veterinary medicine with assumed independent features led us to look once again at this model. The effectiveness of NB despite violation of the independence assumption is still open for discussion. In this study we try to develop a bound relating dependency level of features and the classification error of Naïve Bayes. As dependency between more than two features is difficult to define and express analytically, we consider a simple two class two feature example problem. Using simulations we established empirical bounds measured by Yules Q-statistic between calculable error and error related to the true distribution.

1 Introduction

The Naïve Bayes (NB) classifier has proved to be a useful classification tool, despite the independence assumption imposed upon the features. NB has been shown in numerous studies to perform well even when the features obviously violate the assumption [1, 2, 3, 4, 5, 6].

A non-traditional dataset originating from the veterinary medicine domain motivated us to look at Naïve Bayes again. BSE in cattle and Scrapie in sheep are important notifiable neurodegenerative diseases. They are currently of global concern with the first case of BSE being diagnosed in the USA in December 2003 [7]. The datasets concerning BSE and Scrapie are non-traditional in the sense that they consist of probability estimates of a feature (clinical sign) being present given a certain class (disease). These probability estimates were given by field experts. Since only the marginal probabilities were estimated features were assumed to be independent. Each of the two datasets contains 200+ features and 50+ diseases. We had no information about any dependencies amongst the features, therefore it made sense to try NB first. The NB model has had success in the medical domain in the past, [8, 9]. Medical experts often prefer the NB to other algorithms due to its explanation potential. It appears that the decision made by NB follows a path similar to their own decision making process [9].

In this study we took the NB model and tried to find out how much the calculable error of an assumed independent distribution differed from the true (Bayes) error.

S. Singh et al. (Eds.): ICAPR 2005, LNCS 3686, pp. 28–34, 2005.

2   Types of Error

Consider the two feature, two class problem outlined in Table 1. In this example, ω_1 can be class Scrapie and ω_2, class non-Scrapie. We can think of x_1 as the feature (sign) Licking and Biting and x_2 as the feature (sign) Nibbling reflex. Features may either be present (having the value 1) or absent (having the value 0). The entries in the table are the class-conditional probabilities for the respective combination of signs. For example $a = P(x_1 = 0, x_2 = 0|\omega_1)$. We call this table "the dependent distribution" because x_1 and x_2 are not assumed to be independent given any of the two classes. We note that $a + b + c + d = 1 = e + f + g + h$. We also assume that $P(\omega_1) = P(\omega_2) = \frac{1}{2}$, i.e., both classes (diseases) are equally prevalent.

The true Bayes error for the problem in Table 1 is

$$E_1 = \frac{1}{2}\{\min\{a, e\} + \min\{b, f\} + \min\{c, g\} + \min\{d, h\}\} \tag{1}$$

Table 1. The dependent distribution

ω_1	$x_1 = 0$	$x_1 = 1$
$x_2 = 0$	a	b
$x_2 = 1$	c	d

ω_2	$x_1 = 0$	$x_1 = 1$
$x_2 = 0$	e	f
$x_2 = 1$	g	h

If x_1 and x_2 were *assumed to be* independent their joint distribution will be as shown in Table 2 (called "independent distribution"). We are interested in finding out how the classification error will be affected if we use Table 2 instead of Table 1. The independent distribution can be obtained from the dependent distribution by using the calculations in Table 2. However, the dependent distribution cannot be recovered from Table 2.

Table 2. The independent distribution

ω_1	$x_1 = 0$	$x_1 = 1$
$x_2 = 0$	$A = (a + b)(a + c)$	$B = (a + b)(b + d)$
$x_2 = 1$	$C = (a + c)(c + d)$	$D = (b + d)(c + d)$

ω_2	$x_1 = 0$	$x_1 = 1$
$x_2 = 0$	$E = (e + f)(e + g)$	$F = (e + f)(f + h)$
$x_2 = 1$	$G = (e + g)(g + h)$	$H = (f + h)(g + h)$

Denote by E_2^* the error made if we use the independent distribution. There is no easy way of expressing this error analytically because it will depend on whether or not the true (Bayes) classifier (Table 1) and NB make the same decision. For example, if $(a > e$ and $A > E)$ or $(a < e$ and $A < E)$, then E_2^* will have $\min\{a, e\}$ as the first error term in the brackets. If the opposite holds, $E_2^\star$ will have $\max\{a, e\}$ as the first error term. If we assume that Table 2 gives the true distribution, our (incorrect) estimate of the error will be E_2

$$E_2 = \frac{1}{2}\{\min\{A, E\} + \min\{B, F\} + \min\{C, G\} + \min\{D, H\}\} \qquad (2)$$

The best way to see the difference in these two errors is to consider an example. Consider the problem outlined in Tables 3 and 4. Let $\mathbf{x} = [0, 0]^T$ be the case submitted for classification.

Table 3. The dependent distribution

ω_1	$x_1 = 0$	$x_1 = 1$
$x_2 = 0$	0.4	0.2
$x_2 = 1$	0.1	0.3

ω_2	$x_1 = 0$	$x_1 = 1$
$x_2 = 0$	0.3	0.1
$x_2 = 1$	0.5	0.1

Table 4. The independent distribution calculated from Table 3

ω_1	$x_1 = 0$	$x_1 = 1$
$x_2 = 0$	0.3	0.3
$x_2 = 1$	0.2	0.2

ω_2	$x_1 = 0$	$x_1 = 1$
$x_2 = 0$	0.32	0.08
$x_2 = 1$	0.48	0.12

In the dependent distribution $\mathbf{x} = [0, 0]^T$ would be classified as class ω_1 with the error of this decision being $P_1(\text{error}, \mathbf{x} = [0, 0]^T) = \min\{0.4, 0.3\} \times \frac{1}{2} = 0.15$.

However, when we look at the independent distribution modelled from the dependent distribution we see that $\mathbf{x} = [0, 0]^T$ would be classified as class ω_2. According to the independent distribution $P_2(\text{error}, \mathbf{x} = [0, 0]^T) = \min\{0.3, 0.32\} \times \frac{1}{2} = 0.15$. However, there is a mistake according to the true distribution, i.e., $P_2^\star(\text{error}, \mathbf{x} = [0, 0]^T) = 0.4 \times \frac{1}{2} = 0.2$, so $E_2^\star$ is increased.

In the non-traditional dataset we can only calculate E_2 as we have no other information available. In this study we wanted to look at the relationship between the error we can calculate and the error we were making in relation to the true distribution.

3   Experiments

3.1   Empirical

We generated 10,000 pairs of random matrices of dependent classes as in Table 1. We then calculated the the independent distributions from these as in Table 2. We took our measure of dependence to be Yules Q statistic. Q_1 depicts the level of dependency in class ω_1 and is calculated as

$$Q_1 = \frac{ad - bc}{ad + bc} \qquad (3)$$

The Q-statistic varies from -1 to 1. A value of -1 means the two features are negatively dependent, i.e., the two features always take opposite values; 1 indicates that the features are positively dependent, i.e., the two features always take the same value. A Q-value of zero means the two features are completely independent of one another.

For this simulation we restricted the distribution of class ω_2 so that the two features were independent given class ω_2. This meant that $Q_2 = \frac{eh-fg}{eh+fg} = 0$ throughout and the subtables for ω_2 in Tables 1 and 2 were identical. Only class ω_1 had a true dependent distribution and an independent distribution calculated from this.

The 10,000 points $(Q_1, E_2^\star - E_2)$ are plotted in figure 1. The figure shows that

- $E_2^\star - E_2$ can be both positive and negative. This indicates that we can overestimate or underestimate the error of NB. However, from this simulation it is not clear in what situations over- or underestimation occurs.
- When Q_1 is zero, so is the difference between E_2 and $E_2^\star$. This is expected as this is when the features are conditionally independent for both classes and so NB is known to be optimal.
- The relationship between E_2 and $E_2^\star$ is symmetrical about 0. As the encoding is arbitrary we can interchange the values of 0 and 1 for the features, which is the explanation for the symmetrical pattern.
- As the value of Q_1 reaches out to ± 1 the maximum difference in the two error values increases but there are also points plotted along the zero line indicating that there is no way of knowing exactly what the difference will be even if we know the Q-value of the two features.
- The figure has a pronounced shape which implies there is the possibility of finding a bound on the error difference.

3.2   Bounds

There has been work done on the bounds of the probability of error of NB [10,11]. These bounds have been shown to be arbitrarily tight to the probability of error of the NB model [11]. However, these studies did not look into how the level of dependence of features affects the probability of error of NB. The bounds proposed here are looking at the difference of the two types of error related to the level of dependency of the features.

From the simulations we are able to place an *empirical* bound onto the difference between E_2 and $E_2^\star$. Bound 1 given in equation (4) encloses 95% of the points depicted, in Fig. 1.

$$B_1 = \pm \frac{(Q_1 + Q_1{}^3)}{20} \tag{4}$$

The bound is plotted with a solid line in Fig. 1. Taking the bound one step further we can give a second bound that encompasses 100% of the 10,000 error

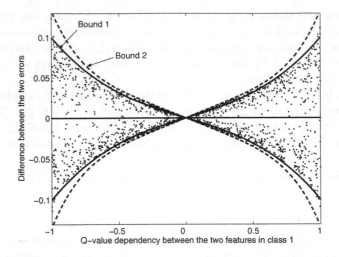

Fig. 1. Scatterplot of the 10,000 data points $(Q_1, E_2^\star - E_2)$ randomly generated as described

differences, see Fig. 1. The formula for the second bound is given below

$$B_2 = \pm \frac{(Q_1 + Q_1^{5})}{15}. \tag{5}$$

This bound is plotted with the dashed line in Fig. 1. These bounds show us that if we know that the two features are independent in one class and we know the Q-value of the features dependency in the second class then it is likely that we will be within $\pm B_2$ of the true error committed on the true distribution.

3.3   Real Data

Using four data sets from the UCI data repository (SPECT, Wine, Thyroid and Glass) [1] it was possible to find pairs of features replicating the conditions required by the empirical simulations.

- Only binary data was considered. Any data sets that were not binary were converted by using the Gini criterion as a splitting function.
- All problems were transformed into two class ones. If the data set contained more then two classes only class one and class two were used.
- The classes were assumed to be equiprobable.
- Our scenario requires that the two features were independent in one class. We looked at all pairs of features in the two classes and selected all pairs that were indeed independent in one class i.e. had a Q-value of zero in one of the two classes.

---

[1] Data sets available at: http://www.ics.uci.edu/~mlearn/MLRepository.html

Plotted in Fig. 2 all the pairs of features that were independent in one class. The x-axis is their Q-value in the second class and the y-axis is $E_2^\star - E_2$. Also plotted is the bound B_2 shown as the dashed line. From Fig. 2 we can see that all the points fit within the bound. All feature pairs that were independent in both classes had no difference in their values of E_2 and $E_2^\star$ as expected and are all plotted on (0,0). This strengthens our findings during the empirical study but cannot serve as a proof of this bound.

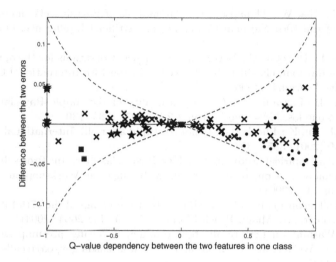

Fig. 2. Scatterplot of the pairs of features found in the real data analysis fitting the requirements of the study. Bound B_2 is also plotted. SPECT - •, Wine - ×, Thyroid - ■, Glass - ★.

4   Conclusion

In this study we looked again at the probability of error when using the NB classifier. By considering the dependency between features we investigated the difference in the error we can calculate from an independent distribution and the true error.

We derived two empirical bounds on the difference in the errors. If two features are independent in one class and we have the Q-value of their dependence for the other class then $E_2^\star - E_2$ is contained within $\pm B_2$.

This finding was then strengthened by the use of real data sets from the UCI repository. The difference of E_2 and $E_2^\star$ of all feature pairs fitting the requirements of the simulation fell within the constraint of bound B_2. This is an initial study which opens up interesting questions. The next step would be to look for bounds for more than two features and also for the cases of dependent distribution (not fixing $Q_2 = 0$).

Acknowledgments

This study was possible thanks to Dr. P. Cockcroft, School of Veterinary Medicine, University of Cambridge; who kindly gave us access to his data sets on BSE and Scrapie.

References

1. Langley, P., Iba, W., Thompson, K.: An analysis of Bayesian classifiers. In: Proceedings of the 10th National Conference on Artificial Intelligence. (1992) 399 – 406
2. Domingos, P., Pazzani, M.: Beyond independence: Conditions for the optimality of the simple Bayesian classifier. In: Proceedins of the 13th International Conference on Machine Learning. (1996)
3. Domingos, P., Pazzani, M.: On the optimality of the simple Bayesian classifier under zero-one loss. Machine Learning **29** (1997) 103 – 130
4. Hand, D.J., Yu, K.: Idiot's Bayes - not so stupid after all? International Statistical Review **69** (2001) 385 – 398
5. Rish, I.: An empirical study of the Naïve Bayes classifier. In: Proceedings of the International Joint Conference on Artificial Intelligence, Workshop on "Empirical Methods in AI". (2001)
6. Zhang, H.: The optimality of Naive Bayes. In: Proceedings of the 17th International FLAIRS conference, Miami Beach Florida, 17 - 19 May 2004 (2004)
7. USDA: Washington firm recalls beef products following presumptive BSE determination. Website article (2003) http://www.fsis.usda.gov/oa/recalls/prelease/pr067-2003.htm.
8. Titterington, D.M., Murray, G.D., Murray, L.S., Spiegelhalter, D.J., Skene, A.M., Habbema, J.D.F., Gelpke, G.J.: Comparison of discrimination techniques applied to a complex data set of head injured patients. Journal of the Royal Statistical Society. Series A (General) **144** (1981) 145 – 175
9. Kononenko, I.: Inductive and Bayesian learning in medical diagnosis. Applied Artificial Intelligence **7** (1993) 317 – 337
10. Hashlamoun, W.A., Varshney, P.K., Samarascoriya, V.N.S.: A tight upper bound on the Bayesian probability of error. IEEE Transactions on Pattern Analysis and Machine Intelligence **16(2)** (1994) 220 – 224
11. Avi-Itzhak, H., Diep, T.: Arbitrarily tight upper and lower bounds on the Bayesian probability of error. IEEE Transactions on Pattern Analysis and Machine Intelligence **18(1)** (1996) 89 – 91

Effective Probability Forecasting for Time Series Data Using Standard Machine Learning Techniques

David Lindsay[1] and Siân Cox[2]

[1] Computer Learning Research Centre
davidl@cs.rhul.ac.uk
[2] School of Biological Sciences,
Royal Holloway University of London, Egham, Surrey, TW20 OEX, UK
s.s.e.cox@rhul.ac.uk

Abstract. This study investigates the effectiveness of probability forecasts output by standard machine learning techniques (Neural Network, C4.5, K-Nearest Neighbours, Naive Bayes, SVM and HMM) when tested on time series datasets from various problem domains. Raw data was converted into a pattern classification problem using a sliding window approach, and the respective target prediction was set as some discretised future value in the time series sequence. Experiments were conducted in the online learning setting to model the way in which time series data is presented. The performance of each learner's probability forecasts was assessed using ROC curves, square loss, classification accuracy and Empirical Reliability Curves (ERC) [1]. Our results demonstrate that effective probability forecasts can be generated on time series data and we discuss the practical implications of this.

1 Introduction

Probability forecasting has become an increasingly popular doctrine in the machine learning community [1], [2]. Probability forecasting is a practically useful generalisation of the standard pattern classification problem where the aim is to estimate the conditional probability (otherwise known as a probability forecast) of a possible label given an observed object. The problem of making *effective* probability forecasts is well studied [1], [2], [3], [4]. Dawid (1985) gives two simple criteria for describing how effective probability forecasts are:

1. **Reliability** - The probability forecasts "should not lie". When a probability $\hat{p}$ is assigned to an event, there should be roughly $1 - \hat{p}$ relative frequency of the event *not* occurring.
2. **Resolution** - The probability forecasts should be practically useful and enable the observer to easily rank the events in order of their likelihood of occurring. This criterion is more related to classification accuracy and ROC area [1].

Effective probability forecasting is a desirable goal, especially in cost-sensitive decision making domains such as medical and financial applications [1], [2]. However, probability forecasting studies mainly consider data which is roughly drawn from *i.i.d.*

S. Singh et al. (Eds.): ICAPR 2005, LNCS 3686, pp. 35–44, 2005.

probability distributions. We wanted to know whether traditional machine learning algorithms could give effective probability forecasts with time series (or non-*i.i.d.*) data. We have tested several machine learning algorithms on time series datasets from varying problem domains (text, industrial, financial and meteorological). The raw data was converted into a pattern classification problem using a sliding window approach (of size w), and the respective target prediction was set as some discretised future value in the raw time series sequence. Experiments were conducted in the online learning setting, where the learner is continually updated with data, thus modelling the natural way in which time series data is presented. The results demonstrate the ability of the learners to output effective probability forecasts with time series data in addition to revealing the misleading nature of classification accuracy.

2   Pattern Classification of Time Series Data

Our notation will extend upon the commonly used supervised learning approach to pattern classification. Nature outputs information pairs called *examples*. Each example $z_i = (\mathbf{x}_i, y_i) \in \mathbf{X} = \mathbf{Z} \times \mathbf{Y}$ consists of an *object* $\mathbf{x}_i \in \mathbf{X}$ and its *label* $y_i \in \mathbf{Y} = \{1, 2, \ldots, |\mathbf{Y}|\}$. All experiments given in this paper test n training examples in the online learning setting, as this seems a natural method of testing time series data - the true value will often become available with some time lag. With online learning, each object $\mathbf{x}_i, i > 1$ is considered in turn as a 'trial' in the online process and a prediction (in our problem a set of probability forecasts) is made by the learner Γ using the previous $i - 1$ training examples $\Gamma(z_1, \ldots, z_{i-1})$. At the end of each trial the learner Γ is updated with the true label y_i of the ith object, and the process is repeated until all data is tested. In this study we will consider a sequence of n probability forecasts for the $|\mathbf{Y}|$ possible labels output by a learner Γ. Let $\hat{P}(y_i = j \mid \mathbf{x}_i) = \hat{p}_{i,j}$ represent the estimated conditional probability of the jth label matching the true label y_i for the ith object tested.

2.1   Conversion of Time Series Data Using a Sliding Window Approach

Our raw multi-variate time series data $s_{i,j}$ has k attribute readings over m time epochs

$$
\begin{matrix}
s_{1,1} & s_{1,2} & \cdots & s_{1,k} \\
s_{2,1} & s_{2,2} & \cdots & s_{2,k} \\
\vdots & \vdots & \vdots & \vdots \\
s_{m,1} & s_{m,2} & \cdots & s_{m,k}
\end{matrix}
$$

Usually the values s_{ij} are continuous, but can also be discrete (as with the text dataset used in this study). We convert the raw time series data into a pattern classification problem using a simple sliding window technique. Firstly, we choose some window size w, and generate a set of $n = m - w$ training examples, each with kw attributes

$$
\mathbf{x}_i = \langle s_{i-w,1}, \ldots, s_{i,1}, \ldots, s_{i-w,k}, \ldots, s_{i,k} \rangle \tag{1}
$$

Secondly, we choose a time shift $f \geq 1$ of how far into the future we would like to predict for a chosen attribute $1 \leq j \leq k$. Now to generate the corresponding label y_i for each object we can use a variety of methods depending on the task:

1. $y_i = s_{i+f,j}$ shift to the next value (as with text data)

2. $y_i = C\left(\frac{s_{i+f,j} - s_{i,j}}{s_{i,j}} \times 100\right)$ shift in a percentage change (as with finance, meteorological and industrial data)

where C is a discretising function breaking the continuous real value into a set of intervals. In our experiments we used a simple equal width discretisation. Let s_{low} and s_{up} be the minimum and maximum percentage deviations of the chosen attribute

$$s_{\text{low}} = \operatorname*{argmin}_{1 \leq i \leq n} \left\{ \frac{s_{i+f,j} - s_{i,j}}{s_{i,j}} \times 100 \right\} \text{ and } s_{\text{up}} = \operatorname*{argmax}_{1 \leq i \leq n} \left\{ \frac{s_{i+f,j} - s_{i,j}}{s_{i,j}} \times 100 \right\}$$

If we let $s_{\text{range}} = \frac{|s_{\text{up}} - s_{\text{low}}|}{|\mathbf{Y}|}$ be the absolute size of each discrete bin interval between s_{up} and s_{low} and include the infinite boundary cases then the discretising function C maps a real value onto $|\mathbf{Y}|$ disjoint bins

$$C : \mathbb{R} \rightarrow \left\{ (-\infty, s_{\text{low}} + s_{\text{range}}), [s_{\text{low}} + s_{\text{range}}, s_{\text{low}} + 2s_{\text{range}}), \ldots, [s_{\text{up}} - s_{\text{range}}, \infty) \right\}$$

In our experiments s_{up} and s_{low} are estimated from the training data, which can be seen as 'cheating' as the learner is indirectly biased by the data before classication has taken place. However it is reasonable to assume that for practical applications, a basic knowledge of the range of deviation in the data will be known apriori, or at least can be estimated from historical data. Of course any other discretisation method could be used, such as one that encodes some semantic meaning about the sequence deviation.

3   Reliable Probability Forecasting: A Machine Learning Perspective

To calculate the reliability of a finite number of probability forecasts a method of discretising forecasts must be used. For predicted probabilities $\hat{p}_{i,j}$ of each class $j \in \mathbf{Y}$ we define a set of '*bins*' (disjoint sub-intervals) B_j, for example one possible bin choice would be to choose k equal width bins $B_j = \left\{ [0, \frac{1}{k}), [\frac{1}{k}, \frac{2}{k}), \ldots, [\frac{k-1}{k}, 1] \right\}$. Let $n_b^j = \sum_{i=1}^{n} \mathbb{I}_{\{\hat{p}_{i,j} \in b\}}$ count the number of forecasts $\hat{p}_{i,j}$ for class $j \in \mathbf{Y}$ that fall within bin interval $b \in B_j$. There are many possible choices of bin sizes, however we aim to specify bin sizes which encompass enough forecasts (make n_b^j as large as possible for each bin) to obtain practically useful estimates.

Once the sets of bins $B_j, j \in \mathbf{Y}$ have been defined, we can define reliability by calculating various statistics from the individual bins $b \in B_j$. Reliability ensures that for each bin of forecasts with predicted values $\approx \hat{p}$, the frequency of this label *not* occurring in that bin is $\approx 1 - \hat{p}$. To obtain a practically useful estimate of the predicted value represented by each bin we use the average predicted probability $\phi_n^j(b)$ for each bin interval b

$$\phi_n^j(b) = \frac{\sum_{i=1}^n \mathbb{I}_{\{\hat{p}_{i,j} \in b\}} \hat{p}_{i,j}}{n_b^j} \tag{2}$$

The empirical frequency $\rho_n^j(b)$ of each bin b calculates the proportion of predictions in that bin that had true class $y_i = j$

$$\rho_n^j(b) = \frac{\sum_{i=1}^n \mathbb{I}_{\{y_i=j\}} \mathbb{I}_{\{\hat{p}_{i,j} \in b\}}}{n_b^j} \tag{3}$$

To determine whether a bin b contains enough forecasts to be practically useful to gather the $\rho_n^j(b)$ and $\phi_n^j(b)$ statistics, an extra weighting term $\nu_n^j(b) = \frac{n_b^j}{n}$ is used. An example of the bin statistics for the Naive Bayes probability forecasts of class label $y_i = (-25\%, -15\%]$ on the pH chemical dataset are given below

Set of bins b in $B_{(-15\%,-25\%]}$						
Bin Interval b	$\phi(b)$	n_b	$\rho(b) \times n_b$	$\rho(b)$	$\nu(b)$	Colour
(0-0.052]	0.013	1068	156	0.146	0.536	■
(0.052-0.113]	0.069	178	41	0.230	0.089	
(0.113-0.879]	0.538	221	64	0.289	0.111	
(0.879-1]	0.959	510	205	0.402	0.256	■

These statistics show that the Naive Bayes learner is unreliable since the predicted probability $\phi(b)$ and empirical frequency $\rho(b)$ are divergent from one another. Every learner's performance in the reliability criterion can be categorised using the functions $\rho_n^j(b)$, $\phi_n^j(b)$ and $\nu_n^j(b)$. Using them intuitively defines reliability. A learner is *well calibrated* (reliable) if its forecasts $\{\hat{p}_{1,1}, \ldots, \hat{p}_{1,|\mathbf{Y}|}, \ldots, \hat{p}_{n,1}, \ldots, \hat{p}_{n,|\mathbf{Y}|}\}$ and a fixed specification of bins $B_j, j \in \mathbf{Y}$ satisfy

$$R(\Gamma, n) = \sum_{j=1}^{|\mathbf{Y}|} \sum_{b \in B_j} \nu_n^j(b) |\rho_n^j(b) - \phi_n^j(b)| \approx 0 \tag{4}$$

3.1 Assessment of Probability Forecasts

At present the most popular techniques for assessing the quality of probability forecasts are *square loss* $\frac{1}{n} \sum_{i=1}^n \sum_{j=1}^{|\mathbf{Y}|} (\mathbb{I}_{\{y_i=j\}} - \hat{p}_{i,j})^2$ [5] and *ROC curves* [6]. The square loss function assesses a combination of reliability and resolution [4]. The area under the ROC curve is commonly used as a measure of the usefulness of the probability forecasts; the larger the area, the better the forecasts.

The *Empirical Reliability Curve* (ERC) is a visual interpretation of the theoretical definition of reliability (cf. 4) [1]. Unlike the previous methods, the ERC allows visualisation of over- and under-estimation of probability forecasts. For more detail about ERC implementation please refer to [1]. In brief, each coordinate (marked as $\odot$) on

Table 1. Descriptions of the time series data tested and the sliding window conversion parameters used to convert the raw data into an online pattern classification problem. From left to right: name of the dataset, number of examples n, number of raw data attributes k, number of classes/bins $|\mathbf{Y}|$, chosen window size w, chosen time shift f, chosen class index j and general description of the data.

Data Name	No. Expl	No. Att	No. Class	Win Size	Time Shift	Class Indx	Description of Data
pHdata	1991	10	4	3	5	3	Simulation data of a pH neutralisation process in a stirring tank. Time is measured in 10 second epochs.
spotexrates	2546	12	4	4	14	8	Spot prices (foreign currency in dollars) and the returns for daily exchange rates of AUD, BEF, CAD, FRF, DEM, JPY, NLG, NZD, ESP, SEK, CHF, GBP against the US dollar.
darwin	1386	8	5	7	1	7	Monthly values of the Darwin SLP series, from 1882 to 1998. This series is a key indicator of climatological patterns.
Emma	4994	6	6	5	1	1	Created from raw text of the famous novel 'Emma'.

the ERC represents the statistics computed for each bin[1] b, and the coordinate $\odot$ is coloured according the weighting of that bin $\nu(b)$ (black = 1, 1 > shades of grey > 0, white = 0). An example of ERC coordinates and their respective colouring can be seen in the table above. The respective ERC plot for the coordinates in the table above can be seen in Figure 1. A reliable classifier will have ERC coordinates $\big(\phi(b), \rho(b)\big)$ close to the diagonal line of calibration $(0,0) \rightarrow (1,1)$ (where predicted probability equals empirical frequency cf. 2, 3). A trend line is predicted from these coordinates using a weighted regression algorithm [8] (where each training example is weighted according to the value $\nu(b)$). This allows the coordinates which relate to a bin containing a large sample of forecasts to have a greater influence on the shape of the curve. An associated reliability score is computed using the absolute deviation of the ERC coordinates from the line of calibration (cf. 4).

4   Experimental Design

We tested the following learning algorithms as provided by WEKA: Naive Bayes, Distance Weighted K-Nearest Neighbours (DW K-NN), Neural Networks, C4.5 Decision Tree using Laplace Smoothing and Linear Kernel Pairwise Coupled Support Vector Machine using Logistic Regression (PC LR SVM) [9]. We also tested the traditional time series analysis technique, Hidden Markov Models (HMM), as provided by the JAHMM project[2]. To ensure that learning was taking place, we also tested two simple classifiers as controls: NegR which outputs forecasts randomly from a uniform distribution,

---

[1] For our ERC we used the Discretize filter provided by WEKA [5] which uses an MDL criterion to optimally define bin interval sizes [7].

[2] http://www.run.montefiore.ulg.ac.be/~francois/software/jahmm/

and ZeroR which outputs mere label frequencies in the training data, i.e. no object $\mathbf{x}_i$ information is used. Window sizes for each dataset were determined by maximising auto-correlation and auto-covariance using correlograms [10]. The shift parameter f (which describes how far into the future we would like to predict) for each dataset was chosen to give an appropriate level of deviation between time epochs (on average approximately $\pm 25\%$ deviation). The learners were tested on 4 time series datasets in the online learning setting, converting the data as detailed in Table 1.

The pHdata, spotexrates and darwin datasets are from the UCR public time series data repository[3]. For the Emma text dataset[4], each of the 6 window attributes represent 28 possible characters (26 alphabet letters, full stop and space characters). Rather than predicting 28 next possible characters, the task was simplified to 6 letter groupings commonly used in cryptography studies

$$\{\{a,t,n,o,r,i,s\},\{b,g,p,y,w\},\{c,d,h,l,f,m,u\},\{j,k,q,v,x,z\},\{e\},\{`\,`,`.`\}\}$$

5   Experimental Results

The results in Table 2 show the various performance scores of probability forecasts output by standard learning algorithms and simple base learners (NegR and ZeroR) when tested on the time series data. Although the error rates across all data are quite high for the standard learners (greater than 18%), they do outperform the error rates of the base learners indicating that learning has taken place on the data. In terms of the reliability score (cf. 4), the base learner ZeroR performed very well. However, this is because ZeroR is effectively estimating the unconditional probability of each label $\Pr(y)$, instead of the conditional probability of each label $\Pr(y \mid \mathbf{x})$ on each object that we desire.

We tested a variety of DW K-NN learners and Table 2 shows the performance scores of probability forecasts when $K = 1$ and for the value of K that gave the best overall performance. When $K = 1$ the error rate is generally lower than for other values of K. This result supports other studies where 1-NN learners were found to be good classifiers on time series data [11]. However, our results show that larger values of K tend to give better performance in terms of reliability, ROC area and square loss. This is because the estimates of the DW 1-NN are too extreme (close to zero or one), whereas averaging over larger K gives more refined estimates.

Figure 1 shows the ERC plots of probability forecasts output by the HMM, C4.5 Decision Tree and Naive Bayes learners when tested on the pHdata chemical dataset. The ERC plots for the Naive Bayes learner show the most obvious pattern of over- and under-estimation of probability forecasts. For example, when Naive Bayes makes a prediction with estimated probability 0.9, the empirical frequency of correct label prediction is only 0.4 (over-estimation). In contrast when a prediction is made with estimated probability 0.1, the empirical frequency of correct label prediction is ≈ 0.28 (under-estimation). Over- and under-estimation of probability forecasts is made by the C4.5 learner although to a much lesser extent. Probability forecasts made by the HMM

---

[3] http://www.cs.ucr.edu/~eamonn/TSDMA/index.html

[4] Raw text downloaded from http://www.gutenberg.org/catalog/

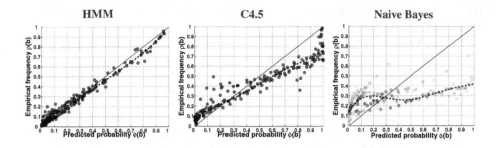

Fig. 1. ERC plots for HMM, C4.5 Decision Tree and Naive Bayes learners tested on the pH chemical dataset. The solid diagonal line represents the *line of calibration*, where predicted probability (*horizontal*) equals observed fraction of correct predictions (*vertical*). Under- and over-estimation of probability forecasts are represented by the reliability curve (dashed line) deviating above and below the line of calibration respectively.

are far more reliable, and this is indicated by the HMM's respective ERC plot which is tight to the line of calibration. These observations are also supported by the reliability scores given on the same dataset in Table 2. The Neural Network and SVM learners display mid-range performance across data but are the slowest and most computationally intensive learners. This is because the particular implementations tested are not incremental by design, and so suffer heavily when tested online.

The Naive Bayes learner's probability forecasts perform the worst of all the standard learners tested, often performing worse than the base learners in terms of reliability and square loss. This is probably because the assumption of independence of attributes is invalid for time series data tested using our sliding window approach. We were unable to get results with the HMM learner on the Emma text dataset because the huge ($28^6 \times 6$) state space made effective computation on the relatively small data sample (4994 examples) impossible. When tested on the pHdata and spotexrates data, the HMM learner gave the most effective probability forecasts of all learners, whilst still attaining the best classification accuracy. The good performance of the HMM on the spotexrates financial data can be expected because many financial engineering models (such as Black Scholes) have long assumed, and empirical studies have also validated that price deviations of financial assets such as stocks and currency exchange rates are roughly Gaussian [12]. However, when tested on the darwin dataset the HMM gives probability forecasts of comparatively poor classification accuracy. Further analysis with the darwin dataset demonstrates the important characteristics of the reliability and resolution criteria [3]. Comparing the performance of the HMM and Naive Bayes on the darwin dataset, HMM exhibits higher error rate (66.5% versus 48.2%) and ROC area (0.64 versus 0.79) but the reliability of the HMM learner is far better than Naive Bayes (0.007 versus 0.024). This observation shows that good reliability does not always follow from good classification accuracy and ROC area, as these scores are often more related to the resolution criteria i.e. the forecasts rank labels effectively. Interestingly, despite these differences between the probability forecasts of Naive

Table 2. Performance of probability forecasts output by various machine learning algorithms when tested on time series data in the online learning setting

Emma Text Data					
Learner	Error	Reliability	ROC Area	Sqr Loss	Time (secs)
NegR	83.68	0.025	0.494	0.891	0.98
ZeroR	59.632	0.004	0.475	0.741	3.56
Naive Bayes	51.502	0.008	0.721	0.663	23.06
DW 1-NN	46.796	0.023	0.736	0.739	15.15
DW 5-NN	50.34	0.007	0.723	0.648	15.44
C4.5	44.393	0.011	0.756	0.636	939
Neural Network	51.343	0.007	0.714	0.649	11458.32
PC LR SVM	52.503	0.014	0.699	0.729	71586.18
HMM	-	-	-	-	-

pH Chemical Data					
Command	Error	Reliability	ROC Area	Sqr Loss	Time (secs)
NegR	75.088	0.028	0.505	0.824	0.59
ZeroR	75.088	0.016	0.612	0.752	0.91
Naive Bayes	52.637	0.052	0.735	0.876	22.97
DW 1-NN	35.610	0.042	0.711	0.706	4.73
DW 9-NN	43.144	0.016	0.808	0.568	4.81
C4.5	31.090	0.029	0.841	0.533	591.67
Neural Network	37.820	0.017	0.841	0.531	6375.66
PC LR SVM	48.016	0.019	0.764	0.633	6999.74
HMM	18.282	0.009	0.914	0.327	18

Spot Exchange Rate Data					
Learner	Error	Reliability	ROC Area	Sqr Loss	Time (secs)
NegR	75.059	0.029	0.500	0.827	1.41
ZeroR	75.334	0.010	0.542	0.752	2.52
Naive Bayes	64.140	0.066	0.621	1.123	370.74
DW 1-NN	27.651	0.032	0.808	0.549	71.22
DW 2-NN	34.289	0.024	0.830	0.516	70.81
C4.5	45.837	0.049	0.722	0.831	16436.29
Neural Network	43.873	0.023	0.777	0.631	221513.64
PC LR SVM	61.626	0.012	0.639	0.723	27658.76
HMM	25.727	0.004	0.876	0.412	213.69

Darwin Climate Data					
Learner	Error	Reliability	ROC Area	Sqr Loss	Time (secs)
NegR	79.870	0.019	0.499	0.867	0.46
ZeroR	83.189	0.008	0.450	0.803	0.66
Naive Bayes	48.268	0.024	0.797	0.762	13.35
DW 1-NN	58.153	0.044	0.642	1.136	3.76
DW 10-NN	47.980	0.008	0.806	0.619	3.5
C4.5	53.463	0.029	0.700	0.885	257.41
Neural Network	47.835	0.007	0.811	0.619	5264.13
PC LR SVM	47.835	0.007	0.806	0.629	4651.11
HMM	66.522	0.007	0.641	0.762	8.08

Bayes and HMM, both learner's forecasts make the same square loss on the `darwin` data. This is because square loss measures a mixture of the resolution and reliability criteria [4]. So in effect the differences between both performance criteria for the HMM and Naive Bayes learners cancel one another out to give the same square loss. The HMM learner's relatively poor performance on the `darwin` data could be attributed to the assumption that the objects x_i are drawn from a multi-variate Gaussian distribution, whereas other standard learning techniques make much weaker *i.i.d* assumptions about the data's probability distribution and so perform better.

6 Discussion

This study has shown that standard learning techniques can be used to generate effective probability forecasts on time series data in the online learning setting. The main findings are as follows:

- Traditional machine learning techniques, namely K-NN, Neural Network, SVM and C4.5, can be a viable alternative to the classical time series analysis technique of HMM, perhaps because they are able to exploit the geometric representation of the problem, and are not restricted to parametric assumptions (such as Gaussian).
- The probability forecasts of the HMM can outperform those of the other standard learning techniques, perhaps when the data windows are roughly Gaussian.
- The Naive Bayes learner demonstrates poor performance for time series data probability forecasting. Despite reasonable classification accuracy, the Naive Bayes probability forecasts are unreliable. This is probably because the Naive Bayes learner invalidly assumes independence of attributes in the sliding window data.

We believe there is much scope for further research. For example, one could investigate various commonly used meta-learning techniques such as Boosting and Bagging which primarily improve classification accuracy [5]. In addition, one could investigate newly developed meta-learning techniques such as Defensive Forecasting [13], which makes no assumption about probability distributions, and Probing [14] which has the useful guarantee that improvement in accuracy is matched with an improvement in reliability. Preliminary research into these meta-learners has indicated significant improvement in accuracy and reliability of probability forecasts. It may also be fruitful to investigate further distance metrics (such as Dynamic Time Warping [15]) and different kernels [9], however this additional complexity would not be suitable unless more efficient incremental versions of algorithms such as SVM and Neural Networks were tested [16]. Further, it may be interesting to investigate a possible correlation between the type of time series data and the performance of the learner used for classification. This may identify learners which are best suited for each data type prior to analysis.

We believe that the methodology presented in this study is practically useful for many time series applications, especially where classification accuracy is not a priority. For example, imagine a financial application where the user may act upon a given prediction by deciding when to invest in a commodity. If the user knows that the predictions have been made by a reliable and reasonably accurate forecaster, they are then in the position to wait for the most ideal situation to invest, i.e. when a desired prediction has a highly emphatic probability forecast.

Acknowledgements

We would like to acknowledge Alex Gammerman, Fionn Murtagh and Volodya Vovk for their helpful suggestions, and in reviewing our work. We would also thank Jean-Marc Francois and Bibi-Rehana Haniff for their advice on Hidden Markov Models and Eamonn Keogh for his expert advice as well as providing the data. Special thanks is given to the support staff at RH: Eddie Howson, Bob Vickers and Adrian Thomas for organising dedicated server time for our experiments. This work was supported by EPSRC (grants GR46670, GR/P01311/01), BBSRC (grant B111/B1014428), and MRC (grant S505/65).

References

1. Lindsay, D., Cox, S.: Improving the Reliability of Decision Tree and Naive Bayes Learners. In: Proc. of the 4th ICDM, IEEE (2004) 459–462
2. Zadrozny, B., Elkan, C.: Transforming Classifier Scores into Accurate Multiclass Probability Estimates. In: Proc. of the 8th ACM SIGKDD, ACM Press (2002) 694–699
3. Dawid, A.P.: Calibration-based empirical probability (with discussion). Annals of Statistics 13 (1985) 1251–1285
4. Murphy, A.H.: A New Vector Partition of the Probability Score. Journal of Applied Meteorology 12 (1973) 595–600
5. Witten, I., Frank, E.: Data Mining - Practical Machine Learning Tools and Techniques with Java Implementations. Morgan Kaufmann, San Francisco (2000)
6. Provost, F., Fawcett, T.: Analysis and Visualisation of Classifier Performance: Comparision Under Imprecise Class and Cost Distributions. In: Proc. of the 3rd ICKDD, AAAI Press (1997) 43–48
7. Fayyad, U., Irani, K.: The attribute selection problem in decision tree generation. In: Proc. of 10th Nat. Conf. on Artificial Intelligence, AAAI Press (1992) 104–110
8. Atkeson, C.G., Moore, A.W., Schaal, S.: Locally Weighted Learning. Artificial Intelligence Review 11 (1997) 11–73
9. Smola, A., Bartlett, P., Schölkopf, B., Schuurmans, C.: Advances in Large Margin Classifiers. MIT Press, Cambridge, MA (1999)
10. Vovk, V., Gammerman, A., Shafer, G.: The Analysis of Time Series: An Introduction. 4th edn. Chapman and Hall, London (1989)
11. Keogh, E., Kasetty, S.: On the Need for Time Series Data Mining Benchmarks: A Survey and Empirical Demonstration. In: Proc. of the 8th ACM SIGKDD, ACM Press (2002) 102–111
12. Hull, J.C.: Options, Futures, and other Derivatives. Fifth edn. Prentice-Hall, Upper Saddle River, NJ (2002)
13. Vovk, V., Takemura, A., Shafer, G.: Defensive Forecasting. In: Proc. of 10th International Workshop on Artificial Intelligence and Statistics, Elecronic publication (2005)
14. Langford, J., Zadronzy, B.: Estimating Class Membership Probabilities Using Classifier Learners. In: Proc. of 10th International Workshop on Artificial Intelligence and Statistics, Elecronic publication (2005)
15. Vlachos, M., Hadjieleftheriou, M., Gunopulos, D., Keogh, E.: Indexing Multi-Dimensional Time-Series with Support for Multiple Distance Measures. In: Proc. of the 9th ACM SIGKDD, ACM Press (2003) 216–225
16. Syed, N., Liu, H., Sung, K.: Incremental Learning with Support Vector Machines. In: Proc. of Workshop on Support Machines at IJCAI-99, Elecronic publication (1999)

A Continuous Weighted Low-Rank Approximation for Collaborative Filtering Problems

Nicoletta Del Buono[1] and Tiziano Politi[2]

[1] Dipartimento di Matematica, Università degli Studi di Bari,
Via E. Orabona 4, I-70125 Bari, Italy
delbuono@dm.uniba.it
[2] Dipartimento di Matematica, Politecnico di Bari,
Via Amendola 126/B, I-70126 Bari, Italy
politi@poliba.it

Abstract. Collaborative filtering is a recent technique that recommends products to customers using other users' preference data. The performance of a collaborative filtering system generally degrades when the number of customers and products increases, hence the dimensionality of filtering database needs to be reduced. In this paper, we discuss the use of weighted low rank matrix approximation to reduce the dimensionality of a partially known dataset in a collaborative filtering system. Particularly, we introduce a projected gradient flow approach to compute a weighted low rank approximation of the dataset matrix.

1 Introduction

Collaborative filtering represents a fundamental issue in the context of data mining and knowledge discovery from data. Techniques related to this field of research (also referred to as social information filtering or recommender system designing) encourage the employment of partial information, collected about a group of users, to predict the unknown tendencies of different users [1,6].

Recommender systems are largely spread in E-Commerce web-sites [7,11]. Particular recommender systems, for instance, are employed to suggest books to customers, based on the pieces of information about books that other customers purchase or like. Another specific recommender system helps customers in the choice of a CD as gift, based on other CDs the recipient has appreciated in the past. More generally, collaborative filtering (CF) can be reviewed as a mechanism for analysing an incomplete dataset (organised into a particular matrix, which will be referred to as rating matrix), in the attempt to determine the values of missing data. Predictions of unknown entries can be performed on the basis of a-priori knowledge or by adapting to the information collected by the passing of time. In the first case the system will rely on the similarity degree which can be assessed among the users' profiles, with the analysis of the answers furnished to a number of preference questionnaires. On the other hand, predictions can be

S. Singh et al. (Eds.): ICAPR 2005, LNCS 3686, pp. 45–53, 2005.

grounded by assuming that whenever two users rate items in a similar way, they share common preferences and, therefore, they will prize new items likewise [6].

Nevertheless, two main problems undermine the performance of a collaborative filtering system: (i) the dimensionality of the filtering dataset, rising with the increasing number of customers and products, (ii) the sparseness of the matrix dataset. It should be noted that the term sparseness is not used here in the matrix linear algebra sense, but it indicates that many entries in the rating matrix are unknown, since the items are not rated by all users.

Singular Value Decomposition (SVD) is a powerful tool, frequently used in the more general context of Latent Semantic Indexing to reduce the dimensionality of the rating matrix and to identify latent factors in the data [2,4,12]. The employment of the SVD technique has been suggested in [3] and [6] for computing a low rank matrix approximation to be used as a pre-processing step in a recommender system. Particularly, the CF problem is treated as a classification task (with two classes) and the original rating matrix is discretised into a binary matrix, whose dimensionality is reduced by the SVD. This pre-processing stage is then used to obtain suitable input data for a feed-forward neural network [3] or a clustering algorithm [6], in order to learn the preferences and to predict rate values for new items. In a web mining context, SVD is adopted to reduce the dimensionality of a Boolean matrix, to recommend web pages [8]. In [10], low rank matrix approximation is computed via SVD, using the MovieLens dataset, after a pre-processing step where missing values are firstly replaced by the average values of available users' ratings of movies. A further application of SVD can be found in [9], where a SVD-based collaborative filtering is applied to ensure users' privacy and to provide accurate predictions in a E-Commerce application.

In this paper, we propose a peculiar approach, based on a weighted low rank approximation of the dataset matrix, to solve the two previously delineated problems connected to a recommender system. Particularly, in order to deal with missing entries in the dataset matrix, we suggest the use of a weight matrix with the employment of two-values weights for referring to each observed and unobserved entry. Moreover, to address the dimensionality issue, we adopt a projected gradient flow approach to compute a continuous weighted low rank approximation. The paper is organized as follows. In the next section, we formalise the problem in a linear algebra context, briefly reviewing some concepts connected with the singular value decomposition, and introducing the weighted low rank approximation associated to a CF problem. Then, we suggest the employment of the projected gradient flow system to obtain a solution for the weighted low rank problem. In section 4, we report the obtained results related to a particular numerical example. Finally, some conclusive remarks and guidelines for future work are sketched in section 5.

2   Weighted Low Rank Approximation of the Rating Matrix

A collaborative filtering problem can be modelled within the framework of a general problem of matrix approximation as follows. Let us represent an initial dataset as a matrix $A \in \mathbb{R}^{n \times m}$, with rows corresponding to users, columns referring to items and each element a_{ij} indicating the rating assigned by the user i to the item j. The problem consists in deriving an approximation matrix A_k, with reduced rank k, which can be used to properly predict missing entries in A. Low rank matrix approximations of $A \in \mathbb{R}^{n \times m}$, with respect to the Frobenius and the Euclidean norm, can be easily obtained using the Singular Value Decomposition (SVD):

$$A = USV^{\top} = \sum_{i=1}^{r} \sigma_i \mathbf{u}_i \mathbf{v}_i^{\top},$$

where $U \in \mathbb{R}^{n \times n}$, $V \in \mathbb{R}^{m \times m}$ are orthogonal matrices ($U^{\top}U = I_n$ and $V^{\top}V = I_m$, being I_p the identity matrix of dimension p), $S \in \mathbb{R}^{n \times m}$ is a diagonal matrix with nonnegative entries $\sigma_1 \geq \sigma_2 \geq \ldots \geq \sigma_r > 0$ and r is the rank of A ($r \leq \min\{m, n\}$). For any k between 1 and r we can formalise the low rank approximation problem as:

$$\min_{\text{rank}(X)=k} \|A - X\| = \|A - A_k\| \tag{1}$$

where $A_k = \sum_{i=1}^{k} \sigma_i \mathbf{u}_i \mathbf{v}_i^{\top}$ is the $k-$th truncated SVD of A and $\| \cdot \|$ denotes the Euclidean or the Frobenius norm. Even if the SVD plays a relevant role in solving (1), it requires to deal with a complete target matrix A. Unfortunately, this scenario seldom occurs in the collaborative filtering context, where the majority of the elements in A are unobserved, since each user typically rates only a reduced subset of the ensemble of items. In order to reduce the dimensionality of the rating matrix A, by computing its best rank k approximation, we proceed by reformulating the CF problem as a *weighted* low rank approximation task. Particularly, we define the nonempty subset of indexes

$$\emptyset \neq N \subset \{(i,j) \in \mathbb{N} \times \mathbb{N} | 1 \leq i \leq m, 1 \leq j \leq n\},$$

such that, for any $(i,j) \in N$, the corresponding element a_{ij} of the rating matrix A is known. Additionally, we define the following pseudonorm:

$$\|A\|_* = \sum_{(i,j) \in N} a_{ij}^2. \tag{2}$$

The problem (1) is equivalent to finding a rank k matrix X_k, such that

$$\|A - X_k\|_* = \min_{\text{rank}(X)=k} \|A - X\|_*. \tag{3}$$

A weight matrix $W \in \mathbb{R}^{m \times n}$ can be adopted to weight the matrix A by means of the coefficient values 1 and 0, corresponding to the known and unknown elements of A, respectively:

$$w_{ij} = \begin{cases} 1 & (i,j) \in N \\ 0 & (i,j) \notin N \end{cases} \tag{4}$$

Consequently, it can be easily proved that (2) is equivalent to: $\|A\|_* = \|W \circ A\|_F$ and (3) corresponds to:

$$\|A - X_k\|_* = \|W \circ (A - X_k)\|_F = \min_{\text{rank}(X)=k} \|W \circ (A - X)\|_F,$$

where $W \circ B$ denotes the entrywise multiplication of W with B.

The weighted low rank approximation task associated to a CF problem can be summarised as follows. Given the rating matrix $A \in \mathbb{R}^{n \times m}$ (describing the CF problem) and the non negative weight matrix $W \in \mathbb{R}_+^{n \times m}$ defined in (4), find the rank k matrix X_k, that minimises the weighted Frobenius distance:

$$\min_{\text{rank}(X)=k} \|W \circ (A - X)\|_F = \|W \circ (A - X_k)\|_F \tag{5}$$

Therefore, we will refer to the matrix X_k as the weighted rank k approximation of A.

3   The Gradient Flow Approach

Once the CF problem has been reformulated as a weighted low rank approximation task, we can compute the matrix X_k by adopting the projected gradient flow approach, whose limiting solutions stand as an approximation of X_k. Here we briefly summarise the main steps of the gradient flow technique, addressing the reader to [5] for a complete description of the overall approach. By representing the matrix X in (5) via the parameters (U, S, V) of its singular value decomposition, we have to minimise the following functional:

$$F(U, S, V) := \langle W \circ (A - USV^\top), W \circ (A - USV^\top) \rangle, \tag{6}$$

subject to the conditions that $U \in \mathcal{St}_{n,k}$, $S \in \mathbb{R}^k$, and $V \in \mathcal{St}_{m,k}$, where $\mathcal{St}_{p,k}$ (for $p = n, m$) denotes the Stiefel manifold (the group of all real matrices of dimension $p \times k$ with orthonormal columns), and $\langle \cdot, \cdot \rangle$ indicates the Frobenius inner product on matrices[1]. The main steps of the projected gradient flow technique are:

1. Compute the gradient of the objective function F in the space $\mathbb{R}^{n \times k} \times \mathbb{R}^k \times \mathbb{R}^{m \times k}$, that is $\nabla F(U, S, V) = (\frac{\partial F}{\partial U}, \frac{\partial F}{\partial S}, \frac{\partial F}{\partial V})$, where

$$\frac{\partial F}{\partial U} = (W \circ (A - USV^\top))VS, \quad \frac{\partial F}{\partial V} = (W^\top \circ (A^\top - VSU^\top))US$$

$$\frac{\partial F}{\partial S} = \text{diag}_k\left(U^\top(W \circ (A - USV^\top))V\right);$$

2. Evaluate the projections $\mathcal{P}_{\mathcal{St}_{n,k}}(\frac{\partial F}{\partial U})$ onto the tangent space of $\mathcal{St}_{n,k}$ and $\mathcal{P}_{\mathcal{St}_{m,k}}(\frac{\partial F}{\partial V})$ onto the tangent space of $\mathcal{St}_{m,k}$;

---

[1] We use the notation $S \in \mathbb{R}^k$ for the diagonal matrix in $\mathbb{R}^{n \times m}$ whose first k diagonal entries correspond to the values in S.

3. Solve the dynamical system:

$$\frac{dU}{dt} = -\mathcal{P}_{St(n,k)}\left(\frac{\partial F}{\partial U}\right), \quad \frac{dS}{dt} = -\frac{\partial F}{\partial S}, \quad \frac{dV}{dt} = -\mathcal{P}_{St(m,k)}\left(\frac{\partial F}{\partial V}\right). \quad (7)$$

At the end of the overall process, the approximation X_k is equal to the product of the limiting solutions of (7), that is: $X_k = U_\infty S_\infty V_\infty^\top$.

When the limiting solution of (7) has been reached, the two resultant matrices

$$U_\infty S_\infty^{1/2} \qquad S_\infty^{1/2} V_\infty^\top$$

can be used to compute the recommendation score for any user i and item j. It should be pointed out that the dimension of $U_\infty S_\infty^{1/2}$ is $n \times k$ and the dimension of $S_\infty^{1/2} V_\infty^\top$ is $k \times m$. Hence, to compute the prediction we simply need to evaluate the dot product of the i-th row of $U_\infty S_\infty^{1/2}$ and the j-th column of $S_\infty^{1/2} V_\infty^\top$.

4   Numerical Examples

The aim of this section is to present a twofold evaluation of the proposed continuous weighted low rank approximation technique (CWLRA). On the one hand, we are going to demonstrate the effectiveness of our approach by considering a matrix reconstruction example. On the other hand, we will report the results of the experimental evaluation of our technique when applied on a well known collaborative filtering task.

Matrix reconstruction example. Assigned a matrix $A \in \mathbb{R}^{n \times m}$, with $n = 100$ and $m = 25$, we want to find its best weighted low rank-k approximation, exploiting a weight matrix W, with growing number of entries equal to 0 (i.e., when the cardinality of N is increased).

We solved the dynamical system in (7) using the `ode113` Matlab solver; the projections on the Stiefel manifolds have been obtained by the modified Gram-Schmidt algorithm. Figure 1 plots the errors on the data matrix. In the figure, the solid line indicates the maximum error on the unknown entries of A, evaluated by $\max_{(i,j) \in N} |a_{ij} - x_{ij}|$. The dashed line shows the error behaviour on the known entries of A, computed by $\max_{(i,j) \in N^c} |a_{ij} - x_{ij}|$, being N^c the complementary set of N, with respect to the set $K = \{(i,j)|1 \leq i \leq n, 1 \leq j \leq k\}$). Finally, the dash-dotted line points out the maximum modulus of a selected element of the matrix $A - W \circ X$.

It can be observed that, even if the error values appear to oscillate at the beginning of the integration, when $|N| \geq 6$ the error on known data converges to zero. This peculiar behaviour demonstrates that the proposed approach is able to predict known data with a good degree of accuracy. Moreover, the error on unknown data tends to stabilise, while the global error reaches a limit point independently from the number of missing data. This characteristic behaviour occurred also in a number of additional experimental simulations we have carried on. (which have not been reported here, due to limited space).

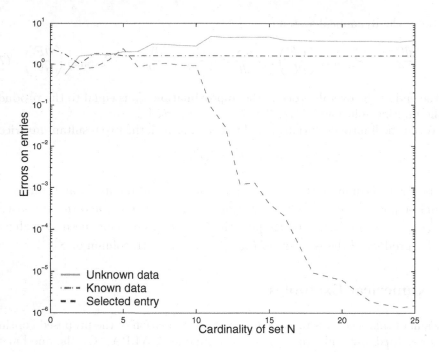

Fig. 1. Error values obtained by employing the computed weighted low rank approximation X_k

Collaborative filtering task. We illustrate the behaviour of the proposed approach to deal with a well known benchmark dataset for CF: the Jester dataset [6]. The Jester dataset collects the ratings (continuous values in the interval [-10,10]) that users assigned at some jokes of a set of one hundred jokes (the average number of ratings per user is equals to 46). We analysed a subset of 1000 Jester data, randomly selected, containing the core set of 10 jokes (rated by all users) and extended with other 10 jokes.

We compared the results obtained using the continuous weighted low rank approximation (CWLRA) technique with those obtained using a SVD factorization of the data matrix, where unobserved values were replaced with zeros [12]. We use the Normalized Mean Absolute Error (NMAE) for comparing the numerical recommendation scores against the actual user ratings for the user-rate pairs in the dataset. Denoting by x_{ij} the prediction for how the user i will rate the item j, the MAE for the user i is

$$MAE_i = \frac{1}{N^c} \sum_{j=1}^{N^c} |a_{ij} - x_{ij}|.$$

The MAE for the entire set of users is the mean MAE on the total number of users (i.e., $MAE = 1/n \sum_{i=1}^{n} MAE_i$) and the Normalized Mean Absolute Error is

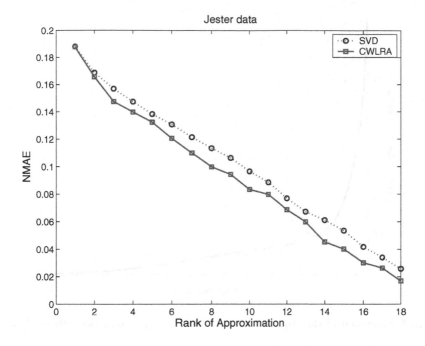

Fig. 2. NMAE as a function of the rank of the approximation

$$NMAE = \frac{MAE}{\max_{ij} a_{ij} - \min_{ij} a_{ij}}.$$

Figure 2 shows the behaviour of the NMAE when the rank of the approximation matrix increases for both the SVD technique (dotted line) and the CWLRA
method (solid line). It should be noted that the results indicate that the CWLRA
method performs better than the computation of the truncated SVD on a modified (filled with zeros) data matrix. Figure 3 illustrates the behaviour of the
objective function (6) during the numerical solution of Jester CF problem for
fixed rank $K = 8$; a similar behaviour occurred all values of rank of the approximation matrix.

5 Conclusive Remarks

In this paper we have proposed a continuous method to compute a weighted low
rank approximation of a rating matrix, in the context of a collaborative filtering
problem. Actually, the employed technique significantly differs from other kinds
of approaches present in literature. In fact, the majority of the suggested mechanisms commonly deal with missing data by replacing them with zero values
[1] or average values [6,10]. Successively, the computation of the truncated SVD
is performed with a priori fixed rank, in order to reduce the problem dimensionality. Our approach, instead of modifying the matrix dataset, introduces a

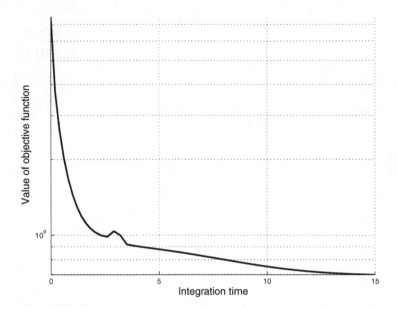

Fig. 3. Behaviour of the objective function for the approximation with $rank = 8$

weight matrix to deal with missing values. Moreover, bearing in mind that we adopted a continuous technique, based on the solution of a dynamical system, additional feature information can be obtained for the approximation matrix: this represents a clear advantage when comparing our approach with other standard methods.

The obtained results are encouraging for further investigations; the application of the developed technique over real-world datasets represents one of the topics for our ongoing research activity. Furthermore, future work could be addressed to extend the proposed approach in order to obtain weighted low rank approximation of a data matrix possessing normalised data (namely, with rows or columns possessing unitary 2-norm).

References

1. Azar, Y., Fiat, A., Karlin A.R., McSherry, F., Saia, J.: Spectral analysis of data. Proceedings of the 33th ACM Symposium on Theory of Computing, 2001.
2. Berry, M.W., Dumais, S.T., O'Brian, G.W.: Using linear algebra for intelligent Information Retrieval. SIAM Review, 37(4), 573-595, 1995.
3. Billsus, D., Pazzani, M.J.: Learning collaborative information filters. Proceedings of 15th International conference on Machine Learning, 1998.
4. Chu, M.T., Del Buono, N., Lopez, L., Politi, T.: On the Low Rank Approximation of Data on the Unit Sphere. To appear on SIAM Journal on Matrix Analysis, 2004.
5. Del Buono, N., Politi, T.: A continuous technique for the Weighted Low-Rank Approximation Problem, Lecture Notes in Computer Science Series (Springer-Verlag), 3044 Vol. II, 988-997, 2004.

6. Goldberg, K., Roeder, T. Gupta, D., Perkins, C.: Eigentaste: a constant time collaborative filtering algorithm. Information Retrieval, 4, 133-151, 2001.
7. Linden, G., Smith, B., York, J.: Amazon.com Recommendations - Item to Item Collaborative Filtering. IEEE Internet Computing, 76–80 (2003).
8. Pryor, M.: The effects of singular value decomposition on collaborative filtering. Dartmounth College CS, Technical Report, PCS-TR98-338, 1998.
9. Polat, H., Du, W.: SVD-based collaborative filtering with privacy. To appear on Proceedings of SAC05, 2005.
10. Sarwar, B.M., Karypis, G., Konstan, J.A., Riedl, J.T.: Application of dimensionality reduction in recommender system- A case study.
11. Schafer, J. B., Konstan, J., Riedl, J.: Recommender systems in E-Commerce. Proceedings of ACM E-Commerce 1999 Conference (1999).
12. Srebro, N., Jaakkola, T.: Weighted Low-Rank Approximations. Proceedings of Twentieth International Conference on Machine Learning (ICML-2003), Washington DC (2003).

GP Ensemble for Distributed Intrusion Detection Systems

Gianluigi Folino, Clara Pizzuti, and Giandomenico Spezzano

ICAR-CNR, Via P.Bucci 41/C, Univ. della Calabria,
87036 Rende (CS), Italy
{folino, pizzuti, spezzano}@icar.cnr.it

Abstract. In this paper an intrusion detection algorithm based on GP ensembles is proposed. The algorithm runs on a distributed hybrid multi-island model-based environment to monitor security-related activity within a network. Each island contains a cellular genetic program whose aim is to generate a decision-tree predictor, trained on the local data stored in the node. Every genetic program operates cooperatively, yet independently by the others, by taking advantage of the cellular model to exchange the outmost individuals of the population. After the classifiers are computed, they are collected to form the GP ensemble. Experiments on the KDD Cup 1999 Data show the validity of the approach.

1 Introduction

The extensive use of Internet and computer networks, besides the known advantages of information exchange, has provided also an augmented risk of disruption of data contained in information systems. In the recent past, several cyber attacks have corrupted data of many organizations creating them serious problems. The availability of *Intrusion Detection Systems(IDS)*, able to automatically scan network activity and to recognize intrusion attacks, is very important to protect computers against such unauthorized uses and make them secure and resistant to intruders. The task of an IDS is, in fact, to identify computing or network activity that is malicious or unauthorized. Most current Intrusion Detection Systems collect data from distributed nodes in the network and then analyze them centrally. The first problem of this approach is a security and privacy problem due to the necessity of transferring data. Moreover, if the central server becomes the objective of an attack, the whole network security is quickly compromised.

In this paper an intrusion detection algorithm (*GEdIDS, Genetic programming Ensemble for Distributed Intrusion Detection Systems*), based on the ensemble paradigm, that employs a Genetic Programming-based classifier as component learner for a distributed Intrusion Detection System (*dIDS*), is proposed.

To run genetic programs a distributed environment, based on a hybrid multi-island model [2] that combines the island model with the cellular model, is used. Each node of the network is considered as an island that contains a learning component, based on cellular genetic programming, whose aim is to generate a

S. Singh et al. (Eds.): ICAPR 2005, LNCS 3686, pp. 54–62, 2005.

decision-tree predictor trained on the local data stored in the node. Every genetic program, however, though isolated, cooperates with the neighboring nodes by collaborating with the other learning components located on the network and by taking advantage of the cellular model to exchange the outmost individuals of the population.

A learning component employs the ensemble method AdaBoost.M2 [7] thus it evolves a population of individuals for a number of rounds, where a round is a fixed number of generations. Every round the islands import the remote classifiers from the other islands and combine them with their own local classifier. Finally, once the classifiers are computed, they are collected to form the GP ensemble. In the distributed architecture proposed, each island thus operates cooperatively yet independently by the others, providing for efficiency and distribution of resources. This architecture gives significant advantages in scalability, flexibility, and extensibility. To evaluate the system proposed, experiments using the network records of the KDD Cup 1999 Data [1] have been performed. Experiments on this data set point out the capability of genetic programming in successfully dealing with the problem of distributed intrusion detection.

Genetic Programming for intrusion detection has not been explored very much. Some proposals can be found in [11,9,8].

The paper is organized as follows. The next section introduces the genetic programming based ensemble paradigm for distributed IDS and describes the distributed programming environment proposed. Section 3 presents the results of experiments.

2   GP Ensembles for dIDS

Ensemble is a learning paradigm where multiple component learners, also called classifiers or predictors, are trained for a same task by a learning algorithm, and then combined together for dealing with new unseen instances. Ensemble techniques have been shown to be more accurate than component learners constituting the ensemble [4,10], thus such a paradigm has become a hot topic in recent years and has already been successfully applied in many application fields.

In this paper such a paradigm has been adopted for modelling distributed intrusion detection systems and the suitability of genetic programming (GP) as component learner has been investigated. The approach is based on the use of cooperative GP-based learning programs that compute intrusion detection models over data stored locally at a site, and then integrate them by applying a majority voting algorithm. The models are built by using the local audit data generated on each node by, for example, operating systems, applications, or network devices so that each ensemble member is trained on a different training set.

The GP classifiers cooperate using a multi-island model to produce the ensemble members. Each node is an island and contains a GP-based learning component extended with the boosting algorithm AdaBoost.M2 [7] whose task is to build a decision tree classifier by collaborating with the other learning components located on the network. Each learning component evolves its population for

Given a network constituted by P nodes, each having a data set S_j
For j = 1, 2, ..., P (for each island in parallel)
    Initialize the weights associated with each tuple
    Initialize the population Q_j with random individuals
end parallel for
For t = 1,2,3, ..., T (boosting rounds)
    For j = 1, 2, ..., P (for each island in parallel)
        Train cGP on S_j using a weighted fitness
        according to the weight distribution
        Compute a weak hypothesis
    end parallel for
    Exchange the hypotheses among the P islands
    Update the weights
end for t

    Output the hypothesis

Fig. 1. The GEdIDS algorithm using AdaBoost.M2

a fixed number of iterations and computes its classifier by operating on the local data. Each island may then import (remote) classifiers from the other islands and combine them with its own local classifiers to form the GP ensemble.

In order to run GP ensembles a distributed computing environment is required. We use dCAGE (distributed Cellular Genetic Programming System) a distributed environment to run genetic programs by a multi-island model, which is an extension of [6]. An hybrid variation of the classic multi-island model, that combines the island model with the cellular model, has been implemented.

The Island model is based on subpopulations, that are created by dividing the original population into disjunctive subsets of individuals, usually of the same size. Each subpopulation can be assigned to one processor and a standard (panmictic) GP algorithm is executed on it. Occasionally, migration process between subpopulations is carried out after a fixed number of generations. The hybrid model modifies the island model by substituting the standard GP algorithm with a cellular GP (cGP) algorithm [6]. In the cellular model each individual has a spatial location, a small neighborhood and interacts only within its neighborhood. The main difference in a cellular GP, with respect to a panmictic algorithm, is its decentralized selection mechanism and the genetic operators (crossover, mutation) adopted.

In dCAGE, to take advantage of the cellular model of GP, the cellular islands are evolved independently, and the outmost individuals are asynchronously exchanged so that all islands can be thought as portions of a single population. dCAGE distributes the evolutionary processes (islands) that implement the detection models over the network nodes using a configuration file that contains the configuration of the distributed system. dCAGE implements the hybrid model as a collection of cooperative autonomous islands running on the various hosts within an heterogeneous network that works as a peer-to-peer system. Each island, employed as a peer IDS, is identical to each other except one that is supposed to be the *collector* island. The collector island is in charge of collecting the GP classifiers from the other nodes, handling the fusion of results on behalf of the other peer IDS, and to redistribute the GP ensemble for future predictions to all network nodes. Respect to the other islands, it has some more duties for administration. The configuration of the structure of the processors is based on a ring topology.

The pseudo-code of the algorithm is shown in figure 1. Each island is furnished with a cGP algorithm enhanced with the boosting technique AdaBoost.M2, a population initialized with random individuals, and operates on the local audit data weighted according to a uniform distribution. The selection rule, the replacement rule and the asynchronous migration strategy are specified in the cGP algorithm. Each peer island generates the GP classifier iterating for a certain number of iterations necessary to compute the number of boosting rounds. During the boosting rounds, each classifier maintains the local vector of the weights that directly reflect the prediction accuracy on that site. At each boosting round the hypotheses generated by each classifier are stored and combined in the collector island to produce the ensemble of predictors. Then the ensemble is broadcasted to each island to locally recalculate the new vector of the weights. After the execution of the fixed number of boosting rounds, the classifiers are used to evaluate the accuracy of the classification algorithm for intrusion detection on the test set.

Genetic programming is used to inductively generate a GP classifier as a decision trees for the task of data classification. Decision trees, in fact, can be interpreted as composition of functions where the function set is the set of attribute tests and the terminal set are the classes. The function set can be obtained by converting each attribute into an attribute-test function. For each attribute A, if $A_1, \ldots A_n$ are the possible values A can assume, the corresponding attribute-test function f_A has arity n and if the value of A is A_i then $f_A(A_1, \ldots A_n) = A_i$. When a tuple has to be evaluated, the function at the root of the tree tests the corresponding attribute and then executes the argument that outcomes from the test. If the argument is a terminal, then the class name for that tuple is returned, otherwise the new function is executed. The fitness is the number of training examples classified in the correct class. The *Cellular genetic programming* algorithm (*cGP*) for data classification has been proposed in [5] and it is described in figure 2. At the beginning, for each cell, the fitness of each individual is evaluated. Then, at each generation, every tree undergoes one of the genetic operators (reproduction, crossover, mutation) depending on the probability test. If crossover is applied, the mate of the current individual is selected as the neighbor having the best fitness, and the offspring is generated. The current tree is then replaced by the best of the two offspring if the fitness of the latter

```
Let p_c, p_m be crossover and mutation probability
for each point i in the population do in parallel
    evaluate the fitness of t_i
end parallel for
while not MaxNumberOfGeneration do
    for each point i in the population do in parallel
        generate a random probability p
        if (p < p_c)
            select the cell j, in the neighborhood of i,
            such that t_j has the best fitness
            produce the offspring by crossing t_i and t_j
            evaluate the fitness of the offspring
            replace t_i with the best of the two offspring
            if its fitness is better than that of t_i
        else
        if ( p < p_m + p_c) then
            mutate the individual
            evaluate the fitness of the new t_i
        else
            copy the current individual in the population
        end if
        end if
    end parallel for
end while
```

Fig. 2. The algorithm cGP

is better than that of the former. The evaluation of the fitness of each classifier is calculated on the entire training data. After the execution of the number of generations defined by the user, the individual with the best fitness represents the classifier.

3   System Evaluation and Results

3.1   Data Sets Description

Experiments over the KDD Cup 1999 Data set [1] have been performed. This data set comes from the 1998 DARPA Intrusion Detection Evaluation Data and contains a training data consisting of 7 weeks of network-based intrusions inserted in the normal data, and 2 weeks of network-based intrusions and normal data for a total of 4,999,000 connection records described by 41 characteristics. The main categories of intrusions are four: Dos (Denial Of Service), R2L (unauthorized access from a remote machine), U2R (unauthorized access to a local superuser privileges by a local unprivileged user), PROBING (surveillance and

Table 1. Class distribution for training and test data for KDDCUP 99 dataset

	Normal	Probe	DoS	U2R	R2L	Total
Train	97277	4107	391458	52	1126	494020
Test	60593	4166	229853	228	16189	311029

probing). However a smaller data set consisting of the 10% the overall data set is generally used to evaluate algorithm performance. In this case the training set consists of 494,020 records among which 97,277 are normal connection records, while the test set contains 311,029 records among which 60,593 are normal connection records. Table 1 shows the distribution of each intrusion type in the training and the test set.

3.2   Performance Measures

To evaluate our system, besides the classical accuracy measure, the two standard metrics of *detection rate* and *false positive rate* developed for network intrusions, have been used. Table 2 shows these standard metrics. Detection rate is computed as the ratio between the number of correctly detected intrusions and the total number of intrusions, that is $DR = \frac{\#TruePositive}{\#FalseNegative+\#TruePositive}$. False positive (also said false alarm) rate is computed as the ratio between the number of normal connections that are incorrectly classifies as intrusions and the total number of normal connections, that is $FP = \frac{\#FalsePositive}{\#TrueNegative+\#FalsePositive}$. These metrics are important because they measure the percentage of intrusions the system is able to detect and how many misclassifications it makes. To visualize the trade-off between the false positive and the detection rates, the ROC (Receiving Operating Characteristic) curves are also depicted. Furthermore, to compare classifiers it is common to compute the area under the ROC curve, denoted as AUC [3]. The higher the area, the better is the average performance of the classifier.

Table 2. Standard metrics to evaluate intrusions

		Predicted label	
		Normal	Intrusions
Actual	Normal	True Negative	False Positive
Class	Intrusions	False Negative	True Positive

3.3   Experimental Setup

The experiments were performed by assuming a network composed by 10 dual-processor 1,133 Ghz Pentium III nodes having 2 Gbytes of memory. Both the training set of 499,467 tuples and the test set of 311029 tuples have been equally partitioned among the 10 nodes by picking them at random. Each node thus

Table 3. Main parameters used in the experiments

Name	Value
max_depth_for_new_trees	6
max_depth_after_crossover	17
max_mutant_depth	2
grow_method	RAMPED
selection_method	GROW
crossover_func_pt_fraction	0.7
crossover_any_pt_fraction	0.1
fitness_prop_repro_fraction	0.1
parsimony_factor	0

contains 1/10 of the instances for each class. On each node we run *AdaBoost.M2* as base GP classifier with a population of 100 elements for 10 rounds, each round consisting of 100 generations. The GP parameters used where the same for each node and they are shown in table 3. All the experiments have been obtained by running the algorithm 10 times and by averaging the results. Each ensemble has been trained on the train set and then evaluated on the test set.

3.4 Results and Comparison with Other Approaches

The results of our experiments are summarized in table 4, where the confusion matrix obtained on the test set by averaging the 10 confusion matrices coming from 10 different executions of *GEdIDS* is showed. The table points out that the prediction is worse on the two classes U2R and R2L. For this two classes, however, there is a discrepancy between the number of instances used to train each classifier on every node and the number of instances to classify in the test set. Table 5 compares our approach with the first and second winner of the KDD-99 CUP competition and the linear genetic programming approach proposed by Song et al. [11]. The table shows the values of the standard metrics described

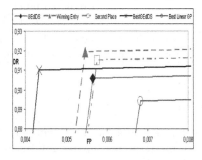

Fig. 3. ROC curves

Table 4. Confusion matrix (averaged over all tries) for dCage. Rows show the true class, columns the predicted ones.

	Normal	Probe	DoS	U2R	R2L
Normal	60250.8	200.2	110.8	15.4	15.8
Probe	832.8	2998.4	263.6	26.4	44.8
DoS	7464.2	465.0	221874.8	19.2	29.8
U2R	139.6	45.2	17.2	11.8	14.2
R2L	15151.4	48.6	232.4	173.8	582.8

Table 5. Comparison with kdd-99 cup winners and other approaches

Algorithm	Detection Rate	FP Rate	ROC Area
Winning Entry	0.919445	0.005462	0,956991
Second Place	0.915252	0.005760	0,954746
Best Linear GP - FP Rate	0.894096	0.006818	0,943639
Avg GEdIDS	0.905812	0.005648	0.950082
Best GEdIDS - FP Rate	0.910165	0.004340	0.952912

above. In particular we show the detection rate, the false positive rate, and the ROC area of these three approaches and those obtained by *GEdIDS*. For the latter we show both the average values of the 10 executions and the best value with respect to the false positive rate. From the table we can observe that the performance of *GEdIDS* is comparable with that of the two winning entries and better than Linear GP. In fact, the average and best *GEdIDS* detection rates are 0.905812 and 0.910165, respectively, while those of the first two winners are 0.919445 and 0.915252. As regard the false positive rate the average value of *GEdIDS* 0.005648 is lower than the second entry, while the best value obtained 0.004340 is lower than both the first and second entries. Thus the solutions found by *GEdIDS* are very near to the winning entries, and, in any case, overcome those obtained with linear GP. These experiments emphasizes the capability of genetic programming to deal with this kind of problem. Finally figure 3 shows an enlargement of the ROC curves of the methods listed in table 5 and better highlights the results of our approach.

4   Conclusions

A distributed intrusion detection algorithm based on the ensemble paradigm has been proposed and the suitability of genetic programming as a component learner of the ensemble has been investigated. Experimental results show the applicability of the approach for this kind of problems. Future research aims at extending the method when considering not batch data sets but data streams that change online on each node of the network.

Acknowledgments

This work has been partially supported by projects CNR/MIUR legge 449/97-DM 30/10/2000 and FIRB Grid.it (RBNE01KNFP).

References

1. The third international knowledge discovery and data mining tools competition dataset kdd99-cup. In *http://kdd.ics.uci.edu/databases/kddcup99/kddcup99.html*, 1999.
2. E. Alba and M. Tomassini. Parallelism and evolutionary algorithms. *IEEE Transaction on Evolutionary Computation*, 6(5):443–462, October 2002.
3. A. P. Bradley. The use of the area under the roc curve in the evaluation of machine learning algorithms. *Pattern Recognition*, 30(7):1145–1159, 1997.
4. Leo Breiman. Bagging predictors. *Machine Learning*, 24(2):123–140, 1996.
5. G. Folino, C. Pizzuti, and G. Spezzano. A cellular genetic programming approach to classification. In *Proc. Of the Genetic and Evolutionary Computation Conference GECCO99*, pages 1015–1020, Orlando, Florida, July 1999. Morgan Kaufmann.
6. G. Folino, C. Pizzuti, and G. Spezzano. A scalable cellular implementation of parallel genetic programming. *IEEE Transaction on Evolutionary Computation*, 7(1):37–53, February 2003.
7. Y. Freund and R. Schapire. Experiments with a new boosting algorithm. In *Proceedings of the 13th Int. Conference on Machine Learning*, pages 148–156, 1996.
8. Wei Lu and Issa Traore. Detecting new forms of network intrusion using genetic programming. In *Proc. of the Congress on Evolutionary Computation CEC'2003*, pages 2165–2173. IEEE Press, 2003.
9. Srinivas Mukkamala, Andrew H. Sung, and Ajith Abraham. Modeling intrusion detection systems using linear genetic programming approach. In *17th International Conference on Industrial and Engineering Applications of Artificial Intelligence and Expert Systems, IEA/AIE 2004*, pages 633–642, Ottawa, Canada, 2004.
10. J. Ross Quinlan. Bagging, boosting, and c4.5. In *Proceedings of the 13th National Conference on Artificial Intelligence AAAI96*, pages 725–730. Mit Press, 1996.
11. D. Song, M.I. Heywood, and A. Nur Zincir-Heywood. A linear genetic programming approach to intrusion detection. In *Proceedings Of the Genetic and Evolutionary Computation Conference GECCO 2003*, pages 2325–2336. LNCS 2724, Springer, 2003.

Clustered Trie Structures for Approximate Search in Hierarchical Objects Collections

R. Giugno, A. Pulvirenti, and D. Reforgiato Recupero

Computer Science Department, University of Catania,
Viale A. Doria, 6, 95125 Catania, Italy
{giugno, apulvirenti, diegoref}@dmi.unict.it

Abstract. Rapid developments in science and engineering are producing a profound effect on the way information is represented. A new problem in pattern recognition has emerged: new data forms such as trees representing XML documents and images cannot been treated efficiently by classical storing and searching methods. In this paper we improve trie-based data structures by adding data mining techniques to speed up range search process. Improvements over the search process are expressed in terms of a lower number of distance calculations. Experiments on real sets of hierarchically represented images and XML documents show the good behavior of our patter recognition method.

1 Introduction

Many applications dealing with XML documents and images describe each object by the hierarchy of its features and the relations among its components [16]. For example, in [9,6] in order to answer queries such as "find all the pages containing the title adjacent to an image", trees describing the geometrical features of document pages are used (figure 1). Trees are also used in connection with applications on images archives [11,1,12], XML documents [18] and natural language processing problems such as information retrieval in digital libraries [9]. An high level tree data model is less sensitive to distortion

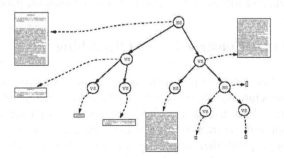

Fig. 1. A MXY-tree [6] describing a document page. Notice that at root level the trie describes the whole page whereas at node level the trie represents the different page paragraphs.

S. Singh et al. (Eds.): ICAPR 2005, LNCS 3686, pp. 63–70, 2005.

and noise problems. Consequently, the continuously emerging searching techniques on trees [2,16] can efficiently treat advanced pattern matching queries on such collections. More precisely, searching in such structures may require finding exact or "approximate" tree and subtree matchings. One way to be precise on approximate searching between trees is by defining distance functions [15,18,16] which assign costs to the operations used to change one tree into the other. Depending on the tree structure, keytree searching problems have complexity ranging from linear (P) to exponential (NP-complete) on the tree size. In order to improve the query processing time, considerable research efforts [10,3,4,7,8,13,16] have arisen to combine several sophisticated data structures with approximate tree matching algorithms that work over generic metric spaces (FQ-tree [3], VP-tree [7], MVP-tree [4], M-tree [8]). Another interesting technique is due to Oflazer [13]. In such model, to retrieve trees whose distance from a given query is below a given threshold, each tree is, in turn, coded as a sequence of paths from its root to every leaf. The sequences of paths describing all the trees are compactly stored in a trie. The trie data structure naturally provides a way to interrogate the database with error tolerant and approximate queries. A strategy is to traverse the trie from the root to the leaves. In this search it is possible to do an early pruning of a branch as soon as it is seen that it leads to trees whose distance from the query exceeds the given threshold. In [10] authors showed that the Oflazer distance is a metric and they speeded up the above searching method using the triangular property of the distance function in connection with a saturation technique. This unfortunately required a quadratic preprocessing time in order to compute all pairwise distances. In this paper we improve the trie-based data structures in [13] by adding a clustering information to speed up the approximate search process. This approach avoids the computation of all pairwise distances and thus it is suitable for applications in which such preprocessing time is not admissible. We compare our proposed method with the method in [13]. Experiments on real data sets of hierarchically represented images and XML documents show the good behavior of our data structure.

The paper is organized as follows. In section 2 we review the Trie data structure construction together with the searching process. In section 3 we describe our new advanced *Clusterd Trie structure* with its advanced searching strategy. In section 4 we present experimental results and comparisons. Section 5 ends the paper.

2   Trie Structure Construction and Searching

We use a *Vertex List Sequence* [13] as a data structure to represent each tree in the collection. This structure is a sequence of lists. There are as many lists in this sequence as leaves in the tree. Each list contains the ordered sequence of vertices in the unique path from the root to the corresponding leaf (see figure 2 for an example of two tries from DB_Stamps [10] with their relative vertex list sequences). The set of vertex list sequences representing the whole database of objects is converted into a trie structure [13,10]. Such a data structure will compress redundancies in the prefixes of the vertex list sequences to achieve a compact data structure.

Following [13,10], the distance between two trees is defined, taking into account structural differences and label differences. Let C be the cost for every different label

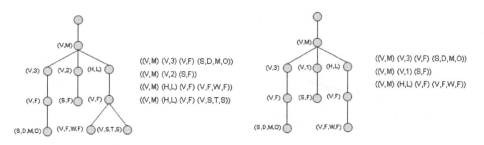

Fig. 2. An example of two tries from DB_Stamps [10] with the associate *Vertex List Sequences*

and S the cost for a structural difference. The distance between two trees is the minimum cost of leaves or branches insertions, deletions, or leaf labels changes necessary to change one tree into the other. The two trees of figure 2, with C=1 and S=2 have a distance of 3 because they differ for a label name ($(V, 2)$ in the first trie and $(V, 1)$ in the second) plus the structural difference of the leaf node labelled (V, S, T, S). The setting for an appropriate value for C and S depends on the given application. Then, given an input trie T, a query object Q and a similarity threshold t we want to retrieve from T all the trees that match Q according to the parameters C and S. Standard searching within a trie corresponds to traversing a path starting from the root node, to one of the leaves, so that the concatenation of the labels on the arcs along this path matches the input query tree. To efficiently perform the search, paths in the trie that lead to no solutions have to be pruned early so that the search is bound to a very small portion of the data structure. The search proceeds, depth first, down the trie, computing the similarity distance between subsequences of the query and the partial sequences obtained chaining together the labels of the nodes of the trie that have been visited so far. Such a distance is formalized in the concept of cutoff distance (see [13,10] for details).

3   Clustered Trie Structure

In our newly proposed method, before constructing the trie, we perform a preliminary clustering of the input objects. We use the metric distance between trees given in [13,10] and the cluster method given in [5]. The information concerning the distribution of the elements in clusters are added in the trie. We call *Clustered Trie Structure* the trie data structure merged with the cluster information. This allows us to slightly modify the trie search process and to improve the performance during the searching. More precisely, in order to cluster, we use the *Antipole Clustering* of bounded radius σ [5,14] which is an indexing scheme designed to support range search queries and nearest neighbor search queries in general metric spaces. The Antipole clustering performs by a recursive top down procedure starting from a given finite set of points S (in our case a set of trees) and checking at each step if a given splitting condition Φ is satisfied. If this is not the case then splitting is not performed and the given subset is a cluster and a centroid having distance approximatively less than σ from every other node in the cluster is computed [14]. Otherwise if Φ is satisfied then a pair of points $\{A, B\}$ of

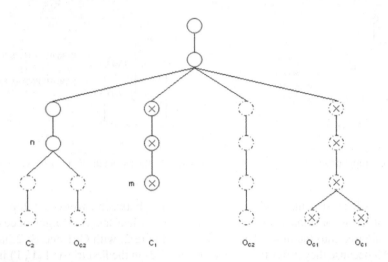

Fig. 3. Advanced Search Method. Case 1 - During the visit of the trie, the paths starting at node n are pruned because the distance between a given query and the partial object (ending at node n) exceeds the threshold. Moreover, since C_2 is a centroid, the triangle inequality is applied to the query and elements in C_2 cluster are pruned (elements with blank dashed nodes are those pruned by this step). Case 2 - The visit reaches the node m, which is a leaf, and its distance with the query is within the threshold. Then, the object C_1 is part of the output set. The triangle inequality is also applied to the centroid C_1. Elements of cluster C_1 satisfying such property are inserted into the output set (elements with $\times$ nodes are those in the output set). Note that all the dashed nodes are not visited by the method because either they are pruned or are inserted into the output set.

S called Antipole pair is generated and it is used to split S into two subsets S_A and S_B ($A \in S_A$ and $B \in S_B$) obtained by assigning each point of S to the subset containing its closest endpoint of the Antipole pair $\{A, B\}$. Once the collection of trees is clustered, the centroids of such clusters are marked and the trie is constructed by inserting first centroids of the clusters and next all the remains elements.

3.1 The Advanced Search Method

In this section we will show an optimization of the basic standard search algorithm discussed previously. This search is applied to the *Clustered Trie Structure* seen in subsection 2. Such an optimization is obtained by applying the triangle inequality property of the used metric distance [10] into each obtained cluster. The improved search algorithm proceeds as follows: a depth first search of the trie is performed starting from the root of the trie until one of the two following cases arises:

1. When the visit reaches a node n, the similarity threshold t is exceeded: the trie is hence pruned at node n and the search backtracks along other paths. If the node

refs to a centroid, the pruning may be also performed on the branches of the trie that refer to its cluster by using the triangle inequality.

2. A leaf m is reached within the similarity threshold: then, the corresponding object M is part of the output set. In this case, if the object M is a centroid, a pruning of its cluster is performed by applying the triangle inequality to all the unvisited portions of the trie which corresponds to the cluster elements. Those elements which satisfy the triangle inequality are inserted into the output set.

An example of both situations is shown in figure 3.

By using such an algorithm we are able to prune, with respect to the standard basic search, some branches of the trie earlier, avoiding distances computation and then speeding up the running time.

4 Experiments

In this section we compare our method with a technique proposed in [13]. We report the results of several comparisons performed on real databases of different sizes up to 5000 elements. We implemented both methods in standard ANSI C (GNU gcc compiler V3.3.1) and we run all experiments on a PC Pentium III 900 Mhz with Linux Operating System. We refer to [10] for comparison of the method in [13] with other distance based index structures [3,4,7,8]. Concerning such a comparison, in [10] the authors report good results.

4.1 Databases

We used the following three databases:

- The database DB_Stamps [10] is a collection of 300 stamps. It has been obtained with the help of a human expert. The expert's intervention provided two actions: first to help in choosing a suitable tree structure to store relevant features of each stamp in the collection; second to obtain the actual trees describing the items. Observe that this step could be, in principle, completely automated provided that good heuristic feature extraction techniques are available.

- The database DB_Docs is provided by a research group of the University of Florence, Italy [9]. It contains 363 elements. Each element which is represented by a MXY tree [6], describes structural and semantic properties of scanned old documents.

- The third database DB_XML [17] obtained by taking a relational biological database (such as ensembl, biosql and chado databases) and converting the results of the queries into XML format, performing the decomposition of the results into normalized entities. We used the Swissprot database generating 5000 XML trees where each tree contains on average eight leaves.

4.2 Preprocessing and Results

The proposed approach requires a preprocessing phase to build the clustered-trie data structure. The preprocessing may be done off line only once. This implies that the proposed approach is best suited when many queries have to be performed on a database where no upgrades are needed. In both cases, the time to construct the data structures is linear on the average to size of the databases. We notice that in our approach the amount of computation resources required by any query is mainly due to distance computations. In figure 4 we report the performance of proposed method compared with the one in [13] by measuring average number of distances computation per query using various threshold values. We set the cost S of removing a branch of a tree equal to two and the cost C of changing the label of a leaf equal to one. We also tried different values for S and C and in all the cases the results were similar to the ones here showed. Table 1 reports the number of clusters discarded during a search and the number of elements returned. Searching with success requires an average time of 0.20 seconds for the first two databases and 20 seconds for the XML databases. We note that, in all the experiments, searching with the largest threshold allows a number of branch substitutions equal at most to half of the average size of the trees in the databases. Classical methods

Table 1. Results of the performed tests. P_C is the number of pruned clusters. *Results* is the number of elements returned form the range search query. R_c is the radius of the clusters in the clustered-trie data structure.

NameDB	Threshold	P_C	$Results$	R_c
DB_Stamps	1	88%	1,27	3
	14	50%	59,83	13
DB_Docs	1	98%	0,53	3
	14	87%	60,27	6
DB_XML	1	96%	0,53	2
	4	64%	2563	2

and strategies of clustering analysis can be applied to tuned the method in order to use the best cluster radius according to the specific input set (these analysis is out of the purpose of the paper). Experiments show that the improvement becomes relevant when the threshold grows. For small thresholds the two methods have almost the same behavior, this is due to the pruning step based on the cutoff distance. We recall that during a partial visit of a branch we discard its subtree if the partial distance between the query and the portion of visited branch exceeds the threshold value. For larger thresholds the impact of the pruning step based on the cutoff distance is low effective.

On the other hand our proposed method first visits the centroid of a cluster and applies the triangle inequality in order to prune earlier all the elements of the cluster. We can notice that a cluster may be spread in *several subtrees* of the trie. This aspect makes the technique more effective with respect to the one proposed by [13]. We can also notice that the worst case running time to process a query in both approaches is linear in the dimension of the database.

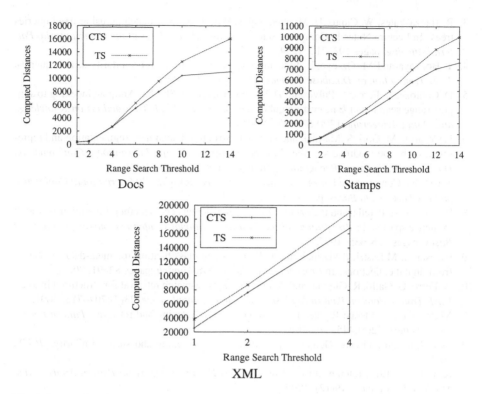

Fig. 4. Comparison between the proposed method (named CTS) and the method (TS) in [13]. Figure reports the average computed distances per queries for DB_Docs, Stamps and XML database, for different threshold values.

5 Conclusions and Future Work

In this paper we showed how a preliminary clustering of the data in a collection of structured objects such as images and documents speeds up the performance of a trie-based searching technique. In order to use the best clustering radius for the given input set, methods of clustering analysis are under study. Future research directions will also investigate on secondary memory management of the clustered trie data structure, its parallel implementation and its exention to k-nearest searching.

References

1. Abdullah A. Al-Shaher and Edwin R. Hancock. Arabic character recognition using structural shape decomposition. *CAIP*, pages 478–486, 2003.
2. S. Amer-Yahia, S. Cho, L. V. S. Lakshmanan, and D. Srivastava. Minimization of tree pattern queries. In *Proceedings of the ACM SIGMOD International Conference on Management of Data*, pages 497–508, 2001.

3. R. Baeza-Yates, W. Cunto, U. Manber, and S. Wu. Proximity matching using fixed-queries trees. In Lecture Notes In Computer Science, editor, *Proceedings of the Combinatorial Pattern Matching*, pages 198–212, 1994.
4. T. Bozkaya and M. Ozsoyogl. Indexing large metric spaces for similarity search queries. *ACM Transaction on Database Systems*, 24(3):361–404, 1999.
5. D. Cantone, A. Ferro, A. Pulvirenti, D. Reforgiato, and D. Shasha. Antipole indexing to support range search and k-nearest neighbor metric spaces. *IEEE Transactions on Knowledge and Data Engineering*, 17(4):535–550, 2005.
6. F. Cesarini, M. Gori, S. Marinai, and G. Soda. Structured document segmentation and representation by the modified x-y tree. In *Proceedings of the Fifth International Conference on Document Analysis and Recognition , Munich*, pages 563–566, 1999.
7. T. Chiueh. Content-based image indexing. In *Proceedings of the 20 International Conference on Very Large Data Bases*, pages 582–593, 1994.
8. P. Ciaccia, M. Patella, and P. Zezula. M-tree: An efficient access method for similarity search in metric spaces. In *Proceedings of the 23th International Conference on Very Large Data Bases*, pages 426–435, 1997.
9. F.Cesarini, M.Lastri, S.Marinai, and G.Soda. Page classification for meta-data extraction from digital collections. In *Proceedings of the DEXA, Munich*, pages 82–91, 2001.
10. A. Ferro, G. Gallo, R. Giugno, and A. Pulvirenti. Best-match retrieval for structured images. *IEEE Transactions on Pattern Analysis and Machine Intelligence*, 23(7):707–718, 2001.
11. Xiaoyi Jiang and Horst Bunke. Preface:image/video indexing and retrieval. *Pattern Recognition Letters*, 22(5):445–446, 2001.
12. Tyng-Luh Liu and Davi Geiger. Approximate tree matching and shape similarity. *ICCV*, 1999.
13. K. Oflazer. Error-tolerant retrieval of trees. *IEEE Transactions on Pattern Analysis and Machine Intelligence*, 19(12), 1997.
14. C. Di Pietro, V. Di Pietro, G. Emmanuele, A. Ferro, T. Maugeri, E. Modica, G. Pigola, A. Pulvirenti, M. Purrello, M. Ragusa, M. Scalia, D. Shasha, S. Travali, and V. Zimmitti. Anticlustal: Multiple sequences alignment by antipole clustering and linear approximate 1-median computation. *IEEE Computer Society Bioinformatics Conference 2003 (CSB2003), Stanford CA*, 2003.
15. D. Shasha, J. T. L. Wang, H. Shan, and K. Zhang. Atreegrep: Approximate searching in unordered trees. Submitted, 2002.
16. D. Shasha, J.T-L Wang, and R. Giugno. Algorithmics and applications of tree and graph searching. *Proceeding of the ACM Symposium on Principles of Database Systems (PODS)*, pages 39–52, 2002.
17. http://www.fruitfly.org/cgi-bin/sql2xml/sql2xml.pl.
18. http://www.alphaworks.ibm.com/aw.nsf/faqs/xmltreediff.

On Adaptive Confidences for Critic-Driven Classifier Combining

Matti Aksela and Jorma Laaksonen

Neural Networks Research Centre,
Laboratory of Computer and Information Science,
P.O.Box 5400, Fin-02015 HUT, Finland
matti.aksela@hut.fi, jorma.laaksonen@hut.fi

Abstract. When combining classifiers in order to improve the classification accuracy, precise estimation of the reliability of each member
classifier can be very beneficial. One approach for estimating how confident we can be in the member classifiers' results being correct is to
use specialized critics to evaluate the classifiers' performances. We introduce an adaptive, critic-based confidence evaluation scheme, where each
critic can not only learn from the behavior of its respective classifier,
but also strives to be robust with respect to changes in its classifier.
This is accomplished via creating distribution models constructed from
the classifier's stored output decisions, and weighting them in a manner
that attempts to bring robustness toward changes in the classifier's behavior. Experiments with handwritten character classification showing
promising results are presented to support the proposed approach.

1 Introduction

In an attempt to improve pattern recognition performance, several approaches
can be taken. One approach often found beneficial is to combine the results of
several classifiers in the hope that the combination will outperform its members.
The basic operation of a committee classifier is to take the outputs of a set of
member classifiers and attempt to combine them in a way that improves accuracy. As a committee merely combines the results produced by its members,
the member classifiers have a significant effect on the performance. The member
classifiers' two most important features that affect the committee's performance
are (i) their individual error rates, and (ii) the correlatedness of the errors between the members. The more different the mistakes made by the classifiers are,
the more beneficial the combination of the classifiers can be [1].

In order to combine the results as effectively as possible, one approach is to
obtain some measure of how confident we can be that a classifier is making a
correct prediction. Therefore, for combining classifiers in an intelligent way, it
would obviously be beneficial to be able to estimate the reliability of each classifier in an accurate fashion. One way of accomplishing this is to use a separate
classification unit to decide whether the classifier is correct or not. Schemes with

S. Singh et al. (Eds.): ICAPR 2005, LNCS 3686, pp. 71–80, 2005.

separate experts for evaluating the classifiers' reliability are often called *critic-driven models*, as each classifier can be assigned a critic that provides a measure of confidence in the classifier's decisions. Critics themselves are specialized classifiers providing an estimate of how likely it is expected that the classifier in question is correct. This prediction can be made based either only on the output of the classifier or also on its inputs.

In pattern recognition the problem is basically to estimate the posterior probabilities of the classes for a given input sample and to choose the class most probably correct. However, estimating the true posterior probabilities is far from simple in many cases, for example when using prototype-based classifiers. The critic-driven approach is one method of attempting to produce accurate estimates on how likely the classifiers are to be correct and thus which label should be chosen. Also, it may be necessary to combine different types of classifiers, and thus it is advantageous to pose as few requirements for the classifiers as possible.

Critic-driven approaches to classifier combining have been investigated previously, e.g. in a situation where the critic makes its decision based on the same input data as the classifier [2] and in a case where scaling schemes and activation functions for critics were examined [3]. Most critic-driven schemes are static in the sense that they have no memory of previous input samples. In general, however, a classifier's performance tends to be similar in similar situations, for example for samples belonging to the same class. It could thus be beneficial to take advantage of this by incorporating also information on the classifier's prior performance into the critic.

Conversely, also the classifier's performance may change in time – for example in the case of handwriting recognition the writer's style may change due to a different situation or we may encounter an entirely new writer. Therefore the critic scheme should be capable of robustness also under changing conditions, a trait somewhat suppressed by the desire to use all collected information for the predictions. Thus it should be advantageous to find a balance between the impact of the older and the more recent samples. In this paper, we propose an approach where information on the classifier's prior performance is used for the critic's decisions while incorporating a weighting scheme to focus on the most recent samples in order to improve the robustness.

2   Adaptive Class-Wise Confidences

In most critic-driven schemes the critic bases its decisions on only the current sample and the member classifiers' outputs. However, the approach presented in this paper is based on our Class-Confidence Critic Combining (CCCC) scheme [4], a model that attempts to learn continuously from the behavior of the critic's associated classifier. Each critic gives its estimate on the classifier's accuracy based not only on the classifier's output to the sample at hand, but also the classifier's prior performance in similar situations. This enables the critic to adapt to the performance of the particular classifier it has been assigned to.

Learning the classifier's performance will inevitably also make the critic less robust to changes in the classifier's behavior, as its evaluations are now by definition based also on the collected knowledge on the classifier's prior performance. To counter this, we in this paper expand the CCCC model further by introducing weighting schemes to make the critics emphasize the most recent inputs. The original model is also modified to produce confidence estimates for all classes from each classifier for each input sample. The output of the combiner is then selected from this array of confidences using one of the standard methods to be presented in Section 2.5.

2.1 The Proposed Approach

In this paper we use prototype-based classifiers that calculate the input's distance from the nearest prototype of each class. Each member classifier produces a vector of distances, with one distance for each class, where a smaller distance indicates a better match. These distances are stored in the critics which attempt to model the distances' distributions. Based on the current and modeled prior distances for the same class, a confidence value is calculated by the critic. These values are then used in deciding the committee's final output. If the classifier is not based on distances, any measure that decreases as similarity increases can be used, or a native confidence-measure may be transformed into a distance for the distribution estimates. For example, if we have a confidence measure $t \in [0, 1]$, we may simply use $1 - t$ as the distance.

Each time a new input sample is processed, it is classified by all member classifiers who calculate the smallest distance to each class. Each classifier's critic then produces an estimate on that classifier's correctness based on the classifier's prior behavior for the same class and that particular input's normalized distances. The final output of the committee is decided from the classifier outputs and confidences obtained from the critics by using one of the combination rules discussed below. After the correctness of the classification has been established, the input is incorporated into the respective critic's distribution model to refine the estimate of the input data distribution. Additionally, weights are used for the distance values stored in the distribution models and adjusted in order to strive for more robust behavior with respect to changes in the classifier's performance.

2.2 Normalizing Distance Values Obtained from Classifiers

As such the distances produced by the member classifiers can be over a wide range of numerical values and have no confidence interpretation on their own. The distances computed in the classifiers may be normalized as follows.

Let there be K classifiers each calculating some type of distances and deciding its output based on minimizing that distance measure. Let there be C pattern classes. Each classifier may have $p \geq C$ prototypes to which the distance is calculated, but at least one prototype for each class. Now let x be the input sample, k the classifier index, $k \in [1, \ldots, K]$, and c the class index, $c \in [1, \ldots, C]$.

From each classifier for each input we may find the shortest distance to the nearest prototype of each class, $d_c^k(x) \in [0, \infty]$.

In practice the classifiers should not produce an infinite distance if matching to any prototype of that class is possible. But for example in the stroke-based handwritten character classifiers used in the experiments presented in this paper, the distance is defined to be infinite if the number of strokes is different in the input and the prototype. Thus also the case of infinite distance should be taken into account. We can now define normalized distances to be used in the critics' distributions as

$$q_c^k(x) = \begin{cases} \frac{d_c^k(x)}{\sum_{i=1}^{C} \hat{d}_i^k(x)} & \text{, if } d_c^k(x) \text{ is finite} \\ 1 & \text{, otherwise} \end{cases} \tag{1}$$

$\hat{d}_c^k(x)$ used in the summation equals $d_c^k(x)$ if $d_c^k(x)$ is finite and is otherwise zero. However, if the distance to only one class is finite, the normalized distance for that class is defined to be zero. If the distances are close to another, $q_c^k(x)$ becomes relatively large, but if the distance to the nearest prototype is much smaller than the others, the normalized value is notably smaller for that class. If the prototype matches the input sample exactly, also the normalized distance equals zero.

2.3   Distribution Types

In order to obtain confidences for decisions on previously unseen x, the values $q_c^k(x)$ must be modeled somehow. The approach of gathering previous values into distribution models from which the value for the confidence can be obtained as a function of $q_c^k(x)$ has been chosen for this task.

One key point in the effectiveness of a scheme based on confidence values calculated from distribution models is in the ease of creating and modifying the models. The amount of data that is obtained from each distribution is quite limited, and in a real situation may vary greatly between distributions due to the fact that some classes occur more frequently than others. The methods should therefore be capable of producing reliable estimates even with small amounts of data. Kernel-based distribution estimates fulfill these requirements and two such schemes have been experimented with and are explained below.

Let us first shorten the notation by using $z \equiv q_c^k(x)$. The confidence obtained from the distribution i then stands as $p^i(q_c^k(x)) = p_c^i(z)$. The distribution model i contains N_i previously collected values $z_j^i, j = 1, \ldots, N_i$. The weight assigned to each sample of the critic is denoted with $w_i(z_j^i)$. The distribution index i runs over the distributions for each class c in each member classifier k.

Triangular kernel distribution model estimate: This distribution estimate uses a triangular kernel function,

$$p_{\text{Tri}}^i(z) = \frac{1}{\sum_{j=1}^{N_i} w_i(z_j^i)} \sum_{j=1}^{N_i} w_i(z_j^i) \max\{0, (b - |z - z_j^i|)\} \tag{2}$$

defined by the kernel bandwidth b, which is given as a parameter.

Gaussian kernel distribution model estimate: The distribution is estimated with a Gaussian function as the kernel. The kernel bandwidth b is used as the variance for the Gaussian.

$$p^i_{\text{Gauss}}(z) = \frac{1}{\sum_{j=1}^{N_i} w_i(z^i_j)} \sum_{j=1}^{N_i} w_i(z^i_j) e^{-\frac{(z-z^i_j)^2}{2b}}. \tag{3}$$

Both distribution models are initialized so that while no data points have been collected, the confidence is evaluated as if one value of zero distance had been obtained. The kernel bandwidth b is optimized from the training data.

2.4 Adapting the Critics

It is assumed that information on the correctness of earlier decisions can be obtained and is available for adapting the critics. The first phase of adaptation is the modification of the distribution models. The $q^k_c(x)$ values received from the member classifiers are incorporated into the corresponding critic's distribution model if the classifier was correct. In practice this is done by appending the new $q^k_c(x)$ value to the list of values for that distribution model.

Additionally, a weight is assigned with each distance value stored in the critic's distribution model to facilitate emphasizing newer inputs. For the second phase of adaptation, the weights will be modified to obtain more robust behavior. Three approaches to adjusting the weights of the sample points have been experimented with, and a constant weighting scheme is used for reference.

Constant weights: The weight for each sample is constant, $w_i(z^i_j) = 1$ for all z^i_j. Weighting the samples equally is used as a reference to examine the benefits of the proposed true weighting schemes.

Class-independent weights: The weights are initially set in an increasing order by using an increasing counter (sample index) $n(z^i_j)$ scaled with a suitable constant. If known beforehand, the total number of test samples N can be used for the scaling factor to obtain the weights

$$w_i(z^i_j) = \frac{n(z^i_j)}{N}. \tag{4}$$

These weights do not depend on the distribution model the sample is inserted into, so within each distribution there can be large differences in the weights.

Class-dependent weights: For each distribution, the weights are scaled linearly every time a new sample is inserted. As a result, each sample has weight equal to the ratio of its index $n_i(z^j_i)$ in that particular distribution model and the total number of samples in that model, N_i,

$$w_i(z^i_j) = \frac{n_i(z^i_j)}{N_i}. \tag{5}$$

This results in the first sample having the smallest weight of $1/N_i$ and the most recent sample having the weight $N_i/N_i = 1$.

Decaying weights: When a new sample is inserted to the distribution model, the weights are recalculated to decrease in accordance with a decay constant λ so that

$$w_i(z_i^j) = \max\{0, 1 - \lambda(N_i - n_i(z_j^i))\}. \tag{6}$$

Effectively the inverse of the decay constant λ states how many previous samples the distribution "remembers" at any given time point, with the newest samples being given the most weight.

2.5   Final Confidence and Decision Mechanisms

As the committee now has label information from the member classifiers and the corresponding confidence values from the critics to work with, a scheme is needed for combining them into the final result. The output of the committee is the label of the class that it decides the input most likely belongs to. This label should be deduced from the available information in an optimal manner.

The overall confidence $u_c^k(x)$ given by the critic k for the input x belonging to class c is obtained from the corresponding distribution model by weighting the confidence estimate with a running evaluation of the classifiers' overall correctness rate. This rate $p(\text{classifier } k \text{ correct})$ is obtained by tracking how many times classifier k has been correct so far and dividing that by the total number of samples classified. Hence the overall confidence is

$$u_c^k(x) = p_c^k(q_c^k(x)) \cdot p(\text{classifier } k \text{ correct}). \tag{7}$$

For the input sample x the decision schemes take the confidences in each label $u_c^k(x)$ from the critics and attempt to form the best possible decision. As the decision mechanisms, especially in the beginning, do not have very much information to work on, a default rule is needed. The default rule here is to use the result of the classifier ranked to be the best on the member validation database. This default rule is applied if no critic suggests a result or several results have exactly the same confidence value.

The effectiveness of the decision scheme is naturally a very important factor in the overall performance of a classifier combination method. It has been often found that in a setting where confidences for all labels can be obtained and the most likely one should be chosen, four basic ways of combining confidences, the *sum, product, min* and *max* rules, can be very effective in spite of their simplicity [5]. Also in this work these decision schemes are used.

Product rule: For each label, the confidences of the critics are multiplied together, and then the label with the greatest total confidence is chosen,

$$c(x) = \arg\max_{j=1}^{C} \prod_{k=1}^{K} u_j^k(x). \tag{8}$$

Sum rule: For each label, the confidences of the critics are summed together, and then the label with the largest resulting confidence is selected,

$$c(x) = \arg\max_{j=1}^{C} \sum_{k=1}^{K} u_j^k(x). \tag{9}$$

Min rule: For each label, the smallest confidence from a critic is discovered, and then the label with the largest minimum confidence is chosen,

$$c(x) = \arg\max_{j=1}^{C} \min_{k=1}^{K} u_j^k(x). \tag{10}$$

Max rule: For each label, the largest confidence from a critic is discovered, and then the label with the largest maximum confidence is selected,

$$c(x) = \arg\max_{j=1}^{C} \max_{k=1}^{K} u_j^k(x). \tag{11}$$

2.6 An Example of the Committee's Operation

In order to illustrate the operation of the combination scheme, let us review an example. In this example we shall use the triangular kernel distribution model estimate and the *sum* rule for determining the final output, and modify the weights according to the *decaying* weights scheme. Now let there be K classifiers and C classes.

For the input sample x each classifier k outputs C values $d_c^k(x)$, with each value corresponding to the distance to the nearest prototype of class c in classifier k. Then this batch of distances is normalized as in equation (1) to obtain again C values $q_c^k(x)$ for each classifier $k = 1, \ldots, K$.

Now the normalized distances are examined by the respective critics, who calculate their confidence values from their distribution models of existing data points as in equation (2). This results in a set of C confidence values $p_c^k(q_c^k(x))$ for every classifier. These confidence values are further adjusted in the critic by weighting them with the respective classifier's correct classification rate in accordance with equation (7). The final output is then selected using the *sum* rule of equation (9) from the final confidences $u_c^k(x)$ obtained from the critics.

For updating the distributions it is assumed that information on the correctness of the result can be obtained after the classification. Now for each of the K classifiers and their respective critics, if that particular classifier was correct, the normalized distance for the correct class is stored into that classifier's critic's distribution model. Furthermore, the weight corresponding to each collected normalized distance value in the distribution model is updated in accordance to equation (6). After the distributions of all the classifiers that were correct have been updated, the next input sample can be processed.

3   Experiments

The committee experiments were performed using a total of six different classifiers. The used data was online handwritten characters written one-by-one. The collection and preprocessing is covered in detail in [6]. All letters, upper and lower case, and digits were used in the experiments.

Table 1. Recognition error rates of the member classifiers

Classifier	Distance measure	Accuracy
1	DTW-PP-MC	79.98%
2	DTW-PL-MC	79.22%
3	DTW-PP-BBC	78.82%
4	DTW-PL-BBC	77.72%
5	DTW-NPP-MC	79.17%
6	DTW-NPP-BBC	77.68%

The member classifiers were based on stroke-by-stroke distances between the given character and prototypes. Dynamic Time Warping (DTW) was used to compute one of three distances, the point-to-point (PP), the normalized point-to-point (NPP), or point-to-line (PL) [6]. The PP distance simply uses the squared Euclidean distance between two data points as the cost function. In the NPP distance the distances are normalized by the number of matchings performed. In the PL distance the points of a stroke are matched to lines interpolated between the successive points of the opposite stroke.

All character samples were scaled so that the length of the longer side of their bounding box was normalized and the aspect ratio kept unchanged. The centers of the characters were moved to the origin. For this we used two different approaches: the center of a character was defined either by its 'Mass Center' (MC) or by its 'Bounding Box Center' (BBC) [6].

The data formed three independent databases consisting of different writers. Database 1 consists of 9961 characters from 22 different writers, which were written without any visual feedback. The pressure level thresholding of the measured data into pen up and pen down movements was set afterwards individually for each writer. The *a priori* probabilities of the classes were somewhat similar to that of the Finnish language. Databases 2 and 3 were collected with a program that showed the pen trace on the screen and recognized the characters online. They both contain data from eight different writers and a total of 8077 and 8047 characters, respectively. The minimum writing pressure for detecting and displaying pen down movements was the same for all writers. The distribution of the character classes was approximately even.

Database 1 was used for forming the initial user-independent prototype set for the DTW-based member classifiers. The prototype set for the DTW-based classifiers consisted of seven prototypes per class. Database 2 was used for evaluating the values for the necessary numeric parameters for the committee and determining the performance rankings of the classifiers. Database 3 was used as a test set. The configurations and corresponding error rates of the member classifiers are shown in Table 1. For experiments with the triangular kernel, the decay parameter was set to $\lambda = 0.26$ for the *product*, *sum* and *max* rules and $\lambda = 0.08$ for the *min* rule. Similarly, for the Gaussian kernel the decay parameter was set to $\lambda = 0.27$ and $\lambda = 0.10$, respectively. The kernel bandwidth for both kernel types was in all cases set to $b = 0.4$.

The data is ordered so that all samples from one writer are processed before moving on to the next writer. The adaptive critics are not reinitialized in between writers, so the committee works in a writer-independent fashion.

4 Results

This preliminary set of experiments examines the two applied distribution types used in conjunction with the four weighting schemes presented. The results obtained with all the four combination rules are shown in Table 2. As can be seen, the triangular kernel performs slightly better than the Gaussian kernel for all the combinations. The results using the constant weighting scheme of both the triangular and Gaussian kernel functions seem somewhat disappointing, as they are outperformed by the respective member classifiers. This was however to be expected, as the distribution models strive to model all input data, and with several different subjects providing the data, the distribution estimate collected from the previous writers may well be suboptimal for the new writer.

Robustness with respect to the changing environment is clearly much better obtained by including the use of the weighting schemes. With the *product* and *sum* rules, the accuracies obtained with the most effective *decaying* scheme clearly outperform those of the member classifiers. The *class-independent* and *class-dependent* schemes perform on roughly the same level. In all cases they provide improvement over the situation where no weighting scheme is used, but less than the *decaying* weights scheme. The *decaying* approach is clearly the most effective one, suggesting that using only a subset consisting of $1/\lambda$ most recent samples is more effective in modeling the classifier's performance than using all available data for the distribution estimates. In these experiments $1/\lambda$ corresponded to between five and ten most recent samples.

It can also be noted that the two most effective combination methods are the *product* and *sum* rules. This may be due to the fact that the confidences, while not always satisfying the properties of being a valid probability, share many of the characteristics of probability values and are modeled in a similar fashion. Although the classifiers are hardly independent, the *product* rule seems to be effective. Furthermore the *product* and *sum* rules take the results from all the classifiers and their confidences from the critics into account when making the decision. The *min* rule can also be seen as trusting the most doubtful critic, which, although risk minimization in a way, clearly is not the optimal scheme. Also the *max* rule trusts a single critic, the most confident one, for the decision making process. In its greediness this appears to be the least beneficial strategy.

5 Conclusions

This paper has presented a scheme for calculating adaptive confidence values from member classifiers' distance values. A distribution model that estimates the member classifiers' performance based not only on their performance for the

Table 2. Experiment results with different weighting schemes and decision methods

Distribution model	Weight scheme	Product rule	Sum rule	Min rule	Max rule
Triangular kernel	Constant	75.07%	74.36%	73.89%	72.28%
Triangular kernel	Class-independent	76.07%	75.22%	74.70%	72.40%
Triangular kernel	Class-dependent	76.06%	75.85%	74.65%	72.98%
Triangular kernel	Decay	81.48%	80.78%	78.65%	73.48%
Gaussian kernel	Constant	70.61%	69.91%	71.88%	71.69%
Gaussian kernel	Class-independent	71.13%	70.21%	72.01%	71.34%
Gaussian kernel	Class-dependent	71.50%	70.86%	71.90%	72.06%
Gaussian kernel	Decay	80.60%	79.92%	78.38%	73.31%

sample being processed, but also learning from prior samples was suggested. In the presented approach a weighting scheme forcing the distributions to emphasize most recent samples is used to enhance robustness. This two-staged adaptive confidence evaluation scheme should provide an effective balance between learning from prior samples and being robust with respect to changes in the member classifiers' performances.

Some preliminary results were presented using two types of kernel functions for constructing the distance distribution models and three weighting schemes to provide robustness. These were applied in an experiment where handwritten character data was recognized. The results clearly showed that the applied scheme can improve upon the results of the member classifiers and that the weighting greatly enhances performance. Especially the decaying weights scheme, where the weight of the samples in the distribution model decays linearly to zero, was found to be effective. This suggests that estimating the confidence of a classifier based on a subset consisting of the newest prior results may be a very effective strategy.

References

1. Kuncheva, L., Whittaker, C., Shipp, C., Duin, R.: Is independence good for combining classifiers. In: Proceedings of the 15th ICPR. Volume 2. (2000) 168–171
2. Miller, D., Yan, L.: Critic-driven ensemble classification. IEEE Transactions on Signal Processing **47** (1999) 2833–2844
3. Hongwei Hao, Cheng-Lin Liu, H.S.: Confidence 'evaluation for combining classifiers. In: Proceedings of International Conference on Document Analysis and Recognition. (2003) 755–759
4. Aksela, M., Girdziušas, R., Laaksonen, J., Oja, E., Kangas, J.: Methods for adaptive combination of classifiers with application to recognition of handwritten characters. International Journal of Document Analysis and Recognition **6** (2003) 23–41
5. Kittler, J., Hatef, M., Duin, R., Matas, J.: On combining classifiers. IEEE Transactions on Pattern Analysis and Machine Intelligence **20** (1998) 226–239
6. Vuori, V., Laaksonen, J., Oja, E., Kangas, J.: Experiments with adaptation strategies for a prototype-based recognition system of isolated handwritten characters. International Journal of Document Analysis and Recognition **3** (2001) 150–159

The RW2 Algorithm for Exact Graph Matching

Marco Gori, Marco Maggini, and Lorenzo Sarti

Dipartimento di Ingegneria dell'Informazione,
Università degli Studi di Siena,
Via Roma, 56-53100 Siena, Italy
{marco, maggini, sarti}@dii.unisi.it

Abstract. The RW algorithm has been proposed recently to solve the exact graph matching problem. This algorithm exploits Random Walk theory to compute a topological signature which can be used to match the nodes in two isomorphic graphs. However, the algorithm may suffer from the presence of colliding signatures in the same graph, which may prevent the procedure from finding the complete mapping between the matching nodes. In this paper we propose an improved version of the original algorithm, the RW2 algorithm, which progressively expands the node signatures by a recursive visit of the node descendants and ancestors to disambiguate the colliding signatures. The experimental results, performed on a benchmark dataset, show that the new algorithm attains a better matching rate with almost the same computational cost as the original one.

1 Introduction

Graph–based techniques have emerged as a powerful tool for data representation in pattern recognition. Due to the growing interest in graphical structures, many efforts have been spent for devising algorithms which are able to process graphs with low computational costs. Among the problems related to structural pattern recognition, graph matching, both exact and approximate, plays a crucial role. While approximate graph matching tries to determine the degree of similarity of two labelled input graphs finding the optimal match between the nodes in the two graphs, the exact graph matching requires the mapping between the nodes of the two input graphs to be structure preserving, in the sense that if two nodes in the first graph are linked by an edge, then they should be mapped to two nodes in the second graph which are also linked by an edge. Even if exact graph matching seems to be less interesting in pattern recognition applications with respect to approximate matching, however it represents an important problem, also from a theoretical point of view. Graph isomorphism is a more stringent version of exact matching, since it requires to determine a bijective exact matching. The method proposed in this paper is designed to solve this particular problem. Actually, from the computational complexity point of view, the graph isomorphism problem lies in the *limbo* between the P and NP classes. In fact, so far, neither a polynomial algorithm was devised nor proof has been given that the problem belongs to the class of NP problems.

S. Singh et al. (Eds.): ICAPR 2005, LNCS 3686, pp. 81–88, 2005.

As recently reported in [1], the exact matching methods proposed in the literature can be classified into two main categories, tree search algorithms and other techniques. Most of the exact matching algorithms are based on a tree search with backtracking (see e.g. [2,3]). The basic idea is that a match (initially empty) is iteratively expanded by adding to it new pairs of matching nodes, which satisfy some conditions that ensure the compatibility with the constraints imposed by the particular matching in which we are interested. In particular, a very efficient algorithm, known as VF, was proposed in [4].

Among the other techniques, the most interesting one is Nauty [5], that is based on group theory. Nauty guarantees, in the average case, impressive performances, although it has been verified that on a benchmark dataset provided by the IAPR-TC–15 [6], Nauty is outperformed by the above mentioned VF algorithm and by RW algorithm [7]. Other interesting approaches are based on spectral theory applied to the graph adjacency matrix. The pioneering method based on spectral analysis was proposed by Umeyama [8], while in [9,10] an hybrid approach for approximate graph matching, based both on spectral techniques and random walks, was proposed.

In this paper we present an improvement of the polynomial RW algorithm presented in [7,11]. The RW algorithm exploits Markovian Random Walks in order to compute a sort of topological signature for each node of an input graph. Isomorphic graphs share the same node signatures, except for a permutation. Determining the permutation of the topological signatures is equivalent to compute the mapping between the nodes of the pair of isomorphic graphs. Unfortunately, if sets of nodes in a graph share the same signatures, the RW algorithm is not able to determine entirely the mapping between a pair of isomorphic structures. The aim of this paper is to present an extended version of the RW algorithm, which performs a recursive visit of the input graph, starting each visit from the nodes with colliding signatures, if any, in order to enrich the topological signatures of these nodes and to obtain not colliding representations. Some experiments on the IAPR–TC–15 benchmark dataset were carried out in order to evaluate if the new version of the RW algorithm is able to reconstruct the mapping on a larger subset of the dataset, and to compare the computational cost of the extended algorithm with respect to the original one.

The paper is organized as follows. In the next section the Markovian Random Walk used to compute the node signatures is defined. In Section 3 the new algorithm is described, and in Section 4 the experimental results are reported. Finally, in Section 5, the conclusions are drawn.

2   Graph Isomorphism Using Random Walks

A Random Walk (RW) model can be exploited to determine a set of topological signatures for each node belonging to a given graph. A Random Walk is described by a probabilistic model which computes a sequence of probability distributions over the set of vertices of a graph G, such that $x_p(t)$ represents the probability of being in vertex p at time t. This probability distribution is described by the real vector $\boldsymbol{x}(t) = [x_1(t), ..., x_N(t)]$, where N is the number of vertices in G. The probabilities are updated at each time step by considering a set of actions that can be performed. In particular, we assume that the RW can move from node q to node p following the arc (q, p), if it exists, or jumping directly to p. Thus, the probability distribution is updated using the following equation:

$$x_p(t+1) = \sum_{q \in G} x(p|q, j) \cdot x(j|q) \cdot x_q(t) + \sum_{q \in pa(p)} x(p|q, l) \cdot x(l|q) \cdot x_q(t) , \quad (1)$$

where $x(p|q, j)$ and $x(p|q, l)$ are the probabilities of moving from the node q to the node p by performing a jump or by traversing the arc (q, p) respectively, $x(l|q)$ and $x(j|q)$ represent the bias between the two possible actions, i.e. jump and follow an arc, and $pa(p)$ represents the set of the parents of node p. These parameters affect the RW behavior and can be chosen in order to compute different topological signatures. Moreover, since they represent probabilities, their values must be normalized such that $\forall q \in G$: $\sum_{p \in G} x(p|q, j) = 1$, $\sum_{p \in ch(q)} x(p|q, l) = 1$, being $ch(q)$ the set of the children of node q, and $x(j|q) = 1 - x(l|q)$.

Considering the update equations for each node in the graph, the RW model can be described in matrix form as

$$\boldsymbol{x}(t+1) = (\boldsymbol{\Sigma} \cdot \boldsymbol{D}_j)' \boldsymbol{x}(t) + (\boldsymbol{\Delta} \cdot \boldsymbol{D}_l)' \boldsymbol{x}(t) = \boldsymbol{T} \cdot \boldsymbol{x}(t) , \quad (2)$$

where $\boldsymbol{\Sigma}, \boldsymbol{\Delta} \in I\!R^{N,N}$ collect the probabilities $x(p|q, j)$ and $x(p|q, l)$ respectively; $\boldsymbol{D}_j, \boldsymbol{D}_l \in I\!R^{N,N}$ are diagonal matrices, whose diagonal values are the probabilities $x(j|q)$ and $x(l|q)$. The entry (p, q) of matrix $\boldsymbol{\Delta}$ is not null only if the corresponding entry of the graph adjacency matrix $\boldsymbol{A}$ is equal to 1, i.e. if the vertices p and q are linked by an arc. Finally, the transition matrix is $\boldsymbol{T} = (\boldsymbol{\Sigma} \cdot \boldsymbol{D}_j + \boldsymbol{\Delta} \cdot \boldsymbol{D}_l)'$.

The signatures of the graph are obtained considering the steady state $\boldsymbol{x}^*$ of the Markov chain defined in equation (2). In fact, since the matrix $\boldsymbol{T}$ is obtained by adding non–negative matrices, then also the transition matrix $\boldsymbol{T}$ is strictly positive. Thus, the resulting Markov chain is irreducible and, consequently, it has a unique stationary distribution (see e.g. [12]) given by the solution of the equation $\boldsymbol{x}^* = \boldsymbol{T}\boldsymbol{x}^*$, where $\boldsymbol{x}^*$ satisfies $\boldsymbol{x}^{*'}I\!I = 1$, being $I\!I$ the N–dimensional vector whose entries are all 1s.

We consider a RW for which the action bias probabilities $x(j|q)$ and $x(l|q)$ are independent of the node q. Thus, we define a parameter $d \in (0, 1)$, called *damping factor*, such that $x(l|q) = d$ and $x(j|q) = 1 - d$. This parameter can be varied in order to compute different topological signatures $\boldsymbol{x}(d)$. Moreover, we choose the other parameters considering uniform probability distributions. The target for a jump is selected using a uniform probability distribution over all the N nodes of the graph, i.e. $x(p|q, j) = 1/N$, $\forall p \in G$. Finally, we assume that all the arcs from node q have the same probability to be traversed, i.e. $x(p|q, l) = 1/|ch(q)|$. The probability of reaching p following a link cannot be computed for vertices without outgoing arcs (*sink* nodes). For these nodes, the RW behavior is set such that $x(j|p_{sink}) = 1$ and $x(l|p_{sink}) = 0$.

Considering these choices, equation (2) can be rewritten as

$$\mathbf{x}(t+1) = \frac{(1-d)}{N} \cdot \mathbb{1} + d \cdot \mathbf{W} \cdot \mathbf{x}(t), \tag{3}$$

where $\mathbf{W} = (\boldsymbol{\Theta}\mathbf{A})'$ being $\mathbf{A}$ the adjacency matrix of the graph and $\boldsymbol{\Theta}$ the diagonal matrix whose (p, p) element is equal to $1/|ch(p)|$. The topological signatures computed using equation (3) can be used to determine the map between two isomorphic graphs. In particular, the following proposition holds (see [7,11]).

Proposition 1. *If two graphs $\boldsymbol{G}_1$ and $\boldsymbol{G}_2$ are isomorphic then $\boldsymbol{x}_1^\star = \boldsymbol{P}\boldsymbol{x}_2^\star$, where $\boldsymbol{P}$ is the permutation matrix such that $\boldsymbol{A}_1 = \boldsymbol{P}\boldsymbol{A}_2\boldsymbol{P}'$, being $\boldsymbol{A}_1, \boldsymbol{A}_2$ the adjacency matrices of $\boldsymbol{G}_1$ and $\boldsymbol{G}_2$, respectively.*

Unfortunately, the vice–versa is not always true. In fact, there exists a class of not isomorphic graphs which, for any value of the damping factor, share the same RW steady state except for a permutation of the entries. The graphs belonging to this class have a common isomorphic and regular subgraph [7].

3   The RW2 Algorithm

The RW algorithm proposed in [7] is based on the RW signatures described in the previous section, and its behavior can be summarized as follows:

1. Given two graphs $\boldsymbol{G}_1$ and $\boldsymbol{G}_2$, the algorithm computes two topological signatures for each graph using a randomly chosen damping factor d; the first one is obtained considering the steady state of the RW, while the second one is computed taking as input the reverse graph obtained changing the direction of each arc in the original graph.
2. If the two graphs share the same signatures, except for a permutation of the entries, and the signatures are all distinct, then the permutation matrix P which represents the map between the two graphs is univocally determined. The matrix P is the only candidate for matching the pair of graphs, because of Proposition 2.1. If the steady state vectors do not share the same signatures, then the graphs are not isomorphic.
3. If the two graphs share the same signatures, except for a permutation of the entries, but the signatures are not all distinct, then an additional signature is computed by using another value of d and this signature is appended to the previous one. The enriched signatures are used to repeat the checks of points 2 and 3.
4. The algorithm is halted if the two graphs are not isomorphic or if the isomorphism is determined, or, finally, if N distinct values for d are chosen (see [7] for a theoretical motivation of this choice).

The RW algorithm is not always able to determine the matrix P when the two graphs are isomorphic, but, as shown by the experimental results reported in [7,11], this situation is very rare. This limitation is due to the presence of colliding signatures. In fact, if two or more nodes in a graph share the same topological signatures, those nodes cannot be matched directly and all their permutations are to be checked. This situation can be due to the particular choice of the parameter d or to symmetries in the graphs.

The new version of the RW algorithm (RW2) tries to overcome this limitation. In fact, the computation of the RW steady state determines a set of equivalence classes $C = \{c_{s_1}, ..., c_{s_M}\}$, where the equivalence class i collects all the nodes sharing the same value s_i for the signature, i.e. the nodes having colliding signatures such that $c_{s_i} = \{v \in V | x_v^* = s_i\}$. In order to discriminate between the nodes in the same equivalence class, their topological signatures can be incrementally expanded by considering recursively the signatures of their neighbors. For each node in the class c_{s_i}, a breadth–first visit of the input graph is performed and the minimum spanning tree which has the selected node as root is progressively built. Anytime a level of the spanning tree is added (level l of the spanning tree collects the nodes l "steps" away from the starting node), the signatures of the nodes belonging to the new level are appended to the signature of the root node. At each step only the nodes which continue to have colliding signatures are furtherly considered for the signature expansion. The breadth–first visit is halted when the minimum spanning tree is completely built or when all the nodes have distinct signatures. In this last situation, we are able to determine the permutation matrix P, applying the same procedure to the second graph, and matching the pairs of nodes which share the same (expanded) signatures in the two graphs.

The experimentation showed that in the RW2 algorithm only one value for the damping factor is sufficient to remove the collisions among the signatures, in the cases when this is actually possible. The use of at most N distinct values for d was introduced in the original RW algorithm to overcome the generation of ambiguous signatures due to the particular choice of this parameter. In fact the use of different values for d allows us to give different weight to the contribution of the layers at a given distance from the current node. The signature enrichment procedure introduced in RW2 seems to play the same role, avoiding the need to compute the steady state of the RW model for more values of d. This effect was observed in the experiments but we have not found any theoretical result to explain it, yet.

The RW2 algorithm is polynomial since the computation of the topological signatures [13], the enrichment of the signatures (if it is needed), and the reconstruction of P can be performed in polynomial time. However, also the RW2 algorithm is not always able to determine entirely the matrix P, but as shown in the results reported in the next section, its "matching rate" is better than that of the original RW algorithm.

4   Experimental Results

The RW2 algorithm was evaluated on a subset of the TC-15 graph dataset [6]. The TC-15 dataset contains 18200 pairs of isomorphic graphs divided in five classes: randomly connected graphs, bounded valence graphs, irregular bounded valence graphs, regular meshes, and irregular meshes. The subset we chose to perform the experiments contains random graphs. The RW2 algorithm has been evaluated in order to determine how the topological regularity and the dimension of the graphs affect the isomorphism determination. The results are evaluated using a "Matching Rate" that is equal to $|C|/N$. Since in [14] the performance of five different exact matching algorithms are compared with respect to their execution time, showing that the VF algorithm outperforms all other methods on the TC-15 dataset, we have also compared the execution time of the RW and RW2 algorithms with respect to the VF algorithm[1]. Notice that the VF algorithm is always able to determine entirely the matching between a pair of input graphs, while the RW and RW2 algorithms are not. However, this comparison seems to be significant, since the time performances of the RW and RW2 algorithms are significantly better that those of the VF algorithm.

In the randomly connected graphs belonging to the dataset, the arcs connect vertices without any structural regularity. These graphs are generated by choosing a value r which represents the probability that an arc is present between two distinct vertices. The arcs are added until the desired arc density is reached. If the resulting graph is not connected, appropriate arcs are added to generate a connected graph. The class collects 3000 pairs of graphs generated using $r \in \{0.01, 0.05, 0.1\}$. The size of the graphs ranges from 20 to 1000 vertices.

The results are reported in Table 1, and show that the matching rate increases both with the number of vertices and with the arc density. The results show that the enrichment procedure introduced in the RW2 algorithm allows us to increase the matching rate w.r.t. the original RW algorithm.

Table 1. Matching rate on randomly connected graphs from the TC-15 dataset: comparison between RW and RW2 algorithms. For the datasets containing graphs with 400 or more nodes, both RW and RW2 show an average matching rate equal to 100%.

Number of Nodes	Arc presence probability					
	0.01		0.05		0.1	
	RW	RW2	RW	RW2	RW	RW2
20	88.9%	94.85%	98.15%	99.325%	99.925%	100%
40	90.105%	95.885%	99.94%	99.97%	100%	100%
60	92.9%	97.29%	100%	100%	100%	100%
80	95.18%	98.12%	100%	100%	100%	100%
100	97.765%	99.27%	100%	100%	100%	100%
200	99.805%	99.945%	100%	100%	100%	100%

---

[1] We have implemented the VF algorithm using the VFLIB (available at *http://amalfi.dis.unina.it/graph/*).

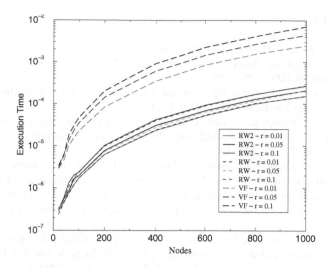

Fig. 1. Execution times on randomly connected graphs from the TC-15 dataset: comparison between the RW, RW2, and VF algorithms for different arc densities

The comparisons of the execution times[2] reported in the plot of figure 1 show that both the RW and RW2 algorithms outperform the VF algorithm. Moreover, the enrichment of the signatures in the RW2 algorithm does not influence significantly the performance of the proposed technique. Thus, even if the RW and RW2 algorithms are not always able to determine the complete matching, the very efficient way in which the computation and the enrichment of the signatures are implemented, allows us to find the eventual matching very fast. Moreover, the performance difference between the algorithms of the RW family and the VF algorithm increases with the size of the graph.

5 Conclusions

In this paper we presented an improved version of the RW algorithm to reduce the problem of colliding signatures. The collisions are disambiguated by enriching the node signatures using a recursive visit of the node neighborhood. The experiments show that the enriched signatures are almost always able to determine the complete mapping between nodes. Moreover, in all successful cases just one value of the dumping parameter is actually needed, whereas all failures are likely due to symmetries in the graph causing collisions that cannot be eliminated by using more values of the dumping factor. Finally the computational cost of the new algorithm is not significantly different from that of the original one.

---

[2] The experiments were run on a PC–IBM with a Pentium 4 2GHZ CPU and 512MB of RAM.

References

1. Conte, D., Foggia, P., Sansone, P., Vento, M.: Thirty years of graph matching in pattern recognition. International Journal of Pattern Recognition and Artificial Intelligence **18** (2004) 265–298
2. Ullmann, J.: An algorithm for subgraph isomorphism. Journal of the Association for Computing Machinery **23** (1976) 31–42
3. Schmidt, D., Druffel, L.: A fast backtracking algorithm to test directed graphs for isomorphism using distance matrices. Journal of the Association for Computing Machinery **23** (1976) 433–445
4. Cordella, L., Foggia, P., Sansone, C., Vento, M.: Evaluation performance of the VF graph matching algorithm. In: Proceedings of the 10th International Conference on Image Analysis and Processing, IEEE Computer Society Press (1999) 1172–1177
5. McKay, B.: Practical graph isomorphism. Congressus Numerantium **30** (1981) 45–87
6. Foggia, P., Sansone, C., Vento, M.: A database of graphs for isomorphism and sub–graph isomorphism benchmarking. In: Proceedings of the 3rd IAPR TC–15 International Workshop on Graph–based Representation. (2001)
7. Gori, M., Maggini, M., Sarti, L.: Graph matching using random walks. In: Proceedings of the International Conference on Pattern Recongition. Volume 3., Cambridge, UK (2004) 394–397
8. Umeyama, S.: An eigen–decomposition approach to weighted graph matching problem. IEEE Transactions on Pattern Analysis and Machine Intelligence **10–5** (1988) 695–703
9. Robles-Kelly, A., Hancock, E.: String edit distance, random walks and graph matching. International Journal of Pattern Recognition and Artificial Intelligence **18** (2004) 315–327
10. Robles-Kelly, A., Hancock, E.: Graph edit distance from spectral seriation. IEEE Transactions on Pattern Analysis and Machine Intelligence **27** (2005) 365–378
11. Gori, M., M. Maggini, M., Sarti, L.: Exact and approximate graph matching using random walks. IEEE Transactions on Pattern Analysis and Machine Intelligence (2005) – To appear.
12. Seneta, E.: Non-negative matrices and Markov chains. Springer-Verlag (1981)
13. Bianchini, M., Gori, M., Scarselli, F.: Inside pagerank. ACM Transactions on Internet Technology **5** (2005)
14. Foggia, P., Sansone, C., Vento, M.: A performance comparison of five algorithms for graph isomorphism. In: Proceedings of the 3rd IAPR TC–15 International Workshop on Graph–based Representation, IEEE Computer Society Press (2001)

Making Use of Unelaborated Advice to Improve Reinforcement Learning: A Mobile Robotics Approach*

David L. Moreno[1], Carlos V. Regueiro[2], Roberto Iglesias[1], and Senén Barro[1]

[1] Dpto. Electrónica y Computación, Universidad de Santiago de Compostela,
15782 Santiago de Compostela, Spain
{dave, rober, senen}@dec.usc.es
[2] Departamento Electrónica y Sistemas, Universidad de A Coruña,
15071 A Coruña, Spain
cvazquez@udc.es

Abstract. Reinforcement Learning (RL) is thought to be an appropriate paradigm for acquiring control policies in mobile robotics. However, in its standard formulation (*tabula rasa*) RL must explore and learn everything from scratch, which is neither realistic nor effective in real-world tasks. In this article we use a new strategy, called Supervised Reinforcement Learning (SRL), that allows the inclusion of external knowledge within this type of learning. We validate it by learning a *wall-following* behaviour and testing it on a Nomad 200 robot. We show that SRL is able to take advantage of multiple sources of knowledge and even from partially erroneous advice, features that allow a SRL agent to make use of a wide range of prior knowledge without the need for a complex or time-consuming elaboration.

1 Introduction

Reinforcement Learning (RL) is an interesting strategy for the automatic resolution of tasks in different domains, among them is robotics [1]. RL only requires a measurement of the system's level of behaviour, so-called reinforcement. Nevertheless, *tabula rasa* RL has strong limitations. RL assumes that the environment as perceived by the system is a Markov Decision Process (MDP), there must not be any *perceptual aliasing*; i.e, the agent cannot consider two situations to which it has to respond with different actions as being equal. Another limitation is the exploitation/exploration dilemma; i.e., deciding between attempting new actions or using previously acquired knowledge. These problems become even more evident in real applications, especially in mobile robotics [2].

However, there usually exists prior knowledge on the task that can be used to improve the learning process, as the RL agent does not start from scratch. In this paper, we show how the Supervised Reinforcement Learning (SRL) [3]

* This work was supported by Xunta de Galicia's project PGIDIT04TIC206011PR. David L. Moreno's research was supported by MECD grant FPU-AP2001-3350.

S. Singh et al. (Eds.): ICAPR 2005, LNCS 3686, pp. 89–98, 2005.

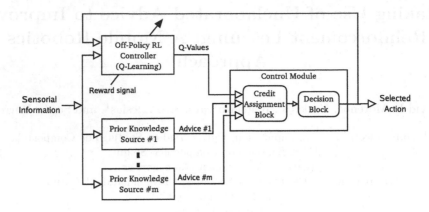

Fig. 1. SRL block diagram

strategy is robust against prior knowledge that is not totally correct (or even completely wrong) and how it is able to use more than one source of knowledge at the same time. These features allow the SRL model to take advantage of prior knowledge without the need for a complex or time-consuming elaboration.

2   Embedding Prior Knowledge: SRL

The objective of the SRL model is to establish a working framework for the development of systems that integrate prior knowledge into RL processes through focusing the exploration of an RL algorithm. SRL comprises several basic blocks (Figure 1): the *Reinforcement Learning Module*, the *Prior Knowledge Sources* (PKSs), and the *Control Module*, which regulates the knowledge transfer between them.

The *RL module* houses the RL algorithm. For this application we use the *Q-Learning* algorithm [4]. The PKSs supply their advice (recommended actions) for the current state of the RL Module, $s(t)$, in the form of a vector of utilities **u** which contains a value $u(s(t), a_i) \in [0, 1]$ for each action a_i. These *u-values* represent an easy way to project knowledge about what actions should be explored first: the more the execution of action a_i is recommended for the current state, the higher $u(s(t), a_i)$ is.

2.1   Control Module

The *Control Module* has the task of amalgamating the utilities that the PKSs supply for each action with the information learnt to date (Figure 1). This module gives priority to the knowledge transfer over exploration of new actions and its design is divided in two blocks: the *Credit Assignment Block* and the *Decision Block*. In general, the operation of the *Control Module* can be divided in the consecutive execution of three different stages.

First, all the advice coming from the PKSs is joined together in the exploitation policy (which is thus called as following it implies exploiting the knowledge that is stored in the PKSs). On the other hand, an exploration policy is also obtained, that recommends executing any of the actions that are not advised by the PKSs.

In a second stage, the policy which seems to be the best one to follow (exploration or knowledge-exploitation) has to be selected. Finally, a decision has to be taken about which action, of those suggested by the previously selected policy, has to be executed.

The first two stages are carried out in the *Credit Assignment Block*, while the third one takes place in the *Decision block*.

Exploitation and Exploration Policies. If $s(t)$ represents the current state of the system, the exploitation policy, $w(s(t))$, represents a normalised sum of all the pieces of advice coming from the PKSs:

$$w(s(t), a_i) = \frac{\sum_{j=1}^{m} u_j(a_i)}{\max_k\{\sum_{j=1}^{m} u_j(a_k)\}}, \forall i = 1, \ldots, n \ , \tag{1}$$

where n is the number of actions and m is the number of PKSs. The higher the value of $w(s(t), a_i)$, the more the execution of action a_i is recommended according to the PKSs.

On the other hand, the exploration policy, $e(s(t))$, recommends the execution of those actions not being advisable by the PKSs:

$$e(s(t), a_i) = 1 - w(s(t), a_i), \quad i = 1, \ldots, n \ . \tag{2}$$

Selection of the Suitable Policy. The next step of the credit assignment block is to decide which of the two policies (knowledge-exploitation or exploration) must be followed. The selected policy, $g(s(t))$, is determined according to the following criterion:

$$g(s(t)) = \begin{cases} e(s(t)) & \text{if } \Omega(e(s(t))) - \delta > \Omega(\mathbf{u_j}(s(t))) \ \forall j = 1, \ldots, m \ , \\ w(s(t)) & \text{otherwise} \ , \end{cases} \tag{3}$$

the function $\Omega(x(s(t)))$, where $x(s(t))$ is $e(s(t))$, or one of the utility vectors $u_j(s(t)), \forall j = 1, \ldots, m$, represents the compatibility of the exploration policy or the utility vectors with those *Q-values* learned for the current state:

$$\Omega(x(s(t))) = \max_i \left\{ x(s(t), a_i) \cdot \left[Q(s(t), a_i) - \min_a Q(s(t), a) \right] \right\} \ . \tag{4}$$

According to equations 3 and 4, it is important to notice that the exploration policy, $e(s(t))$, is chosen whenever its compatibility $\Omega(e(s(t)))$ is higher, with a margin δ, to that of all the suggestions $u_j(s(t))$ in the current state. In this situation, the suggestions either have no information to supply, or are recommending actions that are poorly evaluated by the experience that has been accumulated in the RL module. In this case the advice is considered as not trustworthy and the system must search for new alternatives: it must explore. The parameter δ, the *exploration threshold*, is a positive value that makes possible to regulate the tolerance that the system will have with bad advisors.

Final Action Selection. One of the most classical and popular strategies in RL algorithms to select the final action to be executed is the Softmax algorithm [5]. This strategy, based on the Boltzman distribution, selects an action $a^\star$ with probability:

$$\Pr(s(t), a^\star) = \frac{e^{Q(s(t), a^\star)/T}}{\sum_{i=1}^{n} e^{Q(s(t), a_i)/T}}, \tag{5}$$

where the temperature $T > 0$ controls the amount of randomness.

In our case, we want to adapt equation 5, in order that not just the *Q-values* are taken into account, but also the strategy, $g(s(t))$, selected in the previous stage. According to $g(s(t))$ there is a subset of recommended actions which execution has priority. This prioritization is combined with the learnt Q-values through a last decision vector, $h(s(t))$:

$$h(s(t), a_i) = \frac{g(s(t), a_i) \cdot [Q(s(t), a_i) - \min_a Q(s(t), a)]}{\Omega(g(s(t)))}, \quad \forall i = 1, \ldots, n \ . \tag{6}$$

As this vector reinforces the execution of those actions with a high *Q-value* and which are also recommended by $g(s(t))$, its inclusion on expression 5 seems to be suitable in our case. In this way, an action $a^\star$ is selected with a new value of probability:

$$\Pr(s, a^\star) = \frac{e^{h(s(t), a^\star)/T(s(t))}}{\sum_{j=1}^{n} e^{h(s(t), a_j)/T(s(t))}} \ . \tag{7}$$

It is important to notice that in order to compensate the difference in the probability of every state $s(t)$, each one of them has its own temperature, $T(s)$, which is exponentially decreased every time the state is visited.

3   Application

We have chosen the *wall-following* behaviour as the task to be learnt, as it is one of the most used in mobile robotics [6]. For our experiments we have used a simple state representation [7] that uses only information from the robot's ultrasonic sensors, which are divided into four groups and their measures are discretized according to the values shown in Figure 2. The fifth state variable is the relative *orientation* between the robot and the wall that is discretized into

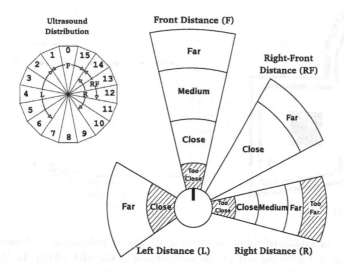

Fig. 2. State representation (except orientation), see text for details

4 values. There are 188 possible states, although not all are met during task development in a normal environment.

A crucial element of the RL system is the reward function. Figure 2 shows the values of the state variables (shaded) that give rise to a negative reward. The actions constitute the final component of the system. In order to simplify the learning process we set a constant linear velocity for the robot (20 cm/s) so that it is only necessary to learn to control the angular velocity. As the maximum angular velocity of the robot is $45°/s$ we have discretized the space of actions as follows:

$$\mathcal{A} = \{-40, -20, -10, -0.3, 0, 0.3, 10, 20, 40\} \quad (°/s) \ . \tag{8}$$

The values used for the parameters of RL and SRL were: 0.2 for the *Learning Rate* (α), 0.99 for the *Discount Factor* (γ) and 0.1 for the *Exploration Threshold* (δ), which is only used on SRL. In our system we have used one PKS, to build it we have used a human expert to decide on the action to be carried out in representative positions within the environment. This PKS (called *Ad-Hoc*) supplies reasonably good advice in 15 states. For each state with advice only an action with maximum utility is recommended. On its own, the advisor is not capable of completely resolving the task.

4   Experimental Results

Our experimental procedure comprises two phases: learning and testing. In the former, the action to be carried out in each *control cycle* (which last about 1/3 of a second) is selected and performed. Learning occurs whenever the state changes

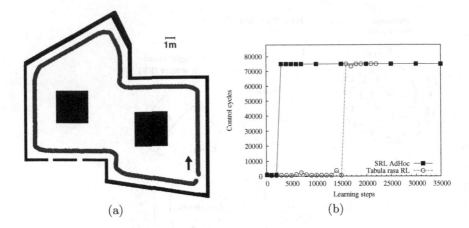

Fig. 3. a) Trajectory of simulated robot performing RL-learnt behaviour. b) Number of failure-free control cycles during test phase, comparison of the SRL (PKS *Ad-Hoc*) and RL *tabula rasa* results. Ten tests were performed for each *Q-value* set.

(*learning step*). The test phase is used to measure the performance of each *Q-value* set and to determine when the learning process has converged. During this phase, the best valued action is always selected and the prior knowledge is never used. Ten tests are performed for each *Q-value* set stored.

As the measurement of the quality of the learnt behaviour we have selected the time before committing an error, as the accumulated reward value does not allow us to discriminate between an error (hitting the wall or travelling away from the wall) and a low performance (e.g. travelling too close to the wall). We consider the task to have been learnt when the system is able to implement it over 75,000 control cycles (about ten laps on the environment) without making an error. Our convergence criterion is that the task should be accurately learnt over five consecutive tests.

4.1   Convergence Time and Erroneous Advice

Both the RL and SRL agents learned on the Nomad 200 simulator and performed on the same environment. Figure 3(a) shows the simulation environment and the trajectory of the simulated robot performing the RL-learnt *wall-following* behaviour. A comparison of the RL and SRL agents' convergence times can be seen in Figure 3(b). The trajectory of the real robot performing the SRL-learnt behaviour is shown in Figure 4.

One of the most interesting characteristics of SRL is its ability to extract beneficial advice from PKSs, and to ignore bad suggestions. In order to verify this we carried out three experiments with increasing amounts of erroneous advice being supplied by the PKSs. In order to construct the erroneous advice we took the advisor *Ad-Hoc* and altered the recommended action to a given percentage

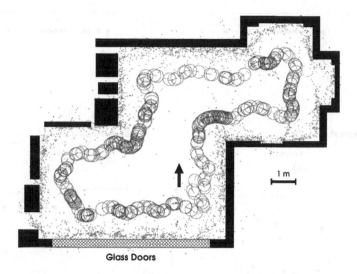

Fig. 4. Trajectory of real robot performing the SRL-learnt *wall-following* behaviour

of those states with advice. We changed the sign of the advised action and increased its values to the maximum (e.g., if in a state the advised action was -0.3, the action 40 would be suggested). The states whose recommended actions were modified were chosen at random.

We carried out three experiments with increasing percentages of erroneous advice: 25, 50 and 100 per cent. In all cases SRL is capable of overcoming erroneous advice and of making use of correct advice to accelerate the learning process. Convergence times were of 9,000, 15,000 and 32,000 learning steps, respectively. A small amount of erroneous advice slows down system convergence; nevertheless, the system still learns faster than *tabula rasa* RL. Figure 5(c) shows the worst situation of all, that in which the advisor supplies no good advice. The system ends up rejecting it, but its convergence time is greater than for *tabula rasa* RL, as SRL initially gives greater priority to the advice than to that which is learnt. The most relevant point of this experiment is that, in spite of the advised action being contrary to the correct one, SRL ends up learning the task.

4.2 Multiple Advice

SRL admits the presence of several PKSs in the *support module*. Taking *Ad-Hoc*, we randomly make 50 per cent of its advice erroneous, using the methodology described above to obtain a new advisor, *Ad-Hoc'₁*. We then construct a second advisor, *Ad-Hoc'₂*, converting all correct advice in *Ad-Hoc'₁* into incorrect advice, and vice versa. In this way, both PKSs supply advice in the same states: one recommends the right action, and the other a wrong one. Figure 5(d) shows how SRL is capable of combining both sources of knowledge and overcoming

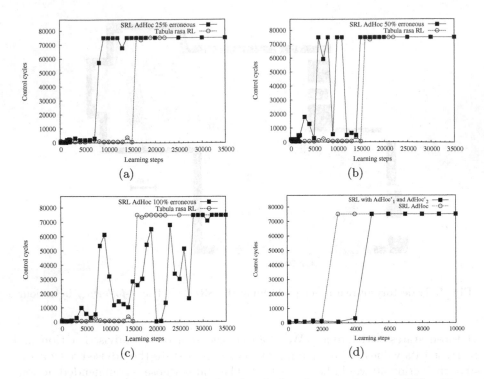

Fig. 5. Number of failure-free control cycles during the test phase: a) SRL with an *Ad-Hoc* advisor with 25% of erroneous advice as opposed to *tabula rasa* RL, b) with 50%, and c) with 100%; d) SRL with an *Ad-Hoc* advisor as opposed to SRL with two *complementary Ad-Hoc* advisors

the erroneous advice that is shared out between the two advisors. SRL succeeds in extracting the good advice for each state, independently of which advisor supplies it, rejecting at the same time the erroneous knowledge.

5   Related Work

Efforts aimed at including prior knowledge into RL in the field of mobile robotics have been based on three techniques. The first is the design of complex reinforcements [8], which has the problem of the specificity of the functions that supply the variable reinforcement, which will probably not be made use of in any other task. Dynamic knowledge-orientated creation of the state space [9] is the second technique. Its main drawbacks are that the stability of the system during learning is not taken into account, and that there is a great dependency on the quality of knowledge: if this is incorrect, the agent cannot learn.

The third and most important approach is the focalization of exploration. There are various approaches, one being to focalize exploration only at the onset of learning [10]. The statistical nature of RL results in there being fluctuations in the learning process, which may lead to the devaluation of an initially-recommended good action. In these cases, the initial focalization of the exploration does not help to stabilize the convergence of the RL algorithm. Another approach is that of Lin [11] and Clouse [12], who propose systems that use external knowledge expressed in the form of sequences of states and actions that are supplied by a teacher. In general, example- or teacher-based systems do not allow the use of knowledge that is not drawn up specifically for the system, incorrect knowledge or various simultaneous teachers.

The latest focalization technique, the one used in SRL, is to employ a control module to regulate the transfer of information between the knowledge sources and the RL algorithm, which is located in the same level. Dixon [13] presents a system of this type, which is similar to SRL. However, his work lacks the generality of SRL, which is designed to allow a broader range of prior knowledge sources and to deal with partial or even erroneous advice. In SRL, the balancing of prior knowledge and learnt information, which is carried out in the *Credit Assignment Block*, allows the designer to use several PKSs without the need for a complex elaboration or any testing of their correctness. Dixon's control modules are far simpler. SRL also contemplates the possibility of the *Control Module* imposing its decisions or assigning overall control to a knowledge source, transforming it into a teacher. Thus, SRL includes the aforementioned approaches at the same time as it permits the use of a much broader set of knowledge sources.

6   Conclusion and Future Work

In this paper we have used a new strategy, called Supervised Reinforcement Learning (SRL), to take advantage of external knowledge within reinforcement learning and we have validated it by learning a *wall-following* behaviour. Thanks to SRL a significant reduction in learning convergence times has been achieved, even using an intuitive prior knowledge source. Thus, just by advising on the best action in 15 of the 188 possible states, there is a 81% reduction in the SLR convergence time with respect to RL (from 16,000 to 3,000 learning steps). We have confirmed the soundness of the proposed mechanism by running the learnt behaviour on a real robot.

We have proved that SRL is capable of overcoming erroneous advice and of making use of any correct advice it may found on the PKSs to accelerate the learning process. A small amount of erroneous advice slows down system convergence and a completely wrong PKS delays it further, but the system always ends up rejecting bad advice. SRL is robust against non-elaborated prior knowledge. Furthermore, we have shown how SRL is capable of combining several sources of prior knowledge, overcoming the erroneous advice that is shared out among them. SRL succeeds in extracting the good advice and rejecting erroneous knowledge. These two features, the ability to overcome erroneous advice and the possibility of using several PKSs, allow the SRL agent to make use of a wide range of PKSs (human knowledge, existing controllers or previously learned information) without the need for a complex or time-consuming elaboration.

References

1. Schaal, S., Atkeson, C.G.: Robot juggling: Implementation of memory-based learning. IEEE Control Systems **14** (1994) 57–71
2. Wyatt, J.: Issues in putting reinforcement learning onto robots. In: 10th Biennal Conference of the AISB, Sheffield, UK (1995)
3. Iglesias, R., Regueiro, C.V., Correa, J., Barro, S.: Supervised reinforcement learning: Application to a wall following behaviour in a mobile robot. In: Tasks and Methods in Applied Artificial Intelligence. Volume 1416., Springer Verlag (1998) 300–309
4. Watkins, C.: Learning from Delayed Rewards. PhD thesis, Cambridge University (1989)
5. Bridle, J.S.: Training stochastic model recognition algorithms as networks can lead to maximum mutual information estimation of parameters. In Touretzky, D., ed.: Advances in Neural Information Processing Systems: Proc. 1989 Conf., Morgan-Kaufmann (1990) 211–217
6. Regueiro, C.V., Rodríguez, M., Correa, J., Moreno, D.L., Iglesias, R., Barro, S.: A control architecture for mobile robotics based on specialists. Volume 6 of Intelligent Systems: Technology and Applications. CRC Press (2002) 337–360
7. Moreno, D.L., Regueiro, C.V., Iglesias, R., Barro, S.: Using prior knowledge to improve reinforcement learning in mobile robotics. In: TAROS 2004, UK (2004)
8. Mataric', M.J.: Reward functions for accelerated learning. In: Int. Conf. on Machine Learning. (1994) 181–189
9. Hailu, G.: Symbolic structures in numeric reinforcement for learning optimum robot trajectory. Robotics and Autonomous Systems **37** (2001) 53–68
10. Millán, J.R., Posenato, D., Dedieu, E.: Continuous-action Q-Learning. Machine Learning **49** (2002) 247, 265
11. Lin, L.J.: Self-improving reactive agents based on reinforcement learning, planning and teaching. Machine Learning **8** (1992) 293–321
12. Clouse, J.A., Utgoff, P.E.: A teaching method for reinforcement learning. In: Machine Learning. Proc. 9th Int. Workshop (ML92), Morgan Kaufmann (1992) 92–101
13. Dixon, K.R., Malak, R.J., Khosla, P.K.: Incorporating prior knowledge and previously learned information into reinforcement learning agents. Technical report, Carnegie Mellon University (2000)

Consolidated Trees: Classifiers with Stable Explanation. A Model to Achieve the Desired Stability in Explanation

Jesús M. Pérez, Javier Muguerza, Olatz Arbelaitz, Ibai Gurrutxaga, and José I. Martín

Dept. of Computer Architecture and Technology, University of the Basque Country,
M. Lardizabal, 1, 20018 Donostia, Spain
{txus.perez, j.muguerza, olatz.arbelaitz, ibai.gurrutxaga,
j.martin}@ehu.es
http://www.sc.ehu.es/aldapa

Abstract. In real world problems solved with machine learning techniques, achieving small error rates is important, but there are situations where an explanation is compulsory. In these situations the stability of the given explanation is crucial. We have presented a methodology for building classification trees, Consolidated Trees Construction Algorithm (CTC). CTC is based on subsampling techniques, therefore it is suitable to face class imbalance problems, and it improves the error rate of standard classification trees and has larger structural stability. The built trees are more steady as the number of subsamples used for induction increases, and therefore also the explanation related to the classification is more steady and wider. In this paper a model is presented for estimating the number of subsamples that would be needed to achieve the desired structural convergence level. The values estimated using the model and the real values are very similar, and there are not statistically significant differences.

1 Introduction

When real world problems are solved using machine learning, the main focus is done in the error (or guess) made by the built model. This aspect is probably the most important, but there are situations where, added to the error rate, the classification made needs to be enclosed with an explanation, so that the decision made by the system in an automatic way can be well-grounded. In real domains such as illness diagnosis, fraud detection in different fields, marketing, etc., the comprehensibility of the classifier is necessary [3], it is not enough to obtain small error rates in the classification. Related to the previous requirements, classifiers can be divided in two groups: without explaining capacity and with explaining capacity. Artificial neural networks, support vector machines, multiple classifiers that due to their complexity do not provide an explanation to the classification, etc. belong to the first group. Among the classifiers with explaining capacity we can mention decision or classification trees (we have selected them for the study presented in this paper), induction rules, etc.

The stability of the given explanation is important when the classifier gives an explanation to the classification made. As Turney found working on industrial applications of decision tree learning, not only to give an explanation but the stability

S. Singh et al. (Eds.): ICAPR 2005, LNCS 3686, pp. 99–107, 2005.
© Springer-Verlag Berlin Heidelberg 2005

of that explanation is of capital importance: "the engineers are disturbed when different batches of data from the same process result in radically different decision trees. The engineers lose confidence in the decision trees even when we can demonstrate that the trees have high predictive accuracy" [9]. Unfortunately, classification trees are unsteady or unstable: trees induced from slightly different subsamples of the same data set are very different in accuracy and structure [4]. The stability of a classifier has been measured by some authors observing if different instances agree in the prediction made for each case of the test set (logical stability or variance) [3], [9]. Nevertheless, since in a decision tree the explanation is given by its structure we need a way to build structurally steady classifiers in order to obtain a convincing explanation (physical stability or structural stability). In paradigms such as bagging and boosting this problem is not solved: some simple classifiers (weak classifiers, mainly classification trees) are combined to make a decision on the whole in order to reduce the error rate, but the solution is based on multiple trees which impedes to give an explanation to the classification. Domingos explained it very clearly in [3]: "while a single decision tree can easily be understood by a human as long as it is not too large, fifty such trees, even if individually simple, exceed the capacity of even the most patient".

We have developed a methodology for building classification trees (Consolidated Trees' Construction Algorithm, CTC) that improves the error rate of standard classification trees and has larger physical or structural stability, see [6]. The algorithm starts extracting several subsamples obtained from the training sample and uses them to build a single classification tree. CTC algorithm is less sensible to the use of subsampling techniques from a structural point of view. Therefore the classification is contributed with a more steady explanation. On the other hand, CTC algorithm is suitable to face class imbalance problems where subsampling techniques (oversampling or undersampling) are used to get through the differences in class distributions.

Consolidated Trees (CT) tend to converge to a common structure as the number of subsamples used to induce the tree increases. The convergence can be modelled so that we can know the number of subsamples we need to use in order to build a CT tree with a certain structural stability. The convergence and the model will be shown in this paper.

The paper proceeds describing CTC algorithm, tat is to say, how a single tree can be built from several subsamples, in Section 2. Details about the experimental methodology and structural measure are described in Section 3. In Section 4 an analysis of the structural stability of CTC algorithm is presented. The model obtained to relate the desired structural stability and the number of subsamples required is presented in Section 5. Finally Section 6 is devoted to show the conclusions and further work.

2 Consolidated Trees' Construction Algorithm

CTC Algorithm uses several subsamples to build a single tree [6] The consensus is achieved at each step of the tree's building process and only one tree is built. The different subsamples are used to make proposals about the feature that should be used to split in the current node. The split function used in each subsample is the gain ratio criterion (the same used by Quinlan in C4.5 [8]). The decision about which feature will be used to make the split in a node of the Consolidated Tree (CT) is accorded among the different proposals by a voting process (not weighted) node by node. Based on this decision, all the subsamples are divided using the same feature. The iterative process is described in Algorithm 1.

Algorithm 1. Consolidated Trees' Construction Algorithm (CTC)

Generate *Number_Samples* subsamples (S^i) from S with *Resampling_Mode* method.
CurrentNode := *RootNode*
for i := 1 to *Number_Samples*
  LS^i := $\{S^i\}$
end for
repeat
  for i := 1 to *Number_Samples*
    CurrentS^i := *First*(LS^i)
    LS^i := LS^i - *CurrentS*^{i}
    Induce the best split $(X,B)^i$ for *CurrentS*^{i}
  end for
  Obtain the consolidated pair (X_c,B_c), based on $(X,B)^i$, $1 \leq i \leq$ *Number_Samples*
  if $(X_c,B_c) \neq$ *Not_Split*
    Split *CurrentNode* based on (X_c,B_c)
    for i := 1 to *Number_Samples*
      Divide *CurrentS*^{i} based on (X_c,B_c) to obtain n subsamples $\{S_1^i, \dots S_n^i\}$
      LS^i := $\{S_1^i, \dots S_n^i\} \cup LS^i$
    end for
  else consolidate *CurrentNode* as a leaf
  end if
  CurrentNode := *NextNode*
until $\forall i$, LS^i is empty

The algorithm starts extracting a set of subsamples (*Number_Samples*) from the original training set. The subsamples are obtained based on the desired resampling technique (*Resampling_Mode*). For example, the class distribution of the original training set can be changed or not, examples can be extracted with or without replacement, different subsample sizes can be chosen, etc.

Decision tree's construction algorithms divide the initial sample in several data partitions. In our algorithm, LS^i contains all the data partitions created from each subsample S^i. When the process starts, the only existing partitions are the initial subsamples.

The pair $(X,B)^i$ is the split proposal for the first data partition in LS^i. X is the feature selected to split and B indicates the proposed branches or criteria to divide the data in the current node. In the consolidation step, X_c and B_c are the feature and branches obtained by a voting process among all the proposals. In the different steps of the algorithm, the default parameters of C4.5 have been used as far as possible.

The process is repeated while LS^i is not empty. The Consolidated Tree's generation process finishes when in the last subsample in all the partitions in LS^i, most of the proposals are not to split it, so, to become a leaf node. When a node is consolidated as a leaf node, the a posteriori probabilities associated to it are calculated averaging the a posteriori obtained from the data partitions related to that node in all the subsamples. Once the consolidated tree has been built, it works the same way a decision tree does.

Previous works [6], [7] show that CT trees have larger discriminating capacity and structural stability than C4.5. Based on these results and due to the lack of space we will present results just for stratified samples without replacement and of 75% of the training set (a single instance of *Resampling_Mode* parameter). For the rest of the parameters, even if more options exist, we have described the ones we are using in this experimentation.

3 Experimental Methodology and Structural Measure

11 databases of real applications from the well known UCI Repository benchmark [1] will be used in order to analyse the structural stability and convergence of Consolidated Trees: *breast-w, hypo, lymph, breast-y, heart-c, soybean large, heart-h, credit-g, glass, liver* and *credit-a*. As validation methodology for the experimentation, we have executed 5 times a 10-fold cross validation [5] with the 11 databases. 1,000 subsamples, that will be used in 20 groups of 50, have been extracted from the training sample in each of the folds of the cross-validation. In each domain the behaviour of CT trees for 5, 10, 20, 30, 40 and 50 subsamples has been analysed in order to evaluate the effect of *Number_Samples* parameter. For each of the folds and value of *Number_Samples* parameter, a CT tree has been built from each group of 50 subsamples. As a consequence, for each value of *Number_Samples*, 1,000 CT trees (5 times, 10 folds, 20 trees) have been built, all of them based on different subsamples (6,000 CT trees for each one of the databases used in the experimentation).

Common, the structural diversity metric, defined in previous works [7], will be used to analyse the physical stability of the induced trees, that is to say, the homogeneity existing in a set of trees. A pair to pair comparison among each possible pair of trees in the group is done. The common nodes among two trees are counted. *Common* is calculated starting from the root and covering the tree, level by level. Two nodes will be considered common nodes, if they coincide in the feature used to make the split, the proposed branches (or stratification) and the position in the tree. When a different node is found the subtree below is not taken into account. The *Common* value of a set of trees is calculated as the average value of all the possible pair to pair comparisons. Somehow, *Common* represents the amount of independent variables that are used in a stable way in the classification, taking into account the significance order. So, this measure quantifies the degree of stable explanation the algorithm is able to contribute with in the classification made. If the complexity of the trees is taken into account and *Common* is normalised in respect to it, *%Common*, the amount of stable explanation relative to all the predicting variables used is captured. *%Common* also indirectly represents the parsimony of the built models (Occam's razor), and besides it allows the comparison of the structural stability of trees induced from different domains, that is, trees that have different complexities (they are situated in different zones of the learning curve).

4 Analysis of the Structural Stability of Consolidated Trees

The aim of the preliminary analysis will be to quantify the stable part of the physical structure in CT trees, that is to say, the average values of *Common* obtained for the 11 databases (5 runs and 10 folds) studied in this paper. These values are shown in Fig. 1. For each database, fold, and value of *Number_Samples* (N_S) parameter, 20 CT trees have been used. Trees built with a certain value of N_S have been compared to all the trees with the same N_S or greater. This comparison is presented in each of the slices in Fig. 1. The comparison among the trees with the same value of N_S appears in the left side of each slice and it shows that as the number of samples used to build the CT trees increases, the common part of the trees also grows: the average value of *Common* is 4.08 when $N\_S = 5$ and it reaches 7.76 when $N\_S = 50$. All the slices have similar shape which shows that, as N_S increases, the similarity of trees built used different numbers of subsamples is even larger than the similarity of trees built with identical number of subsamples. Let's analyse the first slice. Comparing trees built with 5 subamples and trees built with 10 subsamples an average value of *Common* of 4.61 is obtained, which is larger than the value achieved when comparing trees built with 5 subsamples: 4.08. This tendency is maintained in the whole slice: *Common* is 4.83 when the comparison is done among trees built with 5 and 50 subsamples.

It can be observed in Fig. 1 that the same tendency is maintained in all the slices. This means that, from the physical structure point of view, even if the trees are built based on different subsamples, they are more steady as the number of subsamples increases, and they tend to a common structure in each domain. As a consequence we can state that the explanation related to the classification is more steady and wider as the parameter N_S is increased and it seems that an approximation model could be found.

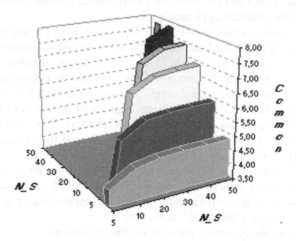

Fig. 1. Average results (11 domains) for *Common* (5x10-fold CVx20CT trees)

Although the increasing tendency is maintained in all the range of values analysed for N_S parameter, there might be a point where it is not worth to increase the number of samples, due to the computational cost of the tree's building process and the error rates achieved. For example, in the experimentation presented in this paper, the average error rate is of 19.72 with a standard deviation $\sigma = 0.15$ and, even if the tendency is to diminish the error rate as N_S parameter increases, there are many databases where the smaller error has been found with 40 subsamples, so it might not be worth increasing N_S. In any case, the error is always smaller than the one achieved with C4.5 trees built based on the same data [6]. If we look to values obtained for %Common, we can observe that, in average, more than half (53.75%) of the structure of CT trees is maintained stable for the range analysed for N_S parameter. This means that at least half of the predictor variables taking part in the classification are maintained steady, providing this way robustness to the given explanation.

5 Estimation of the Number of Subsamples Required to Achieve a Certain Degree of Structural Convergence

In this section we will try to find the number of subsamples required to induce a CT tree so that a fixed structural stability (%Common) is achieved; obtaining solutions with a fixed %Common can be interesting for different uses of classification trees. Therefore, we will try to estimate the value of N_S parameter that needs to be used in each domain in order to achieve the desired %Common.

We want the model that relates %Common to N_S parameter to be robust and with this aim the experimentation (5X10 CV) has been extended to 20 domains: 19 from the UCI repository benchmark [1] (the 11 databases used in previous experimentation and *iris*, *voting*, *hepatitis*, *segment210* (conserving the training/test division of the original data set), *segment2310* (taking into account the whole set of data), *sick-euthyroid*, *vehicle*, *spam*) and one from our environment, *Faithful*, which is a real data application from the electrical appliance's sector.

Based on the results in previous section (%Common is larger when N_S increases and, as a consequence, all the built trees tend to be similar) it seems reasonable to reduce the number of CT trees built in each fold as the number of subsamples increases. We have reduced the computational cost of the experimentation in this section by reducing the number of subsamples generated in each fold to 100 and using them disjointedly to build CT trees. Different number of instances of CTs have been built when varying the parameter N_S: $N\_S = 5$ (20 trees), $N\_S = 10$ (10), $N\_S = 20$ (5), $N\_S = 30$ (3), $N\_S = 40$ (2) and $N\_S = 50$ (2). Even if the computational cost is reduced, the kind of conclusions we can draw, do not vary.

Analysing the evolution of %Common for each one of the 20 databases when the number of samples used to build CT trees (N_S) goes from 5 to 50, we conclude that the values obtained in the experimentation can be approximated with logarithmic regressions with average value for R^2: 0.85.

Equation 1 shows the general expression for the regressions.

$$\%Common_{Domain} = a_{Domain} \cdot Ln(Number\_Samples) + b_{Domain} \qquad (1)$$

The values for coefficients a_{Domain} and b_{Domain} need to be set in order to make this expression practical. We have analysed the characteristics of the domains: size, dimensionality, number of categories of the dependent variable, variable types – discrete, continuous, etc.–- number of missing values, etc., but no correlation has been found with the values of coefficients a and b.

Analysing the values of coefficients a and b for the 20 databases, we find that, for a, the average value is 8.8294 with a standard deviation of $\sigma=4.6629$, whereas, for b, the average value is 25.2345 and the standard deviation $\sigma=25.7777$. Based on these values it seems more critical to find a way of approximating b, whereas for a the average value can be used.

To find a relation among the %*Common* obtained for CT trees and the %*Common* of trees induced based on C4.5 algorithm [8] (even if they are worse in error rate and structural stability they are easier to generate: they have smaller computational cost than CTC algorithm) can probably help in our objective. The analysis made shows that a linear correlation among %*Common* for C4.5 and %*Common* for CT trees exists with an average value of 0.8834 for R^2. So, the relation exists, and as a consequence, it seems that the value of %*Common* of C4.5 trees might be related to coefficients a and b in Equation 1. To try to find this relation we have plotted in Fig. 2 the values of %*Common* of C4.5 trees (axe X) with the values of a and b coefficients (axe Y) for each domain.

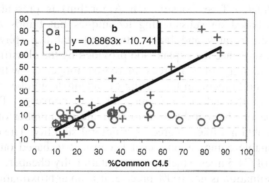

Fig. 2. Relation among %*Common* of C4.5 trees and coefficients a and b in Equation 1

Based on Fig. 2 no relation can be found among the %*Common* of C4.5 trees and coefficient a and, besides, the values of a are similar for every database. Therefore, in any specific domain, the average value obtained for it with the 20 databases, $\hat{a} = 8.8294$, can be used. On the contrary, in the case of coefficient b, the one with greater standard deviation, a relation exists: the points can be approximated with a linear regression. The expression appears in Equation 2 and the values of the coefficients are $C_1 = 0.8863$ and $C_2 = -10.741$.

$$\hat{b}_{Domain} = C_1 \cdot \%CommonC4.5_{Domain} + C_2 \qquad (2)$$

Estimating the value of b based on Equation 2, and using it in Equation 1 together with the value we have fixed for a, in each domain, the %*Common* value achieved for a fixed value of N_S can be predicted. In the experimentation made we have obtained the real values, and, if real and estimated values for each value of N_S parameter are compared, results show that independently of the domain, no significant differences exist (paired t-test with 90% confidence level [2]). The average correlation index among the estimations made and the real values is 0.893.

Based on these results and working out N_S Equation 3 is obtained. As a consequence, we have proposed a model that is able to estimate the number of subsamples that would be needed to achieve the desired structural convergence level ($\%CommonCT_{Desired}$ in Equation 3), with a reasonable computational cost.

$$Number\_Samples = e^{\frac{\%CommonCT_{Desired} - \hat{b}_{Domain}}{\hat{a}}} \tag{3}$$

As an example, the estimation made with this model for *breast-w* database, is that trees built from 34 subsamples or more have to be built if we want to achieve a convergence level (*%Common*) of 90%. This is an acceptable approximation to reality where with 30 subsamples *%Common* is 97.7.

6 Conclusions, Limitations and Further Work

An analysis of the structural stability of classification trees induced using the CTC algorithm (Consolidated Tree Construction Algorithm) is presented in this paper. CTC algorithm is based on subsampling techniques but builds a single tree and achieves smaller error rates and larger structural stability (physical) than C4.5 algorithm. As a consequence the explanation provided by the classifier is steadier.

The structural stability of CT trees increases toogether with the number of subsamples used to build them. In fact, using the adequate number of subsamples the physical structure of the induced CT trees tends to converge. There is a model, presented in this paper, that can be used to estimate the number of subsamples that need to be used to obtain a fixed degree of structural stability. The estimation can be done with a reasonable computational cost because it is based on the study of the common structure of C4.5 trees which are computationally cheaper.

The obtained estimation is not 100% precise, it is an approximation but it is simple and it gives an idea of the number of samples needed to achieve a certain level of convergence which will be enough in most of the applications. On the other hand, the model has not been tried yet with a database that has not been used to obtain it. This could be an experiment to do in the future. The estimation of coefficient a could also be improved in the future, so that the model fits better to reality. On the other hand, more strategies to measure the similarity of trees can be developed.

Acknowledgements

The work described in this paper was partly done under the University of Basque Country (UPV/EHU) project: 1/UPV 00139.226-T-15920/2004. It was also funded by the Diputación Foral de Gipuzkoa and the European Union.

We would like to thank the company Fagor Electrodomésticos, S. COOP. for permitting us the use of their data (Faithful) obtained through the project BETIKO.

The *lymphography* domain was obtained from the University Medical Centre, Institute of Oncology, Ljubljana, Yugoslavia. Thanks go to M. Zwitter and M. Soklic for providing the data.

References

1. Blake, C.L., Merz, C.J.: UCI Repository of Machine Learning Databases, University of California, Irvine, Dept. of Information and Computer Sciences. http://www.ics.uci.edu/~mlearn/MLRepository.html, (1998).
2. Dietterich T.G.: Approximate Statistical Tests for Comparing Supervised Classification Learning Algorithms, Neural Computation, Vol. 10, No. 7, (1998) 1895-1924.
3. Domingos P.: Knowledge acquisition from examples via multiple models. Proc. 14th International Conference on Machine Learning Nashville, TN (1997) 98-106.
4. Drummond C., Holte R.C.: Exploiting the Cost (In)sensitivity of Decision Tree Splitting Criteria, Proceedings of the 17th International Conference on Machine Learning, (2000) 239-246.
5. Hastie T., Tibshirani R. Friedman J.: The Elements of Statistical Learning. Springer-Verlag (es). ISBN: 0-387-95284-5, (2001).
6. Pérez J.M., Muguerza J., Arbelaitz O., Gurrutxaga I., Martín J.I.: Behavior of Consolidated Trees when using Resampling Techniques. Proceedings of the 4th International Workshop on Pattern Recognition in Information Systems (PRIS-2004), Porto, Portugal (2004), 139-148.
7. Pérez J.M., Muguerza J., Arbelaitz O., Gurrutxaga I., Martín J.I.: Analysis of structural convergence of Consolidated Trees when resampling is required. Proceedings of the 3rd Australasian Data Mining Conference (AusDM04), Cairns, Australia (2004), 9-21.
8. Quinlan J.R.: C4.5: Programs for Machine Learning, Morgan Kaufmann Publishers Inc.(eds), San Mateo, California, (1993).
9. Turney P. Bias and the quantification of stability. Machine Learning, 20 (1995), 23-33.

Discovering Predictive Variables When Evolving Cognitive Models

Peter C. R. Lane[1] and Fernand Gobet[2]

[1] School of Computer Science, University of Hertfordshire,
College Lane, HATFIELD AL10 9AB, Hertfordshire, UK
peter.lane@bcs.org.uk
[2] School of Social Sciences and Law, Brunel University,
UXBRIDGE UB8 3PH, Middlesex, UK
fernand.gobet@brunel.ac.uk

Abstract. A non-dominated sorting genetic algorithm is used to evolve models of learning from different theories for multiple tasks. Correlation analysis is performed to identify parameters which affect performance on specific tasks; these are the predictive variables. Mutation is biased so that changes to parameter values tend to preserve values within the population's current range. Experimental results show that optimal models are evolved, and also that uncovering predictive variables is beneficial in improving the rate of convergence.

1   Introduction

Cognitive science aims to devise explanations for the observed behaviour of human or animal participants in different experimental settings. An important component of this science is the construction of computational models which can simulate the observed behaviour. Different classes of models, or *theories*, may be defined based on the underlying representation or learning mechanisms employed. Optimisation with single models on specific tasks has been shown to produce better results than hand optimisation [1,2].

In previous work, we have formalised the process of developing robust computational models, applicable to multiple domains [3,4]. This framework enables us to treat the problem of finding optimal models as one of multi-criteria optimisation, and so apply an evolutionary technique to develop cognitive models. We use a non-dominated sorting genetic algorithm (NDSGA) [5,6] to locate the set of models which are not outperformed on all tasks (taken from categorisation experiments) by any other.

We continue this paper by introducing the psychological data on categorisation in Section 2, describing the classes of models which we explore in Section 3, and then introducing our evolutionary system in Section 4. Section 5 describes our technique of attempting to locate important variables through correlation analysis. Section 6 discusses some experimental results in developing a model of categorisation. The paper is completed with a discussion section and conclusions.

S. Singh et al. (Eds.): ICAPR 2005, LNCS 3686, pp. 108–117, 2005.
© Springer-Verlag Berlin Heidelberg 2005

Table 1. Target behaviours of the 5-4 structure. The first column labels the examples, the second major column gives the probability of responding with category A given that example, and the final column gives the average response time in one experiment. (We show data from two specific experiments, labelled '1ST' and '2ND', and the average data ('AVG') for $P(R_A|E_i)$; classification time was not collected for items E_{10} to E_{16}.)

EXAMPLE	Value	$P(R_A|E_i)$			TIME (S)
		1ST	2ND	AVG	
E1	1 1 1 0	0.78	0.97	0.83	1.11
E2	1 0 1 0	0.88	0.97	0.82	1.34
E3	1 0 1 1	0.81	0.92	0.89	1.08
E4	1 1 0 1	0.88	0.81	0.89	1.27
E5	0 1 1 1	0.81	0.72	0.74	1.07
E6	1 1 0 0	0.16	0.33	0.30	1.30
E7	0 1 1 0	0.16	0.28	0.28	1.08
E8	0 0 0 1	0.12	0.03	0.15	1.13
E9	0 0 0 0	0.03	0.05	0.11	1.19
E10	1 0 0 1	0.59	0.72	0.62	
E11	1 0 0 0	0.31	0.56	0.40	
E12	1 1 1 1	0.94	0.98	0.88	
E13	0 0 1 0	0.34	0.23	0.34	
E14	0 1 0 1	0.50	0.27	0.40	
E15	0 0 1 1	0.62	0.39	0.55	
E16	0 1 0 0	0.16	0.09	0.17	

2 Psychological Data

The problem of categorisation is one of assigning categories to items, and has been widely studied by psychologists and computer scientists for several decades. From a machine learning perspective, the problem is to minimise the error when categorising new examples. However, from a psychological perspective, the problem is much more subtle. Firstly, the aim of a model is to produce a similar pattern of data to that obtained by human participants in the experiment. Secondly, the data to be obtained may be of various kinds: for example, the proportion of correct responses, the time to make a response, or the number of errors during training.

We use data from an experiment called the 5-4 structure, and specifically the experimental data collected by Smith and Minda [7] from thirty earlier studies for proportion of correct responses, and timing data gathered by Gobet et al. [8]. Table 1 summarises some of the psychological data used within this paper. Each example is represented by selecting the binary values for four attributes. The examples are arranged into three groups: the first group (E1-E5) are examples of category A, the second group (E6-E9) examples of category B, and the third group (E10-E16) are known as the 'transfer' examples. During training, the examples of the first two groups are seen and learnt. Finally, all examples, including the training and unseen transfer examples, are presented, and the responses recorded.

Based on the notion of behavioural tests, introduced in [3], we consider the data and how it is compared with the model's performance in each experiment as forming a specific experimental criterion or *task*. The aim of the modeller is to find a model which matches the task as best as possible. For example, a connectionist model may be trained on the examples from the first two groups, and then tested on all the training examples: the output of the model will assign each example into one or other of the categories. By training a collection of models, we can obtain the probability of any given model responding with a category label for each example.

We have described two kinds of experiment against which to judge the behaviour of a computational model: the probability of the model making an error on any given item, and the time to produce a response. Each of these experiments produces, from the model, a behavioural measurement when presented with the examples. There are many ways of quantifying the degree to which a model's performance matches that of the human experimental participants: the degree of match is known as the *fitness* of the model in that experiment. We use two standard techniques: the sum-squared error (SSE) and the average absolute deviation (AAD) of the model's responses from the observed data. SSE is computed by taking the sum of the squared difference between the model's response and the target response across the examples. AAD is the average of the absolute difference between the model's response and the target response.

An individual behavioural test is used to compute the fitness of a model to some behaviour, and thus requires the behavioural data and the fitness comparison method to be specified. Later in the paper, we shall refer to behavioural tests as, for example, 'SSE Time'. This means that we are measuring the deviation of the model's timing responses using the sum-squared error fitness function. Similarly, 'AAD 1st' refers to the average absolute deviation on the first set of data from Table 1 on the probability of responding with category A, etc.

3   Computational Models

We use three different classes of model, all of which are capable of performing the categorisation experiment, and all typical of the kinds of model used within computational modelling. We briefly introduce each class of model below, but first we describe how the models are used within our optimisation technique.

The critical factor behind our technique is that examples of each class can be created by selecting values for the *parameters* which determine how the model performs. For example, the mathematical models have weights, which signify the relative importance of each observed attribute, and connectionist networks have a parameter for the learning rate. The aim of modelling is to find those parameter settings which enable the model to reproduce the observed behaviour (the figures in Table 1) as closely as possible. The novel challenge which we address is to attempt to model multiple kinds of task with each model, and also to locate those models which perform best when compared with the other models.

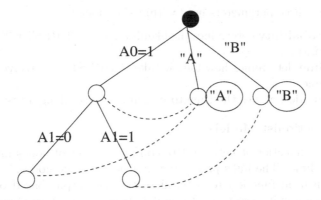

Fig. 1. An example discrimination network, classifying the 5-4 structure

3.1   Mathematical Models

For the class of mathematical models, we use two of the eight models considered by Smith and Minda [7], specifically the *context* model, which uses all previously seen examples to determine its response to a novel example, and the *prototype* model, which forms an overall template for examples of each category, determining its response by how close a novel example fits each template. All of the mathematical models use a formula to provide the probability of responding with category A given certain sets of training data.

The mathematical models are determined by seven parameters:

weights four weights determine the relative significance of each of the attributes defining the examples.
sensitivity is a factor used to scale the response to the observed attribute values.
guessing is a parameter used to capture the fact that people sometimes simply guess a category, without doing any reasoning.
time is used to capture the time required to make a classification.

3.2   Discrimination-Network Models

Discrimination-network models, such as EPAM [8], or CHREST [9], have had a long history within cognitive science. Their strength is in modelling the incremental processes of learning and classification which underpin human behaviour in the categorisation experiments. Fig. 1 illustrates a sample discrimination network, learnt by CHREST when trained on data from the 5-4 experiment. Information is stored as chunks within individual nodes. Tests on the links between nodes are used when sorting a pattern from the root node (the black disc). The dashed links represent *naming links*, which are used by CHREST to associate categories with perceived information. The discrimination network is built up incrementally as the model is given each training example.

There are three parameters used within the model:

learning probability determines the likelihood that CHREST will learn a given training pattern.
reaction time determines how long it takes CHREST to perceive and react to a new example.
sorting time determines the time to match and pass along a test link.

3.3   Connectionist Models

The typical connectionist network [10] comprises a set of nodes interconnected by weighted links. The links pass activation between the nodes, and each node uses an activation function to determine its own output based on the input. We use a single-unit perceptron to model the categorisation experiment, with four input links to capture the four attribute values, and the activation of the perceptron used as the model's output category.

There are four parameters for the perceptron model:

theta is the output threshold value for the perceptron.
eta is the learning rate.
learning probability determines the likelihood that the network will learn a given training pattern.
time is used to capture the time required to make a classification.

4   Multi-criteria Optimisation with a Non-dominated Sorting Genetic Algorithm

We define a space $\mathcal{M}$ of cognitive models by collecting together the abstract space of models of the four theories. Thus, a model, $m \in \mathcal{M}$, will be a specific set of parameter values for one of the classes of theories. Each task described in Section 2 is defined as a function, $f_i(m)$, which produces the fitness of a model for that task. We also assume that we aim to minimise $f_i(m) \geq 0$.

The presence of multiple constraints, f_i, makes the problem a multi-criteria optimisation problem. One of the key challenges is to define 'optimal', because two models may outperform each other on different constraints. Our aim is instead to obtain the *set* of models which are not worse in *all* constraints than any other model. Formally, we say that model m_1 *dominates* model m_2 if:

$$\forall i \bullet f_i(m_1) \leq f_i(m_2) \land \exists j \bullet f_j(m_1) < f_j(m_2)$$

In other words, m_1 does at least as well as m_2 everywhere, but there is at least one constraint in which m_1 does better.

NDSGA is a standard genetic algorithm using the property of non-dominance as a fitness function. Essentially, cross-over is performed across the entire population, but the selection of parents is biased towards those individuals which are not dominated in the current population. We also require the algorithm to maintain populations of each theory type, for the purposes of comparison. Our adapted NDSGA is described in Fig. 2.

1. Four separate, equal-sized populations, are created, each population representing a random collection of models from one of the four theory types.
2. The four populations are pooled, into a set $\mathcal{P}$, and the following four sets are extracted:
   set 1 the non-dominated members of $\mathcal{P}$
   set 2 the non-dominated members of $\mathcal{P} \setminus (\text{set } 1)$
   set 3 the non-dominated members of $\mathcal{P} \setminus (\text{set } 1 \cup \text{set } 2)$
   set 4 the remaining elements, $\mathcal{P} \setminus (\text{set } 1 \cup \text{set } 2 \cup \text{set } 3)$
3. Four new populations are created, each population consisting of models from a single theory, and each of equal size. The populations are constructed by crossover, using relevant individuals from each of the four sets as parents: items in set 1 are twice as likely to be selected as from set 2, etc. Members of set 1 are retained.
4. Mutation is performed with probability *mutate* on the whole population.
5. The process begins again at step 2, until the maximum number of cycles has been reached.

Fig. 2. Modified non-dominated sorting genetic algorithm

5   Discovering Predictive Variables

Each class of model is defined by a set of parameters, or variables, with varying parameter values defining different models. Each model applied to a different task generates a different performance against that task, however it is likely that the importance of each parameter will vary with the task. For instance, timing parameters will clearly be important in tasks measuring response time, but they may also be critical in determining accuracy where training examples are only presented for fixed amounts of time. The question we ask here is whether our genetic algorithm for evolving cognitive models can also identify those parameters which are 'predictive', in the sense of being strongly correlated with performance on specific tasks.

Fig. 3 illustrates the performance of the time parameter for 10,000 instances of the connectionist type of model in two different tasks. As is readily apparent, the performance of the model has a clear global minimum for task 'AAD Time' (on the right), but no such optimum value is apparent for task 'SSE 1st' (on the left). We use a statistical measure, Pearson's product moment correlation, to locate those parameters which take on optimal values in individual tasks. By storing every model tested by the genetic algorithm, along with its fitness on all of the tasks, we test the degree to which any parameter's value correlates with task performance. Specifically, we locate the value of the parameter, p, in the stored models which corresponds to the lowest fitness value for each test. The degree of correlation is then computed between the value of the fitness and the absolute difference of each parameter value from p. A high degree of correlation (> 0.8) means that the parameter acts like the right-hand side of Fig. 3.

The correlated values are used in two ways. Firstly, reporting the correlations is useful in providing additional explanation as to where and why a particular model does well in any given task. Secondly, the stored best values are used to

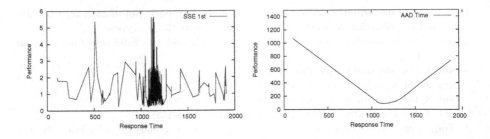

Fig. 3. Graph of performance against parameter value on two tasks

Table 2. Best performance, and model class, on selected individual tasks

Task	Performance	Class
SSE Avg	0.082	Connectionist
AAD Avg	0.057	Connectionist
SSE Time	63814	CHREST
AAD Time	0.069	CHREST

bias mutation of an individual in the population. During mutation of a given parameter, if there is a best value stored, then mutation will pick a new random value near to the best value, in half the cases. In the other half, a new random value is chosen near the current one.

6   Experiments

There were two aims to these experiments: firstly, to confirm that good models were found by the evolutionary process, and secondly, to explore whether the discovery of predictive variables had an impact on the convergence rate. The experiments were run using a population of 200 models, 50 of each type, and allowing training to progress for 500 cycles. The probability of mutation was set at 1.0. Two sets of experiments were run, one with the transfer of bias turned off, and one with it turned on.

Table 2 summarises, for selected tasks, which model type achieved the best performance on that task. Interestingly, different model types do the best on different tasks. The performance measures found here are comparable with the best models in the literature, and, in some cases, exceed the fits obtained. For example, Gobet *et al.* [8] achieved a fit around 0.300 for the 'AAD Time' criterion, whereas our system produced a fit of 0.069. These data confirm that our system discovers useful cognitive models in line with those obtained by practitioners.

Table 3 lists selected correlations detected by the system. The correlations mostly agree with what might be expected. There are strong, and readily apparent, correlations between the different models' timing parameters and the task

Table 3. (Selected) reported parameter values and tasks with high (> 0.8) correlation

Class	Parameter	Value	Task	Correlation
Context	response time	1174.74	SSE Time	0.97
Prototype	response time	1129.73	AAD Time	0.99
Connectionist	response time	1130.62	AAD Time	0.98
CHREST	reaction time	242.52	AAD Time	0.82
CHREST	sorting time	512.60	AAD Time	0.86

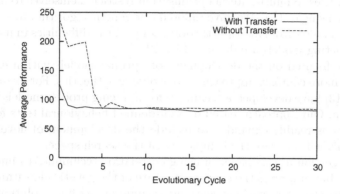

Fig. 4. Convergence rate, with and without transfer of bias, for first 30 learning cycles, of average population performance

involving a timed response. No behavioural effects were apparent for most of the other parameters. Partly, this is a limitation in the current approach, which only seeks a correlation between one parameter's value and the fitness against a task. Most parameters, such as the weights within the mathematical models, work together to produce a behaviour, and more sophisticated analysis techniques are required to locate such dependencies; individually, their correlation with task performance is of the order of 0.3. However, it may also be the case that for many of the parameters, e.g. the learning rate for a connectionist network, there simply is no correlation of the parameter value with the model's performance. This would make the parameter a 'free' variable in the theory; one needed to make the implementation work but which has no explanatory power.

We explored the effect of using the predictive variables discovered by correlation analysis by comparing the rate of convergence of the average performance of the evolving population both with and without the transfer of bias to mutation. Fig. 4 plots the average performance of the entire population of models against training time for a single task, both with and without the transfer of bias. There is a clear improvement in the rate of convergence of the non-dominated set of models when transfer is included. This is readily explained, as the effect of transfer is to reduce the chance that a model will vary from its optimal value.

7 Discussion

Computational modelling is a complex application, to which evolutionary techniques may be profitably applied, as was first proposed by Ritter [1,2]. We have formalised this application as one of multi-criteria optimisation, in which models are drawn from classes of theories, and are optimised against separate experimental criteria. Genetic algorithms are an important technique for solving such multi-criteria problems, and Coello [11] provides a useful summary. Unique to our application is the selection of models from multiple classes of theories, where most approaches to multi-criteria optimisation restrict themselves to individuals drawn from a single class. We have tailored our genetic algorithm to support the evolution of models from multiple theories; some of the difficulties in maintaining useful competing models are discussed in [12].

We have focused on the development of optimal models within established theories. A more complex approach is to evolve new theories. For example, Langley *et al.* [13] have developed a technique for inducing process models from continuous data. Our approach, based on well-defined behavioural tests for specific model types, is readily expanded to include the development of novel theories, although this will increase the complexity of the search space.

Our suggestion in this paper of using correlation techniques to uncover predictive variables improves the convergence rate of the genetic algorithm. Michalski [14] uses stronger machine learning techniques within evolutionary algorithms, suggesting that inductive hypotheses about the performance of specific individuals may be developed. In later work, we intend to extend the range of model types and experiments, and such stronger machine learning techniques may prove beneficial in place of our direct computation of correlations.

8 Conclusions and Further Work

We have described an evolutionary system for developing optimal sets of cognitive models which satisfy multiple experimental criteria. Analysing the evolution of specific model parameters against individual tasks enables the system to pick out optimal values for individual variables. An evolutionary bias in mutation is then employed to guide the system towards these optimal values. Experiments support the value of the technique in locating optimal models across multiple experimental tasks, and also that the identification of predictive variables speeds up the rate of convergence to optimal values.

Further work will focus on widening the range of model types and tasks being explored. As we have argued elsewhere [4], the evolutionary techniques described here are suitable for developing complex models of human behaviour. The selection of predictive variables from the system's own training history will be extended to seek more complex correlations between multiple variables.

Acknowledgements

The authors thank the program chair and three anonymous reviewers for helpful comments which have improved the quality of this paper.

References

1. Ritter, F.E.: Towards fair comparisons of connectionist algorithms through automatically optimized parameter sets. In: Proceedings of the Annual Conference of the Cognitive Science Society, Hillsdale, NJ: Lawrence Erlbaum (1991) 877–881
2. Tor, K., Ritter, F.E.: Using a genetic algorithm to optimize the fit of cognitive models. In: Proceedings of the Sixth International Conference on Cognitive Modeling, Mahwah, NJ: Lawrence Erlbaum (2004) 308–313
3. Lane, P.C.R., Gobet, F.: Developing reproducible and comprehensible computational models. Artificial Intelligence **144** (2003) 251–63
4. Gobet, F., Lane, P.C.R.: A distributed framework for semi-automatically developing architectures of brain and mind. In: Proceedings of the First International Conference on e-Social Science. (2005)
5. Goldberg, D.E.: Genetic Algorithms in Search Optimization and Machine Learning. Reading, MA: Addison-Wesley (1989)
6. Srinivas, N., Deb, K.: Multiobjective optimization using nondominated sorting in genetic algorithms. Evolutionary Computation **2** (1994) 221–248
7. Smith, J.D., Minda, J.P.: Thirty categorization results in search of a model. Journal of Experimental Psychology **26** (2000) 3–27
8. Gobet, F., Richman, H., Staszewski, J., Simon, H.A.: Goals, representations, and strategies in a concept attainment task: The EPAM model. The Psychology of Learning and Motivation **37** (1997) 265–290
9. Gobet, F., Lane, P.C.R., Croker, S.J., Cheng, P.C.H., Jones, G., Oliver, I., Pine, J.M.: Chunking mechanisms in human learning. Trends in Cognitive Sciences **5** (2001) 236–243
10. McLeod, P., Plunkett, K., Rolls, E.T.: Introduction to Connectionist Modelling of Cognitive Processes. Oxford, UK: Oxford University Press (1998)
11. Coello, C.A.C.: An updated survey of GA-based multiobjective optimization techniques. ACM Computing Surveys **32** (2000)
12. Lane, P.C.R., Gobet, F.: Multi-task learning and transfer: The effect of algorithm representation. In: Proceedings of the ICML-2005 Workshop on Meta-Learning. (2005)
13. Langley, P., Sanchez, J., Todorovski, L., Dzeroski, S.: Inducing process models from continuous data. In: Proceedings of the Nineteenth International Conference on Machine Learning, Sydney: Morgan Kaufmann (2002) 347–54
14. Michalski, R.S.: Learning evolution model: Evolutionary processes guided by machine learning. Machine Learning **38** (2000) 9–40

Mathematical Morphology and Binary Geodesy for Robot Navigation Planning

F. Ortiz, S. Puente, and F. Torres

Automatics, Robotics and Computer Vision Group,
Dept. Physics, Systems Engineering and Signal Theory. University of Alicante,
P.O. Box 99, 03080 Alicante, Spain
{fortiz, Santiago.puente, Fernando.torres}@ua.es

Abstract. A new method for obtaining the optimal path to robot navigation in 2-D environments is presented in this paper. To obtain the optimal path we use mathematical morphology in binary worlds and the geodesic distance. The navigation algorithm is based on the search for a path of minimum cost by using the wave-front of the geodesic distance of the mathematical morphology. The optimal path will be the one that minimize the direction changes of the robot. The algorithm of optimal path will be applied in several and complex 2-D environments.

1 Introduction

In path planning for robot navigation it is necessary to know information about the environment in order to efficiently execute the navigation tasks. The information can be represented with different abstraction levels, considering geometric and/or topological information of the environment [1]. In this paper, the approach used considers geometric information of the environment for generating a map and then, using it, the planner computes the optimal navigation path between two points of the environment.

The techniques for path generation in the environment can be divided into three categories: global, local or mixed [2]. These techniques can be used with geometric or topological information.

Global techniques need a global representation of the environment to search for a global solution to the problem, considering the full representation of the environment.

Local techniques only consider some part of the environment for planning the path, i.e., the vicinity of the search point. These techniques are less time-consuming than the global ones, but they can achieve a local minimum instead of the global one.

Mixed techniques combine the ideas of global and local techniques, obtaining good results by using the best characteristics of each technique.

Once the environment representation is made and the search technique is selected, the relevant literature offers different approaches for seeking the best navigation path for going from one point to another. Taking these techniques into consideration, we can emphasize the following ones: fuzzy logic [3], discrete artificial potential fields [4], Voronoi graphs, genetic algorithms [5], graph search optimization paths [6].

Our approach supposes a known world, represented in a 2-D geometric map. This map is transformed into a discrete world representation, using a global approach of

S. Singh et al. (Eds.): ICAPR 2005, LNCS 3686, pp. 118–126, 2005.

the world. By using the geodesic transformations of the mathematical morphology, all the possible paths have been computed. Once these paths are obtained, we propose a method for choosing the best of these, using a branch-and-bound algorithm, taking several rules for choosing the optimal one into consideration.

In other hand, mathematical morphology is a non-linear image-processing approach, which is based on the application of the lattice theory to spatial structures. Mathematical morphology is a powerful image-analysis technique with applications in filtering, enhancement, feature extraction, etc [7,8,9].

In binary morphology, image objects are considered as sets. All the morphological operations are based on the interaction between the original image and another set of a known shape, called the structuring element. In geodesic transformations, the morphological operators applied to an original image involve a second image, known as the mask, which conditions the final results.

This paper is divided into the following sections: In Section 2 geodesic transformations of the mathematical morphology are presented. In Section 3, the rules used for the branch-and-bound algorithm are explained. In Section 4, several examples of the algorithm are shown, computing the best path in the same world and using different starting and finishing points of the trajectory. Finally, our conclusions are presented Section 5.

2 Geodesy and Connectivity in Mathematical Morphology

Geodesic distance was introduced in the framework of image analysis, in 1980, by Lantuéjoul and Beucher [10] and is the base for several morphological operators [11]. The definition of the geodesic distance is as follows:

Let A be a connected set. The geodesic distance $d_A(i,j)$ between two pixels i and j is defined as the length of the shortest path from i to j. C being $= (c_1, c_2, ..., c_n)$ the co-ordinates of the path joining i and j, all of which are included in A:

$$d_A(i, j) = \min\{length(C) \mid c_1 = i, c_n = j \quad and \quad C \subseteq A\} \tag{1}$$

The set A is referred to as the geodesic mask. If A is not connected, the two pixels i and j may belong to two different connected components of A. In this case, the geodesic distance is infinite. The geodesic distance is always greater than or equal to the euclidean distance. In Figure 1, the geodesic distance in a 2-D environment can be seen.

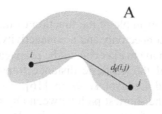

Fig. 1. Geodesic distance in a 2-D environment

If we consider the grouping of pixels, we can calculate the geodesic distance between a point i and any sub-set of pixels X in A. In such a case, $d_A(i,X)$ is the shortest distance between i and any point j in X:

$$d_A(i, X) = \min_{\forall j \in X} \{d_A(i, j)\} \qquad (2)$$

Given that the geodesic distance satisfies the axioms of a metric, the distance between any sub-set X in A from any pixel i in A, is $d_A(i,X)$.

The geodesic distance can be calculated by geodesic dilation. Indeed, the successive thresholds of the geodesic distance function of X in A correspond to the successive geodesic dilations of X in A:

$$\delta_A^{(n)}(X) = \{i \in A \,|\, d_A(i, X) \leq n\} \qquad (3)$$

Where the geodesic dilation of size n of a marker set X with respect to a mask A is obtained by performing n successive geodesic dilations of X with respect to A:

$$\delta_A^{(n)}(X) = \delta_A^{(n)}(X) \left[\delta_A^{(n-1)}(X) \right] \qquad (4)$$

The geodesic dilation of size 1 of the marker set X with respect to the mask A is the intersection of the dilation of the marker set X with the geodesic mask:

$$\delta_A^{(1)}(X) = \delta^{(1)}(X) \cap A \qquad (5)$$

2.1 Distance and Connectivity

Connectivity is classically studied in a topological or a graph-theoretic setting [12,13]. With different types of connectivity between pixels, the results of the geodesic distance will be different. In flat settings, the connectivity will be either 4 or 8. Connectivity 8 allows diagonal movements, while connectivity 4 does not. In Figure 2 we can see the different connectivity [14].

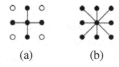

(a)   (b)

Fig. 2. Connectivity graph. 4-connectivity (a) and 8-connectivity (b)

In Figure 3, we show an example of the function of the geodesic distance in a mask from a marker set, using 4-connectivity and 8-connectivity. The grey shades of the mask represent obstacles in which the marker can not be propagated. The marker is a 3x3 set.

The main application of the geodesic distance transformation was described by Verbeek *et al* in [15] and Lengyel *et al* in [16]. By back-tracking the distance propagation, one can find the shortest path between any two points. In this article, we employ the geodesic distance for this very objective. We will move the robot from i to j in a 2-D environment A, by means the wave-front of the geodesic distance $d_A(i,j)$.

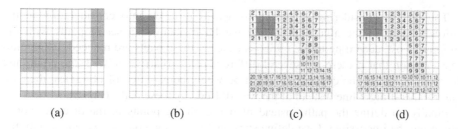

(a)   (b)   (c)   (d)

Fig. 3. Geodesic distance function in a mask (a) from a marker (b), using 4-connectivity and 8-connectivity

In this new application the marker will be a mobile robot and the mask will be the environment of movement. More than one path linking i and j in the environment may have the same minimal length $d_A(i,j)$, so that we will have to use algorithms that reduce the number of paths, based on a given movement cost in following the path.

3   Selecting Reference Points for the Trajectory

To accomplish a trajectory by a robot, the geodesic morphology gives us several possible solutions according to the connectivity of the discrete world used for its computation. Taking that into consideration, one of these trajectories has to be selected to be followed by the robot arm. A branch-and-bound algorithm has been used to select the best path. The characteristics used for determining the optimal path are the followings:

- Minimizing the number of changes of direction. If there are two paths with the same distance but with different amounts of changes in direction, the one with the least changes is selected as a good one (Figure 4.a).
- Selecting the shortest euclidean path, taking into account that two paths with the same number of directional changes can arrive at the same point by different ways. To resolve this problem, a linear movement, either horizontal or vertical, is associated with a weight of 5 and a diagonal movement with a weight of 7. This is because the diagonal distance between two points is √2 and the linear distance, either horizontal or vertical, is 1. The computation of the square-root is more time-consuming than the use of integral numbers. The values 5 and 7 have been selected to minimize the error that arises in that approach. The best path of the two will be the one with the lower weight (Figure 4.b)

The information used for selecting the optimal path with these rules, can give us several paths with the same geodesic and euclidean costs. This happens because we are using a discrete representation of the world. The real world is not discrete, so that continuous movements can be followed, and the directions of these movements are not restricted to eight values.

The fact of considering a continuous world is taken into consideration in the following step of the algorithm. When we have obtained a set of paths with the minimum cost, according to the previous criteria, these paths are reduced, taking into consideration that if there are several different paths to arrive at the same point, and the straight line in the continuous world between them is free of obstacles, then we can use this straight line instead of the way defined by these paths (Fig. 5).

Finally, to define the path, instead of using all the points in the discrete world, which will be impossible if we define straight lines in the continuous world, only the ones in the corners where the trajectory changes direction are stored for controlling the path. Between each pair of points the path has to be a straight line. In Figure 6 we can see a flowchart to the algorithm presented here.

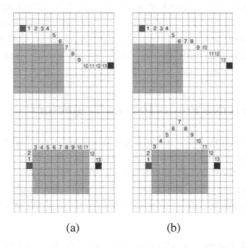

(a)   (b)

Fig. 4. (a) Several paths with the same geodesic distance (8-connectivity) and with different amounts of changes. The best one is the one in the left image. (b) Several paths with the same number of directional changes but with different euclidean costs. The best one is the one in the left image. (8-connectivity).

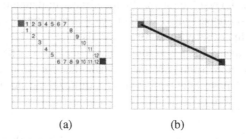

(a)   (b)

Fig. 5. (a) We can see several paths with the same cost for arriving from one point to another. (b) The best continuous path is shown, the straight line. To know whether it is possible to follow this line, the discrete positions marked in the image have to be free of obstacles.

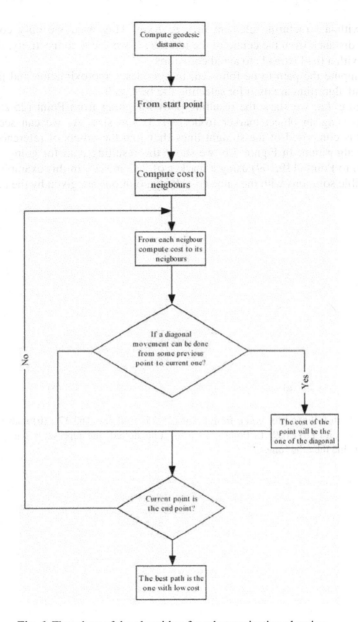

Fig. 6. Flowchart of the algorithm for robot navigation planning

4   Geodesic Paths for Robot Navigation

The examples for testing the algorithm use a 2-D environment, which has been converted to a discrete world. The robot (marker) is represented by a $n \times m$ set ($n \in \mathbb{Z}^2$, $m \in \mathbb{Z}^2$). To reduce calculations, we dilate the obstacles and edges of the environment (mask) with a structuring element of size nxm. This way, we only compute the geodesic distance from the centre of the marker. If we use 8-connectivity, the dilation is made with a $(n+1)$x$(m+1)$ to avoid collisions.

To compute the path to be followed, the geodesic approximation and the branch-and-bound algorithm are used for selecting the best path.

In Figure 7.a, we show the resulting path for going from Point (25,25) to Point (163,17), using an object marker (robot) of 3x3 in size. As we can see, the path followed is composed of the straight lines that join the points of reference obtained from the algorithm. In Figure 7.b, we show the resulting path for going from Point (134,109) to Point (119,109) using a marker of 3x3 in size. In this example, there are two possible solutions with the same cost. Both solutions are given by the algorithm.

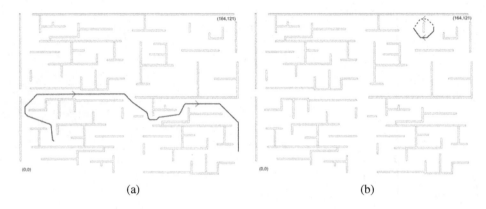

(a)   (b)

Fig. 7. (a) Path followed for going from Point (25,25) to Point (163,17). (b) Path followed for going from Point (134,109) to Point (119,109). The dotted line represents one path and the continuous line the other one.

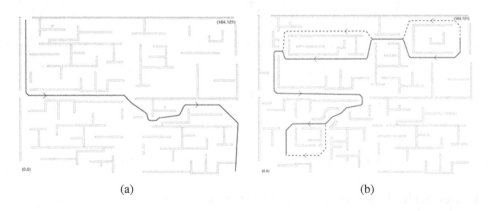

(a)   (b)

Fig. 8. (a) Path followed for going from Point (5,118) to Point (161,2). (b) Path followed for going from Point (154,102) to Point (21,14).

In Figure 8.a, we show the resulting path for going from Point (5,118) to Point (161,2) using an object of 3x3 in size. In this example, there is only one possible path to be followed by the object. In Figure 8.b, we show the resulting path for going from Point (154,102) to Point (21,14), using an object of 3x3 in size. In this example, the environment used for computing the path is different from the one used in the previous examples, but of the same size. There are several paths for going between desired points. The first route is indicated by the continuous line, and another route is represented by the dotted line. These two routes can be combined to gives us the 8 possible routes with the minimum cost. These 8 paths have the same directional changes and the same euclidean cost.

All the examples of the new algorithm for robot navigation has been tested in a Intel Pentium 4 processor 2GHz, and the CPU time (in seconds) has been similar to current algorithms for navigation planning: 0.5-1 seconds in above discrete worlds.

5 Conclusions

In this paper, we have presented a new approach for obtaining the optimal path for robot navigation in 2-D environments. The originality of this algorithm is that it is based on the geodesic distance of the mathematical morphology. We seek the paths with minimum cost. This cost is the number of directional changes in the movement. Taking that into consideration, for paths with the same cost, the one with the minimum euclidean distance in the continuous world is selected.

Several examples of the application of this algorithm have been presented. In these examples, the path obtained for going from the starting to the finishing points have been shown. It can be appreciated that the selected path uses a straight line in the continuous world, between pairs of points, and these lines do not exist in the discrete world. This is an important optimization of the algorithm proposed, which computes the path in the discrete world and gives the solution in the continuous one.

At present, we are working in the adaptation of our algorithm for a fast calculation in discrete and non discrete worlds and in a 3-D navigation. This approach in 3-D can be used for air and submarine navigation, and for computing a robot arm trajectory in a known environment.

References

1. Zavlangas, P., Tzafestas, S.: Integration of topological and metric maps for indoor mobile robot path planning and navigation. Lecture Notes in Artificial Intelligence. Vol. 2308 (2002) 121-130.
2. Reid, M.: Path Planning Using Optically Computed Potencial Fields. In: Proc. IEEE International Conference on Robotics and Automation, Atlanta (1993) 295-300.
3. Meigoli, V. Kamalodin, S., Nikravesh, Y., Talebi, H.: A new global fuzzy path planning and obstacle avoidance scheme for mobile robots. In: Proc. Advanced Robotics, Coimbra (2003) 1296-1301.
4. Zhuang, X., Meng, Q., Yin, B., Wang, H.: Robot path planning by artificial potential fields optimization based on reinforcement learning with fuzzy state. In: Proc. Intelligent Control and Automation, Shanghai, (2002) 1166-1170.

5. Zou, X., Cai, Z., Sun, G.: Non-smooth environment modelling and global path planning for mobile robots. Journal of Central South University of Technology. Vol.10, i.3 (2003) 248-254.
6. Hwang, J., Kim, J., Lim, S., Park, K.: A fast path planning by path graph optimization. IEEE Transactions on Systems Man and Cybernetics Part A-Systems and Humans. Vol.33, i.1 (2003) 121-128.
7. Serra, J.: Image Analysis and Mathematical Morphology. Vol I, Academic Press, London (1981).
8. Serra, J.: Image Analysis and Mathematical Morphology. Vol II: Theoretical Advances, Academic Press, London (1988).
9. Heijmans, H.: Morphological Image Operators. Academic Press, New York (1994).
10. Lantuéjoul, C., Beucher, S.: Geodesic distance and image analysis. Mikroskopie.Vol.37 (1980) 138-142.
11. Soille, P.: Morphological Image Analysis. Principles and Applications, Springer-Verlag, Berlin (1999).
12. Dugundji, J.: Topology. Allyn & Bacon, Boston, MA (1966).
13. Diestel, R.: Graph Theory. Springer-Verlag, New York (1997).
14. Ortiz, F.: Procesamiento Morfológico de Imágenes en Color. Aplicación a la Reconstrucción Geodésica. In: Virtual Library Miguel de Cervantes www.cervantesvirtual.com, Alicante (2002) 121-136.
15. Verbeek, P., Dorst, L., Verwer, B., Groen, F.: Collision avoidance and path finding through constrained distance transformation in robot state space. In: Proc. Intelligent Autonomous Systems, Amsterdam (1986) 634-641.
16. Lengyel, J., Reichert, J., Donald, B., Greenberg, D.: Real-time robot motion planning using rasterizing. Computer Graphics. Vol.24, i.4 (1990) 327-335.

Neural Network Classification: Maximizing Zero-Error Density*

Luís M. Silva[1,2], Luís A. Alexandre[1,3], and J. Marques de Sá[1,2]

[1] INEB -Instituto de Engenharia Biomédica, Lab. de Sinal e Imagem Biomédica,
Campus da FEUP, Rua Dr. Roberto Frias, 4200 - 465 Porto - Portugal
[2] Faculdade de Engenharia da Universidade do Porto - DEEC,
Rua Dr. Roberto Frias, 4200 - 465 Porto - Portugal
[3] IT - Networks and Multimedia Group,
Covilhã - Portugal

Abstract. We propose a new cost function for neural network classification: the error density at the origin. This method provides a simple objective function that can be easily plugged in the usual backpropagation algorithm, giving a simple and efficient learning scheme. Experimental work shows the effectiveness and superiority of the proposed method when compared to the usual mean square error criteria in four well known datasets.

1 Introduction

The work by Príncipe and co-workers [1,2], proposes the use of information measures such as entropy as cost functions for adaptive systems, which are expected to deal better with high-order statistical behaviours than the usual mean square error (MSE). In particular, they proposed the minimization of the error (difference between the output and the target of the system) entropy. The idea is simple. Minimizing the error entropy is equivalent to minimizing the distance between the probability distributions of the target and system outputs [1]. Thus, the system is learning the target variable. The particular application to neural network classification with Rényi's entropy of order $\alpha = 2$ [3,4] and Shannon's entropy [5] has been sucessful. We propose a different but related procedure. The minimization of error entropy is basically inducing a Dirac distribution (the minimum entropy distribution) on the errors. It has been shown that under mild conditions, this Dirac can be centered at zero [3] and thus the error is made to converge to zero. For this reason, we propose to update the weights of a classification neural network by maximizing the error density at the origin. As we will see, this procedure provides a simple objective function and with no need for integral estimation as in other approaches.

* This work was supported by the Portuguese FCT-Fundação para a Ciência e a Tecnologia (project POSI/EIA/56918/2004). First author is also supported by FCT's grant SFRH/BD/16916/2004.

2   The Zero-Error Density Maximization Procedure

Consider a multi-layer perceptron (MLP) with one hidden layer, a single output y and a two-class target variable (class membership for each example in the dataset), t. For each example we measure the (univariate) error $e(n) = t(n) - y(n)$, $n = 1, \ldots, N$ where N is the total number of examples. As discussed above, the minimization of the error entropy induces a Dirac distribution on the errors. It can also be seen that when encoding the classification problem such that $t \in \{-a, a\}$ and $y \in [-a, a]$ for $a > 0$, the induced Dirac distribution must be centered at the origin and thus the error is made to converge to zero [3]. Hence, adapting the system to minimize the error entropy is equivalent to adjusting the network weights in order to concentrate the errors, giving a distribution with a higher peak at the origin. This reasoning leads us to the adaptive criteria of maximizing the error density value at the origin. Formaly,

$$\mathbf{w} = \arg\max_{\mathbf{z}} f(0; \mathbf{z}) \qquad (1)$$

where $\mathbf{w}$ is the weight vector of the network and f is the error density. We denote this principle as Zero-Error Density Maximization (Z-EDM). As the error distribution is not known, we rely on nonparametric estimation using Parzen windowing

$$\hat{f}(e) = \frac{1}{Nh} \sum_{n=1}^{N} K\left(\frac{e - e(n)}{h}\right) \qquad (2)$$

and the Gaussian kernel

$$K(x) = \frac{1}{\sqrt{2\pi}} \exp\left(-\frac{1}{2}x^2\right) . \qquad (3)$$

This is a common and useful choice, because it is continuously differentiable, an essential property when deriving the gradient of the cost function. Hence, our new cost function for neural network classification becomes

$$\hat{f}(0) = \frac{1}{Nh} \sum_{n=1}^{N} \frac{1}{\sqrt{2\pi}} \exp\left(-\frac{1}{2}\frac{e(n)^2}{h^2}\right) . \qquad (4)$$

3   Backpropagating the New Criterion

3.1   Determining the gradient

As we will see, the new criterion can easily substitute MSE in the backpropagation algorithm.

If w is some network weight then

$$\frac{\partial \hat{f}(0)}{\partial w} = \frac{1}{Nh} \sum_{n} \frac{1}{\sqrt{2\pi}} \frac{\partial}{\partial w} \exp\left(-\frac{1}{2}\frac{e(n)^2}{h^2}\right)$$

$$= -\frac{1}{Nh^3} \sum_{n} K\left(\frac{0 - e(n)}{h}\right) e(n) \frac{\partial e(n)}{\partial w} . \qquad (5)$$

Basically one has

$$\frac{\partial \hat{f}(0)}{\partial w} = \sum_n a(n)e(n)\frac{\partial e(n)}{\partial w} \tag{6}$$

with

$$a(n) = -\frac{1}{Nh^3}K\left(\frac{0 - e(n)}{h}\right).$$

For the case of MSE $a(n) = 1, \forall n$. The computation of $\frac{\partial e(n)}{\partial w}$ is as usual for the backpropagation algorithm. Note that the procedure is easily extended for multiple output networks. Taking a target encoding for class $\mathcal{C}_k$ as $[-1, \ldots, 1, \ldots, -1]$ where the 1 appears at the k-th component and using the multivariate Gaussian kernel with identity covariance, the gradient is straightforward to compute

$$\frac{\partial \hat{f}(0)}{\partial w} = -\frac{1}{Nh^{M+2}}\sum_{n=1}^{N}K\left(\frac{\mathbf{0} - \mathbf{e}(n)}{h}\right)\sum_{k=1}^{M}e_k(n)\frac{\partial e_k(n)}{\partial w} \tag{7}$$

where M is the number of output units and $\mathbf{e}(n) = (c_1(n), \ldots, c_M(n))$. Having determined (7) for all network weights, the weight update is given, for the m-th iteration, by the gradient ascent (we are maximizing) rule

$$w^{(m)} = w^{(m-1)} + \eta\frac{\partial \hat{f}(0)}{\partial w}.$$

3.2   Choice of η and h

The algorithm has two parameters that one should optimally set: the smoothing parameter, h, of the kernel density estimator (3) and the learning rate, η. As already seen in previous work [4,5] we can benefit from an adaptive learning rate procedure. By monitoring the value of the cost function, $\hat{f}(0)$, the adaptive procedure ensures a fast convergence and a stable training. The rule is given by

$$\eta^{(m)} = \begin{cases} u\,\eta^{(m-1)} & \hat{f}(0)^{(m)} \geq \hat{f}(0)^{(m-1)} \\ d\,\eta^{(m-1)} \wedge restart & otherwise \end{cases} \quad, 0 < d < 1 \leq u \quad.$$

If $\hat{f}(0)$ increases from one epoch to another, the algorithm is in the right direction, so η is increased by a factor u in order to speedup convergence. However, if η is large enough to decrease $\hat{f}(0)$, then the algorithm makes a *restart* step and decreases η by a factor d to ensure that $\hat{f}(0)$ is being maximized. This *restart* step is just a return to the weights of the previous epoch.

   Although an exhaustive study of the behaviour of the performance surface has not been made yet (this is a topic for future work), we believe that the smoothing parameter h has a particular importance in the convergence success. Just as in the case of entropy, the "dilatation property" mentioned in [2] may also occur. If h is increased to infinity, the local optima of the cost function disappears, letting an unique but biased global maximum to be found. Also note that, as training evolves, it is expected that the errors $\mathbf{e}(n)$ get concentrated around $\mathbf{0}$. Hence, we may benefit from an adaptive rule that starts with a high value of h that is decreased as training evolves. Clearly, this rule should be based on some measure of the local behaviour of the cost function or the gradient. However, we have not yet been successful with this adaptation rule and we postpone this objective as future work. The strategy was then to perform experiments with some fixed h and choose the best ones.

4   Experimental Results

4.1   Convergence Capacity in a Vowel Discrimination Problem

In the first experiment we evaluated the convergence capacity of several MLP's (2, 6 and 10 hidden units) trained using Z-EDM and MSE cost functions, when applied to a vowel discrimination problem. The data, designated PB12, contains 608 examples produced by 76 speakers measuring the first and second formants of the vowels i, I, a and A [6]. The MLP's were trained 100 times with the whole dataset and a convergence success was counted whenever the final training error was below 9%. We varied the number of training epochs, initial learning rate η and smoothing parameter ($h = 2$ and 5) in the case of Z-EDM. Table 1 shows the convergence success rates for Z-EDM and MSE. Below these values, the mean training errors and standard deviations (over the 100 repetitions) are presented. In this Table and in the following, *hid* stands for the number of hidden units.

Table 1. Convergence success rates in 100 repetitions of different MLP's trained with Z-EDM and MSE. Below are the mean training errors and standard deviations.

hid	2		6		10	
epochs	Z-EDM	MSE	Z-EDM	MSE	Z-EDM	MSE
200	71%	6%	100%	87%	100%	90%
	9.54(4.38)	37.9(21.1)	7.31(0.19)	9.78(8.26)	7.22(0.08)	9.11(8.88)
500	96%	21%	100%	97%	100%	99%
	7.61(2.05)	28.6(18.3)	6.62(0.28)	7.77(6.83)	6.58(0.22)	6.61(4.72)
1000	99%	38%	100%	96%	100%	100%
	7.51(2.03)	20.7(14.8)	6.07(0.21)	7.80(7.21)	6.14(0.24)	5.83(0.29)

The results of Table 1 show that the proposed method is clearly more powerful in classifying this dataset. In fact, we encounter already a very good performance for the case of 2 hidden units, while MSE has a global poor performance. By inspecting the training errors and standard deviations, we also find a higher stability of Z-EDM. We've also noted that Z-EDM was not influenced by the initial value of the learning rate, while MSE became very unstable for very high values of η. For 2 hidden units and 200 training epochs, Z-EDM preferred $h = 5$ while for higher training epochs $h = 2$ worked better. This can be related to the smoothness of the performance surface and the dilatation property mentioned

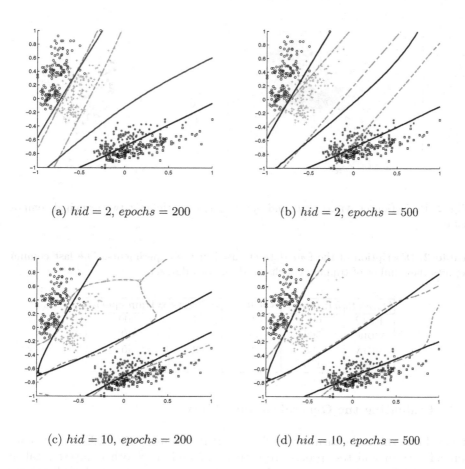

(a) *hid* = 2, *epochs* = 200   (b) *hid* = 2, *epochs* = 500

(c) *hid* = 10, *epochs* = 200   (d) *hid* = 10, *epochs* = 500

Fig. 1. Decision boundaries for PB12. Solid dark line was obtained with Z-EDM and dashed light line with MSE.

earlier. With a small h and consequently a less smoother surface, the number of training epochs (200) may not be sufficient in most cases. This can be surpassed by increasing h at a cost of biasing the optimal solution. Thus the results of Table 1 were obtained with an initial $\eta = 0.5$ and $h = 2$ except for *epochs* = 200 where $h = 5$.

Figure 1 shows decision boundaries obtained with Z-EDM and MSE in different situations. The top figures were obtained with *hid* = 2 and the bottom with *hid* = 10; the left figures used *epochs* = 200 and the right ones *epochs* = 500. The figures show evidence of the stability of Z-EDM and the poor performance of MSE for *hid* = 2. Also, we encounter a higher adaptation of MSE decision lines to the data for *hid* = 10, which can be a drawback in terms of generalization.

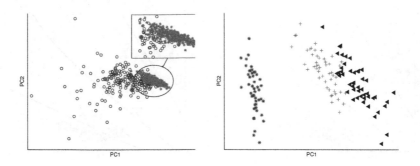

Fig. 2. Projection of WDBC (*left*) and IRIS (*right*) onto the first two principal components

Table 2. Description of the four datasets used in the experiments. The last column reports the number of training epochs used for each dataset.

Datasets	#Instances	#Features	#Classes	#Train epochs
PB12	608	2	4	500
WDBC	569	30	2	40
PIMA	768	8	2	45
IRIS	150	4	3	90

4.2   Evaluating the Generalization Ability

To evaluate the generalization ability of MLP's trained with Z-EDM, we conducted a train and test procedure with PB12 and three other datasets taken from the UCI repository [7]. Two of these datasets are from medical applications. WDBC is concerned with the diagnosis problem between benign and malignant breast cancer and PIMA deals with the diagnostic of diabetes according to the World Health Organization. The fourth dataset is the well known IRIS created by R. A. Fisher. Table 2 gives a brief description of the four datasets.

Figure 2 shows WDBC (left) and IRIS (right) projected onto the first two principal components. As we can see, WDBC has a simple structure and low complexity MLP's should be sufficient to achieve good results. The projection of IRIS shows that one of the classes is linearly separable from the other two, while the latter are not. Note that this is a class structure very similar to the one encountered for PB12 (see Fig. 1).

The following procedure was performed 50 times: divide the data in two subsets, half for training and half for testing; train the network and compute the test set error; interchange the roles of the training and test sets; perform training and test again. The number of training epochs used is reported in Table 2 for each dataset. This procedure was applied to several MLP's varying the number of hidden units from 2 to 20.

Table 3 shows the mean test errors and standard deviations (in brackets). The results of PB12 confirm the previous experiments. For a low number of hid-

Table 3. Test error rates (%), standard deviations (in brackets) and p-values for the Mann-Whitney test of the train and test procedure for several MLP's, trained with Z-EDM and MSE. The right column presents the best results.

PB12	2	5	6	11	20	Best
Z-EDM	8.79(2.64)	7.53(0.56)	7.44(0.58)	7.51(0.58)	7.42(0.47)	7.32(0.53)$\rightarrow hid = 9$
MSE	31.2(13.2)	10.1(5.68)	11.0(7.18)	7.34(0.62)	7.14(0.67)	7.10(0.48)$\rightarrow hid = 15$
p-value	0.000	0.052	0.020	0.227	0.012	0.054

WDBC	2	3	4	5	6	Best
Z-EDM	2.55(0.50)	2.55(0.46)	2.50(0.55)	2.58(0.50)	2.40(0.37)	2.38(0.37)$\rightarrow hid = 18$
MSE	3.11(0.53)	3.18(0.70)	3.25(0.70)	3.08(0.48)	3.17(0.65)	2.99(0.88)$\rightarrow hid = 20$
p-value	0.00	0.00	0.00	0.00	0.00	0.00

PIMA	2	4	6	8	10	Best
Z-EDM	23.5(0.80)	23.2(0.67)	23.4(0.87)	23.3(0.66)	23.5(0.91)	23.2(0.67)$\rightarrow hid = 4$
MSE	24.0(0.91)	23.5(0.78)	23.4(0.95)	23.5(0.86)	23.3(0.88)	23.3(0.88)$\rightarrow hid = 10$
p-value	0.002	0.027	0.761	0.346	0.191	0.569

IRIS	2	6	9	13	19	Best
Z-EDM	4.02(1.32)	4.12(1.26)	3.80(1.04)	4.23(1.26)	4.15(1.13)	3.80(1.04)$\rightarrow hid = 9$
MSE	20.9(15.2)	5.67(4.54)	6.02(6.04)	5.12(3.53)	5.15(3.26)	4.96(2.72)$\rightarrow hid = 18$
p-value	0.000	0.091	0.003	0.265	0.094	0.001

den units, MSE fails to converge in most cases, giving higher mean test errors and standard deviations. From Fig. 3(a), where a more complete set of results is presented, we can see that only for $hid = 11, 19$ and 20, MSE performs equally to Z-EDM. Thus, Z-EDM reveals more stability contrasting with the high dependency of MSE on the number of hidden units. In WDBC, Z-EDM clearly outperforms MSE. All the tested MLP's for this dataset achieved better results than the ones trained with MSE. This behaviour is evident from Fig. 3(b). In what concerns PIMA, Fig. 3(c) shows that the mean test error line for Z-EDM is mostly below the one from MSE, although the differences are not as high as in the previous datasets. For example, with $hid = 6$ both methods achieve the same test error. It was also interesting to evaluate the behaviour of the train and test procedure in the IRIS dataset. As expected, we found similar results as in PB12 (see Fig. 3(d)). For all tested MLP's, MSE had difficulties in finding consistently the best solutions. This was not encountered for Z-EDM, which gave stable results along the various values of hid. e significance of the differences encountered in the results, we performed statistical tests of two types. The parametric t test for two independent samples and the corresponding nonparametric test of Mann-Whitney. The second one, which is a test for the equality in locations of the two samples, is preferable because it does not rely on distributional assumptions and

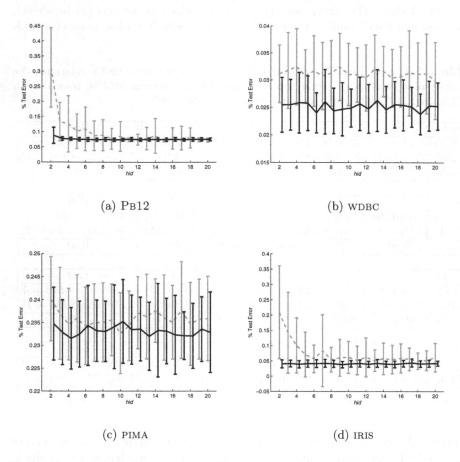

(a) PB12   (b) WDBC

(c) PIMA   (d) IRIS

Fig. 3. Errorbar plot for the test results of the four datasets. Dark solid line was obtained by Z-EDM and light dashed line by MSE. The dark line is slightly shifted to the right for better viewing. Vertical bars from the mean represent one standard deviation.

is more robust to outliers. Nevertheless, the results found for both tests were quite similar. Hence, we opted to show the results for the Mann-Whitney test where p-values below 0.05 show evidence of different locations for the test error distributions coming from MSE and Z-EDM. Except for WDBC, where all the results of Z-EDM are significantly better, we found that when the number of hidden units increases, the differences tend to be not significant. However, we have to take some care while evaluating the p-values, because the existence of (many) strong outliers may lead to wrong conclusions (see for example in PB12 with $hid = 5$). For low complexity MLP's, MSE is clearly outperformed by Z-EDM. This can also be seen on the right column of Table 3, where the best results are presented for each method.

5   Conclusion

We propose a neural network classification method using the error density at the origin as the adaptive criterion. This leads to a simple objective function that can be easily used with the usual backpropagation algorithm. It can be seen as a kind of weighted mean square error, but where information about the error distribution at the origin is taken into account when updating the network's weight vector. The method was evaluated in four datasets and compared to the usual mean square error. We found that Z-EDM was more stable and less dependent on the number of hidden units. The capacity of consistently finding the best solutions was higher for Z-EDM, mainly for small complexity MLP's. It also had a better performance in predicting unseen patterns. Several questions have to be further studied, in particular the relation between the behaviour of the performance surface and the kernel smoothing parameter. This study may also provide insights for an adaptive rule for h. These experiments will be extended to a more general set of benchmark datasets to evaluate generalization ability and to make comparisons with other methods.

References

1. D. Erdogmus and J. C. Principe.: An Error-Entropy Minimization Algorithm for Supervised Training of Nonlinear Adaptive Systems. IEEE Transactions on Signal Processing, 50(7):1780–1786, 2002.
2. D. Erdogmus and J. Principe.: Generalized information potential criterion for adaptive system training. IEEE Transactions on Neural Networks, 13(5):1035–1044, 2002.
3. J.M. Santos, L.A. Alexandre, and J. Marques de Sá.: The Error Entropy Minimization Algorithm for Neural Network Classification. In Int. Conf. on Recent Advances in Soft Computing, Nottingham, United Kingdom, 2004.
4. J.M. Santos, L.A. Alexandre, and J. Marques de Sá.: Optimization of the Error Entropy Minimization Algorithm for Neural Network Classification. In C. Dagli, A. Buczak, D. Enke, M. Embrechts, and O. Ersoy, editors, Intelligent Engineering Systems through Artificial Neural Networks, volume 14, pages 81–86. ASME Press Series, 2004.
5. L.M. Silva, J. Marques de Sá, and L.A. Alexandre.: Neural Network Classification using Shannon's Entropy. In European Symposium on Artificial Neural Networks, 2005.
6. R. Jacobs, M. Jordan, S. Nowlan, and G. Hinton.: Adaptive mixtures of local experts. Neural Computation, 3:79–87, 1991.
7. C.L. Blake and C.J. Merz.: UCI Repository of machine learning databases. University of California, Irvine, Dept. of Information and Computer Sciences, 1998. http://www.ics.uci.edu/~mlearn/MLRepository.html.

Taxonomy of Classifiers Based on Dissimilarity Features

Sarunas Raudys

Institute of Mathematics and Informatics,
Akademijos st. 4, Vilnius 08663, Lithuania
raudys@ktl.mii.lt

Abstract. A great number of linear and nonlinear classification algorithms can follow from a general representation of discriminant function written as a weighted sum of kernel functions of dissimilarity features. Unified look at the algorithms allows obtaining intermediate classifiers which after tuning of the weights and other model's parameters can outperform popular dissimilarity based methods. Simulations with artificial and real world data sets revealed efficiency of single layer perceptron trained in a special way.

Keywords: Classification, dissimilarity, neural networks, single layer perceptron, training.

1 Introduction

A number of research papers utilizing dissimilarity representations while solving classification tasks were increasing during last decades. These classification methods are based on compactness hypothesis [1-3], which claims that similar objects also should be close in their representation. Similarity feature based approach in pattern classification actually goes back to 1951 to pioneering work of Fix and Hoghes [4], however, the similarity based decision making roots can be traced earlier (see. e.g. motivating paper of Attneave [5]).

A great amount of different approaches and methods based on similarity or dissimilarity of objects has been proposed during these decades: kernel discriminant analysis [6, 7], potential functions [1], support vector machines (SVM) [8 - 10], case-based reasoning (CBR) [11 - 13] and many others. Statistical methods such as nearest means (Euclidean distance), certain piecewise-linear classifiers, radial basic functions and learning vector quantization neural networks also could be considered as decision making algorithms working in dissimilarity feature space. For a general introduction into statistical pattern recognition methods see e.g. [3, 14, 15].

Duin and his colleagues [16] made an attempt to segregate dissimilarity-based classification (DSC) algorithms into a separate group. Their attempt gave an impulse to share peculiarities of the algorithms and to take a general look at these groups of methods. The kernel and support vector based methods were reviewed and compared in [9, 10]. Perner [13] performed comparison of DSC and CBR approaches and demonstrated that both approaches relies on the same ideas and concluded that DSC is a special variant of CBR that is influenced by the traditional ideas of pattern recognition. The CBR considers such vital for practitioners problems like the right case description, defining appropriate similarity measures, organization of a large number of cases for efficient retrieval where similar cases are grouped together,

S. Singh et al. (Eds.): ICAPR 2005, LNCS 3686, pp. 136 – 145, 2005.

acquisition and refinements of new cases for entry in the case bases, generalization of specific cases applicable to a wide range of situations. Many of these problems were ignored in statistical pattern recognition so far. Nevertheless, the pattern recognition community investigates important questions of common interest such as fast implementation of search of nearest neighbors [17] and determination of dissimilarity metrics. Mottl *et al.* [18] adopted dissimilarity measure from computational biology and treated the pair-wise similarity measure of two discrete valued vectors as inner product in an imaginary Hilbert space. The dissimilarity is evaluated as the likelihood that two vectors have the same evolutionary origin by way of calculating the so-called alignment score between two sequences. In case of two classes, they suggested to use maximal margin (support vector) classifier. Santini and Jain [19] performed deep analysis of a number of similarity measures. In particular, they investigated similarity measures that exhibit several features that match experimental findings in humans.

It is worth mentioning that CBR systems deal with very large number of case classes, while most theoretical findings in statistical pattern recognition were done in two category case. As a result, learning in CBR systems utilizes narrow gamma of classifiers. Surely, further cooperation of different approaches could create a new quality. An objective of present paper to consider specific questions arising in final decision making performed either in CBS, DCS, SVM or in other pattern recognition methods utilizing the similarity concepts. To get more deep insight and in order to analyze a wider gamma of methods we restrict ourselves to two category situation.

The first classification algorithm based on dissimilarities, the k-nearest neighbor (k-NN) method [3, 4, 12], became very popular. Some authors advocate that the k-NN method outperforms other classifiers if correct number of neighbors and suitable similarity measure are selected (see e.g. [17]). Main problems are: high number training vectors and long computation times. Pekalska and Duin [18] considered a distance matrix directly as new training data. They proposed to use normal density-based linear and quadratic classifiers constructed on this data representation. Pekalska and coauthors [20 - 22] have demonstrated that the classifiers, based on weighted combination of dissimilarities, may be built on all training objects and thereby utilize global rather than local information which often leads to very good performance.

Often dissimilarity based classifiers give good results especially if complex nonlinear decision boundary is required to discriminate pattern classes. Sometimes utilization of dissimilarity concept does not lead to desirable result. Moreover, a number of "dissimilarity methods" became high. So, often an end user meets difficulties in selection of a proper method. As a result, a necessity arises to take unified look at variety on dissimilarity concept based classification methods, elucidate their positive and negative sides, to find new ways of proper use and develop recommendations for end users.

Unfortunately, up to now except recent PhD thesis of Pekalska [21], no systematic analysis of dissimilarity based classification algorithms was performed. In this paper we represented discriminant functions of many known pattern classifiers as weighted sums of kernels of new dissimilarity features. Depending on a way how the weights are calculated and on a kernel scaling, different supplementary linear and nonlinear classification algorithms follow from this representation. Simulation experiments performed with three real world data sets show that new algorithms can compete and often outperform many of known dissimilarity feature based classifiers.

2 Taxonomy of the Algorithms

Consider a case where p-dimensional vectors, $x = [x_1, x_2, ..., x_p]^T$ have to be allocated to one of two pattern classes. Above, superscript "T" denotes transpose operation. Let vectors $x_{j1}, x_{j2}, ... x_{jN_j}$ represent training set of the j-th pattern class. We will restrict our analysis with two category case and Euclidean distance based dissimilarity features (for reviews of a great variety of dissimilarity measures, see e.g. [13, 19 - 22] and references therein)

$$y_{js} = (x - x_{js})^T (x - x_{js})^{1/2}, \tag{1}$$

where s is a current number of vector's, x_{js}, in j-th class representation set.

In many classifiers, distances (1) are transformed by means of some kernel function. Often researcher use and speak well of the Gaussian density based kernel

$$y^*_{js} = \exp(-\alpha\, y^2_{js}), \tag{2}$$

where positive scalar α defines a width of the kernel.

Exponential transformation of the distances makes forming nonlinear decision boundaries easier. If all training set vectors compose representation set, unified expression of discrimiant function is

$$g^*(x) = \sum_{j=1}^{2} \sum_{s=1}^{N_j} w_{js} \exp(-\alpha(x - x_{js})^T (x - x_{js})) + w_0, \tag{3}$$

where w_0, w_{js} ($j = 1, 2,\ \ s = 1, 2, ..., N_j$) are weights. The weights have to be found from training data.

We will show that *representation (3) can result in a number of different classifiers*. In the nearest neighbor classification, previously unseen examples are assigned to the classes of their most similar neighbors in the training set. In k-NN approach, allocation of unknown vector x is performed according to a majority of training vectors of one pattern class. In kernel based (Parzen window approach) methods, potential function classifiers, certain estimate of multivariate density (named also as a potential function in some papers) is calculated for each pattern class

$$p_j(x) = \frac{1}{N_j} \sum_{s=1}^{N_j} \exp(-\gamma(x - x_{js})^T (x - x_{js})). \tag{4}$$

Allocation of unknown vector x is performed according to a sign of discriminant function $g(x) = q_1 p_1(x) - q_2 p_2(x)$, where q_j stands for prior probability of j-th pattern class ($q_1 + q_2 = 1$). If scalar γ is very small, values of $-\alpha(x - x_{js})^T (x - x_{js})$ are small too. Utilization of first term of Taylor expansion results that

$$\exp(-\alpha(x - x_{js})^T (x - x_{js})) \sim -\alpha(x - x_{js})^T (x - x_{js}). \tag{5}$$

Let us consider *two limit cases*. Let $\alpha \to 0$ at first. Inserting (5) into Eq. (3) with $w_{1s}=1$, $w_{2s}= -1$ after some simple algebra we obtain following discriminant function (DF), $g(x)$,

$$g\ (x) \equiv x^T w^{(E)} + w_0^{(E)} + \frac{1}{2}(1 - 1/N)(\text{tr}\ (\hat{\Sigma}_2 - \hat{\Sigma}_1)), \qquad (6)$$

where $w^{(E)} = \hat{\mu}_1 - \hat{\mu}_2$, $\hat{\mu}_j$ and $\hat{\Sigma}_j$ are sample p-dimensional mean vector and $p \times p$ covariance matrix of j-th class in original feature space (we assumed $N_2 = N_1 = N$), $w_0^{(E)} = -\hat{\mu}^T w^{(E)}$, $\hat{\mu} = \frac{1}{2}(\hat{\mu}_1 + \hat{\mu}_2)$.

Note, that $w_0^{(E)}$, $w^{(E)}$ are weights of Euclidean distance classifier (EDC). Ignoring term $\frac{1}{2}(1 - 1/N)(\text{tr}\ (\hat{\Sigma}_2 - \hat{\Sigma}_1))$ in the bias term, we see that in limit case when $\alpha \to 0$, similarity representation of DF (3) gives linear classifier similar to EDC. Let now $\alpha \to \infty$. Then the exponent $\exp(-\alpha c)$ in Eq. (3) approaches zero very fast. For that reason, we can use only first term of Taylor expansion. As a result, in this limit situation, we approach the nearest neighbor classifier, 1-NN. If ⌐⌐167 shaped kernels would be used, in each local sub-area, we could obtain classifiers similar to k-NN ones where kernel width would be associated with number of nearest neighbors, k.

In dissimilarity features approach, the new features corresponding to similar training vectors are highly correlated. Consequently, covariance matrices could become singular. Due to large correlations, the EDC is not appropriate for practical use. On the other hand, the nearest neighbor classifiers suffer from other shortcomings [20]. So, we have to use intermediate α values in practice.

If case of equal among themselves weights, i.e, $w_{21} = , ..., = w_{2N} = -w_{11} = , ... , = -w_{1N}$, we have the Parzen window classifier. In case of unequal weights, however, one can obtain more sophisticated classifiers. In Pekalska and Duin [20] approach, the weights were determined from a theory of standard and regularized discriminant analysis. Their approach originally proposed for similarity features (1) (actually, it is a situation when $\alpha \to 0$) can be used for kernel based dissimilarity features (2) too. In support vector machine (SVM) based methods (for introduction see e.g., [9, 10]), one also often uses kernel based dissimilarity features (2). In the SVM approach, the representation set is reduced essentially: only vectors that are closest to discriminant hyperplane are determining the decision boundary.

In addition to statistical and support vector classifiers utilized to find the weights w_0, w_{js}, one can make use of the single layer perceptron (SLP). In principle, while training SLP one can obtain a number of linear classifiers of different complexity. If a) the data centre is moved to a centre of coordinates, $\hat{\mu}$, before training the perceptron and b) one starts training of the perceptron from zero initial weights ($w_0 = 0$, $w = 0$), then after the first training iteration in the batch mode, one obtains the Euclidean distance classifier. In further training, one obtains linear regularized discriminant analysis (LRDA), standard linear Fisher classifier or the Fisher with the pseudo-inverse of covariance matrix if sample size is very small. In subsequent training, one can obtain robust classifier, a minimum empirical error classifier and approach the support vector machine [15, 23, 24]. Which classifier will be obtained practically depends on training conditions and most importantly on stopping moment. A variety of supplementary classifiers could be obtained if SLP would be trained in the dissimilarity feature space.

If representation set is large, the covariance matrix could become singular. In principle, eigen-values of the covariance matrix could differ in billions of times. For that reason, training of the perceptron becomes very slow (see Chapter 4 in [15]). To speed up training process and to reduce the generalization error it is worth to perform whitening transformation of the data

$$Z = T (y^* - \hat{\mu}),\qquad(7)$$

where $T = \Lambda^{-1/2} \Phi^T$, Λ, Φ^T are eigen-values and eigenvectors of $\hat{\Sigma} = \frac{1}{2}(\hat{\Sigma}_1 + \hat{\Sigma}_2)$ and y^* is dissimilarity feature vector composed of components y^*_{js} defined by Eq. (2).

One can make use of a variety of statistical methods to improve sample estimate of covariance matrix, $\hat{\Sigma}$. Consequently, a great number of extra classifiers could be derived again.

3 Analysis of 2D Example

A main objective of present paper is to discuss a broad gamma of various nonlinear decision boundaries which could be obtained while using the dissimilarity features. In order to explain parallels and differences between distinct classification rules in Fig. 1abc we present specially constructed artificial two class data where decision boundary is nonlinear. We depict 100+100 2D two class vectors (pluses and circles) that compose both the training and the representation sets. The data vectors of each of two pattern classes are distributed uniformly between two "ellipses" composed of four straight lines and four segments of the circles. There is no margin and no intersection between vectors of opposite categories. Ideal decision boundary is the "ellipse" depicted as thin solid line. Decision boundary of the Parzen window classifier with optimal value of smoothing parameter (it was a small positive constant found from test set (5000+5000 vectors) estimates of classification error) is depicted as nonlinear weaved bold curve in dots in Fig. 1a. It correspond to $P_{gen} = 0.0566$.

Much better classification results were obtained if linear classifiers based on dissimilarity representation (2) were designed. The LRDA with covariance matrix estimate $S_{RDA} = (1-\lambda)S_{sample} + \lambda I$ (λ – regularization parameter, I – identity matrix) resulted in a smooth decision boundary similar to that depicted in figures 1b and 1c (the closest to 200D hyperplane vectors are marked by large circles). To find optimal regularization constant the test set was used. The generalization error, $P_{gen} = 0.0265$ is twice smaller as that of Parzen window classifier and confirms conclusions obtained by Pekalska and Duin [20]. Standard SVM with soft margin (package LIBSVM, www.csie.ntu.edu.tw/~cjlin/libsvm/) gave smooth decision boundary where three training vectors were misclassified. The generalization error only insignificantly exceeded that of RDA: $P_{gen} = 0.0294$.

Training of the SLP classifier with gradually increasing learning step (see e.g. [15], Chapter 4) after 35000 batch iterations allowed obtaining "SVM" with positive margin. Performance of such SVM, $P_{gen} = 0.0345$ (bold dotted decision boundary in Fig. 1b). A histogram of distribution of distances to discriminant hyperplane in dissimilarity feature space (2) shows that rather wide margin was obtained in similarity feature space (Fig. 2). Decision boundary and support vectors marked by

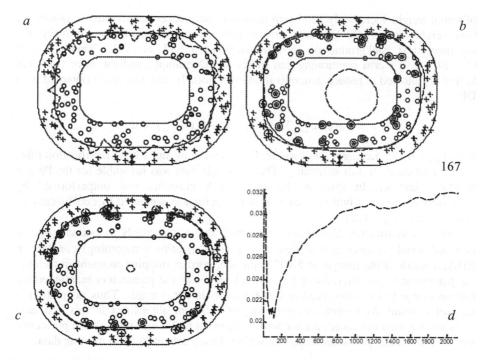

Fig. 1. 100+100 2D two class training vectors (pluses and circles) and decision boundaries of Parzen window classifier (*a*), SVM (*b*), SLP trained in 200D dissimilarity feature space (*c*). Learning curve: generalization error of SLP classifier as a function of a number of training epochs (*d*).

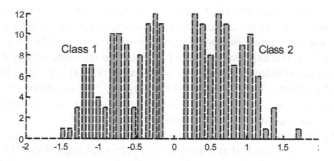

Fig. 2. A histogram of distribution of distances of 200 training vectors to discriminant hyperplane in 200D dissimilarity feature space

large circles around 2D training vectors in original feature space, show, however, that in *original* 2D *feature space, a part of "support vectors" are far away from the nonlinear decision boundary* (Fig. 1*b*).

The best classification performance, however, was achieved after early stopping of the SLP training procedure. After preliminary data transformation performed in order to equalize eigen-values of covariance matrix $\hat{\Sigma}$, $P_{gen} = 0.0205$. Due to good position of initial weight vector only 120 batch iterations were necessary in order to obtain the best weight vector in 200D dissimilarity feature space (see learning curve in Fig. 1d; for theoretical background of such transformation look at [15], Chapter 5, and [23, 24]). It is worth mentioning that in above 2D example, utilization of Gaussian kernel (2) allowed to reduce generalization error notably: the best result obtained with DF

$$g(x) = \sum_{j} \sum_{s} w_{js} (x - x_{js}))^T (x - x_{js})) + w_0$$

was: $P_{gen} = 0.185$, i.e. *eight times worse*. Parzen window and k-NN classification rules suffer from curse of dimensionality. Therefore, 2D data was favorable for the Parzen window classifier. In spite of this factor, PW classifier was outperformed by classifiers where contribution of each training vector was weighted by coefficients w_{js} ($s = 1, 2, ..., N$, $j = 1, 2$).

In all classification algorithms considered above, one needs to estimate one or two additional parameters (coefficient α, a regularization (smoothing) constant in RDA, a width of the margin in SVM approach, correct stopping moment in training the perceptron). For determining the optimal values of the parameters mentioned we had very large test set composed of 10000 vectors in our example. Thus, adaptation to test set is small. Moreover, in comparing of different algorithms in this and next sections, we used the same data for testing them. In real world recognition problems, we meet additional problems while evaluating these parameters from training data.

4   Experiments with Three Real World Data Sets

Dissimilarity feature transformation performed before the classifier design is an important step aimed to improve classification accuracy while solving real-world problems. The results obtained with the artificial data do not generalize. Above example was used to illustrate specific characteristics of similarity features based classification algorithms. Below we present comparative simulation experiments with real world data. The experiments were performed in the same way as described above. We formed dissimilarity features as $y_{js} = (x - x_{js})^T (D+\varepsilon I)^{-1} (x - x_{js})$ in order to avoid numerical problems. Here **D** stands for a diagonal matrix composed of variances of input variables, x_{js}, **I** stands for $r \times r$ identity matrix and $\varepsilon = 0.001$. Afterwards, variances of all the dissimilarity features were normalized to 1.

The Satellite data (S) consists of 15787 8-dimensional spectral feature vectors from 25 small crop fields that belong to two pattern classes; 4384+3555 vectors were allocated for classifiers design and 4242+3606 vectors were used for testing. The experiments were repeated many times with different randomly chosen training sets composed of design set vectors. In Table 1 we presented mean generalization errors when standard classifiers in original feature space were used. Notation "NormEDC" stands for nearest mean classifier (EDC) when for determining the prototypes of the classes and for decision making, the input features were normalized according to their standard deviations. SLP was trained according to suggestions presented in [15]. MLP was trained according to Levenberg-Marquardt optimization. For each number of hidden units, h, five random initializations and subsequent training sessions were

Table 1. Mean generalization errors of standard classifiers in original feature spaces (Satellite (S) and character recognition (Jch) tasks)

Training set size	LRDA	NormEDC	Fisher DF	SLP	MLP $h=5$	MLP $h=20$
S^1 4384+4242	0.159 (0.16)	0.215	0.160	0.137	0.045	0.029
S^{10} 219+177	0.162 (0.02)	0.215	0.162	0.139	0.083	0.068
S^{10} 87+71	0.163 (0.13)	0.215	0.163	0.143	0.111	0.098
Jch^{100} 100+100	0.053 (0.55)	0.124	0.057	0.056	0.081	0.069

performed. In all experiments, optimal number of training epochs and the best initialization were selected according to test set estimates (4242+3606 vectors). Left column's superscripts indicate a number of experiments; average values of optimal regularization parameter of LRDA, λ, are given between the brackets.

In analysis of similarity concept based classifiers, we used geographic coordinates of each pixel to form representation set composed of $r = 50$ pixels which formed 50 dissimilarity features. Three kernels: a) y_{js} defined by Eq. (1) – D^1, b) $(y_{js})^2 - D^2$ (no kernel), and c) y_{js} defined by Eq. (2) with $\alpha=5$ and $\alpha=2$ (near optimal value), were considered. Results presented in upper three rows of Table 2 show that classification accuracy depends crucially on selection of the kernel type. Utilization of SLP trained in whitened 50-dimensional feature space appeared useful and outperformed regularized discriminant analysis in case of utilization of all three kernels.

Table 2. Generalization errors of classifiers based on the similarity features

Training set size	LRDA (D^2)	LRDA (D^1)	LRDA $\exp\alpha5$	LRDA $\exp\alpha2$	SLP (D^2)	SLP (D^1)	SLP $\exp\alpha5$	SLP $\exp\alpha2$
S^1 4384+4242	.159 (.23)	.052 (.001)	.100 (.001)	.074 (.001)	.132	**.041**	.055	.048
S^{10} 219+177	.161 (.28)	.060 (.001)	.102 (.001)	.082 (.001)	.137	**.055**	.070	.064
S^{10} 87+71	.164 (.34)	.069 (.002)	.105 (.001)	.088 (.001)	.142	**.067**	.079	.075
Jch^{100}100+100	.040 (.95)	.050 (.001)	.070 (.001)	.050 (.001)	.050	**.035**	.040	.035

Similar Japanese characters (Jch) recognition task was one of pairs of similar handwritten characters considered in [25] (see an example, 間 間 ; 196 features).

To test competing classifiers, LRDA, SLP and MLP, we have chosen 28 first principal components of the data. In dissimilarity feature analysis, we selected every 7th vector of the design set (28 new dissimilarity features). In each of 100 experiments, we randomly selected 100+100 vectors for training and the same amount for testing. Like in previous examples, we found that classification accuracy depends on kernel type. Consequently, specially trained SLP is a good choice in designing weighted sums of kernel representations of the dissimilarity features.

Machine Vibration Data. In experiments with 200-dimensional spectral data (good and defected electro motors) we obtained similar conclusions: in series of 100 experiments performed with random splits of the data into 100+100 vectors for training and 100+100 ones for test, the SLP classifier trained in 50D similarity feature space (Eq. (2)) the SLP classifier resulted in P_{gen}= 0.048, while LRDA gave P_{gen}= 0.061. Again, the best result obtained in original feature space was much worse.

In all cases except one (MLP with 20 hidden units and very large training set in the experiments with the Satellite data), specially trained SLP outperformed MLP, SLP and LRDA in original feature spaces (compare tables 1 and 2).

5 Conclusions

In this paper we represented discriminant functions of many known pattern classifiers as the weighted sums of kernels of new dissimilarity features. Depending on a way how the weights are calculated and on kernel scaling (parameter α in Eq. (3)), a great amount of linear and nonlinear classification algorithms follows from this representation. The unified look allows obtaining intermediate classifiers with good generalization properties. We found that selection of the kernel's type while forming dissimilarity features is very important step in designing the classifiers. Selection of functional kernel is problem dependent. Utilization of data whitening and subsequent training of the single layer perceptron allows obtaining the classifiers that can compete and often outperform LRDA, MLP and many other dissimilarity feature based classifiers.

Acknowledgments

The author thanks Dr. Krystyna Stankiewicz, Masakazu Iwamura and Robert Duin for providing real-world data sets for the experiments.

References

[1] Arkedev A.G., Braverman E.M. *Computers and Pattern Recognition*: Thompson, Washington, DC, 1966.

[2] Duin R.P.W. Compactness and complexity of pattern recognition problems. In: Perneel C., editor. *Proceedings of the Int. Symposium on Pattern Recognition. In Memoriam Pierre Devijver, Brussels*, B, February 12. Brussels: Royal Military Academy; p. 124–128, 1999.

[3] Duda R.O., Hart .P.E. and Stork D.G. *Pattern Classification and Scene Analysis*. 2nd edition. John Wiley, New York, 2000.

[4] Fix E, Hodges JLJr. Discriminatory analysis, nonparametric discrimination: consistency properties. *Report No. 4, Project 21-49-004*. USAF School of Aviation Medicine, Randolph Field, TX, 1951.

[5] Attneave F. Dimensions of similarity, *Am. J. Psychology*, 63: 516-556, 1950.

[6] Parzen E. On estimation of probability function and mode. *Annals of Mathematical Statistics,* 33:1065–76, 1962.

[7] Wolverton C.T. and Wagner T.J. Asymptotically optimal discriminant functions for pattern classification. *IEEE Transactions on Information Theory*, IT-15:258–65, 1969.

[8] Boser B., Guyon, I. and Vapnik V. A training algorithm for optimal margin classifiers. In *Proceedings of the Fifth Annual Workshop on Computational Learning Theory*, Pittsburgh: ACM, pp. 144–152, 1992.

[9]   Scholkopf B., Burges CJC. and Smola A.J. *Advances in Kernel Methods: Support vector learning.* MIT Press, Cambridge, MA, 1999.

[10] Cristianini N. and Shawe-Taylor J. *An Introduction to Support Vector Machines and Other Kernel-based Learning Methods.* Cambridge Univ. Press, Cambridge, UK, 2000.

[11] Aha, D.W., Kibler, D., Albert, M.K. Instance-based learning algorithms. *Machine Learning,* 6, 37–66, 1991.

[12] Althoff, K.-D. Case-based reasoning. In: S. K. Chang (Ed.), *Handbook on Software Engineering and Knowledge Engineering.* Vol. 1 (Fundamentals), World Scientific, 549-588, 2001.

[13] Perner P. Are case-based reasoning and dissimilarity-based pattern recognition two sides of the same coin? *Lecture Notes in Artificial Intelligence.* Springer-Verlag, 2123: 35-51, 2001.

[14] Schuermann J. *Pattern Classification: A unified view of statistical and neural approaches,* John Wiley, New York, 1996.

[15] Raudys S. *Statistical and Neural Classifiers: An integrated approach to design.* Springer, London, 2001, 2001.

[16] Duin R.P.W., Pekalska E., and De Ridder D. Relational discriminant analysis. *Pattern Recognition Letters,* 20 (11-13): 1175-1181, 1999.

[17] Paclik P. and Duin R.P.W. Dissimilarity-based classification of spectra: computational issues. *Real-Time Imaging,* 9: 237–244, 2003.

[18] Mottl, V. Seredin, O. Dvoenko, S. Kulikowski, C. Muchnik, I. Featureless pattern recognition in an imaginary Hilbert space. *Proc. 16th International Conference on Pattern Recognition.* Vol. 2: 88- 91, 2002.

[19] Santini S. and Jain R. Similarity measures. *IEEE Trans. on Pattern Analysis and Machine Intelligence,* 21(9): 871-883, 1999.

[20] Pekalska E. and Duin R.P.W. Dissimilarity representations allow for building good classifiers. *Pattern Recognition Letters,* 23(8): 943–56, 2002.

[21] Pekalska E. *Ph. D. Thesis.* Delft University of Technology, 2005.

[22] Pekalska E., Paclik P. and Duin R.P.W. A generalized kernel approach to dissimilarity based classification. *Journal of Machine Learning Research,* 2: 175–211, 2002.

[23] Raudys S. Evolution and generalization of a single neurone. I. SLP as seven statistical classifiers. *Neural Networks,* 11: 283–96, 1998.

[24] Raudys S. How good are support vector machines? *Neural Networks,* 13, 9-11, 2000.

[25] Raudys S. and Iwamura M. Structures of covariance matrix in handwritten character recognition. *Lecture Notes in Computer Science,* Springer-Verlag. LNCS, 3138: 725-733, 2004.

Combination of Boosted Classifiers Using Bounded Weights

Hakan Altınçay and Ali Tüzel

Computer Engineering Department, Eastern Mediterranean University,
Gazi Mağusa, KKTC Mersin 10, Turkey
{hakan.altincay, ali.tuzel}@emu.edu.tr

Abstract. A recently developed neural network model that is based on bounded weights is used for the estimation of an optimal set of weights for ensemble members provided by the AdaBoost algorithm. Bounded neural network model is firstly modified for this purpose where ensemble members are used to replace the kernel functions. The optimal set of classifier weights are then obtained by the minimization of a least squares error function. The proposed weight estimation approach is compared to the AdaBoost algorithm with original weights. It is observed that better accuracies can be obtained by using a subset of the ensemble members.

1 Introduction

Ensemble of weak classifiers is considered as alternative approach to develop strong classifiers in pattern classification problems. Boosting is an iterative ensemble design technique which takes into account the classification results of the previous classifiers to construct additional ones. AdaBoost is the most popular boosting algorithm which can be considered as applying a steepest descent search to minimize an exponential loss function [1,2]. The resultant classifiers are combined using weighted majority voting to generate a joint decision.

Recently, a new neural network model is proposed for classification purposes [3]. This approach is mainly inspired from the bounded weight nature of the support vector machines (SVM) [4,5]. The main idea is to estimate the weights in the links by minimizing the least squared error between the network outputs and the target vectors, leading to a quadratic optimization problem. However, similar to the SVM model, the magnitude of the weights are restricted to $0 \le \alpha_i \le C$. In fact, the bound constraint in SVM approach is one of the sources of its strengths since the learning capability can be controlled and hence overfitting can be avoided. Because of this, it is expected that putting bounds on the link weights may also improve the performance of neural network classifiers. The main advantage of the bounded model is that the kernel function used should not necessarily satisfy Mercer's condition as in SVM. Experimental results have shown that bounding the link weights in neural networks could provide better results compared to SVM [3].

Least squared error minimization with bound constraint on the solution has several advantages from ensemble design point of view. Firstly, its convex

S. Singh et al. (Eds.): ICAPR 2005, LNCS 3686, pp. 146–153, 2005.

quadratic objective function provides global optimal solution. Since arbitrary kernels can be used, it can be used for the weighted combination of classifiers where each classifier is considered as a different kernel function. Moreover, the bounded weight nature reduces the risk of overfitting. The weights having zero values may be expected to correspond to redundant classifiers and hence an implicit classifier subset selection is performed.

In this study, the framework described above is modified for the estimation of an optimal set of weights for ensemble members obtained using AdaBoost algorithm. The ensemble members are defined to be the kernel functions where the solution of least squares error provides the optimal set of classifier weights. The proposed approach is compared to the original weights generated by AdaBoost algorithm and it is observed that better combined accuracies can be obtained by using a subset of the ensemble members.

2   Least Square Error Minimization with Bounded Weights

Consider a two-class classification problem with the training set $S = \{(\bar{x}_n, y_n)\}$, $n = 1, \ldots, N$ where $\bar{x}_n \in \Re^d$ denotes the nth input sample and $y_n \in \{+1, -1\}$ is its label. Consider a three layer neural network where each hidden node represents the kernel function $\kappa(\bar{x}_n, \bar{x})$ defined for the nth training sample. The output is the weighted linear combination of the kernel function outputs where the label of $\bar{x}_n$ denoted by y_n is either $+1$ or -1. α_n is the weight of the nth kernel and b is the decision threshold to be computed by solving the following convex quadratic programming problem [6]:

$$\min_{\alpha, b} \frac{1}{2} \sum_{i=1}^{N} \left\| y_i - \left(\sum_{n=1}^{N} \alpha_n y_n \kappa(x_n, x_i) + b \right) \right\|^2 \tag{1}$$

$$\text{subject to} \ 0 \leq \alpha_n \leq C, \ n = 1, 2, \ldots, N$$

where C is a constant upper bound. The discriminant function is defined as,

$$g(\bar{x}) = \sum_{n=1}^{N} \alpha_n y_n \kappa(\bar{x}_n, \bar{x}) + b. \tag{2}$$

During testing an unseen data $\bar{x}_t$, if $g(\bar{x}_t) \geq 0$, $\bar{x}_t$ is assigned to the first class. Otherwise, the second class is selected.

The objective function of the optimization problem defined in Eq. (1) can be formulated in matrix form as,

$$J = \frac{1}{2} \|\bar{y} - (KY\bar{\alpha} + b\bar{e})\|^2 \tag{3}$$

where $\bar{y} = [y_1, y_2, \ldots, y_N]^T$, $\bar{\alpha} = [\alpha_1, \alpha_2, \ldots, \alpha_N]^T$, $X = [\bar{x}_1, \bar{x}_2, \ldots, \bar{x}_N]$, $Y = diag(y_1, y_2, \ldots, y_N)$, $\bar{e}$ is an N-dimensional column vector of ones and,

$$K = \begin{bmatrix} \kappa(\overline{x}_1, \overline{x}_2) & \cdots & \kappa(\overline{x}_1, \overline{x}_N) \\ \vdots & \ddots & \vdots \\ \kappa(\overline{x}_N, \overline{x}_1) & \cdots & \kappa(\overline{x}_N, \overline{x}_N) \end{bmatrix} \tag{4}$$

The objective function is also equivalent to,

$$J = \frac{1}{2}\widehat{\alpha}^T \widehat{K}^T \widehat{K}\widehat{\alpha} - \widehat{q}^T \widehat{\alpha} \tag{5}$$

where $\widehat{\alpha} = [\overline{\alpha}^T, b]$ which is an $N + 1$ dimensional vector, $\widehat{K} = [KY|\overline{e}]$ and $\widehat{q}^T = [\overline{y}^T KY, \overline{y}^T \overline{e}]$. This is a convex quadratic optimization problem since the term $\widehat{K}^T \widehat{K}$ is positive semi-definite. The main difference from SVM's is that the Mercer's condition is not required in the kernel function selection. Hence, this approach is more flexible.

Consider the discriminant function given in Eq. (2). The kernel function located at the nth training sample together with its label, $y_n \kappa(\overline{x}_n, \overline{x})$ can be considered as a primitive classifier. The classifier always provides support for the class to which $\overline{x}_n$ belongs. Then, the discriminant function can be considered as a multiple classifier systems implemented as a weighted linear combination of N classifiers. Since some of the weights may come out to be zero after training, this approach can be considered to implicitly apply classifier subset selection.

Remembering that the Mercer's condition is not required to be satisfied, this framework can be considered for the weighted linear combination of classifiers. In this study, a slightly modified formalism is proposed for this purpose. The resultant system is used for the weighted linear combination of the classifiers which are obtained using AdaBoost algorithm.

3   Linear Classifier Combination with Bounded Weights

Let $f_m(.)$ denote the mth classifier in an ensemble of M classifiers. Using weighted linear combination rule, the combined discriminant function can be defined as [7],

$$g(\overline{x}) = \sum_{m=1}^{M} \alpha_m f_m(\overline{x}) + b. \tag{6}$$

where α_m are the classifier weights. Applying the least squares error minimization technique with bounded weights, the optimization problem can be defined as,

$$\min_{\alpha, b} \frac{1}{2} \sum_{n=1}^{N} \left\| y_n - \left(\sum_{m=1}^{M} \alpha_m f_m(\overline{x}_n) + b \right) \right\|^2 \tag{7}$$

$$\text{subject to } 0 \le \alpha_m \le C, \; m = 1, 2, \ldots, M$$

In matrix form, the objective function can be written as,

$$J = \frac{1}{2}\|\overline{y} - (K\overline{\alpha} + b\overline{e})\|^2 \tag{8}$$

where $\bar{y}$, $\bar{\alpha}$ and $\bar{e}$ are defined as before and,

$$
K = \begin{bmatrix} f_1(\overline{x}_1) & f_2(\overline{x}_1) & \cdots & f_M(\overline{x}_1) \\ f_1(\overline{x}_2) & f_2(\overline{x}_2) & \cdots & f_M(\overline{x}_2) \\ \vdots & & \ddots & \vdots \\ f_1(\overline{x}_N) & f_2(\overline{x}_N) & \cdots & f_M(\overline{x}_N) \end{bmatrix} \tag{9}
$$

We can expand J further as,

$$
\begin{aligned}
J &= \frac{1}{2}\left(\bar{y} - (K\bar{\alpha} + b\bar{e})\right)^T \left(\bar{y} - (K\bar{\alpha} + b\bar{e})\right) \\
&= \frac{1}{2}\bar{\alpha}^T K^T K\bar{\alpha} - \bar{y}^T(K\bar{\alpha} + b\bar{e}) + b\bar{e}^T K\bar{\alpha} + \frac{1}{2}b\bar{e}^T\bar{e}b + \frac{1}{2}\bar{y}^T\bar{y} \\
&= \frac{1}{2}\widehat{\bar{\alpha}}^T \widehat{K}^T \widehat{K}\widehat{\bar{\alpha}} - \bar{q}^T\widehat{\bar{\alpha}} + \frac{1}{2}\bar{y}^T\bar{y}
\end{aligned} \tag{10}
$$

where, $\bar{\alpha}$ is defined as before and $\widehat{K} = [K|\bar{e}]$ and $\bar{q}^T = [\bar{y}^T K, \bar{y}^T\bar{e}]$. Since $\bar{y}^T\bar{y}$ is constant, it can omitted. The optimization problem becomes,

$$
\frac{1}{2}\widehat{\bar{\alpha}}^T \widehat{K}^T \widehat{K}\widehat{\bar{\alpha}} - \bar{q}^T\widehat{\bar{\alpha}} \tag{11}
$$
$$
\text{subject to} \quad 0 \le \alpha_m \le C, \ m = 1, 2, \ldots, M
$$

where $\widehat{\bar{\alpha}} = [\bar{\alpha}^T, b]$. It should be noted that the resultant formalism is almost identical to BNN where the training sample dependent kernel function is replaced by arbitrary classifiers. In this case, $\widehat{K}$ is an $N \times M$ matrix where it was an $N \times N$ before. The solution of this optimization problem provides optimal classifier weights and the threshold, b. Due to the positive semi-definite matrix $\widehat{K}^T\widehat{K}$, this is a convex quadratic problem where the classifiers can be arbitrarily selected without having to satisfy any condition.

4   Experiments

In order to investigate the effectiveness of the proposed weighted linear combination technique, combination of classifiers obtained using AdaBoost is considered. The performances obtained using the weights computed by the proposed technique are compared with the original weights of AdaBoost for three different base classifiers namely, normal densities based quadratic discriminant classifier (QDC), nearest mean classifier (NMC) and neural networks (NNET) [8]. In the simulations, 2-class problems are considered with $C = 1$ and $f_m(\overline{x}_t)$ is set as $+1$ and -1 when the most likely class is first and second, respectively. The experiments are performed on an artificial and eight real datasets. Using an artificial data set including 2-dimensional pattern samples, it is aimed at visualization of the advantages of the proposed framework by inspecting the selected classifiers. The 'Highleyman' classes are generated using the PRTOOLS toolbox [9]. One thousand samples are generated where half of the data are used for training and

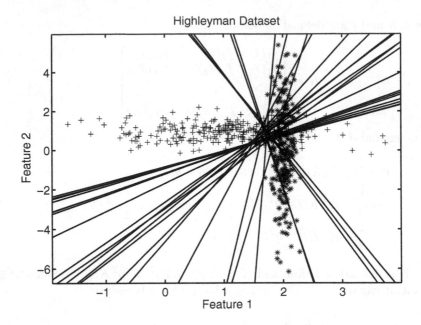

Fig. 1. An ensemble of 25 NMC classifiers on highleyman dataset

the remaining half for testing. The AdaBoost algorithm is run to generate an ensemble of 25 NMC type classifiers [8]. The classifiers developed by AdaBoost and the subset of the classifiers selected by the proposed algorithm are illustrated on the scatter plot of the training data in Figures 1 and Figures 2 respectively.

As it is seen in the figure, the proposed algorithm selected a small subset of the available classifiers. The figure shows that, although the classifiers are trained on resampled data sets, some of the resultant classifiers may be correlated. Since the NMC type classifier is a weak one, it is less sensitive to changes in the training set providing decision boundaries that may be close to some others leading to correlated decisions. The classifiers getting nonzero weights are different from each other, leading to fusion of less correlated decisions.

The experiments are also conducted on eight IAPR TC-5 benchmarking datasets[1] obtained from the UCI machine learning repository. The experiments are repeated for first ten splits (tra_0_0 to tra_1_4 for training and tst_0_0 to tst_1_4 for testing) and the average accuracies are computed.

The experimental results are presented in Tables 1 and 2. In Table 1, the second column provides the number of training and test samples used. The average accuracies over ten simulations for the NMC type base classifier are given in the third column. Following two columns list the accuracies achieved by AdaBoost and the proposed technique (referred as BBoost) respectively for an ensemble size of $M = 25$ classifiers. Table 2 presents the results obtained for

---

[1] http://algoval.essex.ac.uk/tc5/

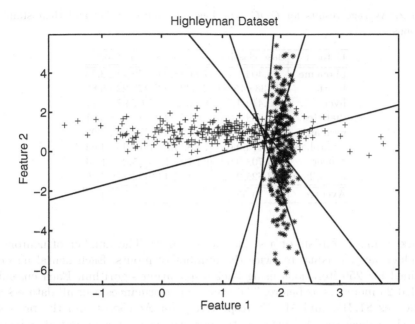

Fig. 2. The subset of classifiers selected by the proposed algorithm

Table 1. Average results for NMC type base classifiers (in %) and their standard deviations

Dataset	train-test count	Base classifier	AdaBoost	BBoost
phoneme	4323-1081	71.53±2.5	72.01±2.33	75.34±2.36
heart	216-54	63.70±8.29	69.07±5.86	70.19±5.27
liver	276-69	55.07±5.16	60.44±6.19	65.80±6.27
diabetes	614-154	62.97±5.32	70.71±4.55	73.44±3.01
german	800-200	63.30+4.49	63.75±4.79	68.10±2.77
australian	552-138	65.52±1.25	65.79±1.26	69.06±2.87
cancer	546-137	95.70±1.86	95.79±1.83	96.25±1.89
ionosphere	281-70	72.81±5.89	85.04±3.77	85.47±4.09
Average	N/A	68.83	72.83	75.46

QDC type base classifier. As it can be seen in the tables, the bounded weighting technique improves accuracies with the use of smaller number of classifiers.

The main observation is that the improvement is more significant in the NMC base classifier case. Since the weights are bounded, subset of the ensemble members are generally used. When averaged on all datasets, 9 and 15 classifiers are used for NMC and QDC type base classifiers respectively. As it was described above, it is reasonable to have fewer NMC classifiers selected since they are expected to be more correlated compared to the QDC ensemble.

The proposed scheme is also evaluated for comparatively less stable classifiers. For this purpose, neural networks are used as base classifiers. Each neural

Table 2. Average results for QDC type base classifiers (in %) and their standard deviations

Dataset	Base classifier	AdaBoost	BBoost
phoneme	78.40±1.47	79.03±2.08	79.71±2.26
heart	68.33±15.99	77.22±5.99	77.22±5.86
liver	58.41±4.59	65.07±5.53	67.97±6.17
diabetes	73.05±2.23	74.54±3.47	73.83±3.91
german	70.80±4.47	71.65±3.37	72.00±2.92
australian	61.17±9.08	74.83±6.61	79.36±4.64
cancer	94.70±0.93	94.79±1.73	95.06±1.74
ionosphere	83.62±2.72	91.18±4.00	92.74±2.36
Average	73.56	78.54	79.74

network member consists of a single hidden layer. The number of neurons in the hidden layer is equal to twice the number of inputs. Each neural network is trained for 250 iterations using backpropagation algorithm. Each ensemble included 25 members as before. The average performances over all datasets are obtained as 81.31% and 81.46% respectively for AdaBoost and the proposed approach, respectively. In this case, the performance is not improved as before. This is in fact what we expected. Since the NNET type classifiers are less stable, the resultant ensemble members are not correlated as in NMC case. In other words, AdaBoost algorithm does not generate redundant members as in NMC case. This is also evident from the number of classifier selected by the proposed approach. When averaged over all datasets, 21 out of 25 classifiers are selected.

There are other parameters that also affect the performance of AdaBoost. For instance, the size of datasets is important for boosting. It is already known that AdaBoost performs better for larger datasets [10]. We observed that the proposed technique has a potential also in the case of small datasets. Although the improvement is not significant, the proposed technique improved the accuracies provided by AdaBoost for the small datasets, "liver" and "ionosphere" more than 1%. This is reasonable since samples drawn from small datasets may not give a good representation for the real distribution of the data. Hence, some ensemble members may provide misleading information. The proposed approach may be useful in discarding such poor members.

5   Conclusions

In this study, a least squares error minimization approach with bounded weights is developed for the estimation of an optimal set of weights for ensemble members obtained using AdaBoost algorithm. It is observed that the proposed approach may provide better combined accuracies by using a subset of the ensemble members. Since the gain in accuracy is different for NMC, QDC and neural networks, the proposed approach should be further investigated for the relation between the base classifier and the gain in the classification accuracy. In particular, the effect of capability and stability of the base classifier on the performance of the proposed approach should be studied. Its effectiveness in selection of classifier subsets obtained by approaches other than AdaBoost should also be studied. The extended version of bounded neural networks is also available for multi-classes case [3]. As a further research, extension of the proposed technique into multi-classes case will be considered.

References

1. R. E. Schapire. A brief introduction to boosting. In *Proceedings of the Sixteenth International Joint Conference on Artificial Intelligence*, 1999.
2. R. E. Schapire. The boosting approach to machine learning: An overview. *MSRI Workshop on Nonlinear Estimation and Classification, Berkeley, CA*, 2002.
3. Y. Liao, S. C. Fang, and H. L. W. Nuttle. A neural network model with bounded-weights for pattern classification. *Computers and Operations Research*, 31:1411–1426, 2004.
4. C. J. C. Burges. A tutorial on support vector machines for pattern recognition. *Data Min. Knowl. Discov.*, 2(2):121–167, 1998.
5. B. Scholköpf. Statistical learning and kernel methods. *Technical Report, MSR-TR-2000-23, Microsoft Research*, 2000.
6. D. G. Luenberger. *Introduction to linear and nonlinear programming*. Addison-Wesley Pub. Co., 1973.
7. S. Hashem. Optimal linear combinations of neural networks. *Neural Networks*, 10(4):599–614, 1997.
8. R. O. Duda, P. E. Hart, and D. G. Stork. *Pattern Classification*. John Wiley and Sons, 2000.
9. R. P. W. Duin. PRTOOLS (version 4.0). A Matlab toolbox for pattern recognition. *Pattern Recognition Group, Delft University, Netherlands*, 2004.
10. M. Skurichina and R. P. W. Duin. Bagging, boosting and the random subspace method for linear classifiers. *Pattern Analysis and Applications*, 5:121–135, 2002.

Prediction of Commodity Prices
in Rapidly Changing Environments

Sarunas Raudys[1,2] and Indre Zliobaite[1]

[1] Dept. of Informatics, MIF, Vilnius University
[2] Institute of Mathematics and Informatics, Naugarduko st. 24, Vilnius 03225, Lithuania
raudys@ktl.mii.lt, indre.zliobaite@mif.vu.lt

Abstract. In dynamic financial time series prediction, neural network training based on short data sequences results to more accurate predictions as using lengthy historical data. Optimal training set size is determined theoretically and experimentally. To reduce generalization error we: a) perform dimensionality reduction by mapping input data into low dimensional space using the multilayer perceptron, b) train the single layer perceptron classifier with short sequences of low-dimensional input data series, c) each time initialize the perceptron with weight vector obtained after training with previous portion of the data sequence, d) make use of useful preceding historical information accumulated in the financial time series data by the early stopping procedure.

Keywords: Classification, changing environments, commodity prices, forecasting, neural networks.

1 Introduction into the Problem

A characteristic feature of current research in economic and financial data mining methods is the environment dynamics. The changes in nature of the data could be minor fluctuations of the underlying probability distributions, steady trends, random or systematic, rapid substitution of one prediction task with another [1]. Therefore, in the financial prediction task, the algorithms ought to include the means to reflect the changes, to be able to operate in new, unknown environments, to adapt to sudden situational changes, to compete with other methods and to survive [2]. If only very short training sequences could be allowed, one needs to reduce both dimensionality of the data (number of features) and complexity of the training algorithm. Moreover, one needs to find ways to utilize useful information accumulated in the times series history. The objective of the present research was to analyze the theoretical background of factually unsolvable financial time series prediction task, suggest methods to achieve more reliable forecasting results in situations when the environment is changing permanently. We reformulate forecasting task as pattern classification problem in present paper.

2 Originality and Contribution

Analysis of financial time series usually is aimed to predict the price of a given security in the future. What is foremost important in this task is the direction of the

S. Singh et al. (Eds.): ICAPR 2005, LNCS 3686, pp. 154–163, 2005.

price change (i.e. price goes up or down as compared to today's value) prior to absolute value of the prediction error (i.e. by how many basis points the price changes). Wrong direction of prediction for investors concerned might bring significantly higher losses than absolute error in predicted security price, having right predicted direction of the price change. We analyze financial time series prediction task delimiting it to classification task, i.e. classifying the trading days into days of increase and days of decrease in security price as compared to previous periods. Such delimitation serves for two purposes: 1) analysis gets less complicated and the results achieved are more convenient to interpret; 2) analysis gives more value added in practice. We solve financial time series prediction task using classification tools, therefore, hereinafter we refer to *prediction when addressing the task* and we refer to *classification when discussing the tools.*

To develop classification and prediction algorithms aimed to work in changing environments *training of the algorithms ought to be based on short learning sequences.* For that reason, the algorithms should operate in a low dimensional space and be able to make use of partially correct historical information accumulated in the past. To achieve this goal, data is mapped into a low dimensional space by wrapper approach based multilayer perceptron MLP training firstly considered in [3]. To have simple nonlinear classifier for final classification, second order polynomial features are formed in the new feature space and single layer perceptron (SLP) based classifier is trained starting from weight vector obtained from previous portion of the time series data. To save useful information extracted from previous data the training process is terminated far beyond minimum of current cost function is obtained.

By analysis of training dynamics of standard Fisher linear discriminant function a presence of minimum of training sequence length is demonstrated theoretically if data is changing stochastically. The optimal length of training data is inversely proportional to intensity of data changes. Usefulness of the suggested forecasting methodology is demonstrated by solving commodity prices forecasting task from stock market statistical data.

3 Researches in the Field and Competing Techniques

Financial forecasting task is of great interest to businessmen as well as to academicians. The reader is referred to excellent review [4], where several methods for forecasting using artificial neural networks (ANN) are compared, to a large extent applicable to various financial time series. The environment dynamics and researches in the field performed already for several years are considered in [2, 4, 5]. Kuncheva [1] states classification of possible changes in the class descriptions into: random noise, random trends, random substitutions and systematic trends and suggests employing different strategies for building the classifier, depending on the type of changes.

To select small number of attributes for training and performing forecasting task, typically one uses fixed length of historical data and reduces the number of input factors. Moody states and experimentally shows in [6], that there exists an optimum for training window length at which the test error is minimal. Sliding window approach for commodity price prediction was used by Kohzadi et al [7] back in 1996. Fieldsend and Singh in [8] give a novel framework for implementing multi-objective optimization within evolutionary neural network domain for time series prediction. Sound results were achieved. Unfortunately, due to huge profit opportunities involved in financial time series forecast, a disclosure of the most significant findings in this field is limited particularly.

4 Proposed Method

In this section we present theoretical foundation of pattern classification algorithm used to forecast non-stationary financial time series based on the knowledge that excessive increase in training set length can deteriorate generalization error when algorithm is applied to forecast future data.

4.1. Training Set Length and Generalization Error

Consider a p-dimensional two category classification problem. Suppose the classes are Gaussian with different mean vectors μ_1, μ_2 and common pooled $p \times p$ covariance matrix $\Sigma = ((\sigma_{sr}))$. Then asymptotically optimal is Fisher linear discriminant function $g_F(X) = X^T \hat{w}_F + w_0$, where $\hat{w}_F = \hat{\Sigma}^{-1}(\hat{\mu}_1 - \hat{\mu}_2)$, $w_0 = -\frac{1}{2}(\hat{\mu}_1 + \hat{\mu}_2)^T \hat{w}_F$ are weighs of sample based DF and classification of vector X is performed according to a sign of discriminant function (DF). Sample mean vectors and covariance matrix are estimated from training set vectors: $\hat{\mu}_1 = \frac{1}{N} \sum_{j=0}^{N-1} X_{2j+i}$ and $\hat{\mu}_2 = \frac{1}{N} \sum_{j=0}^{N-1} X_{2j+2}$;

$\hat{\Sigma} = \frac{1}{n} \sum_{i=1}^{2} \sum_{j=0}^{N-1} (X_{2j+i} - \hat{\mu}_i)(X_{2j+i} - \hat{\mu}_i)^T$, where N is number of training vectors in one class, $n=2N$; (for a general introduction into statistical pattern recognition see e.g. [9, 10]). If true distributions of input vectors, X, are Gaussian with parameters μ_1, μ_2, Σ, expected probability of misclassification (mean generalization error) is [10]

$$\bar{\varepsilon}_N^F \approx \Phi\{ -\frac{1}{2} \delta (T_M T_\Sigma)^{-\frac{1}{2}}\} \tag{1}$$

where $\delta^2 = (M)^T \Sigma^{-1}(M)$ is a squared Mahalanobis distance, the term $T_M = 1 + 4p/(\delta^2 n)$ arises due to inexact sample estimation of the mean vectors of the classes and the term $T_\Sigma = 1 + p/(n-p)$ arises due to inexact sample estimation of the covariance matrix. $M = (m_1, ..., m_s)^T = \mu_1 - \mu_2$.

Equation (1) shows that with an increase in training set size, N, the mean generalization error decreases monotonically. If Fisher discriminant function is utilized to classify time series data which is varying in time, in estimation of parameters μ_1, μ_2, Σ, the changes of the data cause additional inaccuracies that are accumulating with time. Thus, with an increase in training set size the mean generalization error decreases at first, reaches a minimum and then starts increasing. We will demonstrate this phenomenon theoretically.

The simplest, a random drift, model of changes in distribution of X will be considered. Here we suppose that after each time moment, dependent Gaussian random variables, $\varsigma_{2j+i} = \sum_{s=0}^{j} \xi_{2s+i}$ $(i=1,2; j=0,1,\ldots, N\text{-}1\}$ are added to components of X_{ij}. Random contributions, ξ_{2s+i}, are accumulating. Then

$$\hat{\Sigma}^* = \frac{1}{n}\sum_{i=1}^{2}\sum_{j=0}^{N-1}(X_{2j+i} - \hat{\mu}_i + (I\varsigma_{2j+i} - \frac{1}{N}\sum_{j=0}^{N-1}\varsigma_{2j+i}))(X_{2j+i} - \hat{\mu}_i + (I\varsigma_{2j+i} - \frac{1}{N}\sum_{j=0}^{N-1}\varsigma_{2j+i}))^T,$$

where I stands for p-dimensional column vector composed of ones, $I = (1, 1, \ldots , 1)^T$, and $\xi_{2s+i} \sim N(0,\alpha^2))$ are independent.

After tedious combinatory and statistical analysis utilizing first terms of Taylor series expansion we obtain expectations of the means and covariance matrix with respect to random variables ς_{2j+i}:

$$E\hat{\mu}_i^* = \hat{\mu}_i , \ E\hat{\Sigma}^* = \hat{\Sigma} + E \times \beta, \tag{2}$$

where E stands for $p \times p$ matrix composed of ones and $\beta = \frac{1}{3} \alpha^2 N$.

Using (2), for small α and N we find effective Mahalanobis distance

$$\delta^* = (M^T(\Sigma + E\beta)^{-1}M)(M^T(\Sigma + E\beta)^{-1}\Sigma(\Sigma + E\beta)^{-1}M)^{-1/2} \tag{3}$$

Then for small α and N we become aware that we can use Eq. (2) to calculate approximate values for mean generalization error. Note that both the effective Mahalanobis distance and mean generalization error depend on all components of vector M and matrix Σ. In Fig. 1a we present a graph $\bar{\varepsilon}_N^F$ as a function of N. Theoretical calculations confirm that statistical dependence between subsequent components of multivariate time series deteriorates generalization performance and diminishes optimal training length.

4.2 Effect of Initialization and Early Stopping in Changing Environments

Analysis performed in previous sections has shown that the length of learning sequence should be short, if time series characteristics are changing all the time. It means that in training we ought to reject old data. Old data, however, may contain information, which if correctly used may appear useful. One of the ways to save previously accumulated data in situations of permanent environmental changes is to start training from the weight vector obtained with previous portion of the data, which is not precise enough to be used for training. It was demonstrated that information contained in the initial weight vector can be saved if training of the perceptron is stopped in a right time [11]. Due to the fact that in perceptron training we can obtain a sequence of the classifiers of varying complexity [10], and due to unknown accuracies of initial and final weight vectors the early stopping should be performed in empirical way.

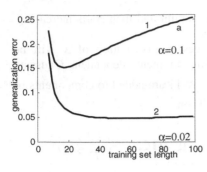 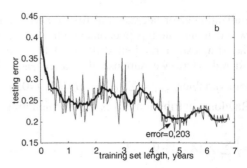

Fig. 1. Generalization errors $\overline{\varepsilon}_N^F$, as a function of training set length N; a) the Fisher classifier, theoretical result ($p=10$; $m_s=1.2526$; $\sigma_{ss}=1$; $\sigma_{1s}=0.1$, if $r=1$; $\sigma_{rs}=0$, if $r>1$), b) MLP, pork price forecasting task. The bold line in graph b denotes approximation of the results.

4.3 Two Stage Pattern Classification Algorithm

The algorithm used to solve commodity price forecasting task is divided into to stages (see Table 1).

Table 1. The algorithm proposed

1st stage	Step 1	Data preparation (get TR and TE)
	Step 2	Initial dimensionality reduction using MLP (TR,TE->TR$_3$,TE$_3$)
2nd stage	Step 3	Derivation of polynomial features (TR$_3$, TE$_3$ ->TR$_9$,TE$_9$)
	Step 4	SLP training on TR$_9$, testing on TE$_9$ using "sliding window" approach

In the next section the underlying considerations of stage 1 and stage 2 are discussed. Data preparation and the flow of training and testing are presented further in Simulation experiment, Section (5).

The First Stage of the Algorithm. In order to design functional classification algorithm capable to work well if trained on very short non-stationary time series data, we have to reduce the number of features at first, as large dimensionality of input vectors increases a need of training samples.

For dimensionality reduction we selected multilayer perceptron (MLP) used to map the data into a low dimensional space as suggested in [3]. This simple feature extraction (FE) method performs linear feature extraction with nonlinear performance criterion. The *r new features*: $z_1, z_2, ..., z_r$, are linearly weighted sums, $z_s = \sum_{j=1}^{p} w_{sj} x_j$, ($s = 1, 2, ..., r$) of p inputs $(r < p)$ calculated in r hidden neurons. The new extracted feature space depends on minimization criterion used in training, i.e. on complexity of decision boundary. Thus, the number of hidden units of MLP (the number of new features, *r*) are affecting the complexity of feature extraction procedure. In spite of simplicity, it is very powerful feature extraction method, which allows make use of discriminatory information contained in all input features. Nevertheless, in finite training sample size situations, one cannot use many hidden units. After several preliminary experiments performed with data TR (every second vector was used for training the MLP classifier and remaining vectors of TR were used for validation, determining the number of training epoch, tL), three hidden neurons we selected, i.e. $r = 3$. Then the MLP, trained on all data TR, after tL epochs produced new *three-dimensional data sets*, TR_3 and TE_3. It was the first stage of the algorithm.

The Second Stage of the Algorithm. Following to the considerations presented in previous sub-section, in the second stage of the algorithm, we have to apply adaptive classifier capable to make use of the historical data series information accumulated in starting weight vector w_{start}. Analysis showed that in the second stage, non-linear boundary would be preferable.

The MLP classifier can be easily trapped into bad local minima. Therefore, we have chosen SLP classifier, performed in polynomial 2nd order feature space derived from TR_3 and TE_3: instead of three features, z_1, z_2, z_3, we used nine new ones: z_1, z_2, z_3, $(z_1)^2$, $(z_2)^2$, $(z_3)^2$, z_1z_2, z_1z_3, z_2z_3. Thus the SLP classifier was trained in *9-dimensional (9D) space*.

To save possibly useful information contained in starting *10-dimensional* weight vector w_{start}, we had to stop training early, much earlier as minimum of cost function was obtained. In practical application of this approach, in the first iteration cost function to be minimized could be very large. Therefore, each new training session started from *scaled* initial weight vector $\kappa \times w_{start}$, where parameter κ was determined from a minimum of cost function estimated from the testing set after recording current test results.

5 Simulation Experiment

Data Used. To demonstrate usefulness of the algorithm described above we analyzed a real word *5-dimensional* financial data recorded in a period from November 1993 till January 2005. The price of Pork Bellies was chosen as forecasting target.

The following variables were used as inputs for the algorithm: x_1 - spring wheat, x_2 - raw cane sugar (as other eatable commodities), x_3 - gold bullion (as alternative currency), x_4 - American Stock Exchange (AMEX) oil price index (supposed to be influential for eatable commodity prices due to transportation and techniques) and x_5 - Pork Bellies price. Input data vectors were formed using four days price history of each of the presented variables. *20-dimensional* data matrix X was split into training data TR (first 1800 days history), and testing data TE (last 1100 days).

As we are dealing with financial variables, a highlight from theory in finance needs to be addressed. The Efficient Market Hypothesis [12] states that in efficient market, the prices reflect all the information available from the market. Thus, statistically significant forecast can be made only in situations where either the market is not efficient enough in terms of information flow or the problem solving method is unexpected for other participants. Therefore, we constructed original index,

$$Y_t = (B_{t+2} + B_{t+1})/(B_t + B_{t-1}) - (B_{t+1} + B_t)/(B_{t-1} + B_{t-2}),$$

where B_t is Pork Bellies price at day t, formed from historical data. Y_t is our forecasting target. Such index was not used by other researchers.

We formulate forecasting task as pattern classification problem. We calculate Y_t for all training data TR at first. Then we selected two threshold values Y_{max} and Y_{min} in a way that Y_{max} is *the smallest* value from 25% of *the highest* Y_t, which were calculated from training set TE. Similarly, Y_{min} is *the highest* value from 25% of *the smallest* Y_t, which were calculated from training set TE. This way we split training, as well as testing data (using the same thresholds, but the thresholds were determined only from training data) into categories C_1 ($Y_t > Y_{max}$), C_2 ($Y_t < Y_{min}$) and $C_{average}$ – the remaining 50% of data. The first two classes C_1 and C_2, were used to develop classification algorithm based on the theory presented above.

Experimental Design. *9-dimensional* two category training and test data sets, TR_9 and TE_9, composed of 450+450 and 199+184 vectors were obtained as described in Section 4.3. Tuning of number of hidden units, MLP initialization, optimal number of MLP training epochs were determined on pseudo-validation sets formed from each training subset by means of colored noise injection [10, 13].

Data TE_9 was split into 44 not intersecting blocks composed of 25 consecutive days. In each iteration, the training subset consisted of L days ending one day before a current testing block starts (depending on L, the training blocks intersected by 0 - 98%; at optimum, L_{opt}=210, we had 88% intersection). Note that in our analysis we skipped the middle pattern class, $C_{average}$. Therefore, numbers of vectors from classes C_1 and C_2, in each single block were different. Training was performed if there were more than two vectors of each class (C_1 and C_2) in current training block. If no testing vectors were in the current testing block training, we used κ=1.

For training the neural network and testing the results, *sliding window* approach was used. The SLP was trained on *one subsequent 9-dimensional block* data, starting from weight vector w_{start} obtained after training with data of preceding block. The supplementary training before testing on each of 44 testing blocks was essential, as the environment affecting commodity prices is changing very rapidly and training takes place on short history.

Results. The core simulations presented in this paper are stated in Table 2. In Fig. 1b we have an influence of training set length (in years) on average generalization error in classifying 383 two category (C_1 and C_2) *20D* test vectors of TE by means of MLP with 3 hidden units: 1100 days for testing (44 blocks, 25 days each).

We see several minima in smoothed graph (bold curve). Presence of several minima is caused by the fact that real world environmental changes do not follow simplified assumptions used for derivation analytical formula in Section 4.1. The first convex sector in averaged graph does not lead to optimum test error, as there is a lack of training vectors as compared to number of features (p=20) at that short training window.

Table 2. Simulation experiments

Fisher classifier, theoretical result	Fig. 1a	Generalization error, as a function of training set length. Experiment with artificial data.
MLP, pork price forecasting task	Fig. 1b	Generalization error, as a function of training set length. Experiment on real data repeated 179 times having different training set length L (20…1800). Without feature reduction. MLP used for training.
SLP, pork price forecasting task	Fig. 2a	Generalization error, as a function of training set length. Experiment on real data repeated 179 times having different training set length L (20…1800). Feature reduction using MLP. Final training using SLP.
SLP, Pork price forecasting task	Fig. 2b	Generalization error, as a function of number of training epochs. Experiment on real data repeated 150 times having different number of training epochs (1…150). Feature reduction using MLP. Final training using SLP.

In Fig. 2 we have similar graphs performed with SLP in *9-dimensional* feature space. In this case, *two types of experiments* were performed with the test set data TE_9:

a) for fixed number of training epochs, t^*, (this number was evaluated during additional experiments), training window, L, vas varying in interval [20, 1800] days;

b) training window length, L, was fixed (it was also obtained in additional experiments) and a number of training epochs varied in interval [1, 150].

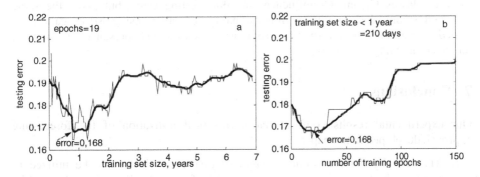

Fig. 2. Classification error as a function of window size (in years) (a) and number of epochs (b). The bold line denotes approximation of the results.

We see that minimum of the generalization error in *9D* space (16.8%) is notable lower as that in original, *20D* feature space (20.3%). It means that our strategy to extract features and to train SLP in low-dimensional space with initialization with previous weight vector and early stopping appeared fruitful. We have to admit that in experiments with SLP in *9D* space the test data participated in determining optimal length of training history and the number of training epochs. It is a shortcoming. The only consolation is that in nowadays, the world trade market changes so rapidly that any optimality parameters determined six years ago does not fit nowadays. Similar gains in accuracy were obtained in forecasting experiments of oil and sugar prices.

6 Implementation

In dynamic financial time series prediction, neural network training based on short data sequences results to more accurate result as using lengthy historical data. The optimal training set size is stated theoretically and experimentally. To reduce generalization we suggest:

  a) to map the data into low dimensional space using multilayer perceptron,
  b) to make final forecasting by single layer perceptron classifier trained in low dimensional space,
  c) to initialize SLP with weight vector obtained after training the perceptron with previous portion of data sequence,
  d) save useful preceding historical information by early stopping.

  The proposed methodology was tested on other financial time series as well, in particular, for oil and sugar forecasting task, the gains achieved supported the conclusions made here. As the objective of the present research is to analyze methodological issues rather than detailed presentation of the results with other commodity price forecasting are left out of the scope of the paper.

  When forming initial data classes C_1 and C_2, 50% of the data was omitted due to business reasons. Experiments showed that taking all the data or different percentage of it to classes C_1 and C_2 influences absolute testing error, but gives the same principal results, i.e. gain from dimensionality reduction, using MLP, gain from scaled preservation of information from previous trainings, presence of optimum lengths of learning sequences.

7 Conclusions

Our experimental results confirm that in practical utilization of the forecasting approach developed in the paper:

  1. Theoretically derived optimal length of time series history to be utilized to improve the forecasting algorithm do not fit for practical application since the parameter of the environmental changes are not known and since in real world, the changes follow much more complicated laws (if they could exist in reality).
  2. Forecasting strategy consisting on a) reduction the number of features by utilizing long history of multivariate financial data series by means of MLP based linear feature extraction, b) training the SLP classifier in low-dimensional polynomial features space starting from scaled previous weight vector and early stopping could become a good alternative to existing forecasting methods. In each practical application, optimal length of times series history and optimal perceptron's training times have to be determined from latest data history and additional non-formal end users information. Possible solution to find these values is to solve several forecasting problems simultaneously.

Acknowledgements

The authors thank Dr. Aistis Raudys for sharing the Matlab codes, useful and challenging discussions in the field.

References

[1] Kuncheva, L. Classifier ensembles for changing environments. *Lecture Notes in Artificial Intelligence*, Springe-Verlag, 3077: 1-15, 2004.

[2] Raudys S. Survival of intelligent agents in changing environments. *Lecture Notes in Artificial Intelligence*, Springer-Verlag, 3070: 109-117, 2004.

[3] Raudys, A. and J.A. Long (2001). MLP based linear feature extraction for nonlinearly separable data. *Pattern Analysis and Applications*, 4(4): 227-234, 2001.

[4] Huang, W., Lai, K.K., Nakamori, Y., Wang S. Forecasting foreign exchange rates with artificial neural networks: a review. *Int. J. of Inf. Technology & Decision Making*, 3(1) 1: 145-165, 2004.

[5] Yao, X. (1999). Evolving ertificial neural networks. *Proceedings of IEEE*, 87(9): 1423-1447.

[6] Moody, J., *Economic Forecasting: Challenges and Neural Network Solutions. International Symposium on Artificial Neural Networks*, Hsinchu, Taiwan, 1995.

[7] N. Kohzadi, M. Boyd, B. Kermanshahi and I. Kaastra . A comparison of artificial neural network and time-series models for forecasting commodity prices. *Neurocomputing*, 10: 169-181, 1996.

[8] Feldsend, J.E. and Singh, S. Pareto evolutionary neural networks. *IEEE Trans. Neural Networks*, 16(2): 338-353, 2005.

[9] Duda, P.E., Hart, R.O. and Stork, D.G. *Pattern Classification*. 2nd ed. Wiley, NY, 2000.

[10] Raudys, S. (2001). *Statistical and Neural Classifiers: An integrated approach to design.* Springer, NY, 2001.

[11] Raudys S. and Amari S. Effect of initial values in simple perception. In *Proc.* 1998 *IEEE World Congress on Comput. Intelligence*, IEEE Press, Vol. IJCNN'98: 1530–1535, 1998.

[12] Fama, E.F. Efficient capital markets: A review of theory and empirical work. *Journal of Finance*, 25: 383-417, 1970.

[13] Skurichina M., Raudys S., Duin R.P.W. K-nearest neighbors directed noise injection in multilayer perceptron training, *IEEE Trans. on Neural Networks*, 11: 504–511, 2000

Develop Multi-hierarchy Classification Model: Rough Set Based Feature Decomposition Method

Qingdong Wang, Huaping Dai, and Youxian Sun

National Key Laboratory of Industry Control Technology, Zhejiang University,
Hangzhou,310027, P.R. China
qdwang@iipc.zju.edu.cn

Abstract. Model development on high dimension database is very difficult. This paper presents a new rough set based machine learning method, named feature decomposition method, to discover concept hierarchies and develop a multi-hierarchy model of database. For the databases which we are familiar with, the feature group can be selected by experience of expert. When dealing with the databases without any background knowledge, a new criterion based on rough set is presented to select the features to form a feature group. According to some measures of rough set theory, the objects defined on the proposed feature group are labeled by a new intermediate concept. The concept hierarchies of the database have specific meaning, which increased the transparency of data mining process and enhance the comprehensibility of the model. Each feature group and the corresponding intermediate concept compose the structure of the database. Finally rule induction can be processed on the intermediate concepts. The algorithm presented is verified by datasets from UCI. The results show that the multi-hierarchy model established by feature decomposition method can get high classification accuracy and have better comprehensibility.

Keywords: Rough set, feature decomposition, classification, rule induction, concept hierarchy.

1 Introduction

The volume of data is growing at an unprecedented rate, both in the number of features and instances. Data mining resolves the problem by offering tools for discovery of patterns, associations, rules and structures in data. Fayyad et al. [1] claim that the explicit challenges for the KDD research community is to develop method that facilitate the use of data mining algorithms for real-world databases. Researchers realize that in order to use these tools effectively, an important part is pre-processing. In many applications, the number of features can be quite large, many of which can be irrelevant and redundant. Feature transformation and feature selection are some frequently used techniques in data pre-processing. Feature transformation is a process through which a new set of features is created. Feature selection is different from feature transformation in that no new features will be generated, but only a subset of original features is selected and feature space is reduced.

S. Singh et al. (Eds.): ICAPR 2005, LNCS 3686, pp. 164–171, 2005.
© Springer-Verlag Berlin Heidelberg 2005

In the last decade, feature selection has enjoyed increased interest by the data mining community. Consequently many feature selection algorithms have been processed, some of which have reported remarkable accuracy improvement [2]. Mining of transformed data sets exhibits about the same classification accuracy with the increased transparency and lower complexity of the developed models. The most useful form of transformation of data sets is decomposition [3]. There are two forms of decomposition in space, feature set decomposition [3, 4, 5] and object set decomposition [6].

Besides concentrating on individual features, the feature set can be divided into some feature groups by intermediate concepts, which are based on aggregation of the original features. A more mathematically exact method for binding the aggregate conceptual directions of a data set is Principal Component Analysis (PCA) [7]. But PCA is suitable for continuous attributes. For discrete attributes, feature relevance decomposition becomes especially important in data sets. Zupan et al. [5] presented a function decomposition approach for machine learning. According to the approach, the new concept was formed based on function decomposition method, an approach originally developed to assist in the design of digital circuits. But the approach is limited to consistent datasets and nominal features. Butine [8] used the concept aggregation to classify free text documents into predefined topics. Kusiak [9] introduced feature bundling method, which is of particular interest in temporal data mining as relationships are formed among features rather than their values. The relationships among features tend to be more stable in time comparing to the relationships among feature values.

Although many learning methods attempt to either extract or construct features, both theoretical analyses and experimental studies indicate that many algorithms are inscrutable to the users [10]. Moreover these methods do not attempt to use all the relevant features and ignore the relationship between the condition attributes. In this paper, we present a new feature decomposition method to discover concept hierarchies and construct the intermediate concept. This method based on rough set theory of Pawlark [11] facilitates the creation of a multi-attributes model. For the data sets which we are familiar with, feature group can be selected by skill and experience of expert. When dealing with the databases without any background knowledge, a new criterion based on rough set theory is presented to select the features to form a feature group. According to some measures of rough set theory, the instances defined on the proposed feature group are labeled by a new intermediate concept. The concept hierarchies of the database have specific meaning, which enhances the comprehensibility of the model. Each feature group and the corresponding intermediate concept compose the structure of the database.

The rest of the paper is organized as follows. Section 2 introduces some basic concept of Rough Set Theory. Section 3 presents the feature decomposition algorithm in detail. Section 4 the method is experimentally evaluated on several dataset coming from UCI depository and the last section contains final conclusions.

2 Preliminaries

The rough set methodology was introduced by Pawlak [11] in the early 1980s as a mathematical tool to deal with uncertainty. It tries to extend capabilities as the treatment of incomplete knowledge, the management of inconsistent pieces of information, knowledge reduction preserving information, etc. The framework is especially suitable to the determination of attribute-value relationships in attribute-value systems [12]. Here, we only introduce the basic notation from rough set approach used in the paper.

Formally, the decision table is defined as 4-tuple $DT=(U, A, V, f)$, where U is a non-empty finite set of objects and A is a non-empty finite set of condition C and decision D attributes, such that $C \cup D = A$ and $C \cap D = \varnothing$. V is a non-empty finite set of attribute values. Each attribute $a \in A$ can be viewed as a function that maps elements of U into a set V_a. f is an information function, $f : U \times A \to V$.

Let $IND(P) \in U \times U$ denote an indiscernibility relation defined for a non-empty set of attributes $P \subseteq A$ as:

$$IND(P) = \left\{ (x, y) \in U \times U : \underset{q \in P}{\forall} f(x,q) = f(y,q) \right\}$$

If $(x, y) \in IND(P)$ we will say that x and y are *P-indiscernible*. Equivalence classes of the relation $IND(P)$ are referred to as *P-elementary sets*. In the rough set approach the elementary sets are the basic building blocks of our knowledge about reality. The family of all equivalence classes of $IND(P)$, i.e., the partition determined by P, will be denoted by U/P.

For every subset $X \subseteq U$ we define the lower approximation $B_*(X)$ and the upper approximation $B^*(X)$ as follows:

$$B_*(X) = \left\{ x \in U : [x]_B \subseteq X \right\}, B^*(X) = \left\{ x \in U : [x]_B \cap X \neq \varnothing \right\}.$$

Another important issue in data analysis is discovering dependencies between attributes. Let D and C be subsets of A. we will say that D depends on C in a degree k, denoted $C \Rightarrow_k D$, if $k = \gamma(C,D) = \dfrac{|POS_C(D)|}{|U|}$, where $POS_C(D) = \underset{X \in U/D}{\bigcup} C_*(X)$

called a positive region of the partition U/D with respect to C, is the set of all elements of U that can be uniquely classified to blocks of the partition U/D, by means of C. The coefficient k expresses the ratio of all elements of the universe, which can be properly classified to blocks of the partition U/D, employing attributes C and will be called *the degree of the dependency*.

3   Feature Decomposition Algorithm

Feature decomposition methodology can be considered as effective strategy for changing the representation of a learning problem. By using some intermediate concept instead of single complex features, several sub problems with different and simple concepts are defined. There are two important issues in feature decomposition problem. One is how to select features to aggregate into a feature group. It might be obtained manually based on expert's knowledge on a specific domain or induced without human interaction by some criterion. Another is how to define the intermediate concept. In this section, we first present an approach to select features to aggregate into one feature group. Information gain is computed to measure the significance of feature groups and get rid of the redundant features. The intermediate concepts are discovered according to two measures based on Rough Set theory. After that, the description of algorithm is presented to express the process of feature decomposition and model development.

3.1 Feature Selection

Feature decomposition methodology attempts to select the features by some criteria to form a feature group and label the union by a new intermediate concept. About the model of the transformed dataset, the classification accuracy does not decrease significantly, and the comprehensibility will be increased. The first step in feature decomposition method is to obtain the structure of databases. The original features should be selected to aggregate into feature groups. There are several methods for the selection of features.

For a given dataset, if we are familiar with the background knowledge, supervised selection can process under the human interaction. The dataset can be examined by expert, who accessed each feature group according to practical meaning, the cost of data collection, the requirement for real time, etc. The expert selects the features to form the feature groups that seemed most meaningful and comprehensible. When dealing with the databases without any background knowledge, we utilize the basic ideas in rough set, and set up an approach to select the features.

An information system is defined as $S = (U, C \cup D, V, f)$, where C is the condition attribute set. D is the decision attribute set. $B \in C$. For random $a_i \in C$ and $a_i \notin B$, we define the selection criterion as follows.

$$k = \frac{\gamma(B + \{a_i\}, D) - \gamma(B, D)}{card(V_{a_i})} = \frac{card(POS_{B + \{a_i\}}(D)) - card(POS_B(D))}{card(U) \cdot card(V_{a_i})}$$

where $card(U)$ means the number of instances in set U, $card(V_{a_i})$ means the number of values of attribute a_i. If the selection criterion k gets the maximum value for a_i, we aggregate a_i into the set B to form a feature group. That is to say, union of B and a_i can get the maximum enhancement of classification power.

3.2 Discovering the Intermediate Concept

Our algorithm is a heuristic method with rough set measures to find the optimal partition of the intermediate concept c_i. The following are the measures based on rough set theory:

1) Consistency measure: Given the feature group G_i and the corresponding intermediate concept c_i, We define a criteria based on the degree of dependency for partition evaluation as below.

$$J_1 = \left| \frac{\gamma(C, D) - \gamma(R \cup \{c_i\}, D)}{\gamma(C, D)} \right| \leq \delta$$

where $\gamma\{R\cup\{c_i\},D\}$ expresses the degree of dependency between attributes $R\cup\{c_i\}$ and D, R means the remains features except feature group $G_i\square$ δ is a user given threshold. This definition expresses that $DT_{new}=\{U,R\cup\{c_i\}\cup D,V,f\}$ has the approximate degree of the dependency compared with DT. The threshold δ suits the characteristic of consistency measure because real-world data is usually noisy and if δ is set to 0 strictly then it may happen that "good" features are filtered away. Especially, when $J_{c1}=0$, the new decision system has the same classification performance as the original datasets.

2) Min-value measure: Generally, the number of the equivalence classes divided by the proposed feature group is $n=\prod_{i=1}^{k}card(a_i)$ where a_i is the feature which is selected into the feature group. By merging the equivalence classes under the restriction of consistency measure J_1, we minimize the number n and get Min-value measure J_2.

$$J_2 = \min(Card(V_{c_i}))$$

The smaller $card(V_{c_i})$, the simpler structure of dataset will get.

3.3 Description of the Algorithm

Algorithm. Feature decomposition method based on rough set.
Given: A N-case data set T containing m-dimensional features, denoted by $DT=(U, C\cup D)$.
Step 1: Select the features to form some feature groups G_1, G_2,..., G_k.
Step 2: Compute the significance of feature groups by the information theory.
Step 3: According to the significance sequence of feature groups, compute the intermediate features c_1, c_2, ..., c_k by the two measures J_1, J_2 of rough set theory
Step 4: Attribute reduct, induce the rule sets by rough set theory on data set $DT_i=(U, G_i\cup c_i)$, $i=1,2...,k$
Step 5: Induce the rule set by rough set theory on data set $DT'=(U, \{c_1,c_2,...,c_k\}\square D)$ to obtain the multi-hierarchy model.

4 Experiments

The data sets used in the experiment are real-world datasets coming from *UCI Machine Learning Repository* [14]. The experiment was conducted on six benchmark datasets. Table 1 describes the character of datasets where N is the number of instances, M is the size of the condition feature set, and M' is the number of the intermediate concepts obtained by our algorithm. We carry out experiment to examine (1) whether multi-hierarchy classification model can achieve high classification accuracy compared with some classical approaches and (2) whether the multi-hierarchy structure is more comprehensible and easy to interpret.

The experimental procedure is to run feature decomposition method to get the intermediate concepts, compare the performance with two different classifiers, in

Table 1. Datasets and comparative results of classification accuracy (%)

Dataset character				Classification accuracy (%)		
Name	N	M	M'	C4.5	RSES	Our algorithm
Car	1728	6	3	100	100	100
Australian	690	15	4	84.5	86.5	88.3
Monk-1	124	6	2	100	88.7	100
Monk-2	169	6	2	64.8	73.6	89.5
Monk-3	122	6	2	94.4	94.7	85.8
Tictactoe	958	9	3	84.1	80.7	87.03

which C4.5 is a standard machine learning algorithms and has been widely used to compare with newly-proposed algorithms, while RSES directly generates decision rules according to rough set analysis. Table 1 indicates that the multi-hierarchy model can achieve high classification accuracy. For example, on *Car, Australian, Monk-1, Monk-2,* and *Tictactoe,* our algorithm all gets better performance. On *Monk-3,* it can be seen that the accuracy for our algorithm can achieve 85.8% and the result is inferior to other algorithm to a certain extent. This is because *Monk-3* has noisy data. The existence of noisy data can take an impact on the process of discovering intermediate concepts. This impact becomes more serious in the small feature set, for example the feature group G_i. And as a result it makes the classification rate of our algorithm on *Monk-3* lower than other methods.

Next the comprehensibility of model can be illustrated with the example of *monk* and *car* datasets. The intermediate concept of every dataset and the corresponding feature group are listed in table 2. The features in bracket are the original features. The features outside the bracket are the new intermediate concepts which were created by the algorithm.

The MONK's problems rely on an artificial robot domain, in which robots are described by six different attributes. Detailed descriptions of Monk's problems can be found in [15].*Monk-1* and *Monk-2* have no noise data, but *Monk-3* has 5% noise in the training set. In the six attributes $x_1, x_2, ..., x_6, x_1$ means head shape, and x_2 means body shape. So we can define a new intermediate concept c_1 which means shape factor of the instances by the experience. The new concept which has specific meaning can be discovered by our algorithm. The value number of the new concept c1 on *Monk-1, Monk-2* and *Monk-3* is 2, 3 and 5 respectively. For example, the two values of the intermediate concept on *Monk-1* are easy to interpret. If the body shape is the same as the head shape, the intermediate concept is equal to 1. If not, it is equal to 2. And our algorithm maps 9 combinations of x_1 and x_2 to only two values of the intermediate concept. For *Monk-2,* because of the complexity of the features combination, most of the algorithm cannot produce excellent classification rules. But the accuracy obtained by our algorithm on *Monk-2* reaches to 89.5%, which is obviously higher than the classification accuracy of other algorithms. The reason for increased classification accuracy with the intermediate concept might due to the fact the associations among features and decisions are stronger than those built on single feature. In fact, the intermediate concept can be looked as the target of the regression function defined on a subset of features.

Car dataset is a famous database to test the performance of classification model. It has six attributes. The first two attributes, buying price (a_1) and the price of the

Table 2. The intermediate concept and the corresponding feature group

Dataset	Feature Group
Car	price factor (a_1, a_2) comfort factor (a_3, a_4, a_5) safety factor (a_6)
Monk	shape factor (a_1, a_2) appearance factor (a_3, a_4, a_5, a_6)
Austrilian	k_1(a_1, a_2, a_9, a_{11}) k_2(a_3, a_4, a_8, a_{10}) k_3(a_5, a_7, a_{15}) k_4(a_6, a_{12}, a_{13}, a_{14})
Tictactoe	top factor (a_1, a_2, a_3) meddle factor (a_4, a_5, a_6) bottom factor (a_7, a_8, a_9)

maintenance (a_2), can be aggregated into one intermediate concept, named *price factor*. Like *price factor*, we can get *comfort factor* (a_3, a_4, a_5) and *safety factor* (a_6), then develop multi-hierarchy model on these intermediate concepts with high comprehensibility. For example, a customer can evaluate a car by the three combined criterion instead of concentrating on individual features. And another advantage of our algorithm is that it can break down a high dimensionality problem into several manageable problems. For example, there are 1728 instances in *car* dataset. After run our algorithm on *car* dataset, it maps 16 combinations of a_1 and a_2 to only 4 values of *price factor*. And transform 36 combinations of a_3, a_4 and a_5 to 3 values of *comfort factor*. It can be seen that the number of objects in each subset of multi-hierarchy model is much less than in original dataset. So the rule set obtained by our algorithm is much simpler and with high support degree.

In a word, the multi-hierarchy model developed by our algorithm is not only a powerful model with high performance but also a simple model with better comprehensibility.

5   Conclusions and Future Work

In the most data mining applications, the model is developed on the original data set. In this paper a new method based on rough set named feature decomposition algorithm was introduced. Two measures based on rough set theory are presented to discover the new intermediate concept. And rule induction is processed by rough set theory to build the multi-hierarchy model. Experiments on datasets coming from UCI are made to evaluate the performance of the method. The results show that the multi-hierarchy model established by feature decomposition method can get high classification accuracy and have better comprehensibility. We conclude the main advantages of the feature decomposition method as follows:

1. This method can enhance the transparency of the model. The model developed by the feature decomposition method expresses the hierarchy structure of database

clearly. And the rule sets induced by the method are comprehensible. So in decision support tasks, reasons for the decision can be clearly identifiable.

2. Feature decomposition method can break down a high dimensionality problem into several manageable problems. It makes the problem more feasible by reducing its dimensionality.

3. This method can deal with the inconsistent dataset, process appropriate treatment of noise in data. Though the existence of noise degrades the performance of model, the results of experiment verify that this method can extract the structure of dataset and build multi-hierarchy model.

Acknowledgement

This work was supported by National Natural Science Foundation of China (60304018) and China 973 Plan under grant number 2002CB312200.

References

1. Fayyad, U., Piatesky-Shapiro, G., and Smyth P., *From Data Mining to Knowledge Discovery: An Overview*, Advances in Knowledge Discovery and Data Mining, pp1-30, MIT Press, 1996
2. Liu and H. Motoda, Feature Selection for Knowledge Discovery and Data Mining, Kluwer Academic Publishers, 1998
3. Kusiak, A., Decomposition in Data Mining: An Industrial Case Study, IEEE Transactions on Electronics Packaging Manufacturing, Vol. 23, No. 4, 2000, pp.345-353
4. Maimon, O., and Rokach L., Improving Supervised Learning by Feature Decomposition, FoIK 2002, LNCS 2284, pp.178-196, 2002
5. Zupan, B., Bohanec, M., Demsar, J., and Bratko, I., Feature Transformation by Function Decomposition, IEEE intelligent systems & their applications, 13:38-43, 1998.
6. Chan, P.K. and Stolfo, S.J., A Comparative Evaluation of Voting and Meta-learning on Partitioned Data, Proc. 12th Intl. Conf. on Machine Learning ICML-95, 1995
7. I.T. Jolliffe. Principal Component Analysis, Springer-Verlag, New York, 1986
8. Buntine, W., Graphical models for discovering knowledge, in U.Fayyad, G.Piatetsky-Shapiro, P.Smyth, and R.Uthurusamy, editors, Advances in Knowledge Discovery and Data Mining, pp59-82. AAAI/MIT Press, 1996
9. Kusiak, A., Feature Transformation Methods in Data Mining, IEEE Transactions on Electronics Packaging Manufacturing, Vol. 24, No. 3, 2001, pp.214-221
10. Ridgeway, G., Madigan, D., Richardson, T. and O'Kane, J., Interpretable Boosted Naïve Bayes Classificatioin, Proceedings of the Fourth International Conference on Knowledge Discovery and Data Mining, pp 101-104
11. Pawlak Z. Rough Sets. Theoretical Aspects of Reasoning About Data, Kluwer Academic Publishers, Dordrecht, 1991
12. Ziarko W. Variable Precision Rough Set Model. Journal of Computer and System Sciences, 1993. 46:39~59
13. UCI repository of machine learning databases (1996). http://www.ics.uci.edu/~mlearn/mlrepository.html. Department of information and computer science, university of California.
14. S. B. Thrun, J. Bala, E. Bloedorn, I. Bratko, e tal. The MONK's problems: a performance comparison of different learning algorithms. Technical Reports. Carnegie Mellon University. CMU-CS-91-197. December 1991.

On Fitting Finite Dirichlet Mixture
Using ECM and MML

Nizar Bouguila and Djemel Ziou

Université de Sherbrooke,
Sherbrooke, Qc, Canada J1K 2R1
{nizar.bouguila, djemel.ziou}@usherbrooke.ca

Abstract. Gaussian mixture models are being increasingly used in pattern recognition applications. However, for a set of data other distributions can give better results. In this paper, we consider Dirichlet mixtures which offer many advantages [1]. The use of the ECM algorithm and the minimum message length (MML) approach to fit this mixture model is described. Experimental results involve the summarization of texture image databases.

1 Introduction

Finite mixture models have continued to receive increasing attention over the years [2]. These models are used in various fields such as image processing, pattern recognition, machine learning and remote sensing. For multivariate data attention has focused on the use of Gaussian components. However, for many applications the Gaussian can fail when the partitions are clearly non-Gaussian. In [1], we have demonstrated that the Dirichlet can be a good choice to overcome the problems of the Gaussian. In dimension dim the Dirichlet distribution with parameters $\boldsymbol{\alpha} = (\alpha_1, \ldots, \alpha_{dim+1})$ is given by:

$$p(\boldsymbol{X}|\boldsymbol{\alpha}) = \frac{\Gamma(|\boldsymbol{\alpha}|)}{\prod_{i=1}^{dim+1} \Gamma(\alpha_i)} \prod_{i=1}^{dim+1} X_i^{\alpha_i - 1} \tag{1}$$

where $\sum_{i=1}^{dim} X_i < 1$, $|\boldsymbol{X}| = \sum_{i=1}^{dim} X_i$, $0 < X_i < 1 \ \forall i = 1 \ldots dim$, $X_{dim+1} = 1 - |\boldsymbol{X}|$, and $|\boldsymbol{\alpha}| = \sum_{i=1}^{dim+1} \alpha_i$, $\alpha_i > 0 \quad \forall i = 1 \ldots dim + 1$. This distribution is the multivariate extension of the 2-parameter Beta distribution. The mean of the Dirichlet distribution is given by:

$$\mu_i = E(X_i) = \frac{\alpha_i}{|\boldsymbol{\alpha}|} \tag{2}$$

A mixture with M components is defined as : $p(\boldsymbol{X}|\Theta) = \sum_{j=1}^{M} p(\boldsymbol{X}|\boldsymbol{\alpha}_j)p(j)$ where $p(j)$ $(0 < p(j) < 1$ and $\sum_{j=1}^{M} p(j) = 1)$ are the mixing parameters and $p(\boldsymbol{X}|\boldsymbol{\alpha}_j)$ is the Dirichlet distribution. The symbol Θ refers to the entire set of parameters to be estimated: $\Theta = (\boldsymbol{\alpha}_1, \ldots, \boldsymbol{\alpha}_M, p(1), \ldots, p(M))$, where $\boldsymbol{\alpha}_j$ is the

S. Singh et al. (Eds.): ICAPR 2005, LNCS 3686, pp. 172–182, 2005.

parameters vector of the jth component. The EM algorithm is a popular method for iterative maximum likelihood (ML) estimation of finite mixture distributions. This algorithm, however, is unattractive when the M-Step is complicate [2]. This is the case of the Dirichlet mixture. Indeed, the M-Step involves the inverse of the $(dim + 1) \times (dim + 1)$ Fisher information matrix which is not easy to compute especially for high-dimensional data. In this paper, we introduce another approach based on the ECM algorithm which replace a complicated M-step of the EM algorithm with several computationally simpler CM-Steps [3]. The determination of the number of components is based on the MML approach. The rest of the paper is organized as follows. Section II, discusses the basic concepts of the EM algorithm and proposes the ECM algorithm as a method to overcome the problems of the EM in the case of Dirichlet mixtures. In Section III, we present the MML approach for the selection of the number of clusters. Section IV is devoted to experimental results, and Section V ends the paper with some concluding remarks.

2 ML Estimation of a Dirichlet Mixture Using ECM

We consider now ML estimation for a M-component mixture of Dirichlet distributions. Given the set of independent vectors $\mathcal{X} = \{X_1, \ldots, X_N\}$, the log-likelihood corresponding to an M-component mixture is:

$$L(\Theta, \mathcal{X}) = log \prod_{i=1}^{N} p(X_i|\Theta) = \sum_{i=1}^{N} log \sum_{j=1}^{M} p(X_i|\alpha_j)p(j) \qquad (3)$$

It's well-known that the ML estimate: $\hat{\Theta}_{ML} = argmax_{\Theta}\{L(\Theta, \mathcal{X})\}$ which can not be found analytically. The maximization defining the ML estimates is subject to the constraints $0 < p(j) \leq 1$ and $\sum_{j=1}^{M} p(j) = 1$. Obtaining ML estimates of the mixture parameters is possible through EM and related techniques [2]. The EM algorithm is a general approach to maximum likelihood in the presence of incomplete data. In EM, the "complete" data are considered to be $Y_i = \{X_i, Z_i\}$, where $Z_i = (Z_{i1}, \ldots, Z_{iM})$ with $Z_{ij} = 1$ if X_i belongs to class j and $Z_{ij} = 0$ otherwise. The relevant assumption is that the density of an observation X_i given Z_i is given by $\prod_{j=1}^{M} p(X_i|\alpha_j)^{Z_{ij}}$. The resulting *complete-data log-likelihood* is:

$$L(\Theta, \mathcal{Z}, \mathcal{X}) = \sum_{i=1}^{N} \sum_{j=1}^{M} Z_{ij} log(p(X_i|\alpha_j)p(j)) \qquad (4)$$

The EM algorithm produces a sequence of estimates $\{\Theta^t, t = 0, 1, 2 \ldots\}$ by applying two steps in alternation (until some convergence criterion is satisfied):

1. **E-step:** Compute $\hat{Z}_{ij}$ given the parameter estimates from the initialization:

$$\hat{Z}_{ij} = \frac{p(X_i|\alpha_j)p(j)}{\sum_{j=1}^{M} p(X_i|\alpha_j)p(j)} \qquad (5)$$

2. **M-step:** Update the parameter estimates according to:

$$\hat{\Theta} = argmax_\Theta L(\Theta, \mathcal{Z}, \mathcal{X}) \tag{6}$$

The quantity $\hat{Z}_{ij}$ is the conditional expectation of Z_{ij} given the observation $\boldsymbol{X_i}$ and parameter vector Θ. The value Z_{ij}^* of $\hat{Z}_{ij}$ at a maximum of Eq. 4 is the conditional probability that observation i belongs to class j (the *posterior* probability); the classification of an observation X_i is taken to be $\{k/Z_{ik}^* = max_j Z_{ij}^*\}$, which is the Bayes rule. The EM algorithm has been shown to monotonically increase the log-likelihood function. When we maximize Eq. 6, we obtain:

$$p(j)^{(t)} = \frac{1}{N} \sum_{i=1}^{N} \hat{Z}_{ij}^{(t-1)} \tag{7}$$

However, we do not obtain a closed-form solution for the $\boldsymbol{\alpha}_j$ parameters. We therefore use the Fisher scoring method to estimate these parameters [1]. The inconvenient of this approach is that it involves the inverse of the $(dim + 1) \times (dim + 1)$ Fisher information matrix which is not easy to compute especially for high-dimensional data. One of reasons of the popularity of the EM algorithm is that the M-step involves only complete-data ML estimation. But, if the M-Step is complicated as in the case of the Dirichlet mixture, the EM algorithm becomes less attractive. In many cases, however, the ML estimation is simpler if maximization is undertaken conditional on some functions of the parameters. For this goal, Meng and Rubin [3] introduced an algorithm called ECM which replaces a complicated M-step of the EM algorithm with several computationally simpler CM-Steps. As a consequence the ECM converges more slowly than the EM in terms of number of iterations, but can be faster in total computer time. Another important advantage of the ECM is the preservation of the convergence properties of the EM, such as its monotone convergence. Now, we focus on the use of this algorithm for the estimation of Dirichlet mixture.

By substituting Eq. 2 in Eq. 1, the Dirichlet distribution can be written as the following:

$$p(\boldsymbol{X}||\boldsymbol{\alpha}|, \boldsymbol{\mu}) = \frac{\Gamma(|\boldsymbol{\alpha}|)}{\prod_{i=1}^{dim+1} \Gamma(\mu_i|\boldsymbol{\alpha}|)} \prod_{i=1}^{dim+1} X_i^{\mu_i|\boldsymbol{\alpha}|-1} \tag{8}$$

where $\boldsymbol{\mu} = (\mu_1, \ldots, \mu_{dim+1})$. By this reparameterization, the parameters of the Dirichlet mixture to estimate will be $\xi = (\boldsymbol{\mu}_1, \ldots, \boldsymbol{\mu}_M, |\boldsymbol{\alpha}_1|, \ldots, |\boldsymbol{\alpha}_M|, p(1), \ldots, p(M))$. This set of parameters can be divided into three subsets $\xi_1 = (|\boldsymbol{\alpha}_1|, \ldots, |\boldsymbol{\alpha}_M|)$, $\xi_2 = (\boldsymbol{\mu}_1, \ldots, \boldsymbol{\mu}_M)$, and $\xi_3 = (p(1), \ldots, p(M))$. Then, the different parameters ξ_1, ξ_2 and ξ_3 can be calculated independently. The likelihood for ξ_1 alone is:

$$p(\mathcal{X}|\xi_1) \propto \prod_{i=1}^{N} \left[\sum_{j=1}^{M} p(j) \frac{\Gamma(|\boldsymbol{\alpha}_j|)}{\prod_{l=1}^{dim+1} \Gamma(\mu_{jl}|\boldsymbol{\alpha}_j|)} \prod_{l=1}^{dim+1} X_{il}^{\mu_{jl}|\boldsymbol{\alpha}_j|-1} \right] \tag{9}$$

For the estimation of $|\alpha_j|$, we use a Newton-Raphson method:

$$|\alpha_j|^{(t)} = |\alpha_j|^{(t-1)} - \left(\frac{\partial^2 logp(\mathcal{X}|\xi_1^{(t-1)})}{\partial^2|\alpha_j|}\right)^{-1}\frac{\partial logp(\mathcal{X}|\xi_1^{(t-1)})}{\partial|\alpha_j|} \qquad (10)$$

The likelihood for ξ_2 alone is:

$$p(\mathcal{X}|\xi_2) \propto \prod_{i=1}^{N}\left[\sum_{j=1}^{M}p(j)\prod_{l=1}^{dim+1}\frac{X_{il}^{\mu_{jl}|\alpha_j|-1}}{\Gamma(\mu_{jl}|\alpha_j|)}\right] \qquad (11)$$

By maximizing $p(\mathcal{X}|\xi_2)$ taking into account the constraint $\sum_{l=1}^{dim+1}\mu_{jl} = 1$, we obtain:

$$\mu_{jl}^{(t)} = \frac{\mu_{jl}^{(t-1)}\sum_{i=1}^{N}p(\boldsymbol{\mu_j}^{(t-1)}|\boldsymbol{X_i})\left(log(X_{il}) - \Psi(\mu_{jl}^{(t-1)}|\alpha_j|^{(t)})\right)}{\sum_{l=1}^{dim+1}\left[\mu_{jl}^{(t-1)}\sum_{i=1}^{N}p(\boldsymbol{\mu_j}^{(t-1)}|\boldsymbol{X_i})\left(log(X_{il}) - \Psi(\mu_{jl}^{(t-1)}|\alpha_j|^{(t)})\right)\right]} \qquad (12)$$

Then, on the iteration t of the ECM algorithm, the E-Step is the same as given above for the EM algorithm, but the M-Step is replaced by three CM-Steps, as follows:

- **CM-Step1:** Calculate $\xi_1^{(t)}$ using Eq. 10 with ξ_2 fixed at $\xi_2^{(t-1)}$ and ξ_3 fixed at $\xi_3^{(t-1)}$.
- **CM-Step2:** Calculate $\xi_2^{(t)}$ using Eq. 12 with ξ_1 fixed at $\xi_1^{(t)}$ and ξ_3 fixed at $\xi_3^{(t-1)}$.
- **CM-Step2:** Calculate $\xi_3^{(t)}$ using Eq. 7 with ξ_1 fixed at $\xi_1^{(t)}$ and ξ_2 fixed at $\xi_2^{(t)}$.

3   MML Approach for the Determination of the Number of Clusters

3.1   MML Principle

Let us consider a set of data $\mathcal{X}$ controlled by a mixture of distributions with vector of parameters ξ. According to information theory [4], the optimal number of clusters of the mixture is that which requires a minimum amount of information, measured in nats, to transmit $\mathcal{X}$ efficiently from a sender to a receiver. The message length is defined as minus the logarithm of the posterior probability.

$$MessLen = -log(P(\xi|\mathcal{X})) \qquad (13)$$

The MML principle has strong connections with Bayesian inference, and hence uses an explicit prior distribution over parameter values. Wallace [5] and Baxter [6] give us the formula for the message length for a mixture of distributions:

$$MessLen \simeq -log(h(\xi)) - log(p(\mathcal{X}|\xi)) + \frac{1}{2}log(|F(\xi)|) - \frac{N_p}{2}log(12) + \frac{N_p}{2} \quad (14)$$

where $h(\xi)$ is the prior probability, $p(\mathcal{X}|\xi)$ is the likelihood, and $|F(\xi)|$ is the Fisher information, defined as the determinant of the Hessian matrix of minus the log-likelihood of the mixture. N_p is the number of parameters to be estimated and is equal to $M(dim+3)$ in our case. The estimation of the number of clusters is carried out by finding the minimum with regards to ξ of the message length $MessLen$. We will determine the expression of MML for a Dirichlet mixture.

3.2   Fisher Information for a Mixture of Dirichlet Distributions

Fisher information is the determinant of the Hessian matrix of the logarithm of minus the likelihood of the mixture. The Hessian matrix of a mixture leads to a complicated analytical form of MML which cannot be easily reproduced. We will approximate this matrix by formulating two assumptions, as follows. First, it should be recalled that (ξ_1, ξ_2) and ξ_3 are independent because any prior idea one might have about (ξ_1, ξ_2) would usually not be greatly influenced by one's idea about the value of the mixing parameter vector ξ_3. Furthermore, we assume that ξ_1 and ξ_2 are also independent. The Fisher information is then [6]:

$$F(\xi) \simeq F(\xi_1)F(\xi_2)F(\xi_3) \quad (15)$$

where $F(\xi_3)$ is the Fisher information with regards to the probability of the mixture. $F(\xi_1)$ and $F(\xi_2)$ are the Fisher information with regards to the vectors ξ_1 and ξ_2. In what follows we will compute each of these separately. For $F(\xi_3)$, it should be noted that the mixing parameters satisfy the requirement $\sum_{j=1}^{M} p(j) = 1$. Consequently, it is possible to consider the generalized Bernoulli process with a series of trials, each of which has M possible outcomes labeled first cluster, second cluster, ..., M^{th} cluster. The number of trials of the jth cluster is a multinomial distribution of parameters $p(1), p(2), \ldots, p(M)$. In this case, the determinant of the Fisher information matrix is [6]:

$$F(\xi_3) = \frac{N}{\prod_{j=1}^{M} p(j)} \quad (16)$$

For $F(\xi_1)$ and $F(\xi_3)$, we assume that the components of ξ_1 and ξ_2 are independent, then:

$$F(\xi_1) = \prod_{j=1}^{M} F(|\boldsymbol{\alpha}_j|) \quad (17)$$

$$F(\xi_2) = \prod_{j=1}^{M} F(\boldsymbol{\mu}_j) \quad (18)$$

let us consider the jth cluster $\mathcal{X}_j = (\boldsymbol{X}_l, \ldots, \boldsymbol{X}_{l+nj-1})$ of the mixture, where $l \leq N$, with parameters $|\boldsymbol{\alpha}_j|$ and $\boldsymbol{\mu}_j$. The choice of the jth cluster allows us to simplify the notation without loss of generality. The Hessian matrix when we consider the vector $\boldsymbol{\mu}_j$ is given by:

$$H(\boldsymbol{\mu}_j) = \frac{\partial^2}{\partial \mu_{jk_1} \partial \mu_{jk_2}}(-logp(\mathcal{X}_j|\boldsymbol{\mu}_j)) \tag{19}$$

where $k_1 = 1 \ldots dim + 1$ and $k_2 = 1 \ldots dim + 1$. Straight forward manipulations give us the determinant of the matrix $H(\boldsymbol{\mu}_j)$:

$$F(\boldsymbol{\mu}_j) = n_j^{dim+1}|\boldsymbol{\alpha}_j|^{2(dim+1)} \prod_{k=1}^{dim+1} \Psi'(\mu_{jk}|\boldsymbol{\alpha}_j|) \tag{20}$$

By substituting Eq. 20 in Eq. 18 we obtain:

$$F(\xi_2) = \prod_{j=1}^{M} \left(n_j^{dim+1}|\boldsymbol{\alpha}_j|^{2(dim+1)} \prod_{k=1}^{dim+1} \Psi'(\mu_{jk}|\boldsymbol{\alpha}_j|) \right) \tag{21}$$

Now we determine the Fisher information when we consider $|\boldsymbol{\alpha}_j|$. The second derivative is given by:

$$-\frac{\partial^2 logp(\mathcal{X}_j||\boldsymbol{\alpha}_j|)}{\partial^2|\boldsymbol{\alpha}_j|} = n_j \left(-\Psi'(|\boldsymbol{\alpha}_j|) + \sum_{k=1}^{dim+1} \mu_{jk}^2 \Psi'(\mu_{jk}|\boldsymbol{\alpha}_j|) \right) \tag{22}$$

and represent the Fisher information. By substituting Eq. 22 in Eq. 17, we obtain:

$$F(\xi_1) = \prod_{j=1}^{M} n_j \left(-\Psi'(|\boldsymbol{\alpha}_j|) + \sum_{k=1}^{dim+1} \mu_{jk}^2 \Psi'(\mu_{jk}|\boldsymbol{\alpha}_j|) \right) \tag{23}$$

Finally the complete Fisher information for the mixture is found by substituting Eq. 16, Eq. 21 and Eq. 23 in Eq. 15.

3.3   Prior Distribution $h(\xi)$

The performance of the MML criterion is dependent on the choice of the prior distribution $h(\xi)$. Several criteria have been proposed for the selection of prior $h(\xi)$. Following Bayesian inference theory, the prior density of a parameter is either constant on the whole range of its values or the value range is split into cells and the prior density is assumed to be constant within each cell. Since ξ_1, ξ_2 and ξ_3 are independent, we have:

$$h(\xi) = h(\xi_1)h(\xi_2)h(\xi_3) \tag{24}$$

We will now define the three densities $h(\xi_1)$, $h(\xi_2)$, and $h(\xi_3)$. The vector ξ_3 has M dependent components; i.e. the sum of the mixing parameters is one. Thus, we omit one of these components, say $p(M)$. The new vector has $(M - 1)$ independent components. We treat the $p(j)$, $j = 1 \ldots M - 1$ as being the parameters of a multinomial distribution. With the $(M-1)$ remaining mixing parameters, $(M-1)!$ possible vectors can be formed. Thus, we set the uniform prior density of ξ_3 to [6]:

$$h(\xi_3) = \frac{1}{(M-1)!} \tag{25}$$

For $h(\xi_2)$, since $\boldsymbol{\mu}_j$, $j = 1 \ldots M$ are assumed to be independent:

$$h(\xi_2) = \prod_{j=1}^{M} h(\boldsymbol{\mu}_j) \tag{26}$$

Using the same approach as for the vector ξ_3, we set the uniform prior density of $\boldsymbol{\mu}_j$ to:

$$h(\boldsymbol{\mu}_j) = \frac{1}{dim!} \tag{27}$$

Indeed, $\sum_{k=1}^{dim+1} \mu_{jk} = 1$. By substituting Eq. 27 in Eq. 26, we obtain:

$$h(\xi_2) = \frac{1}{dim!^M} \tag{28}$$

For $h(\xi_1)$, since $|\boldsymbol{\alpha}_j|$, $j = 1 \ldots M$ are assumed to be independent:

$$h(\xi_1) = \prod_{j=1}^{M} h(|\boldsymbol{\alpha}_j|) \tag{29}$$

We will now calculate $h(|\boldsymbol{\alpha}_j|)$. In the absence of other knowledge about the $|\boldsymbol{\alpha}_j|$, we use the principle of ignorance by assuming that $h(|\boldsymbol{\alpha}_j|)$ is locally uniform over the ranges $[0, e^3|\hat{\boldsymbol{\alpha}}_{pop}|]$ (in fact, we know experimentally that $|\boldsymbol{\alpha}_j| < e^3|\hat{\boldsymbol{\alpha}}_{pop}|$, where $|\hat{\boldsymbol{\alpha}}_{pop}|$ is the estimated parameter when we consider the entire population. We choose the following uniform priors in accordance with Ockham's razor (a simple priors which give good results) [7]:

$$h(|\boldsymbol{\alpha}_j|) = \frac{e^{-3}}{|\hat{\boldsymbol{\alpha}}_{pop}|} \tag{30}$$

By substituting Eq. 30 in Eq. 29, we obtain

$$h(\xi_1) = \prod_{j=1}^{M} \frac{e^{-3}}{|\hat{\boldsymbol{\alpha}}_{pop}|} = \frac{e^{-3M}}{|\hat{\boldsymbol{\alpha}}_{pop}|^M} \tag{31}$$

By substituting Eq. 31, Eq. 28 and Eq. 25 in Eq. 24, we obtain:

$$h(\xi) = \frac{e^{-3M}}{|\hat{\boldsymbol{\alpha}}_{pop}|^M (M-1)! dim!^M} \tag{32}$$

The expression of MML for a finite mixture of Dirichlet distributions is obtained by substituting Eq. 32 and Eq. 15 in Eq. 14.

3.4 Estimation and Selection Algorithm

The algorithm of selection and estimation is thus as follows:

Algorithm
For each candidate value of M:

1. Initialization
2. E-Step: Compute the *posterior* probabilities:
   $$\hat{Z}_{ij} = \frac{p(X_i|\alpha_j)p(j)}{\sum_{j=1}^{M} p(X_i|,\alpha_j)p(j)}$$
3. CM-Steps:
   (a) **CM-Step1:** Calculate $\xi_1^{(t)}$ using Eq. 10 with ξ_2 fixed at $\xi_2^{(t-1)}$ and ξ_3 fixed at $\xi_3^{(t-1)}$.
   (b) **CM-Step2:** Calculate $\xi_2^{(t)}$ using Eq. 12 with ξ_1 fixed at $\xi_1^{(t)}$ and ξ_3 fixed at $\xi_3^{(t-1)}$.
   (c) **CM-Step2:** Calculate $\xi_3^{(t)}$ using Eq. 7 with ξ_1 fixed at $\xi_1^{(t)}$ and ξ_2 fixed at $\xi_2^{(t)}$.
4. If the convergence test is passed, terminate, else go to 2.
5. Calculate the associated criterion $MML(M)$ using Eq. 14.
6. Select the optimal model M^* such that: $M^* = \arg\min_M MML(M)$

details about the initialization algorithm can be found in [1]. The convergence test can involve the stabilization of the parameters or the likelihood function.

4 Experimental Results

The application concerns the summarization of image databases. Interactions between users and multimedia databases can involve queries like "Retrieve images that are similar to this image". A number of techniques have been developed to handle pictorial queries. Summarizing the database is very important because it simplifies the task of retrieval by restricting the search for similar images to a smaller domain of the database. Summarization is also very efficient for browsing. Knowing the categories of images in a given database allows the user to find the images he or she is looking for more quickly. Using mixture decomposition, we can find natural groupings of images and represent each group by the most representative image in the group. In other words, after appropriate features are extracted from the images, it allows us to partition the feature space into regions that are relatively homogeneous with respect to the chosen set of features. By identifying the homogeneous regions in the feature space, the task of summarization is accomplished. For the experiment, we used the *Vistex* gray-level texture database obtained from the MIT Media Lab. In our experimental framework, each of the 512×512 images from the *Vistex* database was divided into 64×64 images. Since each 512×512 "mother image" contributes 64 images to our database, ideally all of the 64 images should be classified in the same class. In the experiment, six homogeneous texture groups, "Bark", "Fabric", "Food",

Fig. 1. Sample images from each group. (a) Bark, (b) Fabric, (c) Food, (d) Metal, (e) Sand, (f) Water.

Table 1. Number of clusters found by three criteria (MML, MDL and AIC)

Number of clusters	MML	MDL	AIC
1	-12945.10	-12951.40	-12974.90
2	-12951.12	-13001.52	-13019.12
3	-12960.34	-13080.37	-13094.23
4	-13000.76	-13206.73	-13225.57
5	-13245.18	**-13574.98**	**-13591.04**
6	**-13765.04**	-13570.09	-13587.64
7	-13456.71	-13493.50	-13519.50
8	-13398.16	-13387.56	-13405.92
9	-13402.64	-13125.41	-13141.95
10	-13100.82	-13001.80	-13020.23

Table 2. Confusion matrix for image classification by a Dirichlet mixture

	Bark	Fabric	Food	Metal	Sand	Water
Bark	250	0	0	0	6	0
Fabric	0	248	8	0	0	0
Food	0	9	375	0	0	0
Metal	0	0	0	250	0	6
Sand	4	0	0	0	380	0
Water	3	0	0	7	2	372

"Metal", "Water" and "Sand" were used to create a new database. A database with 1920 images of size 64×64 pixels was obtained. Four images from each of the Bark, Fabric and Metal texture groups and 6 images from Water, Food and Sand were used. Examples of images from each of the categories are shown in Fig. 1. In order to determine the vector of characteristics for each image, we used the cooccurrence matrix introduced by Haralick et al. [8]. For relevant representation of texture, many cooccurrences should be computed, each one considering a given neighborhood and direction. In our application, we have considered the following four neighborhoods: $(1; 0)$, $(1; \frac{\pi}{4})$, $(1; \frac{\pi}{2})$, and $(1; \frac{3\pi}{4})$. For each of these neighborhoods, we calculated the corresponding cooccurrence matrix, then derived from it the following features: Mean, Energy, Contrast, and Homogeneity [9]. Thus, each image was represented by a $16D$ feature vector. By

Table 3. Confusion matrix for image classification by a Gaussian mixture

	Bark	Fabric	Food	Metal	Sand	Water
Bark	240	0	0	3	8	5
Fabric	0	236	12	0	4	4
Food	0	12	365	4	0	3
Metal	0	2	2	242	4	6
Sand	8	2	0	0	370	4
Water	5	1	0	10	5	363

applying our algorithm to the texture database using MML and other different selection selection criteria such that MDL and AIC [2], only the MML criterion found six categories (see Table 1). In what follows we use the selection found by the MML. The classification was performed using the Bayesian decision rule after the class-conditional densities were estimated. The confusion matrix for the texture image classification is given in Table 2. In this confusion matrix, the cell $(classi, classj)$ represents the number of images from $classi$ which are classified as $classj$. The number of images misclassified was small: 45 in all, which represents an accuracy of 97.65 percent. From Table 2, we can see clearly that the errors are due essentially to the presence of macrotexture, i.e., the texture at large scale, (between Fabric and Food for example) or because of microtexture, i.e., the texture at pixel level (between Metal and Water for example). Table 3 shows the confusion matrix for the Gaussian mixture.

5   Conclusion

In this paper, we have proposed a new method based on the ECM algorithm to estimate the parameters of a Dirichlet mixture. The ECM algorithm replaces a complicated M-step of the EM algorithm with several computationally simpler CM-Steps. The number of clusters is determined using an MML-based approach. From the experimental results, we can say that the Dirichlet distribution offers strong modeling capabilities.

Acknowledgement

The completion of this research was made possible thanks to the the Natural Sciences and Engineering Research Council of Canada, Heritage Canada and Bell Canada's support through its Bell University Laboratories R&D program.

References

1. N. Bouguila, D. Ziou and J. Vaillancourt. Unsupervised Learning of a Finite Mixture Model Based on the Dirichlet Distribution and its Application. *IEEE Transactions on Image Processing*, 13(11):1533–1543, November 2004.
2. G.J. McLachlan and D. Peel. *Finite Mixture Models*. New York: Wiley, 2000.

3. X. L. Meng and D. B. Rubin. Maximum Likelihood Estimation via the ECM Algorithm: a General Framework. *Biometrika*, 80(2):267–278, 1993.
4. C.E. Shannon. A Mathematical Theory of Communication. *Bell System Tech.*, 27:379–423, 1948.
5. C. S. Wallace and D. L. Dowe. MML clustering of multi-state, Poisson, von Mises circular and Gaussian distributions. *Statistics and Computing* , 10(1):73–83, 2000.
6. R. A. Baxter and J. J. Olivier. Finding Overlapping Components with MML. *Statistics and Computing*, 10(1):5–16, 2000.
7. W. Jeffreys and J. Berger. Ockham's Razor and Bayesian Analysis. *American Scientist*, 80:64–72, 1992.
8. R. M. Haralick, K. Shanmugan and I. Dinstein. Texture features for image classification. *IEEE Transactions on Systems, Man and Cybernetics*, 8:610–621, 1973.
9. M. Unser. Sum and Difference Histograms for Texture Classification. *IEEE Transactions on Pattern Analysis and Machine Intelligence*, 8(1):118–125, 1986.

Disease Classification from Capillary Electrophoresis: Mass Spectrometry

Simon Rogers[1], Mark Girolami[1], Ronald Krebs[2], and Harald Mischak[2]

[1] Bioinformatics Research Centre, Department of Computing Science,
University of Glasgow, Glasgow, UK
srogers@dcs.gla.ac.uk
http://www.dcs.gla.ac.uk/~srogers
[2] Mosaiques Diagnostics and Therapeutics AG, Feodor-Lynen-Str. 21,
D-30625 Hannover, Germany

Abstract. We investigate the possibility of using pattern recognition techniques to classify various disease types using data produced by a new form of rapid Mass Spectrometry. The data format has several advantages over other high-throughput technologies and as such could become a useful diagnostic tool. We investigate the binary and multi-class performances obtained using standard classifiers as the number of features is varied and conclude that there is potential in this technique and suggest research directions that would improve performance.

1   Introduction

In recent years, microarrays have enabled researchers to measure the expression of entire genomes simultaneously. Some work has been undertaken to investigate how well classifiers built using microarray data can discriminate between healthy and diseased samples and samples of differing diseases and disease stages [1]. However, although this is very interesting from a feature (i.e. gene) selection perspective, as a general diagnostic tool, it is unlikely to prove useful. There are several underlying reasons for this. Firstly, the cost of microarray analysis and the time required to perform the analysis are both currently prohibitive. Secondly, the mRNA levels measured by a microarray only give a partial picture of the proteomic activity inside the cell. Finally, samples have to be very localised. For example, to diagnose a bladder cancer, a sample of bladder tissue would be required. This is obviously a highly invasive procedure.

In this paper, we consider a new form of biological data (introduced in [2,3,4]) generated using Mass Spectrometry (MS) and assess whether it has potential as a diagnosis tool, using various pattern classification techniques. This data can be obtained very rapidly and inexpensively, suggesting that it may be well suited for a diagnostic purpose. Also, the data is collected from a urine sample. This is easily obtained and therefore can potentially be used to diagnose any disease that will cause a change in the particle content of the urine.

The remainder of the paper is set out as follows. In the next section, we introduce the data generation process. In section 3, we discuss the data pre-filtering

S. Singh et al. (Eds.): ICAPR 2005, LNCS 3686, pp. 183–191, 2005.

and pre-processing and briefly mention the classification algorithms used. In sections 4 and 5 we present results and conclusions.

2   CE/MS Data Generation

Recently, a new MS approach has been investigated that couples capillary electrophoresis (CE) directly to MS enabling detailed analysis to be available quickly (< 1 hour) and directly from a suitable (e.g. urine) sample [2,4,3]. Traditionally MS has been used to identify individual proteins but typically cannot be performed on a sample consisting of various proteins. Separation techniques exist to isolate individual proteins from such a mixture but these tend to be highly labour intensive and therefore expensive and slow. Here, the CE takes a complex sample of particles (in this case, the particles can be anything that might be found in the urine, not necessarily complete proteins) and by applying a charge differential along the capillary, separates the various particles in time. The output is connected directly to the MS realising a mass profile that evolves with time. This data is then analysed by *Mosaiques Visu* software that detects and outputs intensity values at the unique mass/time peaks (for details, see [4]). The separation in time means that only a small fraction of the particles are applied to the MS at any particular time. If the sample was applied directly to the MS without this stage it would be far more difficult to distinguish between individual particles.

This method has many possible diagnostic advantages over microarrays. Firstly, the analysis is quick and non-invasive. Secondly, in the case of using urine samples, there is the potential to be able to diagnose any diseases that would result in a variation of the products found in urine. However, there are drawbacks to this method. Firstly, the data produced is of a very high dimension ($\sim 30,000$ features) and the number of available samples is relatively small. Secondly, as no real control is imposed on the sample being analysed, it is possible that amongst this high number of features, there will be many due to other, spurious factors.

To date, there has been some research focused on the potential of MS proteomics data as a diagnostic tool, but using serum rather than urine. For example, Lilien *et al* [5] use Principal Components Analysis and a linear discriminant to distinguish between the MS spectra of serum samples from patients with various tumours. Similarly, Wagner *et al* [6] use supervised techniques to try and create protein profiles from MS analysis of serum samples. These approaches are all based on identifying whole, specific proteins whereas CE/MS can detect a much wider range of particles.

3   Method

3.1   Data

The data set we shall use consists of analysis of 632 samples that come from one of 22 separate classes from individuals with various renal diseases, cancers and diabetes as well as samples from healthy individuals. We have performed binary classification with a variety of algorithms on a large number of pairs of classes from this set, however, in this investigation, we will concentrate on a group of five classes - Bladder Cancer (BLA - 47 instances), Renal Cancer (REN - 25 instances), Prostate Cancer (PCA - 8 instances), Benign Prostate (PB - 12 instances) and Healthy (NK - 41 instances). The total number of features is 28378. The inclusion of benign prostate samples is interesting as clinical differentiation between individuals with prostate cancer and those with a benign growth is challenging and the two different conditions require vastly different treatment.

3.2 Feature Pre-filtering

This particular form of data has several important characteristics. Firstly, although the total number of possible features is very high ($\sim 30,000$), in each sample, only a small proportion of these values are non-zero, indicating that this particular particle was not present or, and this distinction may be important, not detected. Therefore, if we call our N (samples) $\times M$ (features) dataset $\mathbf{X}$, the vast majority of the x_{ij} values are zeros. Secondly, those values that are present take values over a very large range (see figure 1(a) (top)). To overcome this second problem, we have adopted a log transform[1]. Figure 1(a) shows the binary classification performances for a wide range of pairwise comparisons and algorithms with and without this transform. We can see that in the vast majority of problems the log transform improves performance (all points below the $y = x$ line).

The first problem is not quite so straightforward to address. As an initial step, we perform a simple pre-filtering. For a given classification problem (i.e. 2 or more classes), we only keep features that appear (i.e. are non-zero) in at least $\rho\%$ of the data samples for one or more of the classes. Note that we do not force the feature to be present $\rho\%$ of the time across *all* of the classes as it is possible that both presence and absence of a feature as well as presence with varying magnitude could be indicative of changing condition. We will investigate the effect of varying ρ in more detail later.

3.3 Classifiers

In this investigation we limit ourselves to two main classes of classifiers. Naive Bayes classifiers (NB) and Support Vector Machines (SVM's). This is by no means a complete list but serves as a reasonable starting point. Due to limitations of space, a description of these algorithms is omitted, readers are referred to [7,8] for more details. When using a NB classifier, it is necessary to determine the parametric form of the density function that will be used for each feature. Here, we have considered the following four (defining the data matrix $\mathbf{X}$ as before, and

---

[1] specifically, $\log(x_{ij} + 1)$, where the additive term ensures that our zero values remain at zero.

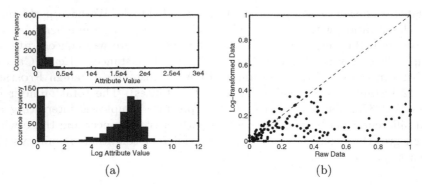

Fig. 1. Log transformation example (left, note that due to the wide range of values, many bars on the top histogram are too small to be visible) and binary classification errors with and without log transform

indexing the individual features with $m = 1 \ldots M$ and the classes $c = 1 \ldots C$ and defining the $M \times 1$ data vector $\mathbf{x}$)

- **Gaussian:** Each class is defined by an M dimensional Gaussian distribution with diagonal covariance. i.e. $p(\mathbf{x}|c, \mu_c, \sigma_c) = \prod_{m=1}^{M} \mathcal{N}(x^m|\mu_c^m, \sigma_c^m)$
- **Binomial:** The data is transformed to a binary representation (i.e. value present (non-zero) or absent (zero). Each class is then defined by an M dimensional vector of probabilities $\mathbf{p}_c$, where $p_c^m = P(x_m = 1|c)$. i.e.
  $p(\mathbf{x}|c, \mu_c, \sigma_c) = \prod_{m=1}^{M}(p_c^m)^{x_m}(1 - p_c^m)^{1-x_m}$
- **Multinomial:** As binomial but with > 2 possible states. Each class is now defined by M vectors of state probabilities.
- **Exact-zero Gaussian:** This density is intended to capture more accurately the characteristics of the particular problem. For each class, we now have an M dimensional vector of probabilities as in the binomial case. If the value is non-zero we then assume it can be modelled by a Gaussian. Therefore, defining the indicator variable t_m which is 1 iff x_m is non-zero, $p(\mathbf{x}|c, p_c, \mu_c, \sigma_c) = \prod_{m=1}^{M}(p_c^m \mathcal{N}(x^m|\mu_c^m, \sigma_c^m))^{t_m}(1 - p_c^m)^{1-t_m}$.

In all cases, we define the prior distributions for each class to be the proportion of training instances from that class and use the standard maximum likelihood solutions for the parameter values.

As an alternative to the probabilistic, Naive Bayes classifiers, we consider the SVM. To use an SVM, a kernel function must be defined. We use a linear kernel (a simple dot product in the input space) after normalising each feature to have zero mean and unit variance. It may be the case that there are other more suitable kernels that could be used however, due to the high dimensionality, this is a reasonable starting point. In addition to this, we set the margin parameter C to infinity (i.e. a hard-margin).

Multi-class SVM Classifiers. Naive Bayes classifiers are naturally multi-class. SVM's however can not be naturally generalised to the multi-class setting.

However, various tree based heuristics can be used to split the problem down into a set of binary decisions. We experiment with two of these here

- **Directed Acyclic Graph (DAG):** In a C class problem, the DAG SVM [9] formulates the problem as a tree with $(C(C-1)/2)$ nodes. At each node, an SVM is trained between two of the classes in the problem. When testing a point, we start by assuming that the point could belong to any of the C classes. It then moves through the tree and at each SVM, one class is removed from the possible solution until only one class remains. For example, in a 3 class problem, we might train our first classifier on class 1 versus class 3. When testing, if the test point is classified as 1, we remove 3 from the list of possible solutions and move on to the classifier between 1 and 2.
- **Divide-by-2 (DB2):** The DB2 [10] classifier operates by repeatedly splitting the C class problem into binary problems. For example, in a four class problem, the first classifier might split the data into the meta-classes (1,2) and (3,4). If a test point is classified as belonging to the first class, it is then applied to a classifier between 1 and 2 etc.

In either of these systems, the particular form of SVM can vary between nodes. Presently, we have kept them all the same (linear kernel, $C = \infty$) but employing different ones for different classifications is an obvious next step. This is particularly promising for the DB2 model where it may be sensible to have different classifiers built from different features at different levels in the hierarchy. We will discuss this further below.

4   Results

4.1   Binary Classification

Initially, we have investigated the pairwise classification performance between relevant pairs of classes in the dataset. This has been performed for many pair wise combinations but due to space limitations we will only consider those belonging to the cancer subset here. Table 1 shows the results for our five binary classifiers (SVM and 4 different NB). Each value is the best leave-one-out (LOO) performance obtained when varying ρ, the feature filtering threshold. In some cases the best performance was obtained for several different values of ρ. In these cases, we have shown the minimum and maximum ρ values. We can see from the table that generally, the performance is reasonably good with low values LOO error. The highest errors obtained are for the classification between Prostate Cancer and benign Prostate with a minimum of 10% LOO error (= 2 data points). This is to be expected, partly due to the difficulty of the problem and partly due to the fact that there is such a small number of samples in each class (8 and 12 in PCA and PB respectively). We also note that no-one classifier out-performs the others although the best performance can generally be found from an SVM or Naive Bayes with Gaussian or binomial densities. This is especially interesting as it suggests that in some cases, the magnitude of a value (if it is non-zero) does not improve performance whereas in other cases it does. The relatively poor performance of the exact-zero system seems to suggest that it is not necessary to use both presence and absence information and magnitude information at once. It is worth mentioning that the exact-zero mixture requires the fitting of considerably more parameters than the individual Gaussian and Binomial models and it will be interesting to see if this error rate can be improved as more data becomes available. The results are promising and suggest that discrimination is possible using CE/MS data. However, further validation will be acquired through a planned blind test.

Table 1. Binary LOO performances (errors are percentages and the value in brackets is value for ρ for this particular level of performance)

Classes	Class Sizes	SVM	NB Gauss	NB Bin	NB Mult	Exact Zero
PCA v PB	8 v 12	15.00 (90)	15.00 (95)	10.00 (90)	20.00 (55→ 95)	25.00 (15 → 85)
PCA v NK	8 v 41	0.00 (85)	4.08 (25→95)	0.00 (95)	12.24 (50→85)	6.12 (40→95)
PB v NK	12 v 41	1.80 (65)	0.00 (10→20)	5.66 (30→85)	11.32 (55→75)	7.55 (70)
REN v NK	25 v 41	1.52 (75)	1.52 (20)	6.06 (15)	7.58 (35→50)	6.06 (10→65)
BLA v NK	47 v 41	2.30 (65)	4.55 (15→20)	7.95 (5)	6.82 (45→80)	3.41 (10→15)

4.2   Multi-class Classification

Binary classifications are interesting but are limited from a diagnostic point of view. One of the possible benefits from CE/MS data from urine samples is that any number of different diseases could be identified. Therefore, we turn our attention to multi-class schemes. As discussed above, the various Naive Bayes classifiers can be naturally expanded to a multi-class scenario. For the SVM's, we have used two tree based approaches, DAG and DB2. For DB2, it is necessary to define the hierarchy - i.e. how we want to perform the successive partitions of the C classes to create a series of binary problems. In this example, we have decided on a hierarchy that is sensible from a clinical point of view. The hierarchy is shown in figure 2(a). At the top level, we split NK from everything else (i.e. healthy versus unhealthy). If the point is classified as unhealthy, we perform a further split into (PCA, PB) v (BLA, REN) and then perform a standard binary classification on whichever of these pairs is chosen. The best results (again, as ρ is varied) can be seen in table 2. A plot of the number of features retained against ρ can be seen in figure 2(b). In this case, we see that the two SVM schemes out-perform the various Naive Bayes classifiers. This may be due to the fact that the two SVM schemes do a series of more simple binary classifications, rather than one more complicated multi-class one (as is the case with the Naive Bayes). The DB2 SVM defines a hierarchy over the possible diseases and so is able to classify at varying levels of abstraction. For example, at the top level of the tree (i.e. simply classifying between healthy and unhealthy) there is just one misclassification (a NK) and at the next level (ignoring the 1 wrong NK from the previous classification), there are no errors, suggesting that the errors all occur in the most specific, lowest level. This shows the power of a possible hierarchical

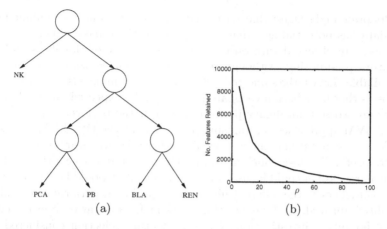

(a)   (b)

Fig. 2. Example hierarchy (left) and number of features retained in multi-class problem as ρ is increased (right)

Table 2. Multi-Class LOO errors. Top line shows overall percentage, second line shows ρ with the actual number of features in brackets, lower rows give the percentage of errors in each particular class / absolute number of errors in each class.

	SVM(DAG)	SVM(DB2)	NB Gauss	NB Bin	NB Mult	Exact Zero
All (%)	6.02	6.77	14.29	19.55	24.81	16.45
ρ (No.Feat)	10 (5411)	25 (2366)	40 (1287)	25 (2366)	60 (720)	60 (720)
PCA	0 / 0	37.5 / 3	37.5 / 3	25.0 / 2	100 / 8	50.0 / 4
PB	33.3 / 4	16.7 / 2	41.67 / 5	41.67 / 5	91.7 / 11	66.7 / 8
REN	12.0 / 3	8.0 / 2	8.0 / 2	16.0 / 4	32.0 / 8	20.0 / 5
BLA	2.1 / 1	0.0 / 0	6.4 / 3	12.8 / 6	2.0 / 1	2.0 / 1
NK	0.0 / 0	4.9 / 1	14.6 / 6	22.0 / 9	12.2 / 5	9.8 / 4

approach - currently we are unable to reliably classify between PCA and PB but it appears that we can reliably classify that something is either PCA or PB from other classes. The ability to visualise the decreasing certainty as we move down the hierarchy is a great bonus to such an approach; something that is lacking from a flat structure such as a simple Naive Bayes classifier.

5 Conclusions and Future Work

In this paper, we have described CE/MS a new rapid, high-throughput form of proteomic data and have performed several simple experiments to try to give some indication of the diagnostic capabilities of the data. The data has several advantages over other similar data formats such as microarray data. It can be produced very rapidly in a non-invasive manner (normally through analysis of a urine sample) and has the potential to be able to diagnose many diseases. However, like microarray data, it is noisy and the number of features is far greater than the number of collected samples - this latter problem is likely to improve as the data generation process is a fraction of the cost of a microarray experiment and obtaining samples for analysis is much more straightforward.

Results presented suggest that pattern recognition techniques combined with CE/MS data has potential as a diagnostic tool. In these basic experiments low LOO errors were observed with only a very basic choice of classifiers and very crude feature pre-filtering. Although results have been presented for only 5 of the 22 available classes, the same general level of performance is observed across other subsets that have been investigated. As might be expected, multi-class performances are worse than binary performances but the performance of the two tree-based SVM approaches is promising. Particularly, the DB2 SVM enables us to classify in a hierarchical manner, revealing where the errors are made and giving a more useful diagnosis. Such an approach isn't limited to SVM classifiers - any particular classifier could be used at each node and it would be expected that by tuning classifiers to the different hierarchical problems performance could be considerably improved. i.e. by extracting relevant features at each level. This is something for future investigation. The only feature selection considered here has involved the initial pre-filtering step. Examining the results, we see that the value of ρ for best performance varies dramatically. This suggests that for some problems, there are a small number of features that are consistently varied for the different diseases, whereas for others, we are obtaining useful information from features that are very rarely present. In these latter cases, it should be remembered that it is possible that several different features could correspond to the same particle that has undergone some small change. This suggests that performance may be improved by combining features or developing more applicable kernel functions, possibly including the mass and time information available for each detected peak. It may also be beneficial to include extra meta-data in the decision making process. This could be clinical history or more general observations (e.g. the gender of the individual). This would be particularly interesting in the hierarchical classifier as different meta-data could be incorporated at each level.

Of the multi-class methods investigated, the hierarchical SVM methods look to be the most promising and it is likely that the performances considered presented here could be improved by careful selection of the classifier at each level. This would involve more careful selection of kernels and a more rigorous feature selection stage.

Finally, the diseases that have been investigated so far have all been chosen due to the fact that they are very likely to produce a change in the urine profile. It would be interesting in future work to investigate whether or not such techniques could be used to diagnose diseases without such an obvious effect or in other testing circumstances.

References

1. Alizadeh, A., Eisen, M., Davis, R., et al.: Different types of diffuse large b-cell lymphoma identified by gene expressing profiling. Nature **403** (2000) 503–511
2. Kolch, W., Neususs, C., Pelzing, M., Mischak, H.: Capillary electrophoresis: Mass spectrometry as a powerful tool in clinical diagnosis and biomarker discovery. Mass Spectrometry Reviews (2005 (in press))

3. Kaiser, T., Wittke, S., Just, I., et al.: Capillary electrophoresis coupled to mass spectrometer for automated and robust polypeptide determination in body fluids for clinical use. Electrophoresis **25** (2004) 2044–2055
4. Weissinger, E., Wittke, S., Kaiser, T., et al.: Proteomic patterns established with capillary electrophoresis and mass spectrometry for diagnostic purposes. Kidney International **65** (2004) 2426–2434
5. Lilien, R.H., Farid, H., Donald, R.: Probabilistic disease classification of expression-dependent proteomic data from mass spectrometry of human serum. Journal of Computational Biology **10** (2003) 925–946
6. Wagner, R., Naik, D., Pothen, A., et al.: Computational protein biomarker prediction: a case study for prostate cancer. BMC Bioinformatics **5** (2004)
7. Mitchell, T.: Machine Learning. McGraw-Hill (1997)
8. Shawe-Taylor, J., Cristianini, N.: Kernel Methods for Pattern Analysis. Cambridge University Press (2004)
9. Platt, J., Cristianini, N., Shawe-Taylor, J.: Large margin DAG's for multiclass classification. Advances in Neural Information Processing Systems **12** (2000) 547–553
10. Vural, V., Dy, J.: A hierarchical method for multi-class support vector machines. In: Proceesings of the 21st International Conference on Machine Learning. (2004)

Analyzing Large Image Databases with the Evolving Tree

Jussi Pakkanen and Jukka Iivarinen

Laboratory of Computer and Information Science,
Helsinki University of Technology,
P.O. Box 5400, FI-02015 HUT, Finland
{jussi.pakkanen, jukka.iivarinen}@hut.fi

Abstract. Analyzing large image databases is an interesting problem that has many applications. The entire problem is very broad and contains difficult subproblems dealing with image analysis, feature selection, database management, and so on. In this paper we deal with efficient clustering and indexing of large feature vector sets. Our main tool is the *Evolving Tree*, an unsupervised, hierarchical, tree-shaped neural network. It has been designed to facilitate efficient analysis and searches of large data sets. Comparison to other similar methods show a favorable performance for the Evolving Tree.

1   Introduction

The Self-Organizing Map (SOM) is a widely used tool in various data analysis tasks [1]. However, some of its intrinsic features make it unsuitable for analyzing very large scale problems. Almost all operations on SOM start by locating the best matching unit (BMU) among all the nodes. This operation scales linearly according to the map size. Analyzing huge data sets requires very large maps. While operating on these maps is not usually intractable, it can be extremely slow. Another drawback is that the map size must be chosen beforehand. While there are some heuristics for this, experimenting with different sized maps is quite time-consuming.

There have been several different approaches to solve these problems. They can be roughly divided into two different groups. The first ones are flat systems that grow during training [2,3]. The second group tries to build efficient search structures to make operations faster. Combinations of these are also quite popular [4,5,6].

In this paper, we present and analyze the Evolving Tree [7,8] (ETree) and its behaviour. The Evolving Tree is a new kind of self-organizing neural network that has been designed to scale to very large problems. In Section 2 we give an overview of the Evolving Tree's algorithms and architecture. This is followed by the novel improvements and analysis in Section 3. In Section 4 we subject our system to several qualitative and quantitative experiments. Then we discuss the results and conclude the paper.

S. Singh et al. (Eds.): ICAPR 2005, LNCS 3686, pp. 192–198, 2005.

2 The Evolving Tree

In this section we briefly describe the ETree algorithm. We refer the readers interested in details on the basic algorithm to [8]. We also describe a new method for controlling the complexity of the system.

The basic building blocks of ETree are the same as in the SOM. We have nodes which have prototype vectors. We also use the same training formulas as the SOM. The difference is that we use these blocks in a very different way. The basic idea is to use a tree topology as opposed to the grid structure of SOM. The leaf nodes are used for data analysis while the trunk nodes maintain an efficient search tree to the leaf nodes. This makes SOM's most time consuming operation, finding the best matching unit (BMU), very fast. This is illustrated in the left image in Figure 1.

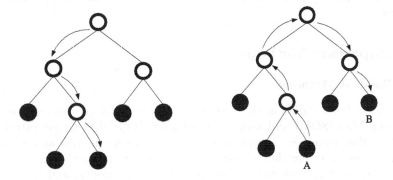

Fig. 1. Fundamental operations of the Evolving Tree. The left image shows how the BMU is found. The right image demonstrates how the tree distance between two nodes is calculated.

Another important part of SOM is the neighborhood. It measures the distance between two nodes along the grid. We use an equivalent metric called the *tree distance*. It computes the distance between two leaf nodes along the the search tree, as can be seen in the second image in Figure 1. With these we can train the tree structure in the same way as in the SOM. The last missing piece for a working system is a method for growing the tree. This is done by counting how many times each node has been the BMU. When this value reaches a pre-specified threshold, we split the node. The leaf node is transformed into a trunk node by giving it some child nodes.

2.1 Weight Decay

Training is usually performed for some amount of epochs and then stopped. Selecting the epoch count beforehand is very difficult, and usually leads to nonoptimal trees. To get around this, we used an algorithm based on weight decay [9].

As mentioned above, every node contains a counter, which says how many times it has been the BMU. After every epoch these counters are multiplied by a constant that is less than one. This inhibits the growth by decreasing the amount of splits. When we discover that the tree has grown only very little, such as under 5%, in one epoch we stop the training.

2.2   K-Means Adjustment

A known property of the SOM is that if the neighborhood function does not go to zero the resulting map will be nonoptimal. One way to optimize the leaf node locations in ETree, while still maintaining ETree's non-supervised nature, is to use a variant of k-means clustering after the training process. First all training vectors are mapped to leaf nodes using the established BMU search. Then the leaf nodes are moved to the center of mass of their respective data vectors. This procedure is repeated a few times to obtain the final leaf node locations.

3   Comparison Methods

3.1   Classical Methods

We used two classical data analysis methods as a basis for comparison, the Self-Organizing Map (SOM) and k-means clustering. Both of them are established algorithms that have been successfully applied in both scientific and industrial settings. These two were chosen because they work in a roughly similar way as ETree. That is, they both have a bunch of prototype nodes, and the training consists of finding BMUs for training vectors and then updating the nodes. The difference between these methods and ETree is that they are nonhierarchical and flat. This makes training and querying slow on large databases.

3.2   Tree-Shaped Methods

We also compared ETree against two other tree-shaped neural network systems. These experiments show the relative performance of modern neural systems. All three systems use a tree structure that is grown during training.

S-Tree. The first test system is the *S-Tree* by Campos and Carpenter [10]. It is noticeably more complex than ETree, having complicated rules for splitting and pruning of nodes. It does not have a concept of the neighborhood function as in SOM and ETree.

CNeT. The other algorithm we tested was *CNeT* by Behnke and Karayiannis [11]. It resembles ETree more than S-Tree. In fact, if you remove the neighborhood function from ETree and remove some portions from CNeT's training algorithms, they reduce to almost the same algorithm. The main difference between these two is the neighborhood function, which CNeT lacks, and the slightly simpler architecture and training formulas of ETree.

4 Experiments

We have used two different kinds of data sets for our experiments. The first data set consists of three different MPEG-7 features [12]: *Homogeneous texture, Edge histogram,* and *Color structure.* These were calculated for 1300 paper surface defect images [13]. This gives us moderate sized databases with dimensions of around 20. The exact dimension varies somewhat depending on the descriptor being used. This data set represents actual industrial data. There are a total of 14 different classes which are fuzzy and overlapping and therefore extremely difficult to classify.

The other data set consists of 18 000 handwritten digits. The digit images were normalized to 32 × 32 pixels with 256 gray scale values. 38 principal components were obtained using PCA and used for our experiments [14]. This is a relatively large classified database. Comparing results to the MPEG-7 feature databases shows how different algorithms scale to larger problems.

All classification results have been obtained by using majority voting and ten fold cross validation.

4.1 Comparison to Classical Methods

We compared ETree to SOM and k-means using two different data sets. The first one is edge histogram, which was used with the paper defect data set. The second one contained the handwritten digits. We tested both the classification accuracy as well as training time. The amount of clusters in k-means was chosen to be approximately the same as the corresponding ETree had leaf nodes.

In Figure 2(a) we have classification results for both datasets. The classification rates for edge histogram are quite similar for all methods. SOM is clearly the worst while k-means obtains the best result. ETree is very close to the performance of k-means. An interesting result is that k-means adjusted ETree is slightly worse than the regular algorithm. Results with the other MPEG-7 data sets were very similar.

The large digit database shows the benefits of k-means adjustment. Regular ETree algorithm achieves only 82% classification rate, which is slightly worse than SOM's. Using the k-means adjustment raises it noticeably to 87%. Plain k-means obtains the extremely good result of 93% which is an expected result. Since the cluster centers in k-means are not constrained by the search tree, they can move more freely.

This freedom does come with a heavy cost, though, as we can see in Figure 2(b). It lists the running times of ten cross-validated training/query rounds. Both versions of ETree take only two minutes, while k-means take up to a half hour. SOM takes almost five times as long as either ETree algorithm. We used the batch version of SOM that is much faster than the iterative SOM. A thing to note is that k-means and SOM were done in Matlab, whereas ETree is a light weight C++ program. Most CPU time in all algorithms is spent calculating the Euclidean distances between vectors at which Matlab is very good. Thus we can conclude that the bulk of the differences stem from differences in the algorithms.

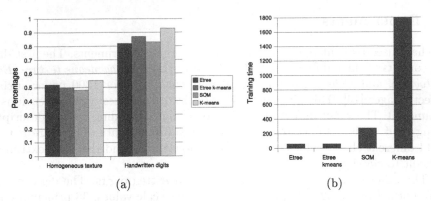

(a)   (b)

Fig. 2. Comparison of ETree to SOM and k-means clustering. (a) The classification percentages and (b) the training times for the handwritten digits.

Another interesting thing to note is that the k-means adjustment to the Evolving Tree adds only a negligible fraction to the total training time, yet it improves classification percentages noticeably. K-means adjustment seems to be useful in large databases, because fine tuning node locations becomes harder and harder as a data set grows.

ETree is clearly the fastest of the algorithms but it still obtains very good classification results. It cannot reach the performance level of k-means, but runs in a fraction of the time. This is a very acceptable trade-off in many cases.

4.2   Comparison to Other Tree-Shaped Neural Systems

All three systems are based on competitive learning and have a tree structure that grows as the training progresses. Algorithmically S-Tree is the most complicated one and ETree is the simplest one. We examined both the classification performance and the training time. All the algorithms were coded in C++, so CPU time comparisons are fair. Because of the similarities we list the times relative to ETree's time.

Figure 3 shows the classification percentages for the different algorithms. The results are consistent with earlier experiments. All three hierarchical systems outperform the regular SOM. ETree and CNeT are very evenly matched, and their performance is very close to k-means (see Figure 2(a)). S-Tree has the worst performance of the hierarchical methods, but the margin is quite small. On the larger digit database, k-means enhanced ETree is clearly the best, CNeT holds second place with S-Tree very close to it.

Figure 4 shows the training times for the different algorithms. The times have been normalized by dividing them with ETree's training time. In all cases S-Tree is the slowest one. CNeT is slower on MPEG-7 data but slightly faster than ETree on the digit database. Overall ETree and CNeT seem to have very similar time complexity.

The Evolving Tree has arguably the best total performance in almost all the tests which makes it the preferred algorithm of the three, especially in these kinds of classification and clustering tasks.

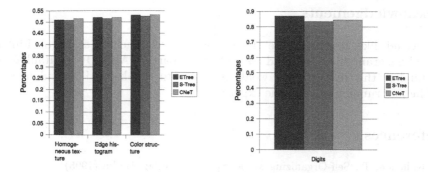

Fig. 3. Classification percentages for ETree, S-Tree, and CNeT using MPEG-7 features and handwritten digits

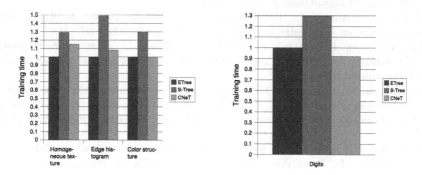

Fig. 4. Normalized training times for ETree, S-Tree, and CNeT using MPEG-7 features and handwritten digits

5   Conclusions

We have examined the Evolving Tree algorithm and compared it against several different algorithms. We have found that ETree's performance is very good against the classical methods, SOM and k-means, but it runs several times faster. ETree performs also favorably against the modern tree-shaped neural systems, S-Tree and CNeT. Its performance is consistently better. It is also roughly as fast, and in many cases faster than them.

These results indicate that ETree could be effectively utilized in large scale problems, that classical methods have not been able to adequately solve. Further research includes evaluating the algorithm with a database of millions or tens of millions of vectors.

We have created a software package that implements the Evolving Tree algorithm. It can be freely downloaded from our web page http://www.cis.hut.fi/research/etree/.

Acknowledgements

We would like to thank the Technology Development Centre of Finland (TEKES's grant 40102/04) and our industrial partner ABB Oy (J. Rauhamaa) for funding this research. We would also like to thank Petri Turkulainen for coding and running some of the experiments.

References

1. Kohonen, T.: Self-Organizing Maps. Springer-Verlag, Berlin (1995)
2. Fritzke, B.: Growing cell structures — a self-organizing network for unsupervised and supervised learning. Neural Networks **7** (1994) 1441–1460
3. Bruske, J., Sommer, G.: Dynamic cell structure learns perfectly topology preserving map. Neural Computation **7** (1997) 845–865
4. Hodge, V.J., Austin, J.: Hierarchical growing cell structures: TreeGCS. IEEE Transactions on Knowledge and Data Engineering **13** (2001) 207–218
5. Luo, F., Khan, L., Bastani, F., Yen-Ling, I., Zhou, J.: A dynamically growing self-organizing tree (DGSOT) for hierarchical clustering gene expression profiles. Bioinformatics (2004)
6. Dittenbach, M., Rauber, A., Merkl, D.: Recent advances with the growing hierarchical self-organizing map. In: Proceedings of the 3rd Workshop on Self-Organizing Maps. Advances in Self-Organizing Maps, Lincoln, England, Springer (2001) 140–145
7. Pakkanen, J., Iivarinen, J.: A novel self-organizing neural network for defect image classification. In: Proceedings of IJCNN 2004, Budapest, Hungary (2004) 2553–2556
8. Pakkanen, J., Iivarinen, J., Oja, E.: The Evolving Tree — a novel self-organizing network for data analysis. Neural Processing Letters **20** (2004) 199–211
9. Bishop, C.: Neural Networks for Pattern Recognition. Oxford University Press (1995)
10. Campos, M., Carpenter, G.: S-TREE: self-organizing trees for data clustering and online vector quantization. Neural Networks **14** (2001) 505–525
11. Behnke, S., Karayiannis, N.: Competitive neural trees for pattern classification. IEEE Transactions on Neural Networks **9** (1998) 1352–1369
12. Manjunath, B.S., Salembier, P., Sikora, T., eds.: Introduction to MPEG-7: Multimedia Content Description Interface. John Wiley & Sons Ltd. (2002)
13. Pakkanen, J., Ilvesmäki, A., Iivarinen, J.: Defect image classification and retrieval with MPEG-7 descriptors. In Bigun, J., Gustavsson, T., eds.: Proceedings of the 13th Scandinavian Conference on Image Analysis. LNCS 2749, Göteborg, Sweden, Springer-Verlag (2003) 349–355
14. Holmström, L., Koistinen, P., Laaksonen, J., Oja, E.: Comparison of neural and statistical classifiers — theory and practice. Research Reports A13, Rolf Nevanlinna Institute, Helsinki (1996)

A Sequence Labeling Method Using Syntactical and Textual Patterns for Record Linkage

Atsuhiro Takasu

National Institute of Informatics,
2-1-2 Hitotsubashi, Chiyoda-ku, Tokyo 101-8430, Japan
takasu@nii.ac.jp

Abstract. Record linkage is an important application area of text pattern analysis. In this paper we propose a new sequence labeling method that can be used to extract entities from a string for record linkage. The proposed method combines a classifier and a Hidden Markov Model (HMM) to utilize both syntactical and textual information from the string. We first describe the model used in the proposed method and then discuss the parameter estimation for this model. The proposed method incorporates a classifier for handling textual information and integrates the classifier with the HMM statistically by estimating the error probability of the classifier. We applied the proposed method to the bibliographic sequence labeling problem, in which bibliographic components are extracted from reference strings. We compared the proposed method with other methods that use textual or syntactical information alone and showed that the proposed method outperforms them.

1 Introduction

Record linkage is an important application area of text pattern analysis. Record linkage refers to the problem of integrating the information in various heterogenous resources focusing on entities. Many kinds of entities are handled in this study, such as vital records (e.g., [10]), and bibliographic records (e.g., [4,15]). Recent advances in scientific literature search systems such as CiteSeer[7] and Google Scholar[13] have attracted many researchers and have led to new record linkage techniques focusing on the bibliographic information as an entity. They include bibliographic matching methods [8,6], and field extraction [14].

The record linkage problem is categorized into four components [2]: (1) duplicate record detection; (2) record integration; (3) reference identification; and (4) co-reference extraction. Duplicate record detection finds records that represent the same entities and merges them into a single record. For bibliographic information, a bibliographic database sometimes contains multiple records representing the same item, and the duplicate record detection must deal with the problem of finding those records and unifying them to make the database clean. The underlying technique of duplicate record detection is record matching. Because a record usually consists of a set of fields, record matching consists of measuring the similarity of values of corresponding fields between the objective

S. Singh et al. (Eds.): ICAPR 2005, LNCS 3686, pp. 199–208, 2005.
© Springer-Verlag Berlin Heidelberg 2005

records and combining the field similarities into a record similarity. For processing efficiency, a blocking technique that roughly separates the database into groups of similar records is important [1].

Record integration is the same problem as duplicate record detection, but it is applied to multiple databases. For bibliographic information, record integration is used to merge multiple bibliographic databases, e.g., the DBLP and Science Citation Index databases. In this problem we confront various kinds of heterogeneity, such as schema mismatches and higher discrepancies in field value representation, so that a more robust matching technique is required.

Reference identification consists of finding the record in a database corresponding to an entity appearing in the text. Finally, for bibliographic information, the problem is to extract entities in articles and find corresponding records in a bibliographic database. Entity extraction and entity matching are the key techniques of this problem.

Co-reference extraction refers to the problem of finding entities in text and merging them into groups, each of which refers to the same entity. For bibliographic information, this problem consists of finding a set of articles that refer to the same article. Though the purpose of co-reference extraction is the same as that of reference identification, it requires the extraction of entities without a well-maintained entity database. Therefore, from the technical point of view, it requires a more powerful entity extraction [18] technique, such as information extraction as is used in the literature of natural language processing.

As described above, the record linkage problem relies on various textual pattern analysis techniques. In this paper we discuss a sequence labeling problem [3] that assigns a label to each token in a sequence and we propose a new labeling method that uses both syntactical and textual features of the token sequence. The sequence labeling problem can be applied to the field extraction step of the reference identification problem. In this paper, we apply the proposed method to bibliographic component extraction from reference strings and show experimental results.

2   Sequence Labeling Based on Score Vectors

Before discussing the proposed method, let us define our notation. A vector is denoted as a bold letter x or by listing its components, such as $(x_1, x_2, \cdots, x_n)$. A sequence is denoted as a bold letter $\mathbf{x}$ or by listing its components as $< x_1, x_2, \cdots, x_l >$. For a sequence $\mathbf{x}$, $|\mathbf{x}|$ denotes the length of the sequence. In sequence labeling, a sequence $< t_1, t_2, \cdots, t_n >$ of tokens is represented with a sequence $< x_1, x_2, \cdots, x_n >$ of feature vectors. For a set Σ of labels, the sequence labeling assigns a label in Σ to each feature vector. To solve this problem, both the syntactical structure of the sequence and the features of tokens represented by a feature vector should be used. Therefore, we must construct a model representing both kinds of information.

A Hidden Markov Model (HMM) [12] is often used for sequence labeling. In this paper, we assume that each state in the HMM corresponds to one label in

Σ, and we denote the corresponding label of a state q as $\psi(q)$. An HMM has an initial probabilities, state transition probabilities and output probabilities as parameters. For a state q, the initial probability, denoted as $\pi(q)$, defines the probability that the transition starts from the state q. For states q_i and q_j, the transition probability, denoted as $\tau(q_i, q_j)$, defines the probability that a transition from state q_i to state q_j occurs. The probability $\tau(q_i, q_j)$ denotes the likelihood a label $\psi(q_i)$ is followed by a label $\psi(q_j)$. For a state q and a feature vector $\boldsymbol{x}$, the output probability, denoted as $P_q(\boldsymbol{x})$, defines the probability that the state emits the feature vector $\boldsymbol{x}$. The probability density function P_q stands for the probability distribution of the label $\psi(q)$ in the feature vector space.

The probability that the HMM M produces a feature vector sequence $\mathbf{x} \equiv\ <\boldsymbol{x}_1, \boldsymbol{x}_2, \cdots, \boldsymbol{x}_n>$ by a state transition $\mathbf{q} \equiv <q_1, q_2, \cdots, q_n>$ is expressed by:

$$P(\mathbf{x}, \mathbf{q} \mid M) = \pi(q_1) \prod_{i=1}^{n} P_{q_i}(\boldsymbol{x}_i) \prod_{i=1}^{n-1} \tau(q_i, q_{i+1}) . \tag{1}$$

In this setting, for a feature vector sequence $\mathbf{x}$, the sequence labeling is solved by finding the following most likely path:

$$\mathbf{q}_{max} \equiv \underset{\mathbf{q}}{\operatorname{argmax}} P(\mathbf{X}, \mathbf{q} \mid M) . \tag{2}$$

Note that each state corresponds to a label and we can obtain a label for a feature vector $\boldsymbol{x}_i$ in $\mathbf{x}$ as $\psi(q_i)$ using the corresponding q_i in $\mathbf{q}_{max}$.

The parameters of the HMM are derived from training data based on the maximum likelihood (ML) estimation, in which only positive examples are used. However, negative examples are also useful for sequence labeling. To incorporate the information from negative examples, we propose a sequence labeling method that combines a classifier and an HMM. For each label c let us assume that there exists a classifier that assigns a score $\phi_c(\boldsymbol{x})$ to a feature vector $\boldsymbol{x}$. Using those classifiers, a feature vector $\boldsymbol{x}$ is converted to a score vector $(\phi_{c_1}(\boldsymbol{x}), \phi_{c_2}(\boldsymbol{x}), \cdots, \phi_{c_{|\Sigma|}}(\boldsymbol{x}))$, where $|\Sigma|$ stands for the number of classes. We call this vector a *score vector* of the feature vector. By using a classifier, we obtain a score vector sequence $<\mathbf{s}_1, \mathbf{s}_2, \cdots, \mathbf{s}_l>$ from a feature vector sequence $<\boldsymbol{x}_1, \boldsymbol{x}_2, \cdots, \boldsymbol{x}_l>$. For a score vector $\boldsymbol{s}$ and a class c, let $\sigma(\boldsymbol{s}, c)$ denote the score for the class c of the score $\boldsymbol{s}$. The score vector represents the likelihood of the feature vector belonging to each class and information about negative examples is incorporated in the learning process of the classifier. Hereafter we consider a score vector sequence instead of a feature vector sequence.

For each state q and a score vector $\boldsymbol{s}$, let us introduce a random variable X_q for the score of the label of q and use the probability $Pr(X_q \leq \sigma(\boldsymbol{s}, q))$ as the output probability of $\boldsymbol{s}$ at state q. This probability has a high value when the score of the classifier is high, so it is suitable for score vector processing. Using this probability, the probability that the HMM M produces a score vector sequence $\mathbf{s} \equiv< \boldsymbol{s}_1, \boldsymbol{s}_2, \cdots, \boldsymbol{s}_n >$ with a state transition $\mathbf{q} \equiv <q_1, q_2, \cdots, q_n >$ is expressed as:

$$Q(\mathbf{s}, \mathbf{q} \mid M) \equiv \pi(q_1) \prod_{i=1}^{n} Pr(X_{q_i} \leq \sigma(\mathbf{s}_i, q_i)) \prod_{i=1}^{n-1} \tau(q_i, q_{i+1}) \ . \tag{3}$$

Then, the sequence labeling for a score vector sequence $\mathbf{s}$ is solved by finding the following state transition of the HMM:

$$\mathbf{q}_{max} \equiv \underset{\mathbf{q}}{\operatorname{argmax}}\, Q(\mathbf{s}, \mathbf{q} \mid M) \ . \tag{4}$$

This optimization problem has the same structure as the most likely path of the HMM, and it can be solved by dynamic programming techniques.

The parameters of the HMM are estimated from training data which are labeled as score vector sequences. For a reference string, let $\mathbf{s}$ be a score vector sequence obtained from the reference string. Parameters of the HMM for the scored sequence can be estimated by the expectation maximization (EM) technique [12] by maximizing the following value

$$\frac{1}{Q(\mathbf{s}|M)} \sum_{\mathbf{q}} Q(\mathbf{s}, \mathbf{q}|M) \log Q(\mathbf{s}, \mathbf{q}|\hat{M}) \tag{5}$$

where $Q(\mathbf{s} \mid M)$ stands for $\sum_{\mathbf{q}} Q(\mathbf{s}, \mathbf{q} \mid M)$ and $\hat{M}$ stands for the model with modified parameters. For reasons of space, we skip the derivation of the algorithm and show only the parameter modification step in each EM algorithm.

Initial probability: For a score vector sequence $\mathbf{s}$ for training, let $I(|\mathbf{s}|, q)$ be the set of paths with length $|\mathbf{s}|$ that start at the state q. Then the modified initial probability of state q_i is given by:

$$\frac{\sum_{\mathbf{q} \in I(|\mathbf{s}|, q_i)} Q(\mathbf{s}, \mathbf{q} \mid M)}{Q(\mathbf{s} \mid M)} \ . \tag{6}$$

Transition Probability: For a path $\mathbf{q}$, let $c_{ij}(\mathbf{q})$ denote the number of transitions from state q_i to q_j in $\mathbf{q}$. Similarly, let c_i denote the number of transitions from state q_i to any state. Then the modified state transition probability from state q_i to q_j is given by:

$$\sum_{\mathbf{q}} Q(\mathbf{s}, \mathbf{q}|M) \frac{c_{ij}(\mathbf{q})}{c_i(\mathbf{q})} \ . \tag{7}$$

Score Distribution: We discuss the estimation of score distribution in the next section.

3   Bibliographic Reference Labeling

3.1   The Bibliographic Reference Labeling Problem

We have been developing a bibliographic record linkage system where the references of several kinds of bibliographic databases and article archives are integrated. This section discusses the application of the proposed method to a bibliographic reference labeling problem.

This problem involves decomposing a reference string into substrings and assigning a label to each substring. For instance, when given the following reference string:

T. Okada, A. Takasu and J. Adachi: "Bibliographic Component Extraction", LNCS 3232, pp. 501-512, 2004.

then the problem is to extract the following tagged strings:

<author>T. Okada</author>,
<author>A. Takasu</author> and
<author>J. Adachi</author>:
"<title>Bibliographic Component Extraction</title>",
<conf>LNCS 3232</conf>,
pp. <page>501-512</page>,
<year>2004</year>.

3.2   Bibliographic Reference Labeling Procedure

This problem is the first step in bibliographic reference identification and co-reference extraction. Extracted bibliographic components are used to find the corresponding records in a bibliographic database or other reference strings in articles. We apply (1) segmentation, (2) feature vector construction, (3) conversion of feature vector to score vector, and (4) labeling, to extract bibliographic components.

Segmentation: A reference string is segmented into subfields using delimiters [11]. As delimiters, we use punctuation such as commas, quotation marks, periods, and strings specific to reference strings such as 'vol.', 'no.', 'pp.' and 'ed.' This method sometimes causes over-segmentation; however, over-segmented substrings are merged in the labeling phase. In this phase, we obtain the following segmented reference.

T. Okada /,/ A. Takasu / and / J. Adachi /: " / Bibliographic Component Extraction / ", / LNCS 3232/ , pp. / 501-512 / , / 2004 /.

Token Feature Construction: To apply classification, subfields must be represented with feature vectors. We use the vector space model for information retrieval [5], where words appearing in subfields are used as features and their frequency is used as the feature value. To handle delimiters and numerical words such as year published, we use the heuristics described in [11]. As a result, each substring in a reference is converted to a feature vector.

Score Vector Conversion: Any classifier that gives a score can be used in this process. We use Support Vector Machines (SVM). For each label in Σ, an SVM classifier is constructed from training data. In the SVM, a feature vector x is projected into the weight vector w in the feature space, and the score is the distance from the decision boundary given by $K(x, w) - b$, where K stands for the kernel used for the SVM and b stands for the decision boundary.

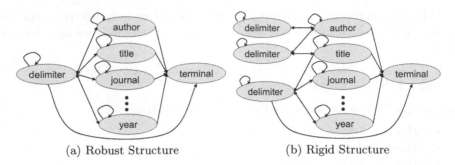

(a) Robust Structure   (b) Rigid Structure

Fig. 1. Graphical Structures of HMMs

Bibliographic Sequence Labeling: When constructing the HMM, we assign one and only one class to each state, as depicted in Figure 1. As shown in the figure, each state has a transition to itself. This transition is used to handle over-segmentation in the segmentation step. When given a score vector sequence **s**, we obtain the most likely state transition by solving the optimization problem (4). The most likely state transition assigns a class (i.e., label) to each score vector in the sequence **s**.

3.3   HMM Construction

Having described the parameter estimation of the HMM in the previous section, in this section we focus on the HMM's graphical structure and the probability estimation of the score vector, which is specific to bibliographic sequence labeling.

The graphical structure of the HMM defines the syntactical structure of the reference strings. Currently we use two kinds of structure. The first is a robust HMM that assumes that the delimiters and bibliographic components appear alternately. The robust HMM accepts reference strings in which bibliographic components appear in any order as long as they are separated by delimiters. The other HMM is a rigid one [14] in which the order of the bibliographic components is defined by the HMM structure. For example, a rule such as "title must be located after authors" is represented by the graphical structure of the HMM. The rigid HMM is more powerful in its ability to modify the classification error in score vector construction; however, it fails to analyze reference strings in which the order of bibliographic components is different from the expected one.

For the probability $P(X_q \leq \phi_q(\boldsymbol{x}))$ in (3), we use the following exponential distribution:

$$\begin{array}{ll} \beta e^{-\lambda_{q,1}(\alpha-x)} & x \leq \alpha \\ 1 - (1-\beta)e^{-\lambda_{q,2}(x-\alpha)} & x > \alpha \end{array} \tag{8}$$

where $\lambda_{q,1}$, $\lambda_{q,2}$, α and β are parameters for the distributions. All classes have score distributions of the form (8), but the values of the parameters are different.

The parameter α stands for the margin boundary for the positive example in SVM. That is, the score of the positive support vectors falls into α. β is a value in the range of $(0,1)$ and stands for the ratio of tokens with scores less than the positive margin boundary. These two parameters are determined by the SVM used for the score calculation. Equation (8) is a heuristic function and we have not validated it analytically; however, the expression (8) is easy to handle mathematically and it fits well with the score function in our preliminary experiment.

In contrast, the parameters $\lambda_{q,1}$ and $\lambda_{q,2}$ are estimated by maximum likelihood estimation. First, the following density function is derived from (8):

$$
\begin{array}{ll}
\lambda_{q,1}\beta e^{-\lambda_{q,1}(\alpha-x)} & x \leq \alpha \\
\lambda_{q,2}(1-\beta)e^{-\lambda_{q,2}(x-\alpha)} & x > \alpha
\end{array} \quad . \tag{9}
$$

For each label c, let us consider the set $\mathbf{P}$ of score vectors that belong to the label c, that is, the set of positive example in the training data. Suppose S is the set of c-components of the score vector in $\mathbf{P}$, where c-component stands for the component corresponding to the label c. Let S_p be $\{v|v \in S, v > \alpha\}$ and S_n be $\{v|v \in S, v \leq \alpha\}$. Then, the parameters $\lambda_{q,1}$ and $\lambda_{q,2}$ are derived by the following formula:

$$
\operatorname*{argmax}_{\lambda_{q,1}} \sum_{s \in S_n} [\log \lambda_{q,1} + \log \beta - \lambda_{q,1}(\alpha - s)]
$$

$$
\operatorname*{argmax}_{\lambda_{q,2}} \sum_{s \in S_p} [\log \lambda_{q,2} + \log (1 - \beta) - \lambda_{q,2}(s - \alpha) \ .]
$$

By solving these optimization problems, parameters for the exponential distribution are obtained as:

$$
\lambda_{q,1} = \frac{1}{\alpha - avg(S_n)} \qquad \lambda_{q,2} = \frac{1}{avg(S_p) - \alpha} \tag{10}
$$

where $avg(S)$ stands for the average of the set S of values. As shown in (10), parameters of the exponential distributions are easily estimated by calculating the average of positive score vectors.

By (10) and the output of SVM training, all parameters of the score distribution for all states are obtained, and consequently, the HMM is obtained. By finding the most likely path of the HMM, we can obtain the bibliographic label for each substring in the reference string.

4   Experimental Results

We applied the proposed bibliographic sequence labeling method to reference strings in academic articles. In this experiment, we used 312 articles in the Transactions of the Institute of Electronics, Information and Communication Engineers, published in 2000. This data set contains both Japanese and English

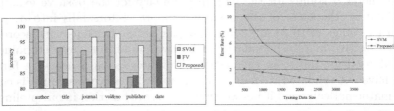

(a) Performance of Three Methods   (b) Error w.r.t. Training Data Size

Fig. 2. Performance Comparison

articles. We extracted 4814 reference strings from the reference sections of these articles. Then, we decomposed each reference string into substrings and assigned a label to each substring manually. In the feature vector construction, we applied a stemmer for English references, and we applied the Mecab morphological analyzer [9] to extract words from Japanese references. For the SVM we used TinySVM [17]. In the preliminary experiment, we did not observe any differences in accuracy among the kernel functions, and adopted the linear kernel because it requires fewer parameters.

In this experiment our goal was to assign one of the following six labels to each subfield: 'author', 'title', 'journal', 'vol&no', 'publisher' and 'date'. In this experiment, we first compared the proposed method with two other methods. One is SVMs without the HMM, where only textual information is used. We applied SVMs to the feature vector of each substring and assigned the label for which the SVM had the highest score, i.e., a one-versus-rest SVM classifier [16]. These results have the label 'SVM' in Figure 2. The other model is the combination of the HMM and a feature vector sequence. In this method we omitted the score vector conversion in the proposed method and used feature vectors as they are. These results have the label 'FV' in Figure 2. Figure 2 (a) shows the performance of the three methods when applying 5-fold validation, using the manually prepared 4814 reference strings. We can derive two implications from this graph. First, as shown in the figure, 'FV' has the worst performance of the three methods. This means that textual information can be used by SVMs more effectively than syntactical information can be used by HMMs. Second, the performance can be improved by using both textual and syntactical information and the proposed method is useful in utilizing this information.

Although the proposed method improves the accuracy, the improvement against SVM is moderate. Figure 2 (b) compares the error rate of the proposed method with SVM with respect to the training data size. The X axis of the figure stands for the number of references used for learning SVM and parameter estimation in the proposed method. As shown in this figure, performance of the SVM is deteriorated for smaller training data. Let us consider, for example, author name in the bibliographic entity extraction problem. It will be difficult for SVM to know that "Takasu" is author's name if "Takasu" is not included in training data. Therefore, SVM requires large amount of training data for handling this problem. On the other hand, from the syntactical information, we can guess it is author's name because it is located before article title even if we don't know the person. This kind of information can be obtained from small amount of training data. Therefore, the proposed method has advantage especially for smaller training data. The cost of constructing training data is labor intensive work, and this advantage of the proposed method is effective in the bibliographic entity extraction problem.

5   Conclusions

In this paper we proposed a new sequence labeling method. The proposed method combines SVMs and HMMs to utilize both syntactical and textual information. The main idea is to convert a feature vector sequence into a score vector sequence using an SVM. We focused on the parameter estimation of the HMM for handling a score vector. We applied the proposed method to the bibliographic sequence labeling problem and showed that the proposed method improves the labeling accuracy.

We used a heuristic probability function for score vector distribution. One future task is to analyze the score vector distribution from both experimental and theoretical points of view, and to derive a suitable function. The graphical structure of the HMM will affect the performance. We plan to study this effect and develop an induction method for the graphical structure.

References

1. A. Aizawa and K. Oyama. "A First Linkage Detection Scheme for Multi-Source Information Integration". *Proc. of Intl. Workshop. on Challenges in Web Information Retrieval and Integration* (WIRI 2005), pp. 31–40, 2005.
2. A. Aizawa, K. Oyama, A. Takasu, and J. Adachi. Techniques and research trends in record linkage studies. *IEICE Tran. on Inforamtion and Systems*, Vol. J88-D-I, No. 3, pp. 576–589, 2005.
3. Y. Altun and T. Hofmann. "Large Margin Methods for Label Sequence Learning". *Proc. of 8th European Conf. on Speeach Communication and Technology*, 2003.
4. F. H. Ayres, J. A. W. Huggill, and E. J. Yannakoudakis. The universal standard bibligraphic code (USBC): its use for clearing, merging and controlling large databases. *Program - Automated Library and Information Systems*, Vol. 22, No. 2, pp. 117–132, 1988.
5. R. Baeza-Yates and B. Ribeiro-Neto. *Modern Infromation Retrieval*. Addison Wesley, 1999.
6. M. Bilenko and R. J. Mooney. "Adaptive Duplicate Detection Using Learnable String Similarity Measures". *Proc. of 9th ACM International Conference on Knowledge Discovery and Data Mining* (SIGKDD2003), pp. 39–48, 2003.
7. CiteSeer.IST: Scientific Leterature Digital Library: http://citeseer.ist.psu.edu/cs
8. Andrew K. McCallum, Kamal Nigam, Jason Rennie, and Kristie S eymore. Automating the construction of internet portals with machine learning. *Information Retrieval*, Vol. 3, No. 2, pp.127–163, 2000.

9. Mecab: http://chasen.org/ taku/software/mecab/
10. H. B. Newcombe, J. M. Kennedy, S. J. Axford, and A. P. James. Automatic linkage of vital records. *Science*, Vol. 130, No. 3381, pp. 954–959, 1959.
11. T. Okada, A. Takasu, and J. Adachi. "Bibliographic Component Extraction Using Support Vector Machines and Hidden Markov Models". *Proc. European Conf. on Research and Advanced Technology for Digital Libraries* (ECDL2004), LNCS 3232, pp. 501–512, 2004.
12. Lawrence. R. Rabiner. "A Tutorial on Hidden Markov Models and Selected Applications in Speech Recognition". *Proceedings of the IEEE*, Vol. 77, No. 2, pp. 257–286, 1989.
13. Google Scholar: http://scholar.google.com
14. A. Takasu. "Bibliographic Attribute Extraction from Erroneous References Based on a Statistical Model". *Proc. of ACM-IEEE Joint Conference on Digital Libraries* (JCDL2003), pp. 49–60, 2003.
15. A. Takasu, N. Katayama, and et. al. "Approximate Matching for OCR-Processed Bibliographic Data". *Proc. of 13th Internationa Conference on Pattern Recognition*, pp. 175–179, 1996.
16. D. M. J. Tax and R. P. W. Duin. "Using Two-Class Classifiers for Multiclass Classification". *Proc. of Infl. Conf. on Pattern Recognition*, 2002.
17. TinySVM: http://chasen.org/ taku/software/TinySVM/
18. Dan Roth Xin Li, Paul Morie. Semantic integration over text: From ambiguous names to identifiable entities. *AI Magazine*, Vol. 26, No. 1, pp. 45–58, 2005.

Recognition Tasks Are Imitation Games

Richard Zanibbi[1], Dorothea Blostein[2], and James R. Cordy[2]

[1] Centre for Pattern Recognition and Machine Intelligence,
1455 de Maisonneuve Blvd. West, Suite GM 606, Montreal, Canada, H3G 1M8
`zanibbi@cenparmi.concordia.ca`
[2] School of Computing, Queen's University, Kingston, ON, Canada, K7L 3N6
{`blostein, cordy`}`@cs.queensu.ca`

Abstract. There is need for more formal specification of recognition tasks. Currently, it is common to use labeled training samples to illustrate the task to be performed. The mathematical theory of games may provide more formal and complete definitions for recognition tasks. We present an imitation game that describes a wide variety of recognition tasks, including the classification of isolated patterns and structural analysis. In each round of the game, a set of 'players' try to match the interpretation of an input produced by a set of 'experts.' The 'playing field' on which experts and players operate is a set of interpretations generated from legal sequences of 'moves' for a round. The expert and player moves transform interpretations, and select interpretations for output. The distance between interpretations in the playing field is defined by a distance metric for interpretations, and the game outcome by a ranking function on distance values observed for players' interpretations. We demonstrate how this imitation game may be used to define and compare recognition tasks, and clarify the evaluation of proposed solutions.

1 Introduction

Many recognition tasks have strong similarities to the children's game 'pin the tail on the donkey.' In that game, players are shown a picture of a donkey without a tail. Players take turns being blindfolded, turned around several times, and then trying to place a pin with an attached tail at the proper location on the donkey. At the end of the game, the adult running the game awards prizes to the children based on the closeness of their pins to the 'proper' location. At first, 'pin the tail on the donkey' might seem like a strange analogy for recognition tasks, but consider the following similarities.

1. **The basic task is to choose points in space.** In the game, a pin is used to pick a physical location within a room. In recognition, interpretations of an input or object are selected from a space of possible interpretations. The relative distances ('closeness') of interpretations within the space are determined by a distance metric (e.g. classification risk or edit distance).
2. **Goal points are determined by an *expert opinion*.** In the game, the adult chooses the optimal tail location(s) using his or her understanding

S. Singh et al. (Eds.): ICAPR 2005, LNCS 3686, pp. 209–218, 2005.

of donkey anatomy. For recognition tasks, one or more experts use their understanding of the problem domain to select goal interpretations.

3. **From an initial point in space, the goal point(s) must be *guessed*.**[1]
In each turn of the game, a blindfolded and disoriented player must make a sequence of guesses as to how to move and eventually place their tail from their starting position. Similarly, despite ambiguities in an input's content or introduced by noise, a recognition protocol must make a sequence of guesses about which interpretations should be considered and/or selected as goal interpretations, starting from some initial interpretation (e.g. 'reject').

4. **Guesses are ranked using their *distance* from the expert opinion.**
In the game, tail locations are ranked using the distance from pins to the location(s) chosen by the adult. For recognition tasks, recognition protocols are ranked using a function of the distances from guessed to goal interpretations within the interpretation space (e.g. the minimum, mean, or median of interpretation space distances observed for a test sample).

We propose that like the players in 'pin the tail on the donkey,' recognition systems evaluated against expert opinions are engaged in an *imitation game* where players producing responses that are *closest* to that of an expert opinion are deemed most successful. The outcome of such a game depends directly upon expert opinion(s), how 'closest' is defined, and the interpretation spaces used by experts and players, which may not coincide. As an extreme example, if the donkey picture in 'pin the tail' is placed too high for a child to reach the goal tail location(s), the child might not consider these locations (i.e. they do not exist in the child's interpretation space). Similarly, the use of different domain models by experts and recognition algorithms may produce 'holes' in the algorithms' interpretation spaces, which may prevent goal interpretations from being considered.

Using an imitation game to define recognition tasks places evaluation within problem definitions, as opposed to treating evaluation as a validation of proposed solutions for more abstract problems. This is similar to how Turing avoided directly considering whether machines "think" by instead considering outcomes of his own famous imitation game [3]. As an example, consider classifying images of handwritten digits (0..9) when the cost (risk) of classification error is fixed, as opposed to when it is not (i.e. solutions try to 'recognize digits,' in the absence of an explicit evaluation scheme). In the second scenario, an evaluation mechanism may be chosen *after* solutions for recognizing digits have been defined. This compares solutions unfairly, particularly if they are designed assuming different evaluation schemes. Evaluation in the context of an imitation game is more meaningful, because the assumptions ('rules') under which solutions ('players') operate are explicit and uniform. Imagine telling children after placing two tails each in 'pin the tail on the donkey' that they will be ranked by *mean* rather than *minimum* pin distance.

---

[1] Interesting discussions pertaining to guessing in pattern recognition have recently been provided by Kanatani [1] and Oommen and Rueda [2].

Games and game theory [4,5] have of course been used previously in the pattern recognition literature. For example, game-theoretic models have been used for combining modules in vision systems [6,7] and for modeling sequential prediction problems [8]. Based on earlier work for monitoring aircraft engines [9], Pau has modeled Bayesian classification using two-player games, with a classifier deciding the *a posteriori* probabilities of input patterns, and a teacher providing the *a priori* class probabilities to the classifier [10]. A zero-sum game between the teacher and classifier is used for the worst-case, when the teacher tries to maximize the number of errors made by the classifier. A bi-matrix game in which teacher and classifier may cooperate is also examined. Optimal strategies for player and teacher in each game (equilibrium solutions) are presented.

We are taking a different tack here, as we present a class of games for defining and comparing various recognition problems, including the classification of isolated patterns and structural analysis. The purpose of our imitation game is to compare recognition strategies rather than optimize a single one, as in Pau's game. For this discussion we assume that all our 'players' produce a sequence of decisions, leaving issues pertinent to parallelism for the future. We use examples from 'pin the tail on the donkey,' digit recognition, and table cell detection for illustration.

2   Rules of the Game: Interpretation Models

Let's consider a simple mathematical model for the rules of 'pin the tail on the donkey,' ignoring the vertical position of pins. We will model the position of a child and their tail-pin as a line segment in R^2, with the child at one point, and the pin at the other. The distances between pins will be defined by their Euclidean distance in R^2. During a turn, a child may do the following actions: move forward, turn varying amounts clockwise or counter-clockwise, and push their tail-pin forward to fix it in a wall. We model these actions as transforming the location of the line segment; walking forward translates the segment, turning rotates the 'pin' point around the 'child' point, and a successful push of the pin into the wall fixes the pin location, and ends the turn. If a child pushes their pin into empty space, the line segment remains unchanged, and the turn continues.

We now have rules for the game, defining the legal space of guesses (*interpretation space*) from the sequences of moves that a child may perform during their turn(s). If we include failed pin pushes (pushes into air), the locations a child may attempt to place the pin includes all the space in R^2 on and between the walls of the room. This provides us with a *generative* model of the interpretation space, in which a player's turn is described by a sequence of model operations (an *operation sequence*), and the interpretation(s) selected as a result.

In our imitation game for recognition tasks, we will define an *interpretation model* as a 7-tuple m:

$$m = (D^m, i_0^m, T^m, \Gamma^m, apply^m, \delta^m, time_{max}^m). \tag{1}$$

D^m is a set of problem domain inputs, the set of elements for which interpretations are constructed by the model. For a model m, the interpretation for all inputs $d \notin D^m$ is defined as the initial interpretation, i_0^m. For a classification model, i_0^m would be 'reject', and for a structural recognition model, i_0^m might be the empty set.

T^m is a set of *model operations* that transform interpretations, the set of possible moves for the game. In 'pin the tail on the donkey,' these were the actions to move and push the tail-pin. Generally speaking, operations in T^m create, delete, classify, segment, and relate entities in interpretations [11]. T^m must include an *accept* operation which marks interpretations for output (as a final 'guess' in the game), and may also include a *reject* transform for reversing an *accept* operation. Model operations in T^m may alter *sets* of interpretations, such as for representing the combination of interpretations by a classifier ensemble, or for generating alternatives. As a simple example, T^m for an interpretation space of digit strings might contain *accept*, *reject*, and a set of string edit operations (e.g. replace a digit, transpose a digit, insert a digit, delete a digit).

Γ^m defines legal sequences of moves (the model operations, T^m) similar to a string grammar with start symbol i_0^m and terminals T^m. Restrictions on the maximum number of moves in a turn may be enforced by defining Γ^m such that legal transformation sequences have a finite maximum length k. All turns begin with the initial interpretation (i_0^m) as the default interpretation. The legal interpretation sequences are denoted L^m (the *language of model m*), and the interpretation space I^m is obtained using the function $apply^m$.

$$L^m = \Gamma^m(i_0^m, T^m) \tag{2}$$

$$I^m = i_0^m \cup \bigcup_{s \in L^m} apply^m(s, i_0^m) \tag{3}$$

$$apply^m(s \notin L^m, i_0^m) = i_0^m \tag{4}$$

The function $apply^m$ constructs interpretations by applying an operation sequence to the initial interpretation (i_0^m); illegal sequences are mapped to i_0^m.

We define one property of operation sequences, the *interpretation history* (h^m). The interpretation history of an operation sequence $s \in L^m$ is the set of unique interpretations generated over the course of applying the sequence.

$$h^m(s = \{t_1 \in T^m, .., t_{|s|} \in T^m\}) = \bigcup_{i=1}^{|s|} apply^m(t_1..t_i) \tag{5}$$

An interpretation history contains all unique interpretations constructed during recognition. This is the set of pin locations for a turn in 'pin the tail on the donkey.'

The distance function δ^m defines a distance between interpretations (this was Euclidean distance in R^2 for our simple 'pin the tail' model). Consider our example of the digit string interpretation space employing string edit operations in T^m. δ^m could be defined as a 'string edit distance,' the minimum number of operations needed to transform one digit string to another. Alternatively, if we use the numerical difference of the numbers represented by the strings for δ^m, the operation sequences and distances are less directly related.

The final component of an interpretation model m is $time_{max}^m$, the maximum time in which an operation sequence may be generated. Unbounded recognition time may be represented using ∞. For both 'pin the tail on the donkey' and real-world recognition problems, $time_{max}^m$ is finite, and relatively small.

3   Playing Recognition Games

We now define an imitation game for recognition tasks, in which a sample of a problem domain is taken, experts define goal interpretations for the sample elements, and players try to guess the expert interpretations. Experts and players use two models, differing only in the time for choosing interpretations. This allows for differences between human 'experts' (e.g. using a GUI to create interpretations) and algorithms under real-time constraints.

For a recognition game, the *recognition game parameters* (g) are an 8-tuple:

$$g = (m^g, time_e^g, time_p^g, \mu^g, n, \phi^g, E^g, P^g). \tag{6}$$

m^g is an interpretation model as described in the previous section. $time_e^g$ is the maximum time allowed for creating expert interpretation(s), and $time_p^g$ is the maximum duration of a player's turn. μ^g is a sampling function returning a list of q elements from a set (e.g. $\mu^g(D^{m^g}, q) = (d_1 \in D^{m^g}, ..., d_q \in D^{m^g})$); n is the sample size used in the game. ϕ^g is a ranking function, defining an ordering for sets of values (e.g. ranking by minimum distance, as in 'pin the tail on the donkey').

Players are represented as functions called *recognition strategies* $(P^g = \{\alpha_1, .., \alpha_p\})$ returning legal operation sequences from the language of m^g (for $i \in \{1..p\}$, $\alpha_i(d \in D^{m^g}, m^g, time_p^g) = s \in L^{m^g}$). The *expert protocol* defining goal interpretations is defined as a pair $E^g = (A^e, \beta)$, where $A^e = \{\alpha_1^e, .., \alpha_j^e\}$ is another set of recognition strategies using time restriction $time_e^g$, and β is a function combining the expert opinions (operation sequences) to produce goal interpretation(s) for an input $d \in D^{m^g}$ $(\beta(\alpha_1^e(d), ..., \alpha_j^e(d)) = \{i_1, ..., i_m\})$.

Given recognition game parameters g, a *recognition game* is an imitation game that proceeds as follows:

1. The game input set D^g is defined by

$$D^g = \mu(D^{m^g}, n) = \{d_1..d_n\} \subseteq D^{m^g}. \tag{7}$$

2. The goal interpretation set $(I_{d_k}^e)$ for each input $d_k \in D^g, k = 1..n$ is defined using the expert protocol E^g

3. A series of 'rounds' is played, one for each $d_k, k = 1..n$. In each round:

   (a) Each player in P^g is given d_k and produces an operation sequence $(\{s_{d_k}^{\alpha_1} .. s_{d_k}^{\alpha_p}\})$, from which the guessed interpretation sets $(\{I_{d_k}^{\alpha_1} .. I_{d_k}^{\alpha_p}\})$ are obtained using $apply^{m^g}$ (e.g. $I_{d_k}^{\alpha_1} = apply^{m^g}(s_{d_k}^{\alpha_1})$)

(b) Each player is scored using the distance function of the game interpretation model (δ^{m^g}), to produce a set of distances $\Delta_{d_k}^{\alpha_x}$ between each guessed interpretation in $I_{d_k}^{\alpha_x}$ and each goal interpretation in $I_{d_k}^e$:

$$\Delta_{d_k}^{\alpha_x} = \bigcup_{i^\alpha \in I_{d_k}^{\alpha_x},\ i^e \in I_{d_k}^e} \delta^{m^g}(i^\alpha, i^e), \forall x = 1..p, \forall k = 1..n \qquad (8)$$

4. The player ranking is determined by applying the ranking function ϕ^g to the scores from each round (e.g. $\phi^g(\{\Delta_{d_1}^{\alpha_1}..\Delta_{d_n}^{\alpha_p}\}) = (\alpha_3, \{\alpha_p, \alpha_1\}, ...)$, where α_3 wins, and α_p and α_1 are tied for second place).

4   Decision Making in Recognition Strategies

In our game, the experts and players are modeled by functions called *recognition strategies*, which return operation sequences for a problem domain input. Operation sequences represent decisions made within a series of *decision spaces* containing the alternatives for each decision; applying a model operation implies that some inference has been made regarding the appropriate model instance(s) for an input [11]. Sequential decision making can be represented by a decision tree, flow chart, or similar representation [4,5]. We will consider properties of a decision tree representation for decision making in our recognition games.

Consider Figure 1, which presents a single turn for 'pin the tail on the donkey.' Shown are the sequence of decisions made ($s = (forward, turn\ 5\ deg.\ CC, push)$), and the set of alternatives considered. At each point in the tree, some alternatives will produce identical pin locations. For example, deciding to turn $30°$ clockwise produces the same pin location as turning $330°$ counter-clockwise. Though not shown in Figure 1, at some points certain moves may not be possible (e.g. moving forward if a chair blocks the child) or even considered. If a child decides after turning that they are at the right location, they may only consider pushing the pin (i.e. the decision space has only one element, 'push pin').

Let us now define the decision spaces, alternative decision sequences, and interpretations considered by a recognition strategy α for an input $d_k \in D^g$. The series of decision spaces encountered by a recognition strategy α may be represented as a list of subsets of the game moves, T^{m^g}:

$$\alpha_{spaces}(\alpha, m^g, d_k \in D^g) = \{\{T_1^\alpha\} \subseteq T^{m^g}, ..., \{T_{|s|}^\alpha\} \subseteq T^{m^g}\} \qquad (9)$$

where $|s|$ is the length of the operation sequence produced by α. The complete set of alternative operation sequences considered for a given series of decision spaces $A = \alpha_{spaces}(\alpha, m^g, d_k \in D^g)$ and the related operation sequence s is then given by:

$$S(A, s) = \{\emptyset\} \cup \bigcup_{i=1}^{|s|} \bigcup_{a \in A_i} cat(s_{[1,i-1]}, a) \subseteq L^{m^g} \qquad (10)$$

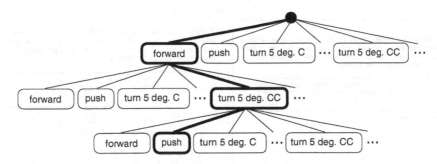

Fig. 1. Decision tree for one child's turn in 'pin the tail on the donkey.' The sequence of moves (*operation sequence*) shown is $s = (forward, turn\ 5\ deg.\ CC, push)$, representing the child stepping forward, turning five degrees counter-clockwise, and then pushing their tail-pin into a wall. '...' is used to represent turning clockwise and counter-clockwise in increments of five degrees (from $10°$ to $355°$).

where *cat* appends elements of a decision space ($a \subset \Lambda_i$) to subsequences of s from lengths 0 ($s_{[1,0]}$) to $|s| - 1$, and $\{\emptyset\}$ is the empty sequence. S also describes the exhaustive set of paths from the root of the corresponding decision tree.

The complete set of points in the game's interpretation space (I^{m^g}) considered by recognition strategy α for input d_k is then given by:

$$C_{d_k}^{\alpha}(A, s) = \bigcup_{y \in S(A,s)} apply^{m^g}(y) \subseteq I^{m^g} \tag{11}$$

Equations (9), (10), and (11) allow us to analyze and compare recognition strategies quantitatively using decision spaces, operation sequences, and interpretations considered. For example, within a recognition game we can determine if a goal interpretation was considered by a strategy, and if so, in which decision space(s) (equivalently, at which nodes in the decision tree).

5   Example: A Table Cell Detection Game

We have previously carried out an informal recognition game for table cell detection [12]. We will now formalize the game, using our imitation game. In the game we compared the detection of table cells within lists of words and lines in segmented tables. We implemented two table recognition algorithms from the literature [13,14] using the Recognition Strategy Language (RSL [12]). RSL formalizes decision making and captures operation sequences that transform interpretations represented as attributed graphs. One of the authors selected five challenging tables from the UW-I technical document corpus [15], and then produced a single set of table cells for each table. Both player algorithms return single interpretations. The distance from the algorithms' cell sets to the author's cell sets were measured using the harmonic mean of cell recall and precision (with a higher mean representing a smaller distance in the interpretation space).

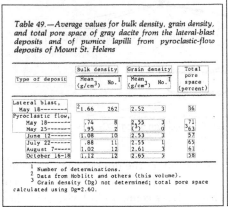

Player 1 (Handley algorithm [13])   Player 2 (Hu et al. algorithm [14])

Expert (author)

Fig. 2. One round of a table cell detection game

The game had a 'best three out of five' outcome, where the algorithm with the highest harmonic mean would 'win' for each round (table). One round from the game is shown in Figure 2, for a table taken from page a038 of the UW-I corpus. For the round shown, Player 2 wins because both recall and precision are higher than for Player 1.

Let us now define the interpretation model m^c for our cell detection game. The problem domain D^{m^c} included the marked tables in the UW-I technical document corpus. The language of m^c was defined with $i_0^{m^c}$ equal to the empty set of cells, T^{m^c} containing basic operations of RSL that modify cell hypotheses (e.g. **accept**, **classify**, **relate**), and Γ^{m^c} defined by legal RSL operation sequences. The function $apply^{m^c}$ was defined by the interpretive layer of the RSL core that generates interpretations from operation sequences. The distance metric δ^{m^c} was 1 - the harmonic mean of cell recall and precision, and $time_{max}^{m^c}$ was roughly 30 minutes, for both the author 'expert' and the algorithms.

The cell detection game parameters g included $m^g = m^c$, with $time_e^g = time_p^g = time_{max}^{m^c}$. Sampling was defined by μ^g (select 'challenging' tables), with sample size $n = 5$. Players were ranked using a 'best three out of five' protocol (ϕ^g). For the expert protocol $E^g = (A^e, \beta)$ the author acted as the sole 'expert', providing 1 interpretation per table ($A^e = \{author\}, \beta =$'choose single best interpretation'). The choice of β was significant, because often the author considered multiple interpretations. Finally, the players P^g were the two table recognition algorithms. One of the algorithms won; complete details may be found elsewhere [12].

The two player algorithms used different models of table structure. To address this, we isolated operations where cell hypotheses may have been affected. In terms of our imitation game, this might have been achieved using a more restrictive model language $\Gamma_r^{m^c} \subset \Gamma^{m^c}$ that included only sequences of RSL operations affecting cell hypotheses. The generated operation sequences could then have been filtered to include only these operations (a well-formed version of such an approach requires further investigation). For both algorithms, we were able to plot cell recall and precision against operation sequence positions affecting cell hypotheses, and observe new metrics describing the recall and precision of the *entire* set of cell hypotheses. The new metrics rely on the interpretation history (see equation (5)), and are called *historical recall* and *historical precision*. Using interpretation histories also simplified error analysis, particularly for locating which operations introduce errors [12].

In the future, we wish to modify RSL to automatically collect the decision space associated with each operation, given an input. We could then use the properties defined in the previous section to further compare recognition strategies. The game model might also provide the basis for a formal semantics of the RSL language.

6   Conclusion

The proposed imitation game defines the variables for a recognition task and the evaluation of proposed solutions explicitly, using an interpretation model and game parameters. This allows recognition tasks to be compared quantitatively in terms of the game variables (e.g. the relative sizes of interpretation spaces), and stipulates the terms of evaluation within the problem definition. Additionally, within the game expert and player moves are transparent, and may be compared using the decision trees they produce for inputs. In the future we are interested in defining general classes of strategies for specific games, as for example Cesa-Bianchi and Lugosi [8] have done, and begin considering *optimal* strategies for recognition games (similar to what Pau has done for Bayesian classification [10]).

For recognition tasks, we commonly talk about *ground-truth* as the set of correct interpretations for a problem domain. However, in practice ground-truth is comprised of interpretations for a sample of a problem domain, produced by experts whose opinions may vary and even conflict [16]. We propose that this type of ground-truth is more accurately understood as expert opinion, as suggested in this paper.

Acknowledgments

We wish to thank Dr. Ching Y. Suen and CENPARMI for providing the resources to write this paper. We also wish to thank Burton Ma, Jeremy Bradbury, Chris McAloney, Dan Ghica, and the anonymous reviewers for their helpful comments. This research was funded by the Natural Sciences and Engineering Research Council of Canada.

References

1. Kanatani, K.: Uncertainty modeling and model selection for geometric inference. IEEE Trans. Pattern Analysis and Machine Intelligence **26** (2004) 1307–1319
2. Oommen, B., Rueda, L.: A formal analysis of why heuristic functions work. Artifical Intelligence **164** (2005) 1–22
3. Turing, A.: Computing machinery and intelligence. Mind (**59**) 433–460
4. Colman, A.: Game theory and experimental games: The study of strategic interaction. Pergamon Press, Oxford, England (1982)
5. Morris, P.: Introduction to Game Theory. Springer-Verlag, New York (1994)
6. Bozma, H., Duncan, J.: A game theoretic approach to integration of modules. IEEE Trans. Pattern Analysis and Machine Intelligence **16** (1994) 1074–1086
7. Chakraborty, A., Duncan, J.: Game-theoretic integration for image segmentation. IEEE Trans. Pattern Analysis and Machine Intelligence **21** (1999) 12–30
8. Cesa-Bianchi, N., Lugosi, G.: Potential-based algorithms in on-line prediction and game theory. Machine Learning **51** (2003) 239–261
9. Pau, L.: An adaptive signal classification procedure. Application to aircraft condition monitoring. Pattern Recognition (**9**) 121–130
10. Pau, L.: Game theoretical signal classification: Application to imperfect or noncooperative learning. Geoexploration **23** (1984) 161–170
11. Zanibbi, R., Blostein, D., Cordy, J.: A survey of table recognition: Models, observations, transformations, and inferences. Int'l J. Document Analysis and Recognition **7** (2004) 1–16
12. Zanibbi, R.: A Language for Specifying and Comparing Table Recognition Strategies. PhD thesis, Queen's University, Kingston, Canada (2004)
13. Handley, J.: Table analysis for multi-line cell identification. In: Proc. Document Recognition and Retrieval VIII (IS&T/SPIE Electronic Imaging). Volume 4307. (2001) 34–43
14. Hu, J., Kashi, R., Lopresti, D., Wilfong, G.: Table structure recognition and its evaluation. In: Proc. Document Recognition and Retrieval VIII (IS&T/SPIE Electronic Imaging). Volume 4307. (2001) 44–55
15. Phillips, I., Chen, S., Haralick, R.: CD-ROM document database standard. In: Proc. Second Int'l Conf. Document Analysis and Recognition, Tsukuba Science City, Japan (1993) 478–483
16. Lopresti, D., Nagy, G.: Issues in ground-truthing graphic documents. In Blostein, D., Kwon, Y.B., eds.: Graphics Recognition: Algorithms and Applications. Volume 2390 of LNCS. Springer, Berlin (2002) 46–66

Use of Input Deformations with Brownian Motion Filters for Discontinuous Regression

Ramūnas Girdziušas and Jorma Laaksonen

Helsinki University of Technology, Laboratory of Computer and Information Science,
P.O. Box 5400, FI-02015 HUT, Espoo, Finland
Ramunas.Girdziusas@hut.fi, Jorma.Laaksonen@hut.fi

Abstract. Bayesian Gaussian processes are known as 'smoothing devices' and in the case of n data points they require $O(n^2) \ldots O(n^3)$ number of multiplications in order to perform a regression analysis. In this work we consider one-dimensional regression with Wiener-Lévy (Brownian motion) covariance functions. We indicate that they require only $O(n)$ number of multiplications and show how one can utilize input deformations in order to define a much broader class of efficient covariance functions suitable for discontinuity-preserving filtering. An example of the selective smoothing is presented which shows that regression with Brownian motion filters outperforms or improves nonlinear diffusion filtering especially when observations are contaminated with noise of larger variance.

1 Introduction

Pattern analysis can be improved if feature extraction and regression stages are not separated, but designed in a systematic way which bridges boundaries between supervised and unsupervised learning. The aim of this work is to justify rather informally one such inference principle known as 'learning by input deformations', e.g. [4], which will be considered in the context of fast one-dimensional Gaussian process (GP) regression.

A central quantity of interest is the covariance matrix (possibly infinite dimensional) postulated for the joint vector of observations and model output at any input location. GP regression then proceeds by estimating the conditional density of the model output given observations. In the simplest case of regression with a single optimal parameter setting this can be interpreted as Tichonov regularization. The optimal model minimizes the sum of squared errors between its output and the observations, and the Euclidean norm of the model output vector, weighted by its inverse covariance matrix which can be diagonal.

Unfortunately, most frequently applied GP models already assume the opposite of this case, namely dense covariance matrices which require $O(n^2) \ldots O(n^3)$ number of multiplications to solve the regression task. Surprisingly, more efficient covariance functions exist that yield filtering in $O(n)$ number of multiplications.

One such example is the synthesis of covariance functions based on the zero-pole diagram of their constrained Laplace transform [6]. However, this can be

S. Singh et al. (Eds.): ICAPR 2005, LNCS 3686, pp. 219–228, 2005.

difficult to accomplish especially in boundary value problems and situations which demand discontinuous or output-dependent covariance functions. We define an alternative class of efficient covariance functions by considering nonlinear input deformations of a particular type of the GP models, known as Wiener-Lévy process or simply Brownian motion.

For this purpose, the GP regression modeling is stated in Section 2. Section 3 shows how to extend the applicability of Brownian motion (bridge) covariance functions by deforming their inputs. Section 4 explains the meaning of deformations in connection to nonlinear diffusion filtering [10]. An example of discontinuous rgeression is analyzed in Section 5, while Section 6 concludes our study.

2   Gaussian Process Regression

The hypothesis that regularities in data can be modeled as a Gaussian random process lies in the core of many regression techniques. Let us assume that the true signal $u(\mathbf{x})$, i.e. the vector of its sample values $\mathbf{u}_{1:n} = \{u_i | i = 1, 2, \ldots, n, u_i \in \mathbb{R}^1\}$ given at any n spatial locations $\mathbf{x}_{1:n} = \{\mathbf{x}_i | i = 1, 2, \ldots, n, \mathbf{x}_i \in \mathbb{R}^d\}$ is normally distributed with zero mean and the covariance matrix $\mathbf{K}_\theta \in R^{n \times n}$. The signal can only be observed imperfectly via observations $\mathbf{y}_{1:n}$:

$$p(\mathbf{y}, \mathbf{u}, \boldsymbol{\theta}) = p(\mathbf{y}|\mathbf{u}, \theta_0)p(\mathbf{u}|\mathbf{x}, \boldsymbol{\theta}) = \mathcal{N}(\mathbf{u}, \theta_0\mathbf{I})\mathcal{N}(\mathbf{0}, \mathbf{K}_\theta) , \tag{1}$$

where $\boldsymbol{\theta}$ denotes the unknown parameters of the GP model including variance of 'an additive noise' θ_0. Given little knowledge about the specific values of optimal model parameters $\boldsymbol{\theta}$, it is often enough to perform the approximate Bayesian regression $p(\mathbf{u}|\mathbf{y}) \approx p(\mathbf{u}|\mathbf{y}, \boldsymbol{\theta}^*)$ with a single parameter setting $\boldsymbol{\theta}^*$:

$$\mathbf{u}^* \equiv E[\mathbf{u}|\mathbf{y}] = \arg\min_{\mathbf{u}} \left(\frac{1}{2\theta_0^*}||\mathbf{u} - \mathbf{y}||^2 + \frac{1}{2}\mathbf{u}^T\mathbf{K}_{\theta^*}^{-1}\mathbf{u} \right) = \mathbf{K}_{\theta^*}(\mathbf{K}_{\theta^*} + \theta_0^*\mathbf{I})^{-1}\mathbf{y} . \tag{2}$$

Eq. (2) gives an intuition behind the Gaussian process regression: regularity properties of the model outputs are controlled by a quadratic form determined by the inverse covariance matrix $\mathbf{K}_\theta^{-1}$. Given an input value $\mathbf{x}$, the prediction is

$$u^*(\mathbf{x}) = \sum_{i=1}^{n} \alpha_i k(\mathbf{x}, \mathbf{x}_i), \quad \alpha_i \equiv [(\mathbf{K}_{\theta^*} + \theta_0^*\mathbf{I})^{-1}\mathbf{y}]_i . \tag{3}$$

The GP model Eq. (1) usefully completes the regularization in Eq. (2) by providing means to compare different covariance models [4]. The optimal model $\boldsymbol{\theta}^*$ can be chosen to maximize the logevidence $\mathcal{B}$ of observations $\mathbf{y}$ [9,4]:

$$-2\mathcal{B}(\boldsymbol{\theta}) = n \ln 2\pi\theta_0 + \ln \det(\mathbf{K}_\theta + \theta_0\mathbf{I}) + \theta_0^{-1}\mathbf{y}^T(\mathbf{y} - \mathbf{u}^*). \tag{4}$$

This criterion is also relevant to the issue of computational complexity because, if suitably modified, it would yield models with sparse inverse covariance $\mathbf{K}_\theta^{-1}$. However, paradoxical situations are very likely when human observes that there exists a model with smaller evidence values and slightly higher error rates, but allows to estimate the regressor thousand times faster. This can be avoided if an efficient model hypothesis space is chosen *a priori*.

3   Input Deformations in Brownian Motion Filters

3.1   Covariance Functions of Brownian Motion

In what follows, we will consider only one-dimensional filtering, i.e. $\mathbf{x} \equiv x \in \mathbb{R}^1$. A variety of useful covariance functions can be derived in the case of boundary value problems, where the combinations of the values of the model output and its derivative are known on the boundary. Without loss of generality we consider the case when $x \in [0, 1]$.

A general source for the efficient covariances lies in the exact Markovian factorization of Eq. (1). One such example is the GP model with the exponential covariance function $k(x_i, x_j) = e^{-\theta_1 |x_i - x_j|}$. When sampling the signal at regular locations $x = h, \ldots, nh$, Eq. (1) then decomposes into [5]

$$p(\mathbf{u}|\mathbf{t}, \gamma) = p(u_1) \prod_{i=2}^{n} p(u_i|u_{i-1}) = p(u_1) \prod_{i=2}^{n} \mathcal{N}(\gamma u_{i-1}, 1 - \gamma^2), \quad \gamma \equiv e^{-\theta_1 h}. \quad (5)$$

Eq. (5) allows direct application of the Kalman filter to estimate the states $u_{1:n}$ in $O(n)$ multiplications because no matrix inverses will be present as the states in the decomposed model Eq. (1) become scalar. One can add that Eq. (4) can be employed to find the optimal value of γ by rewriting $\ln \det(\mathbf{K}_\gamma + \theta_0 \mathbf{I}) = \ln \det(\mathbf{I} + \theta_0 \mathbf{K}_\gamma^{-1}) - \ln \det \mathbf{K}_\gamma^{-1}$ and exploiting the fact that the exponential covariance yields a tridiagonal inverse $\mathbf{K}_\gamma^{-1}$ whose determinant can be evaluated recursively [3].

One of the most studied Gaussian random processes, which admits the particularly simple case of the Markovian decomposition $p(u_i|u_{i-1}) = \mathcal{N}(u_{i-1}, h)$, is known as the Brownian motion or Wiener-Lévy process [5,8]. The covariance function of the corresponding zero mean GP model is

$$k_0(x_i, x_j) = \min(x_i, x_j). \quad (6)$$

This process can intuitively be understood as the 'integrated white noise' as its increments are independent and distributed with zero mean and variance equal to the step size h. If a regression outcome on the boundaries is known, e.g. $u(0) = a$ and $u(1) = b$, then the conditioned Brownian motion can be defined [8] with mean as the linear trend $E[u|x] = a + (b - a)x$ and the covariance function

$$k_1(x_i, x_j) = \min(x_i, x_j) - x_i x_j. \quad (7)$$

This is known as the 'Brownian bridge from a to b on $x \in [0, 1]$'. Eqs. (6) and (7) define 'mildly' non-stationary covariance functions in the sense that a pair of any two points separated by the same distance will have different correlation properties if the points are closer to the boundaries. Zero-correlations can take place only between the boundary and interior points, negative correlations do not exist.

Fig. 1 shows the result of filtering a rectangular pulse in the additive Gaussian noise by employing Brownian bridge from 0 to 0. As the variance of the noise

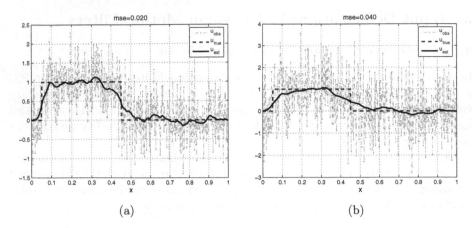

Fig. 1. Regression with the Brownian bridge model when the variance of the additive zero-mean Gaussian noise is (a) $\theta_0 = 0.25$ and (b) $\theta_0 = 1$. As most of conventional GP models, Brownian motion blurs away the edges of a signal.

increases, the effect of blurring away edges strengthens. Similar result could be obtained with any stationary smoothing kernel function. However, whereas the most frequently used Gaussian kernel results in a dense covariance matrix, Brownian motion processes yield tridiagonal inverses which require only $O(n)$ number of multiplications when computing a matrix-vector product in Eq. (2).

3.2   Input Deformations for Discontinuous Regression

The Brownian motion covariance functions not only yield tridiagonal inverse covariance matrices, but this important feature is also preserved when using input deformations. Therefore, the range of applicability of the Brownian motion filters can be considerably expanded. Consider a transformation

$$x \mapsto \tilde{x} = \frac{1}{\sqrt{\int_0^1 p^{-1}(x)dx}} \int_0^x p^{-1}(x)dx, \tag{8}$$

where $p(x) > 0$ is not necessarily continuous, but can be chosen so that $\tilde{x}$ is a bijection. Now let us consider the discrete counterpart of Eq. (8) on the grid $x_0 = 0, x_1, \ldots, x_n, x_{n+1} = 1$ and apply the midpoint approximation

$$\int_0^{x_i} p^{-1}(x)dx \approx \sum_{k=1}^{i} \frac{h_k}{p_{k-1/2}}, \tag{9}$$

where $h_k \equiv x_k - x_{k-1}$ and $p_{k-1/2} \equiv p((x_{k-1} + x_k)/2)$. Eq. (9) would hold with $O(h)$ or $O(h^2)$ accuracy ($h = \max h_k$) depending whether $p(x)$ is continuous by itself or together with its first derivative. In general, we do not even require this to be the case and consider Eq. (9) as the defined input deformation.

We can then build the covariance matrices of size $n \times n$ by sampling the functions given by Eqs. (6) and (7) at $\tilde{x}$ rather than x. The inverse covariance matrix of the Brownian bridge then becomes [11,2]

$$\mathbf{K}_\theta^{-1} = \begin{pmatrix} a_1 + a_2 & -a_2 & & & \\ -a_2 & a_2 + a_3 & -a_2 & & \\ & \ddots & \ddots & \ddots & \\ & & -a_{n-1} & a_{n-1} + a_n & -a_n \\ & & & -a_n & a_n + a_{n+1} \end{pmatrix}, \qquad (10)$$

where $a_i \equiv \frac{1}{h_i} p_{i-1/2}$. When the matrix is of the form in Eq. (10), a matrix-vector product can be computed via Thomas algorithm [10] or by means of dynamic programming in general.

A skeleton of a regression algorithm which uses the Brownian motion filter with input deformations can now be presented:

Define $p(x) \equiv p(x, \boldsymbol{\theta})$ on $x_0 = 0, x_1, \ldots, x_n, x_{n+1} = 1$.

Compute kernel matrix $\mathbf{K}_\theta \in \mathbb{R}^{n \times n}$ in Eq. (10).

Solve for $\mathbf{u}^*$ by optimizing $\boldsymbol{\theta}$ including θ_0:

$$(\mathbf{u}^*, \boldsymbol{\theta}^*) = \arg \min_{\mathbf{u}, \boldsymbol{\theta}} \{ n \ln 2\pi\theta_0 + \ln \det(\mathbf{K}_\theta + \theta_0 \mathbf{I}) + \theta_0^{-1} \mathbf{y}^T (\mathbf{y} - \mathbf{u}^*) \},$$

subject to $\mathbf{u} = (\mathbf{I} + \theta_0 \mathbf{K}_\theta^{-1})^{-1} \mathbf{y}$. \qquad (11)

There is no guarantee that a cost function will have a single optimum w.r.t. all its parameters. We will minimize it w.r.t θ_0 by employing the conjugate gradients algorithm [7,4] for a pre-selected set of values of the remaining parameters $\boldsymbol{\theta}$, finally picking up the global minimum.

4   Connection to Nonlinear Diffusion Filtering

Input deformations can be used to expand the applicability of regression with Brownian covariance matrices within the constraint of efficient $O(n)$ filtering. At this point, however, it remains unclear how to choose the corresponding input deformation. Eq. (8) has an intuitive meaning. Up to a scaling by a diagonal matrix, Eq. (10) is Green's matrix obtained by considering discrete approximation to

$$-\partial_x [p(x) \partial_x u] = y, \quad u(0) = 0, \quad u(1) = 0. \qquad (12)$$

More precisely, the approximate solution to Eq. (12) would be $\mathbf{u} = \mathbf{H}^{-1} \mathbf{K}_\theta \mathbf{y}$ with $\mathbf{H} = \operatorname{diag}[2/(h_1 + h_2), \ldots, 2/(h_n + h_{n+1})]$, see [2]. One can check that the solution to Eq. (12) also minimizes the sum of two Euclidean norms $||u - y||^2 + ||p(x)\partial_x u||^2$. The function $p(x)$ appears as the penalty weight for the norm of the derivative of the solution.

In spite of its clear meaning, the function $p(x)$ can seldom be known and it has to be estimated from observations alone. Section 5 will study a particular parametric form designed to solve an edge-preserving filtering problem shown in

Fig. (1). However, it is important to note that one can choose a different path by employing what is known as *nonlinear diffusion filtering* [10].

According to nonlinear diffusion principle, one assumes that the function $p(x)$ depends on the gradient of the true signal, and it should be small where the gradients are sharp so that the smoothing does not blur the edges. A very rough estimate of the gradient can be made from observations, and the results can be greatly improved by applying a variant of Eq. (12) iteratively.

Table 1 indicates the connection between (i) the regularization in Eq. (12), (ii) the GP regression with the covariance function of Brownian bridge enhanced via input deformations and (iii) nonlinear diffusion filtering. One can see that a single iteration of nonlinear diffusion filtering corresponds to the above-discussed regression with input deformations where the additive noise is spatially variant, e.g. $\theta_0 \equiv \theta_0(x_k) = \tau 2/(h_k + h_{k+1})$, where τ is the time step obtained by approximating the time derivative with the first order divided difference.

Table 1. GP regression model with the Brownian covariance function and its connection to diffusion filtering. The first row clarifies the action of the inverse covariance matrix stated in Eq. (10) in the continuous case. The models in the second row would be equivalent only if a discrete model is considered on the regular grid with spacing h and $\theta_0 = 1/h$. The nonlinear diffusion filtering in the third row starts with $\mathbf{u}_0 = \mathbf{y}$, τ denotes the discrete time step.

Name	Discrete Model	Continuous Space Model
Poisson Equation	$\mathbf{HK}_\theta^{-1}\mathbf{u} = \mathbf{y}$	$-\partial_x(p\partial_x u) = y$
Brownian regression	$(\mathbf{I} + \theta_0\mathbf{K}_\theta^{-1})\mathbf{u} = \mathbf{y}$	$-\partial_x(\tilde{p}\partial_x u) + u = y$
Nonlinear Diffusion	$(\mathbf{I} + \tau\mathbf{HK}_t^{-1})\mathbf{u}_t = \mathbf{u}_{t-1}$	$-\partial_x(p(u)\partial_x u) = -\partial_t u$

An example of nonlinear diffusion filtering is presented in Fig. 2. For moderate variance values, such as $\theta_0 = 0.25$, nonlinear diffusion works rather fine as can be seen in Fig. 2a. There is a problem in automating the optimal choice of the stopping time, but at least such quantity exists. However, increasing the noise level to $\theta_0 = 1$ makes the determination of the optimal stopping time nearly impossible. As the result in Fig. 2b shows, the early stopping based on minimal value of the mean of the sum of the squared errors between the true signal and the diffusion outcome would not solve the problem even if it were possible to evaluate it. In this example, a structure similar to the true signal emerges at about $T = 100$ iterations, this result in shown in Fig. 2c, whereas the steady state decays to the mean value as can be seen in Fig. 2d. Clearly, nonlinear diffusion process gets uncontrollable with the increasing noise variance. Eq. (12) could as well be used as a starting point in regression without any reference to the GP regression. However, such an approach would require continuity assumptions on the function $p(x)$ and making it data-dependent in the nonlinear diffusion way turns out to miss some important information which can be incorporated and adapted into $p(x)$ via the Bayesian GP model.

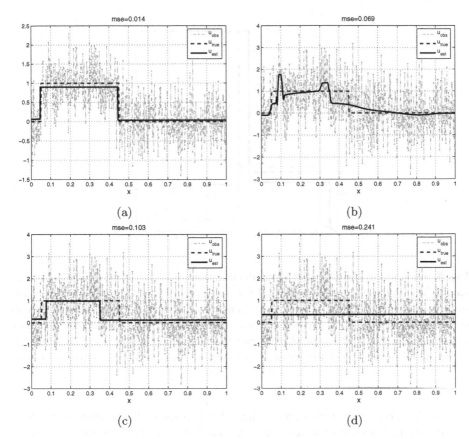

Fig. 2. Typical examples of nonlinear (edge-preserving) diffusion filtering by using $p(u) = 1 - \exp(-c(\frac{\lambda}{||\partial_x u_\sigma||})^s)$, see [1] for more details. (a) $\theta_0 = 0.25$ and (b)–(d) $\theta_0 = 1$. The outcome of the nonlinear diffusion filtering depends on the stopping time: (b) $T = 14$, (b) $T = 100$, (c) $T = 1200$. Here the case (b) corresponds to the minimal achievable mean squared error between the estimated and true signals, (c) indicates the diffusion outcome which preserves the structure of the original signal, while (d) shows the steady state. The examples in (b)–(d) show that nonlinear diffusion filtering works only when $\theta_0 < 1$.

5   Experiment: Regression with Input Deformations

Consider the filtering problem presented in Figs. 1 and 2. Suppose that we know the nature of the signal to be found in the noisy observations, e.g. we can assume that there are two discontinuities whose precise locations have to be discovered. At the first glance this minor information seems to be very marginal because, as can be seen in Fig. 2a, nonlinear diffusion finds the solution without such assumption. On the other hand, the knowledge that there are two discontinuities turns out to be crucial when the noise variance increases from $\theta_0 = 0.25$ to

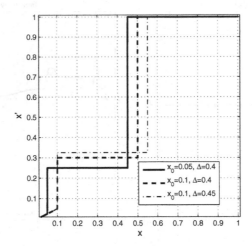

Fig. 3. Example of the adaptive input deformations used with the Brownian bridge model. The shape of a one-dimensional warp is chosen to be the simplest one which could yield edge preserving of two discontinuities when filtering a rectangular pulse in very noisy observations. The optimal input deformation depends on two parameters: location of the first and second discontinuities, denoted by x_0 and $x_0 + \Delta$. They can be estimated from noisy observations by maximizing the logevidence criterion Eq. (4).

$\theta_0 = 1$. Incorporating such a constraint allows to solve the problem reliably in one iteration. Based on the assumption that there are two discontinuities, we are now posed with the problem of finding them. In order to solve this task, we apply input deformation shown in Fig. 3 with the covariance function of the Brownian motion. This transformation is by no mean unique, but it is one of the simplest ways to incorporate two discontinuities *a priori*. The parameter x_0 denotes the location of the first discontinuity whereas $x_0 + \Delta$ stands for the second discontinuity.

Next, we apply the algorithm described in Section 3.2. The variance parameter θ_0 is usually slightly over-estimated, but this does not affect much the optimal location of discontinuities. The dependence of logevidence on two discontinuities can be seen in Fig. 4. The maximum appears to be very close to the true values $x_0 = 0.05$ and $\Delta = 0.4$. If the noise variance is increased up to $\theta_0 = 1.4$, it would become impossible to locate two discontinuities at the same time because the second large maxima would occur at $x_0 = 0.075$ and $\Delta = 0.376$, e.g. the pulse, which is shorter and starts slightly later, would compete with the ideal solution. Most of the local maxima occur on the line $\Delta \approx -x_0 + 0.45$. They correspond to the incorrect estimation of the first discontinuity while preserving the second one. The results of the Brownian motion regression with adapted input deformations are indicated in Fig. 5. The algorithm works even in the case $\theta_0 = 1$ avoiding the difficulties present in nonlinear diffusion.

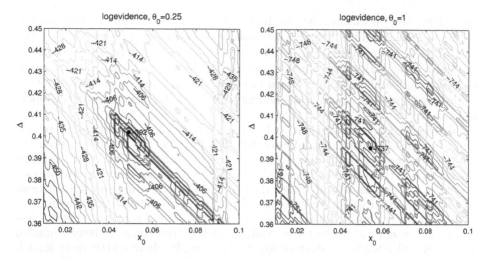

Fig. 4. Adaptive estimation of the input deformation in the case of two unknown discontinuities. The logevidence has multiple maxima, but the global one is close to the optimal values $x_0 = 0.05$ and $\Delta = 0.4$.

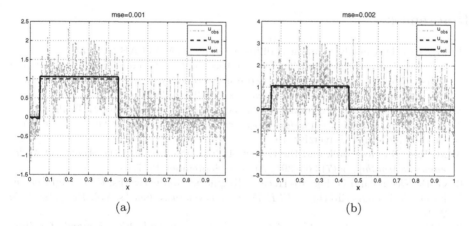

(a)   (b)

Fig. 5. The Brownian bridge regression with evidence maximizing input deformations obtained according to Fig. 4: (a) $\theta_0 = 0.25$ and (b) $\theta_0 = 1$. The edges of the signal are well-preserved, and the problem of the optimal stopping time is avoided. The algorithm works with larger values of variance of additive noise such as $\theta_0 = 1$, whereas the performance of nonlinear diffusion is rather poor, clf. Fig. 2.

6   Conclusions

The covariance functions of the Brownian motion or its variant known as the Brownian bridge are examples of possibly the simplest GP models whose range of applicability can be expanded to filtering of long discontinuous signals in boundary value problems. At the first glance, the use of input deformations seems to be plagued by the need to consider nonlinearities, i.e. one has to define the function that transforms the model inputs which are then passed through the nonlinear model. However, if one considers the Brownian motion kernel, input deformations can be usefully related to the spatial variance model because they indirectly impose the spatially-dependent penalty on the derivatives of the model output.

Currently, the most efficient and general way to solve discontinuous regression problems is based on nonlinear diffusion filtering [10]. However, even the simplest examples indicate that such an approach works only for moderate variance values of the additive noise. As the variance of the noise in the observations increases, it becomes difficult to set the optimal stopping time and the locations of discontinuities can not be determined precisely. This should come as no surprise because nonlinear diffusion operates on the general principle 'smooth less where the rough estimate of the gradient of the true signal is large', applied iteratively. This work has shown a synthetic example where a little more knowledge, such as the number of discontinuities, yields robust filtering. Moreover, the use of the evidence criterion helps in clarifying model assumptions and it indicates how the appearance of local maxima with the increasing variance of noise demands better edge-preserving filters.

References

1. Frederico D'Almeida. Nonlinear diffusion toolbox. MATLAB Central.
2. Q. Fang, T. Tsuchiya, and T. Yamamoto. Finite difference, finite element and finite volume methods applied to two-point boundary value problems. *Journal of Computational and Applied Mathematics*, 139:9–19, 2002.
3. I. M. Gelfand and S. V. Fomin. *Calculus of Variations.* Prentice-Hall, 1963.
4. M.N. Gibbs. *Bayesian Gaussian Processes for Regression and Classification.* Ph.d. thesis, Cambridge University, 1997.
5. M. Kac. Random walk and the theory of Brownian motion. *The American Mathematical Monthly*, 54(7):369–391, August–September 1947.
6. G. De Nicolao and G. Ferrari-Trecate. Regularization networks: Fast weight calculation via Kalman filtering. *IEEE Trans. on Neural Networks*, Vol. 12(2):228–235, 2001.
7. C.E. Rasmussen. *Evaluation of Gaussian Processes and Other Methods for Nonlinear Regression.* Ph.d. thesis, The University of Toronto, 1996.
8. S. E. Shreve. *Stochastic Calculus for Finance II: Continuous Time Models.* Springer, 2000.
9. J. Skilling. Bayesian numerical analysis. In *Physics and Probability*, pages 207–221. Cambridge Univ. Press, 1993.
10. J. Weickert, B. M. ter Haar Romeny, and M. A.Viergever. Efficient and reliable schemes for nonlinear diffusion filtering. *IEEE Trans. on Image Processing*, 7(3):398–410, March 1998.
11. T. Yamamoto. Inversion formulas for tridiagonal matrices with applications to boundary value problems. *Numer. Funct. Anal. and Optimiz.*, 22(3 and 4):357–385, 2001.

Hierarchical Clustering of Dynamical Systems Based on Eigenvalue Constraints

Hiroaki Kawashima and Takashi Matsuyama

Graduate School of Informatics, Kyoto University,
Yoshida-Honmachi Sakyo, Kyoto 6068501, Japan
{kawashima, tm}@i.kyoto-u.ac.jp

Abstract. This paper addresses the clustering problem of hidden dynamical systems behind observed multivariate sequences by assuming an interval-based temporal structure in the sequences. Hybrid dynamical systems that have transition mechanisms between multiple linear dynamical systems have become common models to generate and analyze complex time-varying event. Although the system is a flexible model for human motion and behaviors, the parameter estimation problem of the system has a paradoxical nature: temporal segmentation and system identification should be solved simultaneously. The EM algorithm is a well-known method that solves this kind of paradoxical problem; however the method strongly depends on initial values and often converges to a local optimum. To overcome the problem, we propose a hierarchical clustering method of linear dynamical systems by constraining eigenvalues of the systems. Due to the constraints, the method enables parameter estimation of dynamical systems from a small amount of training data, and provides well-behaved initial parameters for the EM algorithm. Experimental results on simulated and real data show the method can organize hidden dynamical systems successfully.

1   Introduction

Hybrid dynamical systems (hybrid systems) such as switching dynamical systems [6] and segment models [10] have become common models for speech recognition, computer vision, graphics, and machine learning researchers to generate and analyze complex time-varying event (e.g., human speech and motion [3, 12, 2, 9]). They assume that a complex event is consist of dynamic primitives, which is often referred to as phonemes, movemes [3], visemes, motion textons [9], and so on. For instance, a cyclic lip sequence in Figure 1 can be described by simple lip motions (e.g., "open", "close", and "remain closed"). Once the set of dynamic primitives is determined, an observed or generated time-varying pattern can be partitioned by temporal intervals with the labels of primitives.

A hybrid system represents each dynamic primitive by a simple dynamical system, and models transition between dynamical systems by a discrete-event model, such as an automaton and a hidden Markov model. Therefore, the system has a capability of generating and analyzing multivariate sequences that consist of temporal regimes of dynamic primitives.

In spite of the flexibility of hybrid systems, especially for modeling human motion and behaviors such as gestures and facial expressions, the real applications

S. Singh et al. (Eds.): ICAPR 2005, LNCS 3686, pp. 229–238, 2005.

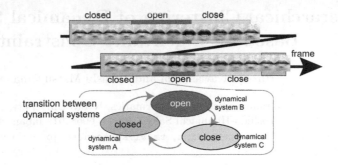

Fig. 1. Example of a lip image sequence modeled by a hybrid system

are often beset with difficulties of parameter estimation due to the paradoxical nature of the estimation problem, as we see in the next paragraph.

This paper proposes a bottom-up approach that estimates a set of dynamical systems based on an agglomerative hierarchical clustering process, which iteratively merges dynamical systems. A constraining method for system eigenvalues (spectra) is proposed to identify stable linear dynamical systems, which are appropriate systems to model human motion, from a small number of training sequences. In this paper, we use only linear dynamical systems to model dynamic primitives and often omit the term "linear". Since the hierarchical clustering method provides approximate parameters of linear dynamical systems comprised in a hybrid system, it successfully initializes refinement process of the overall system such as a maximum likelihood estimation process.

Difficulty of the Parameter Estimation: Let us assume that a large amount of training data (multivariate sequences) is given. Then, the parameter estimation problem requires us to simultaneously estimate temporal partitioning of the training data (i.e., segmentation and labeling) and a set of dynamical systems. The reason is that identification methods of dynamical systems require partitioned and labeled training sample sequences; meanwhile segmentation and labeling methods of the sample sequences require an identified set of dynamical systems. The expectation-maximization (EM) algorithm [5] is a well-known method that solves this kind of paradoxical problems with iterative calculations; however, it strongly depends on the initial parameters and does not converge to the optimum solution, especially if the model has a large parameter space to search. Therefore, the parameter estimation of hybrid systems necessitates an initialization method that searches an appropriate set of dynamical systems (i.e., the number and parameters of dynamical systems) from given training data.

The Assumed Parameter Estimation Scheme: To solve the problem above, we assume a multiphase learning approach (see Figure 2). The first step is a hierarchical clustering process of dynamical systems, which is applied to a comparatively small number of typical sequences selected from given training data set. For the second step, we assume a refinement process for all the system parameters based on a maximum likelihood method via EM algorithm [12]. The method not only refines parameters of dynamical systems but estimates parameters of

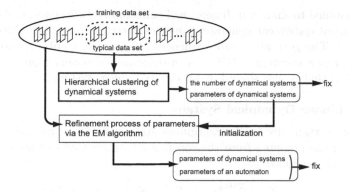

Fig. 2. The assumed parameter estimation scheme of a hybrid system. This paper concentrates on a hierarchical clustering of dynamical systems that works as an initialization process of the EM algorithm.

the automaton that models transition between the constituent linear dynamical systems. This refinement process is applied to all the given training data. Thanks to the estimated parameters in the hierarchical clustering process, the refinement process can be initialized by parameters that are relatively close to optimum compared to randomly selected parameters. As a result, the refinement process converges to the optimal solution successfully. This paper concentrates on the first step of the multiphase approach as an initialization process for the second step (i.e., EM algorithm).

The Advantage of the Hierarchical Clustering: Although several clustering approaches have been proposed to find a set of linear dynamical systems from given training sequences, such as greedy approaches [9], we propose an agglomerative hierarchical clustering method that extracts dynamical systems. The reason is that the method provides useful interfaces, such as the history of model fitting errors in each merging steps, to determine the number of clusters.

In Section 2, we describe a structure of a hybrid system. Section 3 explains the hierarchical clustering method proposed in this paper. We evaluate the method using simulated and real data to verify the expressiveness of the extracted dynamical systems in Section 4.

2   A Hybrid Dynamical System

2.1   System Architecture

A hybrid system is a generative model that can generate multivariate vector sequences by changing (or switching) the activation of constituent dynamical systems [6, 12]. In most case, the dynamical systems are linear. The system has a two-layer architecture. The first layer has a finite state automaton that models stochastic transition between dynamic primitives. The automaton has an ability of generating interval sequences, where each interval is labeled by one of the dynamic primitives. The second layer consists of a set of multiple dynamical systems $\mathcal{D} = \{D_1, ..., D_N\}$. In this paper, all the constituent dynamical systems are assumed to share a n-dimensional continuous state space, and each activated dynamical system can generate sequences of continuous (real valued) state vector $x \in \mathbf{R}^n$. The generated state sequences are mapped to observation sequences of multivariate vector $y \in \mathbf{R}^m$ in a m-dimensional observation space by a linear function that is also shared by all the dynamical systems.

2.2 Linear Dynamical Systems

The state transition in the continuous state space by a linear dynamical system D_i, and the mapping from the continuous state space to the observation space is modeled as follows:

$$x_t = F^{(i)} x_{t-1} + g^{(i)} + \omega_t^{(i)}, \quad y_t = H x_t + v_t, \tag{1}$$

where $F^{(i)}$ is a transition matrix and $g^{(i)}$ is a bias vector. Note that each dynamical system has $F^{(i)}$ and $g^{(i)}$ individually. H is an observation matrix that defines linear projection from the continuous state space to the observation space. $\omega^{(i)}$ and v is a process noise and an observation noise, respectively. We assume that the process noise and the observation noise has Gaussian distribution $\mathcal{N}(0, Q^{(i)})$ and $\mathcal{N}(0, R)$, respectively. The notation $\mathcal{N}(a, B)$ is a Gaussian distribution with an average vector a and a covariance matrix B. As we described in the previous subsection, we assume that all the dynamical systems share a continuous state space to simplify the model and to reduce the parameters. Using the notations above, we can consider the probability distribution functions: $p(x_t | x_{t-1}, d_t = D_i) = \mathcal{N}(F^{(i)} x_{t-1}, Q^{(i)})$ and $p(y_t | x_t, d_t = D_i) = \mathcal{N}(H x_t, R)$, where the variable d_t represents an activated dynamical system at time t.

Calculation of Likelihood in Intervals: Let us assume that a continuous state has a Gaussian distribution at each time t. Then, the transition of the continuous state becomes a Gauss-Markov process, which is inferable in the same manner as Kalman filtering [1]. Therefore, the predicted state distribution under the condition of observations from 1 to $t-1$ is formulated as $p(x_t | y_1^{t-1}, d_t = D_i) = \mathcal{N}(x_{t|t-1}^{(i)}, V_{t|t-1}^{(i)})$ and $p(y_t | y_1^{t-1}, d_t = D_i) = \mathcal{N}(H x_{t|t-1}^{(i)}, H V_{t|t-1}^{(i)} H^{\mathrm{T}} + R)$, where the average vector $x_{t|t-1}^{(i)}$ and covariance matrix $V_{t|t-1}^{(i)}$ are updated every sampled time t. Suppose that the dynamical system D_i represents an observation sequence $y_b^e \triangleq y_b, ..., y_e$, which has a duration length $e - b + 1$, then the likelihood that the system D_i generates the sequence is calculate by the following equation:

$$p(y_b^e | d_b^e = D_j) = \prod_{t=b}^{e} p(y_t | y_1^{t-1}, d_t = D_j), \tag{2}$$

where we assume a Gaussian distribution $N(x_{\mathrm{init}}^{(i)}, V_{\mathrm{init}}^{(i)})$ for the initial state distribution in each interval represented by dynamical system D_i.

In the following sections, we assume the observation matrix and the noise covariance matrix is $H = I$ (unit matrix) and $R = O$ (zero matrix), respectively, to concentrate on extracting dynamic primitives represented by transition matrices. Hence, the parameters to be estimated in a hybrid system become the following.

- the number of dynamical systems N
- the parameters of dynamical systems $\Theta = \{\theta_1, ..., \theta_N\}$, where

  $\theta_i = \{F^{(i)}, g^{(i)}, Q^{(i)}, x_{\text{init}}^{(i)}, V_{\text{init}}^{(i)}\}$ is a parameter set of dynamical system D_i
- the parameters of an automaton that models transition between dynamics

As we described in the Section 1, we concentrate on estimating N and Θ that initializing the EM algorithm. We assume that the parameters of the automaton are estimated and the parameter set Θ is refined by the EM algorithm.

2.3   Generable Time-Varying Patterns by Linear Dynamical Systems

The generable class of time-varying patterns (corresponds to trajectories of points in the state space) from a linear dynamical system can be described by the eigenvalues of the transition matrix. To concentrate on the temporal evolution of the state in the dynamical system, let us assume the bias and the process noise term is zero in Equation (1). Using the eigenvalue decomposition of the transition matrix:

$$F = E\Lambda E^{-1} = [e_1, ..., e_n]\text{diag}(\lambda_1, ..., \lambda_n)[e_1, ..., e_n]^{-1},$$

we can solve the state at time t with initial condition x_0:

$$x_t = F^t x_0 = (E\Lambda E^{-1})^t x_0 = E\Lambda^t E^{-1} x_0 = \sum_{p=1}^{n} \alpha_p e_p \lambda_p^t, \qquad (3)$$

where e_p and λ_p is a corresponding eigenvalue and eigenvector pair. We omit the indices i for simplification. A weight value α_p is determined from the initial state x_0 by calculating $[\alpha_1, ..., \alpha_n]^{\mathsf{T}} = E^{-1} x_0$. Hence, the generable patterns from the system can be categorized by the sign (especially in the real parts) and the norm of the eigenvalues $\lambda_1, ..., \lambda_n$. For instance, the system can generate time-varying patterns that converge to certain values if and only if $|\lambda_p| < 1$ for all $1 \leq p \leq n$ (using the term in control theory, we can say that the system is stable); meanwhile, the system can generate non-monotonous or cyclic patterns if the imaginary parts have nonzero values.

3   Hierarchical Clustering of Dynamical Systems

The goal of the hierarchical clustering process is to estimate the parameters N and Θ by assuming only a small amount of typical training data is given.

Let us assume that a multivariate sequence $y_1^T \triangleq y_1, ..., y_T$ is given as a typical training data (we consider a single training data without loss of generality), then we simultaneously estimate a set of dynamical systems $\mathcal{D}$ (i.e., the number of dynamical system N and the parameter set Θ) with an interval set $\mathcal{I}$ (i.e., segmentation and labeling of the sequence), from the training sample y_1^T. Note that, the number of intervals K is also unknown. We formulate the problem as the search of the linear dynamical system set $\mathcal{D}$ and the interval set $\mathcal{I}$ that maximizes the total likelihood of the training data: $\mathcal{L} = P(y_1^T | \mathcal{I}, \mathcal{D})$. Because the likelihood monotonously increases with an increase in the number of dynamical systems, we need to determine the right balance between the likelihood and the number N. A hierarchical clustering approach provides us an interface, such as the history of model fitting errors in each merging step, to decide the number of dynamical systems.

To identify the system parameters from only a small amount of training data, we need constraints to estimate an appropriate dynamics. In this paper, we concentrate on extracting human motion primitives observed in such as facial motion, gaits, and gestures; therefore, constraints based on stability of dynamics are suitable to find motion that converges to a certain state from an initial pose. The key idea to estimate stable dynamics is the method that constrains on eigenvalues. If all the eigenvalues are lower than 1, the dynamical system changes the state in a stable manner, as we described in Subsection 2.3.

In the following subsections, we first propose a constrained system identification method that constrains an upper bound of eigenvalues in the transition matrices of linear dynamical systems. The method enables us to find a set of dynamical systems that represents only stable dynamics. Second, we describe an agglomerative hierarchical clustering of dynamical systems based on the pseudo distance between two dynamical systems. The algorithm also merges two interval sets labeled by the same dynamical system in each iteration step. Thus, the clustering method solves two problems simultaneously: temporal segmentation and parameter estimation.

3.1   Constrained System Identification

Given a continuous state sequence mapped from an observation space, the parameter estimation of a transition matrix $F^{(i)}$ from the sequence of continuous state vectors $x_b^{(i)}, .., x_e^{(i)}$ becomes a minimization problem of prediction errors. Let us use the notations $X_0^{(i)} = [x_b^{(i)}, ..., x_{e-1}^{(i)}]$ and $X_1^{(i)} = [x_{b+1}^{(i)}, ..., x_e^{(i)}]$, if the temporal interval $[b, e]$ is represented by a linear dynamical system D_i. Then, we can estimate the transition matrix $F^{(i)}$ by the following equation:

$$F^{(i)*} = \arg\min_{F^{(i)}} ||F^{(i)} X_0^{(i)} - X_1^{(i)}||^2 = \lim_{\delta^2 \to 0} X_1^{(i)} X_0^{(i)\mathsf{T}} (X_0^{(i)} X_0^{(i)\mathsf{T}} + \delta^2 I)^{-1}, \quad (4)$$

where I is the unit matrix and δ is a positive real value.

To set a constraint on the eigenvalues, we stop the limit in the Equation (4) before $X_0^{(i)\mathsf{T}} (X_0^{(i)} X_0^{(i)\mathsf{T}} + \delta^2 I)^{-1}$ convergences to the pseudo-inverse matrix of $X_0^{(i)}$. Using Gershgorin's theorem in linear algebra, we can determine the upper bound of eigenvalues in the matrix from its elements. Suppose $f_{uv}^{(i)}$ is an element in row u and column v of the transition matrix $F^{(i)}$. Then, the upper bound of the eigenvalues is determined by $\mathcal{B} = \max_u \sum_{v=1}^{n} |f_{uv}^{(i)}|$. Therefore, we search a nonzero value for δ, which controls the scale of elements in the matrix, that satisfies the equation $\mathcal{B} = 1$ via iterative numerical methods.

3.2   Hierarchical Clustering of Dynamical Systems

The hierarchical clustering algorithm is initialized by partitioning the training sequence into motion and stationary pose intervals, which are simply divided using the scale of the first-order temporal difference of training data. In the

Hierarchical clustering of dynamical systems

```
for i ← 1 to N do
    D_i ← Identify (I_i)
end for
for all pair(D_i, D_j) where D_i, D_j ∈ D
do
    Dist (i, j) ← CalcDistance (D_i, D_j)
end for
while N ≥ 2 do
    (i*, j*) ← arg min_(i, j) Dist (i, j)
    I_i* ← MergeIntervals (I_i*, I_j*)
    D_i* ← Identify (I_i*)
    erase D_j* from D;    N ← N - 1
    for all pair(D_i*, D_j) where D_j ∈ D
    do
        Dist(i*, j) ← CalcDistance (D_i*, D_j)
    end for
end while
```

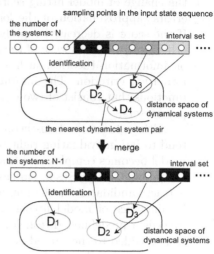

Fig. 3. Identify is a constrained system identification that we described in Subsection 3.1 I_i is an interval set that comprises intervals labeled by D_i. CalcDistance calculates the distance between the two modes defined in Subsection 3.2. MergeIntervals merges two interval set belongs to the nearest dynamical system pair.

first step of the algorithm, a single dynamical system is identified from each interval in the initial interval set. Then, we calculate a pseudo distances for all the dynamical system pairs based on the distance definition in the next paragraph. In the second step, the nearest dynamical systems are merged iteratively based on an agglomerative hierarchical clustering (see Figure 3.2). As a result, all the dynamical systems are merged to one dynamical system. We discuss the determination of the number of dynamical systems in the remaining of this subsection.

Distance Definition between Dynamical Systems: We define a pseudo distance between dynamical systems D_i and D_j as an average of two asymmetric divergences: $Dist(D_i, D_j) = \{KL(D_i||D_j) + KL(D_j||D_i)\}/2$, where each of the divergences is calculated as an approximation of Kullback-Leibler divergence [7]:

$$KL(D_i||D_j) \sim \frac{1}{|I_i|} \sum_{I_k \in I_i} \left\{ \log p(y_{b_k}^{e_k}|d_{b_k}^{e_k} = D_i) - \log p(y_{b_k}^{e_k}|d_{b_k}^{e_k} = D_j) \right\},$$

where $y_{b_k}, ..., y_{e_k}$ is a partitioned sequence by interval I_k. $|I_i|$ is the summation of interval length in the interval set I_i that is labeled by a linear dynamical system D_i. Note that we can calculate the likelihoods based on Equation (2).

Cluster Validation Problem: The determination of the appropriate number of dynamical systems is an important problem in real applications. The problem is often referred to as the cluster validation problem, which remains essentially unsolved. There are, however, several well-known criteria, which can be categorized into two types, to decide the number of clusters. One is defined based on the change of model fitting scores, such as log-likelihood scores and prediction errors (approximation of the log-likelihood scores), during the merging steps. If the score is decreased rapidly, then the merging process is stopped [11]. In other words, it finds *knee* of the log-likelihood curve. The other is defined based on information theories, such as minimum description length and Akaike's information criterion. The information-theoretical criteria define the evaluation functions that consist of two terms: log-likelihood scores and the number of free parameters.

Although information-theoretical criteria work well in simple models, they tend to fail in evaluating right balance between the two terms, especially if the model becomes complex and has a large number of free parameters [8]. Because the problem also arises in our case, we use model fitting scores directly. First, we extract candidates for the numbers of the dynamical systems by finding peaks in difference of model fitting errors between adjacent two steps. If the value exceeds a predefined threshold, then the number of dynamical systems in that step is added to the candidates. We consider that user should finally decide the appropriate number of dynamical systems from the extracted candidates.

4   Experimental Results

For the first evaluation, we used simulated sequences for training data to verify the proposed clustering method, because it provides the ground truth of the estimated parameters. Three linear dynamical systems and their parameters were set manually. The dimension was $n = 2$; therefore each of the system had 2×2 transition matrix. A two-dimensional vector sequence $Y = [y_1, ..., y_L]$ (Figure 4 (b)) was generated as an observation sequence from simulated transition between the dynamical systems based on the activation pattern in Figure 4 (a). The length of the sequence was $L = 100$. We then applied the clustering method proposed in Section 3. Figure 4 (c) shows the overall model fitting error between the original sequence Y and generated sequences $Y^{gen}(N)$ from the extracted N dynamical systems. The error was calculated by the Euclid norm: $Err(N) = ||Y - Y^{gen}(N)|| = \sqrt{\sum_{t=1}^{L} ||y_t - y^{gen}(N)_t||^2}$. Figure 4 (d) shows the results of temporal segmentation partitioned by the extracted dynamical systems in each iteration step. We see that the error increases monotonously with the decrease in the number of dynamical systems. Note that there are several steep slopes in the chart. The steep slopes correspond to the iteration steps in which dynamical system pairs with a long distance were merged. The candidates of the number were determined as $N = 3$ (which corresponds to the ground truth) and $N = 8$ by extracting the steps in which the difference $Err(N - 1) - Err(N)$ exceeds the given threshold. Consequently, the history of model fitting errors helps us to decide the appropriate number of dynamical systems.

For the second evaluation, we applied the clustering method to real video data. A frontal facial image sequence was captured by 60fps camera. Facial feature points were tracked by the active appearance model [4, 13], and eight feature points around the right eye were extracted. The length of the sequence

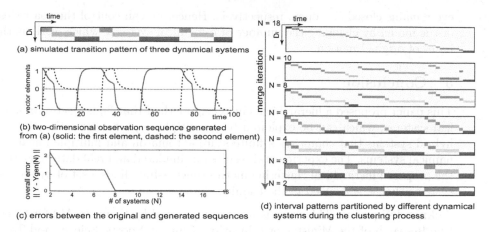

(a) simulated transition pattern of three dynamical systems

(b) two-dimensional observation sequence generated from (a) (solid: the first element, dashed: the second element)

(c) errors between the original and generated sequences

(d) interval patterns partitioned by different dynamical systems during the clustering process

Fig. 4. Clustering results on the simulated sequence generated from three dynamical systems

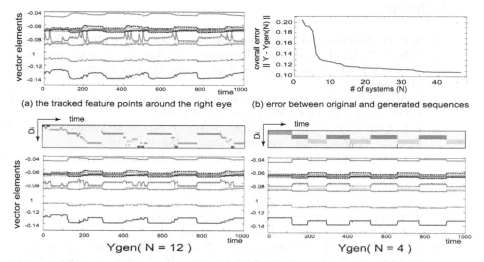

(a) the tracked feature points around the right eye

(b) error between original and generated sequences

(c) interval patterns partitioned by extracted dynamical systems (above) and generated sequences (below)

Fig. 5. Clustering results on the feature point sequence around the right eye during the subject smiled four times

was $L = 1000$. We then applied the clustering method to the obtained 16-dimensional vector sequence that comprised x- and y-coordinates of the feature points (both coordinate coefficients were plotted together in Figure 5 (a)). The candidates of the number of dynamical systems were determined as $N = 3$ and $N = 6$. Figure 5 (b) and (c) shows the error $||Y - Y^{gen}(N)||$ in each step and the generated sequences in the steps of $N = 12$ and $N = 4$. We see that the generated sequence $Y^{gen}(12)$ remains the spikes, which represent eye blinks, appeared in the original sequence; meanwhile, $Y^{gen}(4)$ smoothes out them. For instance, the dominant dynamical systems D_2 and D_3 represents the intervals in which the eye remains closed and open, respectively. Hence, we can control the coarseness of the model by changing the number of dynamical systems, which work as the bases of original motion.

5   Conclusion

This paper proposed a hierarchical clustering method that finds a set of dynamical systems, which can be exploited to a multiphase parameter estimation for hybrid systems that comprises a finite state automaton and multiple linear dynamical systems. The experimental results on simulated and real data show that the proposed hierarchical clustering method successfully finds a set of dynamical systems that is embedded in the training data.

Acknowledgment. This work is in part supported by Grant-in-Aid for Scientific Research of the Ministry of Education, Culture, Sports, Science and Technology of Japan under the contract of 13224051 and 16700175.

References

1. B. D. O. Anderson and J. B. Moor. *Optimal Filtering*. Prentice-Hall, 1979.
2. B. N. A. Blake, M. Isard, and J. Rittscher. Learning and classification of complex dynamics. *IEEE Trans. on Pattern Analysis and Machine Intelligence*, 22(9):1016–1034, 2000.
3. C. Bregler. Learning and recognizing human dynamics in video sequences. *Proc. of Intl. Conference on CVPR*, pages 568–574, 1997.
4. T. F. Cootes, G. J. Edwards, and C. J. Taylor. Active appearance model. *Proc. European Conference on Computer Vision*, 2:484–498, 1998.
5. A. P. Dempster, N. M. Laird, and D. B. Rubin. Maximum likelihood from incomplete data via the em algorithm. *J. R. Statist. Soc. B*, 39:1–38, 1977.
6. Z. Ghahramani and G. E. Hinton. Switching state-space models. *Technical Report CRG-TR-96-3, Dept. of Computer Science, University of Toronto*, 1996.
7. B. H. Juang and L. R. Rabiner. A probabilistic distance measure for hidden markov models. *AT & T Technical Journal*, 64(2):391–408, 1985.
8. D. A. Langan, J. W. Modestino, and J. Zhang. Cluster validation for unsupervised stochastic model-based image segmentation. *IEEE Trans. on Image Processing*, 7(2):180–195, 1998.
9. Y. Li, T. Wang, and H. Y. Shum. Motion texture: A two-level statistical model for character motion synthesis. *SIGGRAPH*, pages 465–472, 2002.
10. M. Ostendorf, V. Digalakis, and O. A. Kimball. From hmms to segment models: A unified view of stochastic modeling for speech recognition. *IEEE Trans. Speech and Audio Process*, 4(5):360–378, 1996.
11. D. K. Panjwani and G. Healey. Markove random field models for unsupervised segmentation of textured color images. *IEEE Trans. on Pattern Analysis and Machine Intelligence*, 17(10):939–954, 1995.
12. V. Pavlovic, J. M. Rehg, and J. MacCormick. Learning switching linear models of human motion. *Proc. of Neural Information Processing Systems*, 2000.
13. M. B. Stegmann, B. K. Ersboll, and R. Larsen. FAME - a flexible appearance modelling environment. *Informatics and Mathematical Modelling, Technical University of Denmark*, 2003.

An Optimally Weighted Fuzzy k-NN Algorithm

Tuan D. Pham[1,2]

[1] Bioinformatics Applications Research Centre
[2] School of Information Technology, James Cook University,
Townsville, QLD 4811, Australia
tuan.pham@jcu.edu.au

Abstract. The nearest neighbor rule is a non-parametric approach and has been widely used for pattern classification. The k-nearest neighbor (k-NN) rule assigns crisp memberships of samples to class labels; whereas the fuzzy k-NN neighbor rule replaces crisp memberships with fuzzy memberships. The membership assignment by the conventional fuzzy k-NN algorithm has a disadvantage in that it depends on the choice of some distance function, which is not based on any principle of optimality. To overcome this problem, we introduce in this paper a computational scheme for determining optimal weights to be combined with different fuzzy membership grades for classification by the fuzzy k-NN approach. We show how this optimally weighted fuzzy k-NN algorithm can be effectively applied for the classification of microarray-based cancer data.

1 Introduction

Methods for pattern classification have been applied for solving many important problems which can be either abstract (conceptual classification) or concrete (physical classification). Methodologies and techniques for machine learning and recognition have been extensively studied by many researchers from many different disciplines. However, there is still no unifying theory that can be applied to all kinds of pattern recognition problems. Most techniques for pattern classification and recognition are problem-oriented. Among many approaches for pattern classification including linear discriminant analysis, Bayesian classifier, Markov chains, hidden Markov models, neural networks, and support vector machines, the k-nearest neighbor decision rule, which is a procedure for deciding the membership of an unknown sample by a majority vote of the k-nearest neighbors, is one of the most popular classification methods chosen for solving many practical problems in many disciplines ranging from image, text, speech, to life and natural sciences [1,3,7,8,15,16,17].

Although the k-nearest neighbor (k-NN) rule is a suboptimal procedure, it has been shown that with unlimited number of samples the error rate for the 1-NN rule is not more than twice the optimal Bayes error rate [6], and as k increases this error rate asymptotically approaches the optimal rate [5]. Given the advantages of the k-NN classifier, it has been pointed out that the assumption of equal weights in the assignment of an input vector to class labels can reduce

S. Singh et al. (Eds.): ICAPR 2005, LNCS 3686, pp. 239–247, 2005.

the accuracy of the k-NN algorithm, particularly when there is a strong degree of overlap between the sample vectors. To overcome this problem, the fuzzy k-NN algorithm [13] assigns a fuzzy membership for an unknown sample $\mathbf{x}_u$ to class label y, denoted as μ_{yu}, as a linear combination of the fuzzy membership grades of k nearest samples:

$$\mu_{yu} = \frac{\sum_{i=1}^{k} c_i \, \mu_{yi}}{\sum_{i=1}^{k} c_i} \tag{1}$$

where the denomination is used as the normalization, μ_{yi} is the fuzzy membership which assigns labeled sample $\mathbf{x}_i$ to class label y, and c_i is the weight that is inversely proportional to the distance between $\mathbf{x}_i$ and the unknown sample $\mathbf{x}_u$

$$c_i = \frac{1}{d_{iu}^p} \tag{2}$$

in which the distance $d_{iu} = ||\mathbf{x}_u - \mathbf{x}_i||$, and the exponent $p = \frac{2}{q-1}$, where q is an integer variable.

Based on (2), different choices of p in terms of q will lead to different values for μ_{yu}. In other word, μ_{yu} can be estimated by using different exponent weights that are inversely proportional to any power of the distance function. Equation (1) is also known to be the general inverse distance estimator which offers flexibility in estimating the fuzzy membership of the unknown sample $\mathbf{x}_u$ to y class with respect to its k nearest neighbors. When p approaches 0, the inverse distance estimate approaches the average of the fuzzy membership grades of the labeled samples. When p approaches ∞, the inverse distance estimate tends to have a strong bias to the closest neighbor of $\mathbf{x}_u$. The choice of the inverse distance exponent p is arbitrary, and the most conventional choice for p is 2, which also means $q = 2$.

After the assignment of the fuzzy membership grades of an unknown vector $\mathbf{x}_u$ to all class labels, the fuzzy k-NN classifier assigns $\mathbf{x}_u$ as belonging to the class label whose fuzzy membership for $\mathbf{x}_u$ is maximum.

It can be seen that the determination of the set of weights $\{c_i\}$ is based on some arbitrary distance measure between the unkown sample and its neighbors, and not based on any optimal criterion. In this paper, we apply the method of ordinary kriging to determine an optimal set of weights for the fuzzy k-nearest neighbor decision rule. Kriging is known as the best linear unbiased estimator [10]. This estimation is *linear* because the estimates are the weighted linear combinations of the available data; it is *unbiased* because it imposes a condition that the mean error is equal to zero; it is *best* because its aim is to minimize the error variance. Other estimation methods can be linear or theoretically unbiased. However, the distinguishing feature of kriging is that its formulation is based on the minimization of the error variance. We present the proposed optimally weighted fuzzy k-NN algorithm in the following section.

2 Optimally Weighted Fuzzy k-NN Algorithm

As the conventional k-NN computes the fuzzy membership grade of an unknown sample by linearly combining the fuzzy membership grades of the neighbor samples using the weighting coefficients obtained from a metric measure, the approach presented herein is to determine the set of optimal weighting coefficients in terms of a statistical measure.

Using a similar expression of the conventional fuzzy k-NN algorithm, the proposed k-NN algorithm assigns the fuzzy membership for an unknown sample $\mathbf{x}_u$ to class label y as an optimally weighted linear combination of the fuzzy membership grades of k nearest samples:

$$\mu_{yu} = \sum_{i=1}^{k} w_i \, \mu_{yi} \tag{3}$$

where μ_{yu} and μ_{yi} have been previously defined, $\{w_i, i = 1, \ldots k\}$ are the optimal weights which indicate the relationship between $\mathbf{x}_i$ and $\mathbf{x}_u$, and to be determined. It is noted that the normalization is not needed in (3) because $\sum_{i=1}^{k} w_i = 1$.

The set of optimal weights expressed in (3), which quantify the relationships between the unknown and available samples can be equivalently derived from the estimate of the value of the unknown sample $\mathbf{x}_u$, which results in the set of optimal weights for the linear combination of the available samples:

$$\hat{\mathbf{x}}_u = \sum_{i=1}^{k} w_i \, \mathbf{x}_i \tag{4}$$

where $\hat{\mathbf{x}}_u$ is the estimate of $\mathbf{x}_u$, and $\mathbf{x}_i, \ldots, \mathbf{x}_k$ are available sample data.

There are different approaches for determining the weights to the available or neighbor data with respect to the unknown value, and different approach leads to different computational scheme. One particular approach for computing these weights optimally is to minimize the average error of estimation. Let r_j denote the error between any particular estimated $\hat{\mathbf{x}}_j$ value and the true value $\mathbf{x}_j$:

$$r_j = \hat{\mathbf{x}}_j - \mathbf{x}_j \tag{5}$$

then the average error, denoted as r_a, of k estimates is

$$r_a = \frac{1}{k} \sum_{j=1}^{k} r_j \tag{6}$$

However, minimizing r_a is unrealistic because the true values $\mathbf{x}_1, \ldots, \mathbf{x}_k$ are not known. One possible solution to this problem is the use of ordinary kriging computational scheme that considers the unknown values as the outcome of a random process and solves the problem by statistical procedures. In other words, it is not possible to minimize the variance of the actual errors, but it is possible to minimize the variance of the modeled error which is defined as the difference between the random variables modeling the estimate and the true value. As the result of statistical and analytical analysis, kriging computes the set of optimal weights by solving the following system of equations:

$$\mathbf{C}\,\mathbf{w} = \mathbf{D} \tag{7}$$

where

$$\mathbf{C} = \begin{bmatrix} C_{11} & \cdots & C_{1k} & 1 \\ \cdot & \cdots & \cdot & \cdot \\ \cdot & \cdots & \cdot & \cdot \\ \cdot & \cdots & \cdot & \cdot \\ C_{k1} & \cdots & C_{kk} & 1 \\ 1 & \cdots & 1 & 0 \end{bmatrix}$$

$$\mathbf{w} = \begin{bmatrix} w_1 & \cdots & w_k & \beta \end{bmatrix}^T$$

and

$$\mathbf{D} = \begin{bmatrix} C_{1u} & \cdots & C_{ku} & 1 \end{bmatrix}^T$$

where C_{ij} is the covariance of $\mathbf{x}_i$ and $\mathbf{x}_j$, $w_1, \ldots, w_k$ are kriging (optimal) weights, and β is a Lagrange multiplier.

The values of the kriging weights can be obtained by solving

$$\mathbf{w} = \mathbf{C}^{-1}\,\mathbf{D} \tag{8}$$

where $\mathbf{C}^{-1}$ is the inverse of the covariance matrix $\mathbf{C}$.

It is known that the solution of a kriging system can result in negative weights that should be avoided in order to ensure the robustness of the estimation. One can adopt a simple and effective procedure for correcting negative weights which was proposed by Journel and Rao [12]. This method determines the largest negative weight and adds an equivalent positive constant to all weights which are then normalized:

$$w_i^* = \frac{w_i + \alpha}{\sum_{i=1}^{k}(w_i + \alpha)}, \; \forall i \tag{9}$$

where w_i^* is the corrected weight of w_i and

$$\alpha = -\min_i w_i \tag{10}$$

The derivation of the kriging system expressed by (7) can be shown in that the probabilistic model employed by kriging is a stationary random function that consists of several random variables, one for each of the available values and one for the unknown value. Let $V(\mathbf{x}_1), \ldots, V(\mathbf{x}_k)$ be the random variables for k samples $\mathbf{x}_1, \ldots, \mathbf{x}_k$ respectively; and $V(\mathbf{x}_u)$ be the random variable for $\mathbf{x}_u$. These random variables are assumed to have the same probability distribution, and the expected value of the random variables at all locations is $E\{V\}$. Thus, the estimate of $\mathbf{x}_u$ is also a random variable and expressed by a weighted linear combination of the random variables at k locations:

$$\hat{V}(\mathbf{x}_0) = \sum_{i=1}^{k} w_i \, V(\mathbf{x}_i) \tag{11}$$

Thus, the error of estimation is

$$R(\mathbf{x}_u) = \sum_{i=1}^{k} w_i V(\mathbf{x}_i) - V(\mathbf{x}_0) \tag{12}$$

The expected value of the error of estimate is

$$E\{R(\mathbf{x}_u)\} = \sum_{i=1}^{k} w_i E\{V(\mathbf{x}_i)\} - E\{V(\mathbf{x}_0)\} \tag{13}$$

Based on the assumption that the random function is stationary, (13) becomes

$$E\{R(\mathbf{x}_u)\} = \sum_{i=1}^{k} w_i E\{V\} - E\{V\} \tag{14}$$

To satisfy the unbiased condition, $E\{R(\mathbf{x}_u)\}$ must be set to zero:

$$E\{R(\mathbf{x}_u)\} = 0 = \sum_{i=1}^{k} w_i E\{V\} - E\{V\} \tag{15}$$

which leads to

$$E\{V\} \sum_{i=1}^{k} w_i = E\{V\} \tag{16}$$

Therefore

$$\sum_{i=1}^{k} w_i = 1 \tag{17}$$

The variance of the random variable $V(\mathbf{x}_u)$ is given by

$$Var\{\sum_{i=1}^{k} w_i V_i\} = \sum_{i=1}^{k} \sum_{j=1}^{k} w_i w_j Cov\{V_i V_j\} \tag{18}$$

Given that $R(\mathbf{x}_u) = \hat{V}(\mathbf{x}_u) - V(\mathbf{x}_u)$ and using (18), the error variance is defined as

$$Var\{R(x_u)\} = Cov\{\hat{V}(x_u)\hat{V}(x_u) - 2Cov\{\hat{V}(x_u)V(x_u)\} \\ + Cov\{V(x_u)V(x_u)\} \tag{19}$$

which can be written as

$$\sigma_R^2 = \sigma^2 + \sum_{i=1}^{k}\sum_{j=1}^{k} w_i w_j C_{ij} - 2\sum_{i=1}^{k} w_i C_i \tag{20}$$

which defines the variance of error as a function of $w_1, \ldots, w_k$.

An optimal choice for the kriging weights is to minimize σ_R^2. This can be done by the Lagrangean method:

$$\sigma_R^2 = \sigma^2 + \sum_{i=1}^{k}\sum_{j=1}^{k} w_i w_j C_{ij} - 2\sum_{i=1}^{k} w_i C_i + 2\beta(\sum_{i=1}^{k} w_i - 1) \tag{21}$$

where β is a Lagrange multiplier.

After differentiating (21) with respect to all w_i and β and setting each one to zero, we obtain

$$\sum_{j=1}^{k} w_j C_{ij} + \beta = C_{iu}, \ \forall i = 1, \ldots, k \tag{22}$$

Expressions (22) and (17) define the ordinary kriging system of equations expressed in (7), which is represented in the form of matrix notation. If the data are spatially related then the covariance can be calulated as [10]

$$C(h) = \frac{1}{N(h)} \sum_{(i,j)|h_{ij}=h} \mathbf{x}_i \mathbf{x}_j - (\frac{1}{n}\sum_{k=1}^{n} \mathbf{x}_k)^2 \tag{23}$$

in which the covariance is a function the lag distance h, $N(h)$ is the number of pairs that $\mathbf{x}_i$ and $\mathbf{x}_j$ are separated by h, and n is the total number of data.

Alternatively, the covariance function $C(h)$ can be replaced by the variogram function, denoted as $\gamma(h)$, which is half the average squared difference between the paired data values:

$$\gamma(h) = \frac{1}{2N(h)} \sum_{(i,j)|h_{ij}=h} (\mathbf{x}_i - \mathbf{x}_j)^2 \tag{24}$$

It can be noted that the computation of the kriging weights that are used to make inference about the fuzzy membership grade of an unknown sample with respect to a particular class is not restricted to the sense that the data are spatially related. Both conventional and spatial covariance values can be used in the computation of the kriging system to derive the set of optimal weights for the proposed fuzzy k-NN algorithm.

3 Results

The proposed approach was used to study an important problem of gene expression microarrays. Microarray-based measure of gene expressions is one of the most recent breakthrough technologies in experimental molecular biology [4]. The utilization of microarrays allows simulateneous study and monitoring of tens of thousands of genes. One of the most useful quantitative analyses and interpretations of microarray-based data is diseased-state classification [14,18].

The data used to test the proposed algorithm is the microarray-based hereditary breast cancer data which were first studied by Hedenfalk et $al.$ [9]. The data consist of 22 cDNA microarrays with 3226 genes. The twenty-two breast tumor samples were collected from the biopsy specimens of 7 patients with germ-line mutations of BRCA1, 8 patients with germ-line mutations of BRCA2, and 7 patients with sporadic cases. The ratio data was truncated from below at 0.1 and above at 20. Log of the ratio data were used to classify BRCA1, BRCA2, and sporadic. The microarray data can be represented in matrix notation as $\mathbf{X} = [\mathbf{x}_{ij}]$, $i = 1,\ldots,N$, $j = 1,\ldots,M$, where N and M are the numbers of tumor samples and genes respectively.

To determine the fuzzy membership grades for sample data, the fuzzy c-means algorithm (FCM) [2] was applied to partition the data set into three fuzzy prototypes according to the three classes. The FCM performs the partition based on the following objective function

$$J_m = \sum_{i=1}^{N}\sum_{y=1}^{c}(\mu_{yi})^m d_{yi}^2 \tag{25}$$

where

$$d_{yi}^2 = ||\mathbf{x}_i - \mathbf{v}_y||_A^2 = (\mathbf{x}_i - \mathbf{v}_y)^T A(\mathbf{x}_i - \mathbf{v}_y) \tag{26}$$

in which c is the number of clusters or fuzzy prototypes, m is the weighting exponent, $1 \leq m < \infty$, $\mathbf{v} = (\mathbf{v}_1, \mathbf{v}_2, \ldots, \mathbf{v}_c)$, the vector of cluster centers, $\mathbf{v}_y = (v_{y1}, \ldots, v_{yM})$, $|| \ ||_A$ is the A-norm which is positive-definite $(M \times M)$ weight matrix, and if A is the identity matrix then it becomes the Euclidean norm.

The FCM tries to minimize J_m by iteratively updating the partiton matrix using the following equations:

$$\mathbf{v}_y = \frac{\sum_{i=1}^{N}(\mu_{yi})^m \mathbf{x}_i}{\sum_{i=1}^{N}(\mu_{yi})^m} \tag{27}$$

where $1 \leq y \leq c$.

$$\mu_{yi} = \frac{1}{\sum_{z=1}^{c}(d_{yi}/d_{zi})^{2/(m-1)}} \tag{28}$$

where $1 \leq i \leq N$, and $1 \leq y \leq c$.

Ten subsets of the cancer data set were randomly selected, each consists of 22 tumor samples and 100 genes, to test the proposed method and compare its results with those obtained by the k-NN and fuzzy k-NN algorithms. The leave-one-out method was used to evaluate the classification performances of the k-NN (KNN), fuzzy k-KNN (FKNN), and the optimally weighted fuzzy k-NN (OWFKNN) classifiers. The numbers of nearest neighbors for the classification were: $k=$ 5, 10, and 15. The weighting exponent m expressed in (25) was taken

Table 1. Average classification results (%) obtained by KNN, FKNN, and OWFKNN

k	KNN	FKNN	OWFKNN
5	89.5	91.4	92.9
10	90.5	92.4	94.4
15	91.9	94.8	97.7

to be 2, and for the FKNN, the parameter p defined in (2) was also 2. The fuzzy prototypes obtained from the FCM were used as the mean values for calculating the covariances which were included in the computation of the kriging system.

For $k= 5$, the total average percentage of classification accuracy for the KNN, FKNN, and OWFKNN are 89.5%, 91.4%, and 92.9% respectively. For $k= 10$, the total average percentage of classification accuracy for the KNN, FKNN, and OWFKNN are 90.5%, 92.4%, and 94.4% respectively. For $k= 15$, the total average percentage of classification accuracy for the KNN, FKNN, and OWFKNN are 91.9%, 94.8%, and 97.7% respectively. It can be seen that the OWFKNN outperformed the other two classifiers in all test cases. The results are shown in Table 1 and can be seen that the classification results for all algorithms become better when k is increased. The performance of OWFKNN was particularly improved when more nearest neighbor samples were considered in the sense of statistical correlation. On the computational aspect of each algorithm, the KNN is the simplest and fastest method, whereas the OWFKNN requires the most computational effort which is due to the computations of fuzzy prototypes, covariance matrix, and kriging system of equations.

4   Conclusions

We have presented an optimal fuzzy k-NN algorithm based on the concept of kriging which tries to determine the weights of the labeled vectors in such a way that the error variance is minimized and subjected to unbiasedness. In addition to the optimal choice of the weighting parameters used to infer the fuzzy class memberships of an unknown sample, this optimally weighted fuzzy k-NN algorithm can particularly be useful for classifying data which are spatially correlated; whereas conventional k-NN, fuzzy k-NN, and other extended versions [3] of k-NN algorithms do not handle this type of problem. The proposed classifier was tested with microarray breast cancer data and found to be superior to both k-NN and conventional fuzzy k-NN algorithms.

References

1. Ablavsky, V., and Stevens, M.R.: Automatic feature selection with applications to script identification of degraded documents, Proc. 7th Int. Conf. Document Analysis and Recognition (ICDAR 2003), **2** (2003) 750-754.
2. Bezdek, J.C. (1981) *Pattern Recognition with Fuzzy Objective Function Algorithms*. Plenum Press, New York.

3. Baoli, L., Qin, L. and Shiwen Y.: An adaptive k-nearest neighbor text categorization strategy. ACM Trans. Asian Language Information Processing, **3** (2004) 215-226.
4. Brazma, A. and Vilo J.: Gene expression data analysis. Federation of European Biochemical Societies Letters, **480** (2000) 17-24.
5. Cover, T.M. and Hart, P.E.: Nearest neighbor pattern classification. IEEE Trans. Information Theory, **13** (1967) 21-27.
6. Duda, R. and Hart, P.: *Pattern Classification and Scene Analysis*. John Wiley & Sons, New York, 1973.
7. Ginneken, B.V., and Loog, M: Pixel position regression - application to medical image segmentation. Proc. 17th International Conference on Pattern Recognition (ICPR'04) **3** (2004) 718-721.
8. Guo, G., Wang, H., Bell, D., Bi, Y., and Greer, K: An kNN model-based approach and its application in text categorization. Lecture Notes in Computer Science, **2945** (2004) 559-570.
9. Hedenfalk, I., Duggan, D., Chen, Y., Radmacher, M., Bittner, M., Simon, R., Meltzer, P., Gusterson, B., Esteller, M., Kallioniemi, O., Wilfond, B., Borg, A. and Trent, J.: Gene-expression profiles in hereditary breast cancer. The New England Journal of Medicine, **344** (2001) 539-548.
10. Isaaks, E.H. and Srivastava, R.M.: *An Introduction to Applied Geostatistics*. Oxford University Press, New York, 1989.
11. Journel, A.G. and Huibregts, C.J,: *Mining Geostatistics*. Academic Press, London, 1978.
12. Journel, A.G. and Rao, S.E.: Deriving conditional distribution from ordinary kriging. Stanford Center for Reservoir Forcasting, Stanford University Report No. 29 (1996) 25 p.
13. Keller, J.M., Gray, M.R. and Givens, J.A.: A fuzzy k-nearest neighbor algorithm. IEEE Trans. Systems, Man and Cybernetics, **15** (1985) 580-585.
14. Nguyen, D.V. and Rocke, D.M.: Tumor classification by partial least squares using microarray gene expression data. Bioinformatics, **18** (2002) 39-50.
15. Paliwal, K.K. and Rao, P.V.S.: Application of k-nearest-neighbor decision rule in vowel recognition. IEEE Trans. Pattern Analysis and Machine Intelligence, **5** (1983) 229-231.
16. Tokola, T., Pitknen, J., Partinen, S., and Muinonen, E.: Point accuracy of a nonparametric method in estimation of forest characteristics with different satellite materials. International Journal of Remote Sensing, **17** (1996) 2333-2351.
17. Troyanskaya,O., Cantor,M., Sherlock,G., Brown,P., Hastie,T., Tibshirani,R., Bostein,D. and Altman,R.B.: Missing value estimation methods for DNA microarrays. Bioinformatics, **17** (2001) 520-525.
18. Zhou, X., Liu, K.Y. and Wong, S.T.C.: Cancer classification and prediction using logistic regression with Bayesian gene selection. J. Biomedical Informatics, **37** (2004) 249-259.

A Tabu Search Based Method for Minimum Sum of Squares Clustering

Yongguo Liu[1,2], Libin Wang[1], and Kefei Chen[1]

[1] Department of Computer Science and Engineering, Shanghai Jiaotong University,
Shanghai 200030, P. R. China
{liu-yg, wang-lb, chen-kf}@cs.sjtu.edu.cn
[2] State Key Laboratory for Novel Software Technology, Nanjing University,
Nanjing 210093, P. R. China

Abstract. In this article, the metaheuristic algorithm, tabu search, is proposed to deal with the clustering problem under the criterion of minimum sum of squares clustering. The presented method integrates four moving operations and mutation operation into tabu search. Its superiority over local search clustering algorithms and another tabu clustering approach is extensively demonstrated for artificial and real life data sets.

1 Introduction

The clustering problem is a fundamental problem that frequently arises in a great variety of fields such as pattern recognition, machine learning, data mining, and statistics. In clustering analysis, objects to be studied are generally denoted by points in m-dimensional Euclidean space and the objective is to group these objects into different clusters such that a certain similarity measure is optimized. In this paper, we focus on the minimum sum of squares clustering problem stated as follows: Given N objects in R^m, allocate each object to one of K clusters such that the sum of squared Euclidean distances between each object and the center of its belonging cluster for every such allocated object is minimized. This problem can be mathematically described as follows:

$$\min_{W,C} J(W,C) = \sum_{i=1}^{N} \sum_{j=1}^{K} w_{ij} \parallel \mathbf{x}_i - \mathbf{c}_j \parallel^2 \tag{1}$$

where $\sum_{j=1}^{K} w_{ij} = 1$, $i = 1, \ldots, N$. If object $\mathbf{x}_i$ is allocated to cluster C_j whose cluster center is $\mathbf{c}_j$, then w_{ij} is equal to 1; otherwise w_{ij} is equal to 0. In Equation 1, N denotes the number of objects, K denotes the number of clusters, $X = \{\mathbf{x}_1, \ldots, \mathbf{x}_N\}$ denotes the set of N objects of m attributes,

S. Singh et al. (Eds.): ICAPR 2005, LNCS 3686, pp. 248–256, 2005.

$C = \{C_1, \ldots, C_K\}$ denotes the set of K clusters, and $W = [w_{ij}]$ denotes the $N \times K$ $0-1$ matrix. Cluster center $\mathbf{c}_j$ is calculated as follows:

$$\mathbf{c}_j = \frac{1}{n_j} \sum_{\mathbf{x}_i \in C_j} \mathbf{x}_i \qquad (2)$$

where n_j denotes the number of objects belonging to cluster C_j. It is known that the clustering problem is a nonconvex program which possesses many locally optimal values, resulting that its solution often falls into these traps. Many clustering approaches have been developed [1]. Among them, K-means algorithm is a very important one as it is a typical iterative hill-climbing method. This method is proven to fail to converge to a local minimum under certain conditions [2]. In [3], another iterative method, a breadth-first search technique, for the clustering problem is reported. According to this algorithm, an alternative approach based on a depth-first search is proposed [4]. In [5], two algorithms based on hybrid alternating searching strategies are presented to overcome the drawbacks of either a breadth-first search or a depth-first search in the clustering problem. It has been proved that these four algorithms called Moving 1, Moving 2, Moving 3, and Moving 4 in [6], respectively, own stronger convergence states than K-means algorithm. They have the same time complexity as K-means algorithm [5]. Moreover, these moving methods can get much better clustering results sooner than K-means algorithm [5,6]. In [7], the genetic algorithm is applied to deal with the clustering problem. But this algorithm needs up to 10000 iterations to attain the correct result. Even so, it cannot reach the best results in many cases. It is seen that only adding the number of iterations is not a good way. Tabu search is a metaheuristic technique that guides the local heuristic search procedures to explore the solution space beyond the local optimality [8], which has been successfully applied to image processing, pattern recognition, etc. In [9], tabu search is proposed to deal with the clustering problem, called TABU-Clustering in this paper. It encodes the solution as a string. After the specified number of iterations, the best solution obtained is viewed as the clustering result. To efficiently use tabu search in various kinds of applications, researchers combine it with the local descent approach. In [10], Nelder–Mead simplex algorithm, a classical local descent algorithm, and tabu search are hybridized to solve the global optimization problem of multiminima functions. Since moving methods with better performance than K-means algorithm are simple and computationally attractive, we propose to combine four moving methods with tabu search, called MT-Clustering including MT-Clustering 1, MT-Clustering 2, MT-Clustering 3, and MT-Clustering 4, to explore the proper clustering result. Moreover, mutation operation, a genetic operator used in the genetic algorithm, is adopted to establish the neighborhood of tabu search in this paper.

The remaining part of this paper is organized as follows: In Section 2, MT-Clustering algorithm and its components are extensively described and analyzed. In Section 3, performance comparisons between our algorithm and others are conducted on different data sets. Finally, the conclusions are drawn in Section 4.

2 MT-Clustering Algorithm

Tabu search is a metaheuristic method that guides the local heuristic search procedures to explore the solution space beyond the local optimality. It is introduced by
Fred Glover specifically for combinatorial problems. The basic elements of tabu
search are described in the following [9]. The detail introduction to tabu search can be
found in [8].

♦ **Configuration** denotes an assignment of values to variables. That is, it is a solution to the optimization problem to be solved.
♦ **Move** denotes a specific procedure for getting a trial solution that is feasible to
  the optimization problem and related to the current configuration. That is, a new
  solution (a neighbor) can be generated by some perturbation on the current configuration.
♦ **Neighborhood** denotes the set of all neighbors, which are the "adjacent solutions" that can be reached from the current configuration. It also includes
  neighbors that do not satisfy the feasible conditions defined.
♦ **Candidate subset** denotes a subset of the neighborhood. It is to be examined
  instead of the whole neighborhood, especially for huge problems where the
  neighborhood include many elements.
♦ **Tabu restrictions** are constraints that prevent the chosen moves to be reversed
  or repeated, which play a memory role for the search by making the forbidden
  moves as tabu. The tabu moves are stored in the tabu list.
♦ **Aspiration criteria** denote rules that determine when the tabu restrictions can be
  overridden, thus removing a tabu classification otherwise applied to a move. If a
  certain move is forbidden by the tabu restrictions then the aspiration criteria,
  when satisfied, can make this move allowable.

```
Begin
    set parameters and the current solution X_c at random
    while (not termination-condition) do
        perform moving operation
        use mutation operation to generate the neighborhood
        select the proper neighbor of X_c as the new current solution
        update the tabu list and the best solution X_b
    end
    output solution X_b
end
```

Fig. 1. General description of MT-Clustering algorithm

Figure 1 gives the general description of MT-Clustering. It is seen that its most
procedures observe the architecture of tabu search. Based on the structure of tabu
search, MT-Clustering algorithm gathers the global optimization property of tabu
search and the local search capability of four moving approaches together. Besides
main procedures of tabu search, MT-Clustering algorithm integrates two operations:
moving operation and mutation operation. Moving operation uses one of four moving
methods to modulate the distribution of objects belonging to different clusters and to

improve the similarity between objects and their centroids. Meanwhile, mutation operation provides a good neighborhood for tabu search to avoid getting stuck in local optima and to find the proper result. In this article, we let the length of solutions equal to the size of objects the same as that in [9], which is suitable for computing the objective function value and comparing with the TABU-Clustering. That is, the value of the ith element of the string denotes the cluster number assigned to the ith element, where $i = 1, \ldots, N$. For instance, a clustering partition, (x_1, x_3) (x_2, x_5, x_8) (x_4, x_6, x_7), can be represented by $(1\ 2\ 1\ 3\ 2\ 3\ 3\ 2)$.

2.1 Moving Operation

Corresponding to four moving methods, we get four different moving operations. Firstly, we describe the change in the sum of squared Euclidean distances and cluster centers after an object moves from its belonging cluster to another one. For cluster C_j, its sum of squared Euclidean distances is given by:

$$J_j = \sum_{x_i \in C_j} \| \mathbf{x}_i - \mathbf{c}_j \|^2 \tag{3}$$

If object $\mathbf{x}_i$ belonging to cluster C_j is reassigned to cluster C_k, cluster centers are moved accordingly, J_j will decrease by ΔJ_{ij}

$$\Delta J_{ij} = n_j \| \mathbf{x}_i - \mathbf{c}_j \|^2 / (n_j - 1) \tag{4}$$

and J_k will increase by ΔJ_{ik}

$$\Delta J_{ik} = n_k \| \mathbf{x}_i - \mathbf{c}_k \|^2 / (n_k + 1) \tag{5}$$

After such a move, the new total sum of squared Euclidean distances is updated by:

$$J' = J - \Delta J_{ij} + \Delta J_{ik} \tag{6}$$

and new cluster centers of C_j and C_k will become:

$$\mathbf{c}'_j = (n_j \mathbf{c}_j - \mathbf{x}_i) / (n_j - 1) \tag{7}$$

$$\mathbf{c}'_k = (n_k \mathbf{c}_k + \mathbf{x}_i) / (n_k + 1) \tag{8}$$

Based on above descriptions, four moving operations are described as follows: Given solution $\mathbf{X}_c = x_1, \ldots, x_i, \ldots, x_N$, $i = 1, \ldots, N$, where N is the number of objects.

Moving Operation 1:

Object $\mathbf{x}_i$ belonging to cluster C_j is reassigned to cluster C_k, $k = 1, \ldots, K$, iff

$$\min(\Delta J_{ik}) < \Delta J_{ij}, \; j = 1, \ldots, K, \text{ and } j \neq k \tag{9}$$

Then cluster centers $\mathbf{c}_j$ and $\mathbf{c}_k$ are updated based on Equations 7 and 8, respectively.

Moving Operation 2:

Object $\mathbf{x}_i$ belonging to cluster C_j is reassigned to cluster C_k, $k = 1, \ldots, K$, iff

$$\Delta J_{ik} < \Delta J_{ij}, \; j = 1, \ldots, K, \text{ and } j \neq k \tag{10}$$

Then cluster centers $\mathbf{c}_j$ and $\mathbf{c}_k$ are updated based on Equations 7 and 8, respectively.

Moving Operation 3:

Object $\mathbf{x}_i$ belonging to cluster C_j is reassigned to cluster C_k, $k = 1, \ldots, K$, if the number of iterations is odd then run moving operation 1, else run moving operation 2.

Moving Operation 4:

Object $\mathbf{x}_i$ belonging to cluster C_j is reassigned to cluster C_k, $k = 1, \ldots, K$, if the number of iterations is even then run moving operation 1, else run moving operation 2. After moving operation, the modified solution is viewed as the current solution.

2.2 Neighborhood Creation

In this paper, mutation operation, a genetic operator in the genetic algorithm, is adopted to establish the neighborhood of tabu search. It is stated as follows: Given the current solution $\mathbf{X}_c = x_1, \ldots, x_i, \ldots, x_N$, $x_i = j$, $j = 1, \ldots, K$, the mutation probability p_m, and the size of the neighborhood N_t, for $i = 1, \ldots, N$, draw a random number $p_i \sim u(0,1)$. If $p_i < p_m$, then $x_i^t = x_i$, $t = 1, \ldots, N_t$; otherwise $x_i^t = k, k = 1, \ldots, K, k \neq j$. Here, the mutation probability is used to moderate the shake-up on the current solution and create a neighbor. The higher the value of this parameter, the less shake-up is allowed and, in consequence, the more similar the neighbor to the current solution, and vice versa. Determination of the mutation probability is the process of seeking the balance between exploration and exploitation. In this article, we use variable mutation operation in order to keep a good tradeoff between exploration and exploitation in the neighborhood. That is, the mutation probability reduces with the increase of the number of iterations. In [11], extensive researches have been conducted and the best parameter settings for the genetic algorithm are given. Among them, the mutation probability is recommended to be in the range [0.005, 0.01]. We choose the terminal probability to be 0.005. Then different original probabilities are compared as shown in Figure 2.

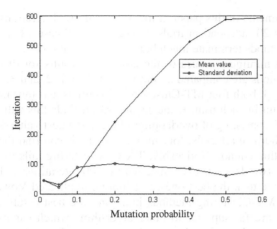

Fig. 2. Comparison of different original mutation probabilities

It is found that, the mean value and the standard deviation of iterations where the clustering result is obtained vary with the original mutation probability. When this value is equal to 0.05, the best performance is achieved. So, we choose the original one to be 0.05.

3 Experimental Evaluation

Firstly, we analyze the time complexities of algorithms employed in this paper. K-means algorithm and four moving methods have the same time complexity $O(KmN)$. The time complexity of TABU-Clustering is $O(GN_tmN)$, where G is the number of iterations. Four moving operations take the same time $O(KmN)$. Neighborhood creation takes $O(N_tmN)$ time. Hence, the time complexity of MT-Clustering algorithm is $O(N_tGmN + KGmN)$. In most cases, K is a small number, then the time complexity is $O(GN_tmN)$, which is the same as that of TABU-Clustering.

Performance comparisons between MT-Clustering algorithm and other techniques are conducted in Matlab on an Intel Pentium III processor running at 800MHz with 128MB real memory. Five data sets are considered for the purpose of conducting the experiments, two artificial data sets (Data_2_3, Data_2_5) and three real life data sets (Vowel, German Towns, and British Towns). Data_2_3 is a two dimensional data set having 200 nonoverlapping objects where the number of clusters is 3. Data_2_5 is a two dimensional data set having 250 overlapping objects where the number of clusters is five. Vowel consists of 871 Indian Telugu vowel sounds having three features and six classes [12]. Other two classical data sets, German Towns [13] and British Towns [14], are considered for different number of clusters the same as those of [5]. In MT-Clustering algorithm, both the size of the neighborhood N_t and the size of the tabu list T are chosen to be 20, which are recommended in [9] by computer simulations. For all experiments in this paper, a maximum of 1000 iterations is fixed. Each experiment includes 20 independent trials. However, in all cases, K-means algorithm and four moving methods terminate much before 1000 iterations.

The average and minimum values of the clustering results for different data sets obtained by ten methods are compared as shown in Tables 1, 2, and 3, respectively. In Table 1, for Data_2_3, both four MT-Clustering algorithms and four moving methods find the optimal result in each trial; K-means algorithm finds this value in only some cases. For Data_2_5 consisting of overlapping objects, the best value is found by four MT-Clustering methods in all trials; four moving methods can find the optimal value but K-means algorithm cannot. Noticeably, TABU-Clustering fails to attain the optimal values for these two data sets even once within 1000 iterations and its best values obtained are far worse than the best ones. For more complicated Vowel, except MT-Clustering 1, other MT-Clustering methods both find the best result, and these MT-Clustering methods are far superior to other algorithms which cannot find the best value in all trials. Here, it is seen that moving operation 3 and moving operation 4 with hybrid moving strategies help tabu search to obtain better results than moving operation 1 and moving operation 2 with single moving strategy.

Table 1. Results of different clustering algorithms for Data_2_3, Data_2_5, and Vowel

Algorithm	Data_2_3 Avg (min)	Data_2_5 Avg (min)	Vowel Avg (min)
K-means	2073.15(827.08)	488.71(488.06)	32606098.87(30706183.98)
Moving 1	827.08(827.08)	488.56(488.02)	32121416.60(30742706.05)
Moving 2	827.08(827.08)	488.43(488.02)	31322897.26(30690653.07)
Moving 3	827.08(827.08)	488.49(488.02)	31716824.69(31370642.87)
Moving 4	827.08(827.08)	488.36(488.02)	31843867.22(30742706.05)
TABU-Clustering	9524.01(8410.43)	2582.60(2485.43)	247501537.17(243149977.91)
MT-Clustering 1	827.08(827.08)	488.02(488.02)	30726507.58(30723649.03)
MT-Clustering 2	827.08(827.08)	488.02(488.02)	30842773.00(30686238.38)
MT-Clustering 3	827.08(827.08)	488.02(488.02)	30710294.63(30686238.38)
MT-Clustering 4	827.08(827.08)	488.02(488.02)	30719492.19(30686238.38)

Table 2 shows the results obtained by ten methods for German Towns of different number of clusters. All MT-Clustering methods can both find the optimal results. But for other approaches, with the increase of the number of clusters, their ability to attain the best value greatly degrades. When $K \geq 8$, K-means algorithm and moving methods except Moving 4 cannot find the best values in each trial.

In Table 3, it is seen that the proposed method is still superior to other algorithms. In all cases, four MT-Clustering methods can obtain better results than others. However, as other approaches, with the increase of the number of clusters, their ability to attain the best values also degrades. MT-Clustering 2 is the best one of all methods.

For all data sets, the performance of the proposed algorithm is obviously superior to that of other approaches. It is surprising that the performance of TABU-Clustering with the same time complexity as MT-Clustering algorithm is found to be the poorest in all cases. Meanwhile, we also find that it can still obtain improved results if more generations are executed. But, as aforementioned, we do not think that it is a good way to attain the best result by only adding the number of iterations.

Table 2. Results of different clustering algorithms for German Towns

Algorithm	Four Clusters Avg (min)	Six Clusters Avg (min)	Eight Clusters Avg (min)	Ten Clusters Avg (min)
K-means	59333.36(49600.59)	34250.86(30535.39)	26503.21(21744.50)	20627.46(17160.61)
Moving 1	50217.47(49600.59)	32330.11(30535.39)	23192.85(21507.15)	17878.08(17187.99)
Moving 2	50557.34(49600.59)	32636.08(30535.39)	23512.16(21499.99)	17817.20(16711.94)
Moving 3	49600.59(49600.59)	32274.68(30535.39)	23120.22(21776.33)	17926.72(17060.13)
Moving 4	52121.22(49600.59)	32388.20(30535.39)	22922.68(21483.02)	17814.06(16505.67)
TABU-Clustering	71837.59(64312.22)	55019.44(46476.18)	47343.00(44578.01)	40400.33(35086.87)
MT-Clustering 1	49600.59(49600.59)	31433.64(30535.39)	21496.82(21483.02)	16426.59(16307.96)
MT-Clustering 2	49600.59(49600.59)	31134.22(30535.39)	21498.02(21483.02)	16396.93(16307.96)
MT-Clustering 3	49600.59(49600.59)	31502.50(30535.39)	21497.18(21483.02)	16387.59(16307.96)
MT-Clustering 4	49600.59(49600.59)	31870.77(30535.39)	21496.69(21483.02)	16381.11(16307.96)

Table 3. Results of different clustering algorithms for British Towns

Algorithm	Four Clusters Avg (min)	Six Clusters Avg (min)	Eight Clusters Avg (min)	Ten Clusters Avg (min)
K-means	193.41(180.91)	156.88(147.85)	128.18(117.67)	113.38(100.34)
Moving 1	187.91(180.91)	148.84(144.61)	128.23(121.33)	115.06(105.35)
Moving 2	182.59(180.91)	147.11(144.32)	123.23(115.04)	108.39(102.94)
Moving 3	183.71(180.91)	147.68(144.32)	126.98(116.09)	110.09(102.94)
Moving 4	182.31(180.91)	147.03(144.32)	125.89(117.22)	107.42(102.94)
TABU-Clustering	205.23(193.58)	170.79(163.05)	150.49(144.48)	135.36(129.52)
MT-Clustering 1	180.91(180.91)	144.32(144.32)	119.30(116.00)	107.42(102.94)
MT-Clustering 2	180.91(180.91)	144.03(141.46)	116.42(113.50)	97.92(92.73)
MT-Clustering 3	180.91(180.91)	144.32(144.32)	116.55(116.00)	100.64(93.06)
MT-Clustering 4	180.91(180.91)	144.03(141.46)	118.29(116.00)	100.90(93.06)

4 Conclusions

In this paper, a tabu search based method for the minimum sum of squares clustering problem, called MT-Clustering, is proposed. Based on the structure of tabu search, MT-Clustering algorithm gathers the global optimization property of tabu search and the local search capability of four moving methods together. Moreover, we adopt variable mutation operation to establish the neighborhood and determine a proper initial mutation probability by computer simulations. Performance comparisons between MT-Clustering algorithm and other algorithms are conducted on artificial data sets and real life data sets. As a result, our approach and TABU-Clustering own the same time complexity, but our approach obtains much better performance for experimental data sets than TABU-Clustering and other local search clustering algorithms. Meanwhile, how to improve the stability of the proposed algorithm in complicate cases will be the subject of future publications.

Acknowledgements

This research was partially supported by National Natural Science Foundation of China (#90104005, #60273049) and State Key Laboratory for Novel Software Technology at Nanjing University.

References

1. Jain, A.K., Dubes, R.: Algorithms for clustering data. Prentice-Hall, New Jersey (1988)
2. Selim, S.Z., Ismail, M.A.: K-means-type algorithm: generalized convergence theorem and characterization of local optimality. IEEE Transactions on Pattern Analysis and Machine Intelligence. 6 (1984) 81-87
3. Duda, R.O., Hart, P.E.: Pattern classification and scene analysis. Wiley, New York (1972)
4. Ismail, M.A., Selim, S.Z., Arora, S.K.: Efficient clustering of multidimensional data. In: Proceedings of 1984 IEEE International Confference on System, Man and Cybernetics, Halifax. (1984) 120-123
5. Ismail, M.A., Kamel, M.S.: Multidimensional data clustering utilizing hybrid search strategies. Pattern Recognition. 22 (1989) 75-89
6. Zhang, Q.W., Boyle, R.D.: A new clustering algorithm with multiple runs of iterative procedures. Pattern Recognition. 24 (1991) 835-848
7. Murthy, C.A., Chowdhury, N.: In search of optimal clusters using genetic algorithms. Pattern Recognition Letters. 17 (1996) 825-832
8. Glover, F., Laguna, M.: Tabu search. Kluwer Academic Publishers, Boston (1997).
9. Al-sultan, K.S.: A tabu search approach to the clustering problem. Pattern Recognition. 28 (1995) 1443-1451
10. Chelouah, R., Siarry, P.: A hybrid method combining continuous tabu search and Nelder–Mead simplex algorithms for the global optimization of multiminima functions. European Journal of Operational Research. 161 (2005) 636-654
11. Schaffer, J.D., Caruana, R.A., Eshelman, L.J., Das, R.: A study of control parameters for genetic algorithms. In: Proceedings of the 3rd International Conference on Genetic Algorithms. Morgan Kaufmann (1989) 51-60
12. Pal, S.K., Majumder, D.D.: Fuzzy sets and decision making approaches in vowel and speaker recognition. IEEE Transactions on System, Man and Cybernetics. SMC-7 (1977) 625-629
13. Spath, H.: Cluster analysis algorithms. Wiley, Chichester (1980)
14. Chien, Y.T.: Interactive Pattern Recognition. Marcel-Dekker, New York (1978)

Approximation of Digital Circles
by Regular Polygons

Partha Bhowmick[1] and Bhargab B. Bhattacharya[2]

[1] Computer Science and Technology Department,
Bengal Engineering and Science University, Shibpur, Howrah, India
partha@becs.ac.in
[2] Center for Soft Computing Research,
Indian Statistical Institute, Kolkata, India
bhargab@isical.ac.in

Abstract. [1] In this paper, we show that an ideal regular (convex) polygon corresponding to a digital circle is possible for some of the digital circles, especially for the ones having smaller radii. For a circle whose ideal regular polygon is not possible, an approximate polygon, tending to the ideal one, is possible, in which the error of approximation can be controlled by the number of vertices of the approximate polygon. These (ideal or approximate) polygonal enclosures of digital circles have several applications in approximate point set pattern matching. We have reported the conditions under which an ideal regular polygon definitely exists corresponding to a digital circle, and the conditions under which the existence of an ideal regular polygon becomes uncertain. Experimental results have been given to exhibit the possibilities of approximation and the tradeoff in terms of error versus approximation.

1 Introduction

Polygonal approximation is a common, useful, and efficient representation of a digital curve in discrete domain [4,9,10,12,13,16,17,19]. The overall objective is to approximate a given digital curve by a polygonal chain satisfying certain optimality criterion (e.g., global approximation error is minimized, or local error is within some predefined threshold). Such approximations reduce significantly the storage requirements and facilitate the extraction and processing of desired features from a given set of digital curves.

A large number of methods have been proposed for approximating a digitized curve into a list of line segments. To cite a few, there is an algorithm by Pikaz and Dinstein [8], meant for optimal polygonal approximation of digital curves, where, given a value for the maximal allowed distance between the approximation and the curve, the algorithm finds an approximation with minimal number of vertices. Perez and Vidal [7] proposed another algorithm in which the number

---

[1] This work is funded in part by the CBIR Project, Computer Science and Technology Department, Bengal Engg. & Sc. University.

S. Singh et al. (Eds.): ICAPR 2005, LNCS 3686, pp. 257–267, 2005.

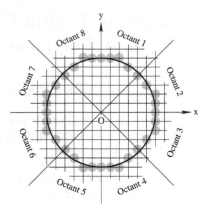

Fig. 1. A real circle, $\mathcal{C}^R(\mathsf{O}, 6)$, and the corresponding digital circle, $\mathcal{C}^Z(\mathsf{O}, 6)$

of segments is fixed a priori, and the error criterion is the sum of the square of Euclidean distance from each point of the contour to its orthogonal projection onto the corresponding line segment. Salotti [11] has proposed some improvements to make the algorithm more efficient, in particular for a large number of segments. Nevertheless, the engrossing problem of polygonal approximation of digital curves has been tackled by many other different classes of methodologies, such as Hough Transform [5], Hopfield Neural Networks [3], Genetic Algorithms [15], Particle Swarm Optimization [18], etc. Reviews on different polygonal approximation methods are available in several literatures [10].

Apart from irregularly shaped curves, the common geometric primitives, e.g. lines, polygons, circles, etc., have found wide applications in various fields, and their weird and challenging nature in the discrete domain have drawn immense research interest in recent years. The construction, properties, and characterization of digital circles make a very interesting area of research, and there exist several works on digital circles and related problems [6,14]. Described in this paper is a novel work that explores and exhibits the subtleties and various possibilities on approximation of a digital circle by a regular (convex) polygon.

It may be mentioned here that approximation of a given digital circle by a suitable regular polygon has significant applications for *approximate* matching of point sets (e.g. fingerprint matching) on two-dimensional plane, using circular range query [1]. The process becomes faster and efficient if we can find a suitable regular polygon in R^2 (meant for polygonal range query) corresponding to the given digital circle, such that all the grid points lying on and inside the digital circle should lie on and inside the polygon, and vice versa.

2   Approximation of a Digital Circle by a Real Polygon

A circle $\mathcal{C}^R(q, r)$ lying in the real plane R^2, $q \in \mathsf{R}^2$ and $r \in \mathsf{R}^+$ being its center and radius respectively, can be realized in the discrete domain Z^2 by a digital circle

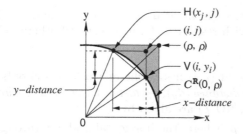

Fig. 2. *x-distance* and *y-distance* of a grid point (i, j) from $C^R(O, \rho)$

$C^Z(\alpha, \rho)$ having center α and radius ρ, where, α is the nearest grid point in Z^2 corresponding to q in R^2, and ρ is the nearest integer corresponding to r. [2] Now, for the digital circle $C^Z(\alpha, \rho)$, if we consider the center α as the origin (point of reference) of the local coordinate system in Z^2, then it may be observed that the set of grid points, enumerated w.r.t. α, representing the circle $C^Z(\alpha, \rho)$, will be always independent of α and will be depending only on its radius ρ. Hence, we can draw the digital circle $C^Z(\alpha, \rho)$ centered at any point $\alpha \in Z^2$, provided $C^Z(O, \rho)$ is known, where, $O = (0,0)$.

2.1   Generation of a Digital Circle

The conversion of a digital circle $C^Z(O, \rho)$ from the corresponding real circle $C^R(O, \rho)$ is done using the property of 8-axes symmetry (Fig. 1) of digital circles [2]. In order to obtain the complete circle $C^Z(O, \rho)$, therefore, generation of the first octant, $C^{Z,I}(O, \rho)$, suffices.

While generating $C^{Z,I}(O, \rho)$, decision is taken to select between east pixel (E: $(i+1, j)$) or south-east pixel (SE: $(i+1, j-1)$), standing at the current pixel (i, j), depending on which one between E and SE is nearer to the point of intersection of the next ordinate line (i.e., $x = i + 1$) with the real circle $C^R(O, \rho)$. In the case of ties, since any one between E and SE may be selected, we select SE. It is interesting to note that such a tie is possible only if there is any computation error, the reason being as follows.

Let a tie occurs when the real circle $C^R(O, \rho)$ has $(i, j+\frac{1}{2})$ as the corresponding point of intersection in the first octant with the vertical grid line $x = i$. That is, $(i, j) \in Z^2$ lies on $C^{Z,I}(O, \rho)$ according to the tie-resolution policy. Since the point $(i, j + \frac{1}{2})$ lies on $C^R(O, \rho)$, we have $i^2 + (j + \frac{1}{2})^2 = \rho^2$, or, $\rho^2 - (i^2 + j^2 + j) = \frac{1}{4}$, which is impossible, since $\rho^2 \in Z$, and $(i^2 + j^2 + j) \in Z$. Hence we observe the following fact on tie:

Fact 1. *A tie occurs only if there is a computation error.*

Now, in relevance with the regular polygonal enclosure of $C^Z(O, \rho)$, we make the following definitions, whose underlying significances are apparent in Fig. 2.

---

[2] For simplicity of notations, $C^Z(\alpha, \rho)$ is also used in this paper to denote the set of grid points (pixels) constituting the digital circle with center at α and radius ρ in an appropriate context.

Definition 1. A point (x, y) (in Z^2 or in R^2, as the case may be) is said to be *lying in the first octant* (with respect to O, unless stated otherwise) if and only if $0 \leq x \leq y$.

Definition 2. If a grid point (i, j) lies in the first octant and outside $C^R(O, \rho)$, then its *x-distance* (d_x), *y-distance* (d_y), *xy-distance* (isothetic distance, $d_\perp$), and *radial distance* (d_r) from $C^R(O, \rho)$ are given by:

$$d_x = \begin{cases} i - x_j, & \text{if } j \leq \rho, \text{ where, the horizontal grid line through } (i, j) \\ & \text{intersects } C^R(O, \rho) \text{ in the first quadrant at } (x_j, j); \\ \infty, & \text{otherwise.} \end{cases}$$

$$d_y = \begin{cases} j - y_i, & \text{if } i \leq \rho, \text{ where, the vertical grid line through } (i, j) \\ & \text{intersects } C^R(O, \rho) \text{ in the first quadrant at } (i, y_i); \\ \infty, & \text{otherwise.} \end{cases}$$

$$d_\perp = \min (d_x, d_y).$$
$$d_r = \sqrt{i^2 + j^2} - \rho.$$

It may be noted that Defs. 1 and 2 can be easily extended for a point lying in one of the remaining seven octants. It may be also noted that, if a grid point (i, j) lies in any one of octants 1, 8, 4, and 5, then $d_\perp = d_y$, and if it lies any one of the other four octants, namely octants 2, 7, 3, and 6, then $d_\perp = d_x$.

Hence, from the principle of construction of $C^Z(O, \rho)$, if any grid point (i, j) lies on $C^Z(O, \rho)$ but outside $C^R(O, \rho)$, then it must have isothetic distance strictly less than $\frac{1}{2}$ grid unit from $C^R(O, \rho)$. This, in turn, ensures that any grid point, lying outside $C^R(O, \rho)$ in any octant with isothetic distance not less than $\frac{1}{2}$ grid unit from $C^R(O, \rho)$, does not lie on $C^Z(O, \rho)$. And this is true as well if the center of the circle is at any grid point α instead of O. This analysis is captured in the following fact:

Fact 2. *Any grid point, not lying on $C^Z(\alpha, \rho)$ and lying outside $C^R(\alpha, \rho)$, must have xy-distance at least $\frac{1}{2}$ grid unit from $C^R(\alpha, \rho)$.*

2.2   Enclosing Circle

Let $\mu(i_\mu, j_\mu)$ be a grid point, lying on $C^{Z,I}(O, \rho)$, such that μ has maximum y-distance from $C^R(O, \rho)$, the corresponding y-distance being δ_μ, where, $0 < \delta_\mu < \frac{1}{2}$ for $\rho \geq 2$ (for $\rho = 1$, $\delta_\mu = 0$). If $(\rho_\mu, \theta_\mu) \in R^2$ be the corresponding polar coordinates of μ, then $\rho_\mu = \sqrt{i_\mu^2 + j_\mu^2} = \rho + \epsilon_\mu$, and $\theta_\mu = \tan^{-1}(j_\mu/i_\mu)$, where, $\epsilon_\mu(< \delta_\mu \sin \theta_\mu)$ is the radial distance of μ from $C^R(O, \rho)$, as shown in Fig. 3.

The next point of curiosity about a digital circle is that, whether μ (the grid point, lying on $C^{Z,I}(O, \rho)$, having maximum y-distance from $C^R(O, \rho)$) is unique or not. To uncover the fact about the uniqueness of μ, let us consider two distinct grid points, μ_1 and μ_2, lying on $C^{Z,I}(O, \rho)$, such that each of them has maximum y-distance, δ_μ, from $C^R(O, \rho)$. Let $\mu_1 = (i_1, j_1)$ and $\mu_2 = (i_2, j_2)$. Since in the first octant, no two grid points lying on $C^{Z,I}(O, \rho)$ can have same abscissa, $i_1 \neq i_2$. Therefore, w.l.g., let $i_1 > i_2$, which implies $j_2 > j_1$, since both μ_1 and μ_2 lie on $C^{Z,I}(O, \rho)$. It may be observed that, since $\mu_1 = (i_1, j_1)$ and $\mu_2 = (i_2, j_2)$ have same y-distance from $C^R(O, \rho)$, $j_1 = j_2$ is not possible. Let the corresponding

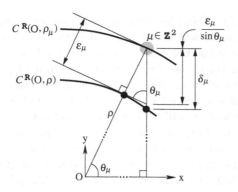

Fig. 3. Isothetic distance, $\delta_\mu \left(0 < \delta_\mu < \frac{1}{2}\right)$, and radial distance, $\epsilon_\mu \left(\frac{1}{\sqrt{2}}\delta_\mu < \epsilon_\mu < \delta_\mu\right)$, for $\mu \in C^{Z,I}(O, \rho),\ \rho \geq 2$

points of intersection of $C^R(O, \rho)$ with the vertical grid lines $x = i_1$ and $x = i_2$ in the first octant be (i_1, y_1) and (i_2, y_2) respectively, where, $y_1, y_2 \in R^+ \setminus Z^+$. Since (i_1, y_1) and (i_2, y_2) lie on $C^R(O, \rho)$, we have:

$i_1^2 + y_1^2 = i_2^2 + y_2^2 = \rho^2$, and $j_1 - y_1 = j_2 - y_2 = \delta_\mu$.

So, $y_2^2 - y_1^2 = i_1^2 - i_2^2$, or, $y_2^2 - y_1^2 \in Z^+$ [since $i_1^2 - i_2^2 \in Z^+$],

or, $(y_1 + y_2)(y_2 - y_1) \in Z^+$, or, $(y_1 + y_2)((j_2 - \delta_\mu) - (j_1 - \delta_\mu)) \in Z^+$,

or, $(y_1 + y_2)$ is rational [since $(j_2 - j_1) \in Z^+$].

Now $y_1^2 = \rho^2 - i_1^2 = a$ (say) is an integer but not a perfect square, since $y_1 \notin Z$. Similarly, $y_2^2 = \rho^2 - i_2^2 = b$ (say) is also a non-square integer. Therefore, $y_1 + y_2 = \sqrt{a} + \sqrt{b}$ would be irrational. The reason is as follows. Let $\sqrt{a} + \sqrt{b}$ be a rational number c, if possible. Then, $a = (c - \sqrt{b})^2 = c^2 + b - 2c\sqrt{b}$. Since the product of any rational number and any irrational number is always irrational, $2c\sqrt{b}$ is irrational. Hence a becomes irrational, which is a contradiction.

Therefore, $y_1 + y_2$ can never be rational, thereby contradicting our assumption that μ is not unique. Hence μ is a unique grid point in the first octant that lies on $C^{Z,I}(O, \rho)$, and has maximum y-distance from $C^R(O, \rho)$. The result obtained, along with its proof (with minor modifications), can be applied equally well for any arbitrary center $\alpha \in Z^2$. Thus we establish the following fact:

Fact 3. *In the set of grid points lying on $C^{Z,I}(\alpha, \rho)$, there is a unique grid point that has maximum y-distance from $C^R(\alpha, \rho)$.*

Now consider the circle $C^R(O, \rho_\mu)$ passing through μ (Fig. 3). Since μ lies in the first octant, we have $0 < i_\mu \leq j_\mu$ (for $\rho \geq 2$, $i_\mu \neq 0$), whence $45^0 < \theta_\mu < 90^0$, which implies $\frac{1}{\sqrt{2}}\delta_\mu < \epsilon_\mu < \delta_\mu$. Let $\nu(i_\nu, j_\nu)$, $\nu \neq \mu$, be any grid point that lies on $C^{Z,I}(O, \rho)$. Let δ_ν be the y-distance of ν from $C^R(O, \rho)$, and (ρ_ν, θ_ν) be the corresponding polar coordinates of ν, such that, $\rho_\nu = \sqrt{i_\nu^2 + j_\nu^2} = \rho + \epsilon_\nu$, and $\theta_\nu = \tan^{-1}(j_\nu/i_\nu)$, where, $\epsilon_\nu = \delta_\nu \sin \theta_\nu$. Therefore, ν would lie on or inside $C^R(O, \rho_\mu)$ if and only if $\epsilon_\nu \leq \epsilon_\mu$. Further, since μ is the grid point having maximum y-distance from $C^R(O, \rho)$, we have $\delta_\nu < \delta_\mu$ ($\delta_\nu \neq \delta_\mu$, since from Fact 3, the grid point in the first octant with maximum y-distance is unique). Hence, if ν be such that in spite of δ_ν being less that δ_μ, θ_ν is so high compared to θ_μ that ϵ_ν is higher than ϵ_μ, then ν will not lie on or inside $\mathcal{C}^R(O, \rho_\mu)$.

On the contrary, if we call the point $\breve{\mu}$ as the grid point lying on $\mathcal{C}^{Z,I}(O, \rho)$, for which $\epsilon_{\breve{\mu}}$ is maximum in $\{\epsilon_\nu \mid \nu$ lies on $\mathcal{C}^{Z,I}(O, \rho)\}$, then it may happen that there exists some grid point $\nu' = (i', j')$ lying in the first octant and outside $\mathcal{C}^{Z,I}(O, \rho)$ for which $\epsilon_{\nu'}$ will not exceed $\epsilon_{\breve{\mu}}$ (although $\delta_{\nu'}$ is not less than $\frac{1}{2}$, as stated in Fact 2), and therefore, ν' will be lying on or inside $\mathcal{C}^R(O, \rho_{\breve{\mu}})$.

Considering the above artefacts, it can be inferred that, for the set of all digital circles, we can not find the canonical solution for the set of *enclosing circles*, for all possible values of ρ, so that for each $\mathcal{C}^Z(O, \rho)$, no grid points but those lying on or inside $\mathcal{C}^Z(O, \rho)$ would lie on or inside $\mathcal{C}^R(O, \rho')$. And no less importantly, we can, for sure, construct the circle $\mathcal{C}^R(O, \rho_\mu)$, passing through μ, where, μ has maximum y-distance from $\mathcal{C}^R(O, \rho)$, so that all grid points, lying on or inside $\mathcal{C}^Z(O, \rho)$, would lie on or inside $\mathcal{C}^R(O, \rho_\mu)$, if and only if μ has maximum radial distance (ϵ_μ) from $\mathcal{C}^R(O, \rho)$ among all grid points lying on $\mathcal{C}^{Z,I}(O, \rho)$. This is stated in Fact 4, where the center of the digital circle, in general, is considered at $\alpha \in \mathbb{Z}^2$, with obvious justification.

Fact 4. *If there exists a grid point μ lying on $\mathcal{C}^{Z,I}(\alpha, \rho)$, such that $\delta_\mu = max \{\delta_\nu \mid \nu$ lies on $\mathcal{C}^{Z,I}(\alpha, \rho)\}$ and $\epsilon_\mu = max \{\epsilon_\nu \mid \nu$ lies on $\mathcal{C}^{Z,I}(\alpha, \rho)\}$, then $\mathcal{C}^R(\alpha, \rho_\mu)$ is the smallest enclosing circle for $\mathcal{C}^Z(\alpha, \rho)$. If no such grid point μ exists, then the existence of an enclosing circle for $\mathcal{C}^Z(\alpha, \rho)$ becomes uncertain.*

2.3   Enclosing Polygon

Definition 3. A regular polygon in $\mathbb{R}^2$ that encloses the digital circle $\mathcal{C}^Z(\alpha, \rho)$, such that each point in $\mathbb{Z}^2$ lying on or inside the digital circle $\mathcal{C}^Z(\alpha, \rho)$ lies on or inside the polygon, and no point in $\mathbb{Z}^2$ lying outside the digital circle $\mathcal{C}^Z(\alpha, \rho)$ lies on or inside the polygon, is defined as a *regular polygonal enclosure*, $\mathcal{E}^R(\mathcal{C}^Z(\alpha, \rho))$, for the digital circle $\mathcal{C}^Z(\alpha, \rho)$.

Let there exists a grid point μ lying on $\mathcal{C}^{Z,I}(O, \rho)$, such that $\delta_\mu = max \{\delta_\nu \mid \nu$ lies on $\mathcal{C}^{Z,I}(O, \rho)\}$ and $\epsilon_\mu = max \{\epsilon_\nu \mid \nu$ lies on $\mathcal{C}^{Z,I}(O, \rho)\}$. Then, from Fact 4, $\mathcal{C}^R(O, \rho_\mu)$ is the smallest enclosing circle for $\mathcal{C}^Z(O, \rho)$. Hence, any grid point ν', lying in the first octant and outside $\mathcal{C}^Z(O, \rho)$, will lie outside $\mathcal{C}^R(O, \rho_\mu)$.

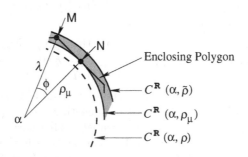

Fig. 4. Generation of the enclosing polygon of $\mathcal{C}^Z(\alpha, \rho)$

Let μ' be a grid point, lying in the first octant and outside $C^R(O, \rho_\mu)$, such that $\epsilon_{\mu'} = \min \{\epsilon_{\nu'} = \rho_{\nu'} - \rho \mid \nu' \in Z^{+2}$ lies in the first octant and outside $C^R(O, \rho_\mu)\}$, where, $(\rho_{\nu'}, \theta_{\nu'})$ and $(\rho_{\mu'}, \theta_{\mu'})$ are the polar coordinates of ν' and μ' respectively. Let $\tilde{\rho} = \max \{\rho' \mid \rho_\mu < \rho' < \rho_{\mu'}\}$, i.e., $\tilde{\rho} = \lim_{h \to 0+} (\rho_{\mu'} - h)$. Therefore, any grid point in the first octant that lies outside $C^R(O, \rho_\mu)$ will also lie outside $C^R(O, \tilde{\rho})$. Furthermore, $C^R(O, \tilde{\rho})$ will be the largest enclosing circle of $C^Z(O, \rho)$ in accordance with our consideration of $\tilde{\rho}$. And this argument holds true as well if we shift the center of the digital circle from O to α.

Thus, all grid points, lying on and inside $C^R(\alpha, \rho_\mu)$, will be lying strictly inside $C^R(\alpha, \tilde{\rho})$; and all grid points, lying strictly outside $C^Z(\alpha, \rho_\mu)$, will be lying strictly outside $C^R(\alpha, \tilde{\rho})$. As a result, there will be no grid point that would lie on or inside $C^R(\alpha, \tilde{\rho})$ and outside $C^R(\alpha, \rho_\mu)$. Hence, we can construct a regular polygon in R^2, centered at α and having n vertices, such that the following conditions are simultaneously satisfied:

(**c1**) each vertex lies on or inside $C^R(\alpha, \tilde{\rho})$;
(**c2**) each edge touches $C^R(\alpha, \rho_\mu)$;
(**c3**) each edge subtends same angle 2ϕ $(= \frac{2\pi}{n}$ rads.) at α.

Let N be the point of contact of one edge of the enclosing polygon with $C^R(\alpha, \rho_\mu)$ (Fig. 4). Let M be one of the two vertices adjacent to N, and λ is the distance of M from α. So, the angle subtended at α by the line segment MN is ϕ, which implies $\cos \phi = \rho_\mu / \lambda$. Since M should lie inside the annular region between $C^R(\alpha, \rho_\mu)$ and $C^R(\alpha, \tilde{\rho})$, including the circumference of $C^R(\alpha, \tilde{\rho})$, λ should not exceed $\tilde{\rho}$. Hence, $\cos \phi$ should be at least $\rho_\mu / \tilde{\rho}$, which implies ϕ should be strictly less than $\cos^{-1}(\rho_\mu / \rho_{\mu'})$, since $\tilde{\rho} < \rho_{\mu'}$. Since there will be no grid point that will lie in the annular space between $C^R(\alpha, \rho_\mu)$ and $C^R(\alpha, \tilde{\rho})$ (including the circumference of $C^R(\alpha, \tilde{\rho})$), this polygon, therefore, would be a *regular polygonal enclosure* for the digital circle $C^Z(\alpha, \rho)$. The following fact is enumerated to conclude these findings.

Fact 5. *If there exists a grid point μ lying on $C^{Z,I}(\alpha, \rho)$, such that $\delta_\mu = \max \{\delta_\nu \mid \nu \text{ lies on } C^{Z,I}(\alpha, \rho)\}$ and $\epsilon_\mu = \max \{\epsilon_\nu \mid \nu \text{ lies on } C^{Z,I}(\alpha, \rho)\}$, then there always exists some $\mathcal{E}^R(C^Z(\alpha, \rho))$ corresponding to $C^Z(\alpha, \rho)$.*

Now, subject to the conditions **c1**, **c2**, and **c3**, if 2ϕ be the angle subtended at α by each edge of the regular polygonal enclosure for the digital circle $C^Z(\alpha, \rho)$, where, $\phi < \cos^{-1}(\rho_\mu / \rho_{\mu'})$, then the minimum number of edges n_{min} of the corresponding polygon can be obtained using Fact 6.

Fact 6. *Minimum number of edges n_{min} of $\mathcal{E}^R(C^Z(\alpha, \rho))$ is given by $n_{min} = \lfloor \pi / \phi_0 \rfloor + 1$, where, $\phi_0 = \cos^{-1}(\rho_\mu / \rho_{\mu'})$, μ' being the grid point with properties discussed above.*

3   Experiments and Results

Exhaustive procedural checking has revealed that each of the digital circles with radii from 1 to 10 has a grid point μ that has simultaneously maximum isothetic distance and maximum radial distance, which implies that each of the digital circles with radii from 1 to 10 has regular polygonal enclosure. To cite a few

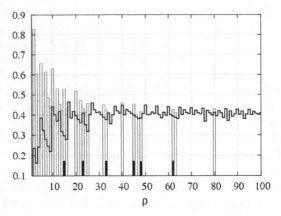

Fig. 5. ϵ_μ (thick-lined silhouette) and $\epsilon_{\mu'}$ (thin-lined bars) plotted against ρ. $\epsilon_{\mu'}$ has been shown only when it exceeds ϵ_μ. A regular polygonal enclosure exists for a digital circle with radius ρ if $\epsilon_{\mu'} > \epsilon_\mu$.

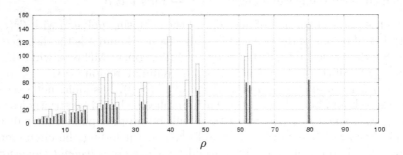

Fig. 6. $2k_{min}$ (thick line) and n_{min} (thin bars) plotted against the radii (ρ) of the digital circles having regular polygonal enclosures

more, for $\rho = 12 - 16, 20 - 25, 32, 33, 40$, the corresponding digital circles possess regular polygonal enclosures. The chance of existence of such a grid point μ, therefore, goes on decreasing as the radius of the circle gets increasing.

The thin-lined vertical bars, shown in Fig. 5, indicate the circles for which regular polygonal enclosures exist. But it should be noticed that for each of these circles having regular polygonal enclosures, it is not always true that $\exists\ \mu \in \mathcal{C}^{z,I}(\alpha, \rho)$, s.t. $\delta_\mu = \max\{\delta_\nu \mid \nu$ lies on $\mathcal{C}^{z,I}(\alpha, \rho)\}$ and $\epsilon_\mu = \max\{\epsilon_\nu \mid \nu$ lies on $\mathcal{C}^{z,I}(\alpha, \rho)\}$ are valid simultaneously (vide Sec. 2.2 and Fact 5). The fillets on the axis of ρ within the thin-lined bars, shown in Fig. 5, indicate the circles for which $\exists\ \mu \in \mathcal{C}^{z,I}(\alpha, \rho)$, s.t. $\delta_\mu < \max\{\delta_\nu \mid \nu$ lies on $\mathcal{C}^{z,I}(\alpha, \rho)\}$ and $\epsilon_\mu = \max\{\epsilon_\nu \mid \nu$ lies on $\mathcal{C}^{z,I}(\alpha, \rho)\}$. And those without fillets indicate the circles for which $\exists\ \mu \in \mathcal{C}^{z,I}(\alpha, \rho)$, s.t. $\delta_\mu = \max\{\delta_\nu \mid \nu$ lies on $\mathcal{C}^{z,I}(\alpha, \rho)\}$ and $\epsilon_\mu = \max\{\epsilon_\nu \mid \nu$ lies on $\mathcal{C}^{z,I}(\alpha, \rho)\}$.

Theoretically, if there exists a regular polygonal enclosure for a digital circle $\mathcal{C}^z(\alpha, \rho)$, then minimum number of edges n_{min} for the polygon is given by $n_{min} =$

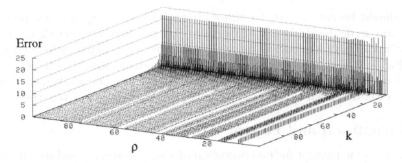

Fig. 7. Error (%) plotted against ρ and k shows that higher values of $k = n/2$ drastically reduce the error

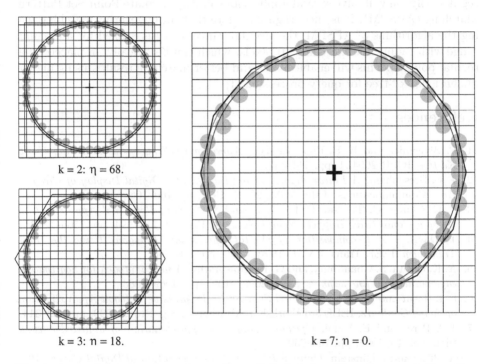

k = 2: η = 68.

k = 3: η = 18.

k = 7: η = 0.

Fig. 8. Approximate regular polygons for a digital circle with radius $\rho = 8$. For other values of k, the corresponding errors are: $\eta(k = 4, 5) = 8$, $\eta(k = 6) = 4$, $\eta(k \geq 7) = 0$.

$\lfloor \pi/\phi_0 \rfloor + 1$. While testing with different digital circles, however, it is found that if a digital circle $\mathcal{C}^{\mathbb{Z}}(\alpha, \rho)$ possesses a regular polygonal enclosure with minimum number of vertices n_{min}, then in practice, it can be often enclosed by a regular polygonal enclosure with lesser number of vertices. This is illustrated in Fig. 6, where the thin-lined vertical bars denote the values of n_{min}, and the thick vertical lines represent the corresponding minimum even number of vertices, $2k_{min}$, found experimentally, that the regular polygonal enclosures should possess, satisfying Def. 3.

It should be noted that, for those circles with $\epsilon_{\mu'} \not> \epsilon_\mu$, regular polygonal enclosures, as defined in Def. 3, are certainly not possible. For such digital circles, approximate polygonal enclosures with low errors are possible, which is evident from Fig. 7. In Fig. 8, a few approximate polygons have been shown for a sample digital circle with radius $\rho = 8$ to demonstrate the rapid convergence of the polygon towards ideal situation with increase in its number of vertices.

4   Conclusion and Future Works

This work is an attempt for manifestation of some interesting and useful properties of digital circles, based on which an approximate (if not ideal) regular polygon of a digital circle can be obtained. As discussed in Sec. 1, the approximate regular polygon will have several applications in Approximate Point Set Pattern Matching (APSPM). It is encouraging to note that, in fingerprint matching using the algorithms of APSPM, the circular range query is evoked for a circular range with radius not exceeding 10 pixels, which can be "ideally" replaced by a regular polygon. This will enable faster and better matching results, using higher dimensional kd-tree for query processing [1].

References

1. P. Bhowmick and B. B. Bhattacharya, *Approximate Fingerprint Matching Using Kd-tree*, ICPR 2004, 1, pp. 544–547.
2. J. E. Bresenham, *A Linear Algorithm for Incremental Digital Display of Circular Arcs*, Communications of the ACM, 20(2), 1977, pp. 100–106.
3. P. C. Chung, C. T. Tsai, E. L. Chen, and Y. N. Sun, *Polygonal Approximation Using a Competitive Hopfield Neural Network*, PR, 27(11), 1994, pp. 1505–1512.
4. T. Davis, *Fast Decomposition of Digital Curves into Polygons Using the Haar Transform*, IEEE Trans. PAMI, 21(8), 1999, pp. 786–790.
5. A. K. Gupta, S. Chaudhury, and G. Parthasarathy, *A new approach for aggregating edge points into line segments*, PR, 26(7), 1993, pp. 1069–1086.
6. P. I. Hosur and K. K. Ma, *A Novel Scheme for Progressive Polygon Approximation of Shape Contours*, IEEE 3rd Workshop Multimedia Sig. Proc., 1999, pp. 309–314.
7. J.C. Perez and E. Vidal, *Optimum polygonal approximation of digitized curves*, PRL, vol. 15, 1994, pp. 743–750.
8. A. Pikaz and I. Dinstein, *Optimal Polygonal Approximation of Digital Curves*, PR, 28(3), 1995, pp. 373–379.
9. I. Debled-Rennesson, S. Tabbone, and L. Wendling, *Fast Polygonal Approximation of Digital Curves*, Proc. ICPR, 2004, pp. 465–468.
10. P. L. Rosin, *Techniques for Assessing Polygonal Approximation of Curves*, IEEE Trans. PAMI, 19(6), 1997, pp. 659–666.
11. M. Salotti, *Improvement of Perez and Vidal Algorithm for the Decomposition of Digitized Curves into Line Segments* Proc. ICPR 2000, pp. 878–882.
12. M. Sarfraz, M. R. Asim, and A. Masood, *Piecewise Polygonal Approximation of Digital Curves*, Proc. 8th Intl. Conf. Information Visualisation (IV'04), 2004, pp. 991–996.
13. K. Schröder and P. Laurent, *Efficient Polygon Approximations for Shape Signatures*, Proc. ICIP, 1999, pp. 811–814.

14. M. Worring and A. W. M. Smeulders, *Digitized Circular Arcs: Characterization and Parameter Estimation*, IEEE Trans. PAMI, 17(6), 1995, pp. 587–598.
15. P. Y. Yin, *Genetic Algorithms for Polygonal Approximation of Digital Curves*, IJPRAI, 13, 1999, pp. 1061–1082.
16. P. Y. Yin, *A Tabu Search Approach to Polygonal Approximation of Digital Curves*, IJPRAI, 14(2), 2000, pp. 243–255.
17. P. Y. Yin, *Ant Colony Search Algorithms for Optimal Polygonal Approximation of Plane Curves*, PR, 36, 2003, pp. 1783–1797.
18. P. Y. Yin, *A discrete particle swarm algorithm for optimal polygonal approximation of digital curves*, J. Visual Comm. Image Reprsn., 15(2), 2004, pp. 241–260.
19. Y. Zhu and L. D. Seneviratne, *Optimal Polygonal Approximation of Digitised Curves*, IEE Proc. Vis. Image Signal Process., 144(1), 1997, pp. 8–14.

A Novel Feature Fusion Method Based on Partial Least Squares Regression

Quan-Sen Sun [1,2], Zhong Jin [3,2], Pheng-Ann Heng [4], and De-Shen Xia [2]

[1] School of Science, Jinan University, Jinan 250022, China
qssun@126.com
[2] Department of Computer Science,
Nanjing University of Science &Technology, Nanjing 210094, China
deshen_x@263.net
[3] Centre de Visió per Computador, Universitat Autònoma de Barcelona, Spain
zhong.jin@cvc.uab.es
[4] Department of Computer Science and Engineering,
The Chinese University of Hong Kong, Hong Kong
pheng@cse.cuhk.edu.hk

Abstract. The partial least squares (PLS) regression is a new multivariate data analysis method. In this paper, based on the ideas of PLS model and feature fusion, a new non-iterative PLS algorithm and a novel method of feature fusion are proposed. The proposed method comprises three steps: firstly, extracting two sets of feature vectors with the same pattern and establishing PLS criterion function between the two sets of feature vectors; then, extracting two sets of PLS components by the PLS algorithm in this paper; and finally, doing feature fusion for classification by using two strategies. Experiment results on the ORL face image database and the Concordia University CENPARMI database of handwritten Arabic numerals show that the proposed method is efficient. Moreover, the proposed non-iterative PLS algorithm is superior to the existing iterative PLS algorithms on the computational cost and speed of feature extraction.

1 Introduction

The Partial least squares (PLS) regression is a novel multivariate data analysis method developed from practical application in real word. PLS regression originally developed by Wold has a tremendous success in chemometrics and chemical industries for static data analysis[1-4]. The robustness of the generated model also makes the partial least squares approach a powerful tool for regression analysis, dimension reduction and classification technique, being applied to many other areas such as process monitoring, marketing analysis and image processing and so on[5-7]. PLS regression provides a modeling method between two sets of data (PLS of single dependent variable is referred as PLS1, PLS of Multi-dependent variable is referred as PLS2), during the regression modeling, both the reduction of primitive data and the elimination of redundant information (i.e. noise) could be achieved. PLS model is effective because it integrates the multivariate linear regression (MLR)[8], principal component analysis (PCA)[9] and canonical correlation analysis (CCA)[10] together naturally while it is

S. Singh et al. (Eds.): ICAPR 2005, LNCS 3686, pp. 268–277, 2005.
© Springer-Verlag Berlin Heidelberg 2005

convenient for the analysis of the multi-dimensional complexity system. PLS method has received a lot of attention and interest in recent years.

In PLS modeling methods, a classical algorithm that the nonlinear iterative partial least squares(NIPALS) was given by Wold[11]. Based on this, several different modification versions of the iterative PLS methods have been given[12]. However, in practical applications the iterative PLS modeling may suffer from overfitting or local minima. In this paper based on the idea of PLS model, we present a new PLS modeling method under the orthogonal constraint. It is a non-iterative PLS algorithm. Furthermore, a novel method of feature fusion is proposed and it has been used in the application of pattern classification.

The rest of this paper is organized as follows. In Section 2, the "classical" PLS algorithm is described and a new PLS modeling method is presented. Then both of their performances have been analyzed. In Section 3, based on PLS model a new method of feature fusion for pattern classification is proposed. In Section 4, we show our experiment results using the ORL face image database and the Concordia University CENPARMI database of handwritten Arabic numerals. Finally, conclusions are drawn in Section 5.

2   PLS Modeling Method

Let A and B be two groups of feature sets on pattern sample space Ω, any pattern sample $\xi \in \Omega \subset R^N$, whose two corresponding feature vectors are $x \in A \subset R^p$ and $y \in B \subset R^q$, respectively. Given two data matrices $X \in R^{p \times n}$ on sample space A and $Y \in R^{q \times n}$ on sample space B, where n is total number of samples. Further we assume centered variables; i.e. the columns of X^T and Y^T are zero-mean. Let $S_{xy} = XY^T$ and $S_{yx} = S_{xy}^T$ ($(1/n-1)S_{xy}$ denotes between-set covariance matrix of X and Y). Note the zero mean does not impose any limitations on the methods discussed since the mean values can easily be estimated.

2.1   The Classical PLS Algorithm (C-PLS)

Basically, the PLS method is a multivariable linear regression algorithm that can handle correlated inputs and limited data. The algorithm reduces the dimension of the predictor variables (input matrix, X) and response variables (output matrix, Y) by projecting them to the directions (input weight α and output weight β) that maximize the covariance between input and output variables. This projection method decomposes variables of high collinearity into one-dimensional variables (input score vector t and output score vector u). The decomposition of X and Y by score vectors is formulated as follows:

$$X = \sum_{i=1}^{d} t_i p_i^T + E = TP^T + E \; ; \; Y = \sum_{i=1}^{d} u_i q_i^T + F = UQ^T + F .$$

Where p_i and q_i are loading vectors, and E and F are residuals.

More precisely, PLS method is to find a pair of directions (weight vectors) α_k and β_k such that

$$\{\alpha_k ; \beta_k\} = \arg \max_{\alpha^{\mathrm{T}}\alpha = \beta^{\mathrm{T}}\beta = 1} \mathrm{Cov}(X^{\mathrm{T}}\alpha, Y^{\mathrm{T}}\beta) = \arg \max_{\alpha^{\mathrm{T}}\alpha = \beta^{\mathrm{T}}\beta = 1} \alpha^{\mathrm{T}} S_{xy}\beta \qquad (1)$$

for $k = 1, 2, \cdots, d$.

The nonlinear iterative partial least squares (NIPALS) algorithm[11] is an iterative process, which can be formulated as follows.

Let $E_0 = X^{\mathrm{T}}$, $F_0 = Y^{\mathrm{T}}$ and $h = 1$ of initialization, and randomly initialize u_1 of the Y-score vector. Iterate the following steps until convergence:

$$1.\ \alpha_h^{\mathrm{T}} = u_h^{\mathrm{T}} E_{h-1} /(u_h^{\mathrm{T}} u_h) \qquad 4.\ \beta_h^{\mathrm{T}} = t_h^{\mathrm{T}} F_{h-1} /(t_h^{\mathrm{T}} t_h)$$

$$2.\ \alpha_h = \alpha_h / \| \alpha_h \| \qquad 5.\ \beta_h = \beta_h / \| \beta \|$$

$$3.\ t_h = E_{h-1}\alpha_h \qquad 6.\ u_h = F_{h-1}\beta_h$$

After the convergence, by regressing E_{h-1} on t_h and F_{h-1} on u_h, the loading vectors $p_h = E_{h-1}^{\mathrm{T}} t_h / t_h^{\mathrm{T}} t_h$ and $q_h = F_{h-1}^{\mathrm{T}} u_h / u_h^{\mathrm{T}} u_h$ can be computed.

Although NIPALS algorithm is efficient and robust and it can be extended to some nonlinear PLS models[13]. It may lead the uncertain results that the score vector u_1 of initialization is random. A faster and more stable iterative algorithm has proposed by Höskuldsson[2] and Helland[14]. As mentioned above, it can be shown that the weight vector α_h also corresponds to the first eigenvector of the following eigenvalue problem

$$E_{h-1}^{\mathrm{T}} F_{h-1} F_{h-1}^{\mathrm{T}} E_{h-1}\alpha_h = \lambda_h^2 \alpha_h \qquad (2)$$

The X-scores t are then given as

$$t_h = E_{h-1}\alpha_h \qquad (3)$$

Similarly, eigenvalue problems for the extraction of u_h and β_h estimates can be derived.

The above iterative algorithm gives out a linear PLS modeling method. It is called the classical PLS algorithm(C-PLS) in this paper. C-PLS exist the facts that $t_i^{\mathrm{T}} t_j = 0$ for $i \neq j$, $t_i^{\mathrm{T}} u_j = 0$ for $j > i$ and $\alpha_i^{\mathrm{T}}\alpha_j = \beta_i^{\mathrm{T}}\beta_j = 0$ for $i \neq j$.

2.2   A New Non-iterative PLS Modeling Method(NI-PLS)

In this section, we present a new PLS modeling method based on idea of PLS model. Under the object criterion function (1), we give out a theorem based on the following the orthogonal constraint

$$\alpha_k^{\mathrm{T}}\alpha_i = \beta_k^{\mathrm{T}}\beta_i = 0 \qquad (4)$$

for all $1 \leq i \leq k$, where $k = 1, 2, \cdots, d$.

Theorem. Under the criterion (1), the number of effective weight vectors, which satisfy constraint (4), is r (r is the number of non-zero eigenvalues of matrix $S_{xy}S_{yx}$) pairs at most. $d(\leq r)$ pairs of weight vectors are composed of vectors which are selected from the eigenvectors corresponding to the first d maximum eigenvalues of eigenequations (5) and (6)

$$S_{xy}S_{yx}\alpha = \lambda^2\alpha \tag{5}$$

$$S_{yx}S_{xy}\beta = \lambda^2\beta \tag{6}$$

and satisfying

$$\begin{cases} \alpha_i^T\alpha_j = \beta_i^T\beta_j = \delta_{ij} \\ \alpha_i^T S_{xy}\beta_j = \lambda_i\delta_{ij} \end{cases} \quad (i, j = 1,2,\cdots,d) \tag{7}$$

Where λ_i^2 is the non-zero eigenvalue corresponding to eigenvectors α_i and β_i, and

$$\delta_{ij} = \begin{cases} 1 & i = j \\ 0 & i \neq j \end{cases}$$

Proof. Using Lagrange Multiplier Method to Transform Eq.(1):

$$L(\alpha,\beta) = \alpha^T S_{xy}\beta - \frac{\lambda_1}{2}(\alpha^T\alpha - 1) - \frac{\lambda_2}{2}(\beta^T\beta - 1) \tag{8}$$

where λ_1 and λ_2 are Lagrange multipliers. Let

$$\partial L(\alpha,\beta)/\partial\alpha = S_{xy}\beta - \lambda_1\alpha = 0 \tag{9}$$

$$\partial L(\alpha,\beta)/\partial\alpha = S_{yx}\alpha - \lambda_2\beta = 0 \tag{10}$$

Multiplying both side of Eq.(9) and (10) by α^T and β^T respectively, considering the constraint in Eq.(1), we obtain $\alpha^T S_{xy}\beta = \lambda_1\alpha^T\alpha = \lambda_1$ and $\beta^T S_{yx}\alpha = \lambda_2\beta^T\beta = \lambda_2$. Since $S_{yx}^T = S_{xy}$, so $\lambda_1 = \lambda_1^T = (\alpha^T S_{xy}\beta)^T = \beta^T S_{yx}\alpha = \lambda_2$. We can infer that to obtain the maximum value of λ is the same as to maximize the criterion function (1) under constraint (4). Let $\lambda_1 = \lambda_2 = \lambda$, then eigenequations (5) and (6) can inferred via Eq.(9) and Eq.(10). Since both $S_{xy}S_{yx}$ and $S_{yx}S_{xy}$ are symmetric matrices, and $rank(S_{xy}S_{yx}) = rank(S_{yx}S_{xy}) \leq rank(S_{xy})$, so that the two eigenequations (5) and (6) have the same non-zero eigenvalues. Let $\lambda_1^2 \geq \lambda_2^2 \geq \cdots \geq \lambda_r^2 > 0$ ($r \leq rank(S_{xy})$), then the r pairs of eigenvectors corresponding to them are orthonormal, respectively, namely $\alpha_i^T \alpha_i = \beta_i^T \beta_i = \delta_{ij}$ .Due to $\alpha_i^T S_{xy} \beta_j = \alpha_i^T S_{xy} (\lambda_j^{-1} S_{yx} \alpha_j) = \lambda_j^{-1} \alpha_i^T (\lambda_j^2 \alpha_j) = \lambda_j \delta_{ij}$, then conclusion (7) is true. □

The above theorem gives out a laconic algorithm that can solve the orthogonal weight vectors. Supposing all d pairs of weight vectors are $\{\alpha_i; \beta_i\}_{i=1}^d$, which make up of two projection matrices $W_x = (\alpha_1, \alpha_2, \cdots \alpha_d)$ and $W_y = (\beta_1, \beta_2, \cdots \beta_d)$. By two linear transformations $z_1 = W_x^T x$ and $z_2 = W_y^T y$, we can obtain two sets of PLS components from the same pattern so as to achieve the goal of dimensionality reduction.

From Eq.(9) and (10), we know that we only need to solve one among each pair of weight vectors, the other one can be solved by Eq.(9) or Eq. (10). Generally, we can choose the low-order matrix $S_{xy} S_{yx}$ or $S_{yx} S_{xy}$ to find its eigenvalues and eigenvectors, such that the computational complexity can be lowered.

2.3 The Algorithm Analysis

As mentioned above, two algorithms are based on the constraints of orthogonal weight vectors. The NI-PLS can't assure that the extracted PLS components are uncorrelated; but it are uncorrelated that the PLS components are extracted by the classical PLS in Section 2.1. Theoretically, from the angle of pattern classification, the performance of the latter classification is better than that of the former one; but in terms of the angle of algorithm's complexity, the former one is better. We could say that both of them have its own advantage over the other, so we should use either one of them depending on the nature of problems in real situation.

Moreover, from the process of solving the above weight vectors, we could see that it is valid for either big sample size or small sample size to adapt the PLS modeling method in reducing the dimension. Since no matter the PLS model is singular or not, it cannot be affected whether for the total scatter matrix of the training sample. Due to this property, the PLS model has better performance in general. It overcomes some modeling difficulties which will appear when the high-dimensional small sample size problems are processed with those methods such as the Fisher discrimination analysis[15], CCA and MLR.

3 Feature Fusion Strategy and Design of Classifier

3.1 Classification Based on Correlation Feature Matrix

Through the above PLS algorithms, we can acquire two sets of PLS components on sample space A and sample space B respectively. Each a pair of PLS components constitute the following matrix:

$$M = [z_1, \ z_2] = [W_x^T x, \ W_y^T y] \in R^{d \times 2} \tag{11}$$

Where matrix M is called the correlation feature matrix of feature vectors x and y (or pattern sample ξ). The distance between any two correlation feature matrices $M_i = [z_1^{(i)},\ z_2^{(i)}] \square M_j = [z_1^{(j)},\ z_2^{(j)}]$ is defined as

$$d(M_i, M_j) = \sum_{k=1}^{2} \| z_k^{(i)} - z_k^{(j)} \|_2 \tag{12}$$

where $\| \cdot \|$ represents the vector's Euclidean distance.

Let $\omega_1, \omega_2, \cdots, \omega_c$ be the c known pattern classes, and assume that $\xi_1, \xi_2, \cdots, \xi_n$ are the all training samples and their corresponding correlation feature matrixes are $M_1, M_2, \cdots, M_n$. For any testing sample ξ, its correlation feature matrix $M = [z_1,\ z_2]$. If $d(M, M_l) = \min_j\ d(M, M_j)$ and $\xi_j \in \omega_k \square$then $\xi \in \omega_k$.

3.2 The Quadratic Bayesian Classifier

By the following linear transformation (13), we can extract a new d-dimension combination feature of each a pair of PLS components, and is used in classification.

$$z = \begin{pmatrix} W_x \\ W_y \end{pmatrix}^{\mathrm{T}} \begin{pmatrix} x \\ y \end{pmatrix} \tag{13}$$

In d-dimension combination feature space, we use the quadratic Bayesian classifier to classification. The quadratic Bayesian function is defined as:

$$g_i(x) = \frac{1}{2}\ln\left|\sum_l\right| + \frac{1}{2}(x - \mu_l)^{\mathrm{T}} \sum_l^{-1} ((x - \mu_l)) \tag{14}$$

where μ_l and $\sum_l$ denote the mean vector and the covariance matrix of class l, respectively. The classifying decision-making based on the discriminant function will be $x \in \omega_k$ if sample x satisfies $g_k(x) = \min_l g_l(x)$.

4 Experiments and Analysis

4.1 Experiment on ORL Face Image Database

Experiment is performed on the ORL face image database (http://www.cam-orl.co.uk). There are 10 different images for 40 individuals. For some people, images were taken at different times. And the facial expression (open/closed eyes, smiling/nonsmiling) and facial details (glasses/no glasses) are variables. The images were taken against a dark homogeneous background and the people are in upright, frontal position with tolerance for some tilting and rotation of up to 20^0. Moreover, there is some variation in scale of up to about 10%. All images are grayscale and normalized with a resolution of 112×92. Some images in ORL are shown in Fig.1.

Fig. 1. Ten images of one person in ORL face database

In this experiment, we use the five images of random extraction of each person for training and the remaining five for testing. Thus, the total amount of training samples and testing samples are both 200.

We use the original face images to make the first training sample space $A = \{x \mid x \in R^{10304}\}$. Performing the cubic wavelet transformation on each original image, low-frequency sub-images with 28×23 resolution are obtained, and the second training sample space $B = \{y \mid y \in R^{168}\}$ are constructed; then, combining those two groups of features, and using the two PLS algorithm in Section 2, we can obtain the d pairs of weight vectors, respectively. By using the linear transformation $z_1 = W_x^T x$ and $z_2 = W_y^T y$, the feature vectors of the two sets can be reduced to d-dimensional(d varies from 1 to 168) discrimination feature vectors; finally, we proceed to classify according to the feature fusion strategy and classifier in Section 3.1, and the classification results and once time(s) in ten experiments are shown in Table 1.

Table 1 shows that the two kinds of PLS methods all reach the better classification result, and their average recognition rates is above 95%. But seeing from the feature and classification, the computational time for C-PLS is 2.7 multiple one for NI-PLS. This matches the result of complexity analyzes of the algorithm in Section 2.3. Compared with the C-PLS, adopting the NI- PLS algorithm, not only dose it not reduce the recognition rate, but also the feature extraction speed is raised consumedly.

Table 1. Comparison of the recognition accuracy (%) of two PLS methods at ten experiments

Methods	1	2	3	4	5	6	7	8	9	10	Average	time(s)
NI-PLS	95.5	95.5	96.5	92.5	95.0	94.0	96.5	94.5	96.0	94.5	95.0	189
C-PLS	95.0	95.5	92.0	97.0	96.0	94.0	96.5	95.0	94.5	95.5	95.1	516

Note: in this table, the recognition rates are obtained when the number the PLS component all are chosen to 50. The time mean that once experiment is accomplished as the number of PLS component varying from 1 to 168, and without taking into account of wavelet transformation.

4.2 Experiment on CENPARMI Handwritten Numerals Database

The goal of this experiment is to test the validity of the algorithm with the big sample proposed in this paper. The Concordia University CENPARMI database of handwritten Arabic numerals, popular in the world, is adopted. In this database, there are 10 class, i.e. 10 digits (from 0 to 9), and 600 sample for each. Some images of original samples are shown in Fig. 2. Hu et al.[16] had done some preprocessing work and extracted four kinds of features as follows:

- X^G: 256-dimensional Gabor transformation feature,
- X^L: 21- dimensional Legendre moment feature,
- X^P: 36-dimensional Pseudo-Zernike moment feature,
- X^Z: 30-dimensional Zernike moment feature.

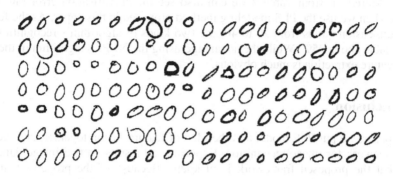

Fig. 2. Some images of digits in CENPARMI handwritten numeral database

In this experiment, we use the first 300 images of each class for training and the remaining 300 for testing. Thus, the total amount of training samples and testing samples are both 3000. Combine any two features of the above four features in the original feature space, using two algorithms described in Section 2 and according to the feature fusion strategy and classifier in Section 3.2, we can obtain classification results that is shown in Table 2. For the sake of further comparing algorithm's performance, in Table 3, as the combination feature's dimension varies form 1 to 40, we also list the classification error rates and the classification time of two kinds of algorithms when we combine Gabor feature and the Legendre feature.

Table 2. Comparison of classification error rates of two methods under different feature combination

Methods	X^G-X^L	X^G-X^P	X^G-X^Z	X^L-X^P	X^L-X^Z	X^P-X^Z
NI-PLS	0.0403(31)	0.0863(33)	0.0850(30)	0.0857(33)	0.0863(30)	0.1910(27)
C-PLS	0.0407(30)	0.0870(34)	0.0847(30)	0.0857(35)	0.0857(30)	0.1853(27)

Note: the value in ( ) denotes the dimension of combination feature(i.e. the number the PLS component) as the optimal result is achieved.

Table 3. Comparison of classification error rates and time(s) of two methods on the Gabor feature and the Legendre feature

Dimension	1	5	10	15	20	25	30	35	40	Time(s)
NI-PLS	0.755	0.204	0.110	0.060	0.045	0.043	0.040	0.043	0.044	256
C-PLS	0.757	0.264	0.113	0.063	0.045	0.041	0.040	0.043	0.043	587

From Table 1 and Table 2, we can see that, the two kinds of algorithm all take the similar classification results when the four sets of features are combined differently. Among them the classification error rate by combining the Gabor feature and the Legendre feature is lower than those of others, and the optimal recognition rate is up to 96%. Moreover, from Table 2 we can also see the classification error rate drops quickly when we use the PLS modeling method proposed in this paper. As the feature vector dimension varies from 20 to 40, the two kinds of algorithm's recognition rate are all stable above 95%. While the advantage using the NI-PLS algorithm in the time of the feature extraction is much obvious.

5 Conclusion

In this paper, we put PLS model and the idea of feature fusion together, and create a new framework for image recognition. Experimental results on two image databases show that the proposed framework is efficient. Because of the property that PLS works no matter whether the total covariance matrix of the training samples is singular or not, the proposed method is more suitable for processing small sample size classification problems in high dimensional spaces. From the comparison of the two PLS modeling methods mentioned in this paper, the proposed non-iterative PLS(NI-PLS) is superior to the classical PLS algorithm(C-PLS) at the complexity of algorithm and the speed of feature extraction.

Acknowledgements

We wish to thank National Science Foundation of China under Grant No. 60473039. This work was supported by the Research Grants Council of the Hong Kong Special Administrative Region under Earmarked Research Grant (project no. CUHK 4223/04 E).

References

1. Wold, S., Martens, H., Wold, H.: The Multivariate Calibration Problem in Chemistry Solved by the PLS Method. Proceedings of Conference on Matrix Pencils, Lecture Notes in Mathematics, Springer, Heidelberg (1983) 286–293
2. Höskuldsson, A.: PLS Regression Methods. Journal of Chemometrics 2 (1988) 211–228
3. Yacoub, F., MacGregor, J.F.: Product Optimization and Control in the Latent Variable Space of Nonlinear PLS Models. Chemometrics and Intelligent Laboratory Systems 70 (2004) 63–74

4. Barker, M., Rayens, W.S.: Partial Least Squares for Discrimination. Journal of Chemometrics 17 (2003) 166-173
5. Kesavanl, P., Lee, J.H., Saucedo, V., Krishnagopalan, G.A.: Partial Least Squares (PLS) Based Monitoring and Control of Batch Digesters. Journal of Process Control 10 (2000) 229–236
6. Chin, W.W.: The Partial Least Squares Approach for Structural Equation Modeling. In: Marcoulides, G.A.(Ed.), Modern Methods for Business Research, Lawrence Erlbaum Associates, London (1998) 295–336
7. Nguyen, D.V., Rocke, D.M.: Tumor Classification by Partial Least Squares Using Microarray Gene Expression Data. Bioinformatics 18 (2002) 39–50
8. Turk, M., Pentland, A.: Face Recognition Using Eigenfaces. Proc. IEEE Conf, On Computer Vision and Pattern Recognition 3 (1991) 586–591
9. Izenman, A.J.: Reduced-Rank Regression for the Multivariate Linear Model. Journal of Multivariate Analysis 5 (1975) 248–264
10. Hotelling, H.: Relations between Two Sets of Variates. Biometrika 28 (1936) 321–377
11. Wold, H.: Non-Linear Iterative Partial Least Squares(NIPALS) Modeling. Some Current Developments. In: Krishnaiah, P.R. (Ed.), Multivariate Analysis, Academic Press, New York (1973) 383–407
12. Wold, S., Trygg, J., Berglund, A., Antti, H.: Some Recent Developments in PLS Modeling. Chemometrics and Intelligent Laboratory Systems 58 (2001) 131–150
13. Tang, K.L., Li, T.H.: Comparison of Different Partial Least-Squares Methods in Quantitative Structure–Activity Relationships. Analytica Chimica Acta 476 (20030 85–92
14. Helland, I.S.: On the Structure of Partial Least Squares. Comm. Statist. Simulation Comput. 17 (1988) 581–607
15. Belhumeur, P.N., Hespanha, J., Kriegman, D.J.: Eigenfaces vs. Fisherfaces: Recognition Using Class Specific Linear Projection. IEEE Trans. Pattern Anal. Machine Intell. 19 (1997) 711–720
16. Hu, Z.S., Lou, Z., Yang, J.Y., Liu, K., Suen, C.Y.: Handwritten Digit Recognition Basis on Multi-Classifier Combination. Chinese J. Computer 22 (1999) 369–374

Combining Text and Link Analysis for Focused Crawling

George Almpanidis and Constantine Kotropoulos

Aristotle University of Thessaloniki, Department of Infomatics,
Box 451, GR-54124 Thessaloniki, Greece
{galba, costas}@aiia.csd.auth.gr
http://www.aiia.csd.auth.gr

Abstract. The number of vertical search engines and portals has rapidly increased over the last years, making the importance of a topic-driven (focused) crawler evident. In this paper, we develop a latent semantic indexing classifier that combines link analysis with text content in order to retrieve and index domain specific web documents. We compare its efficiency with other well-known web information retrieval techniques. Our implementation presents a different approach to focused crawling and aims to overcome the limitations of the necessity to provide initial training data while maintaining a high recall/precision ratio.

1 Introduction

In contrast with large-scale engines such as Google [1], a search engine with a specialised index is more appropriate to services catering for specialty markets and target groups since it has more structured content and offers a high precision [2]. The main goal of this work is to provide an efficient topical information resource discovery algorithm when no previous knowledge of link structure is available except that found in web pages already fetched during a crawling phase. We propose a new method for further improving targeted web information retrieval (IR) by combining text with link analysis and make novelty comparisons against existing methods.

2 Web Information Retrieval – Text and Link Based Techniques

The expansion of a search engine using a *web crawler* is seen as a task of *classification* requiring automatic categorisation of text documents into specific and predefined categories. The visiting strategy of new web pages usually characterises the purpose of the system. Generalised search engines that seek to cover as much proportion of the web as possible usually implement a *breadth-first* (BRFS) algorithm [3]. The BRFS policy uses a simple FIFO queue for the unvisited documents and provides a fairly good bias towards high quality pages without the computational cost of keeping the queue ordered [4]. Systems on the other hand that require high precision and targeted information must seek new pages in a more intelligent way. The crawler of such a system *focused* or *topic-driven* crawler is assigned the task of automatically classifying crawled pages to existing category structures and simultaneously discovering web information related to

S. Singh et al. (Eds.): ICAPR 2005, LNCS 3686, pp. 278–287, 2005.

the specified domain while avoiding irrelevant regions of the web. A popular approach for focused resource discovery is the *best-first search* (BSFS) algorithm where two URL queues are maintained; one containing the already visited links (from here on **AF**) and another having the, yet unvisited, references of the first queue, also called *crawl frontier* (from here on **CF**) [5]. The challenging task is periodically reordering the links in the CF efficiently. The importance metrics can be either interest driven where the classifier for document similarity checks the text content and popularity/location driven where the importance of a page depends on the hyperlink structure of the crawled document.

Although the physical characteristics of web information is distributed and decentralized, the web can be viewed as one big virtual text document collection. In this regard, the fundamental questions and approaches of traditional IR research (e.g. term weighting, query expansion) are likely to be relevant in web IR [6]. The language independent vector space model (VSM) representation of documents has proved effective for text classification [7]. This model is described with indexing terms that are considered to be coordinates in a multidimensional space where documents and queries are represented as binary vectors of terms resulting to a term-document two-dimensional $m \times n$ matrix A where m is the number of terms and n is the number of documents in the collection.

Contrary to text-based techniques, the main target of link analysis is to identify the *importance* or *popularity* of web pages. This task is clearly derived from earlier work in bibliometrics academic citation data analysis where *impact factor* is the measure of importance and influence. More recently, link and social network analysis have been applied to web IR to identify authoritative information sources [8]. Here, the impact factor corresponds to the ranking of a page simply by a tally of the number of links that point to it, also known as *backlink* (BL) count or *in-degree*. But BL can only serve as a rough, heuristic-based, quality measure of a document, because it can favour universally popular locations regardless of the specific query topic. PageRank (PR) is a more intelligent connectivity-based page quality metric with an algorithm that recursively defines the importance of a page to be the weighted sum of its backlinks' importance values [9]. An alternative but equally influential algorithm of modern hypertext IR is HITS, which categorises web pages to two different classes; pages rich and relevant in text content to the user's query (*authorities*) and pages that might not have relevant textual information but can lead to relevant documents (*hubs*) [10]. Hubs may not be indexed in a vertical engine as they are of little interest to the end user, however both kind of pages can collaborate in determining the visit path of a focused crawler.

Latent semantic indexing (LSI) is a concept-based automatic indexing method that models the semantics of the domain in order to suggest additional relevant keywords and to reveal the "hidden" concepts of a given corpus while eliminating high order noise [11]. The attractive point of LSI is that it captures the higher order "latent" structure of word usage across the documents rather than just surface level word choice. The dimensionality reduction is typically computed with the help of Singular Value Decomposition (SVD), where the eigenvectors with the largest eigenvalues capture the axes of the largest variation in the data. In LSI, an approximated version of A, denoted as $A_k = U_k S_k V_k^T$, is computed by truncating its singular values, keeping only the $k = rank(A_k) < k_0 = rank(A)$ larger singular values. Unfortunately, for matrix decompositions such as SVD in dynamic collections, once an index is created it will be obsolete when new data (terms and documents) is inserted to the system. Adding new pages or modifying existing ones also means that the corpus index has to be regenerated for both the recall and the crawling phase. Depending on the indexing technique followed, this can be a computationally intensive procedure. But there are well-known relatively inexpensive methods that avoid the full reconstruction of the term-document matrix [12]. *Folding-in* is based on the existing latent semantic structure and hence new terms and documents have no effect on the representation of the pre-existing terms and documents. Furthermore, the orthogonality in the reduced k-dimensional basis for the column or row space of A (depending on inserting terms or documents) is corrupted. *SVD-updating*, while more complex, maintains the orthogonality and the latent structure of the original matrix.

3   Focused Crawling

3.1   Related Works in Focused Crawling

Numerous techniques that try to combine textual and linking information for efficient URL ordering exist in the literature. Many of these are extensions to PageRank and HITS. An extension to HITS where nodes have additional properties and make use of web page content in addition to its graph structure is proposed in [13] as a remedy to the problem of nepotism. An improvement to HITS is probabilistic HITS (PHITS), a model that has clear statistical representations [14]. An application of PageRank to target seeking crawlers improves the original method by employing a combination of PageRank and similarity to the topic keywords [15]. The URLs at the frontier are first sorted by the number of topic keywords present in their parent pages, then by their estimated PageRanks. A BSFS crawler using PageRank as the heuristic is discussed in [16]. In [17] an interesting extension to probabilistic LSI (PLSI) is introduced where existing links between the documents are used as features in addition to word terms. [18] proposes supervised learning on the structure of paths leading to relevant pages to enhance target seeking crawling. A link-based ontology is required in the training phase. Another similar technique is reinforcement learning [19] where a focused crawler is trained using paths leading to relevant goal nodes. The effect of exploiting hypertext features such as segmenting Document Object Model (DOM) tag-trees of a web document and combining this information with HITS is studied in [20]. Keyword-sensitive crawling strategies such as URL string analysis and other location metrics are investigated in [21]. An intelligent crawler that can adapt online the queue link-extraction strategy using a self-learning mechanism is discussed in [22]. Work on assessing different crawling strategies regarding the ability to remain in the vicinity of the topic in vector space over time is described in [23]. Various approaches to combine linkage metrics with content-based classifiers have been proposed in [24] and [25]. [26] uses tunnelling to overcome some of the limitations of a pure BSFS approach.

3.2   Hypertext Combined Latent Analysis (HCLA)

The problem studied in this paper is the implementation of a focused crawler for target topic discovery, given unlabeled (but known to contain relevant sample documents) textual data, a set of keywords describing the topics and no other data resources. Taking into account these limitations many sophisticated algorithms of the Sect. 2, such as HITS and context graphs, cannot be easily applied. We evaluate a novel algorithm called *Hypertext Content Latent Analysis* or **HCLA** from now onwards that tries to combine text with link analysis using the VSM paradigm. Unlike PageRank, where simple eigen-analysis on globally weighted adjacency matrix is applied and principal eigenvectors are used, we choose to work with a technique more comparable with HITS. While the effectiveness of LSI has been demonstrated experimentally in several text collections yielding an increased average retrieval precision, its success in web connectivity analysis has not been as direct. There is a close connection between HITS and LSI/SVD multidimensional scaling [27]. HITS is equivalent to running SVD on the hyperlink relation (source, target) rather than the (term, document) relation to which SVD is usually applied. Our main assumption is that terms and links in an expanded matrix are both considered for document relevance. They are seen as *relationships*. In the new space introduced, each document is represented by both the terms it contains and the similar text and hypertext documents. This is an extension of the traditional "bag-of-words" document representation of the traditional VSM described in Sect. 2. Unlike [17], we use LSI instead of PLSI. The proposed representation, offers some interesting potential and a number of benefits. First, text only queries can be applied to the enriched relationships space so that documents having only linking information, such as those in CF, can be ordered. Secondly, the method can be easily extended for the case where we also have estimated content information for the documents in CF. This can be done using the anchor text or the neighbour textual context of the link tag in the parent's html source code, following heuristics to remedy for the problem of context boundaries identification [16]. Moreover, we can easily apply local weights to the terms/rows of the matrix, a common technique in IR that can enhance LSI efficiency. While term weighting in classic text IR is a kind of linguistic favouritism, here it can also been seen as a method of emphasizing either the use of linking information or text content. An issue in our method is the complexity of updating the weights in the expanded matrix, especially when a global weighting scheme is used. For simplicity, we do not use any weighting scheme. The steps of our method are described as follows.

Let A be the original term-document representation while $\begin{pmatrix} L_{[m \times a]} \\ G_{[a \times a]} \end{pmatrix}$ and $\begin{pmatrix} O_{[m \times b]} \\ R_{[a \times b]} \end{pmatrix}$ are the new document vectors projected in the expanded term-space having both textual (submatrices $L_{[m \times a]}$ and $O_{[m \times b]}$) and connectivity components ($G_{[a \times a]}$ and $R_{[a \times b]}$).

- With a given text-only corpus of m documents and a vocabulary of n terms we first construct a term-document matrix $A_{m \times n}$ and perform a truncated SVD $A_k = SVD(A, k) = U_k S_k (V_k)^T$. Since this is done during the offline training phase we can estimate the optimum k.

- After a sufficient user-defined number of pages (a) have been fetched be the crawler, we analyse the connectivity information of the crawler's current web graph and insert $a = |AF|$ new rows as "terms" (i.e. documents from AF) and $a + b = |AF| + |CF|$ web pages from both AF and CF as "documents" to the matrix. The SVD-updating

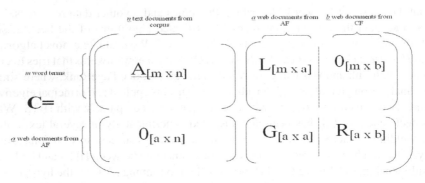

Fig. 1. Expanded connectivity matrix in HCLA. Matrix C is [(m + a) x (n + a + b)]. AF=Already Visited links, CF=Crawl Frontier docs

technique helps avoiding the reconstruction of the expanded index matrix. Because G and R in Fig. 1 are typically sparse, the procedure is simplified and the computation is reduced. For inserting $t = a$ terms and $d = a + b$ documents we append

$$D_{[(m+a)\times(a+b)]} = \begin{pmatrix} L_{[m\times a]} & 0_{[m\times b]} \\ G_{[a\times a]} & R_{[a\times b]} \end{pmatrix} \text{ to } B_{[(m+a)\times n]} = \begin{pmatrix} A_{[m\times n]} \\ 0_{[a\times n]} \end{pmatrix} \text{matrix.}$$

- Since we do not have any information of direct relationship between these web pages and the text documents $\{d_i\}$ of the original corpus, we just add a terms/rows with zero elements at the bottom of A_k. This allows the recomputing of SVD with minimum effort, by reconstructing the term-document matrix. If $SVD(B) = U_B S_B (V_B)^T$ the $k - SVD$ of the matrix after inserting a documents, then we have:

$$U_B = \begin{pmatrix} U_{[m\times k]} \\ 0_{[a\times k]} \end{pmatrix}, S_B = S_k, V_B = V_k \tag{1}$$

The above step does not follow the SVD-updating technique since the full term-document matrix is recreated and a $k-$truncated SVD of the new matrix B is recomputed. In order to insert fetched and unvisited documents from the AF and CF queues as columns in the expanded matrix we use an SVD-updating technique to calculate the semantic differences introduced in the column and row space. If we define $SVD(C) = U_C S_C V_C^T$, $F = \left(S_k | U_B^T D \right)$ and $SVD(F) = U_F S_F V_F^T$ then, matrices U_c, S_c and V_c are calculated according to [12]:

$$V_C = \begin{pmatrix} V_B & 0 \\ 0 & I_{[a+b]} \end{pmatrix} V_F, S_C = S_F, U_C = U_B V_F \tag{2}$$

Accordingly, we project the driving original query q in the new space that the expanded connectivity matrix C represents. This is done by appending a rows of zeroes to the bottom of the query vector: $q_C = \begin{pmatrix} q_{[m\times 1]} \\ 0_{[a\times 1]} \end{pmatrix}$. By applying the driving query q_C of the test topic we can to compute a total ranking of the expanded matrix C. Looking at Fig. 1 we deduce that we only need to rank the last $b=|CF|$ columns. The scores of each document in CF are calculated using the cosine similarity measure:

$$\cos \theta_j = \frac{e_j^T V_C S_C (U_C^T q_C)}{||S_C V_C^T e_j||_2 ||q_C||_2} \tag{3}$$

where $|| \cdot ||_2$ is the L_2 norm. Once similarity scores are attributed to documents, we can reorder the CF, select the most promising candidate and iterate the above steps.

4 Implementation – Experimental Results - Analysis

In this work we evaluate five different algorithms. BRFS is only used as a baseline since it does not offer any focused resource discovery. The rest are cases of BSFS algorithms with different CF reordering policies. The 2nd algorithm is based on simple BL count [21]. Here the BL of a document v in CF is the current number of documents in AF that have v as an outlink. The 3rd algorithm (SS1) is based on the Shark-Search algorithm [28]. The 4th algorithm (SS2) is similar to SS1 but the relevance scores are calculated in a pre-trained VSM using a probability ranking based scheme [7]. Since we work with an unlabelled text corpus, we use the topic query to extract the most relevant documents and use them as sample examples to train the system. The 5th algorithm is based on PageRank. Here, no textual information is available, only the connectivity between documents fetched so far and their outlinks. In order to achieve convergence we assume that from nodes with no outlinks we can jump with probability one to every other page in the current web graph. In this application, the exact PR values are not as important as the ranking they induce on the pages. This means that we can stop the iterations fairly quickly even when the full convergence has not been attained. In practice we found that no more than 10 iterations were needed. The 6th algorithm (HCLA) is the one this paper proposes. In the training phase choosing $k=50$ for the LSI of the text corpus (matrix A) yielded good results. For our experiments the WebKB corpus was used [29]. This has 8275 (after eliminating duplicates) web documents collected from universities and manually classified in 7 categories. For algorithms SS1, SS2, HCLA we selected each time three universities for training the text classifier and the fourth for testing. Documents from the "misc" university were also used for HCLA since the larger size of the initial text corpus can enhance the efficiency of LSI. Although the WebKB documents have link information we disregarded this fact in the training phase and choose to treat them only as textual data but for the testing phase we took into account both textual and linking information. The keyword-based queries that drive the crawl are also an indicative description of each category. In each case as seeds we considered the root documents in the "department" category. This entails the possibility of some documents being unreachable nodes in the vicinity tree by any path starting with that seed, something that explains the $<100\%$ final recall values in Fig. 2. We repeated the experiments for each category and for every university. Categories having relatively limited number of documents (e.g. "staff") were not tested. Evaluation tests measuring the overall performance were performed by calculating the average ratio of relevant pages retrieved out of the total ground-truth at different stages of the crawl. Due to the complexity of PR and HCLA we chose to follow a BSFSN strategy, applying the reordering policy every N documents fetched for all algorithms (except BRFS). This is supported by the results of [30] which indicate that explorative crawlers outperform

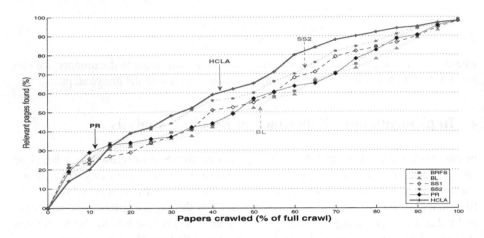

Fig. 2. Algorithm performance for WebKB

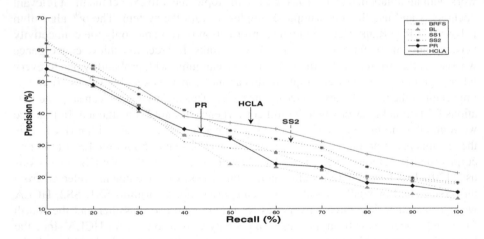

Fig. 3. Recall-Precision graph for category student

their more exploitive counterparts. We experimented with values of $N = 10, 25, 50$. The preprocessing involved fixing HTML errors, converting text encoding, filtering out all external links (outlinks that are not found inside the corpus), stemming and using a word stoplist for both the train and test text documents. The results in Fig. 2 and Fig. 3 depict the superiority of our method especially at higher recall ranges.

We must also consider that in our implementation we didn't use term weighting, which is argued to boost LSI performance [11]. BRFS performance matched or exceeded in some cases SS1 and BL. This can be attributed to the structure of the WebKB corpus and the quality of the seed documents. The unimpressive results of PR justify the assertion that it is too general for use in topic-driven tasks due to its minimal exploitation of the topic context [16], [23]. In a BSFS strategy it is crucial that the time needed for reorganising the crawl frontier is kept at a minimum. In our work, we do not need to recompute the SVD of the highly dimensional matrix C, but perform calculations on the reduced matrices of Sect. 2. Also, we follow a BSFSN algorithm where the reordering of the CF, and consequently the term-document matrix expansion and SVD computation, are performed every N documents fetched. Naturally, the value N has a significant influence in the processing time of the algorithm and the efficiency of the reordering analysis [30]. For the results presented here it is $N=50$. A parameter not well documented is the choice of k (number of important factors) in LSI. While trial and error offline experiments can reveal an optimum value for the text corpus (matrix A), there is no guarantee this will remain optimal for the expanded matrix C.

5 Conclusions

This work has been concerned with a statistical approach to text and link processing. We argue that content- and link-based techniques can be used for both the classifier and the distiller of a focused crawler and propose an alternative document representation where terms and links are combined in an LSI based algorithm. A positive point in our method is that its training is not dependent on a web graph using a previous crawl or an existing generalised search service but only on unlabeled text samples making the problem a case of unsupervised machine learning. Because LSI performance is sensitive to the size of the trained corpus performance can suffer severely when little data is available. Therefore, starting a crawl with a small term-document matrix A is not recommended since at early stages the extra linking-text information from the crawl is minimal. Appending extra text documents in the training phase, even being less relevant to the topics of the current corpus, can enhance the crawling process. At later stages when more information is available to the system we can remove these documents and retrain the model. We also believe that a hybrid strategy where HCLA is facilitated in the early stages of the crawl by a more explorative algorithm can be a practical alternative. Both HCLA and PR methods proved significantly slower requiring more processor power and memory resources. Practically, HCLA was up to 100 times slower than the simple BRFS on some tests and PR performed similarly, something that has been attested by [16]. The dynamic nature of the crawler means that computational complexity increases as more documents are inserted in AF and CF. A solution to the problem is to limit the size of both queues by discarding less authoritative documents at the bottom of the queues during the reordering phase.

References

1. Google Search Technology Online at http://www.google.com/technology/index.html
2. R. Steele, "Techniques for Specialized Search Engines", in Proc *Internet Computing*, Las Vegas, 2001
3. S. Chakrabarti, M. Berg, and B. Dom, "Focused crawling: a new approach to topic-specific Web resource discovery", *Computer Networks*, vol. 31, pp. 1623-1640, 1999.
4. M. Najork and J. Wiener, "Breadth-first search crawling yields high-quality pages", in Proc. 10th *Int. World Wide Web Conf.*, pp. 114-118, 2001.

5. A. Arasu, J. Cho, H. Garcia-Molina, A. Paepcke, and S. Raghavan, "Searching the Web", *ACM Transactions on Internet Technology*, vol. 1, no. 1, pp. 2-43, June 2001.

6. K. Yang, "Combining text- and link-based methods for Web IR", in Proc. 10th *Text Rerieval Conf. (TREC-10)*, Washington 2002, DC: U.S. Government Printing Office.

7. G. Salton, A. Wong, and C. S. Yang, "A vector space model for automatic indexing". *Communications of the ACM*, vol. 18, no. 11, pp. 613-620, 1975.

8. A. Ng, A. Zheng, and M. Jordan, "Stable algorithms for link analysis", in Proc. *ACM Conf. on Research and Development in Infomation Retrieval*, pp. 258-266, 2001.

9. S. Brin and L. Page, "The anatomy of a large-scale hypertextual web search engine", *WWW7 / Computer Networks*, vol. 30, no. 1-7, pp.107-117, 1998.

10. J. Kleinberg, "Authoritative sources in a hyperlinked environment", in *Proc. 9th Annual ACM-SIAM Symposium Discrete Algorithms*, pp. 668-677, Jan. 1998.

11. M. Berry and M. Browne. *Understanding Search Engines: Mathematical Modeling and Text Retrieval*, Philadelphia, PA: Society of Industrial and Applied Mathematics, 1999.

12. G. O'Brien, Information Management Tools for Updating an SVD-Encoded Indexing Scheme. Master's thesis, University of Tennessee, Knoxville, TN. 1994.

13. K. Bharat and M. Henzinger, "Improved algorithms for topic distillation in hyperlinked environments", in Proc.*Int. Conf. Research and Development in Information Retrieval*, pp. 104-111, Melbourne (Australia), August 1998.

14. D. Cohn and H. Chang, "Learning to probabilistically identify authoritative documents", in Proc.17th*Int. Conf. Machine Learning*, pp. 167-174, 2000.

15. P. Srinivasan, G. Pant, and F. Menczer, "Target Seeking Crawlers and their Topical Performance", in Proc. *Int. Conf. Research and Development in Information Retrieval*, August 2002.

16. M. Chau and H. Chen, "Comparison of three vertical search spiders", *Computer*, vol. 36, no. 5, pp. 56-62, 2003.

17. D. Cohn and T. Hoffman, "The Missing Link-A probabilistic model of document content and hypertext connectivity", *Advances in Neural Information Processing Systems*, vol. 13, pp. 430-436, 2001.

18. M. Diligenti, F. Coetzee, S. Lawrence, C. L. Giles, and M. Gori, "Focused crawling using context graphs", in Proc. 26th*Int. Conf. Very Large Databases (VLDB 2000)*, pp. 527-534, Cairo, 2000.

19. J. Rennie and A. McCallum, "Using reinforcement learning to spider the Web efficiently", in Proc.16th*Int. Conf. Machine Learning (ICML99)*, pp. 335-343, 1999.

20. S. Chakrabarti, "Integrating the Document Object Model with hyperlinks for enhanced topic distillation and information extraction", in Proc.10th*Int. World Wide Web Conf*, pp. 211-220, Hong Kong, 2001.

21. J. Cho, H. G. Molina, and L. Page, "Efficient Crawling through URL Ordering", in Proc.7th*Int. World Wide Web Conf.*, pp. 161-172, Brisbane, Australia 1998.

22. C. Aggarwal, F. Al-Garawi, and P. Yu, "Intelligent Crawling on the World Wide Web with Arbitrary Predicates", in Proc. 10th*Int. World Wide Web Conf.*, pp. 96-105, Hong Kong, 2001.

23. F. Menczer, G. Pant, M. Ruiz, and P. Srinivasan. "Evaluating topic-driven web crawlers", in Proc. *Int. Conf. Research and Development in Information*, pp. 241-249, New Orleans, 2001.

24. P. Calado, M. Cristo, E. Moura, N. Ziviani, B. Ribeiro-Neto, and M.A. Goncalves, "Combining link-based and content-based methods for web document classification", in Proc. 12th*Int. Conf. Information and Knowledge Management*, pp. 394-401, New Orleans, USA, November 2003.

25. I. Varlamis, M. Vazirgiannis, M. Halkidi, and B. Nguyen, "THESUS: Effective thematic selection and organization of web document collections based on link semantics", *IEEE Trans. Knowledge & Data Engineering*, vol. 16, no. 6, pp. 585-600, 2004.

26. D. Bergmark, C. Lagoze, and A. Sbityakov, "Focused Crawls, Tunneling, and Digital Libraries", in Proc. 6th European Conf. Research and Advanced Technology for Digital Libraries, pp. 91-106, 2002.
27. S. Chakrabarti, Mining the Web: Discovering Knowledge from Hypertext Data. San Francisco, CA: Morgan Kaufmann Publishers, 2002.
28. M. Hersovici, M. Jacovi, Y. S. Maarek, D. Pelleg, M. Shtalhaim, and S. Ur, "The shark-search algorithm. An Application: tailored Web site mapping", Computer Networks and ISDN Systems, vol. 30, pp. 317-326, 1998.
29. CMU World Wide Knowledge Base and WebKB dataset. Online at http://www-2.cs.cmu.edu/~webkb
30. G. Pant, P. Srinivasan, and F. Menczer, "Exploration versus exploitation in topic driven crawlers", in Proc. 2nd Int. Workshop Web Dynamics, May, 2002.

A Weighting Initialization Strategy for Weighted Support Vector Machines

Kuo-Ping Wu and Sheng-De Wang

Dept. of Electrical Engineering, National Taiwan University,
Taipei 106, Taiwan, R.O.C.
{wgb, sdwang}@hpc.ee.ntu.edu.tw
http://hpc.ee.ntu.edu.tw

Abstract. This paper presents a problem independent weighting strategy for weighted support vector machines (SVMs). SVMs can be applied with a weighting to each training vector to reflect the importance of different classes or training samples. Weightings are often assigned to the two classes inversely proportional to the sample count of each class, or according to a priori knowledge. Such a strategy can be applied to skewed data sets to balance the importance, error contribution and cost between the two classes. In this paper we propose a strategy to give each training pattern a weighting according to their distances to the classifier. The strategy regards the importance of the training patterns to the training process but not the importance of the data to the problem, thus it is suitable for general SVM applications. Experiments show that the performance of the proposed method is competitive to standard SVM while the training processes are even sped up.

1 Introduction

Support vector machines (SVMs)[1] are designed for binary classifying problems[2]. It gives promising results in many pattern recognition applications, such as document categorization, medical diagnosis and prognosis, and so on. Modern implementations of SVM often incorporate fast training algorithms that can reduce the computation time and the consumption of memory. Sequential minimal optimization (SMO)[3] is a widely used technique for the fast training algorithms. These algorithms make it possible to apply SVM on large scale problems.

The SVMs are often used without weightings. That is, no knowledge of the importance of the classes or the data patterns is incorporated. The classes are equally important, and so do the data. However, for a problem where the data distribution is skewed, i.e., most outcomes belong to one class, or for a problem where the cost of miss–classifying is not the same on different classes, it is beneficial to consider adding weightings to data patterns[4]. The weighting factor is in fact in the SVM model and can be introduced for these problems. The weightings for the classes are decided and affect the objective function to be minimized. Research about the imbalanced cases often concerns the penalties of

S. Singh et al. (Eds.): ICAPR 2005, LNCS 3686, pp. 288–296, 2005.

false positives and false negatives that are not equally important[5][6][7]. They give different classes different weightings according to the sample counts or the importance of the classes.

The weighting strategy mentioned above is problem dependent. Since the SVMs find the decision function (or the optimal hyperplane) to separate classes without making use of any problem domain knowledge, the weightings should be chosen according to the model or the data itself. In this work we present a strategy to set the weighting for each training pattern. The strategy gives weightings to data patterns according to their distances to the optimal hyperplane. The concept is based on the observation that a more closer data pattern to the hyperplane contributes more to the determination of the hyperplane itself. However, the optimal hyperplane is unknown to us in the training phase. Thus, we propose using an estimate of the hyperplane to approximate the underlying concept. The estimate of the hyperplane is computed by using a random subset of the given training data. We call this phase of training the stage 1 model. That is, we propose using a small SVM to initialize the whole SVM such that the training process can converge faster. The proposed weighting strategy is obviously not for the sample count balance or the cost reasons, but to reflect the importance of each sample and improve the training process. The remaining part of the paper is organized as follows. The SVM model is stated in section 2, and the proposed method will be introduced in section 3. Section 4 provides the experiment results showing that the proposed strategy maintains the testing performance and improves the training speed.

2   Weighted Support Vector Machine

Given the labeled training patterns (x_i, y_i), $i = 1, \ldots, l$, where $x_i \in R^n$ is the feature vector and $y_i \in \{+1, -1\}$ is the class label of x_i, an SVM looks for the linear discriminant function $f(x) = \omega^T x + b$ which maximizes the minimal distance from the data to the function. The maximized minimal distance is called the margin, and the found function is called the optimal separating hyperplane. Those data close to the hyperplane are the support vectors. An SVM finds the hyperplane by solving the following optimization problem:

$$\min_{\omega, b, \xi} \frac{1}{2} |\omega| + C \sum_{i=1}^{l} \xi_i \tag{1}$$

$$\text{subject to } y_i \left(\omega^T x_i + b \right) \geq 1 - \xi_i, \ \xi_i \geq 0,$$

or its dual problem

$$\min_{\alpha} \frac{1}{2} \alpha^T Q \alpha - e^T \alpha \tag{2}$$

$$\text{subject to } 0 \leq \alpha_i \leq C, \ i = 1, \ldots, l, \ y^T \alpha = 0,$$

where e is the vector of all ones, C is the penalty of error which is positive, Q_{ij} is $y_i y_j \langle x_i, x_j \rangle$ and ξ_i is the relaxation parameter. The decision function of the classifier is

$$sign\left(\sum_{i=1}^{l} y_i\alpha_i \langle x_i, x\rangle + b\right) . \tag{3}$$

$\langle .,.\rangle$ indicates the inner product operation.

When the problem is complicate or is linearly non–separable, a mapping ϕ : $X \rightarrow F$ can be applied on the data first, transforming them into a higher dimensional space called the feature space, and then an SVM can look for the hyperplane in that space. The discriminant function then becomes $f(x) = \omega^T\phi(x)+b$, and the decision function therefore becomes a non-linear classifier

$$sign\left(\sum_{i=1}^{l} y_i\alpha_i \langle \phi(x), \phi(y)\rangle + b\right) \tag{4}$$

in the input space. Kernel functions are introduced to eliminate computations with the mapping function. A kernel function $K(x,y) = \langle \phi(x), \phi(y)\rangle$ gives the inner product value of x and y in the feature space. The most often used kernel functions are radial basis functions: $K(x,y) = \exp\left(-\gamma |x-y|^d\right), \gamma > 0$.

For the weighted case, Eq. (1) can be rewritten to

$$\min_{\omega,b,\xi} \frac{1}{2}|\omega|^2 + s_+C\sum_{i:y_i=+1}\xi_i + s_-C\sum_{j:y_j=-1}\xi_j \tag{5}$$
$$\text{subject to } y_i\left(\omega^T x_i + b\right) \geq 1 - \xi_i, \xi_i \geq 0 .$$

And we can combine the class weightings s_+, s_- with C

$$\min_{\omega,b,\xi} \frac{1}{2}|\omega|^2 + C_+\sum_{i:y_i=+1}\xi_i + C_-\sum_{j:y_j=-1}\xi_j \tag{6}$$
$$\text{subject to } y_i\left(\omega^T x_i + b\right) \geq 1 - \xi_i, \xi_i \geq 0 .$$

A choice of s_+, s_- can be referred to the ratio between the sample numbers of the two classes[8]. However, such choice requires knowledge of the problem, and possibly affects the training results[5].

Rewrite Eq. (5) to

$$\min_{\omega,b,\xi} \frac{1}{2}|\omega|^2 + C\sum_{i=1}^{l} s_i\xi_i \tag{7}$$
$$\text{subject to } y_i\left(\omega^T x_i + b\right) \geq 1 - \xi_i, \xi_i \geq 0,$$

where the weightings can be applied to the samples separately. Instead of using a priori knowledge of samples of the problem, we propose a strategy to decide the weightings by the property of samples in the SVM training process. For a linearly non–separable problem, a kernel function can be incorporated to the weighted models similar to the non-weighted case.

3   Weighting Separate Data

The hyperplane is decided by support vectors. The support vectors are the samples "near" the hyperplane. They "push" the hyperplane to maximize the margin. After the training process, the data with non-zero α values are defined as the support vectors. Therefore, only the support vectors really involve in computing the decision function. If we know a priori which samples would be the support vectors and train the SVM with only those samples, we still have the same classifier that trained with all training patterns, as illustrated with Fig. 1.

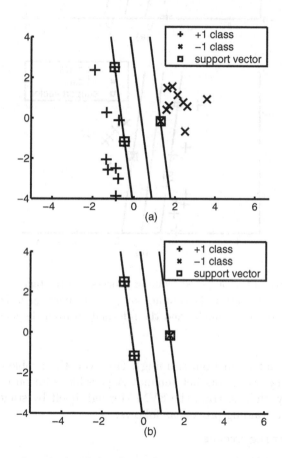

Fig. 1. The hyperplane is decided only with certain samples. (a) An SVM classifier trained with all samples. (b) An SVM classifier trained with samples to be support vectors.

As we can see from the above statement, the support vectors and the hyperplane is dependent on each other. Once the support vectors are known, to train an SVM we can set the weightings of samples other than support vectors to zero

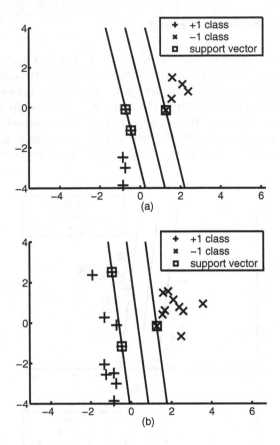

Fig. 2. An example of the proposed training process. (a) In stage 1, a standard SVM classifier is trained with part of the training samples. (b) In stage 2, the weighted SVM is trained with all training samples and the information from the model generated in stage 1.

(or just simply omit them from training). However, the problem is that we do not know the support vectors beforehand. A possible solution is to use a boot-straping strategy, that is, train the SVM through itself by stages. We propose using a two-stage training process as follows:

```
two-stage training process
    stage 1
        select a subset Xs of all sample set X
        train the standard SVM model S1 with Xs
    stage 2
        judge the importance of all samples in X by S1
        weight the samples in X as W according to their importance
        train the weighted SVM model S2 with X and W
```

An example of the training process is illustrated in Fig. 2.

We suggest training the models of the two stages in the same feature space. That is, use the same kernel function and related parameters. The hyperplanes found in different feature spaces may be very different, which means the relative locations or distances of the samples to the hyperplane change. If the two training processes are held in the same feature space, the two classifiers are possible to be similar, that is, the samples near the first hyperplane are more possible to be near the second one. One trick that should be mentioned is about setting the value of the penalty parameter C. If C is different between training processes, even using the same sample set and feature space, the resultant classifiers may be different because the cost of the relaxation term changes. Fig. 3 shows the effects of different training parameters, where different kernel functions and values of C are used. Although the training data sets are the same with those of Fig. 1, Fig. 3 has different resultant SVMS.

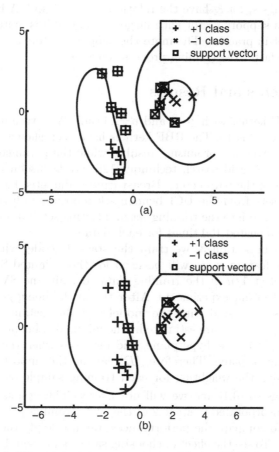

Fig. 3. Effects of different training parameters (a) An SVM classifier trained with same data set of Fig. 1a but a RBF kernel is used instead of a linear kernel. (b) An SVM classifier trained in same feature space but with different penalty parameter C.

Since the training time complexity of the SVMs is experimentally shown about $O(l^2)$[9][10], we expect reducing the sample size would speed up the training process significantly running stage 1. When training an SVM with working set selection methods[9][11], training samples are split into the working set B which contains q samples and the fix set N contains the rest. Eq. (2) can be decomposed into

$$\min_{\alpha} -\alpha_B^T (1 - Q_{BN}\alpha_N) + \tfrac{1}{2}\alpha_B^T Q_{BB}\alpha_b + \tfrac{1}{2}\alpha_N^T Q_{NN}\alpha_N - \alpha_N^T 1 \qquad (8)$$

$$\text{subject to } 0 \le \alpha_i \le C, \, i = 1, \ldots, l, \, \alpha_B^T y_B + \alpha_N^T y_N = 0,$$

where $\alpha = \begin{vmatrix} \alpha_B \\ \alpha_N \end{vmatrix}$, $y = \begin{vmatrix} y_B \\ y_N \end{vmatrix}$ and $Q = \begin{vmatrix} Q_{BB} & Q_{BN} \\ Q_{NB} & Q_{NN} \end{vmatrix}$. The optimization process is iteratively updating the working set B by selecting those samples with non zero α value. When the number of candidates are larger than the working set size q, the samples with larger output error and gradient of α are preferred. α_i and the weightings s_i for ξ_i have the relation $0 \le \alpha_i \le s_i C$. A larger weighting would result in a support vector with large α value, and is more likely to be in the weighting set in practice. The more the support vectors are included in the working set, the faster the training process converges.

4   Experiments and Results

We use the RBF kernel with $\gamma=5$ and $d=2$ in all SVM training process. The penalty parameter C is 10. The RBF kernels have been shown helpful in training the SVMs to generate promising results, while the parameters we used are decided arbitrary. A grid search technique[12] may be useful to find the better combinations of the parameters. Breast-cancer, diabetis, flare-solar, german and heart data sets from the UCI benchmark repository[13] are tested. Each dataset is partitioned into the training set and testing set 100 times[14][15], thus experiments are repeated 100 times for each data set.

As suggested in section 3, we train the stage 1 model with the standard SVM. The same feature space will be used for the weighted SVM in stage 2. We randomly select $1/n$ of the training data to train the SVM in the stage 1. n is 2,3 and 4 in our experiments. After the model being generated, all the training samples are tested with the model and the distance values D from samples to the hyperplane are estimated. The minimal value of $|D|$ is zero when a pattern falls right on the hyperplane, and increases when the pattern locates away from the hyperplane. Therefore, $|D|$ can be the importance index of a pattern. We assign the weighting for each training sample as $\frac{1}{|D|+1}$, thus the weighting belongs to (0,1] and we will not over-weighting a sample. Then the weightings are incorporated to stage 2 of the SVM training process that uses all training data. To compare the performances, we also apply the standard SVM on each data set. To see the effect of choosing samples in stage 1, we also run the proposed method with first 1/3 samples of each data set. The average testing rates of the 100 trials are shown in Table 1, and the total training times are shown in Table 2.

Table 1. Average testing rate (%)

Data set		breast-cancer	diabetis	flare-solar	german	heart
standard svm		74.18	75.75	67.61	76.33	82.33
proposed, $n=$	2(random)	74.56	75.83	67.62	76.27	82.69
	3(random)	74.52	75.78	67.62	76.41	82.55
	4(random)	74.82	75.72	67.63	76.36	82.95
	3(first 1/3)	74.77	75.72	67.62	76.31	83.10

Table 2. Training time (sec.)

Data set			breast-cancer	diabetis	flare-solar	german	heart
standard svm			291.79	768.13	653.72	1376	132.2
proposed, $n=$	2 (random)	stage 1	106.59	291.94	236.07	438.36	54.723
		total	277.66	773.55	676.42	1309.8	145.38
	3 (random)	stage 1	58.197	171.39	139.99	239.05	32.772
		total	235.79	653.41	599.13	1104.1	122.39
	4 (random)	stage 1	37.325	119.07	91.116	165.81	23.964
		total	216.87	595.94	550.14	1033.3	115.34
	3 (first 1/3)	stage 1	58.623	179.1	133.97	241.63	33.476
		total	237.75	666.07	592.6	1114.2	124.02

The results show that in most cases the testing performance of the proposed method maintains, while the proposed method is faster than standard, non-weighted (or, the weighting=1 for each sample) SVM when $n > 2$. In the case $n = 2$, training the stage 1 SVMs may cost more time then which could be saved in stage 2, and the proposed method performs about the same as standard SVM does. When n gets larger, more training time can be saved in stage 1, while training time of stage 2 maintains the same, so the total training time will reduce. According to our experiments, the options of the sample selection in stage 1 does not matter. However, it is still possible that an SVM model generated in stage 1 cannot represent the data well when n is large (too few samples are chosen) or unrepresentative samples are chose.

5   Conclusion

We have proposed a weighting strategy for weighted support vector machines. The concept behind the strategy is the observation that a sample closer to the optimal hyperplane would affect the objective function more than the others. We find such samples by a simpler classifier that is trained with a subset of data patterns. Assigning larger weightings to them in the subsequent training process will emphasize their contribution to the relaxation term of the optimizing objective function, and makes the training process converging faster. That is, we propose to initiate the weightings for a weighted SVM by the information from a previously trained SVM model with a subset of the training samples.

The experiments show that a random subset selected from training samples works fine. There exist some other methods that can also generate a subset from training samples[16][17], which can also be combined with our strategy. The simulation results show that the proposed weighting strategy is effective both in keeping the SVM testing performance and in speeding up the SVM training process.

References

1. Vapnik, V. : Statistical Learning Theory. New York: Wiley (1998)
2. Cortes C., Vapnik, V. : Support vector networks. Machine Learning **20** (1995) 273–297
3. Platt, J.C. : Fast training of support vector machines using sequential minimal optimization. In: Scholkopf, B., Burges, C.J.C.,Smola, A.J. (eds.) : Advanced in Kernel Methods-Support Vector Learning. Cambridge MA: MIT Press (1999)
4. Xu, P., Chan, A.K. : Support vector machines for multi-class signal classification with unbalanced samples. Proc. IJCNN 2003 **2** 1116–1119
5. Li, D., Du, S., Wu, T. : A weighted support vector machine method and its application. Proc. 5th world congress on intelligent control and automation (2004) 1834–1837
6. Lee, K.K., Gunn, S.R., Harris, C.J., Read, P.A.S. : Classification of imbalanced data with transparent kernels. Proc. IJCNN 2001 **4** 2410–2415
7. Veropoulos, K., Cambell, C., Cristianini, N. : Controlling the sensitivity of support machines. Proc. international joint conference on artificial intelligence 1999
8. Morik, K., Brockhausen, P., Joachims, T. : Combining statistical learning with a knowledge-based approach, a case study in intensive case monitoring. Proc. 16th International conference on machine learning (1999) 268-277.
9. Platt, J.C. : Sequetial minimal optimization: A fast algorithm for training support vector machines. Microsoft Research, Technical Report MST-TR-98-14 (1998)
10. Maruyama, K.-I., Maruyama, M.,Miyao H., Nakano, Y. : A method to make multiple hypotheses with high cumulative recognition rate using SVMs. Pattern Recognition **37** (2004) 241–251
11. Osuna, E., Freund, R., Girosi, F. : Training support vector machines:an application to face detection. Proc. CVPR'97 (1997)
12. Hsu, C.-W., Lin, C.-J. : A comparison of methods for multiclass support vector machines. IEEE trans. Neural Networks **13** (2002) 415–425
13. Blake, C. L., Merz, C. J. : UCI Repository of Machine Learning Databases. [Online], Univ. California, Dept. Inform. Comput. Sci., Irvine, CA., (1998) available from World Wide Web: http://www.ics.uci.edu/mlearn/MLRepository.html
14. Ratsch, G., Onoda T., Muller, K.-R. : Soft margins for AdaBoost. Machine Learning **42** (2001) 287–320
15. Benchmark Repository. [Online], available from World Wide Web: http://ida.first.fraunhofer.de/projects/bench/
16. Franc,V., Hlavac,V. : Training set approximation for kernel methods. in Proc. Computer vision winter workshop (2003)
17. Shih, L., Chang, Y.-H., Rennie, J.D.M., Karger, D. : Not Too Hot, Not Too Cold: The Bundled-SVM is Just Right. in Proc. ICML-2002 Workshop on Text Learning (2002)

Configuration of Neural Networks for the Analysis of Seasonal Time Series

T. Taskaya-Temizel and M.C. Casey

Department of Computing, School of Electronics and Physical Sciences,
University of Surrey, Guildford GU2 7XH, UK
{t.taskaya, m.casey}@surrey.ac.uk

Abstract. Time series often exhibit periodical patterns that can be analysed by conventional statistical techniques. These techniques rely upon an appropriate choice of model parameters that are often difficult to determine. Whilst neural networks also require an appropriate parameter configuration, they offer a way in which non-linear patterns may be modelled. However, evidence from a limited number of experiments has been used to argue that periodical patterns cannot be modelled using such networks. In this paper, we present a method to overcome the perceived limitations of this approach by determining the configuration parameters of a time delayed neural network from the seasonal data it is being used to model. Our method uses a fast Fourier transform to calculate the number of input tapped delays, with results demonstrating improved performance as compared to that of other linear and hybrid seasonal modelling techniques.

1 Introduction

In time series analysis, the recognition of patterns is important to facilitate the estimation of future values. This is especially evident for financial time series forecasting, where techniques such as technical and regression analyses have been developed that rely upon the identification of different temporal patterns [1]. In particular, regression analysis, whose application area is not only limited to financial forecasting [2,3,4], relies on the identification of patterns within the series, such as *trend* and *seasonality*. These patterns can be modelled using statistical techniques, such as autoregressive (AR) variants, but constructing such models is often difficult. Whilst the application of neural networks to time series analysis remains controversial [1,4], they appear to offer improved performance, for example, when used in hybrid models [5]. In hybrid models applied to seasonal data series, seasonality is first decomposed using techniques such as linear filters [1]. One such method is the application of the autocorrelation function to determine the input lags used to build a linear AR model, and in particular only the significant lags are selected [3]. However there is evidence to suggest that using just these selected lags does not give optimal linear models [6]. Furthermore, it is unclear whether the lags obtained from the autocorrelation function are sufficient for use in a non-linear model, such as a neural network. In this paper

S. Singh et al. (Eds.): ICAPR 2005, LNCS 3686, pp. 297–304, 2005.

we describe a method to configure a neural network to model time series that exhibit cyclic behaviour using a more appropriate selection of input lags. When applied to a *time delayed neural network* (TDNN), our method demonstrates similar performance compared to more complex hybrid modelling techniques.

In statistics, periodical variations are treated in two different ways. First, if periodical patterns change stochastically during time, one can apply seasonal differencing to eliminate seasonality. Second, if behaviour of the periodicity is deterministic, models can be applied by taking into account the type of the variation, such as whether the changes are in additive or multiplicative form. Linear processes such as AR models, seasonal AR models and the Holt-Winters method are among the few seasonal modelling techniques that have proven successful [1,3]. However, to model non-linear patterns, appropriate non-linear techniques are required, such as neural networks.

Non-linear neural networks are capable of extracting complex patterns in time series successfully to some degree [2,7], although identifying whether non-linear models are required remains difficult [1]. A well-known technique to perform temporal processing is to use memory of past input and activation values within the network to allow it to identify temporal patterns. These TDNN models [7,8] are widely used because of their simplicity, either on their own [9,10,11], or in hybrid models such as with an autoregressive integrated moving average model (ARIMA) [5,12,13]. Nevertheless, it has been reported that such hybrids do not necessarily outperform single models [11].

The limitations of neural networks are essentially an inability to cope with changes in mean and variance [9], which can be attributed, for example, to trend and exponential seasonality. The mean and variance of a series may be stabilised using techniques such as differencing and the Box-Cox transformation [14]. Similarly, it has been argued that periodical patterns should be removed prior to modelling with a neural network [12,13,15,16]. However, it has been shown that certain periodical patterns can be successfully modelled using neural networks if the network is configured appropriately, and the time series preprocessed to stabilise the mean [10].

The studies so far have focused on model selection tools designed for building linear AR models, which are then used to construct a TDNN for cyclic series [9,12]. For example, Cottrell [9] used Akaike and Bayesian Information Criteria to find an optimum TDNN. For the Sunspot data, their results suggested an architecture with four input delays and four hidden neurons. Despite this, the Sunspot data exhibits an 11 year cycle, and results suggest that a minimum lag of 11 is required (for example, [2,17]). In contrast to Cottrell's approach, Zhang and Qi [12] used the autocorrelation function to configure a TDNN. First, they obtained the significant lags within the series based on the an analysis of the autocorrelation. Then they employed these lags to construct the input delays of TDNN. In this paper we describe a method, which is an extension of that described in [10], for selecting the input delays using a fast Fourier transform, comparing our results with Zhang and Qi's.

2   Configuring TDNN for Seasonal Time Series

A TDNN is a variant of the multi-layer perceptron in which the inputs to any node i can consist of the outputs of the earlier nodes generated between the current step t and step $t - d$, where $d \in Z^+$ and $\forall d < t$. Here, the activation function for node i at time t is:

$$y_i(t) = f \left(\sum_{j=1}^{M} \sum_{d=1}^{T} w_{ij}(t - d) y_j(t - d) \right) \tag{1}$$

where $y_i(t)$ is the output of node i at time t, $w_{ij}(t)$ is the connection weight between node i and j at time t, T is the number of tapped delays, M is the number of nodes connected to node i from preceding layer, and f is the activation function, typically the logistic sigmoid. In this paper, we consider the case when we have tapped delays in the input layer only: an IDNN.

In order to set the number of delays in the input layer, we obtain the cycle information from the training data using a fast Fourier transform, which sometimes gives similar results to that of applying the autocorrelation function, but is more convenient to use within a systematic method. In the choice of the cycle information, we consider the dominant cycles within the data as determined by the outliers in the amplitude response. Then we construct the TDNN with the number of tapped delays IL equal to each of the extracted dominant cycles, choosing the final configuration based upon the best performing network on the validation data set. Specifically:

1. Given a time series $\{x_i\}_{i=1}^{K}$, create training $\{x_i\}_{i=1}^{TR}$, validation $\{x_i\}_{i=TR+1}^{VL}$ and test data sets $\{x_i\}_{i=VL+1}^{K}$, where i is the time index.
2. Stabilise the mean of the series by computing the first-order difference.

$$x_{i+1}' = x_{i+1} - x_i \tag{2}$$

   where x' is the stabilised time series.
3. Estimate the number of input tapped delays IL using the dominant cycle information in the differenced series:
   (a) Compute the fast Fourier transform of $\{x_i'\}_{i=1}^{VL}$:

$$X_i = \sum_{j=1}^{VL} x_j' w_{VL}^{(j-1)(i-1)} \tag{3}$$

   where $w_{VL} = e^{(-2\pi i)/VL}$ is a VLth root of unity.
   (b) Let $R_i^0 = |X_i|$ be the amplitude response of X_i.
   (c) Discard the periods that are greater than $VL/2$.
   (d) Set $j = 1$.
       Let S be a set and set $S = \oslash$.
       Let P_i be the set of periods.

While not finished

    Compute the mean μ_{j-1} and standard deviation σ_{j-1} of R_i^{j-1}

    Extract the outliers P_i, where $R_i^{j-1} > \mu_{j-1} + 3\sigma_{j-1}$

    Set $S = S \cup P_i$.

    Set R_i^j be the amplitude response without the outliers P_i.

    Set $j = j + 1$.

    If $P = \oslash$ exit loop.

End while

(e) Set the number of input tapped delays IL to the closest integer value ± 2 [12] for each period within S. The best number will be selected experimentally from these according to the test set performance.

4. Restrict the number of nodes in the output layer to unity and set the hidden layer size

$$H \leq \frac{(IL + 1)}{2} \tag{4}$$

5. Normalize the series using the z-score to improve training in the network.

$$z_i = \frac{x_i' - \bar{x}'}{\sigma_{x'}} \tag{5}$$

Note that the outliers P in Step 3d are rounded integer periods.

3   Experiments and Results

To evaluate this method, we chose four industrial production series (from Federal Reserve Board [18]): consumer goods (starting January 1970), durable goods (starting January 1947), fuels (starting January 1954), and total industrial production (starting January 1947)and five U.S. Census Bureau series [19] (starting January 1992): retail, hardware, clothing, furniture, and bookstore, all ending in December 2001 [12]. Each of these monthly data sets exhibit strong seasonalities that are difficult to predict.

In order to compare the selected architectures using the method described with the performance of a TDNN in general, we conducted a number of experiments with network configurations of $2i : 2j : 1$, where $1 \leq i \leq 33$ and $1 \leq j \leq 16$, a total of 528 networks. Each network was configured to use a hyperbolic tangent activation function for the hidden layer and a linear function for the output layer. Training was performed using the gradient descent algorithm for a maximum of 20,000 epochs, with initial learning rate parameter $\lambda = 0.1$, increased by 1.05% if the training error decreased, otherwise decreased by 0.7%, if the training error increased by over 4%. Each configuration was tested with 30 different random initial conditions to provide an average root mean square error (RMSE). The testing data set was used to determine which was the best architecture once training was complete.

3.1 Results

For each of the selected input delay sizes IL per data set, Table 1 shows the architecture selected by our algorithm and the best performing network within the trials. In five data sets (FR Fuels, FR Total Production, USBC Bookstore, USBC Clothing, USBC Retail, USBC Hardware), we obtained the best performing architectures among our trials using our method. In the other four data sets, the method did not pick up the best performing architecture. However, the selected architectures were amongst the top ten within the trials.

Table 1. The selected method's performances compared with the performances of the best network configuration obtained from the trials. Bold values show the correctly identified TDNN configurations by the algorithm. Our algorithm finds the best configurations on five out of nine data sets.

Data Sets	Selected Model		Best Model	
	Config	RMSE	Config	RMSE
FR Consumer Goods	16:02:01	1.25 ± 0.17	24:02:01	1.07 ± 0.89
FR Durable Goods	14:06:01	3.15 ± 0.51	12:16:01	2.91 ± 0.34
FR Total Production	**42:02:01**	0.99 ± 0.05	**42:02:01**	0.99 ± 0.05
FR Fuels	30:02:01	1.73 ± 0.12	32:02:01	1.64 ± 0.13
USBC Bookstore	**12:02:01**	91.51 ± 10.40	**12:02:01**	91.51 ± 10.40
USBC Clothing	**14:02:01**	378.71 ± 68.29	**14:02:01**	378.71 ± 68.29
USBC Furniture	14:02:01	175.02 ± 10.57	48:02:01	161.09 ± 22.63
USBC Retail	**14:02:01**	634.72 ± 30.97	**14:02:01**	634.72 ± 30.97
USBC Hardware	**12:04:01**	37.97 ± 9.34	**12:04:01**	37.97 ± 9.34

In order to understand whether the method selects near optimum parameters for the TDNN, we compared the results given in Table 1 with those for each of the 528 networks constructed over 30 trials. Figure 1 shows the results for each of these networks for the FR total production data set. Part (a) shows the dominant cycles determined by the method, part (b) shows the performance of each of the 528 networks, with the number of neurons within the hidden layer on the x-axis, and the number of input tapped delays on the y-axis. The shading shows the average RMSE, with the dark areas showing the lowest values. Here we see that the best performing architecture is that with 2 hidden neurons and approximately 42 input tapped delays. This corresponds well with our method, which selects 43 as one of the dominant periods, and with the best performing architecture using 42 input delays, within our bounds of ±2, as suggested by [12]. Apart from FR durable goods, a TDNN has optimal performance when the hidden layer size is set between 2 and 4. We can therefore see that our method provides a way in which the dominant cycle information can be successfully used to construct a near-optimum TDNN to model seasonal time series.

In Table 2, we compared our best fit results among the selected networks with the best fit of the hybrid ARIMA models constructed by Zhang and Qi [12].

(a)   (b)

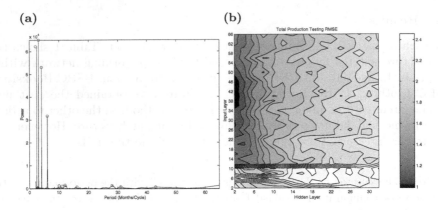

Fig. 1. (a) shows the dominant cycles, which are automatically picked up by the algorithm, (b) shows the error surface of the TDNN on 'Total Production' time series.

Table 2. The best fit selected architectures compared with the performance of TDNN and hybrid architecture constructed by Zhang and Qi [12]. Bold values indicate the minimum RMSE obtained per data set. Our TDNN architectures outperform in all data sets compared to Zhang and Qi's TDNN architectures. They also outperform on five out of nine data sets compared to their hybrid architectures.

Data Sets	Selected Model	RMSE	TDNN [13]	ARIMA-NN[13]
FR Consumer Goods	18:04:01	0.91	1.48	**0.68**
FR Durable Goods	12:02:01	**2.16**	5.98	3.63
FR Total Production	42:02:01	**0.77**	1.62	0.85
FR Fuels	30:04:01	1.30	1.83	**0.81**
USBC Bookstore	12:02:01	**71.10**	170.49	88.74
USBC Clothing	14:02:01	**277.34**	1117.72	315.43
USBC Furniture	14:06:01	144.32	226.68	**99.45**
USBC Retail	14:02:01	**546.58**	1785.77	975.55
USBC Hardware	12:02:01	**19.76**	105.12	49.17

For the hybrid architecture, our TDNN models outperformed in six data sets (FR Durable, FR Total Production, USBC Bookstore, USBC Clothing, USBC Retail, USBC Hardware). Zhang and Qi constructed the networks by taking into consideration the correlation structure of the series. Based on the outcome of their analysis, they considered ten time lags: 1-4, 12-14, 24, 25, and 36, where 12, 24, and 36 months apart are highly correlated. The number of hidden nodes varied between 2 and 14 with an increment of 2. For example, the best neural network configuration for the USBC retail series was 36:12:1, where 36 shows the maximal lagged term. Their findings showed that the input layer should comprise at least a maximal lag of 12 (5 input nodes) for all series. More specifically, they reported that for the detrended data using polynomial fitting, the maximal lags were identified as 13 for USBC retail, 36 for FR consumer goods, and 14 for durable goods. For the detrended and deseasonalised data, they used a maximal lag of 1 for USBC retail, 4 for FR consumer goods and 4 for durable goods. They concluded that they found large discrepancies among the hidden layer nodes. However, we found that neural networks with a small number of hidden nodes perform significantly better than ones with a large number of hidden nodes in our earlier study [11]. We observed that the networks trained with continuous lag information outperform networks without.

4   Conclusion

We have described an algorithm that can be used to find the optimum TDNN configuration for modelling seasonal time series. The method described selects a number of candidate architectures that include those that are the best performing for this technique compared to existing results. The results also demonstrate that a TDNN model can produce comparable performance to other hybrid models. One advantage with this approach is that the performance of an ARIMA neural network hybrid is likely to degrade due to overfitting [11]. In this case it therefore appears that using relatively simple models can improve performance.

One restriction to our method is that it can only model stationary seasonal time series. For example, our method cannot model series in which the amplitude of the cycle increases constantly over time. To be able to model such a series with neural networks, first either the series should be stabilised using an appropriate transformation, or a seasonal AR model should be used. Furthermore, the poorer performance of our method on three data sets (FR Consumer Goods, FR Fuels and USBC Furniture) requires further investigation to determine whether there are any particular characteristics of these that affects our method.

In evaluating the method through comparison of different network configurations, we note that as the number of free parameters increases in the network (the number of input delays and hidden neurons), that the model is likely to overfit both to the training and validation data sets giving poor generalisation. Experimentally this tells us that the input layer size should be set to less than ten percent of the total data set size in order to achieve improved results, but further investigation is required to formalise this. Similarly, we also note that a TDNN generally has optimal performance on the selected data sets when the hidden layer size is set between 2 and 4, commensurate with our previous work [10,11].

Acknowledgements

We are grateful to the Fingrid project [RES-149-25-0028], which provided us a Grid environment comprising 24 machines to run our simulations.

References

1. Chatfield, C.: The Analysis of Time Series. Sixth edn. Texts in Statistical Science. Chapman & Hall, USA (2004)
2. Rementeria, S., Olabe, X.B.: Predicting sunspots with a self-configuring neural system. In: Proceedings of the 8th Conference on Information Processing and Management of Uncertainty in Knowledge-Based Systems (IPMU 2000). (2000)
3. Box, G., Jenkins, G.: Time Series Analysis: forecasting and control. Texts in Statistical Science. Holden Day (1970)
4. Faraway, J., Chatfield, C.: Time series forecasting with neural networks: A comparative study using the airline data. Applied Statistics 47 (1998) 231–250
5. Zhang, G.P.: Time series forecasting using a hybrid ARIMA and neural network model. Neurocomputing 50 (2003) 159–175
6. Rao, T.S., Sabr, M.M.: An introduction to bispectral analysis and bilinear time series models. Lecture Notes in Statistics 24 (1984)
7. Wan, E.: Time series prediction by using a connectionist network with internal delay lines. In Weigend, A., Gershenfeld, N., eds.: Time Series Prediction: Forecasting the Future and Understanding the Past. SFI Studies in the Sciences of Complexity, Addison Wesley (1994) 1883–1888
8. Waibel, A., Hanazawa, T., Hinton, G., Shikano, K., Lang, K.J.: Phoneme recognition using time-delay neural networks. IEEE Transactions on Acoustics, Speech and Signal Processing 37 (1989) 328–339
9. Cottrell, M., Girard, B., Girard, Y., Mangeas, M., Muller, C.: Neural modeling for time series: a statistical stepwise method for weight elimination. IEEE Transactions on Neural Networks 6 (1995) 1355–1364
10. Taskaya-Temizel, T., Casey, M.C., Ahmad, K.: Pre-processing inputs for optimally-configured time-delay neural networks. IEE Electronic Letters 41 (2005) 198–200
11. Taskaya-Temizel, T., Ahmad, K.: Are ARIMA neural network hybrids better than single models? In: Proceedings of International Joint Conference on Neural Networks (IJCNN 2005), Montréal, Canada (2005) to appear
12. Zhang, G.P., Qi, M.: Neural network forecasting for seasonal and trend time series. European Journal of Operational Research 160 (2005) 501–514
13. Tseng, F.M., Yu, H.C., Tzeng, G.H.: Combining neural network model with seasonal time series ARIMA model. Technological Forecasting and Social Change 69 (2002) 71–87
14. Box, G.E.P., Cox, D.R.: An analysis of transformations. JRSS B 26 (1996) 211–246
15. Virili, F., Freisleben, B.: Nonstationarity and data preprocessing for neural network predictions of an economic time series. In: Proceedings of International Joint Conference on Neural Networks (IJCNN 2000), Como, Italy (2000) 5129–5136
16. Nelson, M., Hill, T., Remus, W., O'Connor, M.: Time series forecasting using neural networks: Should the data be deseasonalized first? Journal of Forecasting 18 (1999) 359–367
17. Weigend, A., Gershenfeld, N.A., eds.: Time Series Prediction: Forecasting the Future and Understanding the Past. Addison-Wesley, Reading, MA (1993)
18. Federal Reserve Time Serial Data. http://www.federalreserve.gov/releases/G17/table1_2.htm (2005) Last accessed: Jan.2005.
19. U.S. Census Bureau Monthly Trade and Food Services Data Sets. http://www.census.gov/mrts/www/mrts.html (2005) Last accessed: Jan.2005.

Boosting Feature Selection

D.B. Redpath and K. Lebart

ECE, School of EPS, Heriot-Watt University,
Edinburgh, EH14 4AS, UK
dr2@hw.ac.uk, K.Lebart@hw.ac.uk

Abstract. It is possible to reduce the error rate of a single classifier using a classifier ensemble. However, any gain in performance is undermined by the increased computation of performing classification several times. Here the Adaboost$_{FS}$ algorithm is proposed which builds on two popular areas of ensemble research: Adaboost and Ensemble Feature Selection (EFS). The aim of Adaboost$_{FS}$ is to reduce the number of features used by each base classifer and hence the overall computation required by the ensemble. To do this the algorithm combines a regularised version of Boosting Adaboost$_{Reg}$ [1] with a floating feature search for each base classifier.

Adaboost$_{FS}$ is compared using four benchmark data sets to Adaboost$_{All}$, which uses all features and to Adaboost$_{RSM}$, which uses a random selection of features. Performance is assessed based on error rate, ensemble error and diversity, and the total number of features used for classification. Results show that Adaboost$_{FS}$ achieves a lower error rate and higher diversity than Adaboost$_{All}$, and achieves a lower error rate and comparable diversity to Adaboost$_{RSM}$. However, over the other methods Adaboost$_{FS}$ produces a significant reduction in the number of features required for classification in each base classifier and the entire ensemble.

1 Introduction

Adaboost and ensemble feature selection are two active areas of ensemble research. Adaboost [2] commonly known as Boosting is a successful technique for building ensembles, which influences different classifier decisions using example reweighting. Its operation is closely associated with large margin theory [3]. Ensemble feature selection builds an ensemble using different feature subsets for each base learner. This provides a potentially more active way of promoting decision diversity in comparison to partitioning or reweighting training examples [4]. Unfortunately, the success of building an ensemble this way is less theoretically understood.

Combining these two techniques is of interest for two reasons. Firstly, for cases where Boosting performs poorly or fails. Boosting performs poorly when the selected data and classifier pairing gives inadequate error diversity for ensembling to be beneficial [4]. Boosting fails if the training error is zero using the entire feature description. This happens when training data is small and may not reflect

S. Singh et al. (Eds.): ICAPR 2005, LNCS 3686, pp. 305–314, 2005.

the true class separation of the dataset given sufficient data. Secondly, to reduce the number of features used by each base classifier in the ensemble.

The first examples combining Boosting and feature selection were Boosting *Decision Stumps* [5] and *Domain-Partitioning* [6]. In both cases the motivation was to improve the fitting of the base classifier to match that of a *weak learner*. Boosting using different single features [7] has also been undertaken.

Here Boosting with feature selection is proposed which involves integrating a floating feature search into a soft margin implementation of Boosting. The paper is organised as follows. Section 2 details the new Boosting algorithm. Section 3 details experiments, results and discussion on benchmark data. Section 4 concludes the paper.

2   Boosting with Feature Selection

In its simplest form Adaboost drives down the error rate of an ensembled classifier, by concentrating the classifier at each iteration on examples that are particularly difficult to classify. Control of the learner is achieved by training example weighting. At each iteration a hypothesis is obtained for the base learner using the current weighted training set. These are then updated by placing more weight on training examples that are incorrectly classified. The resulting ensemble decision is then a linear combination of the support for each base hypothesis. This is an approximate large margin classifier, since Boosting has concentrated base learners on the class boundary which has training examples with the highest error.

The success of Boosting depends primarily on the choice of classifier for a given problem. As the method relies on training example weighting or partitioning, the classifier selected must be sensitive to changes in training examples. Many classifiers are relatively stable to training example variation and are unsuitable for Boosting. Unstable classifiers are more successfully Boosted, where classifiers such as decision trees and neural networks are common choices. Clearly a classifier which does not change its decision cannot contribute complementary decisions to an ensemble. Such a condition is know as decision diversity and is the main incentive for classifier ensemble design [4].

Training classifiers using different feature subsets provides an alternative method for introducing variation between decisions. Building ensembles this way has been studied by a number of researchers. Several improvements have been proposed to the inital Random Subspace Method (RSM) [8] based on accuracy refinement [9,10]. Genetic and sequential search methods have also been proposed [11,12,13], but require a careful choice of selection criteria to avoid overfitting. The objective of these methods was to use feature selection for training each base classifier as a means of decorrelating errors between them.

Here Boosting and feature selection are combined. In developing such an algorithm, emphasis must be placed on the intrinsic behaviour of both algorithms. Feature selection is commonly used to identify a set of relevant, non-redundant features. Boosting focuses on individual training examples. Its aggressive operation results in easy to classify examples receiving little weight, the remaining weight placed on hard examples. Weighted examples of this type make it difficult to discriminate between a relevant and an irrelevant feature. However, redundant features can be discriminated. To this end Boosting with feature selection will not perform explicit feature selection and irrelevant features must be removed using pre-processing. A second anticipated problem is overfitting from either small amounts of training data or from noisy examples. This problem is approached using soft margins.

2.1   Soft Margins for Boosting

In this study the Adaboost$_{Reg}$ algorithm is used [1], the functionality of which is centred around the theory of training example *margins* [3]. A training set is defined by $\mathbf{S} = \{S_1, ..., S_M\}$, where $S_m = (\mathbf{x}_m, y_m)$, $\mathbf{x}_m$ is a pattern vector $\mathbf{x} \in \mathbb{R}^N$ and y_m is a class label $y_m \in \{-1, 1\}$. At each Boosting iteration a hypothesis h_t is created by training a base classifier $\mathcal{L}$ according to the training example weighting $\mathbf{w}^{(t)}$ provided by the current Boosting iteration such that $h_t = \mathcal{L}(\mathbf{S}, \mathbf{w}^{(t)})$. As a result Boosting produces an ensemble of weighted hypotheses $\mathcal{H} = \sum_t \alpha_t h_t(\mathbf{x})$, where α_t is the support for each hypothesis. The margin of each training point is then formally defined by

$$\rho_m^{(t)} = \rho(S_m, \boldsymbol{\alpha}^{(t)}) = y_m \sum_t \frac{\alpha_t h_t(\mathbf{x}_m)}{\sum_t \alpha_t}, \tag{1}$$

which is positive only if $\mathcal{H}$ classifies the example correctly. Boosting effectively maximises the smallest example margins, i.e. examples which are most difficult to classify. In turn the smallest training example margin can be used to define the margin of the training set hyperplane

$$\varrho = \arg\min_{m=1:M} \rho_m^{(t)}. \tag{2}$$

Such a definition of a large margin created by Boosting a training set correlates with the good generalisation performance observed [3].

Boosting creates a hard margin decision as it attempts to reduce the training error to zero. This can be a problem in cases where training examples are noisy due to outliers or miss-labelling. These cases can be over emphasised causing overfitting and poor generalisation. To solve this problem the soft margin has been developed for Boosting.

A soft margin approach proposed by Rätsch [1] is to monitor the *influence* of the training examples based on the hypothesis weighting α_r and the example weighting $w_m^{(r)}$, produced over Boosting iterations

$$\mu_m^{(t)} = \sum_{r=1}^t \frac{\alpha_r}{\sum_{r'=1}^t \alpha_{r'}} w_m^{(r)}, \tag{3}$$

which is the average weight of an example computed during the learning process. The rationale is that noisy examples that are hard to classify will have a high average weight, resulting in a high influence $\mu_m^{(t)}$. This can then be used to define the soft margin as follows

$$\bar{\rho}_m^{(t)} = \rho_m^{(t)} + C\mu_m^{(t)}, \tag{4}$$

where C is a regularization factor. Outliers with a consistently small margin receive a high average weighting and influence $\mu_m^{(t)}$. This is then added to the examples margin to increase the soft margin, which on subsequent Boosting iterations decreases the example weight $w_m^{(t)}$. Maximising the soft margin on all examples prevents outliers being forcefully classified by allowing for some errors, which compensates the tendency of Boosting to overweight outlying examples by trading margin and influence

Given this definition of Boosting soft margins, a feature selection algorithm can now be proposed to search for an appropriate set of features on which to train each Boosting base learner.

2.2   Selection Algorithm

Feature selection has many well established selection criteria and selection algorithms. The key decision of which search to use depends on the tolerable level of sub-optimality and criteria monotonicity [14].

Here as a selection criterion the Boosting loss parameter (5) is used, which is based on the exponential sum of individual inverted training margins in their un-normalised form. This is a non-monotonic function as the feature subset size bears no direct relation to the magnitude of the cost.

As a search algorithm the floating search method [15] is used which is simplified for non-monotonic cases. This floating feature search is then conducted for each Boosting iteration to find a hypothesis which best minimises (5).

Using the Adaboost$_{Reg}$ algorithm and integrating the stated floating feature search Adaboost$_{FS}$ takes the form shown in Fig. 1. Adaboost$_{FS}$ begins by selecting a set of features $\mathbf{F}_t = \mathbf{F}^{(k)}$ on which to train each base classifier, where k indicates the feature subset size during the search. Features are assessed using the minimum of (5) as a selection criterion $J(\cdot)$. To calculate this the base classifier is first trained using the feature partitioned training set $\mathbf{S}_{\mathbf{F}_t} = \{(x_{1,f_1}, y_1), ..., (x_{M,f_k}, y_M)\}$, to obtain the hypothesis $h_{t,\mathbf{F}_t} = \mathcal{L}\{S_{\mathbf{F}_t}, \mathbf{w}^{(t)}\}$, which is used to calculate $\rho(S_{m,\mathbf{F}_t}, \boldsymbol{\alpha}^{(t)})$ and hence (5).

The final feature set for each iteration is chosen by inclusion and conditional exclusion stages, both of which seek addition and removal respectively of one feature which minimises the selection criterion (5). Floating search continues until the selection criterion is no longer improved $J(\mathbf{F}^{(k)}) < \Delta(\mathbf{F}^{(k-1)})$, where Δ is an improvement threshold used to control the final accuracy of the search. The selected hypothesis weighting α_t and hypothesis $h_{t,\mathbf{F}_t}$ are then included into the final ensemble. Finally, the new training example weights $w_m^{(t)}$ are updated. On repetition of the algorithm a new feature subset $\mathbf{F}_t$ is found which minimises the cost $J(\cdot)$ using the new example weighting $\mathbf{w}^{(t)}$. The process then continues until the required ensemble size is reached $t = T$.

Input: training set $S = \{(x_{1,1}, y_1), ..., (x_{M,N}, y_M)\}$, number of iterations T, improvement threshold Δ.
Initialise: $w_m^{(1)} = 1/M$ for all $m = \{1, ..., M\}$, empty ensemble $\mathcal{H} = \{\emptyset\}$.
Do for $t = 1, ..., T$,

1. **Select hypothesis:** using the margin loss (5) as a selection criteria $J(\cdot)$ calculated according to **2**.
   Initialise: $k = 0$, feature set $\mathbf{F}^{(k)} = \{\emptyset\}$ and remaining features $\overline{\mathbf{F}}^{(k)} = \{f_1, ..., f_N\}$, where $\mathbf{F}^{(k)} \notin \overline{\mathbf{F}}^{(k)}$.

   1.1 **Forward Inclusion:** of one new feature into the current feature set $\mathbf{F}^{(k)}$ from $\overline{\mathbf{F}}^{(k)}$ according to $j^o = \arg\min_j J(\{\mathbf{F}^{(k)} \cup \overline{\mathbf{F}}_j^{(k)}\})$, where $\forall j = \{1, ..., (N-k)\}$. Set $\mathbf{F}^{(k+1)} = \{\mathbf{F}^{(k)} \cup \overline{\mathbf{F}}_{j^o}^{(k)}\}$, $k = k+1$.

   1.2 **Conditional Exclusion:** of one feature from $\mathbf{F}^{(k)}$ according to $j^o = \arg\min_j J(\{\mathbf{F}^{(k)} \setminus \mathbf{F}_j^{(k)}\})$, where $\forall j = \{1, ..., k\}$.
   if$(J(\{\mathbf{F}^{(k)} \setminus \mathbf{F}_{j^o}^{(k)}\}) < \Delta J(\mathbf{F}^{(k)}))$, then $\mathbf{F}^{(k-1)} = \{\mathbf{F}^{(k)} \setminus \mathbf{F}_{j^o}^{(k)}\}$, $k = k-1$, **goto 1.2**.

   1.3 **Continue:** if $(J(\mathbf{F}^{(k)}) < \Delta J(\mathbf{F}^{(k-1)}))$, then **goto 1.1**, else **goto 3**.

2. **Calculate hypothesis:** using feature partitioned training set $\mathbf{S}_{\mathbf{F}_t}$ and example weighting $\mathbf{w}^{(t)}$.
   2.1 **Train:** $\mathcal{L}$ using $\mathbf{F}_t = \mathbf{F}^{(k)}$ from **1.** to obtain hypothesis $h_{t,\mathbf{F}_t} = \mathcal{L}(\mathbf{S}_{\mathbf{F}_t}, \mathbf{w}^{(t)})$, where $h_t : \mathbf{x} \mapsto \{-1, 1\}$.
   2.2 **Calculate:** hypothesis coefficient and corresponding loss

$$J(\mathbf{F}_t) = \arg\min_{\alpha_t} \sum_{m=1}^{M} \exp\left\{-\frac{1}{2}\left[\rho(S_{m,\mathbf{F}_t}, \boldsymbol{\alpha}^{(t)}) + C\mu_m^{(t)}\right]\sum_{r=1}^{t}\alpha_r\right\}. \tag{5}$$

   if $\alpha_t \leq 0$, then set $T = t - 1$ and **break**.

3. **Include:** hypothesis weighting α_t and hypothesis $h_{t,\mathbf{F}_t}$ to ensemble $\mathcal{H} = \{\mathcal{H} \cup (\alpha_t h_{t,\mathbf{F}_t})\}$.

4. **Update:** training example weights

$$w_m^{(t+1)} = \frac{1}{Z_t}\exp\left\{-\frac{1}{2}\left[\rho(S_{m,\mathbf{F}_t}, \boldsymbol{\alpha}^{(t)}) + C\mu_m^{(t)}\right]\sum_{r=1}^{t}\alpha_r\right\}, \tag{6}$$

   where Z_t is the normalisation factor $\sum_{m=1}^{M} w_m^{(t+1)} = 1$.

5. **Output:** Final hypothesis

$$\mathcal{H}_T(\mathbf{x}) = sign\left(\sum_{t=1}^{T}\alpha_t h_{t,\mathbf{F}_t}(\mathbf{x})\right). \tag{7}$$

Fig. 1. The Adaboost$_{\mathrm{FS}}$ algorithm

3   Performance Assessment and Discussion

The performance of Adaboost$_{FS}$ was assessed using benchmark data and was compared against two other Adaboost feature selection options, one which used all available features and one which used a random selection of features.

Benchmark datasets were selected from the UCI repository [16] for binary learning problems that contained a large number of features. The datasets Wdbc, Sonar and Musk met this specification and were selected for experimentation. Additionally, due to the restricted number of large dimensional binary learning problems, the multiclass dataset Multiple Features (Mfeat) was reformulated as a binary learning problem. The binary problem was to classifying the numbers 0-4 from 5-9. All the features were used in this dataset except the profile correlation and pixel averages. All data was used in the case of Wdbc and Sonar, but the Musk and Mfeat2 were reduced in size (for computational reasons) to 600 training examples using stratified random sampling. Datasets were pre-processed by normalising each feature to zero mean and unit standard deviation. The datasets dimensionalities and class distributions are summarised in Table 1.

Adaboost$_{FS}$ was compared with two other Adaboost feature selection schemes: Adaboost$_{All}$ and Adaboost$_{RSM}$. To compare our method with using all available feature information at each Boosting iteration Adaboost$_{Reg}$ was used. This method was termed Adaboost$_{All}$. To compare our method with an alternative ensemble feature selection strategy for Boosting the Random Subspace Method (RSM) [8] was used. The RSM has no guidance in its choice of subset leading to variable accuracy. However, this lack of accuracy is compensated by base classifier diversity something which the RSM exploits for high dimensional datasets. For direct comparison the RSM method was integrated into Adaboost$_{Reg}$. At each Boosting iteration a randomly size, randomly sampled set of features was sampled without replacement for training the hypothesis. This method was termed Adaboost$_{RSM}$.

Each algorithm was used to Boost an RBF base classifier [1], which had 5 centres, 10 refinement cycles and a regularisation of 10^{-5}. All the algorithms used regularised Boosting. As a comparison the regularisation values $C = \{0, 100\}$ were used for experimentation, the value with lowest test error performance was finally selected. The upper bound for α_t when minimising (5) was 10. For Adaboost$_{FS}$ the threshold Δ was set to 1 to find the lowest possible cost.

Table 1. Summary of benchmark datasets dimensions after rejecting 3% outliers

Dataset Name	Instances	Attributes	Class distribution [1 -1]
Wdbc	551	30	[35 65]
Sonar	201	60	[53 47]
Musk	581	167	[18 81]
Mfeat2	581	193	[50 50]

The performance of the three algorithms was found using ensembles sized $T = 1 \rightarrow 50$ using the following measures. The overall performance of each ensemble was assessed using error rate. To assess ensemble diversity, training error diversity was calculated using the kappa measure [17], where a smaller value represents greater diversity. To assess ensemble error the mean base classifier training error rate was calculated over training decisions. Complexity of the final ensemble was assessed using the total number of features used by each base classifier. Performance was estimated for each algorithm on the benchmark datasets using 10 runs of 2-fold cross validation.

The results ensembling the RBF(5) classifier using Adaboost$_{All}$, Adaboost$_{RSM}$ and Adaboost$_{FS}$ are shown in Figs. 2-5.

Examining the test error for the three methods Fig. 2, Adaboost$_{RSM}$ and Adaboost$_{FS}$ have performed better than Adaboost$_{All}$ for all the datasets Fig. 2.a)-d). There has been a variation in the effectiveness of regularisation for each method. Clearly, regularisation is only necessary if measurements contain outliers. Adaboost$_{All}$ has not performed very well providing only a slight improvement in the performance of a single classifier. Adaboost$_{FS}$ has performed better than Adaboost$_{RSM}$ for the Sonar and Mfeat2 datasets Fig. 2.b),d). The surprisingly good performance of the RSM method using random feature selections cor-

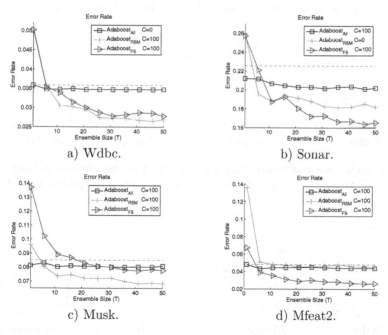

Fig. 2. Test error rate for the four benchmark datasets using Adaboost$_{All}$, Adaboost$_{RSM}$ and Adaboost$_{FS}$ to ensemble the **RBF(5)** base classifer. The dashed line is the mean test error rate of a single classifier.

responds to that found by other ensemble feature selection researchers [8,9,10], which is explained by the increased error diversity.

Examining the training error diversity Fig. 3, Adaboost$_{RSM}$ and Adaboost$_{FS}$ have created much more diverse ensembles than Adaboost$_{All}$ Fig. 3.a)-d). The diversity of Adaboost$_{RSM}$ and Adaboost$_{FS}$ is approximately zero, from the kappa measure that expected by chance. This is expected for Adaboost$_{RSM}$ since the selected feature subsets are completely random. However, it is surprising that Adaboost$_{FS}$ also has this behaviour using more ordered feature selections.

Examining the mean base classifier error rate Fig. 4, Adaboost$_{RSM}$ and Adaboost$_{FS}$ have created ensembles with a higher error rate that Adaboost$_{All}$ in all cases Fig. 4.a)-d). However, this increased classifier error rate has not lead to a greater ensemble error. It has been caused by the Boosting process seeking to minimise the loss (5), which has in turn traded base classifier error rate for increased decision diversity reducing ensemble error. Adaboost$_{All}$ using all the features has not been capable of producing much variation in base classifier accuracy and diversity using weighted examples alone. This explains the poor improvement in ensemble error rate over a single classifier. Adaboost$_{FS}$ and Adaboost$_{RSM}$ using feature selection has been better capable of trading error rate for increased diversity using different features, hence reducing the overall ensemble error. Interestingly, the weaker hypotheses created using feature selection have created better ensembles than a stronger classifier using all available features. This was stated for the original definition of Boosting [2].

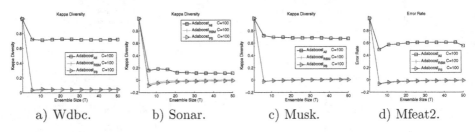

a) Wdbc.   b) Sonar.   c) Musk.   d) Mfeat2.

Fig. 3. Training error diversity (Kappa) for the four benchmark datasets using Adaboost$_{All}$, Adaboost$_{RSM}$ and Adaboost$_{FS}$ to ensemble the **RBF(5)** base classifer

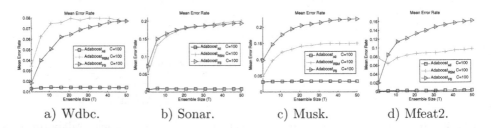

a) Wdbc.   b) Sonar.   c) Musk.   d) Mfeat2.

Fig. 4. Mean base classifier training error rate for the four benchmark datasets using Adaboost$_{All}$, Adaboost$_{RSM}$ and Adaboost$_{FS}$ to ensemble the **RBF(5)** base classifer

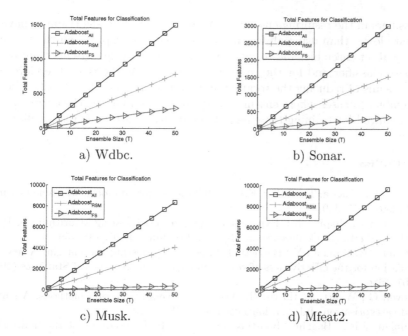

Fig. 5. Total number of features used for classification for the four benchmark datasets using Adaboost$_{All}$, Adaboost$_{RSM}$ and Adaboost$_{FS}$ to ensemble the **RBF(5)** base classifer.

Examining the total number of features used for classification Fig. 5, Adaboost$_{RSM}$ and Adaboost$_{FS}$ have created ensembles with a smaller number of features Fig. 5.a)-d). This is approximately half that of Adaboost$_{All}$ for Adaboost$_{RSM}$ due to the mean random subset size. Adaboost$_{FS}$ has created ensembles considerably smaller than both Adaboost$_{RSM}$ and Adaboost$_{All}$ 5.a)-d).

Adaboost$_{FS}$ using a floating feature search to select suitable features for Boosting has been most effective at reducing ensemble feature dimensionality. Feature selection has removed redundant features focusing each base classifier on feature subsets suitable for reducing the current contribution to the margin loss. Hence, fewer features are required for classification in each base classifier and for the final trained ensemble. The ensemble error rate of Adaboost$_{FS}$ has been slightly higher than Adaboost$_{RSM}$ in two cases, as it has been difficult to improve on this characteristically diverse method. However, Adaboost$_{RSM}$ has produced ensembles with a much larger total number of features. Where applicable, regularisation has improved the performance of the Boosting algorithms by preventing overfitting which has lowered the ensemble error rate, but has slightly increased the total number of features selected for Adaboost$_{FS}$.

4   Conclusion

We have presented Adaboost$_{FS}$, an algorithm which combines Adaboost and feature selection, with the motivation of reducing the number of features used by classifiers in the resulting ensemble. We have shown the algorithm to have an error rate lower than using all features, to compete or improve on the powerful random subset ensemble, and to require significantly less feature information for each base classifier and for the resulting ensemble. Adaboost$_{FS}$ therefore provides a means of reducing the total number of features used for classification by an ensemble. In cases where ensemble test speed is important the extra computation required for training the ensemble using feature selection is justifiable.

References

1. Rätsch, G., Onoda, T., Müller, K.R.: Soft margins for adaboost. Machine Learning **42** (2001) 287–320
2. Freund, Y., Schapire, R.: Experiments with a new boosting algorithm. In Proc. 13th International Conference on Machine Learning (1996) 148–156
3. Schapire, R., Freund, Y., Bartlett, P., Lee, W.: Boosting the margin: A new explanation for the effectiveness of voting methods. The Annuals of Statistics (1998) 1651–1686
4. Brown, G., Wyatt, J., Harris, R., Yao, X.: Diversity creation methods: A survey and categorisation. Information Fusion **6** (2005) 5–20
5. Quinlan, J.R.: Bagging, boosting and c4.5. In Proceedings of the Thirteenth National Conference on Artificial Intelligence (1996) 725–730
6. Schapire, R., Singer, Y.: Improved boosting algorithms using confidence-rated predictions. Machine Learning **37** (1999) 297–336
7. Tieu, K., Viola, P.: Boosting image retrieval. IEEE Conf. on Computer Vision and Pattern Recognition (2000) 228–235
8. Ho., T.: The random subspace method for constructing decision forests. IEEE Transactions on Pattern Analysis and Machine Intelligence **20** (1998) 832–844
9. R. Bryll, R. Gutierrez-Osuna, F.Q.: Attribute bagging: improving accuracy of classifier ensembles by using random feature subsets. Pattern Recognition **36** (2003) 1291–1302
10. Cunningham, P., Carney, J.: Diversity versus quality in classification ensembles based on feature selection. 11th European Conference on machine learning (ECML2000), lecture notes in artificial intelligence (2000) 109–116
11. Guerra-Salcedo, C., Whitley, D.: Feature selection mechanisms for ensemble creation: a genetic search perspective. AAAI-99 (1999)
12. Tsymbal, A., Pechenizkiy, M., Cunningham, P.: Diversity in search strategies for ensemble feature selection. Information Fusion **6** (2005) 83–98
13. *Günter*, S., Bunke, H.: Feature selection algorithms for the generation of multiple classifier systems and their application to handwritten word recognition. Pattern Recognition **25** (2004) 1323–1336
14. Kudo, M., Sklansky, J.: Comparison of algorithms that select features for pattern classifiers. Pattern Recognition **33** (2000) 25–41
15. Pudil, P., Novovivčová, J., Kittler, J.: Floating search methods in feature selection. Pattern Recognition Letters **15** (1994) 1119–1125
16. Blake, C., Merz, C.: UCI repository of machine learning databases (1998)
17. Feiss, J.: Statistical methods for rates and proportions (1981)

Similarity Searching in Image Retrieval with Statistical Distance Measures and Supervised Learning*

Md. Mahmudur Rahman[1], Prabir Bhattacharya[2] and Bipin C. Desai[1]

[1] Dept. of Computer Science, Concordia University, Canada
[2] Institute for Information Systems Engineering, Concordia University, Canada

Abstract. When the organization of images in a database is well described with pre-defined semantic categories, it can be useful for category specific searching. In this work, we investigate a supervised learning approach to associate low-dimensional image features with their high level semantic categories and utilize the category specific feature distribution information in statistical similarity matching. A multi-class support vector classifier (SVC) is trained to predict the categories of query and database images. Based on the online prediction, pre-computed category specific first and second order statistical parameters are utilized in similarity measure functions on the assumption that, distributions are multivariate Gaussian. A high dimensional feature vector would increase the computational complexity, logical database size and moreover, incorporate inaccuracy in parameter estimation. We also propose a fusion (*early, late, and no fusion*) based principal component analysis (PCA) to reduce the dimensionality based on both independent and dependent assumptions of image features. Experimental results on the reduced feature dimensions are reported on a generic image database with ground-truth or known categories. Performances of two statistical distance measures (e.g., Bhattacharyya & Mahalanobis) are evaluated and compared with commonly used Euclidean distance, which show the effectiveness of the proposed technique.

1 Introduction

In recent years, rapid advances in software and hardware technology, availability of the World Wide Web and moreover, digital imaging revolution facilitate the generation, storage and retrieval of large collections of digital images for professional archives to personal use. Effectively and efficiently searching these large image collections poses significant technical challenges. During the last decade, there have been an overwhelming research interests in content-based image retrieval (CBIR) from different communities [7,13]. In a typical CBIR system, a

---

* This work was partially supported by grants from NSERC and ENCS Research Support.

S. Singh et al. (Eds.): ICAPR 2005, LNCS 3686, pp. 315–324, 2005.

user can search the image databases with a visual example image (*query-by-example*) and the system will return an ordered list of images that are perceptually similar to the query image. Currently, most CBIR systems are similarity-based, where similarity between query and target images in a database is measured by some form of distance metrics in feature space [13]. However, CBIR systems generally conduct this similarity matching on a very high-dimensional feature space without any semantic interpretation [7] or paying enough attention about the underlying distribution of the feature space [12]. High-dimensional feature vectors not only increase the computational complexity in similarity matching and indexing, but also increase the logical database size. In the context of image classification, recently supervised machine learning approaches have been applied to classify collection of images into distinguishable perceptual or semantic categories (e.g., indoor-vs.-outdoor, textured-vs.-nontextured etc.) [4]. For many frequently used visual features in CBIR, often their category specific distributions are also available in a database whose perceptual and/or semantic description is reasonably well defined (such as Personal photo collection, Medical images with different modalities etc.). In this case, it is possible to extract a set of low-level features (e.g., colour, texture, shape, etc.) to predict semantic categories of each image by identifying its class assignment using a classifier. Feature descriptors along with their attributes may vary substantially from one category to another. Thus, an image can be best characterized by its feature vector and by exploiting the information of feature distribution from its semantic category. In commonly used geometric similarity measures (e.g., Euclidean), no assumption is made about the distribution of the features and its effectiveness depends on the assumption of a sphere shape distribution of similar images around the query image point in feature space [1]. However, this assumption is not always true in reality. Similarity measures based on empirical estimates of the distributions of features have been proposed in recent years [12]. However, the comparison is most often point wise or statistics of the first order (i.e., mean vector) of the distribution is considered only [6].

This paper is primarily concerned about a principal component analysis (PCA) based dimension reduction and a category based statistical similarity measure technique on the low-dimensional feature space. There are mainly two major contributions in this paper. The first one is to propose a fusion based (*early, late and no fusion*) PCA for dimension reduction of high-dimensional feature vectors on both independent and dependent assumptions of several image features. Secondly, we propose an adaptive statistical similarity matching function based on parameterization (first and second order) of underlying probabilistic distribution of feature space in different image categories. For this, we utilized a multi-class support vector machine (SVM) for online category prediction of query and database images. Hence, category specific statistical parameters in low-dimensional feature space can be exploited by statistical distance measures in real time similarity matching. Feature representation in reduced dimension with different fusion based techniques may affect retrieval performances, which we evaluated through experimantation on a generic image database with known categoreis and ground-truth. We have also evaluated objective comparison results of our adaptive statistical distance measures with precision/recall curves. It showed the effectiveness of our proposed approach through performence improvment compare to commonly used Euclidean distance.

2   Feature Extraction and Representation in PCA Space

The performance of a CBIR system mainly depends on the particular image representation and similarity matching function employed. We have extracted colour, texture and edge features for our image representation at global level. Colour is the most useful low-level feature and its histogram based representation is one of the earliest descriptors, which is widely used in CBIR [13]. For colour feature, a 108 dimensional color histogram is created in vector form on HSV (Hue, Saturation, Value) colour space. In HSV space, the colours correlates well and can be matched in a way that is consistent with human perception. In this work, we uniformly quantized HSV space into 12 bins for hue(each bin consisting of a range of 30°), 3 bins for saturation and 3 bins for value, which results in 108 bins for color histogram. Many natural images of different categories can be distinguished via their homogeneousness or texture characteristics. We extracted texture features from the gray level co-occurrence matrix. A gray level co-occurrence matrix is defined as a sample of the joint probability density of the gray levels of two pixels separated by a given displacement d and angle θ [2]. We obtained four co-occurrence matrices for four different orientations (horizontal 0°,vertical 90 °, and two diagonals 45 ° and 135 °) and normalize the entries [0,1] by dividing each entry by total number of pixels. Higher order features, such as energy, maximum probability, entropy, contrast and inverse difference moment are measured based on each gray level co-occurrence matrix to form a five dimensional feature vector and finally obtained a twenty dimensional feature vector by concatenating the feature vector of each co-occurrence matrix. To reperesent the shape feature on a global level, a histogram of edge direction is constructed. The edge information contained in the images is processed and generated by using the Canny edge detection (with $\sigma = 1$, Gaussian masks of size = 9, low threshold = 1, and high threshold = 255) algorithm [3].The corresponding edge directions are quantized into 72 bins of 5° each. Scale invariance is achieved by normalizing this histograms with respect to the number of edge points in the image. As the dimensions of colour, texture and edge (108+20+72 = 220) feature vectors are high, we need to apply some dimension reduction technique to reduce the computational complexity and logical database size. Moreover, if the training samples used to estimate the statistical parameters are smaller compare to the size of feature dimension, then inaccuracy or singularity may arise for second order (co-variance matrix) parameter estimation.

The problem of selecting most representative feature attributes commonly known as dimension reduction, has been examined by principal component analysis (PCA) [10] in some CBIR systems [13]. The basic idea of PCA is to find m linearly transformed components so that they explain the maximum amount of variances in the input data and mathmetical steps used to describe the method is as follows:

given a set of N feature vectors (training samples) $x_i \in \mathbb{R}^d | i = (1 \cdots N)$, where the mean vector($\mu$) and covariance matrix (C) is estimated as

$$\mu = \frac{1}{N} \sum_{i=1}^{N} x_i \quad \& \quad C = \frac{1}{N} \sum_{i=1}^{N} (x_i - \mu)(x_i - \mu)^T \tag{1}$$

Let ν_i and λ_i be the eigenvectors and the eigenvalues of C, then they satisfy the following:

$$\lambda_i = \sum_{i=1}^{N} (\nu_i^T (x_i - \mu))^2 \tag{2}$$

Here, $\sum_{i=1}^{N} \lambda_i$ accounts for the total variance of the original feature vectors set. Now, PCA method tries to approximate the original feature space using an m dimensional feature vector, that is using m largest eigenvalues account for a large percentage of variance, where typically $m << min(d, N)$. These m eigenvectors span a subspace, where $V = [v_1, v_2, \cdots, v_m]$ is the $d \times m$-dimensional matrix that contains orthogonal basis vectors of the feature space in its columns. The $m \times d$ transformation V^T transforms the original feature vector from $\mathbb{R}^d \to \mathbb{R}^m$ ones. That is

$$V^T(x_i - \mu) = y_i, i = 1 \cdots N \tag{3}$$

where $y_i \in \mathbb{R}^m$ and kth component of the y_i vector is called the kth principal component (PC) of the original feature vector x_i. So, the feature vector in the original $\mathbb{R}^d$ space for query and database images can be projected on to the $\mathbb{R}^m$ space via the transformation of V^T [10].

In CBIR, PCA has been mainly employed to reduce the dimensions of a single feature vector or a composite feature vector combined with various features. In this work, we take a fusion based approach of PCA to account for all types of combinations based on both independent and dependent assumptions of feature space. Here, we describe it as *early fusion, late fusion & no fusion*. In case of *early fusion*, we consider to form a composite feature vector from the three feature types described above before applying any PCA based dimension reduction technique. Let f_c, f_t, and f_e be the global colour, texture and edge feature vector respectively of an image. Now the composite feature vector is formed by simple concatenation of each individual feature vector as $F_{earlyfusion} = (f_c + f_t + f_e)$, where the dimension of $F_{earlyfusion}$ is the sum of individual feature vector dimension. Now we apply PCA to this high-dimensional composite feature vector to convert it in a low-dimensional feature vector in PCA space and called it $F_{earlyfusion}^{PCA}$. After getting this reduced feature vector, we will use it in consequent parameter estimation and similarity matching functions. In case of *late fusion*, we assume independent assumption of feature space for PCA and apply it in each feature space seperately to lower the dimension first. So, here we reduce dimension of each color, texture, and edge feature seperately as $f_c \to f_c^{PCA}$, $f_t \to f_t^{PCA}$, and $f_e \to f_e^{PCA}$ and finally combine them to form a joint feature vector as $F_{\text{latefusion}}^{\text{PCA}} = (f_c^{\text{PCA}} + f_t^{\text{PCA}} + f_e^{\text{PCA}})$ for subsequent analysis. We took a totally independent assumption of feature space in case of *no fusion*. Instead of combining the feature vectors lately as in late fusion strategy, the reduced features will be used seperately as $F_{\text{nofusion}}^{\text{PCA}} = (f_c^{\text{PCA}}, f_t^{\text{PCA}}, f_e^{\text{PCA}})$ in subsequent parameter esimations and similarity matching functions described in later sections.

3   Statistical Distance Measures

Statistical distance measure, defined as the distances between two probability distributions, finds its uses in many research areas in pattern recognition, information theory, and communication. It captures correlations or variations between attributes of the feature vectors and provides bounds for probability of retrieval error of a two way classification problem. Recently, CBIR community also adopted statistical distance measures for similarity matching [1,12]. In this scheme query image q and target image t are assumed to be in different classes and their respective density as $p_q(\boldsymbol{x})$ and $p_t(\boldsymbol{x})$, both defined on $\mathbb{R}^d$. When these densities are multivariate normal, they can be approximated by mean vector μ and covariance matrix C as $p_q(\boldsymbol{x}) = N(\boldsymbol{x}; \mu_q, C_q)$ & $p_t(\boldsymbol{x}) = N(\boldsymbol{x}; \mu_t, C_t)$ where,

$$N(\boldsymbol{x}; \mu, C) = \frac{1}{\sqrt{(2\pi)d|C|}} \exp^{-\frac{1}{2}(\boldsymbol{x}-\mu)^T C^{-1}(\boldsymbol{x}-\mu)} \tag{4}$$

here, $\mathbf{x} \in \mathbb{R}^d$ and $|\cdot|$ is matrix determinant [9]. A popular measure of similarity between two Gaussian distributions is the Bhattacharyya distance, which is equivalent to an upper bound of the optimal Bayesian classification error probability [9] [11]. Bhattacharyya distance (D_{Bhatt}) between query image q and target image t in the database is given by:

$$D_{\text{Bhatt}}(q,t) = \frac{1}{8}(\mu_q - \mu_t)^T \left[\frac{(C_q + C_t)}{2}\right]^{-1} (\mu_q - \mu_t) + \frac{1}{2} \ln \frac{\left|\frac{(C_q + C_t)}{2}\right|}{\sqrt{|C_q||C_t|}} \tag{5}$$

where μ_q and μ_t are the mean vectors, and C_q and C_t are the covariance matrices of query image q and target image t respectively. Equation (5) is composed of two terms, the first one being the distance between mean vectors of images, while the second term gives the class separability due to the difference between class covariance matrices. When all classes have the same covariance matrices, the Bhattacharyya distance reduce to the Mahalanobis distance, a widely used similarity measure in CBIR literatures [6,9].

$$D_{\text{Maha}}(q,t) = (\mu_q - \mu_t)^T C^{-1} (\mu_q - \mu_t) \tag{6}$$

However, if inclusion of both query and target covariance matrices is useful, Bhattacharyya distance will outperform Mahalanobis distance [6] as will be shown in observation section.

The above distance measures will work fine for a single feature or combined feature vector as obtained from early or late fusion based techniques. However, for no fusion based technique, we have to calculate the distance measure for each feature seperately and combine them with appropriate weight for a final distance value. For Bhattacharyya distance it will be as follows:

$$D_{\text{Bhatt}}(q,t) = w_1 * D_{\text{Bhatt}}^{color}(q,t) + w_2 * D_{\text{Bhatt}}^{texture}(q,t) + w_3 * D_{\text{Bhatt}}^{edge}(q,t) \quad (7)$$

whereas same will apply for Mahalanobis distance. Here, w_1, w_2, and w_3 are non-negative weighting factor with normalization ($w_1 + w_2 + w_3 = 1$), which need to be selected experimentally as described in section 6.

4   Category Prediction with Multi-class SVM

Statistical distance measures described in previous section, are not adaptive in nature and assume individual feature distributions of query and database images always belong to seperate classes. However, to utilize category specific distribution information in similarity matching, we need some form of classifier based on supervised machine learning technique to predict the categories online. Support vector machine (SVM) is an emerging machine learning technology which has been successfully used in content based image retrieval [4]. Given training data $(\boldsymbol{x}_1, \ldots, \boldsymbol{x}_n)$ that are vectors in some space $\boldsymbol{x}_i \in \mathbb{R}^n$ and their labels $(y_1, \ldots, y_n)$ where $y_i \in (+1, -1)^n$, the general form of the binary linear classification function is

$$g(x) = \boldsymbol{w} \cdot \boldsymbol{x} + b \quad (8)$$

which corresponds to a separating hyper plane

$$\boldsymbol{w} \cdot \boldsymbol{x} + b = 0 \quad (9)$$

where $\boldsymbol{x}$ is an input vector, $\boldsymbol{w}$ is a weight vector, and b is a bias. The goal of SVM is to find the parameters $\boldsymbol{w}$ and b for the optimal hyper plane to maximize the geometric margin $\frac{2}{||\boldsymbol{w}||}$ between the hyper planes, subject to the solution of the following optimization problem:

$$\min_{\boldsymbol{w}, b, \xi} \quad \frac{1}{2}\boldsymbol{w}^T\boldsymbol{w} + C\sum_{i=1}^{n}\xi_i \quad (10)$$

subject to

$$y_i(\boldsymbol{w}^T\phi(\boldsymbol{x}_i) + b) \geq 1 - \xi_i \quad (11)$$

where $\xi_i \geq 0$ and $C > 0$ is the penalty parameter of the error term. Here training vectors $\boldsymbol{x}_i$ are mapped into a high dimensional space by the non linear mapping function $\phi : \mathbb{R}^n \rightarrow \mathbb{R}^f$, where $f > n$ or f could even be infinite. Optimization problem and its solution can be represented by the inner product. Hence,

$$\boldsymbol{x}_i.\boldsymbol{x}_j \rightarrow \phi(\boldsymbol{x}_i)^T\phi(\boldsymbol{x}_j) = K(\boldsymbol{x}_i, \boldsymbol{x}_j) \quad (12)$$

where K is a kernel function. The SVM classification function is given by [4]:

$$f(\boldsymbol{x}) = sign \left(\sum_{i=1}^{n} \alpha_i y_i K(\boldsymbol{x}_i, \boldsymbol{x}) + b \right) \qquad (13)$$

A number of methods have been proposed for extension to multi-class problem to separate L mutually exclusive classes essentially by solving many two-class problems and combining their predictions in various ways [4]. One technique, commonly known as "one-vs.-one" is to construct SVMs between all possible pairs of classes. During testing, each of the $L*(L-1)/2$ classifier votes for one class. The winning class is the one with the largest number of accumulated votes. We use this technique for the implementation of our multi-class SVM by using the LIBSVM software package [5].

5   Parameter Estimation and Online Similarity Matching

To estimate the parameters of the category specific distributions, feature vectors with *early, late and no fusion* based techniques are extracted from N selected training image samples. It is assumed that feature of each category will have distinguishable normal distribution, which is very natural in many specific purpose and natural image databases with predefined categories. Computing the statistical distance measures between two multivariate normal distributions requires first and second order statistics in the form of mean (μ) and covariance matrix (C) or parameter vector $\theta = (\mu, C)$ as described in previous section. Suppose that there are L different semantic categories in the database, each assumed to have a multivariate normal distribution with mean vector μ_i and covariance matrix C_i, for $i \in L$. However, the true values of μ and C of each category usually are not known in advance and must be estimated from a set of training samples N [9]. The μ_i and C_i of each category are estimated as

$$\mu_i = \frac{1}{N_i} \sum_{j=1}^{N_i} \boldsymbol{x}_{i,j} \quad \& \quad C_i = \frac{1}{N_i - 1} \sum_{j=1}^{N_i} (\boldsymbol{x}_{i,j} - \mu_i)(\boldsymbol{x}_{i,j} - \mu_i)^T \qquad (14)$$

where $\boldsymbol{x}_{i,j}$ is sample j from category i, N_i is the number of training samples from category i and $N = (N_1 + N_2 + \ldots + N_L)$. As we estimate the parameters after applying the PCA based dimension reduction technique, the dimension of $\boldsymbol{x}$ will vary based on *early, late, and no fusion* techniques. Moreover, as *no fusion* based technique assumes totally independent assumption, the parameter vector θ will have seperate μ and C for each feature as $\theta = (\mu_i^{color}, C_i^{color}, \mu_i^{texture}, C_i^{texture}, \mu_i^{edge}, C_i^{edge})$ for $i \in L$. Later these parameters will be used independently for each feature in statistical similarity matching. Similarity measure based on the above statistical parameters would perform better if the right categories of query and database images are predicted in real time. Hence, we utilized the multi-class SVM classifier to predict the categories and based on the online prediction, similarity measure functions will be adjusted to accommodate category specific parameters for query and database images. Figure 1, shows the functional process diagram of the proposed similarity matching technique from a query image view point.

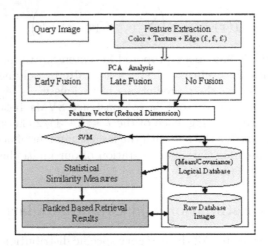

Fig. 1. Process diagram of the proposed similarity matching technique

6   Experimental Setup and Observation

For statistical parameter estimation, PCA and SVM training, we used a labeled database of generic images as training samples. We experimentally selected 15 semantically different categories (Mountain, Beach, Flower, Food, Architecture, etc.) each with 100 images for generating the training samples. However, for actual evaluation of similarity measure functions, we conducted our experiments on the entire database (3000 generic images from the Corel color photo collection) without any labeling. It should be pointed out that some images in one category can have visual features that are very similar to those found in images belong to other categories, hence making image classification and retrieval more difficult.

For SVM training, we used the original feature vector with radial basis kernel function $K(x_i, x_j) = \exp(-\gamma||x_i - x_j||^2), \gamma > 0$. After 10 fold cross validation, we found the best parameters $C = 30$ and $\gamma = .01$ with an accuracy of 80.15% in our current setting and finally trained the whole training set with these parameters. The dimensionality of the feature vector is reduced to $(220 \rightarrow 20)$ for $F_{earlyfusion}^{PCA}$, $(220 \rightarrow 35)$ for $F_{latefusion}^{PCA}$ where, $(f_c^{PCA} \in \mathbb{R}^{16}, f_t^{PCA} \in \mathbb{R}^5, f_e^{PCA} \in \mathbb{R}^{14})$ for $F_{nofusion}^{PCA}$ and accounted for 90.0% of the total variances. In simililariy matching based on *no fusion* technique, we used $w_1 = .6$, $w_2 = .2$, and $w_3 = .2$ for the colour, texture and edge feature which gave the best performance.

For evaluation of the retrieval performance, we used precision-recall metrics. Recall is the ratio of the number of relevant images returned to the total number of relevant images. Precision is the ratio of the number of relevant images returned to the total number of images returned. For experimentation, we selected a set of 10 bench mark queries for each category not included in the database and used *query-by-example* as the search method. Performances of two statistical distance measures, one which utilizes both query and target image category specific parameters (Bhattacharyya) and the other which only utilizes parameters for query cate-

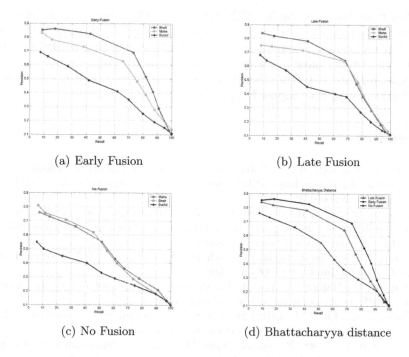

(a) Early Fusion

(b) Late Fusion

(c) No Fusion

(d) Bhattacharyya distance

Fig. 2. Precision-recall curves for distance measures

gory (Mahalanobis), are evaluated and compared with Euclidian distance measure. Figure 2(a), 2(b) & 2(c) presents precision-recall curves of these distance measures based on different fusion techniques. As shown, best performance is always achieved by Bhattacharyya distance measure, whereas Euclidean distance performed poorly in all cases. The result is expected as Euclidean distance does not take into account the variations of its feature attributes in semantic categories. Based on the above observation, we can conclude that distance measures which utilze both the parameters of query and database image categories performed better in a semantically organized database. We obtained better performances in *early* and *late fusion* compare to *no fusion*, as shown in Figure 2(d) for Bhattacharyya distance measure. Hence, it concluded that, features may have some form of dependency in feature space for our generic image database.

7   Conclusion

This paper has examined an image retrieval approach with supervised machine learning and fusion based dimension reduction techniques. High-dimensional feature vectors are projected to a PCA subspace to reduce the dimensionality on both independent and dependent feature space assumptions. Category specific statistical parameters are computed from the feature distributions in reduced feature space on Gaussian assumption and utilized in statistical similarity measure functions based on online SVM prediction. Overall, the assumptions are reasonable, given the constraint that, online similarity matching need to be performed in real time and database is semantically well organized. Performances of different distance measures with three fusion based techniques were evaluated in a generic image database. Experimental results were promising and showed the effectiveness of the proposed approach.

References

1. Aksoy, S., Haralick, R.M.: Probabilistic vs. geometric similarity measures for image retrieval., Proceedings. IEEE Conference on Computer Vision and Pattern Recognition,2(2000) 357–362
2. Aksoy, S., Haralick, R. M.: Texture Analysis in Machine Vision., Chapter Using Texture in Image Similarity and Retrieval,Series on Machine Perception and Artificial Intelligence., World Scientific (2000)
3. Canny, J. : A computational approach to edge detection., IEEE Trans. Pattern Anal. Machine Intell., 8 (1986) 679–698
4. Chapelle, O., Haffner, P., Vapnik, V.: SVMs for histogram-based image classification.,IEEE Transaction on Neural Networks (1999)
5. Chang, C. C., Lin, C.J.: LIBSVM : a library for support vector machines., (2001) Software available at http://www.csie.ntu.edu.tw/ cjlin/libsvm
6. Comaniciu, D., Meer, P., Xu, K., Tyler D.: Retrieval Performance Improvement through Low Rank Corrections., In Workshop in Content-based Access to Image and Video Libraries, Fort Collins, Colorado, (1999) 50–54
7. Eakins John, P.: Towards Intelligent image retrieval., Pattern Recognition, 35 (2002) 3–14
8. Friedman,J.:Regularized Discriminant Analysis., Journal of American Statistical Association, 84 (2002) 165–175
9. Fukunaga,K.: Introduction to Statistical Pattern Recognition., Second ed. Academic Press, (1990)
10. Jain,A.K., Bhandrasekaran, B. : Dimensionality and sample size considerations in pattern recognition practice., Handbook of Statistics, 2 (1987) 835–855
11. Kailath,T.: The divergence and Bhattacharyya distance measures in signal selection.,IEEE Trans. Commun. Technol, COM-15(1967) 52–60
12. Puzicha, J., Buhmann, J., Rubner, Y., Tomasi, C.: Empirical evaluation of dissimilarity measures for color and texture., Intern. Conf. on Computer Vision, (1999)
13. Smeulder, A., Worring, M., Santini, S., Gupta, A., Jain, R.: Content-Based Image Retrieval at the End of the Early Years., IEEE Trans. on Pattern Anal. and Machine Intell., 22, (2000) 1349–1380

Using Patterns to Generate Prime Numbers

Udayan Khurana[*] and Anirudh Koul

Thapar Institute of Engineering and Technology,
Patiala, Punjab State, India-147004
{udayankhurana, anirudhkoul}@gmail.com

Abstract. A new deterministic method to determine primality of any given number is presented in this paper. The underlying principle involves the use of a special series that generates lesser prime numbers till the root of the number under scrutiny. Subsequently, divisibility is performed to check whether the number is prime or not. Such a series characteristically produces all the successive prime numbers along with a few composite numbers as well, the proportion of latter increasing as one moves to higher numbers. This technique is provably more efficient than other deterministic methods that employ division by primes till the root of the number, either by generating those smaller primes or storing them or by simply taking all odd numbers till the square root.

1   Introduction

1.1   Related Work

A Prime number sequence is considered to be a confounded mystery of the most unpredictable nature. Many researchers believe that an algorithm to determine the sequence certainly does not exist. Even primality testing of an integer is a computationally tough problem. Prime numbers are tested to be prime based upon the divisibility by lesser numbers. It is a well known fact that divisibility tests till $\sqrt{n}$ can deterministically establish the prime character of a number n. Many other relations, like $2^n - 1$ being prime for all primes n existed but have been contradicted because they failed to satisfy for certain values of n (In 1536 Hudalricus Regius showed that $2^{11} - 1 = 2047$ is not prime. It is 23×89. Since then, many others have proved this for other values of n). Lucas-Lehmer Test actually checks successfully for the $2^n - 1$ being prime or not. Still, it provides for very few prime numbers. Not all numbers can be tested to be prime or not depending on the above check. So, fundamentally, to generate a prime number or to check for a number to be prime or not without having any knowledge about all the smaller prime numbers, we, inevitably have to perform those divisibility tests. In the year 2002, Agrawal et. al. [AS1] presented the first deterministic algorithm that is polynomial in $log_2 n$ to check the primality of a number. It performs with a complexity $O((log_2 n)^{10.5})$ which is not feasible

---

[*] Part of this work was done while the first author was a project trainee at Tata Research Development and Design Centre, Pune-411013, Maharashtra, India.

S. Singh et al. (Eds.): ICAPR 2005, LNCS 3686, pp. 325–334, 2005.

for the present size of numbers. However, the first of its kind, a deterministic polynomial algorithm is historically the most important development in this field. State of the art algorithms are the probabilistic ones like Solovay-Strassen [SS1] and Miller-Rabin[Ra1] that are used today in cryptographic applications like the RSA. How ever, due to the probabilistic nature, a composite number may very well be reported to be prime. So, to assure a good probability (typically more than 99.99 %), the algorithm is run many times which makes it slow for practical purposes. For e.g., the probability of an erroneous result in the Solovay-Strassen algorithm is $\left(\frac{1}{2}\right)^k$, where k is the number of iterations.

1.2   Motivation

It is known that all prime numbers p, with the exception of 2 and 3 always exist in the form $p = 6n \pm 1$ [We1]. Taking inspiration from the fact, we tried to find sequences in occurrences of prime numbers. In this paper, we present certain experimentally generated sequences [KK2] that are periodic in nature and determine successive "possible" prime numbers with a definite efficiency. Later, we use these sequences to check for large prime numbers and hence we obtain reduction in the number of required divisibility tests (by prime numbers till root of n). Every sequence performs to 100 % efficiency till a certain value (also presented) after which the efficiency goes down gradually but still serves the practical utility. We also argue that storing such sequences is far better (in terms of computation cost) than computing lesser prime numbers (up till $\sqrt{n}$) and then to establish the prime character of n. It is also better than storing the entire lot of prime numbers (till $\sqrt{n}$) because of two reasons. The sequence takes lesser space and can be compressed very effectively. The determination and use of such sequences provides a new field for engineering with prime numbers and indeed leaves a good scope for scientific research.

2   Patterns

2.1   Definition of a Pattern Sequence

The nature of patterns in the text hereon is based upon the difference between successive prime numbers. A pattern sequence h for a prime h, would consist of differences between successive prime numbers that exist between h (actually starting from the prime number next to h) and the product of all prime number less than equal to h which is also the LCM of all the numbers from one to h. For e.g., a pattern sequence 7 consists of differences of all successive prime numbers that are greater than or equal to 11 and less than 221 i.e. $11 + 210$ where $210 (= 2 \times 3 \times 5 \times 7)$ is actually the LCM of all integers from 1 to 7 or more simply the product of all prime numbers from 1 to 7. Pattern sequence 7 is:

$$PS_7 = 2, 4, 2, 4, 6, 2, 6, 4, 2, 4, 6, 6, 2, 6, 4, 2, 6, 4, 6, 8, 4, 2, 4, 2, 4, 8, 6,$$
$$4, 6, 2, 4, 6, 2, 6, 6, 4, 2, 4, 6, 2, 6, 4, 2, 4, 2, 10, 2, 10 \tag{1}$$

The sum of above series is of course 210. In general, sum of PS_h is prime product of h. The prime product of k is defined as the product of all prime numbers less than or equal to h. The pattern sequence and prime product are defined only for prime numbers. Table 1 gives certain prime numbers versus their prime products.

Table 1. Prime Products versus prime numbers

Prime number (h)	Prime Product
3	6
5	30
7	210
11	2310
13	30030
17	510510
19	9699690
23	223092870

2.2   Algorithm to Generate Pattern Sequence h

1. I = h , J = 1, P = 1, sum = 0
2. I = I + 1
3. IF I is Prime, Then GOTO 4 else GOTO 2
4. Last = I
5. I = I + 1
6. IF I is prime, Then GOTO 7 else GOTO 5
7. J = J + 1
8. IF J is prime, GOTO 9 else GOTO 7
9. P = P × J
10. IF J is equal to h, Then GOTO 11 else GOTO 7
11. IF I % 6 = 1, Then GOTO 12 else GOTO 19
12. IF I is Prime, Then GOTO 13 else GOTO 18
13. Diff = I - Last
14. Print Diff
15. Last = I
16. Sum = Sum + Diff
17. IF Sum < P, Then GOTO 18 else GOTO 26
18. I = I + 4
19. IF I is Prime, Then GOTO 20 else GOTO 24
20. Diff = I - Last
21. Print Diff
22. Last = I
23. Sum = Sum + Diff
24. I = I + 2
25. IF Sum < P, Then GOTO 12 else GOTO 26
26. STOP

3 Utility of a Pattern Sequence

3.1 Predicting a Prime Number

Predicting successive prime numbers from Pattern Sequences is like adding the next term in the PS to the present prime number in hand and moving on till requirements are met. For e.g., for Pattern Sequence 7, starting from 11, to obtain the next prime number, we add the first term of the sequence, i.e. 2 to get 13. To get the next prime number, we add the next term i.e. 4 to get 17 and so on until the sequence's last term is added. Then we move back to the first term and carry on in a periodic manner. The sequence is periodic, since after all the terms have been added, the next term to be added is again the first one and the procedure continues repeatedly (cyclic).

For Pattern Sequence 7, i.e. $\{2, 4, 2, 4, 6, 2, 6 \cdots 10, 2, 10\}$

First prime number $= 11$,

The next one $= 11 + 2 = 13$, The next$= 13 + 4 = 17$, The next $= 17 + 2 = 19$, The next$= 19 + 4 = 23$ and so on $\cdots$ up till the last term is added and then back to the first one.

3.2 Usefulness of Pattern Sequences

A Pattern Sequence h is actually a periodic sequence with the period being the length of the particular pattern, which encompasses all the prime numbers that are greater than h. i.e. every big prime number should appear in the list of numbers generated by a pattern sequence by passing through the series periodically. This does not guarantee that every number that is generated by a Pattern Sequence is prime but does guarantee that no prime number is missed from the sequence. Hence, it all comes down to the efficiency of various Pattern Sequences. We define Efficiency(h, N) as the percentage of numbers that are prime from h up to N among the all the numbers generated by the Pattern Sequence h. Greater efficiency is hence desirable from each Pattern Sequence. Efficiency(h, N) for every Pattern Sequence h is 100 % for N less than equal to the square of prime number next to h. For e.g. for the Pattern Sequence 7, efficiency is 100% until $N = 121 (= 11 \times 11$ and 11 is the next prime number to 7). In fact the square is the first prime number to appear in a Pattern Sequence h is the square of prime number next to h. There on, the efficiency gradually decreases for increasing N. Table 2 shows the efficiency of some Pattern Sequence for $N = 100, 1000, 10000, 100000$.

We may infer that, "for the same N, the efficiency is greater for Pattern Sequence h_1 than Pattern Sequence h_2 if $h_1 > h_2$".

4 Application in Determining Prime Numbers

To determine whether a number n is prime or not, one needs to make divisibility tests by all prime numbers till the root of n. So, to check whether a large

Table 2. Efficiency for various values of h and N. It can be seen that N can be scaled in proportion of h.

	EFFICIENCY						
	FOR h Upto N						
h,N	100	300	700	1000	3000	7000	10000	100000
3	73.53	60.40	53.41	50.00	42.96	38.56	36.83	28.72
5	89.28	74.39	66.49	62.31	62.31	48.18	46.02	35.96
7	100.00	86.11	76.69	72.29	62.40	56.14	53.67	41.95
11	100.00	92.54	83.89	79.15	68.58	61.69	59.01	46.15
13	100.00	98.41	90.58	85.64	74.18	66.72	63.89	49.99
17	100.00	100.00	95.42	90.27	78.75	70.81	67.77	53.11
19	100.00	100.00	98.42	94.91	83.33	74.75	71.48	56.06
23	100.00	100.00	100.00	98.25	87.22	78.26	74.79	58.60

number is prime or not, provided a Pattern Sequence h which has a workable Efficiency(h, $\sqrt{n}$) will suit the purpose of obtaining primes till root of n. What is a workable efficiency and what not, is discussed in the next section. For example, determining the prime character of a number of the order of 10^8 (100 million), should require all prime numbers till 10,000 and using the Pattern Sequence 19, we see from Table 2, up till $N = 361$, the efficiency is 100 %. Till 1000 it remains almost 95% and even till 10,000 it is 71 % (There is of course a more gradual decrease). Hence, by making 1.3 times the optimal number of divisions (number of prime numbers till $\sqrt{n}$), we are able to determine the primality of n in lesser steps than the usual method of dividing by all numbers (odd) till the $\sqrt{n}$. Table 3 shows a comparison of the steps taken by the usual method and those by using various Pattern Sequences. So, checking for divisibility of the number n, by every number generated from a suitable Pattern Sequence makes a good way of generating or checking whether a number is prime or not. In the following section we argue how much efficiency is good enough in practice. After that it is argued why is this one a possible alternative to other methods of generating prime numbers.

4.1   How Efficient Is Workable?

We have defined efficiency of a Pattern Sequence till a number N. What figure of the efficiency can be good enough to adopt this method is the natural question that follows. It is also established by now that with increasing value of h, the efficiency also increases for the same value of N. So, greater the value of h, lesser is the time required. An optimal value of h, can be determined experimentally. For example, from Table 3, we see that for a number n of the order 10^9, using even $h = 5$, we obtain results that are twice as good as the normal method. And from Table 2, for values of N (of the order of $\sqrt{n}$), the efficiency is still about 38% only. So, even at 40% efficiency we provide results better than twice as good (i.e. half the time taken). It can be seen that for greater values of h, efficiency is much better and hence the time taken is much lesser. A definite value of h (or efficiency) can be chosen for a definite range of numbers (order of the number) for suitable requirements.

4.2 Usefulness over Others

We now discuss the performance of this technique in comparison to other deterministic ones. We shall provide arguments with respect to the two fundamental and rather nave methods: First, performing divisibility tests by all the odd numbers till $\sqrt{n}$ and second, prior storage of prime numbers till the $\sqrt{n}$ and then performing divisibility tests by all those prime numbers. Comparisons with respect to probabilistic algorithms is not done here as this algorithm is deterministic in nature and provides the results with 100% accuracy, whereas the other category of algorithms ([Ra1],[SS1]) produce the output only with some probability of correctness of the result. Table 3 shows the relative number of steps required by the First one and by our method to test for a number of the order of 10^9 (1 billion). Table 4 actually shows the relative number of steps employed by both the methods. It is clear that our method always outperforms the First one.

Considering the second Method that requires the knowledge of all existing prime numbers till $\sqrt{n}$ is quite impractical in real situations because of two reasons. Firstly, because to test for big prime numbers, lets say, of the order of, 10^{16}, number of primes to be stored is enormously huge (of the order of 10^8 i.e. 100 million) and impossible to always carry along. Also, while running such a program on the computer, the amount of memory required would be too high. That would, in fact require regular swapping in and swapping out which retards the purpose of efficient generation. Also, the limit of n would not known and it is very likely that a prior supposition may fail to perform in all cases (i.e. fall short). The sequence used by our method is in fact much lesser in size and can even be compressed (next subsection) to provide still lesser storage and memory requirements. In such conditions, the method described in this paper moves on the lines of giving a better alternative.

4.3 Performance Analysis

The time complexity of the method can be expressed as the number of divisions performed in the worst case. Hence, maximum number of divisions will be the number of terms generated by the Pattern Sequence till $\sqrt{n}$. If e be the efficiency of the pattern sequence being used and $d(n)$ be the prime density till n,

$$TimeComplexity = O(\sqrt{e \times d(n) \times n}) \qquad (2)$$

The prime number density is a debatable issue for n is general and e is a constant term for a particular pattern sequence being used. For large values of n, $d(n) = \frac{1}{log(n)}$ (see Appendix 6.4) reduces (2) to,

$$TimeComplexity = O(\sqrt{\frac{n}{log(n)}}) \qquad (3)$$

Table 3. Comparison of method First to that of ours using various h for determining a prime number (429467291). Time reported is in milliseconds and the experiment was performed on an IBM R51 Thinkpad with Pentium M and Windows XP as the OS.

Sequence	Time Required (ms)
Odd till $\sqrt{n}$	14.55
Our Method (h)	
5	7.74
7	6.48
11	5.99
13	5.6

Table 4. Comparison of method First to that of ours using various h for determining a prime number (429467291) in terms of the number of required comparison steps

Sequence	No. of Comparisons
Odd till $\sqrt{n}$	32767
Our Method (h)	
5	17477
7	14981
11	13619
13	12573
17	11832
19	11213
23	10730

4.4   Compressing the Sequence

If sequences like those defined in this paper are used for very big values of h, we may have sequences that are indeed large. In, this section we briefly describe a scheme for compression of such sequences. In a typical Pattern Sequence, in most of the sequence, we have combinations of 2, 4, 6 and 8 and also 10 and 12. Let us decide to refer the most frequently occurring combinations like:

2, 4, 2, 4   4, 2, 4, 6   2, 4, 2, 6   4, 2, 4, 2, 4, 6   etc. by some odd integers or any other alias. So, once such a sequence is generated, we can compress the Pattern Sequence in linear time by replacing largest subsequences with the corresponding alias (preferably an odd number) quite in the same manner as general Text Compression by [KK1]. Also, while the prime number testing algorithm is running, it takes care of those few aliases inherently. In fact, on finding an alias control may be shifted over to another function (specific to each alias) that takes as parameter the number and performs the required comparisons. We have only presented the scheme, which should be easy to implement [Ha1].

5   Conclusion and Further Scope

In this paper, a sequential pattern has been used for the fist time to actually deterministically test the prime characteristic of higher numbers. Though there exist It is believed to be a new direction to how we engineer and how we look at prime numbers. Prime numbers are indeed unpredictable in absolute sense, but we have shown that regular patterns exist absolutely for small intervals and work with a slightly decreasing efficiency for values exceeding the particular interval. This should invite interest from applicability as well as scientific needs. The further challenge lies in merging sequence patterns of more than one prime number and coming out with more efficient and in one sense more generic (at least, in the domain of human use) sequences. Another field of research is developing (or extrapolating) higher sequences by using lower ones for the dynamic and automatic growth of sequences. Finally, challenge lies in better utilization and better implementation of such sequences to suit commercial needs.

Acknowledgements

We acknowledge the help and guidance provided by Dr. Sachin Lodha and Ms. Sharada Sundaram. We are thankful to the two anonymous referees whose comments were extremely beneficial while revising the paper. Also, Dr. Maneesha Singh's regular guidance during submission and pre-submission process was a real helping hand.

References

[AS1]   Agrawal, M., Saxena, N. , Kayal, N.: PRIMES in P. August 6, 2002
[CG1]   Conway, J. H.,Guy, R. K.: The Book of Numbers. New York: Springer-Verlag,(1996) p. 130
[Be1]   Berndt, B. C.: Ramanujan's Theory of Prime Numbers. (1994) Ch. 24 in Ramanujan's Notebooks, Part IV. New York: Springer-Verlag
[Ha1]   Havil, J. Gamma: Exploring Euler's Constant. Princeton, NJ: Princeton University Press, (2003)
[KK1]   Khurana U, Koul A: Text Compression and Searchability. (forthcoming)
[KK2]   Koul A., Khurana U: Determination Sequential Patterns in Prime Numbers. Proceedings of The National Conference on Bioinformatics Computing (March 2005)
[LY1]   Lim, C.H., Yung, M.: Information Security Applications, (2005) Vol. 3325
[Na1]   Nagell, T.: General Remarks. The Sieve of Eratosthenes. Introduction to Number Theory. New York: Wiley,(1951) Chapter 15, 51–54
[Ra1]   Rabin: Probabilistic Algorithms for Testing Primality. J. of Num. Th.(1980) 12, 128–138
[SS1]   Solovay and Strassen: A fast Monte-Carlo test for Primality. SIAM. J. of Comp. 6 (1977), 84-85
[Va1]   Tilborg V.: Encyclopedia of Cryptography and Security. H.C.A., (2005)
[We1]   Wells, D.: The Penguin Dictionary of Curious and Interesting Numbers. Middlesex, England: Penguin Books, (1986)

6   Appendix

6.1   More About Sequences and Their Origin

The following text is a direct quote from [KK2], to elucidate more upon the concept and generation of such pattern sequences: With the exception of 2 and 3, all primes are of the form $p = 6n \pm 1$, i.e., $p = 1, 5(mod6)$. The difference in elements in the same row is 2 while that in consecutive diagonal elements is 4, thus forming the sequence 2 4 2 4 2 4 2 4 which is repetitive in nature. The length of this sequence is 2. In this case, the repetitive unit is '2 4' (Sequence I).

$$\underline{2\ 4}\ \underline{2\ 4\ 6\ 2\ 6\ 4}\ \underline{2\ 4\ 2\ 4\ 6\ 2\ 6\ 4}\ \underline{2\ 4\ 2\ 4\ 6\ 2\ 6\ 4}$$

where the repeating sequence is 2 4 2 4 6 2 6 4 (Sequence II). This sequence is of the length 8. Extending this concept further by crossing out all numbers > 7 which are divisible by 7. Close observations of the differences between any two consecutive numbers generate the following sequence, which has 48 elements:

$$2\ 4\ 2\ 4\ 6\ 2\ 6\ 4\ 2\ 4\ 6\ 6\ 2\ 6\ 4\ 2\ 6\ 4\ 6\ 8\ 4\ 2\ 4\ 2\ 4$$
$$8\ 6\ 4\ 6\ 2\ 4\ 6\ 2\ 6\ 6\ 4\ 2\ 4\ 6\ 2\ 6\ 4\ 2\ 4\ 2\ 10\ 2\ 10$$

The elements of the sequence keep repeating periodically after every 48 elements.

6.2   Few Pattern Sequences

Pattern Sequence 5: 4 2 4 2 4 6 2 6.

Pattern Sequence 7 appears in the main text.

Pattern Sequence 11: 4 2 4 6 2 6 4 2 4 6 6 2 6 4 2 6 4 6 8 4 2 4 2 4 14 4 6 2 10 2 6 6 4 2 4 6 2 10 2 4 2 12 10 2 4 2 4 6 2 6 4 6 6 6 2 6 4 2 6 4 6 8 4 2 4 6 8 6 10 2 4 6 2 6 6 4 2 4 6 2 6 4 2 6 10 2 10 2 4 2 4 6 8 4 2 4 12 2 6 4 2 6 4 6 12 2 4 2 4 8 6 4 6 2 4 6 2 6 10 2 4 6 2 6 4 2 4 2 10 2 10 2 4 6 6 2 6 6 4 6 6 2 6 4 2 6 4 6 8 4 2 6 4 8 6 4 6 2 4 6 8 6 4 2 10 2 6 4 2 4 2 10 2 10 2 4 2 4 8 6 4 2 4 6 6 2 6 4 2 4 6 2 6 4 2 4 2 10 2 10 2 6 4 6 2 6 4 2 4 6 8 4 2 4 2 4 8 6 4 6 6 6 2 6 6 4 2 4 6 2 6 4 2 4 2 10 2 10 2 6 4 6 2 6 4 2 4 6 6 8 4 2 4 2 4 8 10 6 2 4 8 6 6 4 2 4 6 2 6 4 6 2 10 2 10 2 4 2 4 6 2 6 6 6 4 6 8 4 2 4 2 4 8 6 4 8 4 6 2 6 6 4 2 4 6 8 4 2 4 2 10 2 10 2 4 2 4 6 2 10 2 4 6 8 4 2 6 4 6 8 4 6 2 4 6 2 6 6 4 6 6 2 6 6 4 2 10 2 10 2 4 2 4 6 2 6 4 2 10 6 2 6 4 2 6 4 6 8 4 2 4 2 12 6 4 6 2 4 6 2 12 4 2 4 8 6 4 2 4 2 10 2 10 6 2 4 6 2 6 4 2 4 6 6 2 6 4 2 10 6 8 6 4 2 4 8 6 4 6 2 4 6 2 6 6 4 6 2 6 4 2 4 2 10 12 2 4 2 10 2 6 4 2 4 6 6 2 10 2 6 4 14 4 2 4 2 4 8 6 4 6 2 4 6 2 6 6 4 2 4 6 2 6 4 2 4 12 2 12.

6.3   Recursive Formulae for Size of Patterns

Consider two Pattern Sequences of successive Prime Numbers h_1 and h_2. Let the number of terms in these sequences be T_1 and T_2 with sum of terms S_1 and S_2 respectively
Number of terms in S_2,

$$T_2 = (h_2 - 1) \times T_1 \tag{4}$$

Sum of terms for Sequence S_2,

$$S_2 = h_2 \times S_1 \tag{5}$$

6.4   Prime Number Density

Prime number density for n is defined as the fraction of primes till n over n. The unpredictable nature of Prime Numbers has invited a lot of debate on the issue prime of density. However, for very large primes, it is believed to be of the order of $\frac{1}{log(n)}$ as proposed by Adrian-Marie Legendre and Carl F.Gauss.

Empirical Study on Weighted Voting Multiple Classifiers

Yanmin Sun, Mohamed S. Kamel, and Andrew K.C. Wong

Pattern Analysis and Machine Intelligence Lab,
Department of Electrical and Computer Engineering,
University of Waterloo,
Waterloo, Ontario, Canada N2L 3G1
{sunym, mkamel, akcwong}@pami.uwaterloo.ca

Abstract. Combining multiple classifiers is expected to increase classification accuracy. Research on combination strategies of multiple classifiers becomes a popular topic. For a crisp classifier, which returns a discrete class label instead of a set of real-valued probabilities respecting to every classes, the often used combination method is majority voting. Both majority and weighted majority voting are *classifier-based* voting schemes, which provide a certain base classifier with an identical confidence in voting. However, each classifier should have different voting priorities with respect to its learning space. This differences can not be reflected by classifier-based voting strategy. In this paper, we propose another two voting strategies in an effort to take such differences into consideration. We apply the AdaBoost algorithm to generate multiple classifiers and vary its voting strategy. Then, the prediction ability of each voting strategy is tested and compared on 8 datasets taken from UCI Machine Learning Repository. The experimental results show that one of the proposed voting strategies, namely *sample-based* voting scheme, achieves better performance in view of classification accuracy.

1 Research Motivation

Combining multiple classifiers is expected to increase classification accuracy [2,7]. Research on combination schemes of multiple classifiers becomes a popular topic. Most reported works in this area focus on classifier fusion with the output of each classifier is scaled to the [0 1] interval [1,9]. In this case, the combining of classifiers is often done using linear combinations of classifier outputs, rank-based combining and voting-based combination. For a crisp classifier, which returns a discrete class label instead of a real-valued probabilities respecting to every classes in a data set, the often used combination method is majority voting, either simple majority or weighted majority voting. Both majority voting and weighted voting are *classifier-based* voting schemes, which mean that a certain base classifier will provide an identical confidence in classifying a set of objects via voting.

By manipulating the training data, such as Bootstrap aggregating (bagging) [3] and AdaBoost [4,6], a set of classifiers are learned with each one concentrating

S. Singh et al. (Eds.): ICAPR 2005, LNCS 3686, pp. 335–344, 2005.

on a specific data space. Hence, each base classifier should have different voting priorities with respect to its specific learning space. However, classifier-based weighting scheme overlooks this difference. Two other weighting strategies for voting can be: 1)*class-based* weighting scheme: for each class label in the data set, a certain base classifier will provide a specific weight, which denotes the prediction confidence respecting to this class label; and 2)*sample-based* weighting scheme: for each sample, a certain base classifier will provide a set of weights with each one indicating the prediction confidence in classifying the sample to a specific class. Given a base classifier, classifier-based weighting scheme provides only one voting weight; class-based weighting scheme provides a vector of voting weights, with each one representing the voting confidence for classifying samples to a specific class; and sample-based weighting scheme provides each testing sample with a vector of voting weights, with each one denoting the predictive confidence for classifying this sample to a particular class.

The learning objective of voting multiple classifiers is to achieve more accurate prediction results. Among these weighting schemes, which one is superior in obtaining the better classification results? In this study, we concentrate on crisp classifier, whose output is a discrete class label instead of a set of real-valued probabilities respecting to every classes in a data set. We adopt AdaBoost algorithm to generate multiple classifiers and vary its voting strategy. Two classification learning algorithms, Naïve-Bayes and an associative classification system, are employed as the base learners in our experiments. The prediction ability of each voting strategy is tested and compared on a representative collection of 8 datasets taken from UCI Machine Learning Repository[10]. This paper is organized as: following the introduction in Section 1, Section 2 introduces related learning algorithms: AdaBoost algorithm; Section 3 presents weighting schemes for voting multiple classifiers; Section 4 explains the two base classification learning systems; Section 5 shows the experiment settings and results; Section 6 highlights the conclusion.

2   AdaBoost Algorithm

AdaBoost (Adaptive Boosting) algorithm introduced by Freund and Schapire [5,6,13,12] is generally considered as an effective boosting algorithm. It weighs each sample reflecting its importance and places most weights on those examples which are most often misclassified by the preceding classifiers. This forces the following learning process to concentrate on those samples which are hard to be correctly classified. The final classification is based on a weighted majority vote of each individual classifier. The weight assigned to each classifier is determined according to classifier's performance on its training set.

AdaBoost algorithm takes as input a training set $\{(\underline{x}_1, y_1), \cdots, (\underline{x}_m, y_m)\}$ where each $\underline{x}_i$ is an n-tuple of attribute values belonging to a certain domain or instance space X, and y_i is a label in a label set $Y = \{c_1, c_2, \cdots, c_K\}$. The Pseudocode for AdaBoost is given as below:

Given: $(\underline{x}_1, y_1), \cdots, (\underline{x}_m, y_m)$ where $\underline{x}_i \in X$, $y_i \in Y = \{c_1, c_2, \cdots, c_K\}$
Initialize $D^1(i) = 1/m$.
For $t = 1, \cdots, T$:
1. Train base learner h_t using distribution D^t
2. Choose weight updating parameter: α_t
3. Update and normalize sample weights:
$$D^{(t+1)}(i) = \frac{D^{(t)}(i)exp(-\alpha_t I[h_t(\underline{x}_i) \neq y_i])}{Z_t}$$
Where, Z_t is a normalization factor.
Output the final classifier: $H(\underline{x}) = sign(\sum_{t=1}^{T} \beta_t h_t(\underline{x}))$

where, for any predicate π, $I[\pi]$ equals 1 if π holds, Otherwise 0. In this study, we apply AdaBoost algorithm to the base classification learning systems whose outputs are discrete class labels.

3   Weighting Schemes for Voting Multiple Classifiers

We vary the weighted voting strategy of AdaBoost algorithm with another three types of voting schemes: majority voting, class-based and sample-based voting schemes.

3.1   Classifier-Based Weighting Schemes

AdaBoost Weighting: In AdaBoost algorithm, the classifier weighting factor β_t is calculated as:

$$\beta_t = \alpha_t = \frac{1}{2}log\frac{1 - err_t}{err_t} \tag{1}$$

where, err_t denoting the weighted training error of the tth classifier:

$$err_t = \frac{\sum\limits_{i,y_i \neq h_t(\underline{x}_i)} D(i)^{(t)}}{\sum\limits_{i} D(i)^{(t)}} \tag{2}$$

β is used as the strength measure respecting to each classifier. This weighted voting scheme is denoted as **Stra1**. The weighted combination of the output of each classifier then becomes:

$$H(\underline{x}) = arg \max_{c_k, k=1 \cdots K} (\sum_{t=1}^{T} \beta_t I[h_t(\underline{x}) = c_k]) \tag{3}$$

Majority Voting: When we set $\beta_t = 1$, the weighted majority voting scheme of Adaboost algorithm becomes to majority voting. Each classifier obtains identical voting weight irrelative to its performance. Majority voting is denoted as **Stra2** in this paper. The output of Equation 3 turns to

$$H(\underline{x}) = arg \max_{c_k, k=1\cdot\cdot K} (\sum_{t=1}^{T} I[h_t(\underline{x}) = c_k]) \qquad (4)$$

3.2   Class-Based Weighting Scheme

Class-based weighting scheme is to provide a classifier with a vector of prediction confidences corresponding to its performances on different classes. The weight associated with each class label is calculated by the training error of this class. Let err_{kt} denote the training error of the tth classifier on class c_k, where $k = 1\cdot\cdot K$:

$$err_{kt} = \frac{\sum\limits_{i,y_i \neq h_t(\underline{x}_i)\&y_i=c_k} D(i)^{(t)}}{\sum\limits_{i,y_i=c_k} D(i)^{(t)}} \qquad (5)$$

then, β_{kt} is the weight of the tth classifier when it predicts an object belonging to class c_k

$$\beta_{kt} = \frac{1}{2} log \frac{1 - err_{kt}}{err_{kt}} \qquad (6)$$

When voting multiple classifier, according to the class label assignment, a specific weight is used for voting. The weighted combination of the output of each classifier then becomes:

$$H(\underline{x}) = arg \max_{c_k, k=1\cdot\cdot K} (\sum_{t=1}^{T} \beta_{kt} I[h_t(\underline{x}) = c_k]) \qquad (7)$$

In the following sections, this class-based weighting scheme will be referred as **Stra3**.

3.3   Sample-Based Weighting Scheme

When a new observation $\underline{x}$ is given, the amount of evidence provided by $\underline{x}$ for c_k being a plausible value of Y can be quantitatively estimated by an evidence measure which is derived from an information-theoretic measure known as the *mutual information*[11]:

$$I(Y = c_k : \underline{x}) = log \frac{P(Y = c_k|\underline{x})}{P(Y = c_k)}$$

Based on the mutual information, the difference in the gain of information when Y takes on the value c_k and when it takes on some other values, given $\underline{x}$, is a measure of evidence provided by $\underline{x}$ in favor of c_k being a plausible value of Y as opposed to other values. This difference, denoted by $W(Y = c_k/y \neq c_k|\underline{x})$, is defined as the *weight of evidence*[11], which has the following form:

$$W(Y = c_k/Y \neq c_k|\underline{x}) = I(Y = c_k : \underline{x}) - I(Y \neq c_k : \underline{x})$$
$$= log\frac{P(Y = c_k|\underline{x})}{P(Y = c_k)} - log\frac{P(Y \neq c_k|\underline{x})}{P(Y \neq c_k)} \tag{8}$$

By applying Bayes formula, Equation (8) can be rewritten equivalently as:

$$W(Y = c_k/Y \neq c_k|\underline{x}) = log\frac{P(\underline{x}|Y = c_k)}{P(\underline{x})} - log\frac{P(\underline{x}|Y \neq c_k)}{P(\underline{x})}$$
$$= log\frac{P(\underline{x}|Y = c_k)}{P(\underline{x}|Y \neq c_k)} \tag{9}$$

The most plausible value of Y is the one with the highest weight of evidence provided by the observation. When weight of evidence is employed as the prediction confidence measure, each classifier will provide a specific prediction confidence for each sample:

$$h_t(\underline{x}) \rightarrow c_k, \qquad \text{with confidence } r_{kt} \qquad 1 \leq k \leq K$$

where r_{kt} denotes the weight of evidence of the tth classifier in predicting the class c_k given the object $\underline{x}$. Then, the weighted combination of the output of each classifier becomes:

$$H(\underline{x}) = arg \max_{c_k, k=1 \cdots K} (\sum_{t=1}^{T} r_{kt} I[h_t(\underline{x}) = c_k]) \tag{10}$$

In view that each base classifier provides a specific prediction confidence in classifying each object to a particular class, this weighting scheme is *sample-based*, which will be referred as **Stra4** in the following sections.

4   The Base Classifier Learning Systems

To test and compare these voting strategies, two distinct kinds of classification systems are specially selected as the base classifiers. One is the well-known and widely used Naïve Bayes classifier which assumes strong independencies among attributes. Another one is a new classification approach in data mining, called *Associative classification*. In this paper, a mathematically well-developed associative classifier, namely High-Order Pattern and Weight of Evidence Rule (HPWR)[14,15] based classification system is selected. The common point of these two classification systems is they both calculate the weight of evidence as the prediction support for classification.

4.1 Naïve Bayes

Naïve Bayesian classification is based on Bayesian Theory assuming independencies among attributes. Given an unlabelled instance $\underline{x}$ containing n attributes $[A_1, \cdots, A_n]$ and target label value c_k, the assumption yields the following model:

$$P(\underline{x}|Y = c_k) = \prod_{m=1}^{n} P(A_m|Y = c_k) \tag{11}$$

Weight of evidence in favor of class label c_k against other values can be formulated as:

$$W(Y = c_k/Y \neq c_k|\underline{x}) = log\frac{P(\underline{x}|Y = c_k)}{P(\underline{x}|Y \neq c_k)}$$

$$= \sum_{m=1}^{n} log\frac{P(A_m|Y = c_k)}{P(A_m|Y \neq c_k)} = \sum_{m=1}^{n} W(Y = c_k/Y \neq c_k|A_m) \tag{12}$$

4.2 Associative Classification—HPWR

Employing residual analysis and mutual information for decision support, HPWR generates classification patterns and rules in two stages: 1) discovering high-order significant events associations using residual analysis in statistics to test the significance of the occurrence of a pattern candidate against its default expectation[15], and 2) generating classification rules with weight of evidence attached to them to quantify the evidence of significant event associations in support of, or against a certain class membership[14].

Suppose $\underline{x}^l$ is a subset of events, and $(\underline{x}^l, Y = c_k)$ is a significant event association of $\underline{x}$ and satisfies $\underline{x}^p \cap \underline{x}^q = \Phi$, $p \neq q$, $1 \leq p, q \leq n$ and $\cup_{l=1}^{n}\underline{x}^l = \underline{x}$. The Equation (9) can therefore be written as:

$$W(Y = c_k/Y \neq c_k|\underline{x}) = log\frac{P(\underline{x}^1|Y = c_k)}{P(\underline{x}^1|Y \neq c_k)} + ... + log\frac{P(\underline{x}^n|Y = c_k)}{P(\underline{x}^n|Y \neq c_k)}$$

$$= W(Y = c_k/Y \neq c_k|\underline{x}^1) + ... + W(Y = c_k/Y \neq c_i|\underline{x}^n)$$

$$= \sum_{l=1}^{n} W(Y = c_k/Y \neq c_k|\underline{x}^l) \tag{13}$$

Thus, the calculation of weight of evidence is to find a proper set of disjoint significant event associations from $\underline{x}$ and to sum each individual weight of evidence provided by the subset of events. For more detail information, please refer to [14].

5 Experimental Evaluation

In this section, we carry on experiments to test and compare their voting accuracies of every voting strategies as described in Section 3.

5.1 Experiment Setups

A representative collection of 8 datasets taken from UCI Machine Learning Repository [10] are used for such evaluation. The description of these datasets are summarized in Table 1. For each data set with missing values, the missing values are treated as having the value "?". The continue attributes of each data set are discretized through the commonly used discretization utility of MLC++ [8] with the default settings. The parameter T governing the number of classifiers generated is set at 10 for these experiments. 5-fold cross-validations are carried out with each data set.

Table 1. Description of Datasets

	Data set	#attr	#Cate.	#rec
1	Auto	25	7	205
2	Chess	36	2	3196
3	Cleve	13	2	303
4	Diabetes	8	2	768
5	Horse	22	2	368
6	Iono	34	2	351
7	Sonar	60	2	208
8	Waveform	21	3	5000

5.2 Experiment Results

Results for Naïve Bayes are reported in Table 2. The results in column labelled **NB** are the average accuracies of the original Naïve Bayes classifiers. **Stra1** stands for the voting strategy of AdaBoost, **Stra2** for majority voting, and **Stra3** for class-based voting, and **Stra4** for sample-based voting. Results in these columns show the average accuracies of the voted classifications, as well as the improvements when the results is compared with those of the original Naïve Bayes classifiers. The best classification result on each data set is marked in bold.

Viewing these experimental results, we find that over these 8 datasets, in most cases the voting results from all these four classifier weighting strategies are better than the original Naïve Bayes classifiers in different degrees. When four voting strategies are compared, Stra1 achieves the best results on 1 dataset, Stra2 on 1 dataset, and Stra4 on 6 datasets. For average classification accuracy, Stra1 improves the performance of Naïve Bayes classifier by 3.9%, Stra2 by 2.9%, Stra3 by 2.7% and Stra4 by 4.3%. For both comparisons, Stra4 gains the best results. When Stra1 is compared with Stra4, i.e., voting strategy of AdaBoost is compared with the sample-based voting strategy, Stra4 slightly over performs Stra1 in viewing of the average classification accuracy.

Results for HPWR are reported in Table 3. The number in column labelled **Order** is the order limitation of association patterns respecting to each dataset. It is stated in [14], high order associations are more complex than the low order

Table 2. Voting Results of Different Weighting Schemes Taking Naïve Bayes as the Base Inducer

Data set	NB Acc%	Stra1 Acc% ↑%		Stra2 Acc % ↑%		Stra3 Acc% ↑%		Stra4 Acc% ↑%	
Auto	74.1	83.8	9.7	**84.0**	**9.9**	79.5	5.4	82.4	8.3
Chess	87.6	94.5	6.9	93.9	6.3	94.3	6.7	**95.2**	**7.6**
Cleve	83.1	84.5	1.4	84.6	1.5	84.6	1.5	**85.3**	**2.2**
Diabetes	70.9	74.6	3.7	70.3	-0.6	71.8	0.9	**74.7**	**3.8**
Horse	78.5	**83.0**	**4.5**	81.1	2.6	80.7	2.2	81.4	2.9
Iono	91.4	93.8	2.4	92.6	1.2	93.8	2.4	**95.5**	**4.1**
Sonar	84.7	85.7	1.0	85.5	0.8	85.3	0.6	**87.7**	**3.0**
Waveform	81.4	83.5	2.1	83.3	1.9	83.1	1.7	**84.0**	**2.6**
Average	81.5	85.4	3.9	84.4	2.9	84.2	2.7	**85.8**	**4.3**

Table 3. Voting Results of Different Weighting Schemes Taking HPWR as the Base Inducer

Data set	Order	HWRC Acc%	Stra1 Acc% ↑%		Stra2 Acc % ↑%		Stra3 Acc% ↑%		Stra4 Acc% ↑%	
Auto	3	67.9	72.4	4.5	76.8	8.9	71.4	3.5	**80.4**	**12.5**
Chess	3	87.8	98.2	10.4	95.0	7.2	97.5	9.7	**98.6**	**11.8**
Cleve	4	78.8	86.6	7.8	83.0	4.2	82.0	3.2	**88.7**	**9.9**
Diabetes	3	75.8	76.2	0.4	76.6	0.8	76.0	0.2	**77.4**	**1.6**
Horse	3	82.9	87.8	4.9	88.3	5.4	85.8	2.9	**88.4**	**5.5**
Iono	2	92.6	**95.7**	**3.1**	93.5	0.9	93.3	0.7	95.1	2.5
Sonar	3	79.0	83.9	4.9	**85.0**	**6.0**	82.6	3.6	83.5	4.5
Waveform	2	75.3	84.6	9.3	**86.6**	**11.3**	83.3	8.0	84.4	9.1
Average		80.0	85.7	5.7	85.6	5.6	84.0	4.0	**87.1**	**7.1**

ones as they describe the properties of the domain more accurately and more specifically than the low order events. The objective of boosting is to generate a more accurate composite classifier by combining moderately inaccurate, or simply "weaker" classifiers. Thus, in order to build "weaker" classifiers, this parameter is set low to ensure the use of low order patterns in this part of experiments. The results in column labelled **HPWR** are the average accuracies of the original HPWR classifiers. Experiment results of various voting strategies are reported using the same template as described for Table 2.

In most cases the voting results from all these four classifier weighting strategies are better than the original HPWR classifiers over these 8 datasets. When four voting strategies are compared, Stra1 achieves the best results on 1 datasets, Stra2 on 2 datasets, and Stra4 on 5 datasets. For average classification accuracy, Stra1 improves HPWR's accuracy by 5.7%, Stra2 by 5.6%, Stra3 by 4.0% and Stra4 by 7.1%. Again, for both comparisons, Stra4 gains the best results. When Stra1 is compared with Stra4, Stra4 performs Stra1 by 1.4% in viewing of the average classification accuracy.

5.3   Sample-Based Weighting Strategy vs. Classifier-Based Weighting Strategy

Traditionally, weighting scheme for weighted voting multiple classifier is classifier-based. It means that a certain component classifier will provide an identical confidence in classifying a set of objects via voting. However, the intention of a boosting algorithm is to force each learning iteration to concentrate on a specific part of the data space by changing the data distribution. It is quite possible that a classifier has different prediction confidences in different data spaces. When classifier-based weighting scheme is adopted, this difference is overlooked in voting.

Strat4 uses the weight of evidence as the voting factor in the final classification. Weight of evidence provides a confidence measure of a classifier in predicting the class label of a particular object. As each classifier will provide a specific prediction confidence for each sample, this weighting scheme is sample-based. Consider that the significant character of a boosting algorithm is that a set of classifiers are learned with each concentrating on a specific data space. Then, it is quite understandable that each classifier should have different voting priorities in respect to its learning space. The sample-based voting strategy reflects the uneven learning concentration across the whole data space of each classifier. The obvious advantage of sample-based weighting scheme takes into account such differences. The overall better experimental result of Stra4 in comparison with that of Stra1 is an convincing empirical support of this advantage.

6   Conclusion

Combining multiple classifiers is expected to increase classification performance evaluated by prediction accuracy. Multiple classifiers are usually combined by majority voting when the outputs of every classifiers are discrete class labels. Both majority voting and weighted majority voting are classifier-based voting schemes. In this paper, we propose another two voting schemes, class-based voting scheme and sample-based voting scheme respectively.

The prediction ability of each voting strategy is then experimentally studied. We adopt AdaBoost algorithm to generate multiple classifiers and vary its weighted voting strategy. Two distinct kinds of classification systems are specially selected as the base inducers as they provide the weight of evidence as the measure of their prediction strength. Experimental results on a representative collection of 8 data sets show that sample-based weighting strategy offers the following advantages: 1) when the weights of evidence furnished by the base learning systems respecting to each classification object are used as prediction confidence measures, this sample-based weighting scheme reflects the distinct learning focus of each classifier on the data space; and 2) experimental results indicate this sample-based weighting scheme is better than that of AdaBoost in view of classification accuracy.

References

1. L. A. Alexandre, A. C. Campilho, and M. Kamel. On combining classifiers using sum and product rules. *Pattern Recognition Letters*, 22(12):1283–1289, October 2001.
2. E. Bauer and R. Kohavi. An empirical comparison of voting classification algoirthm: Bagging, boosting and variants. *Machine Learning*, 36:105–142, 1999.
3. L. Breiman. Bagging predictors. *Machine Learning*, 24(2):123–140, 1996.
4. Y. Freund. Boosting a weak leaning algorithm by majority. *Information and computation*, 121(2):256–285, 1995.
5. Y. Freund and R. E. Schapire. Experiments with a new boosting algorithm. In *Proc. of the Thirteenth International Conference on Machine Learning*, pages 148–156, Morgan Kaufmann, 1996. The Mit Press.
6. Y. Freund and R. E. Schapire. A decision-theoretic generalization of on-line learning and an aplication to boosting. *Journal of Computer and System Sciences*, 55(1):119–139, August 1997.
7. M. Kamel and N. Wanas. Data dependence in combining classifiers. In *Multiple Classifiers Systems, Fourth International Workshop*, pages 11–13, Surrey, UK, June 2003.
8. R. Kohavi, D. Sommerfield, and J. Dougherty. *Data Mining Using MLC++: A machine learning library in C++. Tools with Artificial Intelligence*. IEEE CS Press, 1996.
9. L. I. Kuncheva, J. C. Bezdek, and R. P. W. Duin. Decision templates for multiple classifier fusion: An experimental comparision. *Pattern Recognition*, 34(2):299–314, 2001.
10. P. M. Murph and D. W. Aha. *UCI Repository Of Machine Learning Databases*. Dept. Of Information and Computer Science, Univ. Of California: Irvine, 1991.
11. D. B. Osteyee and I. J. Good. *Information, Weight Of Evidence, The Singularity Between Probability Measures and Signal Detection*. Springer-Velag, Berlin, Germany, 1974.
12. R. E. Schapire and Y. Singer. Boosting the margin: A new explanation for the effectiveness of voting methods. *Machine Learning*, 37(3):297–336, 1999.
13. R. E. Schapire and Y. Singer. Improved boosting algorithms using confidence-rated predictions. *Machine Learning*, 37(3):297–336, 1999.
14. Y. Wang and A. K. C. Wong. From association to classification: Inference using weight of evidence. *IEEE Trans. On Knowledge and Data Engineering*, 15(3):764–767, 2003.
15. A. K. C. Wong and Y. Wang. High order discovery from discrete-valued data. *IEEE Trans. On Knowledge and Data Engineering*, 9(6):877–893, 1997.

Spectral Clustering for Time Series

Fei Wang and Changshui Zhang

State Key Laboratory of Intelligent Technology and Systems,
Department of Automation, Tsinghua University, Beijing 100084, P.R. China
feiwang03@mails.tsinghua.edu.cn
zcs@mail.tsinghua.edu.cn

Abstract. This paper presents a general framework for time series clustering based on spectral decomposition of the affinity matrix. We use the Gaussian function to construct the affinity matrix and develop a gradient based method for self-tuning the variance of the Gaussian function. The feasibility of our method is guaranteed by the theoretical inference in this paper. And our approach can be used to cluster both constant and variable length time series. Further our analysis shows that the cluster number is governed by the eigenstructure of the normalized affinity matrix. Thus our algorithm is able to discover the optimal number of clusters automatically. Finally experimental results are presented to show the effectiveness of our method.

1 Introduction

The recent years have seen a surge of interest in time series clustering. The high dimensionality and irregular lengths of the time sequence data pose many challenges to the traditional clustering algorithms. For example, it is hard for the application of the k-means algorithm [1] since we cannot define the "mean" of time series with different length. Many researchers propose to use hierarchical agglomerative clustering (HAC) for time series clustering [2][3], but there are two main drawbacks of these methods. On one hand, it is difficult for us to choose a proper distance measure when we merge two clusters; on the other hand, it is hard to decide when to stop the clustering procedure, that is, to decide the final cluster number.

Recently Porikli [4] proposed to use HMM parameter space and eigenvector decomposition to cluster time series, however, they didn't give us the theoretical basis of their method, and the variance of the Gaussian function they used to construct the affinity matrix was set empirically, which is usually not desirable.

To overcome the above problems, this paper presents a more efficient spectral decomposition based framework for time series clustering. Our method has four main advantages: (1) it is based on the similarity matrix of the dataset, that is, all it needs are just the pairwise similarities of the time series, so the high dimensionality of time series will not affect the efficiency of our approach; (2) it can be used to clustering time series with arbitrary length as long as the similarity measure between them is properly defined; (3) it can determine the

S. Singh et al. (Eds.): ICAPR 2005, LNCS 3686, pp. 345–354, 2005.

optimal cluster number automatically; (4) it can self-tune the variance of the Gaussian kernel. The feasibility of our method has been proved theoretically in this paper, and many experiments are presented to show its effectiveness.

The remainder of this paper is organized as follows: we analyze and present our clustering framework in Section 2 in detail. In section 3 we will give a set of experiments, followed by the conclusions and discussions in section 4.

2   Spectral Clustering for Time Series

2.1   Theoretical Background

We will introduce the theoretical background of our spectral decomposition based clustering framework in this subsection. Given a set of time series $\{\mathbf{x}_i\}_{i=1}^M$ with the same length d, we form the data matrix $\mathbf{A} = [\mathbf{x}_1, \mathbf{x}_2, \cdots, \mathbf{x}_M]$. A clustering process to $\mathbf{A}$ can result in $\mathbf{B} = \mathbf{AE} = [\mathbf{A}_1, \mathbf{A}_2, \cdots, \mathbf{A}_K]$, where $\mathbf{E}$ is a permutation matrix, $\mathbf{A}_i = [\mathbf{x}_{i1}, \mathbf{x}_{i2}, \cdots, \mathbf{x}_{is_i}]$ represents the i-th cluster, $\mathbf{x}_{ij}$ is the jth data in cluster i, and s_i is the number of data in the i-th cluster.

Now let's introduce some notes. The within-cluster scatter matrix of cluster k is $\mathbf{S}_w^k = \frac{1}{s_k} \sum_{s_i \in k} (\mathbf{x}_{s_i} - \mathbf{m}_k)(\mathbf{x}_{s_i} - \mathbf{m}_k)^T$, where $\mathbf{m}_k$ is the mean vector of the k-th cluster. The total within-cluster scatter matrix is $\mathbf{S}_w = \sum_{k=1}^K s_k \mathbf{S}_w^k$. The total between-cluster scatter matrix is $\mathbf{S}_b = \sum_{k=1}^K s_k (\mathbf{m}_k - \mathbf{m})(\mathbf{m}_k - \mathbf{m})^T$ and the total data scatter matrix is $\mathbf{T} = \mathbf{S}_b + \mathbf{S}_w = \sum_{i=1}^M (\mathbf{x}_i - \mathbf{m})(\mathbf{x}_i - \mathbf{m})^T$, where $\mathbf{m}$ is the sample mean of the whole dataset.

The goal of clustering is to achieve high within-cluster similarity and low between-cluster similarity, that is, we should minimize $trace(\mathbf{S}_w)$ and maximize $trace(\mathbf{S}_b)$. Since $\mathbf{T}$ is independent on the clustering results, then the maximization of $trace(\mathbf{S}_b)$ is equivalent to the minimization of $trace(\mathbf{S}_w)$. So our optimization object becomes

$$\min \quad trace(\mathbf{S}_w) \tag{1}$$

Since $\mathbf{m}_k = \mathbf{A}_k \mathbf{e}_k / s_k$, where $\mathbf{e}_k$ is the column vector containing s_k ones, then $\mathbf{S}_w^k = \frac{1}{s_k} \left(\mathbf{A}_k - \frac{\mathbf{A}_k \mathbf{e}_k}{s_k} \mathbf{e}_k^T \right) \left(\mathbf{A}_k - \frac{\mathbf{A}_k \mathbf{e}_k}{s_k} \mathbf{e}_k^T \right)^T = \frac{1}{s_k} \mathbf{A}_k \left(\mathbf{I}_k - \frac{\mathbf{e}_k \mathbf{e}_k^T}{s_k} \right) \mathbf{A}_k^T$, where $\mathbf{I}_k$ is the identity matrix of order s_k.

Let $\mathbf{J}_k = trace(\mathbf{S}_w^k) = trace \left(\frac{1}{s_k} \mathbf{A}_k \mathbf{A}_k^T \right) - trace \left(\frac{1}{s_k} \frac{\mathbf{e}_k^T}{\sqrt{s_k}} \mathbf{A}_k^T \mathbf{A}_k \frac{\mathbf{e}_k}{\sqrt{s_k}} \right)$. Define $\mathbf{J} = trace(\mathbf{S}_w) = \sum_{k=1}^K s_k trace(\mathbf{S}_w^k)$ and the block-diagonal matrix

$$\mathbf{Q} = \begin{pmatrix} \mathbf{e}_1/\sqrt{s_1} & 0 & \cdots & 0 \\ 0 & \mathbf{e}_1/\sqrt{s_1} & \cdots & 0 \\ \vdots & \vdots & \ddots & \vdots \\ 0 & 0 & \cdots & \mathbf{e}_K/\sqrt{s_K} \end{pmatrix} \tag{2}$$

then $\mathbf{J} = trace \left(\mathbf{BB}^T \right) - trace \left(\mathbf{Q}^T \mathbf{B}^T \mathbf{B} \mathbf{Q} \right)$. Since $\mathbf{B} = \mathbf{AE}$ and $\mathbf{E}$ is a permutation matrix, it's straight forward to show that $trace \left(\mathbf{BB}^T \right) = trace(\mathbf{A}^T \mathbf{A})$,

$trace(\mathbf{Q}^T\mathbf{B}^T\mathbf{B}\mathbf{Q}) = trace(\tilde{\mathbf{Q}}^T\mathbf{A}^T\mathbf{A}\tilde{\mathbf{Q}})$, where $\tilde{\mathbf{Q}} = \mathbf{E}\mathbf{Q}$, that is, $\tilde{\mathbf{Q}}$ is equal to $\mathbf{Q}$ with some rows exchanged. Now we relax the constraint of $\tilde{\mathbf{Q}}$ to $\tilde{\mathbf{Q}}^T\tilde{\mathbf{Q}} = \mathbf{I}$ as in [5]. Then optimizer (1) is equivalent to

$$\max_{\tilde{\mathbf{Q}}^T\tilde{\mathbf{Q}}=\mathbf{I}} H = trace\left(\tilde{\mathbf{Q}}^T\mathbf{A}^T\mathbf{A}\tilde{\mathbf{Q}}\right) \tag{3}$$

which is a constrained optimization problem. It turns out the above optimization problem has a closed-form solution according to the following theorem[5].

Theorem(Ky Fan). *Let* $\mathbf{H}$ *be a symmetric matrix with eigenvalues*

$$\lambda_1 \geqslant \lambda_2 \geqslant \cdots \geqslant \lambda_n$$

and the corresponding eigenvectors $\mathbf{U} = [\mathbf{u}_1, \mathbf{u}_2, \cdots, \mathbf{u}_n]$. *Then*

$$\lambda_1 + \lambda_2 + \cdots + \lambda_k = \max_{\mathbf{X}^T\mathbf{X}=\mathbf{I}} trace(\mathbf{X}^T\mathbf{H}\mathbf{X})$$

Moreover, the optimal $\mathbf{X}^*$ *is given by* $\mathbf{X}^* = \mathbf{U} = [\mathbf{u}_1, \mathbf{u}_2, \cdots, \mathbf{u}_k]\mathbf{R}$, *with* $\mathbf{R}$ *an arbitrary orthogonal matrix.*

From the above theorem we can easily derive the solution to (3). The optimal $\tilde{\mathbf{Q}}$ can be obtained by taking the top K eigenvectors of $\mathbf{S} = \mathbf{A}^T\mathbf{A}$, and the sum of the corresponding largest K eigenvalues of $\mathbf{S}$ gives the optimal H.

The matrix $\mathbf{S}$ can be expanded as

$$\mathbf{S} = \mathbf{A}^T\mathbf{A} = [\mathbf{x}_1, \mathbf{x}_2, \cdots, \mathbf{x}_M]^T[\mathbf{x}_1, \mathbf{x}_2, \cdots, \mathbf{x}_M] = \begin{pmatrix} \mathbf{x}_1^T\mathbf{x}_1 & \mathbf{x}_1^T\mathbf{x}_2 & \cdots & \mathbf{x}_1^T\mathbf{x}_M \\ \mathbf{x}_2^T\mathbf{x}_1 & \mathbf{x}_2^T\mathbf{x}_2 & \cdots & \mathbf{x}_2^T\mathbf{x}_M \\ \vdots & \vdots & \ddots & \vdots \\ \mathbf{x}_M^T\mathbf{x}_1 & \mathbf{x}_M^T\mathbf{x}_2 & \cdots & \mathbf{x}_M^T\mathbf{x}_M \end{pmatrix}$$

Thus the (i, j)-th entry of $\mathbf{S}$ is the inner product of $\mathbf{x}_i$ and $\mathbf{x}_j$, which can be used to measure the similarity between them. Then $\mathbf{S}$ can be treated as the similarity matrix of the dataset $\mathbf{A}$. Moreover, we can generalize this idea and let the entries of $\mathbf{S}$ be some other similarity measure, as long as it satisfies the symmetry and positive semidefinite properties. Since there has been so many methods for measuring the similarities between time series with different length (for a comprehensive study, see [7]), we can drop the assumption at the beginning of this section that all the time series have the same length d and let the entries in $\mathbf{S}$ be some similarity measure that can measure the similarity of time series with arbitrary length. Then our method can be used to cluster time series with any length.

2.2   Estimating the Number of Clusters Automatically

In order to estimate the optimal number of the clusters, we first normalize the rows of $\mathbf{S}$, that is, define $\mathbf{U} = diag(u_{11}, u_{22}, \cdots, u_{MM})$, where $u_{ii} = \sum_{j=1}^{M} \mathbf{S}_{ij}$, then our normalization makes $\mathbf{S}' = \mathbf{U}^{-1}\mathbf{S}$.

In the ideal case, $\mathbf{S}(i,j) = 0$ if $\mathbf{x}_i$ and $\mathbf{x}_j$ belong to different clusters. We assume that the data objects are ordered by clusters, that is $\mathbf{A} = [\mathbf{A}_1, \cdots, \mathbf{A}_K]$, where $\mathbf{A}_i$ represents the data in cluster i. Thus the similarity matrix $\mathbf{S}$ and normalized similarity matrix $\mathbf{S}'$ will become block-diagonal. It can be easily inferred that each diagonal block in $\mathbf{S}'$ has the largest eigenvalue 1 [8]. Therefore we can use the number of repeated eigenvalue 1 to estimate the number of clusters in the dataset. Moreover, Ng et al [9] told us that this conclusion can also be extended to the general cases through matrix perturbation theory.

In practice, since the similarity matrix may not be block-diagonal, we can choose the number of eigenvalues which are most close to 1. Therefore we can predefine a small threshold $\delta \in [0,1]$, and determine the number of clusters by count the eigenvalues λ_i which satisfy $|\lambda_i - 1| < \delta$.

2.3   Constructing the Affinity Matrix

Now the only problem remained for us is to construct a "good" similarity matrix which is almost block-diagonal. We use the Gaussian function $S_{ij} = exp\left(-\frac{d_{ij}^2}{2\sigma^2}\right)$ to construct it like in [4], where d_{ij} is some similarity measure between $\mathbf{x}_i$ and $\mathbf{x}_j$. To distinguish the transformed matrix $\mathbf{S}$ from the previously constructed similarity matrix $\mathbf{D}$, we will call $\mathbf{S}$ affinity matrix throughout the paper. The diagonal elements of the affinity matrix are set to zero as in [9]. A gradient ascent method is used to determine the parameter σ^2.

More precisely, assume the solution of the optimizer (3) is $\tilde{\mathbf{Q}}^* = [\mathbf{q}_1, \cdots, \mathbf{q}_K]$, and $\mathbf{q}_i = (\mathbf{q}_i^1, \cdots, \mathbf{q}_i^M)^T \in \mathbb{R}^M$, where $\mathbf{q}_i^j$ represents the j-th element of the column vector $\mathbf{q}_i$. Then $H = trace\left(\tilde{\mathbf{Q}}^{*T} \mathbf{S} \tilde{\mathbf{Q}}^*\right) = \sum_{i=1}^{K} \mathbf{q}_i^T \mathbf{S} \mathbf{q}_i$, hence

$$H = \sum_{i=1}^{K}\sum_{j=1}^{M}\sum_{k=1}^{M} \mathbf{q}_i^j \mathbf{q}_i^k S_{jk} = \sum_{i=1}^{K}\sum_{j=1}^{M}\sum_{k=1}^{M} \mathbf{q}_i^j \mathbf{q}_i^k exp\left(-\frac{d_{jk}^2}{2\sigma^2}\right)$$

where S_{jk} is the (j,k)-th entry of the affinity matrix $\mathbf{S}$. If we treat H as a function of σ, then the gradient of H is

$$G = \frac{\partial H}{\partial \sigma} = \sum_{i=1}^{K}\sum_{j=1}^{M}\sum_{k=1}^{M} \mathbf{q}_i^j \mathbf{q}_j^k \frac{\partial S_{jk}}{\partial \sigma} = \sum_{i=1}^{K}\sum_{j=1}^{M}\sum_{k=1}^{M} \mathbf{q}_i^j \mathbf{q}_i^k \frac{d_{jk}^2}{\sigma^3} exp\left(-\frac{d_{jk}^2}{2\sigma^2}\right) \quad (4)$$

Inspired by the work in [10], we propose a gradient based method to tune the variance of the Gaussian function. More precisely, we can first give an initial guess of σ, then use G to adjust it iteratively until $\|G\| < \varepsilon$. The detailed algorithm is shown in Table 1.

3   Experiments

In this section, we will give two experiments where we used our spectral decomposition based clustering framework to cluster time series. First we used a

Table 1. Clustering Time Series via Spectral Decomposition

Input: Dataset **X**, Precision ε, Max iteration T, Initialize σ to σ_0, Learning rate α
Output: Clustering results.
1. Choose some similarity metric to construct the similarity matrix **D**.
2. Initialize σ to σ_0, construct the affinity matrix **S**;
3. Calculate **S**$'$ by normalizing **S**, do spectral decomposition on it and find the number of eigenvalues which are closest to 1, which corresponds to the optimal number of clusters K
4. For i=1:T
(a).Solve (3) to achieve $\tilde{\mathbf{Q}}^*$
(b).Compute the gradient according to (4) G^i
(c).If $\|G\| < \varepsilon$, break; else let $\sigma = \sigma + \alpha G^i$
5. Treat each column of the final $\tilde{\mathbf{Q}}^*$ as a new point in $\mathbb{R}^K$ and cluster them into K clusters via kmeans algorithm

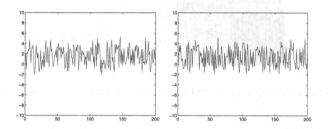

Fig. 1. Samples generated from different HMMs

synthetic dataset generated in the same way as in [2]. This is a two-class clustering problem. In our experiments, 40 time series are generated from each of the 2 HMMs. The length of these time series vary from 200 to 300. We use Both HMMs have two hidden states and use the same priors and observation parameters. The priors are uniform and the observation distribution is a univariate Gaussian with $\mu = 3$ and variance $\sigma^2 = 1$ for hidden state 1, and with mean $\mu = 0$ and variance $\sigma^2 = 1$ for hidden state 2. The transition matrices of them are $A_1 = \begin{pmatrix} 0.6 & 0.4 \\ 0.4 & 0.6 \end{pmatrix}$ and $A_2 = \begin{pmatrix} 0.4 & 0.6 \\ 0.6 & 0.4 \end{pmatrix}$. Fig.1 shows us two samples generated from these two HMMs, the left figure is a time series generated by the first HMM, and the right is generated by the second HMM.

From Fig.1 we cannot easily infer which sample is generated from which HMM. We measure the pairwise similarity of the time series by the BP metric [11] which is defined as follows.

Definition 1 (BP metric). *Suppose we train two HMMs λ_i and λ_j for time series $\mathbf{x}_i$ and $\mathbf{x}_j$ respectively. Let $L_{ij} = P(\mathbf{x}_j|\lambda_i)$ and $L_{ii} = P(\mathbf{x}_i|\lambda_i)$. Then the BP metric between $\mathbf{x}_i$ and $\mathbf{x}_j$ is defined as*

$$L_{BP}^{ij} = \frac{1}{2}\left[\frac{L_{ij} - L_{ii}}{L_{ii}} + \frac{L_{ji} - L_{jj}}{L_{jj}} \right] \qquad (5)$$

The reason why we used the BP metric here is because that it not only considers the likelihood of $\mathbf{x}_i$ under λ_j as usual[2], but also take into account the modeling goodness of $\mathbf{x}_i$ and $\mathbf{x}_j$ themselves. Thus it can be viewed as a relative normalized difference between the sequence and the training likelihoods[11].

After the similarity matrix $\mathbf{D}$ having been constructed by the BP metric, we will come to step 2 in Table 1. The initial variance σ_0 of the Gaussian function is set to 0.1. Fig. 2 shows the normalized affinity matrix and the corresponding top ten eigenvalues, from which we can see that our method is able to discover the correct cluster number 2 automatically.

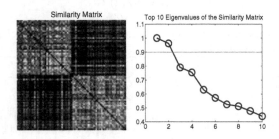

Fig. 2. Affinity matrix and the corresponding top 10 eigenvalues

We use clustering accuracy to evaluate the final clustering results as in [12]. More precisely, if we treat the cluster problem as a classification problem, then the clustering accuracy can be defined as follows.

Definition 2 (Clustering Accuracy). *Let $\{t_i\}$ denote the true classes and $\{c_j\}$denote the clusters found by a cluster algorithm. We then label all the data incluster c_j as t_i if they share the most data objects. Note that the number of clusters need not be the same as the number of classes. Then we can calculate the clustering accuracy η as :*

$$\eta = \frac{\sum_{\mathbf{x}} I(c_i(\mathbf{x}) = t_j(\mathbf{x}))}{M} \tag{6}$$

where $I(\cdot)$ is the indicator function, $c_i(\mathbf{x})$ is the label of the cluster which $\mathbf{x}$ belongs to, $t_j(\mathbf{x})$ is the true class of $\mathbf{x}$, and M is the size of the dataset.

Fig. 3 provides the results of our algorithm after 50 iterations (We don't use the termination condition in Table 1 step 4 (c)). In all these figures the horizontal axis corresponds to the iteration number. The vertical axis of these figures represents the clustering accuracy η in Eq.(6), gradient G in Eq.(4), and the object function value H in Eq.(3).

From Fig.3 we can see that as the iteration procedure goes deeply, the gradient G will become smaller while the the clustering accuracy and the object

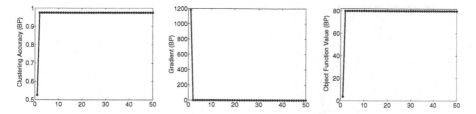

Fig. 3. Experimental results for Smyth's dataset

Table 2. Clustering accuracies on the synthetic dataset

	CHAC	SHAC	AHAC	Our method
BP	0.9500	0.5125	0.9625	0.9725

function value H are increasing. And our algorithm will converge after only two steps in this experiment.

We have compared the clustering accuracies resulted from our approach with the results achieved from hierarchical agglomerative clustering (HAC) methods, since most of the developed time series clustering approaches have adopted an HAC framework[2][3]. In an agglomerative fashion, the HAC method starts with M different clusters, each containing exactly one time sequence. Then the algorithm will merge the clusters continuously based on some similarity measure until the stopping condition is met. There are three kinds of HAC approaches according to the different similarity measure they use to merge clusters[1]. They are *Complete-linkage HAC(CHAC)*, *Single-linkage HAC(SHAC)* and *Average-linkage HAC(AHAC)*, which adopt furthest-neighbor distance, nearest-neighbor distance and average-neighbor distance to measure the similarity between two clusters respectively. In our experiments, the final cluster number of all these HAC methods is set to 2 manually. The final clustering accuracies are shown in Table 2. From which we can see that our algorithm can perform better than HAC methods in this case study.

In the second experiment we use a real EEG dataset which is extracted from the 2nd Wadsworth BCI dataset in BCI2003 competition [13]. According to [13], the data objects can be generated from 3 classes: the EEG signals evoked by flashes containing targets,the EEG signals evoked by flashes adjacent to targets, and other EEG signals. All the data objects have an equal length 144. All the data objects have an equal length 144. Fig. 4 shows an example for each class.

We randomly choose 50 EEG signals from each class. As all the time series have the same length, therefore we can use the Euclidean distance to measure the pairwise distances of the time series. The Euclidean distance between two time series can be defined as follows[14].

Definition 3 (Euclidean distance). *Assume time series $\mathbf{x}_i$ and $\mathbf{x}_j$ have the same length l, then the Euclidean distance between them is simply*

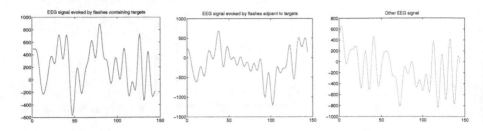

Fig. 4. EEG signals from the 2nd Wadsworth BCI Dataset

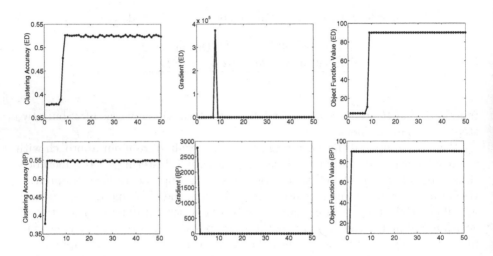

Fig. 5. Experimental results for EEG dataset

$$D_{ij}^{Euc} = \sqrt{\sum_{k=1}^{l}(\mathbf{x}_i^k - \mathbf{x}_j^k)^2} \qquad (7)$$

where $\mathbf{x}_i^k$ refers to the $k-th$ element of $\mathbf{x}_i$.

In our experiments, we adopt both the Euclidean distance (7) and the BP metric (5)to construct the similarity matrix $\mathbf{D}$. And apply the Gaussian function to transform them to the affinity matrices. The initial variance of the Gaussian function is set to 600. The final experimental results of our method after 50 iterations are presented in Fig. 5, where the first row shows the trends of clustering accuracy, gradient, and object function value achieved based on the Euclidean distance (ED) versus iteration, and the second row shows the trends of these indexes achieved based on the BP metric (BP) versus iteration.

Fig.5 shows us that the trend of these indexes are very similar with that in Fig.3. The clustering accuracy and objective function value are increasing and more and more stable with the decreasing of gradient.

We also compared the clustering accuracies achieved from *HAC* methods and our approach, the final cluster number of the *HAC* methods is also set to 3 manually. Table 3 gives us the final results. Each column in Table 3 shows the results of a method. The second and third rows are the final clustering accuracies when we use the Euclidean distance and the BP metric respectively to measure the similarity of pairwise time series in these methods.

Table 3. Clustering results on *EEG* dataset

	CHAC	SHAC	AHAC	Our method
Euclidean	0.4778	0.3556	0.3556	0.5222
BP	0.4556	0.3556	0.4222	0.5444

From the above experiments we can see that for *HAC* methods, if we adopt different distance measures (nearest, furthest, average neighbor distance), the final clustering results may become dramatically different. Moreover, these methods may always fail to find the correct cluster number, which makes us to set this number manually. On the contrary, our spectral decomposition based clustering method can discover the right cluster number automatically, and in most cases the final performance achieved by our approach will be better than *HAC* methods.

4   Conclusion and Discussion

In this paper we present a new spectral decomposition based time series clustering framework. The theoretical analysis guarantees the feasibility of our approach, and the effectiveness of which has been shown by experiments.

The problem remained for us is that the spectral decomposition is time consuming. Fortunately the affinity matrix is Hermitian, and usually sparse. Thus we can use the subspace method like the Lanczos method, Arnoldi method to solve the large eigenproblems [6]. We believe that our approach will be promising and it may have potential usage in many data mining problems.

Acknowledgements

This work is supported by the project (60475001) of the National Natural Science Foundation of China.

References

1. R. O. Duda, P. E. Hart, D. G. Stork. Pattern Classification. 2nd edition. New York, John Wiley & Sons, Inc. 2001.
2. P. Smyth. Clustering Sequences with Hidden Markov Models. In Advances in Neural Information Processing 9 (NIPS'97), MIT Press, 1997. 1997.

3. S. Zhong, J. Ghosh. A Unified Framework for Model-based Clustering. Journal of Machine Learning Research 4 (2003) 1001-1037. 2003.
4. F. M. Porikli. Clustering Variable Length Sequences by Eigenvector Decomposition Using Hmm. International Workshop on Structural and Syntactic Pattern Recognition, (SSPR'04). 2004.
5. H. Zha, X. He, C. Ding, H. Simon, M. Gu. Spectral Relaxation for K-means Clustering. Advances in Neural Information Processing Systems 14 (NIPS'01). pp. 1057-1064, Vancouver, Canada. 2001.
6. G. H. Golub, C. F. Van Loan. Matrix Computation. 2nd ed. Johns Hopkins University Press, Baltimore. 1989.
7. G. Das, D. Gunopulos, and H. Mannila. Finding Similar Time Series, In proceedings of the First European Symposium on Principles of Data Mining and Knowledge Discovery, LNCS 1263, pp. 88-100. 1999.
8. M. Maila and J. Shi. A Random Walks View of Spectral Segmentation. International Workshop on AI and STATISTICS (AISTATS) 2001. 2001.
9. A. Y. Ng, M. I. Jordan, Y. Weiss. On Spectral Clustering: Analysis and an Algorithm. In Advances in Neural Information Processing Systems 14 (NIPS'01), pages 849–856, Vancouver, Canada, MIT Press. 2001.
10. J. Huang, P. C. Yuen, W. S. Chen and J. H. Lai. Kernel Subspace LDA with Optimized Kernel Parameters on Face Recognition. Proceedings of the Sixth IEEE International Conference on Automatic Face and Gesture Recognition (FGR04). 2004.
11. A. Panuccio, M. Bicego, V. Murino. A Hidden Markov Model-based approach to sequential data clustering. Structural, Syntactic and Statistical Pattern Recognition (SSPR'02), LNCS 2396. 2002.
12. F. Wang, C. Zhang. Boosting GMM and Its Two Applications. To be appeared in the 6th International Workshop on Multiple Classifier Systems (MCS'05). 2005.
13. Z. Lin, C. Zhang, Enhancing Classification by Perceptual Characteristic for the P300 Speller Paradigm. In Proceedings of the 2nd International IEEE EMBS Special Topic Conference on Neural Engineering (NER'05). 2005.
14. R. Agrawal, C. Faloutsos, A. Swami. Efficient Similarity Search In Sequence Databases. Proceedings of the 4th International Conference of Foundations of Data Organization and Algorithms (FODO'93). 1993.

A New EM Algorithm for Resource Allocation Network

Kyoung-Mi Lee

Department of Computer Science, College of Information Engineering,
Duksung Women's University, Seoul, Korea
kmlee@duksung.ac.kr
http://namhae.duksung.ac.kr/~kmlee

Abstract. Clustering usually assumes that the number of clusters is known or given. No knowledge of such a priori information is needed to find an appropriate number of clusters. This paper introduces an elliptical clustering algorithm with incremental growth of clusters, which is derived from the batch EM algorithm with a decay factor and a novelty criterion. The proposed algorithm can start with no or a small number of clusters. Whenever unusual data is presented, the algorithm adds a new cluster and finally the number of clusters in the data is obtained after clustering. The usefulness of the proposed algorithm is demonstrated for texture image segmentation and skin image segmentation.

1 Introduction

Given a set of N data, x^n, the objective of clustering is to assign one of a set of J clusters, c_j, such that data within the same cluster has a high degree of similarity, while data belonging to different clusters exhibits a high degree of dissimilarity. Each cluster c_j contains N_j data and can be represented by two parameters: a center, μ_j, and a covariance matrix, Σ_j.

When applied as a clustering algorithm, however, this leads to several practical problems. First of all, these algorithms usually require an assumption that the number of clusters, J, is known. In general, this approach is useful only when the number of clusters is chosen in a correct manner. When no priori information regarding J is available it is desirable that the algorithms be capable of automatically finding an appropriate number of clusters automatically. Another problem is that the choice of distance measures can produce substantial differences in clustering results. Many clustering algorithms using a Euclidean distance are suitable for detecting spherical-shaped clusters; most clusters in real data sets, however, are non-spherical in shape. Therefore, clustering algorithms with Euclidean distance are inappropriate for large and elongated clusters. The third problem is regularization; covariance matrices can become singular or near-singular causing, numerical problems in their inversion. In this paper, we propose a new elliptical clustering algorithm with incremental growth of clusters, by incorporating both a decay factor and a novelty criterion into the EM algorithm. This technique is effective for proper probabilistic and elliptical clustering.

S. Singh et al. (Eds.): ICAPR 2005, LNCS 3686, pp. 355–362, 2005.
© Springer-Verlag Berlin Heidelberg 2005

2 Elliptical Clustering

The batch EM algorithm starts with an initial guess at the maximum likelihood parameters of clusters, $\theta \equiv \{\mu_j, \Sigma_j\}$, and then proceeds to generate successive estimates by iteratively applying the E(Expectation)-step and M(Maximization)-step [3]. Assume that the joint probability for an input data, x^n, and a cluster, c_j, using θ is $P(x^n, j \mid \theta)$. Let $\bar{\theta}$ be the present estimator. The E-step calculates the posterior probability of c_j according to the Bayes rule,

$$P(j \mid x^n, \bar{\theta}) = \frac{P(x^n, j \mid \bar{\theta})}{\sum_{i=1}^{J} P(x^n, i \mid \bar{\theta})}. \tag{1}$$

The M-step updates the parameters as follows:

$$\mu_j = \frac{\sum_{n=1, x^n \in c_j}^{N_j} x^n P(j \mid x^n, \bar{\theta})}{\sum_{n=1, x^n \in c_j}^{N_j} P(j \mid x^n, \bar{\theta})} \quad \text{and} \quad \Sigma_j = \frac{\sum_{n=1, x^n \in c_j}^{N_j} m_j m_j' P(j \mid x^n, \bar{\theta})}{\sum_{n=1, x^n \in c_j}^{N_j} P(j \mid x^n, \bar{\theta})} \tag{2 and 3}$$

where $m_j^n \equiv x^n - \mu_j$ and the prime (') denotes a transpose.

Another method uses on-line or sequential updating of the parameters each time new data x^n is presented [14]. Let $\theta^n \equiv \{\mu_j^n, \Sigma_j^n\}$ be the estimator after x^n. Using the step-wise equation in the on-line EM algorithm, Eq. (2) and (3) can be modified

$$\mu_j^n = \frac{f_j^{n-1}(x) + \eta^n [x^n P_j^n - f_j^{n-1}(x)]}{f_j^{n-1}(1) + \eta^n [P_j^n - f_j^{n-1}(1)]} \quad \text{and} \tag{4}$$

$$\Sigma_j^n = \frac{f_j^{n-1}(m_j m_j') + \eta^n [m_j^n m_j^{n'} P_j^n - f_j^{n-1}(m_j m_j')]}{f_j^{n-1}(1) + \eta^n [P_j^n - f_j^{n-1}(1)]} \tag{5}$$

where $m_j^n \equiv x^n - \mu_j^n$ and $P_j^n \equiv P(j \mid x^n, \theta^{n-1})$. Since a decay factor λ^n is $0 \le \lambda^n \le 1$, a learning parameter $\eta^n = \left(1 + \lambda^n / \eta^{n-1}\right)^{-1}$ is $\frac{1}{n} \le \eta^n \le 1$.

Suppose that the same input data is repeatedly presented, i.e., $x^{n+(t-1)N} = x^n$, and θ is updated at the end of the whole data set iteration, i.e., $n = N$. If λ^n is defined as 0 at $n = (t-1)N+1$ and 1 in all other cases, then η^n is defined by $\eta^n = \eta^{(t-1)N+n} = \frac{1}{n}$. At the beginning of each iteration, $f_j^{n-1}(x)$ is set to $f_j^0(x) + \eta^1 [x^1 P_j^1 - f_j^0(x)] = x^1 P_j^1$. At the end of the whole data set iteration, $f_j^{(t-1)N}(x) = f_j^N(x)$ is satisfied for $\bar{\theta} = \theta^{t-1}$.

Therefore, these will become $f_j^{n-1}(1)$ and $f_j^{n-1}\left(m_j m_j'\right)$. This shows that the on-line EM algorithm with an appropriate choice of λ^n is equivalent to the batch EM algorithm [6].

3 Elliptical Clustering with Incremental Growth

3.1 Proposed Algorithm

To select an appropriate number of clusters in EM, we incorporated resource allocations found to be suitable for on-line modeling of non-stationary processes [4]. The proposed algorithm may start either with a small number of clusters ($J_0 \leq J$) or with no clusters ($J_0 = 0$). If the number of clusters is given, the algorithm initializes J_0 clusters in the same way as the EM or the on-line EM algorithm does. A center, μ_j^0, is set randomly and an initial covariance matrix, Σ_j^0, is set as the unit matrix. If no knowledge of the data is assumed, then the algorithm includes no clusters.

When a new data x^n is incoming, the proposed algorithm first compares it to the centers of existing clusters and finds the nearest cluster using Eq. (1). According to the novelty criterion, the algorithm computes the posterior probability on the nearest cluster, which should be greater than the threshold, ε^n,

$$P\left(nearest \mid x^n, \bar{\theta}\right) > \varepsilon^n. \tag{6}$$

The threshold starts with the largest value, i.e., $\varepsilon^0 = \varepsilon_{max}$, and is multiplied by a decay factor γ for ε^n, $0 < \gamma < 1$, until it reaches the smallest value, ε_{min}. As more data is presented, ε^n is reduced which allows the refinement of the overall clustering. This also provides stability to the overall clustering.

Eq. (6) confirms that the data x^n is not similar to any existing clusters and is currently not appropriately represented by existing clusters. Thus, if Eq. (6) is satisfied, a new cluster is created with x^n and the number of clusters J is increased. The center of the newly allocated cluster is set to the input data and the covariance matrix proportionally is set to the posterior probability on the nearest existing cluster to the new data. A new cluster is set as the following [4]:

$$\mu_{J+1}^n = x^n, \quad \text{and}$$

$$\Sigma_{J+1}^n = \kappa\left(x^n - \mu_{nearest}^n\right)\left(x^n - \mu_{nearest}^n\right)' \tag{7}$$

where $\mu_{nearest}^n$ is the center of the nearest existing cluster from x^n and κ is a constant for a overlap factor which determines the smoothness of the function. This makes the new data more likely to match the newly-created cluster. Thus, after clustering on a set of data, each cluster represents a cluster of data that are near one another in the data space. Eventually, the proposed algorithm finds the desired number of clusters more closely as the data is clustered. This is referred to as an Allocation (A-step).

If the data x^n is similar to one of existing clusters, the M-step updates the center and the covariance matrix using Eq. (4) and (5), respectively. However, updating works only if $\left(\Sigma_j^n\right)^{-1}$ exists. In real world applications, unfortunately, the inverse matrix can become singular. To guarantee non-singularity of $\left(\Sigma_j^n\right)^{-1}$ in the d-dimensional space, the regularized covariance matrix can be defined by the step-wise equation

$$\Sigma_{jR}^n = \Sigma_j^n + \beta \frac{\Sigma_j^{n'}}{d} I_d$$

where β is a constant $(0 < \beta < 1)$ and I_d is a d-dimensional identity matrix. Since computing the inverse formula for $\left(\Sigma_{jR}^n\right)^{-1}$ with each iteration is impractical, the inverse matrix needs to be updated directly. Direct updating can be accomplished by the step-wise equation. The detailed derivation on direct updating of $\left(\Sigma_{jR}^n\right)^{-1}$ is in [7]. The proposed EAM algorithm is summarized in Algorithm I. Note that after adding a new cluster, it is suitable to incorporate some time steps just to update existing parameters and to disallow another allocation.

For some clusters $j = 1, \cdots, J_0$,

Initialize μ_j^0 randomly.

Initialize Σ_j^0 to be a unit matrix.

$\varepsilon^0 = \varepsilon_{\max}$, $J = J_0$, and $n = 0$.

For each data x^n

    E-step: Find the nearest cluster c_j using Eq. (1).

    If Eq. (6) is satisfied, (A-step)

        Allocate a new cluster using Eq. (7).

        $N_{J+1} = N_{J+1} + 1$ and $J = J + 1$.

    else (M-step)

        Update μ_j^n and Σ_j^n using Eq. (4) and $\left(\Sigma_{jR}^n\right)^{-1}$, respectively.

        $N_j = N_j + 1$.

    If $\varepsilon^n > \varepsilon_{\min}$, $\varepsilon^{n+1} = \gamma \varepsilon^n$.

    $n = n + 1$.

Algorithm I. The proposed EAM algorithm

3.2 Discussion

Note that Algorithm I is equivalent to an online EM algorithm with the appropriate choice of J and $\varepsilon_{\max}$. Assume $\varepsilon_{\max}$ is defined as the largest posterior probability on any cluster as long as $\gamma \approx 1$. If the number of initial clusters provided is optimal ($J_0 = J$), this shows that, by simply neglecting the A-step, the proposed EAM algorithm is equivalent to the on-line EM algorithm. If J_0 is a smaller number $(0 \le J_0 < J)$, some μ_j can be initialized to the first similar data to the cluster, c_j, while the on-line EM algorithm initializes all clusters randomly at the beginning of the algorithm. Consequently, the estimator of the EAM algorithm converges to the maximum likelihood estimator.

4 Results

Clustering can be used for image segmentation. In this section, the proposed algorithm is applied to texture image segmentation and skin color region segmentation, and proven more appropriate when compared to the batch EM algorithm.

4.1 Texture Segmentation

The proposed algorithm was tested on Brodatz textures [1] with the pixel spatial resolution set at $N = 256 \times 256$. Fig.1 shows two composite texture images, containing four (the top row in Fig.1) and five (the bottom row in Fig. 1). For texture representation, we used Gabor filters [2] which have optimal localization properties in both the spatial and frequency domains. Each pixel in the image is represented using $d = 24$ Gabor filters (four radial frequencies and six different orientations per frequency). The values used for this experiment were: $\varepsilon_{max} = 0.7$, $\varepsilon_{min} = 0.07$, $\gamma = 0.999$, $\kappa = 0.9$, and $\beta = 0.3$.

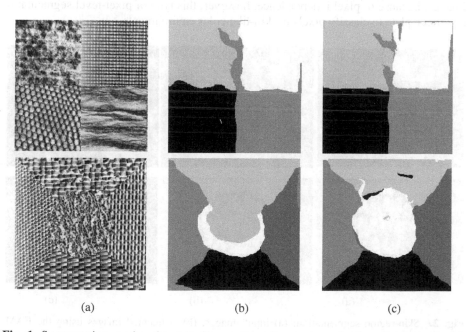

(a)   (b)   (c)

Fig. 1. Segmentation results of texture images (top: four textures, bottom: five texture): (a) original composite textured images, (b) using the batch EM, and (c) using the EAM algorithm

In Fig. 1, segmentation results are shown using the EM (Fig. 1(b)) and the EAM (Fig. 1(c)) algorithm. The EM algorithm assumes a desired number, J, is given, i.e., Fig 1. (b) was obtained with $J = 4$ and $J = 5$, respectively. On the other hand, the EAM algorithm starts without any clusters ($J_0 = 0$), finds an optimal number of clusters, and distinguishes all textures (Fig. 1 (c)). Compared to the EM algorithm in segmentation, the EAM algorithm achieved less segmentation errors with some noisy patches. Since both algorithms depend on the order in which the data is presented, results may differ from one data set to another. Therefore two test images with 20 differently-ordered pixel sets were applied to the algorithms and the images were correctly segmented; on average, at 91.13% using the EAM algorithm and 83.46% using the EM algorithm. In addition, we tested 11 more composite texture images used in Randen's experiments [5]. After all images were repeatedly applied to both algorithms with 20 differently-ordered pixel sets, we achieved averaged results: 71.26% using the EAM algorithm and 67.92% using the EM algorithm.

4.2 Skin-Region Segmentation

In this section, we applied the proposed algorithm to skin-region segmentation. First, an input image was smoothed in the *RGB* color space by a median filter. Each pixel in the smoothed image was then transferred to 6 features (*H, S, V, Y, Cb*, and *Cr*) in the HSV and *YCbCr* color spaces. At a pixel-level, the EAM algorithm was applied to the 6-featured space. Each pixel in clustered regions could be tested using a skin filter [8] to determine whether it was a skin pixel or not. Generally, segmentation at the pixel-level is simple to achieve using a probability model, regardless of complexity in images. Because of pixel independence, however, this type of pixel-level segmentation is incomplete to classify pixels in skin-like color environments.

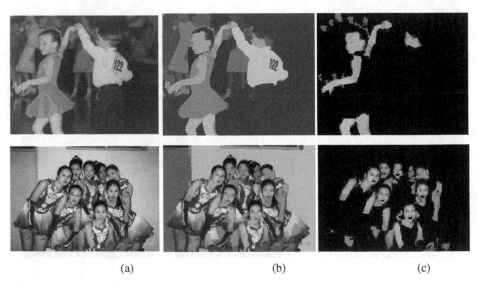

(a)   (b)   (c)

Fig. 2. Skin-region segmentation: (a) input images, (b) segmented images using the EAM algorithm, and (c) skin regions segmented by a skin-likelihood model

Table 1. Segmentation results using Eq.(10) on test images

	Using the batch EM	Using the EAM
Region segmentation	84.65%	90.45%
Skin-region segmentation	86.15%	92.35%

To overcome this problem, we can segment an image at a region-level. A conditional probability of a segmented region R is calculated by averaging probabilities of pixels in the image:

$$P(R \mid skin) = \frac{\sum_{x \in R} P(x \mid skin)}{Area(R)} \tag{8}$$

where $Area(R)$ is the area of the region R, as a number of pixels in the region R, and P (skin|x) the probability of a pixel x in the region R.

Fig. 2 shows the results of the EAM algorithm and skin-region segmentation. We tested 200 images with 20 differently-ordered pixel sets (Table I). The rate of region segmentation is a percentage of correctly clustered pixels and that of skin-region segmentation a percentage of pixels in correctly detected skin-regions. Table I shows that the EAM algorithm with Eq. (8) has better results than the EM algorithm.

5 Conclusion

Clustering is an established method of exploratory data analysis. When the clustering trend has been established for given data, choosing the correct number of clusters becomes important. Regarding the identification of the number of clusters, we proposed an elliptical clustering algorithm with incremental growth of clusters, called the EAM algorithm, and applied this algorithm to texture image segmentation tasks. This paper demonstrated that the proposed algorithm provides better segmentation than the batch EM algorithm.

The proposed algorithm may be improved by adding a pruning strategy for existing clusters whose center is an outlier in the input space and/or whose probability is becoming negligible. This can solve over-clustering problems ($J_0 > J$).

Acknowledgements

This work was supported by Korea Research Foundation Grant No. R04-2003-000-10092-0 (2004).

References

[1] P. Brodatz, *Textures: A photographic album for artists and designers*, Dover publications, NY, 1966.

[2] J.G. Daugman, An information- theoretic view of analogue representation in striate cortex, *Computational neuroscience*, In E.L. Schwartz (Eds.), Cambridge, MIT Press, MA, pp. 403- 424, 1990.

[3]   A.P. Dempster, N.M. Laird and D.B. Rubin, Maximum likelihood from incomplete data via the EM algorithm (with discussion), *Journal of the royal statistical society* B, 39(1):1-38, 1977.

[4]   K.-M. Lee and W.N. Street, An adaptive resource-allocating network for automated detection, segmentation, and classification of breast cancer nuclei, *IEEE Transactions on neural networks*, 14(3):680-687, 2003.

[5]   T. Randen and J.H. Husøy, Filtering for texture classification: a comparative study, *IEEE transactions on pattern analysis and machine intelligence*, 21(4): 291-310, 1999.

[6]   M. Sato and S. Ishii, On-line EM algorithm for the normalized Gaussian network, *Neural computation*, 12(2):407-432, 2000.

[7]   K.-M. Lee, Elliptical clustering with incremental growth and its application to skin color region segmentation, *Journal of KISS*, 31(9):1161-1170, 2004.

[8]   C. Garcia and G. Tziritas, "Face Detection Using Quantized Skin Color Regions Merging and Wavelet Packet Analysis," *IEEE transactions on Multimedia*, 1(3):264-276, 1999.

A Biased Support Vector Machine Approach to Web Filtering

A-Ning Du, Bin-Xing Fang, and Bin Li

Research Center of Computer Network and Information Security Technology,
Harbin Institute of Technology, Harbin 150001, People's Republic of China
{ani, bxfang, libin}@pact518.hit.edu.cn

Abstract. Web filtering is an inductive process which automatically builds a filter by learning the description of user interest from a set of pre-assigned web pages, and uses the filter to assign unprocessed web pages. In web filtering, content similarity analysis is the core problem, the automatic-learning and relativity-analysis abilities of machine learning algorithms help solve the above problems and make *ML* useful in web filtering. While in practical applications, different filtering task implies different userinterest and thus implies different filtering result. This work studies how to adjust the web filtering results to be more fit for the user interest. The web filtering result are divided into three categories: relative pages, similar pages and homologous pages according to different user interest. A Biased Support Vector Machine (BSVM) algorithm, which imports a stimulant function, uses training examples distribution n_+/n_- and a user-adaptable parameter k to deal imbalancedly different classes of the pre-assigned pages, is introduced to adjust the filtering result to be best fit for the user interest. Experiments show that BSVM can greatly improve the web filtering performance.

1 Introduction

The increasing of Internet resources brings up the problem of *information overload, quality enhancement*, which means that people want to read the most interesting messages, and avoid having to read low-quality or uninteresting messages. Web filtering is the activity of classifying a stream of incoming web pages dispatched in an asynchronous way by an information producer to an information consumer[1], which helps people find the most interesting and valuable information and saves Internet users from drowned in information flood.

Recent years, the machine learning (*ML*) paradigm, instead of knowledge engineering and domain experts, becomes more popular in solving the above problem because of its automatically-learning and relativity-analysis abilities. The typical procedure of applying machine learning algorithm to web filtering can be described as follows: a general inductive process automatically builds a web pages filter by learning from a set of pre-assigned pages, namely the characteristics of different categories of user interest, and use this filter to decided

S. Singh et al. (Eds.): ICAPR 2005, LNCS 3686, pp. 363–370, 2005.

whether the web pages is in accord with the user interest. The accurate description of the user interest is the critical precondition of web filtering and the precision and recall ability of the page filter is the main problem of web filtering.

Practically, different web filtering task stands for different user interest. For example, for the task of searching engine, all the web pages with the key words are of the user interest. For the task of harmful information filtering, only the web pages with the user-specified orientation should be selected for the user, though quite a lot more of web pages may be relative to the user interest. Thus, for different filtering tasks which implies different user interest, the filtering result sets are of different size and all these results sets are subsets of the relative web pages set.

This paper studies how to adjust the web filtering results to be more fit for the user interest. The web filtering results are divided into three categories: relative pages, similar pages and homologous pages, each of which is correspondent with a kind of user interest. To achieve more precisely the filtering result, the inductive process is improved so that it can get better precision and recall ability according to the user interest. The improved machine learning algorithm in this paper is based on the Support Vector Machine (SVM) algorithm because that of all the generic machine learning algorithms (*Decision Tree, Rule Induction, Bayesian algorithm* and *SVM*), SVM algorithm has shown to be superior to other machine learning algorithms with the solid foundation of Statistical Learning Theory (*SLT*). The improved algorithm is called Biased Support Vector Machine (BSVM), which imports a stimulant function, uses training examples distribution n_+/n_- and a user-adaptable parameter k to deal imbalancedly different classes of the pre-assigned pages so as to adjust the filtering result to be best fit for the user interest. The remainder of the paper is organized as follows: In Section 2 we briefly examine the work of web filtering and previous machine learning approaches. Section 3 states the problem of user interest, put forward the model of Biased Support Vector Machine and analyzes its efficiency in web filtering. Section 4 closes the paper with our conclusions and future work.

2   Web Filtering and Inductive Constructions

2.1   Web Filtering

Web filtering is the task of assigning a boolean value to each web page vector $d_i \in \mathbb{D}$, where $\mathbb{D}$ is a domain of web pages. A value of $TRUE$ assigned to d_i indicates a decision to page d_i relative to the user interest, while $FALSE$ indicates not. More formally, the task is to approximate the unknown target function $\Psi : \mathbb{D} \rightarrow \{TRUE, FALSE\}$ (which describes how web pages ought to be assigned) by means of a function $\Phi : \mathbb{D} \rightarrow \{TRUE, FALSE\}$ called the filter. How to improve the precision and recall of the filter Φ are the core problems of web filtering, which are also what this paper concentrates on.

The general process of web filtering includes five steps:

1. *user interest acquiring*: acquire many user-assigned web pages as training set

2. *web pages pre-processing*: translate the assigned pages into a set of compact representations of page content. Usually a page d_i is represented as a vector of term weights $d_i = \{w_{1i}, w_{2i}, \cdots, w_{|F|i}\}$, where F is the set of features that occur at least once in at least one document of $\mathbb{D}$, and $0 < w_{ki} < 1$ represents how much feature f_k contributes to the semantics of page d_i

3. *dimensionality reduction*: select feature of high contribution to reduce the size of feature set F

4. *construction of web filters*: build a filter to describe user interest automatically

5. *predict unfiltered web pages*: use the filter to predict an unmarked web page is relative or not

Representation of web pages is the basic step of the process, while the degree of dimensionality reduction is the key infecting factor. What decides the effectiveness of web filters is that the generalization and description ability of web filtering algorithm. The dominant approach is to use *ML* algorithm as filtering algorithms with its high effectiveness.

2.2   Inductive Constructions of Web Filter

The inductive construction of web filter usually consists in the definition of a function $\Phi : \mathbb{D} \to \{TRUE, FALSE\}$ which gives each web page d_i a decision value. General speaking, four general algorithms are often chosen to construct web filter because of their simplicity, flexibility and robustness. A brief analysis of each algorithm is as follows:

- *Decision Trees (C4.5)*: A decision tree[2,3,4] is a graph of nodes connected by arcs with each internal node corresponding to a feature and each arc to a possible value of that feature. Decision tree is easily interpretable by humans and has low computational complexity, which is a quite simple and practical idea in the field of *ML*.
- *Rules Induction (CN2)*: Rule induction methods[5,6] try to find a proper set of *DNF* rules for filtering task such that the error rate on training set is minimal. By use of local optimization techniques, rule induction methods dynamically evaluate rules and revise the covering rule set.
- *Naïve Bayes Algorithms (NB)*: Naïve Bayes algorithm[7] views $\Phi(d_i)$ in terms of $P(c_j|d_i)$ (the probability that the web page d_i belongs to the class c_j) and compute this probability using Bayes' theorem. Naïve Bayes, as a representative probabilistic algorithm, acts well in many applications.

$$\Pr(c_j|d_i) = \frac{p(c_j)\prod_{k=1}^{|F|}p(w_{ki}|c_j)}{p(d_i)} \tag{1}$$

- *Support Vector Machines (SVM)*: Support Vector Machines[8] is a process of finding a surface which separates the positives from the negatives with the widest possible margin among all the surfaces in $|F|$-dimensional space, which is strongly supported by the *Statistical Learning Theory*. SVM acts well in dealing with large scale training set and it has no need of human and machine efforts in parameter tuning.

As is compared in [9], SVM acts well in filtering task with strong robustness and acceptable efficiency and the latter is CN2. While the precondition of NB that omitting the feature dependence reduces its web content analysis ability and the over-fitting problem occurring in the procedure of user interest description makes C4.5 not satisfied.

3   Biased Support Vector Machine for Web Filtering

In this section, we will firstly give a detail analysis of the user interest and filtering result. Then we introduce the model of BSVM and its implementation. At last, we will prove the effect of BSVM in web filtering with experiments and analysis.

3.1   Analysis of User Interest

In practical web filtering applications, the web pages set related to user interest is considerable large. But the users may be interested in only several homologous pages or all the related ones based on the difference of page subjects, writer's viewpoints and expression orientations. So we can divide web filtering tasks into three levels according to the user interest:

- *relativity-filter*: the filtering result contains all the web pages with the same key phrases or key sentences. These web pages express the same subject, but may be not consistent in viewpoint or orientation. Typical applications of relativity-filtering include erotic web pages filtering and hot topic tracing which expect to collect all the web pages related to the topic, regardless of approval or not.
- *similarity-filter*: the filtering result contains all the web pages that hold the same subject, viewpoint and orientation with the user. Typical applications of similarity-filtering include filtering of web pages on racialism or splittism. The similarity-filtering is more strict than relativity-filtering as not only key words or sentences but also orientation is taken into consideration.
- *homology-filter*: the filtering result contains only the web pages with quite a lot of same sentences or paragraphs. The filtering results are almost the same as the user interest, and always this is because that the articles from the official or authoritative website are redistributed by other websites with little modification. An examples of homology-filtering is counting which article is the most reprinted one on the *Bulletin Board Systems*.

We can define the all the filtering results acquired by *ML* algorithms as relative results(R_1) and the filtering results which the *ML* algorithms assign *TRUE* with probability near-to-1 as homologous results(R_k). So the results of similarity-filtering $R_i \in \{R_k \subseteq R_i \subseteq R_1\}$. As is illustrated in the left of Fig.1, most filtering tasks can be described as application of similarity-filtering with different similarity degrees between the web pages acquired and the user interest. To fit the user interest better, we must import adjusting ability into the *ML* algorithms. So the approach proposed in this paper imports a stimulant function, uses training examples distribution n_+/n_- and a user-adaptable parameter k to deals imbalancedly different classes of the pre-assigned pages, so as to be best fit for the user interest. The approach is called Biased Support Vector Machine, and a detailed description and analysis are in the next sections.

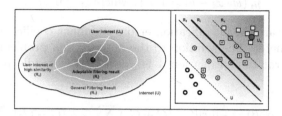

Fig. 1. Analysis and demonstration of filtering result estimation. In the left figure, outside the biggest circles means filtering scope $\mathbb{U}$, the smallest circle means user interest $\mathbb{U}_k$, the biggest circle $\boldsymbol{R}_1$ is the filtering result of general *ML* algorithms as content relativity, the smaller one $\boldsymbol{R}_k$ is the filtering result as content homology. The middle circle $\boldsymbol{R}_i$ means the biased filtering result according to user demand as content similarity. The right is a corresponding demonstration of Biased Support Vector Machine based on the left figure.

3.2   Biased Support Vector Machine Algorithm

In the classical SVM, a penalty function $F = C \cdot \sum \xi_i$ is introduced as additional capacity control function, where the non-negative variable ξ_i is a measure of the misclassification errors and the coefficient C emphasizes the tolerant degree of misclassification error. Consequently the width of the margin decreases with C increasing.

BSVM introduces a stimulant function, $F = C \cdot [(k - 1) \cdot n_- \sum_{y_i=1} \xi_i - n_+ \sum_{y_i=-1} \xi_i]/n$, as the extension of penalty function. In the right figure of Fig.1, the rectangles mean the examples of $y_i = +1$, the circles mean the examples of $y_i = -1$, and those with black dot in them stand for support vectors. Thus we define $n_+ = |\{y_i = +1\}|$ and $n_- = |\{y_i = -1\}|$. The stimulant function uses both training examples distribution n_+/n_- and an user-adaptable parameter k to express the user bias degree of different classes. Together with the effect of penalty function, the bias is described in Equation 2. The width of the margin to the positive side decreases with n_+/n_- or k increasing. Thus BSVM can find a proper separating hyperplane with filtering result R_i between R_1 and R_k.

$$bias = \frac{C + C \cdot (k-1) \cdot n_-/n}{C - C \cdot n_+/n} = \frac{1 + (k-1) \cdot n_-/n}{1 - n_+/n} = \frac{n_+/n + k \cdot n_-/n}{n_-/n} = k + n_+/n_- \quad (2)$$

BSVM is shown as follows. The generalized optimal separating hyperplane is determined by the vector w, that minimizes the functional,

$$\min_{w,b,\xi} \frac{1}{2}\|w\|^2 + C\sum \xi_i + C_1 \sum_{y_i=1} \xi_i - C_2 \sum_{y_i=-1} \xi_i$$

where $C_1 = C \cdot (k-1) \cdot n_-/n, C_2 = C \cdot n_+/n, k \geq 0$ (3)

subject to the constraints of:

$$y_i(w \cdot x_i - b) \geq 1 - \xi_i \text{ where } \xi_i \geq 0, \forall i \quad (4)$$

Here C_1 and C_2 are the classification errors stimulant coefficients, $k \geq 0$ is an adaptable parameter. The solution to the optimization problem of Equation 3 under the constraints of Equation 4 is given by the saddle point of the Lagrangian:

$$L(w,b,\xi,\alpha,\beta) = \frac{1}{2}\|w\|^2 + (C+C_1)\sum_{y_i=1}\xi_i + (C-C_2)\sum_{y_i=-1}\xi_i$$
$$- \sum \alpha_i(y_i[w^T x_i - b] - 1 + \xi_i) - \sum \beta_i \xi_i \quad (5)$$

where α, β are the Lagrange multipliers. The Lagrangian has to be minimized with respect to w, b, ξ and maximized with respect to α, β. The minimum with respect to w, b, ξ of the Lagrangian L is given by

$$\frac{\partial L(\mathbf{w}, b, \xi, \alpha, \beta)}{\partial \mathbf{w}} = \mathbf{w} - \sum_i \alpha_i y_i \mathbf{x}_i = 0 \quad (6)$$

$$\frac{\partial L(\mathbf{w}, b, \xi, \alpha, \beta)}{\partial b} = -\sum_i \alpha_i y_i = 0 \quad (7)$$

$$\frac{\partial L(w, b, \xi, \alpha, \beta)}{\partial \xi_i} = \begin{cases} C + C_1 - \alpha_i - \beta_i & \text{if } y_i = 1 \\ C - C_2 - \alpha_i - \beta_i & \text{if } y_i = -1 \end{cases} = 0 \quad (8)$$

And hence the solution to the problem is given by:

$$\min Q(\alpha) = \frac{1}{2}\sum_{i,j=1}^{n} \alpha_i \alpha_j y_i y_j K(\mathbf{x}_i, \mathbf{x}_j) - \sum_{i=1}^{n} \alpha_i \quad (9)$$

with constraints of:

$$\sum_{i=1}^{n} y_i \alpha_i = 0 \quad (10)$$

and

$$\begin{aligned} 0 \leq \alpha_i \leq C + C_1 \text{ if } y_i = 1 \\ 0 \leq \alpha_i \leq C - C_2 \text{ if } y_i = -1 \end{aligned} \quad (11)$$

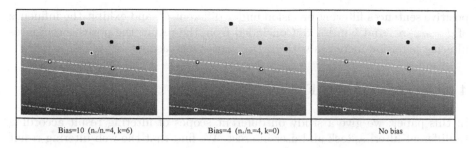

Fig. 2. The simulative training result with different biases. Black spots mean examples of $y_i = +1$, white spots mean those of $y_i = -1$, and spots with circle in them mean support vectors.

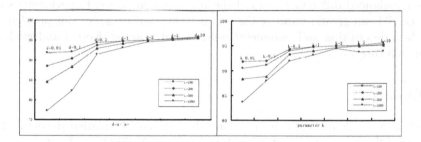

Fig. 3. BSVM filtering efficiency on different k and n_+/n_-. The left figure shows the influence of paramater $d=n_+/n_-$ on the positive sentences filtering precision (k=1). The right figure shows the influence of parameter k on the positive sentences filtering precision $(n_+/n_-=1)$.

3.3   Experiments and Analysis

For easy comprehension of BSVM, we choose Linear function $(C=100)$ as kernel and simulate the training result of different biases under the environment of *Matlab*. As is shown in Fig.2, the separating surface shifts with the parameters of $d = n_+/n_-$ and k changing.

To show the efficiency of BSVM in practical applications, we experiment on benchmark collections of Chinese web pages [1] prepared by *FuDan University*. The collections include 9804 training examples and 9833 evaluating documents, which consist of a set of Chinese newswire stories classified under 20 categories. In this paper, we experiment on a document set made of two related categories(*history* and *politics*) of the benchmark. The document set contains totally 2800 web pages(2000 pages about politics as positives, 800 pages about history as negatives and 1/10 of each as training examples). We compute the

---

[1] The benchmark and a detailed description(in Chinese) are available at http://www.nlp.org.cn/docs/doclist.php?cat_id=16\&type=15.

positive sentences filtering precision under different C, and exhibit the influence of $d = n_+/n_-$ and k in Fig.3. Concluded from the result, the positive sentences filtering precision increases with n_+/n_- and k increasing.

4   Conclusion and Future Work

In this paper, we give a study on different scopes of filtering result according to different filtering task and user interest. We find that the web filtering result can be divided three sets of relative pages set(R_1), similar pages set(R_i) and homologous pages set(R_k) with the relationship of $R_k \subseteq R_i \subseteq R_1$. To adjust the web filtering result to be more fit for the user interest, a Biased Support Vector Machine (BSVM) algorithm in introduced which imports a stimulant function, uses training examples distribution n_+/n_- and a user-adaptable parameter k to deals imbalanced different classes of the pre-assigned pages. Experiments show that BSVM can greatly improve the web filtering performance. But problems of user bias description and parameter self-adaptable are still open and we leave them as future work.

References

1. Belkin, N.J., Croft, W.B.: Information Filtering and Information Retrieval: Two Sides of the Same Coin? Communications of the ACM **35** (1992) 29–38
2. Quinlan, J.R.: Discovering rules by induction from large collections of examples. In Michie, D., ed.: Expert Systems in the Micro-Electronic Age. Edinburgh University Press, Edinburgh (1979) 168–201
3. Quinlan, J.R.: Induction of decision trees. Mach. Learn. **1** (1986) 81–106
4. Quinlan, J.R.: C4.5: programs for machine learning. Morgan Kaufmann Publishers Inc., San Francisco, CA, USA (1993)
5. Chid Apte, F.D., Weiss, S.: Text miningwith decision rules and decision trees. In: Proceedings ofthe Conference on Automated Learning and Discovery, CMU (1998)
6. Clark, P., Niblett, T.: The cn2 induction algorithm. Mach. Learn. **3** (1989) 261–283
7. McCallum, A., Nigam, K.: A comparison of event models for naive bayes text classification. In: AAAI-98 Workshop on Learning for Text Categorization. (1998)
8. Joachims, T.: Text categorization with support vector machines: Learning with many relevant features. In: Proceedings of the European Conference on Machine Learning, Berlin,German, Springer (1998) 137–142
9. Du, A., Fang, B.: Comparison of maching learning algorithms in chinese web filtering. In: proceedings of The third International Conference on Machine Learning and Cybernetics, Shanghai,China (2004) 2521–2526

A New Approach to Generate Frequent Patterns from Enterprise Databases

Yu-Chin Liu[1,2] and Ping-Yu Hsu[1]

[1] Department of Business Administration, National Central University,
Chung-Li, Taiwan 320, R.O.C.
{ycliu, pyhsu}@mgt.ncu.edu.tw
[2] Department of Information Management, TungNan Institute of Technology,
Taipei, Taiwan 222, R.O.C.

Abstract. As data mining techniques are explored extensively, incorporating discovered knowledge into business leads to superior competitive advantages. Most techniques in mining association rules nowadays are designed to solve problems based on de-normalized transaction files. Namely, normalized transaction tables should be transformed before mining methods could be applied, and some previous works have pointed that such data transformation usually consumes a lot of resources. As a result, this study proposes a new method which incorporates mining algorithms with enterprise transaction databases directly.

In addition, in most well-known mining algorithms, the minimum support threshold is used in deciding whether the pattern is frequent or not, and it is crucial to define an appropriate threshold before performing mining tasks. Since setting an appropriate threshold cannot be done intuitively by domain experts or users, they usually set the threshold through trial and error. Usually, while setting different minimum support thresholds, most existing algorithms re-perform all mining procedures. Consequently, it takes a lot of computations. Our new method explores such circumstances and provides ways to flexibly adjust support thresholds without re-doing the whole mining task.

1 Introduction

The idea of extracting knowledge from the data has been explored for a long time. To deem modern data mining and knowledge discovery as scholarly, interdisciplinary fields contributes to the IJCAI-89 Workshop on Knowledge Discovery in Real Databases [5]. Since the database system prevails worldwide, large amounts of data have been accumulating in real-world databases rapidly. Several researchers have proposed algorithms for handling large amounts of data as well as extracting implicit, previously unknown and potentially useful information [1].

Data Mining is application-dependent; in short, different applications need different mining techniques. One of major techniques is mining association rules, which has attracted many researchers in recent years. It has an essential application for Market-Basket Analysis and is used for acquiring strong association

S. Singh et al. (Eds.): ICAPR 2005, LNCS 3686, pp. 371–380, 2005.

rules, shown in the form, "If one buys x, then he/she buys y" [1]. The first step in mining association rules is setting the minimum support and confidence thresholds. After this, based on the predefined support, the frequent patterns are then searched. Finally, all frequent patterns are used to generate rules. Rules that have fulfilled the confidence threshold are called 'association rules'.

Although these mining works emphasize on retrieving knowledge from enterprise databases, researchers seldom focus on developing mining algorithms based on enterprise databases directly. Most well-known algorithms mining association rules based on the De-Normalized Table(Table 1). In short, the normalized transaction databases should be transformed into a de-normalized one before performing mining tasks. This may consume additional computation resources. The aim of this study is to provide a novel algorithm to mine frequent patterns based on original enterprise databases.

Table 1. A Denormalized Table

TID	Items(pid)
1	f a c d g i m p
2	a b c f l m o
3	b f h j o
4	b c k s p
5	a f c e l p m n

It is crucial to select appropriate thresholds for mining association rules. For the support threshold, if it is overestimated, the number of frequent patterns mined may be too few for extracting knowledge. Contrarily, if the support is underestimated, too many possible rules might be discovered. Consequently, it may pay too much computation costs and no valuable knowledge can be discovered. In general, support thresholds might be determined by users or domain experts first, and after, mining algorithms are performed. Based on the mined results, people will re-adjust thresholds as needed, and the appropriate support threshold will be obtained through several trials.

Typically, most well-known algorithms perform such trial-and-errors by redoing the mining procedures completely. The original database is re-scanned, the frequent items are re-calculated as well as the rules are re-generated. Hence, our new method tries not only to integrate the data mining work with enterprise information systems, but it also provides ways to flexibly adjust the minimum support threshold without re-doing whole mining procedures.

Our approach(FPN) consists of two main steps: One is to construct a frequent pattern tree(FPN-tree) containing the complete information for mining frequent patterns. And next, the most-used mining algorithms, such as [4], which are developed for mining frequent patterns based on the FP-tree [4], can be applied to the FPN-tree freely. Hence, this study sets focus on the tree construction phase and proposes an FPTree-like algorithm, FPNTree, which is used for constructing frequent pattern trees from normalized databases directly. The remaining pages are organized as follows: In Section 2, a brief literature review of mining frequent patterns is given. Section 3 introduces the FPNTree Algorithm, in detail. Experimental results in Section 4 show that, while re-adjusting the adequate minimum support thresholds, the FPNTree is more efficient than FPTree(FPGrowth) and ECLAT in 4 real and 2 synthetic datasets. Finally, the conclusion is drawn.

2   Related Work

The problem of mining association rules over basket data was introduced in [1]. Finding such rules provides useful marketing information for target customer-buying behavior. Later, in [2], one of the most famous algorithms, named *Apriori*, proposes to solve this problem based on the de-normalized table(Table 1).

In addition to association rules, mining frequent patterns can be applied to other applications, such as causality, sequential patterns, $\cdots$, etc [4]. There have been many studies, such as [2] [6], adopt the Apriori-like approach to search frequent patterns. The Apriori achieves good performance, but however, it is costly when handling a huge number of candidate sets and becomes tedious as it repeatedly scans the database and pattern matching. So [4] proposes a method to search frequent patterns without candidate generation-and-tests. It first constructs a frequent pattern tree, FP-tree, from de-normalized tables. It is used to store compact and complete information for mining frequent patterns. Second, an FP-tree-based pattern-fragment growth mining method is employed. The FPTree achieves success in an order of magnitude faster than Apriori and is implemented in DBMiner System. Our work, instead of constructing trees from de-normalized tables, provides a novel method to construct the FPN-tree, which is equivalent to the FP-tree, based on normalized transaction databases.

In addition, there are published papers on integrating mining algorithms with relational database systems during past few years. In [6], the aspects of integration are discussed in a purely technical way. Alternatives to speed up the database access of the algorithms are evaluated. [3] presents extended versions of the SQL that directly support data mining. Instead of handling these technical issues, our study aims to mine frequent patterns from original transaction tables of enterprise databases, so it can be further combined with these works.

As pointed in [4], since the support value is usually query-dependent, the difficulty arises when selecting a good minimum support threshold while constructing an FP-tree. [4] recommends selecting a relatively low support, which can satisfy most initial mining queries, in constructing trees. As long as the support is re-set to values higher than the initial one, the FP-tree needn't be re-constructed, and it's easy to re-generate frequent patterns based on the upper portions of the established FP-tree. But however, in case of re-setting to a lower support value, the computation cost for mining frequent patterns with a low support threshold explodes exponentially. Aside from this, under such circumstance, an FP-tree should be re-constructed and the whole mining processes should be re-performed. Hence, the FPNTree proposes a new method to flexibly adjust the minimum support threshold without re-doing the whole mining processes.

3   The FPN Approach

3.1   The Data Preparation

In order to mine frequent patterns directly form normalized enterprize databases, few modifications should be made. Usually, enterprize databases contain two normalized tables related to company sales. One is the Transaction Table used to record products sold of each transaction, and the concatenated primary key may usually be the ProductId(*pid*) and TransactionId(*tid*). The other one is the Product Table which records all products provided in the company. Next, modifications required for these two tables are further explained.

1. Add one index constraint on the *pid* field to the Transaction Table. Therefore, the Transaction Table will keep a logical sequence based on *pid*.
2. In the Product Table, one field named *count* which records the number of transactions of each sold product should be added. So, when one transaction happens, information systems will insert new records to the Transaction Table as well as update the *count* of the Product Table simultaneously.
3. In order to speed the mining process, an index constraint on the *count* field of the Product Table should be added.
4. Another pointer field, represented as *pptr*, is also required in the Product Table for enhancing the efficiency. It links each product in the Product Table to the first tupple of related transaction records. Since the Transaction Table has indexed on the ProductId in first step, all related transactions containing the same ProductId can be retrieved efficiently.

Product Table: Index by Frequency Count / **Logical Sequence of Index view**

pid	count	pptr
3(f)	4	
6(c)	4	
1(a)	3	
2(b)	3	
13(m)	3	
16(p)	3	
12(l)	2	
15(o)	2	
4(d)	1	
5(e)	1	
7(g)	1	
8(h)	1	
9(i)	1	
10(j)	1	
11(k)	1	
14(n)	1	
19(s)	1	

Transaction Table: Index by pid / **Logical Sequence of Index view**

tid	pid
1	1(a)
2	1(a)
5	1(a)
2	2(b)
3	2(b)
4	2(b)
1	3(f)
2	3(f)
3	3(f)
5	3(f)
1	4(d)
5	5(e)
1	6(c)
2	6(c)
4	6(c)
5	6(c)
1	7(g)

3	8(h)
1	9(i)
3	10(j)
4	11(k)
2	12(l)
5	12(l)
1	13(m)
2	13(m)
5	13(m)
5	14(n)
2	15(o)
3	15(o)
1	16(p)
4	16(p)
5	16(p)
4	19(s)

Fig. 1. The modified tables in Enterprize Information Systems

Fig. 1 illustrates these modifications and these two normalized tables are exactly the data input of the FPN Approach. As mentioned, most nowadays algorithms for mining frequent patterns require additional steps to transform original normalized tables(Fig. 1) into the de-normalized flat file(Table 1).

3.2   Preliminaries and Definitions

Example 1 *Let the Transaction and Product Table shown in Fig. 1 be the data sources for mining frequent patterns, and the minimum support threshold $\xi = 3$.*

Definition 1. *A Product Recordset, PR, is define as*

a. $PR = \{\langle pid, count\rangle\}$ *defined to be a projected recordset of the PR recoredset.*
b. *PR is ordered by descending count first and ascending tid secondly.*
c. *freq(PR)=$\{\langle pid, count\rangle \mid count \geq \xi\}$.*
d. *$PR(x) = \{\langle pid, count\rangle \mid pid = x\}$.*
e. *$f(x)$ is a function which returns the value of $PR(x).count$.*
f. *$f^o(x)$ is a function to define the sequence of constructing the FPN-tree.*

For instance, in Example 1, because f,c,a,b,m,p are frequent patterns and $f(f) \geq f(c) \geq f(a) \geq f(b) \geq f(m) \geq f(p)$, the order used to construct the FPN-tree can be defined as $f^o(f) > f^o(c) > f^o(a) > f^o(b) > f^o(m) > f^o(p)$.

Definition 2. *A Transaction Recordset, TR, is define as*

a. *$TR = \{\langle tid, pid\rangle\}$, which is a projected recordset of the Transaction Table.*
b. *TR is ordered by ascending pid first and ascending tid secondly.*
c. *$freq(TR) = \{\langle tid, pid\rangle \mid pid \in freq(PR).pids\}$.*

Definition 3. *A FPN-tree, mined from TR and PR, is defined below.*

a. *It consists of one root, a set of subtrees as the children of the root and a Transaction-Header-Table termed as "THT".*
b. *Other than the root, each node contains two fields: (1)pid and (2)cnt.*
c. *the Transaction-Header-Table consists of two filed: (1)tid and (2)ptr.*

The Transaction-Header-Table, THT, of the FPN-tree is used to point the tree node whose *pid* contained in the transaction. And the path originated form the root to current node represents the *pid*s of the current transaction have been stored in the FPN-tree. Next, it shows the FPN constructs the tree by the descending $f^o(pid)$s and after completing the FPN-tree, it stores the complete information for mining frequent patterns. And then, THT can be deleted.

3.3   The FPNTree Construction

In constructing the tree, FPNTree initially retrieves all $freq(PR).pids$. After this, it first constructs the root as well as establishes the THT and sets all $THT.ptr$s to the root. The initial state of Example 1 is shown in Fig. 2.

Secondly, for each $freq(PR).pid$, the FPNTree retrieves all transactions containing the current $freq(PR).pid$ from TR, i.e. $\sigma_{freq(PR).pid}TR$. After this, FPNTree constructs the tree by updating the $THT.ptr$s of each related transactions (i.e. $\sigma_{freq(PR).pid}TR.tids$). The tree construction principles are summarized as: for each $\sigma_{freq(PR).pid}TR.tid$, FPNTree checks if the current $THT.ptr$ has a child node marked as $freq(PR).pid$. If so, then adds '1' to $THT.cnt$ as well as updates the current $THT.ptr$ to this child node; or else, creates a child node marked as $\langle freq(PR).pid, 1 \rangle$ as well as updates the current $THT.ptr$ to the newly created node. Next, FPNTree is further illustrated by Example 1.

In Step 1, the first tuple of $freq(PR).pid$ is f(i.e. 3), the transactions containing the $pid = f$ are retrieved. As shown in Fig. 2, the $\sigma_{pid=f}TR$ contains fours transactions: tid=1, 2, 3 and 5. Since the first $\sigma_f TR$ is tid=1, the FPNTree checks the $THT.ptr$ with $THT.tid = 1$. Here, the current $THT.ptr$ points to the root and of which there is no child node marked with 'f', FPNTree creates a child node with $\langle pid = f, cnt = 1 \rangle$. At the same time, the current $THT.ptr$ is updated to point this newly created child node. After this, FPNTree proceeds to the second transaction record, i.e. $\sigma_{f,tid=2}TR$. The $THT.ptr$ with $THT.tid = 2$ also points to the root initially, but however there is already a child node marked as f, so the FPNTree adds '1' to the $count$ of the child node as well as updates the current $THT.ptr$ to this child node. The same procedures are repeated with $tid = 3$ and 5. Finally, this child node is marked as $\langle f, 4 \rangle$.

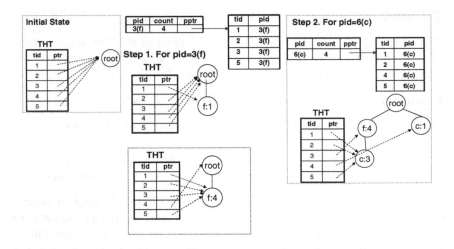

Fig. 2. Initial Step; Step1 $pid = f$ and Step 2 $pid = c$

Next, the second $freq(PR).pid$ is c, and the $\sigma_c TR$ is retrieved(tid=1,2,4 and 5). The current $THT.ptr$ with $THT.tid = 1$ now points to the node: $\langle f, 4 \rangle$, we use $(THT.ptr).pid$ to refer 'f' and $(THT.ptr).cnt$ to indicate '4'. Since there is no child labeled with c, FPNTree creates a child node and marks it as $\langle c, 1 \rangle$. The current $THT.ptr$ with $THT.tid = 1$ is now changed to this new node. Next, as $\sigma_{c,tid=2}TR$, the corresponding $THT.ptr$ points to the $\langle f, 4 \rangle$. At this time, there is already a child node named $\langle c, 1 \rangle$, so we just need to update this child node to $\langle c, 2 \rangle$ and re-point the current $THT.ptr$ to this node. When processing the $\sigma_{c,tid=4}TR$, the corresponding $THT.ptr$ points to the root originally. Since there is no child with $pid = c$ linked to the root directly, a new node is created as $\langle c, 1 \rangle$. Meanwhile, the current $THT.ptr$ is updated to this node. Finally, $\sigma_{c,tid=5}TR$ is processed. So all transactions containing c are stored in the FPN-tree. The

FPNTree algorithm and the constructed FPN-tree is listed in Fig. 3. As reader might observe that excepting the Header Table of the FP-tree, the FPN-tree is equivalent to the FP-tree(The proof has been omitted to save sapce).

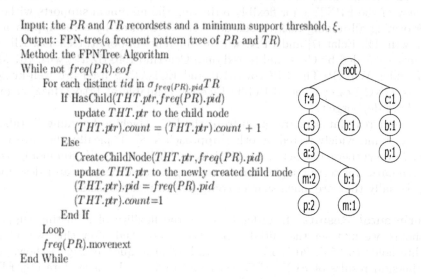

Input: the PR and TR recordsets and a minimum support threshold, ξ.
Output: FPN-tree(a frequent pattern tree of PR and TR)
Method: the FPNTree Algorithm
While not $freq(PR).eof$
    For each distinct tid in $\sigma_{freq(PR).pid}TR$
        If HasChild($THT.ptr, freq(PR).pid$)
            update $THT.ptr$ to the child node
            $(THT.ptr).count = (THT.ptr).count + 1$
        Else
            CreateChildNode($THT.ptr, freq(PR).pid$)
            update $THT.ptr$ to the newly created child node
            $(THT.ptr).pid = freq(PR).pid$
            $(THT.ptr).count=1$
        End If
    Loop
    $freq(PR)$.movenext
End While

Fig. 3. The FPNTree Algorithm

Since the tree is constructed by the descending $f^o(pid)$s, the tree path from the root to the current node shows the frequent $pids$ of the transaction in the descending order. For example, in Fig. 3, the $THT.tid=1$ terminated at the node $\langle p, 2\rangle$. Since the tree is constructed by the order: $f^o(f) > f^o(c) > f^o(a) > f^o(b) > f^o(m) > f^o(p)$, the path from the root to $\langle p, 2\rangle$ registers the transaction($tid=1$) containing $pid = f, c, a, m, p$.

While constructing the tree with ξ, since the PR is ordered by the $f(pid)$s and the TR is ordered by the pid and tid, the FPNTree needn't to scan the whole PR and TR. During the tree construction, FPNTree navigates the $PR.pids$ in the descending order of $f^o(pid)$s. As it reaches the $PR.pid$ whose frequency count lower than ξ, the algorithm will terminate since no frequent items are left.

Property 3.1. Given the TR and PR, as well as ξ, the FPNTree constructs the FPN-tree without scanning the whole TR and PR.

Based on **Property 3.1**, if the new minimum support $\varepsilon > \xi$, the FPN-tree needn't to be re-constructed. The upper portion, whose $f(PR.pid)s \geq \varepsilon$, of the FPN-tree can be used to generate frequent patterns. Contrarily, as the new $\varepsilon < \xi$, since the PR is ordered by the descending $f(PR.pid)$s, FPNTree can continue to construct the FPN-tree with $pids$ whose $f(PR.pid)$s are between ε and ξ. Consequently, no matter the minimum support threshold is increased or decreased, the FPNTree can flexibly construct the tree without re-scan the whole databases as well as re-construct the FPN-tree.

4   Simulation Results

To evaluate the efficiency and effectiveness of the FPNTree, several experiments on various kinds of datasets are tested. Experimental results show the efficiency of the FPNTree for flexibly adjusting the minimum supports without re-performing all mining tasks. Each dataset is evaluated by three approaches, FPGrowth [4], Eclat [7] and FPN, with 6 different support thresholds. All Algorithms are coded by C++ and tested on a Centrino 1.4GHZ PC with 767 MB RAM under WinXP. The FPGrowth1.6 and ECLAT2.9 used are well-known versions, GNC Lesser General Public License, available at http://fuzzy.cs.uni-magdeburg.de/borgelt/.

The four real datasets are taken from the UCI Machine Learning Database Repository and widely tested by other approaches [8]. Typically, they are very dense. Two synthetic datasets, T10I4D10K and T40I10T100K, commonly used as benchmarks for testing algorithms are also generated by generator described in [2]. Usually they are much sparser compared to real datasets.

Experimental Results. In order to show the flexibility of adjusting support thresholds, we first set the initial support to 30%. And after this, we adjust the thresholds to 25%, 20%, 15%, 10% and 5% in sequence. Fig. 4 shows the experimental results of FPN, FPGrowth and ECLAT. In most datasets, FPN outperforms FPGrowth and ECLAT. In general, the computation time used in FPN is 2 to 4 times faster than FPGrowth and ECLAT in real datasets. And on the connect dataset, the computation time required are much less than two other methods. It shows the FPGrowth and ECLAT are 8 to 10 times slower than the FPN. These real datasets are very dense. Especially in the pumsb, there are only 2k items and 500k transactions, however the number of frequent patterns found are 28,096,545 with support=10%. For the synthetic datasets: t10 and t40, which are much sparser than real datasets, the results also present the FPN achieves good performance on both of them. So the flexibility of the FPNTree to adjust minimum support values do save lots of computation costs especially when the support value is low.

Comparison with FPTree. As in Example 1, FPTree first reads the Table 1. And then, FPTree transforms Table 1 into an ordered frequent *pid* projections. Since f, c, a, b, m, p are frequent items, the ordered frequent *pid* projections refer to the left part of Table 2 and all infrequent $l, o, d, e, g, h, i, j, k, n, s$ are deleted. As a result, each transaction contains only the ordered frequent items. And based on the projections, FPTree scans the database by examining one transaction at a time. On the other hand, the FPNTree examines one frequent *pid* at a time, and all *tids* containing the specific *pid* are used to construct the tree. Since the FPTree does not maintain the infrequent *pids*, as decreasing the support values, it needs to re-scan entire database as well as re-construct the tree.

Since different mechanisms of infrequent *pids* recording are embedded in these two algorithms, FPNTree and FPTree inherit significantly different run time complexity. Assume existing a database with I items and a total of S copies

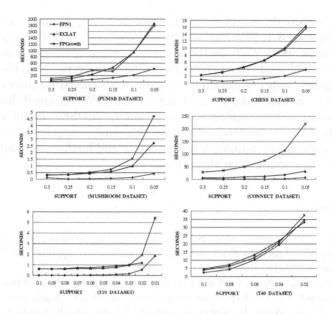

Fig. 4. Comparison of FPN, FPGrowth and ECLAT

Table 2. FPTree vs FPNTree

TID	f c a b m p	l o d e g h i j k n s
1	f c a m p	d g i
2	f c a b m	l o
3	f b	o h j
4	c b p	k s
5	f c a m p	l e n

of these items being sold in numerous transactions. And assume the database is used to build FPN-trees K times with thresholds monotonically decreased. In each iteration of building the FPN-tree, the method only needs to read the transactions associated with the items whose counts fall between thresholds of current and previous iterations. Therefore, the complexity of building the FPN-trees K times with thresholds monotonically decreased is $O(I + S)$, which is independent of the K. On the other hand, the algorithm of FP-tree generation reads the database and decides the counts of items first. And then it sorts the items of each transaction according to the counts. Suppose s is the length of the maximum transaction. In the following iterations, the ordered S is re-read and the FP-trees are re-built with the frequent $pids$. Hence, the complexity of constructing K FP-trees with the given database is $O(K * S + S * log(s))$.

5   Conclusion

Unlike most mining algorithms, the proposed method mines frequent patterns from normalized tables of enterprise databases directly. The FPNTree constructs an FPN-tree to store the complete information for mining frequent patterns. And based on **Property 3.1**, FPNTree only needs to scan the transactions associated with the frequent items. Only in the rare cases, when the support is extremely low, do the method need to scan *tids* associated with every *pid* in the database.

As mentioned, incorporating mining tasks with enterprise information systems would be an interesting research topic. The FPN Approach has room for further research, such as mining sequential data, CLOSET frequent patterns, and maximum frequent patterns. Based on this study, our interest lies in developing new algorithms for mining very dense data and further combining our new method with other well-known algorithms.

References

1. R. Agrawal, I. Imielinski, and A. Swami. Mining association rules between sets of items in large databases. In *Proceedings of International Conference on Management of Data*, pages 207–216, 1993.
2. R. Argawal and R. Srikant. Fast algorithms for mining associations rules. In *Proceedings of International Conference in Very Large Data Bases*, pages 487–499, 1994.
3. J. Han, Y. Fu, W. Wang, K. Koperski, and O. Zaiane. Dmql: A data mining query language. In *In 1996 SIGMOD'96 Workshop on Research Issues on Data Mining and Knowledge Discovery (DMKD'96)*, 1996.
4. J. Han, J. Pei, Y. Yin, and R. Mao. Mining frequent patterns without candidate generation: A frequent-pattern tree approach. *Data Mining and Knowledge Discovery*, 8:53–87, 2004.
5. G. Piatetsky-Shapiro. Knowledge discovery in real databases: A report on the ijcai-89 workshop. *AI Magazine*, 11(5):68–70, 1991.
6. S. Sarawagi, S. Thomas, and R. Agrawal. Integrating association rule mining with relational database systems: Alternatives and implications. In *Proceedings of ACM SIGMOD International Conference in Management of Data*, pages 343–354, 1998.
7. M.J. Zaki. Scalable algorithms for association mining. *IEEE Transactions on Knowledge and Data Engineering*, 12(3):372–390, 2000.
8. M.J. Zaki and K. Gouda. Fast vertical mining using diffsets. In *Proceedings of International Conference on Knowledge Discovering and Data Mining*, 2003.

Consolidated Tree Classifier Learning in a Car Insurance Fraud Detection Domain with Class Imbalance

Jesús M. Pérez, Javier Muguerza, Olatz Arbelaitz, Ibai Gurrutxaga, and José I. Martín

Dept. of Computer Architecture and Technology, University of the Basque Country,
M. Lardizabal, 1, 20018 Donostia, Spain
{txus.perez, j.muguerza, olatz.arbelaitz,
ibai.gurrutxaga, j.martin}@ehu.es
http://www.sc.ehu.es/aldapa

Abstract. This paper presents an analysis of the behaviour of Consolidated Trees, CT (classification trees induced from multiple subsamples but without loss of explaining capacity). We analyse how CT trees behave when used to solve a fraud detection problem in a car insurance company. This domain has two important characteristics: the explanation given to the classification made is critical to help investigating the received reports or claims, and besides, this is a typical example of class imbalance problem due to its skewed class distribution. In the results presented in the paper CT and C4.5 trees have been compared, from the accuracy and structural stability (explaining capacity) point of view and, for both algorithms, the best class distribution has been searched.. Due to the different associated costs of different error types (costs of investigating suspicious reports, etc.) a wider analysis of the error has also been done: precision/recall, ROC curve, etc.

1 Introduction

The application of machine learning to real world problems has to be done considering two important aspects: class distribution affects to classifiers' accuracy and the explanation is very important in some domains.

In real domains such as illness diagnosis, fraud detection in different fields, customer's behaviour analysis (marketing), customer fidelisation, ... it is not enough to obtain high accuracy in the classification, comprehensibility in the built classifier is also needed [2]. The classifying paradigms used to solve this kind of problems need to be able to give an explanation, for example classification trees.

On the other hand, it is very common to find domains where the number of examples for one of the categories (or classes) of the dependent variable is much smaller than for the rest of the classes. These situations are named class imbalance or skewed class distribution.

Classifiers do not behave well when they are trained with very unbalanced data sets. For example, if 99% of the examples in a data set belong to the same class, for a classifier that labels test cases with the majority class, the accuracy will be 99%. Since most classifiers are designed to minimise the error rate, the classifiers built from this kind of data-sets tend to be very simple and nonsense [3].

S. Singh et al. (Eds.): ICAPR 2005, LNCS 3686, pp. 381–389, 2005.
© Springer-Verlag Berlin Heidelberg 2005

Weiss and Provost [8] have shown that each domain has an optimal class distribution to be used for training. Their work shows that, in situations where the class distribution of the training set can be chosen, it is often preferable to use a different distribution to the one expected in reality. So, in environments with skewed class distribution, we can use samples with modified class distribution with the aim of building valid classifiers. Undersampling, eliminating some examples, is the most common strategy to modify the class distribution of a sample. Even if the direct consequence of undersampling is that some examples are ignored, in general, undersampling techniques obtain better results than oversampling techniques (repeating some examples) [3]. In order to avoid the information loss produced by undersampling, multiple classifiers can be built based on subsamples with changed distribution. Most of the information in the original sample can be covered by choosing adequately the number of generated subsamples. Techniques such as bagging and boosting can be a good option in some cases but not in areas where explanation is important. It is clear that "while a single decision tree can easily be understood by a human as long as it is not too large, fifty such trees, even if individually simple, exceed the capacity of even the most patient" [2].

We have developed an algorithm, CTC (Consolidated Tree's Construction Algorithm), that is able to face both problems: several subsamples with the desired class distribution are created from the original training set, but opposite to other algorithms that build multiple trees (bagging, boosting), a single tree is induced, therefore the comprehensibility of the base classifier is not lost.

Fraud detection problems belong to the group of domains where the explanation in the classification is important. We will use the CTC algorithm and compare it to C4.5 [6] in a fraudulent report detection problem from a car insurance company. This is a difficult problem because the experts in the company estimate that the fraud average is in reality higher than 10% or 15% but the fraud examples are very difficult to detect, and, as a consequence, in the data provided by insurance companies this percentage is lower. Many of the examples labelled as not fraudulent in the data-set are actually fraudulent which makes the problem even harder. It is important to provide an insurance company with a tool to know the profile of fraudulent customers or the evidences that make a report suspicious of fraud, in order to investigate them more deeply, so that the fraud does not suppose great financial loss.

The paper proceeds describing how a single tree can be built from several subsamples, CTC algorithm, in Section 2. In Section 3 we describe the main characteristics of the car insurance company's database for fraud detection, and Section 4 contains the methodology used to face the class imbalance problem in the described domain. Section 5 is devoted to describe the results obtained for the fraud detection problem with both algorithms CTC and C4.5. Finally Section 6 is devoted to show the conclusions and further work.

2 Consolidated Trees' Construction Algorithm

Consolidated Trees' Construction Algorithm (CTC) uses several subsamples to build a single tree [4]. The consensus is achieved at each step of the tree's building process and only one tree is built. The different subsamples are used to make proposals about the feature that should be used to split in the current node. The split function used in each subsample is the gain ratio criterion (the same used by Quinlan in C4.5). The decision about which feature will be used to make the split in a node of the Consolidated Tree (CT) is accorded among the different proposals by a voting process (not weighted) node by node. Based on this decision, all the subsamples are divided using the same feature. The iterative process is described in Algorithm 1.

Algorithm 1. Consolidated Trees' Construction Algorithm (CTC)

Generate *Number_Samples* subsamples (S^i) from S with *Resampling_Mode* method.
CurrentNode := *RootNode*
for i := 1 to *Number_Samples*
  $LS^i := \{S^i\}$
end for
repeat
  for i := 1 to *Number_Samples*
    $CurrentS^i := First(LS^i)$
    $LS^i := LS^i - CurrentS^i$
    Induce the best split $(X,B)^i$ for *CurrentS*^{i}
  end for
  Obtain the consolidated pair (X_c, B_c), based on $(X,B)^i$, $1 \le i \le Number\_Samples$
  if $(X_c, B_c) \ne Not\_Split$
    Split *CurrentNode* based on (X_c, B_c)
    for i := 1 to *Number_Samples*
      Divide *CurrentS*^{i} based on (X_c, B_c) to obtain n subsamples $\{S_1^i, ... S_n^i\}$
      $LS^i := \{S_1^i, ... S_n^i\} \cup LS^i$
    end for
  else consolidate *CurrentNode* as a leaf
  end if
  CurrentNode := *NextNode*
  until $\forall i$, LS^i is empty

The algorithm starts extracting a set of subsamples (*Number_Samples*) from the original training set. Based on previous experimentation we find that to use 30 subsamples can be a good trade-off among efficiency and computational cost . The subsamples are obtained based on the desired resampling technique (*Resampling_Mode*). For example, the class distribution of the original training set can be changed or not, examples can be drawn with or without replacement, different subsample sizes can be chosen, etc.

Decision tree's construction algorithms divide the initial sample in several data partitions. In our algorithm, LS^i contains all the data partitions created from each subsample S^i. When the process starts, the only existing partitions are the initial subsamples.

The pair $(X,B)^i$ is the split proposal for the first data partition in LS^i. X is the feature selected to split and B indicates the proposed branches or criteria to divide the data in the current node. In the consolidation step, X_c is the feature obtained by a voting process among all the proposed X. Whereas B_c will be the median of the proposed *Cut* values when X_c is continuous and all the possible values of the feature when X_c is discrete. In the different steps of the algorithm, the default parameters of C4.5 have been used as far as possible.

The process is repeated while LS^i is not empty. The Consolidated Tree's generation process finishes when in the last subsample in all the partitions in LS^i, most of the proposals are not to split it, so, to become a leaf node. When a node is consolidated as a leaf node, the a posteriori probabilities associated to it are calculated averaging the a posteriori obtained from the data partitions related to that node in all the subsamples.

Once the consolidated tree has been built, it works the same way a decision tree does for testing, pruning, etc. This way the explanation of the classifier is not lost even if several subsamples are used to build it.

3   Car Insurance Database for Fraud Detection

One of the factors affecting the price of car insurance policies is the large amount of fraudulent reports that a company is not able to detect is. The company has to assume all the increase in costs produced by this fraud, and, as a consequence, the insurance policies become more expensive. The experts in the companies think that at least 10% or 15% of the produced reports are fraudulent, and, however, about 5% of them is detected. So the databases in insurance companies have the following characteristics: the examples labelled as fraudulent belong to the minority class (class imbalance) and, on the other hand, they are the only 100% reliable data, because among the examples labelled as not fraudulent there are some fraudulent examples that the company has not been able to detect. Therefore the information provided to the algorithm is not correct which makes the machine learning problem difficult to solve.

In order to detect fraud, suspicious reports have to be investigated but this has associated costs: the costs concerning to the investigation itself (staff, resources, etc.), and the cost coming from investigating not fraudulent customers [1]. The company's image can be severely affected by customers that are annoyed when they realise that they are being investigated. When evaluating a report, it will be important for the insurance broker to know the fraud probability assigned to it by the classification system, as well as the factors that have affected to the decision. The explanation given by the classifier about the decision made could be used by the broker to investigate the case. So, if the aim is to have a tool that will help in the detection of fraudulent reports, it is absolutely necessary to use classification paradigms that are able to give an explanation, for example, decision trees. This paper analyses the behaviour of different classifiers with real data from a car insurance company, the kind of problem we have just described. The data-set has 108,000 examples, and just 7.40% of them are fraudulent cases. This database is clearly an example of class imbalance problem with imprecise information for one of the classes.

The database has 31 independent variables that contribute to the report with information of different nature about the accidents: date of the accident (when it happened and when it was communicated), insured person (age, sex, marital status,...), insurance policy and vehicle (fully comprehensive insurance or not, driving experience, kind and use of the vehicle, power,...). When solving this problem with supervised classification, the dependent variable or class will have two categories: fraud and not fraud.

4 Experimental Methodology

The original class distribution of the collected data does not always coincide with the best one to build the classifier when a problem with class imbalance has to be faced. We will make a sweep with different percentages of fraud examples in order to find the class distribution we should use to induce the tree. Based on the methodology proposed by Weiss and Provost in [8] the tried percentages will be 2%, 5%, 7.40% (original distribution), 10%, 20%, 30%, 40%, 50%, 60%, 70%, 80%, 90% and 95%. We will use the methodology described in [8] to compare the CTC algorithm (with parameters mentioned in Section 2) with the well known C4.5 with default settings for the different class distributions. In order to make a fair comparison, the used training sets need to have the same size even if the used percentages are changed. The size of the training set is fixed to 75% of the minority class examples (6,000 examples), so that subsamples of all the mentioned percentages can be used in the experimentation. The remaining 25% of both, majority and minority classes, will be used for test.

Two trees, one with C4.5 and the other with CTC, have been built for each one of the proposed percentages. Even if Weiss and Provost did not prune the trees, the results obtained with pruned trees in this database are substantially better for both algorithms. Therefore, we will present the results obtained by pruning the trees based on the training sample and C4.5 standard pruning. To prune the C4.5 trees we have used the corrector proposed by Weiss and Provost for estimating the a posteriori probability of the leaf nodes, so that they are adapted to the distribution expected in reality. This has to be done because the class distribution of the training set and the class distribution existing in reality (test) do not coincide. Nevertheless, the corrector needs not to be used when pruning CT trees, because the pruning is done with the whole training set (the percentages are the ones expected in reality), and the a posteriori probabilities are corrected due to the backfitting process.

As a validation method, the experimentation has been repeated 10 times.

5 Experimental Results

In the problem described in previous sections, the behaviour of classifiers can not be analysed based just on the error rate. Other aspects will help us to complete our comprehension about the classifier's behaviour: the structural stability of the generated trees and the complexity of the trees will give us information about the quality of the explanation, the ratio among True Positive and False Positive examples (ROC curve) will give us information about the behaviour of the classifier in different environments, etc.

Table 1 shows, for CT trees built with 30 subsamples and C4.5, the error rates and standard deviation (*Error* and σ columns) and the average complexity, measured as the number of internal nodes of the trees (*Compl.* column). The values in different

Table 1. Error, standard deviation and complexity for different class distributions

	C4.5			CT(30)		
	Error	σ	Compl.	Error	σ	Compl.
2%	7.40	0.00	0.0	7.30	0.00	3.8
5%	7.40	0.00	0.0	7.31	0.03	4.8
7.4%	7.40	0.00	0.0	7.31	0.03	4.3
10%	7.40	0.00	0.0	7.31	0.03	4.1
20%	7.57	0.34	12.3	7.30	0.05	5.6
30%	9.75	0.68	103.6	7.30	0.05	6.6
40%	11.70	1.06	183.7	7.31	0.03	6.2
50%	10.87	0.95	141.4	7.29	0.06	7.1
60%	8.69	0.71	44.6	7.30	0.05	6.9
70%	8.59	0.60	28.2	7.30	0.05	7.2
80%	8.27	0.52	17.2	7.29	0.06	6.6
90%	8.59	0.70	11.2	7.29	0.06	6.0
95%	7.40	0.00	0.0	7.31	0.07	3.4

rows are related to different class distributions. Results show that in every case the error is smaller for CTC than for C4.5. The trees built with C4.5 are not able to reduce the error rate of 7.40% that would achieve a trivial classifier that labels all the examples with the majority class (no fraud). The values belonging to average complexity confirm that when the error for C4.5 trees is 7.40% the built trees are trivial classifiers: they are just the root node. For the rest of the percentages, the built classifiers do not make sense because of the achieved error rate and complexity. The values of standard deviation show that the error rates achieved with CT trees are very stable (best values are obtained when class distributions are 50%, 80% and 90%). Besides, all of them are under the threshold of 7.40% (7.30% in average) and with small complexity in average; therefore, giving a simple explanation. These results confirm that CT trees are better situated in the learning curve and also according to the principle of parsimony (Occam's razor).

If we want to evaluate the stability of the explanation given by CT trees we need to measure the structural stability. A structural distance, *Common*, based on a pair to pair comparison among all the trees of the compared set has been defined with this aim. *Common* is calculated starting from the root and covering the tree, level by level. The common nodes among two trees are counted if they coincide in the feature used to make the split, the proposed branches or stratification and the position in the tree [5]. Normalising the *Common* value with the complexity of the tree (as defined before) and making the analysis for the CT trees built for different class distributions, we find a maximum value of 74.27% and minimum of 35.19% being the average 49.36%. So we can say that in average half of the structure of the trees, and, as a consequence the explanation, is maintained. However for C4.5 the %*Common* is in average 10.31%.

The division of the error in false positive (FP) and false negative (FN) is important in this kind of applications. FP quantifies the amount of unnecessary investigations of customers whereas the FN quantifies the fraudulent customers that are not detected. Evidently it is of capital importance not to investigate honest customers in order to achieve a good company image. Even if the objective is to detect all the fraudulent reports, the quantification of the percentage of investigated reports is also important due to the costs and the trouble to the customers it implies. We will analyse these aspects with results of *precision*, *recall* or *sensitivity*, *breakeven point*, ROC curve and AUC [7]. To be brief, we will present results for class distribution of 50% (results for other distributions are similar).

Classification trees can work on a more or less conservative way by modifying the threshold needed to label a node as fraudulent. Table 2 shows results for a wide range of thresholds. When the threshold is 0% (trivial acceptor) all the examples will be classified as fraudulent; this would be the most liberal operation mode. On the other hand, the most restrictive operation mode will be when the threshold is 100%. Only the fraudulent examples belonging to homogeneous nodes would be classified as fraudulent. The adequate threshold is usually selected so that the best trade off among costs of FP and FN is found.

Table 2. Results for conservative and liberal operation mode for CTC and C4.5 algorithms

	CTC			C4.5		
Threshold	Precision	Recall	Reports	Precision	Recall	Reports.
0%	7.41	100.00	27000	7.41	100.00	27000
5%	11.21	89.66	15995	12.20	76.39	12523
10%	12.17	66.53	10932	12.60	70.17	11135
15%	**53.98**	**5.12**	**190**	13.50	41.28	6115
20%	55.03	5.09	185	14.11	22.87	3241
25%	57.84	4.89	169	14.52	16.80	2314
30%	59.92	4.71	157	14.95	14.32	1916
35%	60.31	4.68	155	15.61	13.33	1708
40%	60.64	4.62	152	15.78	11.82	1498
45%	63.23	3.59	113	14.73	10.13	1375
50%	63.39	3.46	109	14.85	9.91	1335
55%	63.51	3.42	108	13.88	8.92	1285
60%	66.43	2.80	84	13.31	8.29	1245
65%	67.05	2.62	78	12.13	7.28	1199
70%	70.11	2.24	64	10.69	6.14	1148
75%	**70.39**	**1.61**	**46**	9.82	5.48	1116
80%	50.00	0.09	3	9.83	5.38	1094
85%	50.00	0.09	3	9.83	5.38	1094
90%	50.00	0.09	3	9.83	5.22	1062
95%	50.00	0.09	3	9.83	5.22	1062
100%	--	0.00	0	--	0.00	0

The *precision* and *recall* are two parameters that can be used to measure the effectiveness of the classifier on basis of the threshold. Examples of conservative and liberal operation mode appear in bold in Table 2. We can observe that if the classifier based on CTC would work in a conservative way (*threshold* 75%), the company would revise only 46 reports (*reports*) and 70.39% of them (*precision*) would be fraudulent (the probability to find a fraudulent report has been increased from 7.4% to 70.39% and the disturbed customers have been very few). If we would like to detect more fraudulent reports, increasing the *recall* but still without disturbing a lot of not fraudulent customers we could decrease the threshold to 15% (liberal example). As a consequence 190 reports would be revised and more than half of them would be fraudulent. Table 2 shows that the trees induced with C4.5 achieve higher *recall* values, but the amount of reports to investigate and the low precision obtained make grow considerably the costs related to investigations and incorrectly revised customers.

A way to find a balance among *precision* and *recall* is to establish a threshold, *breakeven point*, so that both parameters are made equal. Although this is a too general measure for the purposes of our study, CTC clearly beats C4.5; the estimated values are 34.45% for CTC and 15.25% for C4.5. Evidently being the aim the maximisation of *precision* and *recall* when larger the *breakeven point* is, better the behaviour of the algorithm is.

We have also calculated the ROC curves (they are not shown due to lack of space) of both classifiers. The aim is to maximize the TP with a minimum FP, and as a consequence, to maximize the Area Under the ROC Curve (AUC). We have calculated the average AUC for all the analyzed class distributions and the average values obtained are: 68.87% for CTC and 60.71% for C4.5. This indicates that CT trees have better global behaviour than C4.5 trees.

6 Conclusions and Further Work

This paper presents the analysis of the influence of class distribution in a fraud detection problem from a car insurance company for two tree induction algorithms: C4.5 and CTC. The behaviour of both algorithms for different class distributions has been analysed based on the methodology presented in [8]. Thanks to this methodology we have been able to build non trivial C4.5 trees, but results have been better for CT trees. Moreover, both algorithms build a single tree, that is to say, they maintain the explanation in the classification which is essential in real problems of this kind where an explanation added to the classification made is compulsory. The results presented in Section 5 confirm that CT trees behave better than C4.5 trees in many aspects: accuracy, structural stability or explanation, ROC curve, *precision/recall*, etc.

The results obtained in this experimentation could be compared to other strategies that do not lose the explanation even if they use several subsamples to build the classifier. For example the procedure presented in [2], which is able to extract explanation to bagging, and our proposal could be compared. As we mentioned when describing CTC algorithm many parameters can be varied. CTC algorithm can be tested using other base algorithm different to C4.5 such as CHAID, CART, …

Related to the real application, fraud detection in car insurance companies, the difficulty finding fraud examples make us think that the characteristics provided by the experts in the company might not be suitable, so the extraction of more discriminating information could be studied.

Acknowledgements

The work described in this paper was partly done under the University of Basque Country (UPV/EHU), project: 1/UPV 00139.226-T-15920/2004. It was also funded by the Diputación Foral de Gipuzkoa and the European Union. We also want to thank the collaboration of the company ADHOC Synectic Systems, S.A. The *lymphography* domain was obtained from the University Medical Centre, Institute of Oncology, Ljubljana, Yugoslavia. Thanks go to M. Zwitter and M. Soklic for providing the data.

References

1. Chan P.K., Stolfo S.J.: Toward Scalable Learning with Non-uniform Class and Cost Distributions: A Case Study in Credit Card Fraud Detection, Proc. of the 4th Int. Conference on Knowledge Discovery and Data Mining, (1998) 164-168.
2. Domingos P.: Knowledge acquisition from examples via multiple models. Proc. 14th International Conference on Machine Learning Nashville, TN (1997) 98-106.
3. Japkowicz N.: Learning from Imbalanced Data Sets: A Comparison of Various Strategies, Proceedings of the AAAI Workshop on Learning from Imbalanced Data Sets, Menlo Park, CA, (2000).
4. Pérez J.M., Muguerza J., Arbelaitz O., Gurrutxaga I.: A New Algorithm to Build Consolidated Trees: Study of the Error Rate and Steadiness. Advances in Soft Computing, Proceedings of the International Intelligent Information Processing and Web Mining Conference (IIS: IIPWM'04), Zakopane, Poland (2004), 79-88.
5. Pérez J.M., Muguerza J., Arbelaitz O., Gurrutxaga I., Martín J.I.: Analysis of structural convergence of Consolidated Trees when resampling is required. Proc. of the 3rd Australasian Data Mining Conf. (AusDM04), Australia (2004), 9-21.
6. Quinlan J.R.: C4.5: Programs for Machine Learning, Morgan Kaufmann Publishers Inc.(eds), San Mateo, California, (1993).
7. Sebastiani F.: Machine Learning in Automated Document Categorisation. Tutorial of the 18th Int. Conference on Computational Linguistics, Nancy, Francia, 2000.
8. Weiss G.M., Provost F.: Learning when Training Data are Costly: The Effect of Class Distribution on Tree Induction, Journal of Artificial Intelligence Research, Vol. 19, (2003) 315-354.

Missing Data Estimation Using Polynomial Kernels

Maxime Berar[1], Michel Desvignes[1], Gérard Bailly[2], Yohan Payan[3],
and Barbara Romaniuk[4]

[1] Laboratoire des Images et des Signaux, LIS , 961 rue de la houille blanche,
BP 46,38402 St Martin d'Hères cedex, France
{Maxime.Berar, Michel.Desvignes}@lis.inpg.fr
http://www.lis.inpg.fr
[2] Institut de la Communication Parlée (ICP), UMR CNRS 5009,
INPG/U3, 46,av. Félix Viallet, 38031 Grenoble, France
bailly@icp.inpg.fr
[3] Techniques de l'Imagerie, de la Modélisation et de la Cognition (TIMC),
Faculté de Médecine, 38706 La Tronche, France
payan@imag.fr
[4] CreSTIC-LERI, Rue des Crayères, BP 1035 51687 Reims Cedex 2, France
barbara.romaniuk@leri.univ-reims.fr

Abstract. In this paper, we deal with the problem of partially observed objects. These objects are defined by a set of points and their shape variations are represented by a statistical model. We present two models in this paper: a linear model based on PCA and a non-linear model based on KPCA. The present work attempts to localize of non visible parts of an object, from the visible part and from the model, using the variability represented by the models. Both are applied to synthesis data and to cephalometric data with good results.

1 Introduction

DATA compression, reconstruction, estimation and de-noising are common applications of linear Principal Component Analysis (PCA) [1,2] and Kernel PCA [3,4]. In the latter case, this is a non-trivial task as the results provided by Kernel PCA live in some high dimensional feature space. The main problem of KPCA reconstruction and denoising scheme is to retrieve the data in the input space whose image in Kernel Space is known : in fact, every point of the kernel space does not have a pre image in the input space. This is the pre-image problem [3-6].

In this paper, the estimation of a partially observed object in the input space, using a model learned in the feature space F. is addressed. Some part of the observation is known. To solve this problem, spatial relationships between the known part of the observation and the unknown one are represented in a statistical model and used to localize the unknown part. Those relationships are automatically learned in the model. Like in KPCA reconstruction problem, there are two possible approaches to solve this problem.

The first one use an explicit mapping function φ, the second one use Kernel PCA making φ implicit. In the first case estimation consists in computing the inverse of φ

S. Singh et al. (Eds.): ICAPR 2005, LNCS 3686, pp. 390–399, 2005.

(step 2 in Fig. 1) : a global model (polynomial, sigmoid) of the relations is an a-priori knowledge in this case. In the second case the problem is much more complicate (step 5 in Fig. 1).

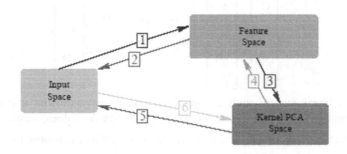

Fig. 1. Three different observations space

The paper is organized as follow : First, the extension of the PCA model to spatial relationship representation and partial object recognition is presented. Next, the KPCA model is described and the extension to partial object localization is given. Polynomial Kernels are detailed and results are illustrated with synthetic and real examples.

2 Linear PCA Model

The extension of the linear PCA model [7] defined here is an elegant way to take into account spatial relations between landmarks and can also estimate the unknown part of the partially visible or occulted model.

Principal Component Analysis is an orthogonal basis transformation, where the new basis is found by diagonalizing the covariance matrix of a dataset.

Let $T_i = (x_{i1}, x_{in}, ..., y_{i1}, y_{in}) \in \mathbf{R}^{2n}$, be the locations of n landmarks. Using PCA, we can write $T_i \approx T + \Phi b$, where T is the mean shape of the pattern, $\Phi = (\phi_1 | ... | \phi_t)$ is a $(n+m) \times (n+m)$ matrix composed with the eigenvectors of the covariance matrix S of the centered data and b is a vector of dimension t : $b = \Phi^t(T_i - \overline{T})$. The dimension t of the vector b is the number of eigenvectors with the largest eigenvalues.

In classical uses of PCA, such as de-noising, $t<n+m$ is chosen by $\sum_{i=1}^{t} \lambda_i \geq 0.95 \sum_{i=1}^{m+n} \lambda_i$. The vector b is then a good approximation of the original dataset and every vector T_i can be represented with the $t_{t<n+m}$ values of the vector b.

Under this hypothesis, if some points (says $t=n$ points) are known, the remaining unknown points can be determined using PCA. Without any approximations, we can write:

$$
\begin{bmatrix} C_1 \\ M \\ C_n \\ X_1 \\ M \\ X_m \end{bmatrix} = \begin{bmatrix} \overline{C}_1 \\ M \\ \overline{C}_n \\ \overline{C}_{n+1} \\ M \\ \overline{C}_m \end{bmatrix} + \begin{bmatrix} \Phi_{1,1} & L & \Phi_{1,n+m} \\ M & O & M \\ \Phi_{n+m,1} & L & \Phi_{n+m,n+m} \end{bmatrix} \begin{bmatrix} b_1 \\ M \\ b_n \\ b_{n+1} \\ M \\ b_{n+m} \end{bmatrix}
$$

This is a linear system with m equations and $n+m$ unknowns that can not be resolved. Since PCA can represent the dataset with $t < n+m$ values, suppose $t <= n$, the unknown vector $(b_1, K, b_n, X_1, K, X_m)$ can be estimated by the following system:

$$
\left\| \begin{bmatrix} C_1 - \overline{C}_1 \\ M \\ C_n - \overline{C}_n \\ -\overline{C}_{n+1} \\ M \\ -\overline{C}_{n+m} \end{bmatrix} - \begin{bmatrix} \phi_{1,1} & L & \phi_{1,t} & 0 & L & L & 0 \\ M & M & M & M & M & M & M \\ \phi_{n,1} & L & \phi_{n,t} & 0 & L & L & 0 \\ \phi_{n+1,1} & L & \phi_{n+1,t} & -1 & 0 & L & 0 \\ M & M & M & 0 & O & M & M \\ M & M & M & M & M & O & 0 \\ \phi_{n+m,1} & L & \phi_{n+m,t} & 0 & L & 0 & -1 \end{bmatrix} \begin{bmatrix} b_1 \\ M \\ b_t \\ X_1 \\ M \\ X_m \end{bmatrix} \right\|^2
$$

In this framework, a linear approximation of spatial relations between known and unknown points are explicitly determined from the eigenvectors of the covariance matrix.

3 KPCA Models

Kernel PCA can be considered as a natural generalization of linear PCA and is very well suited to extract interesting non-linear structures in the data. Closely related to methods applied in Support Vector Machines, it has proved useful for various applications, such as denoising and as a pre-processing step in regressions problems.

3.1 Kernel PCA and Reconstruction

Kernel PCA first map the data from an input space $\mathbf{I}$ into a feature space $\mathbf{F}$ via a (usually non-linear) function and then perform linear PCA on the mapped data. As the feature space $\mathbf{F}$ can be very high dimensional, kernel PCA employs Mercer kernels instead of carrying out the mapping explicitly such as Gaussian kernels $k(x, y) = \exp(-\|x - y\|^2 / c)$ and polynomial kernels $k(x, y) = (1 + x \bullet y)^d$.

Consider data vector x and y in the input space $\mathbf{I} = \mathbf{R}^{2n}$. The non-linear mapping $\Phi : \mathbf{R}^{2n} \to \mathbf{F}$ is defined such that : $\Phi(x) \bullet \Phi(y) \equiv k(x, y)$ where $\bullet$ is the vector dot product in the high dimensional feature space $\mathbf{F}$.

To perform PCA in feature space, we need to find Eigenvalues $\lambda > 0$ and Eigenvectors $V \in \mathbf{F}\backslash\{0\}$ satisfying $\lambda V = CV$ with $C = \left\langle \Phi(x_i)\Phi(x_i)^T \right\rangle$, the covariance matrix computed on the mapped data. Defining the NxN Kernel matrice K: $K_{ij} \equiv \Phi(x_i) \bullet \Phi(x_j)$, the problem becomes :

$$N \lambda \alpha = K \alpha \tag{1}$$

To extract non-linear principal components β_i of a test point $\overset{!}{x}$, the projection onto the k-th component is computed by:

$$\beta_k = (V^k \bullet \Phi(x)) = \sum_{i=1}^{N} \alpha_i^k k(x, x_i) \tag{2}$$

To reconstruct the Φ-image of a vector x from its projections β_k onto the first n principal component in $\mathbf{F}$ (assuming that the Eigenvectors are ordered by decreasing Eigenvalue size), a projection operator P_n is defined by

$$P_n \Phi(x) = \sum_{k=1}^{n} \beta_k V^k \tag{3}$$

When observations are not centered, the centered $\Phi(x)$ are used :

$$\Phi(x) \equiv \Phi(x) - \frac{1}{N} \sum_{i=1}^{N} \Phi(x_i) \quad \forall \overset{r}{x} \in \mathbf{R}^n. \tag{4}$$

In term of dot product, the Gramm matrix replaces the Kernel matrix:

$$K_{ij} = K_{ij} - \frac{1}{N} \sum_{p=1}^{N} K_{ip} - \frac{1}{N} \sum_{q=1}^{N} K_{qj} + \frac{1}{N^2} \sum_{p,q=1}^{N} K_{pq} . \tag{5}$$

3.2 Missing Data Estimation

The problem to solve is the reconstruction of partially unknown examples from the KPCA model and from the known part of the data.

Let $z = (c_1, \mathbf{K}, c_n, x_1, \mathbf{K}, x_m)$ be an example to reconstruct, with the n first coordinates known. The statistical model can be seen as some variability parameters (b in PCA model, β in KPCA model) around a mean shape. Finding the unknown part of x is equivalent to find the shape belonging to the model (i.e. variability parameters) whose first coordinates are given by the known part of x. However we are interested in an estimation in the input space $(x_1, x_2, \ldots x_m)$ rather than in feature space $(\beta_1, \beta_2, \ldots \beta_k)$. So the solution is given by a vector satisfying $P_n \Phi(c) = \Phi(z)$, which is the pre-image with $(x_1, x_2, \ldots x_m, \beta_1, \beta_2, \ldots \beta_k)$ as unknown. Remember that in the classical pre-image, the feature space coordinates $(\beta_1, \beta_2, \ldots \beta_k)$ are known.

When the vector has no pre-image z, the vector z, such as its image is the nearest one to the model, is found by minimizing $\rho(x) = \left\| \Phi(z) - P_N \Phi(c) \right\|^2$, i.e.

$$\rho(x) = \left\| \Phi(z) \right\|^2 - 2\left(\Phi(z) \bullet P_N \Phi(c) \right) + \left\| P_N \Phi(c) \right\|^2 \tag{6}$$

Using equations (2) and (3), kernel notation is introduced to obtain:

$$\rho(x) = k(z,z) - \sum_{k=1}^{N} \left(\sum_{i=1}^{L} \alpha_i^k k(c,x_i) \right) \left(\sum_{i=1}^{L} \alpha_i^k \left(2k(z,x_i) - k(c,x_i) \right) \right) \tag{7}$$

The projection of c and z on the KPCA space are the same :

$$\rho(x) = k(z,z) - \sum_{k=1}^{N} \left(\sum_{i=1}^{L} \alpha_i^k k(c,x_i) \right)^2 \tag{8}$$

This is the general case and minimize $\rho(z)$ depends upon the chosen kernel. This equation can be solved by numerical optimization, but this function presents in general a great number of local minima, sometimes numerically instable. Now, the paper is focused on the polynomial kernels.

3.3 Estimation for Polynomial Kernel

Let pose $z = (c_1, K, c_n, x_1, K, x_m)$ as the known part of z is the known part of x. For polynomial kernels, we have to minimize

$$\rho(z)_{\min} = (1 + x_c \bullet x_c + z_x \bullet z_x)^d - \sum_{k=1}^{N} \left(\sum_{i=1}^{L} \alpha_i^k (1 + x_c \bullet x_{ci} + z_x \bullet x_{xi})^d \right)^2 \tag{9}$$

which is a polynomial of degree 2d with m unknowns. The mapping φ is easily retrieved and is explained using a linear combination of monomial and dot product.

3.3.1 Polynomial Degree One
As the observation must be centered in the Feature space $k(x,y) = (x \bullet y)$. The mapping in this case is linear.

$$\rho(x)_{\min} = (c \bullet c + x \bullet x) - \sum_{k=1}^{N} \left(\sum_{i=1}^{L} \alpha_i^k (c \bullet x_{ci} + x \bullet x_{xi}) \right)^2 = C_{00} + \left\| x \right\|^2 - \sum_{k=1}^{N} \left(C_{0k} + C_{1k} \bullet x \right)^2$$

where $C_{0k} = \sum_{i=1}^{L} \alpha_i^k (c \bullet x_{ci}), C_0 = \left\| c \right\|^2, C_{1k} = \sum_{i=1}^{L} \alpha_i^k x_{xi}$

For an extremum, the gradient has to vanish, which lead to a necessary condition:

$$x = \frac{\sum_{k=1}^{N} C_{0k} C_{1k}}{1 - \sum_{k=1}^{N} \left(C_{1k} \right)^2} = \frac{\sum_{k=1}^{N} \left(\sum_{i=1}^{L} \alpha_i^k \left(c \bullet x_{ci} \right) \right) \left(\sum_{i=1}^{L} \alpha_i^k x_{xi} \right)}{1 - \sum_{k=1}^{N} \left(\sum_{i=1}^{L} \alpha_i^k x_{xi} \right)^2} \tag{10}$$

Not surprisingly, this is the classical PCA solution related in §II.

3.3.2 Polynomial Degree 2

The mapping φ is given by

$$\varphi(x) = (x_1, K, x_n, x_1^2, \sqrt{2}x_1x_{2}, K, \sqrt{2}x_1x_n, x_2^2, \sqrt{2}x_2x_{3}, K, x_n^2).$$

$$\rho(x)_{min} = C_{00} + \left\|\varphi_c(z)\right\|^2 - \sum_{k=1}^{N}\left(C_{0k} + C_{1k} \bullet \varphi_x(z)\right)^2$$

$$with\, C_{0k} = \sum_{i=1}^{L}\alpha_i^k(\varphi_c(z) \bullet \varphi_c(x_i))\; ;\; C_0 = \left\|\varphi_c(z)\right\|^2\; ;\; C_{1k} = \sum_{i=1}^{L}\alpha_i^k\varphi_x(x_i)$$

(11)

The gradient has to vanish:

$$\left(\varphi_x(z) \bullet \varphi_x(z) - \sum_{k=1}^{N}\left(C_{0k}C_{1k} \bullet \varphi_x(z) + \left(C_{1k} \bullet \varphi_x(z)\right)\left(C_{1k} \bullet \varphi_x(z)\right)\right)\right)$$

(12)

Finding the roots of this polynomial
is done by classical numerical method such as newton's one or brent'sone. Note that the solution must be close enough to the mean value of the model and between 3 times the eigenvalue around the mean. This is used as initial value and/or bracketed range. Finding the solution of the general equation (eq 9) give simultaneously the unknown input space data (unknown part of object) and the variability parameter β of the model.

4 Results

4.1 Synthesis Data

In this first experiment, a data set of three points (i.e. six values) is generated (fig 2). Three parameters are needed to perfectly describe these data, i.e. 3 is the theoretical optimal number of variability parameters for PCA and KPCA methods.

1. In these synthesis data, one point is a constant. It should be easily predicted, simply because its mean values is a constant.
2. Variations of a point through the examples are linear. Its trace on the figure 3 is a line. The linear PCA and the Kernel PCA should predict it with a good accuracy.
3. Variations of the last point through the examples are non linear. This point describes an ellipse. A first set uses a full ellipse (on the left side of figure 2) with some indetermination on y-axis when x is known. A second set uses a half ellipse (on the right side of figure 2) where this indetermination is missing.

Independent uniform noise is added to every position.

The PCA and KPCA models are trained on a set of 50 samples. The test set is composed of 300 samples.

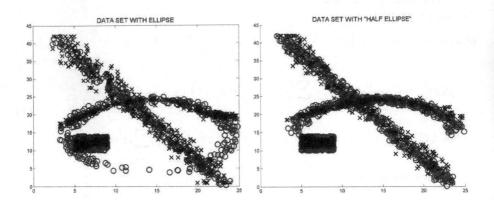

Fig. 2. 3 points with linear and non linear relationships. The circles are initial data set, crosses are reconstructed data set when 3 values are missing in an example.

In this experiment, the last value of each sample is suppressed and this missing data is estimated by our model. First, the value of the function to minimize (eq 12) has a clearly visible minima, and the width of this minima is the width of the added noise

Second, The error of this estimated unknown value is summarized in the table 1, with respect to the number of variability parameters retained, for 3 methods :

1. Polynomial Kernel minimization, described here : variability parameters and pre-image are simultaneously estimated.
2. Explicit second degree polynomial projection with PCA : variability parameters are first estimated, following by pre-image computation
3. Classical PCA

Table 1. Estimation error for a varying number of parameters

Variability Parameters	1	2	3	4	5
Kernel minimization	478.94	54.811	60.662	58.685	56.088
Polynomial function	505.1	598.09	592.09	816.96	962.14
PCA	3504.7	27639	19549	39786	4.952e+005

The results exhibit a large advantage to the non linear method : the non linear aspect of the data is well extracted and represented by these models. Linear PCA cannot deal with such non linear data. The second method, in which the variability parameters are first estimated and then the pre-image computed is less powerful than the use of the kernel trick and the estimation of the variability parameters and the unknown values in one step.

Another Comparison between linear and Kernel PCA can be achieved with the accuracy of the reconstructed points when the number of these reconstructed points. In the previous example, 3 parameters are needed to describe the data. So, 3 values can be retrieved by this method and the number of retained variability parameters varies from 1 to 3 in the linear model, from 1 to 9 in the non linear model. Figure 3 plots the error of global reconstruction when 3 values are missing, with the number of variability parameters used on the x-axis. It becomes clear that non linear method has a large advantage with a better reconstruction error, but with an increased computational cost because more parameters are used.

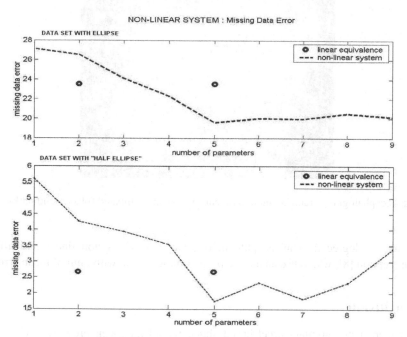

Fig. 3. Reconstruction error for 3 missing points

4.2 Real Data

The goal of cephalometry [2,8] is the study of the skull growth of young children in order to improve orthodontic therapy. It is based on the landmarking of cephalometrics points on tele-radiography, two dimensional X-ray images of the sagittal skull projection (figure 4). These points are used for the computation of features, such as the length or the angle between lines. The interpretation of these features is used to diagnose the deviation of the patient form from an ideal one. It is also used to evaluate the results of different orthodontic treatment. Cephalometric landmarks are linked to the shape of the cranial contour. In this context, the cranial contour is sampled and the landmarks are learned together with the sampled contour [9].

To landmark a new cephalogram, knowing the contour, the unknown part of the model (landmarks) has to be retrieved, with the statistical model and the known part (sampled contour). On these real data, linear PCA and KPCA give the same results, with 4mm of mean error between the real positions of the landmarks and the estimated landmarks. Note that intra-expert variability is about 1mn. This means that the data are non really non linear, or that the non-linearity in these data cannot be represented by a

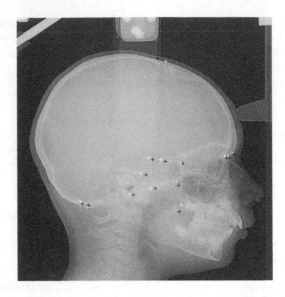

Fig. 4. cephalogram, cranial contour and real (white) and estimated (black) landmarks

polynomial of degree 2. This is quite more than a previous non linear and affine invariant version [8], which use an ad-hoc projection function, with 2mn of mean error.

5 Conclusion

In this paper, a polynomial kernel based shape model has been presented. This non linear model is used to resolve the problem of missing data in an image in a statistical framework. We found equation 12, which can be numerically solve in the general case. Shape parameters and missing data are then estimated. With polynomial kernel, we have to found the roots of a polynomial equation and the solution more robust. The polynomial kernel based model has been compared to classical linear PCA on synthetic and real data. When a non linear relationship exists between data, the kernel model has better accuracy than the linear one, with a larger computational cost.

References

1. T.F.Cootes, G.J. Edwards, C.J.Taylor. Active Appearance Models, IEEE PAMI, Vol. 23 (6), pp. 681-685, 2001.
2. T.J. Hutton, S. Cunningham, P. Hammond. An Evaluation of Active Shape Models for the Automatic Identification of Cephalometric Landmarks. European Journal of Orthodontics, Vol. 22(5), pp. 499-508, 2000.
3. S. Mika, B. Schölkopf, A.J. Smola, K.-R. Müller, M. Scholz, and G. Rätsch. Kernel PCA and de-noising in feature spaces. In M.S. Kearns, S.A. Solla, and D.A. Cohn, editors, Advances in Neural Information Processing Systems 11, pages 536-542. MIT Press, 1999.
4. B. Schölkopf, A. Smola et K. Müller. Non linear component Analysis as a Kernel Eignevalue Problem. Neural Computation, 10(5): 1299-1319, 1998.

5. J. T. Kwok et Ivor W. Tsang. The Pre-Image Problem in Kernel Methods. Proceedings of ICML 2003 : pp.408-415, 2003.
6. Bakir, G.H., J. Weston and B. Schölkopf: Learning to Find Pre-Images. Advances in Neural Information Processing Systems, 16, 449-456. (Eds.) Thrun, S., L. Saul and B. Schölkopf, MIT Press, Cambridge, MA, USA (2004)
7. S. Sclaroff, AP..Pentland, Modal Matching for Correspondence and Recognition. IEEE Transactions on Pattern Recognition and Machine Intelligence, 17(6):545-561, 1995.
8. B. Romaniuk, M. Desvignes, M. Revenu, M.J. Deshayes Linear and Non-Linear Model for Statistical Localization of Landmarks, ICPR, Vol. 4, pp. 393-396, 2002
9. B. Romaniuk, M. Desvignes, "Contour Tracking by Minimal Cost Path Approach. Application to Cephalometry", International Conference on Image Processing ICIP, Singapore, October 2004

Predictive Model for Protein Function Using Modular Neural Approach

Doosung Hwang[1], Ungmo Kim[2], Jaehun Choi[3], Jeho Park[1], and Janghee Yoo[3]

[1] Department of Computer Science, Dankook University,
San 29, Anseo-dong, Cheonan-si, Chungnam, 330-174, Korea
{dshwang, dk_jhpark}@dankook.ac.kr
[2] Department of Computer Engineering, Sungkyunkwan University,
ChunChun-dong 300, JangAn-Gu, Suwon, Kyounggi, 440-746, Korea
umkim@yurim.skku.ac.kr
[3] Electronics and Telecommunications Research Institute,
161 Kajong-Dong, Yuseong-Gu, Daejeon 305-350, Korea
{jhchoi, jhy}@etri.re.kr

Abstract. As interest within bioinformatics has been vastly increased, efforts to predict functional role of proteins have been made using diverse approaches. In this paper, we discuss a protein function prediction method that utilizes protein molecular information including protein interaction data. The proposed method takes the given problem into account as a K-class classification problem and resolves the new problem by using a modular neural network based predictive approach. The simulation demonstrates that the proposed approach predicts the functional roles of Yeast proteins with unknown functional knowledge and is competitive to the other methodologies in KDD Cup 2001 competition.

1 Introduction

The potential value of the knowledge in bioinformatics is perceived as promising by different application areas. This is accelerated with the completion of genome sequencing of several target organisms. Among the various related work, the functional annotation of the proteins is one of the most challenging tasks. Utilizing the bioinformatics information disseminated out in the Internet, the biology related research laboratories published massively new experimental results, updated the existing knowledge and constructed the new relationships between known facts. Accessing the countless available information, various computational methods for protein function prediction have been studied using such data as phenotype [2], gene expression [12], motif [18,19], and protein-protein interaction [5,11,16]. The function prediction methods can be categorized according to training data that they used. Graph-based methods [4,7] utilizes protein-protein interaction data for graph construction while machine learning methods [2,19] take protein-related knowledge as input to the methods. A graph-based approach may fail to define protein function if a protein containing known functions interacts directly or indirectly to proteins without functions. A learning methods

S. Singh et al. (Eds.): ICAPR 2005, LNCS 3686, pp. 400–409, 2005.

generate the prediction rules that are used to assign functional roles to proteins that are not bound with known functions.

In KDD Cup 2001 competition [9], various learning algorithms were challenged for high performance in respect of function prediction problem of Yeast proteins. The performance of neural network was however not comparable to the other methods. Here our main objective is to propose a model for protein function prediction that can be competitive to the winning approaches in the KDD Cup.

In order to fulfill our objective, perceiving the aspect that a protein can contain multiple functions, we take into account the given task as a K-class classification problem. Each decomposed subproblem is then resolved by a modular neural network [3,17]. It is generally accepted that a modular neural network improves learning speed and flexible in data representation over a case of a single neural network for a multi-class classification problem. By doing this, a modular neural network's decomposability makes it possible to accommodate smooth learning of the mapping mechanism implicated in a subproblem. In addition, the training data set used for a modular neural network can be easily understood than the much larger original data set. The class determined by a subproblem covers a local region of the pattern space over which a modular neural network can proceed its learning phase. As a binary or 2-class classifier, a modular neural network with backpropagation algorithm displays fast learning and good generalization. However, in order to exploit the advantages of a modular neural network, the additional works such as the class data analysis, problem decomposition, and output combination are quite demanding.

The rest of this paper is organized as follows. Section 2 describes the application of modular neural networks to multi-class classification problems. The details of data preparation is followed in Section 3. Section 4 shows the experimental results in case of machine learning and graph-based approaches. Finally, a summary of this work is given in Section 5.

2   Modular Neural Network

In oder to resolve the K-class classification problem with a neural network, it is demanded to decompose the K-class problem into a set of 2-class subproblems. Each 2-class subproblem is then addressed by using a modular neural network. Therefore the K modular neural networks are constructed for a K-class problem and a modular network M_i models a classifier for class C_i. As like a neural network requires a training data set, each modular neural network also demands distinct data preparing process that plays a partial role of data preparation for the original K-class classification problem.

A modular neural network needs its own topology according to the complexity of the subproblem and utilizes the training set C_i for its learning. C_i includes the data of positive and negative classes, C_i^+ and C_i^-. During the data preparation of a modular neural network M_i, if a training data e is included in both classes of C_i^+ and C_i^-, the replicated data contained in the negative class should

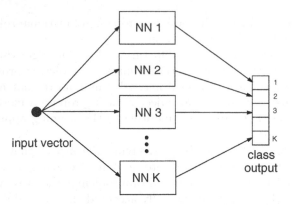

Fig. 1. The structure of the modular neural network for a K-class problem

be eliminated in order to avoid class interference problem. Unlike the general problems that can be resolved nicely by exploiting a neural network, the class interference problem in our case prohibits generation of plausible results that can be achieved with a neural network. We conjecture that e is considered as only in C_i^+ for learning M_i resulting in preventing interference problem. Figure 1 shows the architecture of modular neural networks for a K-class classification problem. An input test data is fed to all modular neural networks and the outputs of the modular neural networks are combined for the global classification. Each modular neural network is constructed by using a multi-layered architecture in order to attack the non-linear classification problem. Our approach is based upon a co-operative modular neural network according to the types of combining modular neural networks [3].

In a learning phase for a modular neural network M_i, the data sizes of C_i^+ and C_i^- are not equal because the set of negative data is the total of all C_k, $k = 1, \cdots, K$, $k \neq i$. This occurs due to that the members of C_i^+ can be detected with little problem, however the identification of members in C_i^- needs to be considered in the scope of the overall data set including the set C_j^+, $j = 1, \cdots, i, i + 1, \cdots, K$. A multi-layered neural network trained on imbalanced data may not learn to discriminate properly between classes [14]. In order to balance the data sizes of positive and negative data, the oversampling strategy of SMOTE[1] [15] is applied to the modular neural network. The synthetic data that are generated by SMOTE enables the modular neural network to learn large and broad decision regions rather than small and specific regions. If the sampling strategies are used in learning a single network for an imbalanced multi-class classification problem, it is difficult to equally approximate the number of present data in classes due to the existence of multiple classes. But if a problem is restricted to two classes, it is easy to equally approximate the number of present data. Therefore, the sampling strategies could be effective in the modular approach of a multi-class classification problem. Figure 2 describes

---

[1] Synthetic Minority Over-sampling Techniques.

the learning phase for a K-class problem. *Sample(E, N)* returns N synthetic data generated from class data E by SMOTE. In order to proceed the learning phase, a training data set needs to be carefully prepared.

Build_Modular_Network(TE, K, α)
> *TE is a set of training data with a pair of $(\mathbf{x}, t)$ where $\mathbf{x}$ is the network input and t is a class data($t \in \{1, \cdots K\}$). K is the number of classes in TE. α is the oversampling rate as a real value.*

for $k = 1, \cdots, K$ do
> 1. prepare the training data E_k for class C_k
>   - add $(\mathbf{x}, 1)$ to E_k^+ if $(\mathbf{x}, t) \in TE$ and $t = k$
>   - add $(\mathbf{x}, 0)$ to E_k^- if $(\mathbf{x}, t) \in TE$ and $t \neq k$
>   - delete $(\mathbf{x}, 0)$ from E_k^- if $(\mathbf{x}, 1) \in E_k^+$ and $(\mathbf{x}, 0) \in E_k^-$
> 2. oversample a set of synthetic data from the subordinate class
>   - if $\mid E_k^+ \mid \ll \mid E_k^- \mid$, add *Sample(E_k^+, $\alpha \times \mid E_k^+ \mid$)* to E_k^+
>   - if $\mid E_k^- \mid \ll \mid E_k^+ \mid$, add *Sample(E_k^-, $\alpha \times \mid E_k^- \mid$)* to E_k^-
>   - update $E_k = E_k^+ + E_k^-$
> 3. learn a modular neural network M_k
>   - construct M_k with hidden neurons
>   - train M_k with E_k

Fig. 2. The learning phase of a modular neural network approach for a K-class problem

3   Data Preparation

Currently, the target database used for our system is KDD Cup 2001 that contains information about 1,243 Yeast proteins. The database consists of four tables: a protein relation and an interaction table for both training and test sets. The databases are filled with unordered category information that has null values in many cases. In the protein relation table, attribute values are observable characteristics of an Yeast protein such as protein(or gene name), essential, class, complex, phenotype, motif, chromosome number, function, and localization. The interaction table explains an interaction between two proteins so that if an interaction exists between two proteins, the fact is represented by interaction type and expression correlation.

As a neural network cannot handle categorical values directly, those values needs to be transformed into vector values. Here, regarding a neural network's performance, we devise an encoding scheme for vector values that has characteristics such as low dimensionality, uniqueness among values, and reasonable distance between two values. In order to satisfy these conflicting requirements, the main ideas are focused on enumeration of attribute values and binary encoding. During enumeration phase, all category information except meaningless values such as "?" or 0 are enumerated. As a result, an unique category information obtains an unique enumeration value. In the next phase, the enumeration values are transformed into binary code by using our own encoding scheme. The point of the encoding scheme is bit-map style transformation considering the largest number for each digit with different weight. This transformation guarantees that at least two bits in the resulted vector implemented as a bitmap are different. As a result, the distance between two points in a vector space is much larger. For instance, a category class has 23 constituent members, the last category member is mapped to a value 23. We here need three bits for the three possible members at the first digit: 0, 1, and 2. Similarly, the second digit needs ten bits from 0 to 9. Hence, the total length of the resulting bitmap has 13 bits and the actual code is encoded as (0 0 1 0 0 0 1 0 0 0 0 0 0) by setting as 1 for the third position among the first three bits and the forth position among the following ten bits. Null attribute values are transformed as all zeros.

By using this coding scheme, it needs 4 bits for 4 essentials, 13 for 23 classes, 16 for 51 complexes, 12 for 12 phenotypes, 23 for 235 motif and 12 for 16 chromosomes. In our analysis over the data, a protein can have multiple values for a single attribute. Moreover, we found 13 functions in the training data and 14 functions in the testing data. For simplicity, we decided to use 13 bits to represent protein functions. If a protein demonstrates multiple functions, more than one bit is set as 1's in its function vector. When the transformation from category information into bit-maps, all tables are linked by protein name.

The prediction of protein function is generally based on the biological fact: two proteins that directly interact are likely to be involved in the similar biological function [6]. By utilizing the hint, the explicit relationship between two interacting proteins is represented in data preparation phase. A binary interaction is simply a symmetrical relationship. However, repeatedly applied transitive relation increases the size of the interaction relation rapidly. In order to avoid a large set of training data caused by representation of transitive relation, we multiply two expression correlations[2] by considering only the transitive relationship within interaction depth 4 and add 1,301 new interactions of expression correlation greater than 0.5. Being based on the abovementioned scheme, the performance of our proposed method is implemented and run experiments.

4   Experiments

We implemented our modular neural network system as a 3-layer feed-forward network with sigmoid activation function. The resilient backpropagation [13] is used for learning a modular neural network. The training and testing data sets are created in a binary vector by the encoding scheme. The size of training data is 6,395 without duplicates. The scheme of training data table consists of essential, class, complex, phenotype, motif, and function in form of binary vector. The negative class data E_i^- data of class f_i is generated with a condition of $function \neq f_i$ and class value 0. Some attributes can found in both the positive and negative set so that it might cause interference within learning phase of a neural network. In order to avoid such a situation, for each data $\mathbf{x} \in E_i^+$ in the

---

[2] An expression correlation is the correlation between gene expression patterns for the two genes. A correlation far from 0 implies that these genes are likely to influence one another strongly.

positive set, the same data in the negative set($\mathbf{x} \notin E_i^-$) is eliminated from the set during the preparation phase.

The class data sets for training are generated from the overall training data set. The number of training data differs highly in positive and negative data. The imbalance rate was 2.14% to 32.1%[3]. The total is not equal because an input vector has multiple target classes and the problem is a K-class classification problem.

The simulation environment uses the necessary parameters: 10^{-3} for the minimum performance gradient, 50.0 for the maximum weight change, 1.2 for increment weight changes and 0.5 for decrement weight change. If the neuron output is greater than 0.95, the output value is 1 . The total error is measured with sum of square error and the performance of each trial is calculated by a confusion matrix. The network topology is $68 \times 82 \times 13$. The training vector does not have chromosome information. The learning rate is set as 0.1 and the *sum-of-square* error is 10^{-3}. The total of iterates is 20,000.

Table 1 is the result of the single neural network approach with 3-fold cross-validation. The 20% of the training data are used for a cross-validation. The performance of the training and cross-validation is around 91.0% while that of the testing set is 87.0%.

Table 1. The simulation results of the single neural network

Experiment	1	2	3
Training(%)	91.7	91.2	91.4
Cross-validation(%)	91.8	91.1	91.3
Testing(%)	87.3	87.0	87.8

Regarding the simulation environments for the proposed modular approach, the topology of a binary classifier is set as $68 \times 16 \times 1$. The learning rate is 0.01 and the error goal is 10^{-3}. The total of iterations is 2,000. As expected, the application of a single neural network took about two hours at learning phase while the modular approach spent one hour for learning.

Table 2 shows the results of the modular neural network approach for each class with 10-fold cross validation. The 100% and 200% of the subordinate classes are added for the simulations. The experiments without sampling shows that the accuracy of the training set is close to 100.0% and that of the cross-validation is ranged within 94.39% to 99.0%. According to this result, we can be sure that the proposed approach is more effective than a single neural network. The sampling strategy improves the accuracy in the training set and the cross-validation set, but not in case of the test set.

Table 3 demonstrates the performance comparing to the other approaches in the competition of the KDD Cup Yeast data. From the results, we can see clearly that the proposed modular neural approach is competitive to the the

---

[3] $\min \left(\mid E_i^+ \mid, \mid E_i^- \mid\right) / \max \left(\mid E_i^+ \mid, \mid E_i^- \mid\right) \times 100$.

Table 2. The simulation results of the modular approach(KDD Cup data)

M	Data set	Original	100%	200%
1	Training	99.60 ± 0.10	99.71 ± 0.10	99.67 ± 0.06
	Cross-validation	94.39 ± 1.43	95.57 ± 0.73	96.89 ± 1.35
2	Training	99.70 ± 0.09	99.79 ± 0.06	99.72 ± 0.09
	Cross-validation	96.99 ± 0.83	97.33 ± 0.94	96.82 ± 1.17
3	Training	99.70 ± 0.10	99.67 ± 0.12	99.78 ± 0.07
	Cross-validation	97.71 ± 0.64	98.11 ± 0.69	98.45 ± 0.61
4	Training	99.23 ± 0.20	99.59 ± 0.09	99.80 ± 0.05
	Cross-validation	97.48 ± 2.99	97.81 ± 0.98	97.99 ± 0.95
5	Training	99.23 ± 0.20	99.37 ± 0.24	99.23 ± 0.29
	Cross-validation	97.48 ± 2.99	97.86 ± 0.95	97.06 ± 1.59
6	Training	99.46 ± 0.09	99.60 ± 0.11	99.62 ± 0.06
	Cross-validation	97.56 ± 0.90	98.03 ± 0.70	98.13 ± 0.76
7	Training	99.78 ± 0.02	99.81 ± 0.03	99.75 ± 0.06
	Cross-validation	99.16 ± 0.45	99.33 ± 0.45	99.42 ± 0.37
8	Training	99.71 ± 0.05	99.63 ± 0.09	99.62 ± 0.10
	Cross-validation	99.16 ± 0.27	99.23 ± 0.55	99.21 ± 0.44
9	Training	99.69 ± 0.09	99.79 ± 0.08	99.82 ± 0.06
	Cross-validation	95.78 ± 0.86	96.88 ± 1.23	96.99 ± 0.62
10	Training	99.61 ± 0.12	99.71 ± 0.08	99.71 ± 0.08
	Cross-validation	96.68 ± 1.02	97.69 ± 0.85	97.43 ± 1.01
11	Training	99.71 ± 0.05	99.73 ± 0.09	99.65 ± 0.10
	Cross-validation	99.42 ± 0.24	99.26 ± 0.23	99.18 ± 0.32
12	Training	99.63 ± 0.09	99.74 ± 0.08	99.79 ± 0.07
	Cross-validation	96.55 ± 0.81	97.71 ± 0.80	97.14 ± 1.17
13	Training	99.68 ± 0.05	99.70 ± 0.10	99.78 ± 0.05
	Cross-validation	99.13 ± 0.22	99.30 ± 0.35	99.46 ± 0.47

Table 3. The performance comparisons(KDD Cup data)

	TP	TN	FP	FN	Acc(%)	no.
SVM	690	4,304	58	282	93.6	381
ICP	654	4,264	90	326	92.2	381
MNN	668	4,295	84	287	93.0	381
2-NN	535	3,265	249	213	89.1	336
χ^2	432	3,234	428	284	83.9	336

SVM(Support Vector Machine) and ICL(Inductive Classification Logic, [10]). Moreover, the modular approach outperforms k nearest–neighbor($k = 2$) and χ^2-statistic methods. The nearest–neighbor [5] and χ^2-statistic [8] approaches utilize protein interaction data only in predicting protein functions. The disadvantages of the nearest–neighbor and χ^2-statistic approaches is that those methods can not assign functions to proteins without interaction with proteins with known functions. The SVM approach was the winner with the accuracy of 93.6 %. The best performance of our approach was 93.0 % with a sampling strategy of 200%.

Table 4. The performance comparisons(MIPS Yeast data)

Test	RS	EQ-1	EQ-2	EQ-3
TP	749	1,258	1,273	1,242
TN	6,205	7,819	8,122	7,917
FP	420	94	97	101
FN	150	1,488	1,547	1,646
Tot	7,524	10,659	11,039	10,906
Acc(%)	92.42	85.16	85.11	83.98

The proposed approach also applied to the training and testing sets created from the MIPS Yeast database [1]. Compared to the KDD Cup data, the MIPS data has the different numbers of categorical values in protein attributes: 4 essentials, 25 classes, 69 complexes, 12 phenotypes, 594 motifs and 17 chromosomes. The number of functional roles is 19. The topology of a modular neural network is $72 \times 24 \times 1$ for each class. The total of proteins is 3,510 and the interaction data is not used. A test set is prepared by random selection(RS), equal selection(ES) per functional class. In equal selection, we prepare three different sets for test data. The 10 % of class data is chosen for test if the data size is greater than 100, otherwise the 25 % of the function class data is selected. The number of proteins is 396 for RS, 561 for ES-1, 581 for ES-2, and 574 for ES-3. The training set for each test set was prepared except its test instance. In the simulation, the accuracy is around 99.0% for the training set and 97.0% for the cross–validation set. The test accuracy is shown in Table 4 by a confusion matrix[4]. The performance is 92.42 % for RS, 85.16 % for EQ-1, 85.11 % for EQ-2, and 83.98 % for EQ-3. The good result of RS might occurs due to the data scarcity of 5 classes. The average of the ER sets is around 85.0 %. The simulation demonstrates that the proposed approach is adaptable in predicting the functions of Yeast protein without interaction data.

5   Conclusion

For the prediction of protein function, various approaches have been discussed and analyzed as the potential value of bioinformatics increases rapidly. The application of a neural network to the prediction of protein function had showed lower performance than the other methods. In this paper, we showed our efforts to apply a neural network to the problem taking advantage of its high quality classification capability. Our approach starts with taking into the problem as a $K-$class classification problem. Each decomposed subproblem is then resolved by applying a modular neural network that can avoid learning interference among classes. The modular neural network in our system were designed based on multi-layered architecture in order to resolve non-linear problem that

---

[4] True Positive(TP), True Negative(TN), False Positive(FP), False Negative(FN), Accuracy(Acc)

contains complex classification boundary. The imbalance problem caused in the modular network was addressed by utilizing oversampling strategy. The experiments demonstrate that the proposed approach can predict the functional roles of Yeast proteins with unknown functional knowledge. More importantly, our approach is competitive to the other methodologies in KDD Cup 2001 data.

We plan to use this framework with the improvement of the modular neural network and apply to other organisms. In real situation, there exists many missing values in protein-related information. This problem was not resolved completely or could not find reasonable compensation method. The method that manipulates the missing values should be studied in respect of biological context and computational model as well for robust protein function classification.

References

1. http://mips.gsf.de/proj/yeast/.
2. A. Clare and R. D. King . Machine learning of functional class from phenotype data. *Bioinformatics*, 18:160–166, 2002.
3. A. J. C. Sharkey. *Combining Artificial Neural Nets: Ensemble and Modular Multi-Net Systems.* Springer, 1999.
4. A. Vazques, A. Flammini, A. Maritan, and A. Vespignani. Global protein function prediction in protein-protein interaction networks. *Nature Biotechnology*, 21(697), 2003.
5. B. Schwikowski, P. Uetz, and S. Fields. A network of protein–protein interactions in yeast. *Nature Biotechnology*, 18(3):1257–1261, December 2000.
6. C. L. Tucker, J. F. Gera, and P. Uetz. Towards an understanding of complex protein networks. *TRENDS in cell biology*, 11(3):102–106, May 2001.
7. P. Uetz *et. al.* A comprehensive analysis of protein–protein interactions in *saccharomyces cerevisiae*. *Nature*, 403(10):623–627, February 2000.
8. H. Hishigaki, K. Nakai, T. Ono, T. Tanigami, and T. Takagi. Assessment of prediction accuracy of protein function from protein–protein interaction data. *Yeast*, 18:523–531, 2001.
9. J. Cheng, C. Hatzis, H. Hayashi, M. A. Krogel, S. Morishita, D. Page, and J. Sese. KDD Cup 2001 report. *SIGKDD Exploration*, 3:47–64, 2001.
10. W. Van Laer. *From Propositional to First Order Logic in Machine Learning and Data Mining - Induction of first order rules with ICL.* PhD thesis, Department of Computer Science, K.U.Leuven, Leuven, Belgium, jun 2002. 239+xviii pages.
11. M. Fellenberg, K. Albermann, A. Zollner, H. W. Mewes, and J. Hani. Integrative Analysis of Protein Interaction Data. In R. Altmann, T.L. Bailey, P. Bourne, M. Gribskov, T. Lengauer, I.N. Shindyalov, L.F. Ten Eyck, and H. Weissig, editors, *Intelligent Systems for Molecular Biology*, volume 8, pages 152–161. AAAI Press, 2000.
12. M. P. S. Brown, W. N. Grundy, D. Lin, N. Cristianini, C. Sugnet, T. S. Furey, M. A. Jr., and D. Haussler. Knowledge-based analysis of microarray gene expression data using support vector machines. *Proceedings of the National Academy of Sciences*, 2000.
13. M. Riedmiller and H. Braun. A direct adaptive method for faster backpropagation learning: The RPROP algorithm. In *Proc. of the IEEE Intl. Conf. on Neural Networks*, pages 586–591, San Francisco, CA, 1993.

14. N. Japkowicz and S. Stephen. The Class Imbalances: A Systematic Study. *IDA Journal*, 6(5):429–449, 2002.
15. N. V. Chawlar, K. W. Bowyer, L. O. Hall, and W. P. Kegelmeyer. SMOTE: Synthetic Minority Over–sampling Techniques. *Journal of Artificail Intelligence Research*, 16:321–357, June 2002.
16. S. Oliver. Guilt-by-association goes global. *Nature*, 403(6770):601–603, 2000.
17. S. Haykin. *Neural Network: A Comprehensive Foundation*. Prentice Hall, 1998.
18. Xiangyun Wang, Diane Schroeder, Dreana Dobbs, and Vasant Honavar. Automated data-driven discovery of motif–based protein function classifiers. *Information Sciences*, 155:1–18, 2003.
19. Xinghua Lu, Chengxiang Zhai, Vanathi Gopalakrishnan, and Bruce G Buchanan. Automatic annotation of protein motif function with Gene Ontology terms. *BMC Bioinfotmatics*, 5(122), 2004.

Using *k*NN Model for Automatic Feature Selection

Gongde Guo[1], Daniel Neagu[1], and Mark T.D. Cronin[2]

[1] Department of Computing, University of Bradford, Bradford, BD7 1DP, UK
{G.Guo, D.Neagu}@Bradford.ac.uk
[2] School of Pharmacy and Chemistry, Liverpool John Moores University, L3 3AF, UK
M.T.Cronin@Livjm.ac.uk

Abstract. This paper proposes a *k*NN model-based feature selection method aimed at improving the efficiency and effectiveness of the ReliefF method by: (1) using a *k*NN model as the starter selection, aimed at choosing a set of more meaningful representatives to replace the original data for feature selection; (2) integration of the Heterogeneous Value Difference Metric to handle heterogeneous applications – those with both ordinal and nominal features; and (3) presenting a simple method of difference function calculation based on inductive information in each representative obtained by *k*NN model. We have evaluated the performance of the proposed *k*NN model-based feature selection method on toxicity dataset Phenols with two different endpoints. Experimental results indicate that the proposed feature selection method has a significant improvement in the classification accuracy for the trial dataset.

1 Introduction

The success of applying machine learning methods to real-world problems depends on many factors. One such factor is the quality of available data. The more the collected data contain irrelevant or redundant information, or contain noisy and unreliable information, the more difficult for any machine learning algorithm to discover or obtain acceptable and practicable results. Feature subset selection is the process of identifying and removing as much of the irrelevant and redundant information as possible. Regardless of whether a learner attempts to select features itself, or ignores the issue, feature selection prior to learning has obvious merits [4]:

(1) Reduction of the size of the hypothesis space allows algorithms to operate faster and more effectively.
(2) A more compact, easily interpreted representation of the target concept can be obtained.
(3) Improvement of classification accuracy can be achieved in some cases.

Feature selection methods are commonly divided into two broad categories: wrapper methods [6] and filter methods [2]. Wrapper methods usually employ a statistical re-sampling technique using the actual target learning algorithm to estimate the accuracy of feature subsets. The wrapper model tends to give superior performance as it finds features better suited to the predetermined learning algorithm. The main problem of this approach is the relative low efficiency especially for large

S. Singh et al. (Eds.): ICAPR 2005, LNCS 3686, pp. 410–419, 2005.
© Springer-Verlag Berlin Heidelberg 2005

datasets and the dependency on the learning algorithm. On the other hand, filters operate independently of any learning algorithm. They make use of all the available training data only when commencing feature selection. When the training data become very large, the filter model is usually a good choice due to its computational efficiency and neutral bias toward any learning algorithm [9]. Many feature selection algorithms [12, 17, 18] have been developed to answer challenging research issues: from handling a huge number of instances, large dimensionality, to deal with data without class labels.

The aim of this study was to investigate an optimised approach for feature selection, termed *k*NNMFS (*k*NN Model-based Feature Selection). This augments the typical feature subset selection algorithm ReliefF [8]. The resulting algorithm was run on toxicity data for phenols to assess the effect of reduction of the training data.

2   Related Work

The basic concept of *Relief* was introduced initially by Kira et al. in 1992 [7]. It is a feature weight-based algorithm inspired by instance-based learning algorithms. It estimates the quality of features according to how well their values distinguish between the instances of the same and different classes that are near each other [9]. The Relief family of algorithms e.g. Relief, ReliefF and RReliefF, are feature subset selection methods that are applied in a pre-processing step before the model is learned, and are amongst the most successful algorithms [10]. The majority of heuristic measures for estimating the quality of the attributes assumes the conditional independence (upon the target variable) of the attributes and is therefore less appropriate in problems which possibly involve much feature interaction. Relief algorithms do not make this assumption. They are efficient, aware of the contextual information, and can estimate the quality of attributes correctly in problems with strong dependencies between attributes [5].

The Relief algorithm works by randomly sampling an instance and locating its nearest neighbour from the same and opposite class. The values of the features of the nearest neighbours are compared to the sampled instance and used to update the relevance scores for each feature. This process is repeated for a user specified number of instances. The rationale is that a useful feature should differentiate between instances from different classes and have the same value for instances from the same class [5]. A major limitation of Relief is that it does not help with reducing redundant features. Another limitation is that it works only for binary classes. These drawbacks are overcome by ReliefF.

ReliefF, an extension of Relief, aims to solve the problem of datasets with multi-class, noisy and incomplete data. It smoothes the influence of noise in the data by averaging the contribution of k nearest neighbours from the same and opposite class of each sampled instance, instead of the single nearest neighbour of Relief, which ensures greater robustness of the algorithm with regards to noise. The user-defined parameter k controls the locality of the estimates. Multi-class datasets are handled by finding the nearest neighbours from each class that are different from the current sampled instance, and weighing their contribution by the prior probability of each class estimated from the training data [8].

One major drawback of ReliefF comes from its limitation to deal with the problem of multi-valued attributes. ReliefF uses a numerical-oriented method as a similarity measurement. It uses a simple strategy to cope with categorical attributes by assigning 0 to the difference function for categorical data with the same value, and 1 for categorical data with different values. This simple strategy cannot, however, measure the contribution of each discrete value of a categorical attribute to the class label appropriately. Another drawback of ReliefF is the setting of a suitable number of instances sampled from the dataset. The number of randomly selected instances, is usually set empirically, or set to the number of the entire dataset.

RReliefF is a further extension of ReliefF to deal with regression problems, where the predicted value is continuous, and therefore nearest hits and misses cannot be used. To solve this difficulty, instead of requiring the exact knowledge of whether two instances belong to the same class or not, a kind of probability that the predicted values of two instances are different is introduced [11]. This probability can be modelled with the relative distance between the predicted (class) values of two instances.

FSSMC (Feature Selection via Supervised Model Construction) [5] is an attempt to deal with the problems facing ReliefF described above. FSSMC chooses a set of more meaningful representatives (the centre of each cluster) to replace the whole dataset and serves as the basis for further feature selection. As the number of chosen representatives can be reduced to only 10 percent of the original datasets on average [3], it is computationally faster than ReliefF. Moreover, it applies a frequency based encoding scheme to transform categorical data to numerical data to cope with the multi-valued attributes.

Although FSSMC improves the computational efficiency compared to ReliefF, there is no significant classification accuracy improvement of FSSMC on most trial datasets [5]. The problem probably is the manner in which it randomly chooses seeds for grouping clusters, thus generating a set of less optimal representatives for feature selection. Moreover, noise in the data will affect the generated representatives both in quality and quantity, e.g. the number of representatives.

The basic idea of kNN model-based classification method (kNNModel) [3] is to find a set of more meaningful representatives of the complete dataset to serve as the basis for further classification. Each chosen representative d_i is represented in the form of $<Cls(d_i), Sim(d_i), Num(d_i), Rep(d_i)>$ which respectively represents the class label of d_i; the similarity of d_i to the furthest instance among the instances covered by N_i; the number of instances covered by N_i; a representation of instance d_i. The symbol N_i represents the area that the distance to d_i is less than or equal to $Sim(d_i)$. kNNModel can generate a set of optimal representatives via inductively learning from the dataset.

3 kNN Model-Based Feature Selection

3.1   A Modified kNN Model

For the convenience of difference function calculation between any two representatives and making use of the generated information in the final model of kNNModel for further feature selection, we make a slight change of the original *k*NNModel by adding some inductive information, e.g. the nearest neighbour and the furthest neighbour covered by a representative. Therefore, each representative in the modified *k*NNModel is represented as $<Cls(d_i), Sim(d_i), Num(d_i), Rep(d_i), Rep(d_{i1}), Rep(d_{i2})>$, in which the additive information such as $Rep(d_{i1})$, $Rep(d_{i2})$ respectively represent the nearest neighbour and the furthest neighbour covered by this representative. As there is not significant change between the original and the modified *k*NNModel, we still called the modified *k*NNModel as *k*NNModel for convention. Moreover, the Heterogeneous Value Difference Metric (*HVDM*) similarity measure [15], instead of the numerical-oriented methods, is used in *k*NNModel to deal with the multi-valued attributes problem.

The modified *k*NNModel method is described as follows:

Algorithm kNNModel
Input: the entire training data D, parameter ε
Output: a set of automatically generated representatives M

1. *For a given similarity measure, create a similarity matrix from a given training set D*
2. *Set to "ungrouped" the tag of all instances and set M=∅*
3. *For each "ungrouped" instance, find its local ε-neighbourhood*
4. *Among all the local ε-neighbourhoods obtained in step 3, find its global ε-neighbourhood N_i. Create a representative $<Cls(d_i), Sim(d_i), Num(d_i), Rep(d_i), Rep(d_{i1}), Rep(d_{i2}))>$ into M to represent all the instances covered by N_i, and then set to "grouped" the tag of all the instances covered by N_i*
5. *Repeat step 3 and step 4 until all the instances in the training set have been set to "grouped"*
6. *Model M consists of all the representatives collected from the above learning process.*

Fig. 1. Pseudo code of the modified *k*NNModel algorithm

In the algorithm above, A *neighbourhood* of a given instance is defined as the set of nearest neighbours around this instance; A *local neighbourhood* is a neighbourhood which covers the maximal number of instances with the same class label; A *local ε-neighbourhood* is a neighbourhood which covers the maximal number of instances with the same class label except for allowing ε exceptions and a *global ε-neighbourhood* is a local ε-neighbourhood which covers the largest number of instances among a set of obtained local ε-neighbourhoods.

3.2 *k*NN Model-Based Feature Selection

A *k*NN model-based feature selection method, *k*NNMFS is proposed in this study. It takes the output of the modified *k*NNModel as seeds for further feature selection. Given a new instance, *k*NNMFS finds the nearest representative for each class and then directly uses the inductive information of each representative generated by *k*NNModel for feature weight calculation. This means the *k* in ReliefF is varied in our algorithm. Its value depends on the number of instances covered by each nearest representative used for feature weight calculation.

The detailed *kNNMFS* algorithm is described as follows:

Algorithm kNNMFS
Input: the entire training data D and parameter ε.
Output: the vector W of estimations of the qualities of attributes.

1. *Set all weights $W[A_i]=0.0$, $i=1,2,...,p$;*
2. *$M:=kNNModel(D, \varepsilon)$; $m=|M|$;*
3. *for $j=1$ to m do begin*
4. *Select representative $X_j=<Cls(d_j)$, $Sim(d_j)$, $Num(d_j)$, $Rep(d_j)$, $Rep(d_{j1})$, $Rep(d_{j2})>$ from M*
5. *for each class C $Cls(d_j)$ find its nearest miss (C) from M;*
6. *for $k=1$ to p do begin*
7.

$$W[A_k] = W[A_k] - \left(diff(A_k,d_j,d_{j1}) + diff(A_k,d_j,d_{j2}) \times \frac{Sim(d_j)}{Num(d_j)}\right)/(2m) +$$

$$\sum_{C \neq Cls(d_j)} \left(\frac{P(C)}{1-P(Cls(d_v))} \times (diff(A_k,d_j,d_{v1}(C)) + diff(A_k,d_j,d_{v2}(C)) \times \frac{Sim(d_v)}{Num(d_v)})/(2m)\right)$$

8. *end;*
9. *end;*

Fig. 2. Pseudo code of the *kNNMFS* algorithm

In the algorithm above, *p* is the number of attributes in the dataset; m is the number of representatives which is obtained from *kNNModel(D, ε)* and is used for feature selection. *diff()* uses HVDM [15] as a different function for calculating the difference between two values from an attribute.

Compared to ReliefF, *kNNMFS* speeds up the feature selection process by focussing on a few selected representatives instead of the whole dataset. These representatives are obtained by learning from the original dataset. Each of them is an optimal representation of a local data distribution. Using these representatives as seeds for feature selection better reflects the influence of each attribute on different classes, thus giving more accurate weights to attributes. Moreover, a change was made to the original difference function to allow *kNNMFS* to make use of the generated information in each representative such as $Sim(d_j)$ and $Num(d_j)$ from the created model of *kNNModel* for the calculation of weights. This modification reduces the computational cost further.

4 Experiments and Evaluation

To evaluate the effectiveness of the newly introduced algorithm *kNNMFS*, we performed some experiments on a dataset of toxicity values for approximately 250 chemicals, all of which contained a similar chemical feature, namely a phenolic group. The toxicity of the phenols was assessed in the ciliated protozoan *Tetrahymena pyriformis*, according to [13]. The full toxicity dataset was reported originally by Cronin et al [1] following experimental measurement of the effects by Schultz and co-workers (College of Veterinary Medicine, University of Tennessee, Knoxville TN, USA). The analysis of the Tetrahymena pyriformis toxicity data allowed for an evaluation of the performance of the *k*NNMFS algorithm and to assess its performance in feature selection for the real-world application of toxicity prediction of the environmental effects of chemicals.

4.1 Toxicity Dataset for Phenols

The hydroxy-substituted aromatic compounds (phenols) form a large and structurally diverse group of chemicals. These are interesting from a toxicological point of view, since the phenols are widely used organic compounds. They elicit a number of toxicities to different species [14]. Thus, there has been much interest in quantitative structure-activity relationships (QSARs) for phenols, due to their ubiquitous presence in the environment and the various toxicities they may have. One of the important tasks in the prediction of the toxicity of phenols using QSAR analysis is the examination of the relevance of the descriptors in the modeling paradigm. This is often a tedious task, considering the large number of descriptors and compounds to be studied. The algorithm proposed in this study is therefore a contribution to the area of analyzing the correlations between chemical descriptors and the development of QSARs. In this study high quality toxicity data for a large number of phenols were collated from a historical source [14], and supplemented by those from further testing, providing data on 250 compounds for the development and validation of QSARs [1]. A total of 173 descriptors were calculated for each compound. These descriptors were calculated to represent the physico-chemical, structural and topological properties that were relevant to toxicity. An explanation of these chemical descriptors and the large variety of software tools used to calculate them is available from [1].

4.2 Evaluation Criteria

An optimal subset is always relative to a certain evaluation criterion. Evaluation criteria can be broadly categorized into two groups based on their dependency on the learning algorithm applied to the selected feature subset. Typically, an independent criterion, as in filter models, tries to evaluate the goodness of a feature, or a feature subset, without the involvement of a learning algorithm in this process. A dependent criterion, as in wrapper models, tries to evaluate the goodness of a feature, or feature subset, by evaluating the performance of the learning algorithm applied on the selected subset.

For the prediction of continuous class values, e.g. the toxicity values in the phenols dataset, dependent criteria: Correlation Coefficient (CC), Mean Absolute Error (MAE), Root Mean Squared Error (RMSE), Relative Absolute Error (RAE), and Root Relative Squared Error (RRSE) are chosen to evaluate the goodness of different feature selection algorithms in the experiments. The evaluation measures for continuous class values prediction are presented in Table 1. For the prediction of discrete classes, e.g. the Mechanism of Action in the Phenols dataset, average classification accuracy and unbiased variance are used as evaluation criteria. The unbiased variance is defined as:

$$s^2 = \frac{1}{n-1} \sum_{i=1}^{n} (x_i - \bar{x})^2$$

where $\bar{x}$ is the sample mean. These evaluation measures are used frequently to compare the performance of different feature selection methods.

Table 1. Evaluation measures for continuous class values prediction

Acronym	Full Name	Equation
CC	Correlation Coefficient	$$r = \frac{n\sum_{i=1}^{n}x_i y_i - \sum_{i=1}^{n}x_i \sum_{j=1}^{n}y_i}{\sqrt{\left[n\sum_{i=1}^{n}x_i^2 - (\sum_{j=1}^{n}x_j)^2\right]\left[n\sum_{i=1}^{n}y_i^2 - (\sum_{j=1}^{n}y_j)^2\right]}}$$
MAE	Mean Absolute Error	$$E_i = \frac{1}{n}\sum_{i=1}^{n}|x_i - y_i|$$
RMSE	Root Mean Squared Error	$$E_i = \sqrt{\frac{1}{n}\sum_{i=1}^{n}(x_i - y_i)^2}$$
RAE	Relative Absolute Error	$$E_i = \frac{\sum_{i=1}^{n}|x_i - y_i|}{\sum_{j=1}^{n}|y_j - \bar{y}|}$$
RRSE	Root Relative Squared Error	$$E_i = \sqrt{\frac{\sum_{i=1}^{n}(x_i - y_i)^2}{\sum_{j=1}^{n}(y_j - \bar{y})^2}}$$

The following terms are used in Table 1 for a set of n data points (x_i, y_i), where x_i represents the predicted value of y_i, y_i is the true class value, and $\bar{y}$ is the average defined by the formula: $\bar{y} = \frac{1}{n}\sum_{j=1}^{n}y_j$.

4.3 Evaluation

[**Experiment 1**]. In this experiment, eight feature selection methods including ReliefF and kNNMFS were performed on the phenols dataset with toxicity as endpoint to choose a set of optimal subsets based on different evaluation criteria. Besides kNNMFS that was implemented in our own prototype, seven other feature selection methods are implemented in the Weka [16] software package.

The experimental results performed on subsets obtained by different feature selection methods are presented in Table 2. In the experiments, a 10-fold cross validation method was used for evaluation. It is obvious that the proposed kNNMFS method performs better than any other feature selection methods evaluated by the linear regression algorithm on the phenols dataset. The performance on the subset after feature selection by kNNMFS using linear regression algorithm is significantly better than that on the original dataset. Compared to ReliefF and other feature selection methods, kNNMFS obtains higher correlation coefficient and lower error rates such as MAE, RMSE, RAE and RRSE.

Table 2. Performance carried out on different subsets after feature selection

FSM	NSF	Evaluated by Linear Regression Algorithm				
		CC	**MAE**	**RMSE**	**RAE**	**RRSE**
GR	20	0.7722	0.4083	0.5291	60.7675%	63.7304%
IG	20	0.7662	0.3942	0.5325	58.6724%	63.1352%
Chi	20	0.7570	0.4065	0.5439	60.5101%	65.5146%
ReliefF	20	0.8353	0.3455	0.4568	51.4319%	55.0232%
SVM	20	0.8239	0.3564	0.4697	53.0501%	56.5722%
CS	13	0.7702	0.3982	0.5292	59.2748%	63.7334%
CFS	7	0.8049	0.3681	0.4908	54.7891%	59.1181%
kNNMFS	35	**0.8627**	**0.3150**	**0.4226**	**46.8855%**	**50.8992%**
Phenols	173	0.8039	0.3993	0.5427	59.4360%	65.3601%

The meaning of the column titles in Table 2 is as follows: FSM – Feature Selection Method; NSF – Number of Selected Features. The feature selection methods studied include: GR – Gain Ratio feature evaluator; IG – Information Gain ranking filter; Chi – Chi-squared ranking filter; ReliefF – ReliefF Feature selection; SVM- SVM feature evaluator; CS – Consistency Subset evaluator; CFS – Correlation-based Feature Selection; kNNMFS – kNN Model-based Feature Selection and Phenols – the original Phenols data set with 173 features.

Table 3. Performance of wkNN algorithm on different phenols subsets

FSM	NSF	10-Fold Cross Validation Using wkNN (k=5)		
		Average Accuracy	**Variance**	**Deviation**
GR	20	89.32	1.70	1.31
IG	20	89.08	1.21	1.10
Chi	20	88.68	0.50	0.71
ReliefF	20	91.40	1.32	1.15
SVM	20	91.80	0.40	0.63
CS	13	89.40	0.76	0.87
CFS	7	80.76	1.26	1.12
kNNMFS	35	**93.24**	**0.44**	**0.67**
Phenols	173	86.24	0.43	0.66

[Experiment 2]. In this experiment, we performed the same feature selection as we did in experiment 1 on the phenols dataset with mechanism of action as endpoint and then carried out classification using Weighted kNN (wkNN) which was implemented in our own prototype.

The experimental results are presented in Table 3. It shows that the proposed kNNMFS method performs better than any other feature selection method on the phenols dataset with mechanism of action as endpoint. The average classification accuracy on the subset after feature selection by kNNMFS using wkNN is higher than that of by any other feature selection methods and that of the original dataset. Compared to the original Phenols dataset, kNNMFS achieves 8.1% improvement in terms of average classification accuracy and has relatively small range of variance.

5   Conclusions

In this paper we present a novel solution to deal with the shortcomings of ReliefF. To solve the problem of choosing a set of seeds for ReliefF, we modified the original kNNModel method by choosing a few more meaningful representatives from the training set, in addition to some extra information to represent the whole training set, and used it as a starter reference for ReliefF. In the selection of each representative we used the optimal but different k, decided automatically for each dataset itself. The representatives obtained can be used directly for feature selection.

Experimental results showed that the performance evaluated on the subsets of the Phenol dataset with different endpoints by kNNMFS is better than that of using any other feature selection methods. The improvement is significant compared to ReliefF and other feature selection methods. The results obtained using the proposed algorithms for chemical descriptors analysis applied in predictive toxicology are encouraging and show that the method is worthy of further research.

Further research is required into investigating the effects of boundary data or centre data of clusters chosen as seeds for kNNMFS.

Acknowledgment

This work was supported partly by the EPSRC project PYTHIA – Predictive Toxicology Knowledge Representation and Processing Tool Based on a Hybrid Intelligent Systems Approach, Grant Reference: GR/T02508/01.

References

1. Cronin, M.T.D., Aptula, A.O., Duffy, J. C. et al.: Comparative Assessment of Methods to Develop QSARs for the Prediction of the Toxicity of Phenols to Tetrahymena Pyriformis, Chemosphere 49 (2002), pp. 1201-1221
2. Fayyad, U.M. and Irani, K.B.: The Attribute Selection Problem in Decision Tree Generation. In Proc. of AAAI-92, the 9th National Conference on Artificial Intelligence (1992), pp. 104-110, AAAI Press/The MIT Press

3. Guo, G., Wang, H., Bell, D. et al.: kNN Model-based Approach in Classification. In Proc. of CoopIS/DOA/ODBASE 2003, LNCS 2888, Springer-Verlag, pp. 986-996 (2003)
4. Hall, M. A.: Correlation-based Feature Selection for Discrete and Numeric Class Machine Learning, In Proc. of ICML'00, the 17th International Conference on Machine Learning, pp. 359 – 366 (2000)
5. Huang, Y., McCullagh, P.J. Black, N.D.: Feature Selection via Supervised Model Construction. In Proc. of the Fourth IEEE International Conference on Data Mining, pp. 411-414 (2004)
6. John, G.H., Kohavi, R. and Pfleger, K.: Irrelevant Feature and the Subset Selection Problem. In W.W. Cohen and Hirsh H., editors, Machine Learning: Proc. of the Eleventh International Conference (1994), pp. 121-129, New Brunswick, N.J., Rutgers University
7. Kira, K. and Rendell, L.A.: A Practical Approach to Feature Selection. Machine Learning (1992): 249-256
8. Kononenko, I.: Estimating attributes: Analysis and Extension of Relief. In Proc. of ECML'94, the Seventh European Conference in Machine Learning (1994), Springer-Verlag, pp.171-182
9. Liu, H., Yu, L., Dash, M. and Motoda, H.: Active Feature Selection Using Classes. In Proc. of PAKDD'03, pp. 474-485 (2003)
10. Robnik, M., Kononenko, I.: Machine Learning, Kluwer Academic Publishers (2003), 53, pp. 23-69
11. Sikonja, M.R. Kononenko, I.: Theoretical and Empirical Analysis of ReliefF and RReliefF. Machine Learning Journal (2003) 53:23- 69
12. Søndberg-Madsen, N., Thomsen, C. and Peña, J. M.: Unsupervised Feature Subset Selection. In Proc. of the Workshop on Probabilistic Graphical Models for Classification (within ECML/PKDD 2003), 71-82 (2003)
13. Scheultz, T.W.: TETRATOX: The Tetrahymena Pyriformis Population Growth Impairment Endpoint – A Surrogate for Fish Lethality. Toxicol. Methods, 7, 289-309 (1997)
14. Schultz, T.W., Sinks, G.D., Cronin, M.T.D.: Identification of Mechanisms of Toxic Action of Phenols to Tetrahymena Pyriformis from Molecular Descriptors. In: Chen, F., Schuurmann, G. (Eds.), Quantitative Structure-Activity Relationships in Environmental Sciences – VII. SETAC Press, Presacola, FL, USA, pp. 329-342 (1997)
15. Wilson, D.R. and Martinez, T.R..: Improved Heterogeneous Distance Functions, Journal of Artificial Intelligence Research (JAIR), 6-1, pp.1-34 (1997)
16. Witten, I.H. and Frank, E.: Data Mining: Practical Machine Learning Tools with Java Implementations, Morgan Kaufmann (2000), San Francisco
17. Biesiada, J. and Duch, W.: Feature Selection for High-Dimensional Data: A Kolmogorov-Smirnov Correlation-based Filter. In Proc. of CORES 2005, the 4th International Conference on Computer Recognition Systems (2005)
18. Sebban, M. and Nock, R..: A Hybrid Filter/Wrapper Approach of Feature Selection Using Information Theory. Pattern Recognition 35(4):835-846 (2002)

Multi-view EM Algorithm for Finite Mixture Models

Xing Yi, Yunpeng Xu, and Changshui Zhang

State Key Laboratory of Intelligent Technology and Systems,
Department of Automation, Tsinghua University, Beijing 100084, P.R. China
yixing@cs.umass.edu, xuyp03@mails.tsinghua.edu.cn,
zcs@mail.tsinghua.edu.cn

Abstract. In this paper, Multi-View Expectation and Maximization (EM) algorithm for finite mixture models is proposed by us to handle real-world learning problems which have natural feature splits. Multi-View EM does feature split as Co-training and Co-EM, but it considers multi-view learning problems in the EM framework. The proposed algorithm has these impressing advantages comparing with other algorithms in Co-training setting: its convergence is theoretically guaranteed; it can easily deal with more two views learning problems. Experiments on WebKB data[1] demonstrated that Multi-View EM performed satisfactorily well compared with Co-EM, Co-training and standard EM.

1 Introduction

Semi-supervised learning, which combines information from both labeled and unlabeled data for learning tasks, has drawn wide attention. Some related research deal with labeled and unlabeled data in problem domains where features naturally divide into different subsets(views)[1][2]. For example, in web-page classification, features can be divided into two disjoint subsets, one concerning words that appear on the page, another concerning words that appear in hyperlinks pointing to that page, etc. Many algorithms have been proposed to utilize this feature division for boosting performance of learning systems such as Co-training[1], Co-EM[3], Co-EMT[2], etc. Blum and Mitchell provided a PAC-style analysis for Co-training, which shows that when the two views are *compatible* and *uncorrelated*, Co-training will successfully learn the target concept with labeled and unlabeled data[1]. Nigam and Ghani demonstrated that when a natural independent split of input features exists, algorithms utilizing this feature split outperform algorithms that do not. They also proposed Co-EM as a probabilistic version of Co-training[3]. Intuitively, Co-EM runs EM algorithm in each view, and before each new EM iteration inter-changes the probabilistic labels generated in each view. However, Co-EM is only a technical design in the view of EM

---

[1] This data is available at http://www.cs.cmu.edu/afs/cs.cmu.edu/project/theo-20/www/data/webkb-data.gtar.gz

S. Singh et al. (Eds.): ICAPR 2005, LNCS 3686, pp. 420–425, 2005.

framework considering the step of interchanging two views' labels, which does not guarantee convergence.

In this paper, we propose Multi-View EM algorithm for finite mixture models, which follows the scheme of feature split and deal with these multi-view learning problems in the EM framework instead of PAC model. The proposed algorithm guarantees convergence.

Section 2 briefly reviews Co-training setting and EM, then describes Multi-View EM for finite mixture models and provides some implements in Gaussian mixture models (GMM) and Naïve Bayes Classifier. Section 3 presents experimental results of comparing Multi-View EM with Co-training, Co-EM and standard EM on WebKB data[1]. Section 4 concludes.

2   Multi-view EM Algorithm for Finite Mixture Models

2.1   The Co-training Setting

The Co-training setting[1] assumes that in real-world learning problems that have a natural way to partition features into two views V_1, V_2, an example x can be described by a triple $[x_1, x_2, l]$, where x_1, x_2 are x's descriptions in two views and l is its label. In this setting, given a learning algorithm L, the sets T and U of labeled and unlabeled samples and the number k of iterations to be performed, Table 1 and Table 2 describe flow charts of Co-training and Co-EM respectively.

Table 1. Flow chart of Co-training

Loop for k iterations or while all samples have been labeled
-Use L, V1(T), V2(T) to create classifiers h_1 and h_2 respectively
-For each class C_i:
-Let E1, E2 be unlabeled examples on which h_1 and h_2 make the most confident predictions for C_i:
-Remove E1, E2 from U, label them according to h_1 and h_2, respectively, and add them to T
-Recreate classifiers h_1 and h_2 with L, V1(T) and V2(T), respectively

Table 2. Flow chart of Co-EM

-Let $S = T \cup U$, h_1 be the classifier obtained by training L on T
Loop for k iterations
-New1=Probabilistically_Label(S, h_1)
-Use L, V2(New1) to create classifier h_2
-New2= Probabilistically_Label(S, h_2)
-Use L, V1(New2) to create classifier h_1
Combine the prediction of h_1 and h_2 by sum rule and then designate labels according to the largest class conditional probability.

2.2   Finite Mixture Models and EM Algorithm

It is said a d-dimensional random variable $x = [x_1, x_2, \cdots, x_d]^T$ follows a k-component finite mixture distribution, if its probability density function can be written as

$$p(x|\theta) = \sum_{m=1}^{k} \alpha_m p(x|\theta_m), \tag{1}$$

where α_m is the prior probability of the mth component and satisfies:

$$\alpha_m \geq 0, \text{ and } \sum_{m=1}^{k} \alpha_m = 1, \tag{2}$$

θ_m is the parameter of the mth density model and $\theta = \{(\alpha_m, \theta_m), m = 1, 2, \cdots, k\}$ is the parameter set of mixture models. For GMM, $\theta_m = \{\mu_m, \Sigma_m\}$.

EM has been widely used in the parameter estimation of finite mixture models. Suppose that one set Z consists of observed data X and unobserved data Y. According to Maximum Likelihood(ML) estimation, the E-Step calculates the complete data expected log-likelihood function defined by the so called Q function[4],

$$Q(\theta, \hat{\theta}(t)) \equiv E[\log p(X, Y|\theta)|X, \hat{\theta}(t)]. \tag{3}$$

The M-Step updates the parameters by

$$\hat{\theta}(t+1) = \arg\max_{\theta} Q(\theta, \hat{\theta}(t)). \tag{4}$$

The EM algorithm performs the E-Step and M-Step iteratively, and the convergence is guaranteed.

2.3   Multi-view EM Algorithm for Finite Mixture Models

For convenience, we describe two-view version of Multi-View EM for finite mixture models in this paper, which can be easily generalised to more views with only slight changes of the corresponding formulas. In this setting, it holds that:

$$\begin{aligned} p(x_1|\theta) &= \sum_{m=1}^{k} \alpha_m \sum_{x_2} p(x_1, x_2|\theta_m) = \sum_{m=1}^{k} \alpha_m p(x_1|\theta_{mV_1}) \\ p(x_2|\theta) &= \sum_{m=1}^{k} \alpha_m \sum_{x_1} p(x_1, x_2|\theta_m) = \sum_{m=1}^{k} \alpha_m p(x_2|\theta_{mV_2}), \end{aligned} \tag{5}$$

where $\{(\alpha_m, \theta_{mV_1}), m = 1, 2, \cdots, k\} = \theta_{V_1}$ and $\{(\alpha_m, \theta_{mV_2}), m = 1, 2, \cdots, k\} = \theta_{V_2}$ denote models' parameter sets of two views respectively.

Multi-View EM fits finite mixture models to observed data according to another criterion instead of ML, which can be formulated as:

$$\begin{aligned} Q''(\theta, \hat{\theta}(t)) &\equiv E[\log(p(X_1, Y|\theta_{V_1})^{w_1} \bullet p(X_2, Y|\theta_{V_2})^{w_2})|X, \hat{\theta}(t)], \\ \hat{\theta}(t+1) &= \arg\max_{\theta} Q''(\theta, \hat{\theta}(t)), \end{aligned} \tag{6}$$

where w_i denotes the weight of the ith view. The M-step of Multi-View EM updates parameters by maximizing the Q'' function. Consider that in the EM parameter estimation of mixture models, for the Q function[4], it holds that:

$$Q(\theta, \hat{\theta}(t)) = \sum_{m=1}^{k} \sum_{i=1}^{N} \ln(\alpha_m) p(m|x^i, \hat{\theta}) + \sum_{m=1}^{k} \sum_{i=1}^{N} \ln(p(x^i|\theta_m)) p(m|x^i, \hat{\theta}), \quad (7)$$

where $p(m|x^i, \hat{\theta})$ denotes the probability that component m generates sample x^i. Therefore in Multi-View EM, it holds that:

$$Q''(\theta, \hat{\theta}(t)) = \sum_{m=1}^{k} \sum_{i=1}^{N} \ln(\alpha_m)(w_1 \cdot p(m|x_1^i, \hat{\theta}_{V_1}) + w_2 \cdot p(m|x_2^i, \hat{\theta}_{V_2})) +$$

$$w_1 \cdot \sum_{m=1}^{k} \sum_{i=1}^{N} \ln(p(x_1^i|\theta_{mV_1})) p(m|x_1^i, \hat{\theta}_{V_1}) + w_2 \cdot \sum_{m=1}^{k} \sum_{i=1}^{N} \ln(p(x_2^i|\theta_{mV_2})) p(m|x_2^i, \hat{\theta}_{V_2}),$$

$$(8)$$

Then M-step of Multi-View EM updates parameters by maximizing this Q'' function which can be formulated as:

$$\alpha_m = \frac{1}{N} \sum_{i=1}^{N} p(m|x^i, \hat{\theta}),$$

$$\hat{\theta}_{mV_1} = \underset{\theta_{mV_1}}{argmax} \sum_{i=1}^{N} \ln(p(x_1^i|\theta_{mV_1})) p(m|x_1^i, \hat{\theta}_{V_1}),$$

$$\hat{\theta}_{mV_2} = \underset{\theta_{mV_2}}{argmax} \sum_{i=1}^{N} \ln(p(x_2^i|\theta_{mV_2})) p(m|x_2^i, \hat{\theta}_{V_2}), \quad (9)$$

$$p(m|x^i, \hat{\theta}) = \frac{w_1}{w_1+w_2} \bullet p(m|x_1^i, \hat{\theta}_{V_1}) + \frac{w_2}{w_1+w_2} \bullet p(m|x_2^i, \hat{\theta}_{V_2}).$$

Nigam *et al.* proposed one scheme to utilize EM for semi-supervised learning [5], which firstly built an initial classifier $\hat{\theta}$ with all the labeled data T, and then began the EM itcrations until convergence:

E-step: for all the unlabeled data $x_u^i \in U$ calculated the probability $p(m|x_u^i, \hat{\theta})$ that each mixture componcnt m (or class m when one class was represented by one component) generated x_u^i; for all the labeled data $(x_l^i, y^i) \in T$, $p(m|x_l^i, \hat{\theta}) = 1$ when $m = y^i$, otherwise $p(m|x_l^i, \hat{\theta}) = 0$.

M-step: re-estimated the classifier $\hat{\theta}$ with all the labeled and unlabeled data x^i and their probability labels $p(m|x^i, \hat{\theta})$.

Different from Co-EM, Multi-View EM has a global optimization objcctive and iterates completely in the framework of EM, which theoretically guarantee its convergence. Following this scheme, Multi-View EM can be utilized for semi-supervised learning similarly.

2.4   Implementation Details of Multi-view EM

For GMM, where $\theta_m = \{\mu_m, \Sigma_m\}$, the updating iterative formulas in the M-step for $\hat{\theta}_{mV_t}$ can be represented by:

$$\mu_{m_t} = \frac{\sum_{i=1}^{N} x_t^i p(m|x_t^i, \hat{\theta}_{V_t})}{\sum_{i=1}^{N} p(m|x_t^i, \hat{\theta}_{V_t})}, \quad \Sigma_{m_t} = \frac{\sum_{i=1}^{N} p(m|x_t^i, \hat{\theta}_{V_t})(x_t^i - \mu_{m_t})(x_t^i - \mu_{m_t})^T}{\sum_{i=1}^{N} p(m|x_t^i, \hat{\theta}_{V_t})}. \quad (10)$$

For Naïve Bayes classifier, consider the simple case which assumes that one class consists of only one component[5]. Then $\theta_m = \{\theta_{a_j|m} = P(a_j|m)\}$, where set $\{a_j\}$ denotes input attributes of all the training samples. The updating iterative formulas of in the M-step for $\hat{\theta}_{mV_t}$ can be represented by [5]:

$$\theta_{a_j|mV_t} = \frac{1 + \sum\limits_{i=1}^{N} Value(x_t^i, a_j) P(m|x_t^i, \hat{\theta}_{V_t})}{n + \sum\limits_{s=1}^{n} \sum\limits_{i=1}^{N} Value(x_t^i, a_s) P(m|x_t^i, \hat{\theta}_{V_t})}, \tag{11}$$

where $Value(x_t^i, a_j)$ denotes the value of attribute a_j in the tth View of sample x^i. In the text classification task, $Value(x_t^i, a_j)$ denotes the frequency of word a_j in the tth View of document x^i.

3 Experiments

In this section, Multi-View EM is compared with Co-training, Co-EM and standard EM for designing Naïve Bayes classifier on WebKB data in semi-supervised learning scenarios.

In WebKB data there are overall 8,282 pages which had been manually classified into 7 categories. In experiment, we firstly extracted page-based view and hyperlink-based view from those pages by the document feature selection method: "χ^2 statistics+Document Frequency Cut"[6]. Here 3000 features were selected to form the page-based view and 2000 features were selected to form the hyperlink-based view. Then we randomly selected 2000 samples from the "course" category and the "others" category to design a 2-category Naïve Bayes classifier for testing performances of Multi-View EM, Co-training, Co-EM and standard EM(using the whole 5000 features) in semi-supervised setting.

In experiment, after removing all the labels of data, we randomly re-labeled part of them then testified performances of Multi-View EM, Co-training, Co-EM and EM with the semi-supervised datasets created. We randomly re-labeled 200, 400, 600 samples in 2000 samples for 20 times respectively and achieved the averages and variations of the accuracy by different semi-supervised learning algorithms. In multi-view EM, weights of page-based view and hyperlink-based view were set by cross-validation, which was set to be 0.2 and 0.8 respectively in experiment. The experimental results are presented in Table 3. It can be observed that Multi-View EM outperformed Co-training Co-EM and EM in this learning task.

Table 3. Semi-supervised learning on WebKB

	labeled samples	Multi-View EM	Co-training	Co-EM	EM
	200	**0.872**±0.013	0.819±0.017	0.861±0.015	0.748±0.011
Accuracy	400	**0.893**±0.006	0.849±0.009	0.881±0.005	0.780±0.008
	600	**0.907**±0.004	0.869±0.009	0.897±0.004	0.808±0.007

In the view of classifier fusion, Multi-View EM can be intuitively regarded as in each round before each EM iteration, independently designing a classifier in each view and then using the weighted sum criterion of classifier fusion[7] to combine probabilistic labels from all the classifiers. Technically, other classifier fusion criteria such as multiplication criterion and max criterion, etc, could be utilized to combine probabilistic labels, but rigidly, in the framework of EM, using these criteria could not guarantee the convergence.

4 Conclusions

In this paper, Multi-View EM for finite mixture models is proposed to handle real-world learning problem which have natural feature splits. As a more proper and close probabilistic version of Co-training than Co-EM, Multi-View EM is completely deduced in the EM framework and its convergence is theoretically guaranteed. Multi-View EM can be intuitively regarded as that it runs EM in each view, and before each new EM iteration combines all the weighted probabilistic labels generated in each view. Future work involves introducing prior knowledge of data and some criteria to automatically choose more appropriate weights of different views and utilizing Multi-View EM for more semi-supervised learning problems. And that active learning could be introduced so that active learning version of Multi-View EM similar to Co-EMT[2] could be designed.

Acknowledgment

This work was supported by National High Technology Research and Development Program of China(863 Program) under contract No.2001AA114190

References

1. Blum, A., Mitchell, T.: Combining labeled and unlabeled data with co-training. In: Proceedings of the 11th Annual Conference on Computational Learning Theory. (1998) 92–100
2. Muslea, I., Minton, S., Knoblock, C.A.: Active + semi-supervised learning = robust multi-view learning. In: Proceedings of ICML-02. (2002)
3. Nigam, K., Ghani, R.: Analyzing the effectiveness and applicability of co-training. In: Proceedings of Information and Knowledge Management. (2000) 86–93
4. Bilmes, J.A.: A gentle tutorial of the em algorithm and its application to parameter estimation for gaussian mixture and hidden markov models. Technical report, ICSI Technical Report TR-97-021,UC Berkeley (1997)
5. Nigam, K., Mccallum, A., Thrun, S., Mitchell, T.: Text classification from labeled and unlabeled documents using em. Machine Learning **39** (2000) 103–134
6. Rogati, M., Yang, Y.: High-performing feature selection for text classification. In: Proceedings of the eleventh international conference on Information and knowledge management, McLean, Virginia, USA (2002)
7. Suen, C.Y.W., Lam, L.: Analyzing the effectiveness and applicability of co-training. In: Proceedings of the First International Workshop on Multiple Classifier Systems (MCS2000), Cagliari, Italy (2000) 52–66

Segmentation Evaluation Using a Support Vector Machine

Sébastien Chabrier[1], Christophe Rosenberger[1], Hélène Laurent[1],
and Alain Rakotomamonjy[2]

[1] LVR - UPRES EA 2078 - ENSI de Bourges - Université d'Orléans,
10 boulevard Lahitolle, 18020 Bourges, France
[2] PSI - FRE CNRS 2645 - INSA de Rouen,
Avenue de l'Université, 76801 Saint Etienne du Rouvray, France
`sebastien.chabrier@ensi-bourges.fr`

Abstract. Segmentation evaluation is a very difficult task even for an
expert. We propose in this article a new unsupervised evaluation criterion
of an image segmentation result. The quality of a segmentation result is
derived without any *a priori* knowledge by taking into account different
evaluation criteria from the literature. We first compare six unsupervised
evaluation criteria on a database composed of synthetic gray level images.
Vinet's measure is used as an objective function to compare the behavior
of the different criteria. We propose in this paper to fuse the best ones
by a support vector machine. We illustrate the efficiency of the proposed
approach through some experimental results.

1 Introduction

Segmentation is a fundamental stage in image processing since it conditions the
quality of interpretation. Many segmentation methods have been proposed in the
literature [9], [4], but it still is difficult to evaluate their efficiency. In order to
make an objective comparison of different segmentation methods or results, some
evaluation criteria have already been defined and some literature is available [12].
This evaluation of a segmentation result obviously makes sense for a given level
of precision (same number of regions or classes ...).

An evaluation criterion can be used for different applications. The first application is the comparison of different segmentation results of a single image.
This enables to compare the behavior of different segmentation methods in order
to choose the most appropriate for a given application. The second application
is to facilitate the choice of the parameters of a segmentation method. Image
segmentation generally needs the definition of some input parameters, which
are usually defined by the user. This task, that is sometimes arbitrary, can be
obtained by determining the best parameters with the evaluation criterion.

Briefly stated, there are two main approaches. On the one hand, there are
evaluation criteria based upon the computation of a dissimilarity measure between a segmentation result and a ground truth (due to the use of synthetic
images or derived by an expert). These methods are of widely use for example

S. Singh et al. (Eds.): ICAPR 2005, LNCS 3686, pp. 426–435, 2005.
© Springer-Verlag Berlin Heidelberg 2005

in medical applications. On the other hand, there are unsupervised evaluation criteria that quantifies the quality of a segmentation result by computing different statistics without any *a priori* knowledge. They are based on the fact that the quality of a segmentation result could be estimated without any information on its interpretation [7]. In [12], a comparative study of evaluation criteria of segmentation results of gray level images is developed. The problem is that most of the tested criteria are not adapted for textured images while most real images are composed of textured regions. In order to solve this problem, Mac Cane [2] showed that it is necessary to use the maximum quantity of different criteria and to combine them. This article deals with this kind of approach.

In the first part of this article, we carry out a comparative study of six unsupervised evaluation criteria. We use a database of synthetic images and Vinet's measure as an objective function. In the second part, we define a new criterion by combining the best criteria in order to improve the quality of evaluation. Finally, we show the efficiency of the proposed method through some experimental results.

2   Developed Method

First, we carry out a comparative study of evaluation criteria from the literature. In order to improve the quality of these evaluation criteria, we then fuse the best ones by using a support vector machine.

2.1   Unsupervised Evaluation Criteria

We selected, from the state of art [12], six unsupervised evaluation criteria of a gray level image segmentation result and one supervised criterion (used as a reference):

- Zeboudj's contrast (Zeboudj) : this contrast takes into account the internal and external contrast of the regions measured in the neighborhood of each pixel.
- Levine and Nazif's inter-class contrast (Inter-regions) [8] : this criterion computes the sum of contrasts of the regions balanced by their surfaces.
- Levine and Nazif's intra-class uniformity (Intra-region) : this criterion computes the sum of the normalized standard deviation of each region.
- Combination of intra-class and inter-class disparity (Intra-inter) : this indicator combines similar versions of the Levine and Nazif inter-class and intra-class contrast.
- Borsotti criterion (Borsotti) [1] : this measure is based on the number, the surface and the variance of the regions.
- Rosenberger's criterion (Rosenberger) [11] : the originality of this method lies in its adaptive computation according to the type of region (uniform or textured). In the textured case, the dispersion of some textured parameters is used and in the uniform case, gray levels parameters are computed.

– Vinet's measure (Vinet) : it is a supervised evaluation criterion. It computes the correct classification rate by comparing the result with a ground truth. Since we work in this study on a database composed of synthetic images, Vinet's measure is used as a point of comparison.

The database used for our tests included 300 synthetic images composed of textured and uniform regions. Each image contained five regions of different types : texture from Brodatz's album [3] or uniform with low noise. One database called $Unif$ was composed of 5 uniform regions, Mix was composed of 2 textured and 3 uniform regions. Finally, Tex only contained textured regions (see Figure 1).

Each image was segmented by the EDISON algorithm which uses the mean shift algorithm [4] and by a classification method (fuzzy K-means) with a number of clusters equal to 5 and with 3 different parameters settings : a 5x5 pixels analysis window and moments of order 1 to 4 (segmentation adapted to uniform images); a 9x9 pixels analysis window, moments of order 1 to 4 and attributes from the cooccurrence matrix (segmentation adapted to slightly textured images); a 15x15 pixels analysis window, moments of order 1 to 4, attributes from the cooccurrence matrix and the normalized autocorrelation (segmentation adapted to strongly textured images).

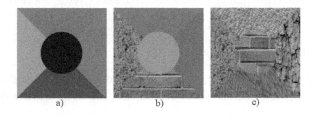

Fig. 1. Example of images for each database a) Unif, b) Mix and c) Tex

We studied the correlation factor of each criterion as an indicator of similarity (see Table 1). It was computed on 1.200 segmentation results (300 images and 4 methods). The absolute value of the correlation factor of two variables is near zero when they are complementary and near 1 when they are similar. The criteria that obtained the best correlation factor with Vinet's measure (ground truth) were Zeboudj, Borsotti and Rosenberger. In order to improve these results, we suggest to fuse these evaluation criteria and to exploit the evaluation of these 1.200 segmentation results in a learning phase.

2.2   Fusion by Support Vector Machine

Let us suppose that we have a set of pairs $\{x_i, y_i\}_{i=1,\cdot\ell}$ with $x_i \in \mathbb{R}^d$ being a vector of d criteria describing the quality of a segmentation result i of a given image and y_i an index quality of segmentation (issued from Vinet's measure associated to the image segmentation result). Our objective is to learn from the knowledge of the training set $\{x_i, y_i\}_{i=1,.\ell}$ a function f that will be able to predict accurately the index quality of a new image segmentation result x. Thus, our idea is to use a supervised learning framework to achieve this goal but also to use this context for the fusion of different criteria and the selection of the most useful.

Table 1. Correlation factors for each evaluation criterion with Vinet's measure

	$Unif$	Mix	Tex
Borsotti	**-0.622**	-0.005	-0.188
Zeboudj	**0.565**	0.252	0.063
Inter-regions	0.200	0.227	0.181
Intra-region	0.158	0.030	0.256
Intra-inter	0.226	0.096	0.010
Rosenberger	**0.471**	**0.306**	**0.313**

This supervised learning problem is a multiclass problem in which the number of classes depends on the cardinality of index quality (for instance, Vinet's measure can be discretized into 10 values yielding thus into a 10-class problem in a d dimension space). The multiclass problem is addressed through a polychotomy based on a *one-against-one* approach [6]. Our baseline machine learning algorithm for each binary problem is a 2-norm Support Vector Machines [5]. Hence, we are looking for a hyperplane in a space $\mathcal{H}$ defined as : $f(x) = \sum_{i=1}^{\ell} \alpha_i^{\star} y_i K(x_i, x) + b$ that maximizes the margin between the hyperplane and the projected data points x_i in $\mathcal{H}$. Hence $\alpha_i^{\star}$ are the solutions to the following optimization problem :

$$\begin{cases} \max_{\alpha_i} \sum_i \alpha_i - \frac{1}{2} \sum_{i,j} \alpha_i \alpha_j y_i y_j (K(x_i, x_j) + \frac{1}{C} \delta_{i,j}) \\ \text{with} \sum_i \alpha_i y_i = 0 \quad 0 \leq \alpha_i \end{cases} \tag{1}$$

where K is the kernel associated to $\mathcal{H}$, $\delta_{i,j}$ is the kronecker symbol and C is a trade-off parameter between the margin width and the number of training examples located beyond the margin.

Note that all the segmentation criteria are implicitly combined through kernel K. In fact, owing to the property of $\mathcal{H}$, the decision function can also be written : $f(x) = \langle \sum_i \alpha_i y_i K(x_i, x) \rangle + b$ which becomes (in the linear case) $f(x) = \langle \sum_i \alpha_i y_i x_i, x \rangle + b$. Thus, each evaluation criterion u of the segmentation results x is weighted by $\sum_i \alpha_i y_i x_i^{(u)}$ in the decision function. If we want to select only a small subset of criteria to fuse, then a criterion selection stage must be performed.

In this work, we have also addressed this problem. In fact, among all the available segmentation criteria, only a small subset of them may be relevant to predict the index quality. Hence, besides learning the decision function f, a criteria selection has been performed. The variable selection algorithm is a backward features ranking algorithm based on the influence of a given criterion on the margin [10] . Hence, each criterion has been weighted by a scaling factor σ. The sensitivity of the margin with regards to a criterion u is related to :

$$| \sum_i \sum_j \alpha_i \alpha_j y_i y_j \frac{\partial K(x_i, x_j)}{\partial \sigma_u} |$$

3   Experimental Result

We have carried out several experiments to assess the performance of our way of merging segmentation criteria. In this first experiment, we have used the same data as in section 2 namely, 1.200 segmentation results. We used 22 criteria (Zeboudj, Borsotti and the Rosenberger's criterion with 20 different parameters). The problem has been turned into a 10 class problem: the quantitative index quality of the segmentation result belongs to $\{1, \cdots, 10\}$. The index quality that we used was Vinet's measure which, for this purpose, was discretized. Our first objective was to rank all the criteria with regards to their relevance in predicting the correct Vinet's measure. We thus ran a *one-against-one* SVM with a variable ranking at each run. SVMs hyperparameters were set to $\sigma = 2$ for the gaussian kernel bandwidth and $C = 10$ so that the number of misclassified segmentation results was low on the training data. Learning and testing sets have been built by randomly splitting each class in 20% and 80% respectively for the learning and testing set. Then, with the random nature due to this, for each binary classifier, 30 trials were performed with different random draws of the learning and testing sets.

The results of this experiment are depicted in Figure 2 (left). We see that learning to predict Vinet's measure of segmentation results is an interesting approach since, with a single criterion, the correct classification rate of segmentation results is around 84%. Combining all criteria by using them as predictive variables yields an improvement of the classification rate. But, the plot also tells us that, by using a small subset of them, we get better results (of the order of 87%) which means that some criteria, instead of bringing information about the segmentation results, tend to make the learning problem harder. Figure 2 (right) gives the average ranking of each criterion over all the trials and binary classifiers. A criterion that was ranked first (most relevant criterion) for all the runs should then get a mean rank of 1. Hence, this plot tells us that the four most relevant variables are the following : 22 (Borsotti's criteria), 1 and 3 (classical Rosenberger's criteria with adaptive parameters and with a fixed set of parameters), and 21 (Zeboudj's criteria).

According to Figure 2 (left), the best performance is achieved when the number of criteria is equal to 7. Hence, we have selected the corresponding criteria and looked for the performance of our criterion fusion approach for different training set size. The experimental setting is similar as above but we did not perform variable ranking, we only used the best criteria. We show in Figure 3 the obtained performances. We can see that even with very few learning examples (1% of the learning set), our criteria fusion method is able to predict accurately (up to 78%) Vinet's measure and that, surprisingly, increasing the size of the training set improves this performance (about 90%). Therefore, just a few segmentation results can be compared to a ground truth and then used as training set.

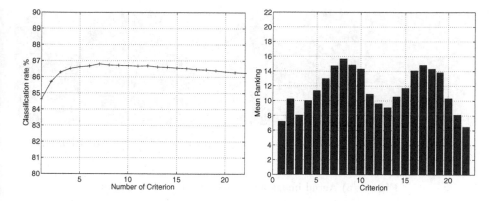

Fig. 2. Results on variables ranking. (left) Performances on the SVMs with regards to the number of evaluation criterion used for predicting Vinet's measure. (right) Mean ranking over all the runs of the 22 different criteria.

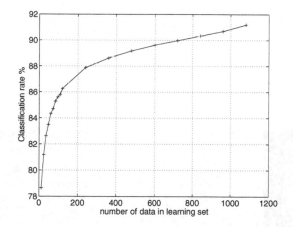

Fig. 3. Classification rate on Vinet's measure prediction according to the number of examples in the training set

We finally illustrate the efficiency of our approach on three real images.

The first image is an aerial one (see Figure 4). The image is composed of textured and uniform regions. The segmentation result that can be visually considered as the best one is EDISON's. Table 2 shows the values for each evaluation criterion. If we focus on Rosenberger's criterion, the value of the EDISON segmentation result is much better than for the others ones. On the contrary, Zeboudj criterion has a bad value for this segmentation result. We used the proposed fusion method using the SVM (the learning process was realized with segmentation results on synthetic images). If we now consider the fusion result, we see that EDISON's segmentation result is correctly preferred.

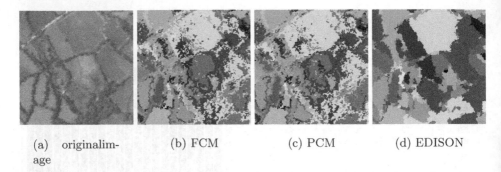

(a) originalim-
age
    (b) FCM
    (c) PCM
    (d) EDISON

Fig. 4. (a) Aerial image and three segmentation results

Table 2. Comparison of three segmentation results of an aerial image of the figure 4 by different evaluation criteria

	FCM	PCM	EDISON
Borsotti	0.0222	0.0297	**0.0155**
Zeboudj	**0.6228**	0.6124	0.5428
Rosenberger	0.6379	0.6328	**0.6973**
SVM	9	9	10

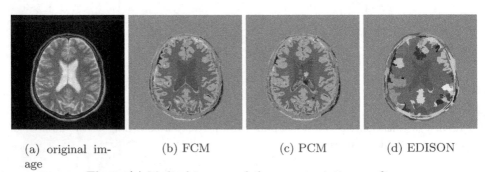

(a) original im-
age
    (b) FCM
    (c) PCM
    (d) EDISON

Fig. 5. (a) Medical image and three segmentation results

The second image is a medical one (see Figure 5). Once again, EDISON's segmentation result is correctly preferred by SVM.

The last image is a radar one (see Figure 6). EDISON's segmentation result seems visually much better than the two other ones. In this case, EDISON's segmentation result is preferred by SVM.

Table 3. Comparison of three segmentation results of a medical image of the figure 5 by different evaluation criteria

	FCM	PCM	EDISON
Borsotti	0.5154	4.4901	**0.0389**
Zeboudj	**0.7279**	0.7184	0.5721
Rosenberger	0.5294	0.5025	**0.6418**
SVM	8	8	9

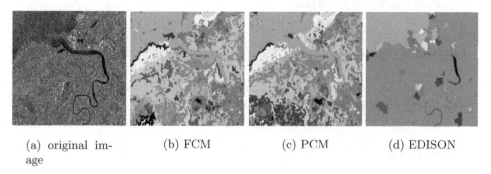

(a) original image   (b) FCM   (c) PCM   (d) EDISON

Fig. 6. (a) Radar image and three segmentation results

Table 4. Comparison of three segmentation results of a radar image of the figure 6 by different evaluation criteria

	FCM	PCM	EDISON
Borsotti	0.1952	0.2793	**0.0293**
Zeboudj	0.1094	**0.1172**	0.0432
Rosenberger	0.4699	0.4677	**0.9074**
SVM	6	6	9

4   Conclusion

Segmentation evaluation is a great challenge and has lots of applications. The first application is the comparison of different segmentation results for a single image. We could compare the behavior of different segmentation methods in order to choose the most appropriate for a given application. The second application is to improve the choice of the parameters of a segmentation method. Image segmentation generally needs the definition of some input parameters, that are usually defined by the user. This sometimes-arbitrary task can be automatic by determining the best parameters by using the proposed evaluation criterion. Another application is the possibility to define new segmentation methods by optimizing these evaluation criteria. Last, an evaluation criterion can be used to fusion several segmentation results of a single image or of the different bands in the multi-components case.

We presented in this paper a comparative study of unsupervised evaluation criteria from the literature. We selected three evaluation criteria for their efficiency : Borsotti's, Zeboudj's and Rosenberger's. The first two are efficient for low textured images while Rosenberger's is more reliable for textured images. We showed the benefit of combining these criteria by a support vector machine to improve their performances.

We are developing new segmentation methods by optimizing this new segmentation evaluation criterion.

Acknowledgments

The authors would like to thank financial support provided by the Conseil Régional du Centre and the European union(FSE).

References

1. M. Borsotti and P. Campadelli and R. Schettini, "Quantitative evaluation of color image segmentation results", Pattern Recognition Letters", (19), 741-747, (1998).
2. B. McCane, "On the evaluation of image segmentation algorithms", Digital Image Computing: Techniques and Applications, (1997).
3. P. Brodatz, Textures "A photographic album for artists and designers", Dover, New York, (1966).
4. D. Comanicu and P. Meer, "Mean shift: A robust approach toward feature space analysis", IEEE Transactions on Pattern Analysis Machine Intelligence, (24), 603-619, (2002).
5. Cristianini and J. Shawe-Taylor, Introduction to Support Vector Machines, Cambridge Univeristy Press, (2000).
6. C.-W. Hsu and C.-J. Lin, "A comparison of methods for multi-class support vector machines", IEEE Transactions on Neural Networks, (13), 415-425, (2002).
7. J. Freixenet, X. Muñoz, D. Raba, J. Marti and X. Cufi, "Yet Another Survey on Image Segmentation: Region and Boundary Information Integration", European Conference on Computer Vision, 408-422, (2002).
8. A.M. Nazif and M.D. Levine, "Low level image segmentation : an expert system", IEEE Transaction on Pattern Analysis and Machine Intelligence, (6), 555-577, (1984).
9. N. Paragios and R. Deriche, "Geodesic Active Regions for supervised texture segmentation", in Proceedings of 7th International Conference on Computer Vision, (1), 926-932, (1999).

10. A. Rakotomamonjy, "Variable selection using SVMbased criteria", Journal of Machine Learning Research, (3), 1357-1370, (2003).
11. C. Rosenberger and K. Chehdi , "Genetic Fusion : Application to multi-components image segmentation", IEEE ICASSP, 2219-2222, (2000).
12. Y.J. Zhang, "A survey on evaluation method for image segmentation", Computer Vision and Pattern Recognition, (29), 1335-1346, (1996).

Detection of Spots in 2-D Electrophoresis Gels by Symmetry Features

Martin Persson and Josef Bigun

School of Information Science, Computer and Electrical Engineering,
Halmstad University, P.O. Box 823
SE-301 18 Halmstad, Sweden
Martin.Persson@ide.hh.se

Abstract. We have implemented an algorithm for detection and segmentation of protein spots in 2-D gel electrophoresis images using symmetry derivative features computed using low level image processing operations. The implementation was compared with a previously published Watershed segmentation and a commercial software. Our algorithm was found to yield segmentation results that were either better than or comparable to the other solutions while having fewer free parameters and a low computational cost.

1 Summary

Two-dimensional gel electrophoresis (2-DE) is a major workhorse in proteomics. 2-DE data comes as spot maps containing a vast number of proteins, requiring automatic image processing for efficient analysis. Quantification of individual proteins and tracing changes in expression between gels require accurate spot detection. Existing spot detection algorithms often require user intervention for setting free parameters and time consuming morphological post processing. We approach the problem by using a set of computationally cheap and robust symmetry derivative features and minimal post processing. A feed forward neural network is used to find decision boundaries in the feature space. The neural network is trained with features extracted from manually segmented 2-DE images. Classification performance is compared with the published non-commercial algorithm of Bettens [1] and one commercial 2DE image analysis program, ImageMaster™ 2D Platinum v5.0 (GE Healthcare, formerly Amersham Biosciences). The result, presented as ROC curves, show that we perform at least as well as both Bettens and Imagemaster in terms of spot detection and segmentation, while using fewer free parameters, and a limited amount of computational resources.

1.1 Originality and Contribution

We propose a set of symmetry derivative features [2, 3] to be used in automatic segmentation of 2DE gel images with minimal post-processing. Symmetry derivatives give immediate information on local shape that otherwise requires time consuming regional processing. In addition to achieving better segmentation performance, this moves the focus of the problem from post processing to basic signal processing.

S. Singh et al. (Eds.): ICAPR 2005, LNCS 3686, pp. 436–445, 2005.

1.2   Introduction to the Problem

2-DE [4] is able to separate thousands of proteins in a sample, presented as image data. Spot detection and segmentation into spot regions is of central importance for the quantitative and differential analysis of proteomics experiments. The segmentation is complicated by the large range of protein concentration affecting spot geometry, overlapping spots, irregular spot shapes, and random noise.

1.3   Alternative Spot Detection Techniques

Most spot detection solutions are closed source, making fair comparison difficult. However many older image analysis packages [5, 6] applied model fitting for direct segmentation of the protein spots. The alternative approach is to use a crude initial segmentation followed by computation of morphological and grey level features of the initial regions for a final decision. The Laplacian of Gaussian (LoG) filter response is a weak feature that has been widely used in segmentation [7, 8, 9]. In recent years the unsupervised Watershed algorithm has also become a popular choice [1, 9] for initial 2DE image segmentation. The second step often consists of iterative model fitting within the regions [1, 8, 10] to closer determine spot properties. Fitting each segmented area to a model is often a computationally expensive step.

1.3.1   Gaussian Fitting

The 2-D Gaussian function is used for smoothing and noise removal in image processing, and is also the most common protein spot model:

$$G(x, y) = \frac{1}{2\pi\sigma^2} e^{-\frac{x^2+y^2}{2\sigma^2}} . \tag{1}$$

As the function is separable smoothing can be done either with a 2-d convolution with the Gaussian kernel above or by convolution with two orthogonal 1d kernels. The reason the Gaussian is a popular in spot modeling is that the Gaussian function corresponds to diffusion from a single point. Most spots are formed under conditions similar to diffusion in the MR/PI directions, making the model appropriate for many spots. The main weaknesses are poor modeling of irregularly (ie non ellipsoid) shaped spots, and in fitting with saturated "flat top" spots. Bettens as well as Rogers [8] address the flat top spot problem by assuming diffusion from a central area rather than a point, representing the spot as a central disc or irregularly shaped region convolved with a Gaussian kernel. In spot detection the Gaussian is used either by iteratively optimizing the parameters around a peak [5] with respect to a residual, or by similar fitting to a segmented region that is expected to contain one spot only.

1.3.2   Second Derivative

The second derivative gives information on the curvature of the local surface. As protein spots appear as dark blobs on a white background they have convex curvature. This weak criterion is widely used in initial spot segmentation [7, 8, 9]. Since the second derivative amplifies noise a smoothing operation is often applied before the derivative is computed. In practice the computation of the second derivative of a gray scale image is often implemented by means of 2-d convolution with the LoG filter

$$LoG(x, y) = \frac{-1}{\pi\sigma^4}\left[1 - \frac{x^2 + y^2}{2\sigma^2}\right]e^{-\frac{x^2+y^2}{2\sigma^2}} .$$ (2)

which in the case of local concave curvature gives a positive value on a spot.

1.3.3 The Watershed Transformation

The watershed transformation (WST) is a powerful segmentation algorithm commonly used for segmentation of locally homogenous grey value images that has been implemented in linear time [11]. The transform finds the reliefs that separates catchment basins around local minima. In 2-DE the WST has been applied to a smoothed grey scale image [1] and to the gradient strength image [9]. The WST tends to over-segment spots when applied to the gradient strength image. Usually heuristics are applied to merge or discard regions in a computationally costly post processing step.

2 Method

2.1 Features and Feature Extraction

For each pixel we generate a feature vector containing the local LoG transform of the original image and the local symmetry derivative response. The LoG captures the concavity of spot surfaces while symmetry derivatives capture spot shape. Symmetry derivatives are powerful textural features that have a wide range of applications in image processing [3, 12, 13]. The main strength of symmetry derivatives is an ability to represent local shape without initial segmentation of the raw image, a costly step in terms of computation. Symmetry Derivatives are differential operators that are based on

$$D_x + iD_y = \partial/\partial x + i\partial/\partial y .$$ (3)

which yields a complex vector field when applied to images. Higher order symmetry derivatives of the n'th order, and their conjugates, are defined as

$$(D_x + iD_y)^n .$$ (4)

$$(D_x - iD_y)^n .$$ (5)

respectively. Applying a differential operator to an image corresponds to convolution with a derivative kernel in the direction of the differential combined with a window function. We choose the discrete Gaussian function and its derivative as window function and derivative kernel respectively. In our application we first use derivative operators to compute the local orientation map of an image

$$Z = (dxf + idyf)^2 .$$ (6)

Further smoothing of Z (convolution with a 2-d window function such as the Gaussian kernel) of the local orientation map directly gives the local Complex moments I_{20} and I_{11}. Complex moments of order $m+n$ are defined as:

$$I_{mn} = \iint (u+iv)^m (u-iv)^n |F(u,v)|^2 \, du \, dv \; . \tag{7}$$

I_{20} can be computed as the local weighted sum of Z, while I_{11} is computed as the weighted sum of the absolute values of Z. The relationship between the magnitude of I_{20} and I_{11} represents how well the local power spectrum of the image f fits to a line, and is widely used for edge and corner detection. Working on the local orientation map in this manner corresponds to using the 0'th order symmetry derivative operator, or linear symmetry. If we instead choose further application of the Symmetry derivative operators before smoothing, then that corresponds to a coordinate transform followed by fitting to a line in the power spectrum. The coordinate transform and the following line fitting also corresponds to fitting certain symmetric shapes to the image in the Cartesian (x,y) coordinate system. We are using the first and second order conjugate symmetry derivatives which corresponds to parabolic and spiral (Circular Symmetry, CS) structures in Cartesian coordinates as shown in *Fig 1*. The choice is motivated by spot geometry. The Local orientation map of spot centers show strong similarity to the Local orientation map of ideal circular patterns, and the spot boundaries analogously have a parabolic structure. As D_x and D_y are implemented with separable 1-D Gaussian filters the computation is very fast.

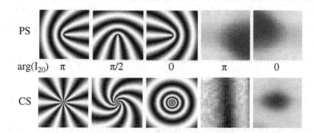

Fig. 1. Phantom and 2-DE patterns in the Cartesian coordinate system and the corresponding argument of the moment I_{20}

We compute two types of moments for the circular and parabolic symmetry derivative responses. For each type we have I_{20} as the local sum of the complex response weighted by a complex filter, and I_{11} as the sum of the magnitudes of the complex responses weighted with the magnitudes of the complex filter. Parabolic symmetry requires 8 1-dimensional convolutions, circular symmetry another 4 (circular symmetry can be computed as applying symmetry derivative operators to parabolic symmetry). Computation of the moments I_{20} and I_{11} requires another 4 convolutions for each type, for a total of 20 1-D convolution operations.

2.2 Classification of Pixels and Final Segmentation of the Gel

We use a Feed forward Artificial Neural Network (ANN) for initial classification of the pixels based upon our feature set. A two layer ANN with a non-linear transfer function is chosen as such a network can find an arbitrarily close approximation of any function or decision boundary [14]. We choose the hyperbolic tangent function as transfer function in the hidden layer and a logistic sigmoid function as transfer function in the output layer. We use two output nodes representing the posterior probabilities of a pixel belonging to the background or a spot core, and let the experiments described in **3.2.1** determine the number of nodes in the hidden layer. Training is done by Resilient Backpropagation [16]. As the neural network ignores spatial continuity in the data the initial result can be expected to be undersegmented. We first split the segmented regions along the local minima of real(I_{20}) for CS. The split segments are then kept if their size is above a certain threshold.

3   Data Set and Experimental Design

3.1   Data Set

Training and validation data were obtained from 8-bit tif images of eight silver stained gels (four human and four E-coli) with varying signal to noise ratio and background intensity. We keep two E-coli gel images as test data. In addition we used an artificially generated gel [16] with known spot positions (available for download at http://www.isbe.man.ac.uk/~mdr/content.php?f=electrophoresis).   The   continuous regions of the smoothed real gel images whose second derivative is above zero were split along the watershed lines of the Laplacian strength image. The resulting regions were manually inspected and if necessary merged with their neighbours. Finally the resulting regions were eroded by two pixels to eliminate the smallest spots and to separate spot cores from boundaries.

The following features were computed from the 8 bit grey value images:

x_1= real(I_{20}) for CS x_2= imag(I_{20}) for CS x_3=I_{11}-|I_{20}| for CS x_4=|I_{20}|/I_{11} for CS
x_5= real(I_{20}) for PS x_6= imag(I_{20}) for PS x_7= |I_{20}|/I_{11} for PS x_8= LoG

Pixels for training of the neural network were selected as follows: For the spot core class all pixels belonging to the identified spot regions in the training set were chosen. For the background class we chose all the pixels within a cityblock distance of two pixels from the spot core, and a number of pixels equal to that of the spot core class were randomly chosen from the remaining pixels.

3.2   Experiments

3.2.1   Selection of Features and Parameters for ANN
Parameter selection is carried out using the Backward Elimination technique. The method consists of two steps. First we set the number of nodes by changing the number of nodes in the hidden layer between 1 to 15 and performing five-fold cross-validation for each configuration. In the second step we use the number of nodes that gave the lowest number of misclassified pixels in a leave one feature out experiments. The feature with the smallest effect on the total error was then removed, and the two steps repeated with the reduced feature set. The result is presented as error bars for the best number of nodes for each number of features.

Table 1. Test parameters

Algorithm	Parameter	Values
Symmetry derivative	Minimum Core size	0 to 19
ImageMaster™ 2D Platinum v5.0[1]	Saliency	1, 2, 5, 10, 20
Bettens Watershed[2]	Minimum watershed size	10 to 160 in increments of 10

[1]Imagemaster also has the parameters *Smooth* and *Min Area* which were held constant at 2 and 5 respectively as experiments showed that they have little effect on the performance on our data set.
[2]Bettens algorithm has one more parameter, *Maximum Grey value in Watershed*, that we after experiments decided to hold constant at 238 for the real images and 250 for the artificial images.

3.2.2 Evaluation of Segmentation Performance

In the final segmentation experiments we feed the neural network with a reduced feature set for an initial classification. Post processing is done by splitting the regions at local minima of the feature x_1, followed by discarding regions of a size smaller than a certain threshold.Using a segment of a real silver stained image containing 215 manually identified spots and an artificial silver stained image with 924 spots we compute precision and recall with respect to spots found. A spot is considered valid if the centre of the segmented region is within the identified core of a real gel, or within 20% of the standard deviation plus three pixels distance of the centre of the artificial spots. As the watershed segmentation only gives target regions for later gaussian fitting we provide an alternative segmentation performance measure as well. For these we consider all segmented regions that overlap precisely one spot center to be valid, giving a measure of the potential improvement from post processing. Precision and recall is computed for our algorithm as well as the Watershed of Bettens and Imagemaster 2D Platinum. The result is presented as ROC curves with respect to different parameters of the algorithms (*Table 1*).

3.2.3 Comparison of Execution Time

As Image Master is a closed source package we choose measure execution time in order to estimate the comparative computational complexity. Cropped versions of a real image with sides 128, 181, 256, 362, 512, 724, and 1024 pixels were used for the experiment. The Symmetry derivative and Watershed based approaches were implemented in Matlab 7.0. For the testing of Image Master we used a downloadable trial version (available at http://www1.amershambiosciences.com/). Execution times for the Matlab implementations were measured using Matlab's internal timer functions, while the execution times for Image Master were measured manually with a stopwatch. All experiments were performed under Windows XP Professional on a AMD Athlon XP2400+ with 1Gb RAM. The results are presented as a log-log plot of execution time vs the number of pixels.

4 Results and Interpretation

4.1 Feature Selection and ANN Parameter Selection

Figure 2 shows the crossvalidation error and standard deviation after removing features. Four features can be removed without hurting performance. The redundant

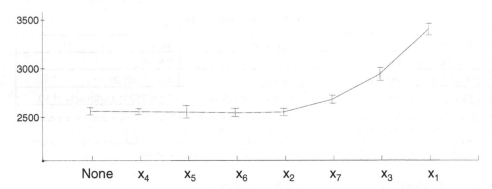

Fig. 2. Crossvalidation error and standard deviation of the average error after successive removal of features. The features were removed in the order listed from left to right.

features represent direction of parabolic symmetry, lack of CS, and swirly spiral patterns. With the exception of the lack of CS this is unsurprising. However one of the remaining features represents relative strength of CS, a property that is related to lack of circular symmetry accordingly. We choose to keep features 1, 3, 7, and 8.

4.2 Segmentation Performance

Fig. 3 shows ROC curves for the experiments. The Symmetry based approach consistently yields a higher recall compared to the most sensitive setting of Image Master, but also finds a higher number of false positives. A large subset of the false positives found by our algorithm are caused by splitting of actual spots. Such false positives are less of a problem than false negatives when the output of a spot detection algorithm is used to find targets for MS analysis, as the goal often is to find low abundance novel proteins. Image Master achieves the highest precision, but never manages to achieve an equal error rate even on the artificial gel. The comparison with the Watershed algorithm shows that we achieve comparable initial segmentation performance except for in the case of the artificial gel. The artificial gel used in the test has an unrealistically low noise level and thus lack the non-spot catchment basins that otherwise cause the Watershed algorithm to oversegment an image. The difference between the segmentation and detection curves shows the potential improvement of the algorithms by further post processing. One low cost improvement could be to compute the spot center using a method that takes pixel values into account rather than computing the center of gravity of the regions only, as done in this paper. Our segmentation approach returns smaller regions, corresponding to spot cores, compared to Watershed and Imagemaster, which would result in computationally cheaper post processing. The results closest to an equal error rate for each algorithm are show in *Table 2*. The results for Image Master are from detection at the most sensitive settings.

4.3 Computational Complexity and Execution Time

Fig 4 shows a log-log plot of execution time vs the number of pixels in the image. A realistic image of size 1024*1024 pixels is segmented in 10s by Image Master, 31s by

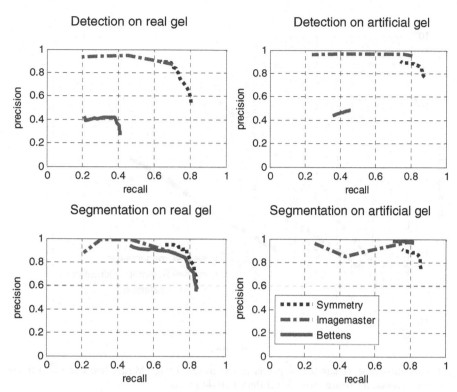

Fig. 3. ROC curves for spot segmentation and detection on real and artificial gels

Table 2. Best performance

	Symmetry Derivatives		Watershed		Imagemaster	
Detection	Precision	Recall	Precision	Recall	Precision	Recall
Real gels	73%	74%	39%	39%	87%	70%
Virtual gel	85%	85%	51%	43%	96%	80%
Segmentation	Precision	Recall	Precision	Recall	Precision	Recall
Real gels	79%	81%	79%	79%	88%	73%
Virtual gel	84%	85%	93%	88%	97%	82%

the Symmetry Derivative based method, and 560s by the Watershed based method. The execution time for the Symmetry derivative approach and Image Master is linear, while our implementation of the Watershed segmentation has much worse performance. The poor performance of the Watershed based segmentation stems from the post processing that is dependent on the number of watersheds as well as their size. Our implementation also uses loops which are very inefficiently implemented in Matlab, further adding to the execution time. Execution times for the Symmetry based approach were on average 3.2 times longer than for Image Master. This difference is remarkably small as Image Master is a compiled, optimised software, while the

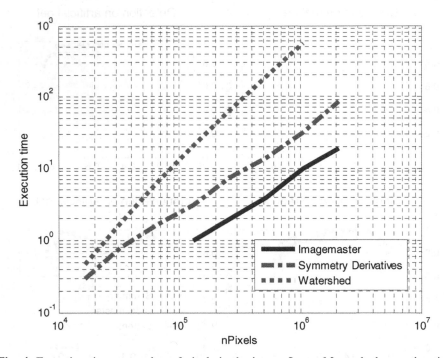

Fig. 4. Execution time vs number of pixels in the image. Image Master had execution times lower than 1 second for images with fewer than 131044 pixels.

Symmetry Derivative based segmentation was implemented in an interpreting language using an inefficient double precision representation of all data.

5 Conclusions

In this paper we have presented a novel 2-DE image segmentation algorithm based upon Symmetry Derivatives and compared it with the widely used Watershed segmentation technique and a commercial software package. We achieve a better equal error rate and a better performance on the recall measure compared to Imagemaster, and comparable or results to a Watershed segmentation approach that relies on significant post processing. The comparison shows that we achieve equivalent or better spot detection compared to the other approaches, using only the basic signal processing operation of one-dimensional convolution and a size criterion.

References

[1]   Bettens E, Scheunders P, Van Dyck D, Moens L, Van Osta P, Electrophoresis, 18, pp 792-798, 1997
[2]   Bigün J., Bigün T., Nilsson K., *IEEE-PAMI* 26, pp 1590-1605, 2004

[3]   Persson M, Bigün J, In J. Bigun and T. Gustavsson, editors, *Scandinavian Conference on Image Analysis*, volume LNCS-2749, pp 520-525. Springer, 2003

[4]   Görg A, Weiss W, Dunn M.J., Proteomics 2004, no 4, pp 3665-3685

[5]   Garrels J.I., J. Biol. Chem. 1989, 264, pp 5269-5282

[6]   Appel R. D., Vargas J. R, Palagi P. M, Walther D. et al Electrophoresis 1997, 18, pp 2724-2734

[7]   Baker M, Busse H, Vogt M, Medical Imaging 2000: Image Processing, San Diego, SPIE, Bellingham 2000, 2979, pp 426-436

[8]   Rogers M, Graham J,. Tonge R.P., Proteomics 2003, no. 6, pp 887-896

[9]   Pleissner, K.-P., Hoffman F, Kriegel K, Wenk C, Wegner S, Sahlström S, Oswald H, Alt H, Fleck E, Electrophoresis 1999, 20, pp 755-765

[10]   Mo X, Wilson R, In J. Bigun and T. Gustavsson, editors, *Scandinavian Conference on Image Analysis*, volume LNCS-2749, pp 430-437. Springer, 2003

[11]   Vincent L, Soille P, IEEE PAMI, 1991, 13, pp 583-598

[12]   Nilsson K, Bigün J, Pattern Recognition Letters, 24, pp 2135-2144, 2003

[13]   Premaratne H. L., Bigün J, Pattern Recognition, 37, pp 2081-2089, 2004

[14]   Funahashi, K, Neural Networks 2 (3), pp 183-192

[15]   Riedmiller M, Braun H, In Proc. of the IEEE Intl. Conf. on Neural Networks, San Francisco 1993

[16]   Rogers M, Graham J,. Tonge R.P., Proteomics 2003, no. 6, pp 879-886

Analysis of MHC-Peptide Binding Using Amino Acid Property-Based Decision Rules

Jochen Supper*, Pierre Dönnes*, and Oliver Kohlbacher

Dept. for Simulation of Biological Systems, WSI/ZBIT, Eberhard Karls University,
Tübingen, Sand 14, D-72076 Tübingen, Germany
doennes@informatik.uni-tuebingen.de

Abstract. The human immune system is a highly complex machinery tuned to recognize specific molecular patterns in order to distinguish self from non-self proteins. Specialized immune cells can recognize major histocompatibility (MHC) molecules with bound protein fragments (peptides) on the surface of other cells. If these peptides originate from virus or cancer proteins, the immune cells can induce controlled cell death. *In silico* vaccine design typically starts with the identification of peptides that might induce an immune response as a first step. This is typically done by searching for specific amino acid patterns obtained from peptides known to be recognized by the immune system. We propose a new method for deriving decision rules based on the physiochemical properties of such peptides. The rulesets generated give insights into the underlying mechanism of MHC-peptide interaction. Furthermore, we show that these rulesets can be used for high accuracy prediction of MHC binding peptides.

1 Originality and Contribution

This study shows how pattern recognition can be applied for analyzing and predicting MHC-binding peptides. We encode the peptides using physiochemical amino acid properties, e.g. hydrophobicity and size, in order to find important determinants of MHC-peptide binding. This gives new insights into the underlying mechanism of MHC-peptide interaction, which is useful in the design of peptide-based vaccines.

2 Introduction

Cancer and infectious disease are two major causes of death in the world today. Both cancer cells and virally infected cells produce proteins that are not present under normal conditions. These and other intracellular proteins are degraded into smaller peptides and some of these are selectively displayed on the cell surface bound to major histocompatibility (MHC) class I molecules. The MHC-peptide complexes serve as a "fingerprint" of the current state of the cell. Cancer cells

---

* These authors contributed equally to this work.

S. Singh et al. (Eds.): ICAPR 2005, LNCS 3686, pp. 446–453, 2005.

and virally infected cells typically display a distinct range of peptides compared to normal cells, which can be recognized by cytotoxic T-cells (CTLs) of the immune system. CTLs can then induce cell death (apoptosis) of such modified self-cells.

During recent years, several clinical experiments have proven that MHC-binding peptides can be used for therapeutic purposes in vaccine development for both cancer [1,2] and viral infection [3,4]. However, the number of MHC-binding candidates from a known protein sequence is about 100-200 fold higher than the number of actual binders [5]. There are up to six different MHC class I variants (alleles) expressed in each individual (human) and there are several hundred known alleles with different binding preferences. This diversity makes humans robust against quickly evolving pathogens that try to avoid recognition. Nonetheless MHC molecules share some common features, the bound peptides usually have a length of eight to ten amino acids and the interaction is typically very strong in a few deep pockets of the MHC molecule. Some MHC alleles have pockets that can accommodate long and charged amino acids, whereas other prefer small hydrophobic ones. The amino acids not buried in the MHC molecules are typically exposed to the extracellular environment, enabling them to interact with T-cells.

In this study we use simple rulesets based on amino acid property descriptors to investigate MHC-peptide binding. The identified properties give an easily interpretable biological explanation of MHC-peptide binding. We are also able to show that only a limited number of positions are crucial for describing the interaction. In comparison with two existing prediction methods, we show that our rule-based approach is well suited for discriminating between binding and non-binding peptides.

A Brief Review of Competing Techniques to Solve the Problem

The first sequence-based prediction methods presented used regular-expression searches for consensus amino acid patterns identified in experimental data, e.g. the *XLXXXXXXV* (*X* indicating any of the 20 amino acids) motif for the MHC allele HLA-A*0201 [6]. The simple motifs have been extended into position-specific scoring matrix (PSSM) methods, where a score is assigned for each amino acid in every position of the peptide. The best known PSSM methods are SYFPEITHI [7] and BIMAS [8]. The SYFPEITHI method assigns scores based on expert knowledge; amino acids contributing strongly to binding are given a score of 10, preferred amino acids are given a lower positive score, and amino acids with a negative contribution are given a negative score. The BIMAS method on the other hand is based on experimentally measured dissociation rates of the MHC-peptide complex. A large number of peptides differing in only one or a few positions were synthesized in order to determine the contribution of individual amino acids to the overall binding energy.

A number of different machine learning methods, such as neural networks [9], hidden Markov models [5], and support vector machines [10], have also been used for prediction. The advantage of these methods is that they use a non-linear description of MHC-peptide interactions, leading to a higher accuracy than the PSSM methods for some alleles. The major drawback of these methods is that they are "black boxes" not giving an easily interpretable description of the underlying mechanism of MHC-peptide binding.

3   The Proposed Method

Here we use decision rules based on amino acid-specific properties for predicting MHC class I binding peptides. The aim of this study is both to find the amino acid properties that describe the interaction in MHC-peptide complexes and to create a prediction method with high accuracy. We test a wide range of different amino acid properties in combination with specific positions of the peptides.

From the rulesets generated we find peptide positions and amino acid properties determining binding of peptides to different MHC alleles. The prediction performance of our method is tested and compared to that of two existing PSSM-based methods.

4   Data Used and Experimental Design

Experimentally verified MHC binding peptides with a length of nine amino acids for the HLA-A*0201, HLA-B*08, and HLA-B*2705 alleles were obtained from the SYFPEITHI database [7]. This gave a total of 246 HLA-A*0201, 45 HLA-B*2705, and 32 HLA-B*08 peptides. A dataset of non-binders was created by extracting peptides from existing proteins in the Ensembl database [11]. The effect of potential binders in the non-binding dataset is considered small since only 1 in 100-200 peptides from a protein typically binds to a certain MHC molecule [5]. The number of non-binders used in the training data was twice the size of known binders for each allele. Duplicate entries were removed from the data sets and peptides were encoded using amino acid-specific properties (e.g. hydrophobicity, side chain volume, and absolute entropy) from the AAindex database [12].

We used the well known C4.5 and C5.0 software packages [13] to create rulesets that consist of unordered collections of if-then rules. We prefer rulesets over decision trees since they give clear descriptions of the rules associated with a certain class. In cases where more than one rule applies, the C5.0 program takes the confidence value of each rule into account to calculate a total vote for each class. Furthermore, there is a default class that is used when none of the rules in the ruleset is applicable. The aim with the sampling procedure is to find amino acid patterns, in terms of biochemical properties describing MHC-peptide binding.

In a preliminary analysis every amino acid property was evaluated separately considering all peptide positions. The rulesets generated were then searched for key positions and subsequent predictions restricted to a limited number of positions showed improved accuracy. Furthermore, different combinations of amino acid properties and position were evaluated. In most cases a limited number of peptide positions and amino acid features gave the best prediction results. The prediction performance was evaluated using Matthews correlation coefficient [14] and fivefold cross-validation. Furthermore, we report the sensitivity, specificity, and total accuracy of our method.

We also compared our method against the existing methods SYFPEITHI and BIMAS by applying the same data and statistics to the online version of these methods.

5   Results and Interpretation

The performance for the best combination of peptide positions and amino acid properties found for each allele is presented in Table 1. The general conclusion from these results is that only a limited number of peptide positions and amino acid properties are needed to describe MHC-peptide binding. This also means that the dimensionality of the classification problem can be reduced, since there is no need to take all sequence positions into account. We also investigated the effect of data splits into training and test sets for the best results, giving only small differences to the results presented here.

Table 1. Results for the best prediction accuracy achieved for each of the studied MHC alleles. The table shows the statistics obtained from fivefold cross-validation, the positions of the peptide considered, and the amino acid properties used.

MHC	MCC	SP	SE	ACC	POS	Properties
HLA-A*0201	0.85	0.95	0.89	0.93	2,4,6,9	BetaStruct,SideChainGyr,BurResidue
HLA-B*2705	0.95	1.0	0.94	0.98	2,9	Stability,LengthSideChain
HLA-B*08	0.92	0.97	0.97	0.97	3,5,9	PosCharge,Stability,AccSurfaceArea

Fig. 1 shows the rulesets generated for HLA-B*2705 (a) and HLA-B*08 (b) using the whole datasets. The rules for HLA-B*2705 are based on two peptide positions and two amino acid properties. The two features found to be the most important for this allele are *Stability* and *LengthSideChain*. *Stability* describes the contribution to protein stability from a certain side-chain [15], whereas the *LengthSideChain* property is a size descriptor for the amino acids [16]. *Stability* is a feature closely correlated to the hydrophobicity of an amino acid, which can be seen in position nine of the peptide where hydrophobic amino acids are preferred. It can be seen from the ruleset that HLA-B*2705 prefers amino acids with long side chains in position two of the peptide. This is important since the "binding pocket" of the MHC molecule is rather spacious. A small amino acid would not be able to fill the pocket. Space-filling effects like this are known to be important for protein structure stability and protein-ligand interaction [17,18].

a.

```
Rule 1: (48/3, lift 2.8)
  Stability_9 > 0.2024793
  LengthSideChain_2 > 0.9756944
  -> class epitope  [0.920]

Rule 2: (84, lift 1.5)
  LengthSideChain_2 <= 0.9756944
  -> class non-epitope  [0.988]

Rule 3: (38, lift 1.5)
  Stability_9 <= 0.2024793
  -> class non-epitope  [0.975]

Default class: non-epitope
```

b.

```
Rule 1: (23/1, lift 2.7)
  PosCharge_3 > 0
  AccSurfaceArea_9 <= 0.2057143
  -> class epitope  [0.920]

Rule 2: (9/1, lift 2.4)
  PosCharge_5 > 0
  Stability_3 > 0.4917355
  -> class epitope  [0.818]

Rule 3: (35, lift 1.5)
  AccSurfaceArea_9 > 0.2057143
  -> class non-epitope  [0.973]

Rule 4: (52/1, lift 1.5)
  PosCharge_3 <= 0
  PosCharge_5 <= 0
  -> class non-epitope  [0.963]

Rule 5: (30/2, lift 1.4)
  PosCharge_3 <= 0
  Stability_3 <= 0.4917355
  -> class non-epitope  [0.906]

Default class: non-epitope
```

Fig. 1. The rulesets created for the HLA-B*2705 (a) and HLA-B*08 (b) alleles. The rules presented here were generated using the whole dataset for each allele.

The small number of features needed to describe the relevant properties for binding gives a very compact model.

The ruleset for HLA-B*08 is based in three amino acid properties. *PosCharge* describes the charge of the amino acids [16] and *AccSurfaceArea* is a measure of how exposed a certain amino acid is to the solvent in known protein structures [19]. The importance of *PosCharge* in position five of the peptide can clearly be seen (Rule 2 and Rule 4). This has been previously described in literature, where the positively charged amino acid lysine has been found in position five of the peptide [20]. *AccSurfaceArea* in position nine is also important for HLA-B*08 (Rule 1 and Rule 3) where small residues are preferred. Figure 2 shows the structure of a HLA-B*08 molecule with a bound peptide (PDB code 1M05). The peptide binding groove is formed by antiparallel β-sheets and two α helices, see Fig. 2a. A cross-section of the MHC-peptide complex can be seen in Fig. 2b. This figure clearly shows how the amino acids in positions 3, 5, and 9 of the peptide are deeply buried in the MHC molecule. These positions are in close contact with the MHC molecule and are crucial for binding, something also captured in the rulesets generated.

The results of the external methods SYFPEITHI and BIMAS can be seen in Table 2. SYFPEITHI performs better than BIMAS for the HLA-A*0201 and HLA-B*08 alleles, but is worse for HLA-B*2705. The ruleset method is better than both SYFPEITHI and BIMAS considering all alleles. The advantage of the ruleset method is that it finds the peptide positions and amino acid properties that best describe MHC-peptide interaction. Thus we can learn the underlying

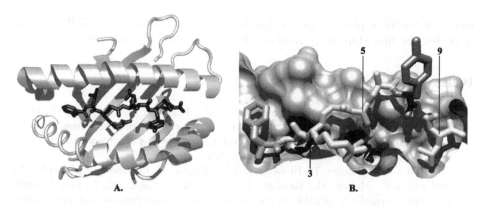

Fig. 2. This figure shows the structure of an HLA-B*08 MHC molecule with a bound peptide. An overview of the structure shows the binding groove formed by β-sheets and two α-helices, as well as the peptide (A). A cross-section of the MHC-peptide complex is also presented (B). Here the positions 3, 5, and 9 of the peptide can be seen to go deeply into the MHC molecule, something also reflected in the rulesets generated for HLA-B*08. The figures were generated using BALLView [21].

Table 2. Prediction performance of the SYFPEITHI and BIMAS methods for the three MHC alleles HLA-A*0201, HLA-B*08, and HLA-B*2705. The measures used for the performance evaluation are Matthews correlation coefficient (MCC), specificity (SP), sensitivity, (SE), and overall accuracy (ACC).

	A*0201				B*2705				B*08			
Method	MCC	SP	SE	ACC	MCC	SP	SE	ACC	MCC	SP	SE	ACC
SYFPEITHI	0.84	0.95	0.88	0.93	0.92	0.96	0.98	0.96	0.83	0.98	0.81	0.92
BIMAS	0.79	0.93	0.86	0.91	0.95	0.97	1.0	0.97	0.79	0.98	0.75	0.91

rules that have to be considered, in terms of MHC-peptide binding, for peptide-vaccine design. Whole proteins or even whole proteomes can easily be analyzed for vaccine candidates using these rules.

6   Conclusions

Decisions rules based on amino acid properties are well-suited to understand and predict MHC-peptide binding for several alleles. The method reveals the patterns recognized by MHC molecules and does so in the form of concise, easily interpretable rules. These rules are found to be meaningful in the context of protein structure and consistent with current immunological knowledge. Interestingly, these simple rulesets outperform methods based on position-specific scoring matrices in predicting MHC binding peptides. The main difference between the two methods lies in the feature encoding. Projecting the peptide sequence into a physiochemical property space obviously yields features well-suited to describe the underlying binding process. The resulting rules are thus quite simple and provide insights into the biophysics of the recognition process. We conclude the rule-based method prove to be a valuable tool in the prediction of MHC binding peptides, the first step in the *in-silico* design of peptide-based vaccines.

References

1. Scheibenbogen, C., Schmittel, A., Keilholz, U., Allgauer, T., Hofmann, U., Max, R., Thiel, E., Schadendorf, D.: Phase 2 trial of vaccination with tyrosinase peptides and granulocyte-macrophage colony-stimulating factor in patients with metastatic melanoma. J. Immunother **23** (2000) 275–281

2. Gjertsen, M.K., Buanes, T., Rosseland, A.R., Bakka, A., Gladhaug, I., Soreide, O., Eriksen, J.A., Moller, M., Baksaas, I., Lothe, R.A., Saeterdal, I., Gaudernack, G.: Intradermal ras peptide vaccination with granulocyte-macrophage colony-stimulating factor as adjuvant: Clinical and immunological responses in patients with pancreatic adenocarcinoma. Int. J. Cancer **92** (2001) 441–450

3. Ourmanov, I., Brown, C.R., Moss, B., Carroll, M., Wyatt, L., Pletneva, L., Goldstein, S., Venzon, D., Hirsch, V.M.: Comparative efficacy of recombinant modified vaccinia virus Ankara expressing simian immunodeficiency virus (SIV) Gag-Pol and/or Env in macaques challenged with pathogenic SIV. J. Virol **74** (2000) 2740–2751

4. De Groot, A.S., Marcon, L., Bishop, E.A., Rivera, D., Kutzler, M., Weiner, D.B., Martin, W.: HIV vaccine development by computer assisted design: the GAIA vaccine. Vaccine **23** (2005) 2136–2148

5. Mamitsuka, H.: MHC molecules using supervised learning of hidden markov models. Proteins: Structure, Function and Genetics **33** (1998) 460–474

6. Falk, K., Rotzschke, O., Stevanovic, S., Jung, G., HG, R.: Allele-specific motifs revealed by sequencing of self-peptides eluted from MHC molecules. Science **351** (1991) 290–296

7. Rammensee, H.G., Bachman, J., Philipp, N., Emmerich, N., Bachor, O.A., Stevanovic, S.: SYFPEITHI: a database for MHC ligands and peptide motifs. Immunogenetics **50** (1997) 213–219

8. Parker, K.C., Bednarek, M.A., Coligan, J.E.: Scheme for ranking potential HLA-A2 binding peptides based on independent binding of individual peptide side-chains. J. Immunol. **152** (1994) 163–175

9. Gulukota, K., Sidney, J., Sette, A., DeLisi, C.: Two complementary methods for predicting peptides binding major histocompatibility complex molecules. J. Mol. Biol. **267** (1997) 1258–1267

10. Dönnes, P., Elofsson, A.: Prediction of MHC class I binding peptides, using SVMHC. BMC Bioinformatics. **3** (2002) 25

11. Hubbard, T., Barker, D., Birney, E., Cameron, G., Chen, Y., Clark, L., Cox, T., Cuff, J., Curwen, V., Down, T., Durbin, R., Eyras, E., Gilbert, J., Hammond, M., Huminiecki, L., Kasprzyk, A., Lehvaslaiho, H., Lijnzaad, P., Melsopp, C., Mongin, E., Pettett, R., Pocock, M., Potter, S., Rust, A., Schmidt, E., Searle, S., Slater, G., Smith, J., Spooner, W., Stabenau, A., Stalker, J., Stupka, E., Ureta-Vidal, A., Vastrik, I., Clamp, M.: The ensembl genome database project. Nucleic Acids Res **30** (2002) 38–41

12. Kawashima, S., Ogata, H., Kanehisa, M.: AAindex: Amino Acid Index Database. Nucleic Acids Res **27** (1999) 368–369

13. Quinlan, R.: C4.5: Programs for Machine Learning. Morgan Kaufmann Publishers, San Francisco, USA (1993)

14. Matthews, B.W.: Comparison of predicted and observed secondary structure of T4 phage lysozyme. Biochim. Biophys. Acta. **405** (1975) 442–451
15. Takano, K., Yutani, K.: A new scale for side-chain contribution to protein stability based on the empirical stability analysis of mutant proteins. Protein Eng **14** (2001) 525–528
16. Fauchere, J.L., Charton, M., Kier, L.B., Verloop, A., Pliska, V.: Amino acid side chain parameters for correlation studies in biology and pharmacology. Int J Pept Protein Res **32** (1988) 269–278
17. Eriksson, A.E., Baase, W.A., Zhang, X.J., Heinz, D.W., Blaber, M., Baldwin, E.P., Matthews, B.W.: Response of a protein structure to cavity-creating mutations and its relation to the hydrophobic effect. Science **255** (1992) 178–183
18. Ishikawa, K., Nakamura, H., Morikawa, K., Kanaya, S.: Stabilization of Escherichia coli ribonuclease HI by cavity-filling mutations within a hydrophobic core. Biochemistry **32** (1993) 6171–6178
19. Janin, J., Wodak, S.: Conformation of amino acid side-chains in proteins. J. Mol. Biol **125** (1978) 357–386
20. DiBrino, M., Parker, K.C., Shiloach, J., Turner, R.V., Tsuchida, T., Garfield, M., Biddison, W.E., Coligan, J.E.: Endogenous peptides with distinct amino acid anchor residue motifs bind to HLA-A1 and HLA-B8. J. Immunol **152** (1994) 620–631
21. Kohlbacher, O., Lenhof, H.P.: BALL–rapid software prototyping in computational molecular biology. Biochemicals Algorithms Library . Bioinformatics **16** (2000) 815–824

Accuracy of String Kernels for Protein Sequence Classification

J. Dylan Spalding and David C. Hoyle

School of Engineering, Computer Science & Mathematics,
University of Exeter, Harrison Building,
North Park Road, Exeter, EX4 4QF, UK
{j.d.spalding, d.c.hoyle}@exeter.ac.uk

Abstract. Determining protein sequence similarity is an important task for protein classification and homology detection. Typically this may be done using sequence alignment algorithms, yet fast and accurate alignment-free kernel based classifiers exist. Viewing sequences as a "bag of words", we test a simple weighted string kernel, investigating the effects of k-mer length, sequence length and choice of weighting. We also extend the kernel to operate on the k-mer frequency representation of a sequence rather than the "bag of words" representation.

Keywords: Protein Classification, Homology, String Kernel.

1 Originality and Contribution

We investigate use of a simple string kernel method for classification of protein sequences. We start from our own implementation of the alignment free Probabilistic Sequence Search Tool (PSST) method previously proposed by Miller and Attwood[1]. The PSST algorithm uses co-occurrence of rare words to determine similarity of two protein sequences. We provide mathematical justification for the heuristic weighting scheme used by Miller and Attwood[1]. We investigate the effect of sequence length on the performance of the algorithm, and adapt it to compare sequences of arbitrary length with no loss of performance. The adapted algorithm performs well with all types of sequences, including families with large and varying numbers of amino acids between conserved regions.

2 Introduction

Classification of protein function is a central task in Bioinformatics, particularly given the large number of uncharacterized proteins from various sequenced organisms. Functional classification is often made by transfer of annotation from homologous sequences, i.e. characterized protein sequences from other organisms that are believed to share an ancestral sequence with the uncharacterized protein. Determination of homology usually proceeds by detection of similarities between protein sequences, as sequence similarity can imply both functional and structural similarity as well as homology. There are many approaches to determining sequence similarity, e.g. local pair-wise methods, such

S. Singh et al. (Eds.): ICAPR 2005, LNCS 3686, pp. 454–460, 2005.

as BLAST[2] and FASTA[3], profile hidden Markov model based methods[4], iterative methods such as PSI-BLAST[5]. Alternative alignment free based methods offer the advantage of speed. Many of these alternative approaches treat sequences as a "bag of words", and consequently are amenable to string kernel treatments[6]. With such kernel methods being based more upon machine learning than biology, there is a need to develop a strategy for efficient estimation of suitable kernel parameter values.

3   Critical Review

The PSST score function S_N described by Miller & Attwood[1], scores sequence similarity according to the co-occurrence of various k-mers. Different k-mers or words are weighted differently,

$$S_N = \sum_i \omega_i N_i^{(1)} N_i^{(2)} . \tag{1}$$

Here $N_i^{(1)}$ and $N_i^{(2)}$ are binary numbers which equal 1 if k-mer i is present in the sequence (1) or (2) respectively, and 0 otherwise. Miller and Attwood[1] set the weight $\omega_i = 1/\rho_i$ where ρ_i is the training sample (database) average of N_i. This choice of weighting rewards rarity of different k-mers, and is a common heuristic approach within text-based information retrieval where rare words are considered more distinctive and therefore discriminative.

The simple score function S_N, given by eq.(1), is just a weighted scalar product between two vectors $\boldsymbol{N}^{(1)} = (N_1^{(1)}, \ldots, N_M^{(1)})$ and $\boldsymbol{N}^{(2)} = (N_1^{(2)}, \ldots, N_M^{(2)})$, with M being cardinality of the set of all possible k-mers. Consequently S_N represents a kernel, similar in form to the Spectrum[7] and the Mismatch[8] kernels used by Leslie *et al.* for Support Vector Machine classification of protein sequences. Both these algorithms derive their scoring weights from the information contained within the sequences to be compared, such as the number of occurrences of word i within a sequence for the spectrum algorithm. The Mismatch kernel extends the idea of the spectrum kernel by setting a limit on the number of words that do not occur in both sequences and which can still be classified as similar.

The variance, $\mathrm{Var}(N_i)$, of the ith feature is given by $\rho_i(1 - \rho_i)$. Choosing a weighting $\omega_i = 1/\mathrm{Var}(N_i)$ is then essentially equivalent to a sensible pre-processing step that produces a set of new features all with unit variance, and one that would be expected to increase classification performance. Since $1/\mathrm{Var}(N_i)$ increases as $\rho_i \to 0$, this provides a heuristic justification for the weighting scheme chosen by Miller and Attwood[1]. A weighting $1/\mathrm{Var}(N_i)$ also up-weights commonly occurring k-mers, i.e. those for which $\rho_i \to 1$. However for this research we will just retain the weighting used by Miller and Attwood. The weighting $\omega_i = 1/\mathrm{Var}(N_i)$ can also be derived in a more principled fashion by optimization of a suitable objective function, but we do not give the details in this short paper.

4   Proposed Method

For this research we take a selected set of classified protein sequences. Classification based upon a "bag of words" treatment of protein sequences will offer advantages over classifiers acting directly upon the primary amino acid sequence only if there is a genuine difference in information (signal) content of the k-mer composition of a sequence in comparison to the information (signal) content encoded in the amino-acid composition. This is actually to be expected since important functional domains within protein typically extend over several amino acids. To determine if such differences do genuinely exist within our test set we calculate the Kullback-Leibler (KL) distance between the observed (sample) k-mer frequencies and a set of theoretical k-mer frequencies produced from a null-model of k-mer generation that operates at the level of individual amino acids. In this case the null-model k-mer frequencies are calculated as the product of the constituent amino acid frequencies. Statistical significance of the calculated KL distance was evaluated by boot-strapping, producing 10000 artificial data sets under the null-model.

As the length L of a sequence increases there will be a corresponding increase in the probability that a particular k-mer i is present in that sequence, with the probability ultimately saturating, i.e. $P(N_i = 1) \rightarrow 1$ as $L \rightarrow \infty$. As a consequence, the lack of positional information in the algorithm means that a local cluster of words in one sequence can match the same words distributed over the entire length of another long sequence[1]. To address this problem Miller & Attwood [1] divided the sequences up into blocks with each block containing 300 amino acids. An alternative approach, closer to the Spectrum kernel, is to replace the binary features N_i by the frequency f_i of k-mer i within the sequence. Therefore we test a second kernel (using an obvious notation),

$$S_f = \sum_i \omega_i f_i^{(1)} f_i^{(2)} . \tag{2}$$

By analogy with the original score function of Miller and Attwood we chose to test the weighting scheme $\omega_i = 1/\pi_i$, where π_i is the population (database) frequency of k-mer i.

5   Experiment Design

Following Miller and Attwood the protein sequences used were taken from the PRINTS[9] database, a database of fingerprints for detecting members of large protein super-families. Each fingerprint is made up of aligned and un-weighted motifs[1]. PRINTS contains over 67,000 sequences, so for speed a smaller database, miniPRINTS, was constructed from 20 families represented in PRINTS, and duplicate sequences were removed. miniPRINTS contains highly divergent super-families as well as some small well-defined families[1].

To investigate the effect of sequence length on the performance of the algorithm, mini-PRINTS was then divided again into three databases of approximately equal size; a database of short, mid-length, and long sequences. The short database contained sequences with less than 355 amino acids, the mid-length database sequences with between 355 and 560 amino acids inclusive, and the long database sequences with more than 560 amino acids.

We implemented the algorithm in Java, removing each sequence within the miniPRINTS database in turn to create a query sequence. This was scored, using (1) or (2), against the remaining sequences in the same miniPRINTS database. The highest score, or top-hit, was used to classify the query sequence. The performance of the kernels in (1) and (2) was investigated for word lengths $k = 3$ and $k = 4$. With a 21 character alphabet (21 amino acids if we include selenocysteine) the number of possible k-mers is 21^k. Beyond $k = 4$, with > 160000, k-mers this raises concerns of computational efficiency. However for large choices of k there is also a significant likelihood that k-mers become specific to a single sequence, with a consequent reduction in recall when querying against the database. A naïve calculation (based upon having $\mathcal{O}(1)$ sequences in the database with at least one k-mer matching the query sequence) indicates that for a typical sequence length L in a database of N sequences we require,

$$k = \frac{\ln NL^2}{\ln 21} \tag{3}$$

Substituting into (3) $L = 450$ and $N = 1000$, appropriate for the mid-length database, gives $k \simeq 6$. This represents an upper estimate on k, beyond which recall is significantly reduced. Consequently we have chosen to primarily perform calculations with $k = 3$ and $k = 4$.

6 Results

For $k = 3$ and $k = 4$ the KL distances between sample k-mer frequencies and the null-model were 0.058 and 0.316 respectively. In both cases this was highly statistically significant ($p < 1/10000$), indicating that as expected the information (signal) content contained within the k-mer composition of a sequence is genuinely different from the information content of its amino-acid composition, and that consequently it is a valid approach to construct a classifier based upon k-mer composition.

Each top-hit score was assigned as being a positive hit of negative hit according to whether it was greater or less than a pre-selected threshold. Variation of the threshold value permits the calculation of a Receiver Operating Characteristic (ROC) curves for each database and each kernel score function. The ROC curves consist of the True Positive Fraction (TPF) plotted against the False Positive Fraction (FPF), with TPF and FPF defined as,

$$\text{TPF} = \frac{\text{True Positives}}{\text{True Positives} + \text{False Negatives}} \;,\; \text{FPF} = \frac{\text{False Positives}}{\text{False Positives} + \text{True Negatives}} \tag{4}$$

The area (AROC) of the resulting curve gives an indication of how well the algorithm under test discriminates between the data and are given in Table 1. For completeness the classification accuracy of the top-hit, irrespective of threshold, is given in Table (2).

Table 1. Area under the ROC curves for each algorithm for each database.

Database	$k = 3$		$k = 4$	
	S_f	S_N	S_f	S_N
Short	0.9395	0.7781	0.9243	0.7916
Mid	0.9525	0.9399	0.9689	0.9633
Long	0.9555	0.8280	0.9800	0.9564

Table 2. Percentage of correct classifications (accuracy) of the algorithm.

Database	$k = 3$		$k = 4$	
	S_f	S_N	S_f	S_N
Short	94.051%	95.795%	95.179%	96.821%
Mid	97.020%	95.889%	96.814%	96.711%
Long	96.107%	68.238%	95.082%	81.455%

The poor results for the kernel S_f operating on short sequences are due to the poor results for the LIPOCALIN and ALPHAHAEM families, with AROC for these families of 0.67 and 0.85 respectively. Since sequences from both of these difficult to classify families are only represented in the short database, this has biased the results for the database of short sequences.

The results show that the accuracy of the kernel S_N, operating on binary features, does reduce as the sequence length increases, whereas the kernel S_f, operating on continuous valued features, performs consistently irrespective of sequence length.

The results for each family within miniPRINTS are listed in Table 3. For comparison we have evaluated each kernel classifier with a uniform weighting scheme $\omega_i \equiv 1, \forall i$. Table 3 shows that for most families, with the exception of FNTYPEIII the kernel S_f performs as well or better than the kernel S_N. The kernel S_f also performs more consistently across the families, with the percentage of correct classifications falling to a minimum of 87.5% for OPSINs, as opposed to 12.24% for classification of GLHYDRLASE3 with S_N. All sequences for GLHYDRLASE3 are contained within the long database, with an average length of 789 amino acids, which could explain the poor performance of S_N with this family. On average, the use of a non-uniform weighting ω_i improves top-hit classification accuracy. Some families, e.g. HEATSHOCK90 and KRINGLE, show marked improvement at the shorter k-mer length $k = 3$. However improvement is by no means consistent across all families - for example GPCRRHODOPSIN and PRION show a small decrease in classification accuracy when using the non-uniform weighting schemes.

7   Conclusion

The quality of these results suggest that even the simplest of kernels may be well suited to the problem of protein sequence classification. The results show that a kernel operating on the k-mer composition, f_i, of a sequence tends to outperform a kernel operating on the binary features N_i. In particular the classification performance S_f is less susceptible to variations in sequence length than S_N. The results also show that the accuracy of all the kernels considered, while related to the discriminatory power given by the AROC values, tends to be greater than the the poorer AROC values suggest. This is because the classification method used is in effect a nearest neighbour classifier, and thus is not effected by how close the two distributions are in general, just the family of the nearest neighbour.

Table 3. Percentage of correct classifications for each family, using the highest scoring sequence to indicate family, with $k = 3$ & $k = 4$ and different choices of weighting scheme ω_i

Family	k=3						k=4					
	$\omega_{i=1}$		$\omega_i = {}^1/\rho_i$		$\omega_i = {}^1/\pi_i$		$\omega_{i=1}$		$\omega_i = {}^1/\rho_i$		$\omega_i = {}^1/\pi_i$	
	S_N	S_f	S_N	S_f	S_N	S_f	S_N	S_f	S_N	S_f	S_N	S_f
ALPHAHAEM	99.33%	99.33%	98.65%	99.33%			99.33%	99.33%	99.33%	99.33%		
BETAAMYLOID	60.00%	100.00%	70.00%	100.00%			100.00%	100.00%	100.00%	100.00%		
CYTOCHROMEF	100.00%	100.00%	100.00%	100.00%			100.00%	100.00%	100.00%	100.00%		
DPTHRIATOXIN	100.00%	100.00%	100.00%	100.00%			100.00%	100.00%	100.00%	100.00%		
EUMOPTERIN	29.79%	91.49%	40.43%	95.74%			87.23%	95.74%	85.11%	95.74%		
FANCINICGENE	50.00%	100.00%	50.00%	100.00%			75.00%	100.00%	100.00%	100.00%		
FNTYPEIII	97.47%	97.59%	97.70%	92.18%			97.01%	89.89%	97.70%	90.57%		
GLHYDRLASE3	8.16%	83.67%	12.24%	97.96%			42.86%	91.84%	48.98%	95.92%		
GPCRRHODOPSN	95.41%	96.91%	93.31%	94.71%			97.01%	95.92%	94.81%	96.31%		
HEATSHOCK90	40.30%	100.00%	53.73%	100.00%			88.06%	100.00%	85.07%	100.00%		
KINESINLIGHT	63.64%	100.00%	63.64%	100.00%			100.00%	100.00%	100.00%	100.00%		
KRINGLE	53.97%	93.65%	66.67%	95.24%			88.89%	96.83%	87.30%	96.83%		
LIPOCALIN	91.21%	94.51%	93.41%	94.51%			97.80%	98.90%	94.51%	96.70%		
NIHGNASESMLL	100.00%	94.12%	100.00%	100.00%			100.00%	100.00%	100.00%	100.00%		
OPSIN	88.75%	84.50%	88.75%	88.75%			91.25%	88.75%	90.00%	88.75%		
PHOTOSYPSAAB	63.41%	100.00%	100.00%	100.00%			100.00%	100.00%	100.00%	100.00%		
PRION	100.00%	100.00%	97.37%	97.37%			100.00%	100.00%	100.00%	100.00%		
RHODOPSIN	91.49%	91.49%	91.49%	91.49%			91.49%	91.49%	89.36%	89.36%		
URICASE	95.83%	95.83%	95.83%	95.83%			95.83%	95.83%	95.83%	95.83%		
ZINCFINGER	68.28%	98.75%	71.86%	98.21%			80.65%	98.03%	80.11%	96.06%		
AVERAGE	**74.85%**	**96.09%**	**79.25%**	**97.07%**			**91.62%**	**97.13%**	**92.41%**	**97.07%**		

Classification performance could potentially be improved by using class labels from all sequences above the threshold rather than just the top-hit, which is equivalent to just a nearest-neighbour classifier. The results from the ROC curves suggest that this type of classifier may perform very well. The use of regular expressions to allow for non-exact matches may also improve results and is something we are currently investigating.

References

1. Miller, C.J., Attwood, T.K.: Psst...the probabilistic sequence search tool. In: 2nd IEEE International Symposium on Bioinformatics and Bioengineering, Washington, IEEE Press (2001) 33–40 ISBN 0769514235.
2. Altschul, S.F., Gish, W., Miller, W., Myers, E.W., Lipman, D.J.: Basic local alignment search tool. Journal of Molecular Biology **215** (1990) 403–410
3. Pearson, W.R.: Rapid and sensitive sequence comparison with fastp and fasta. Methods Enzymol **183** (1990) 63–98
4. Krogh, A., Brown, M., Mian, I., Sjolander, K., Haussler, D.: Hidden markov models in computational biology: Applications to protein modelling. Journal of Molecular Biology **235** (1994) 1501–1531
5. Altschul, S.F., Madden, T.L., Schaffer, A.A., Zhang, J., Zhang, Z., Miller, W., Lipman, D.J.: Gapped BLAST and PSI–BLAST: A new generation of protein database search programs. Nucleic Acids Research **25** (1997) 3389–3402
6. Shawe-Taylor, J., Cristianini, N.: Kernel Methods for Pattern Analysis. Cambridge University Press (2004)

7. Leslie, C., Eskin, E., Weston, J., Noble, W.S.: The spectrum kernel: a string kernel for svm protein classification. In: Proceedings of the Pacific Symposium on Biocomputing, World Scientific (2002) 564–575
8. Leslie, C., Eskin, E., Weston, J., Noble, W.S.: Mismatch string kernels for svm protein classification. In: Advances in Neural Information Processing Systems. (2002)
9. Attwood, T.K., Bradley, P., Flower, D.R., Gaulton, A., Maudling, N., Mitchell, A., Moulton, G., Nordle, A., Paine, K., Taylor, P., Uddin, A., Zygouri, C.: PRINTS and its automatic supplement, prePRINTS. Nucleic Acids Research **31(1)** (2003) 400–402

An Efficient Feature Selection Method for Object Detection

Duy-Dinh Le[1] and Shin'ichi Satoh[1,2]

[1] The Graduate University for Advanced Studies,
Shonan Village, Hayama, Kanagawa, Japan 240-0193
ledduy@grad.nii.ac.jp
[2] National Institute of Informatics,
2-1-2 Hitotsubashi, Chiyoda-ku, Tokyo, Japan 101-8430
satoh@nii.ac.jp

Abstract. We propose a simple yet efficient feature-selection method — based on principle component analysis (PCA) — for SVM-based classifiers. The idea is to select features whose corresponding axes are closest to the principle components computed from a data distribution by PCA. Experimental results show that our proposed method reduces dimensionality similar to PCA, but maintains the original measurement meanings while decreasing the computation time significantly.

1   Introduction

In many object-detection systems, feature selection — which is generally considered as the selection of a smaller subset of features from a large set of features — is one of the critical issues for the following three reasons.

First, there are many ways to represent a target object, leading to a huge input feature set. For example, Haar wavelet features used in [1] are in the order of thousands. However, only small and incomplete training sets are available. As a result, systems will suffer from the curse of dimensionality and overfitting.

Second, a large feature set includes many irrelevant and correlated features that can degrade the generalization performance of classifiers [2,3].

Third, selecting an optimal feature subset from a large input feature set can improve the performance and speed of classifiers. In face detection, the success of systems such as those in [1,4] comes mainly from efficient feature-selection methods.

Most work, however, only focuses on feature-extraction methods, such as principle-component analysis (PCA), linear discriminant analysis (LDA), and independent-component analysis (ICA) [5,6,7], which try to map data from high-dimensional space to lower-dimensional space. This might be because feature-selection methods, such as sequential forward selection (SFS), sequential backward selection (SBS), and sequential forward floating search (SFFS) [8,9], incur very high computational cost.

In this paper, to address these problems, we propose a simple yet efficient feature-selection method for object detection. The main idea is to select features

S. Singh et al. (Eds.): ICAPR 2005, LNCS 3686, pp. 461–468, 2005.

whose corresponding axes are closest to principle components computed by PCA from the data distribution. This is a very naive feature-selection method, but experimental results on different kinds of features show that when working with support vector machine (SVM)-based classifiers, our proposed method has comparable performance, but faster speed, compared to a feature-selection method based on PCA directly.

The rest of the paper is organized as follows: In section 2, feature extraction by PCA is presented. Our feature selection method is introduced in section 3. Experimental results are showed in section 4. Finally, section 5 concludes the paper.

2   Feature Extraction Using PCA

The main steps to extract features using PCA are summarized in the following. The details are given in [5].

Each face image $I(x, y)$ is represented as an $N \times N$ vector Γ_i.

The average face Ψ is computed as: $\Psi = \frac{1}{M} \Sigma_{i=1}^{M} \Gamma_i$ where M is the number of face images in the training set.

The difference between each face and the average face is given as: $\Phi_i = \Gamma_i - \Psi$. A covariance matrix is then estimated as: $C = \frac{1}{M} \Sigma_{i=1}^{M} \Phi_i \Phi_i^T = AA^T$ where $A = [\Phi_1 \Phi_2 ... \Phi_M]$.

Eigenvectors u_i and corresponding eigenvalues λ_i of the covariance matrix C can be evaluated by using a Singular Value Decomposition (SVD) method [5]: $Cu_i = \lambda_i u_i$. Because matrix C is usually very large ($N^2 \times N^2$), evaluating eigenvectors and eigenvalues is very expensive. Instead, eigenvectors v_i and corresponding eigen values μ_i of matrix $A^T A$ ($M \times M$) can be computed. After that, u_i can be computed from v_i as follows: $u_i = Av_i, j = 1, ..., M$.

To reduce dimensionality, only a smaller number of eigenvectors $K(K << M)$ corresponding to the largest eigenvalues are kept. A new face image Γ, after subtracting the mean ($\Phi = \Gamma - \Psi$) can then be reconstructed in eigenspace by the formula: $\tilde{\Phi} = \Sigma_{i=1}^{K} w_i u_i$ where $w_i = u_i^T \Psi$ are coefficients of the projection and can be considered as a new representation of the original face in this eigenspace.

3   The Proposed PCA-Based Feature Selection

The main idea of our naive feature-selection method is to investigate the principle components computed by PCA in the projection space to select corresponding axes in the original space. Selected axes are those closest to these principle components. Specifically, starting from each principle component e_i in the projection space, we try to find the principle axis x_j in the original space closest to e_i. As a result, the jth feature will be selected.

The method is illustrated in Figure 1. According to the data distribution, e_1 and e_2 are principle components sorted by their corresponding eigen values. By using PCA for feature extraction, we can map data from (x_1, x_2) to (y_1, y_2).

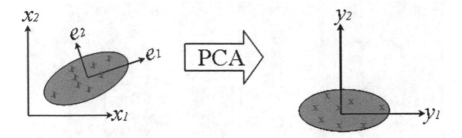

Fig. 1. Feature extraction by using PCA

And by using the proposed feature-selection method, starting from e_1, x_1, which is the closest to e_1, is found. Hence, the first feature, i.e, x_1, will be selected. The proposed algorithm is summarized as follows:

- **Step 1:** Compute principle components $\{e_1, e_2, ..., e_N\}$ from the data distribution by PCA and sort them in the order of the magnitude of eigen values.
- **Step 2:** For each principle component e_i, find the axis x_j that is closest to e_i.
- **Step 3:** Select feature j^{th}.

4   Experimental Results

4.1   Training Data

We demonstrated efficiency of our feature-selection method by building a face detector based on SVM. For training, we used 7,000 face samples and 7,000 non-face samples. Face samples are collected from the Internet, cropped and resized to a size of 24x24 pixels. Non-face samples are generated from 6,278 images with various subjects such as rocks, trees, buildings, scenery, and flowers that contain no faces. Figure 2 shows some of the face and non-face samples. For comparison, 2,450 face samples and 7,000 non-face samples different from the training set were also used.

4.2   Pixel-Based Features

In this experiment, we used the intensity of pixels as features. LibSVM [10] was used to train SVM classifiers with a RBF kernel on selected feature subsets. We compared the performances of SVM classifiers trained on subset features selected by our method and subset features selected from PCA-based feature extraction in which the top-100 and top-200 eigenvectors were used. The results in Figure 3 shows that the performances of the SVM classifiers are comparable, particularly

Fig. 2. Some 24x24 face and non-face samples

when the number of features in each subset is large enough, e.g., 200. However, in terms of speed, the SVM classifier trained on a 200-feature set selected by our method can process 86 patterns per second (PPS) while the SVM classifier trained on the top-200 eigenvectors can only process 80 PPS (i.e., approximately 1.08 times slower).

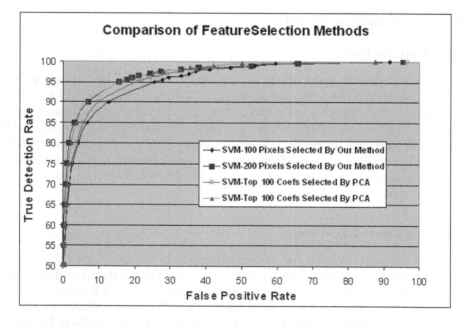

Fig. 3. Performances of SVM classifiers trained on different feature subsets selected from different selection method are comparable when the number of selected features is large enough

Fig. 4. Image of 200 pixels (depicted in white) selected by the proposed selection method

Figure 4 shows 200 pixel features selected by our method. It is easy to see that selected pixels belong to major parts of facial features such as eyes, mouth, and nose.

4.3   Haar Wavelet Features

In recent face detectors based on AdaBoost [1,11,12], Haar wavelet features are used quite extensively because they are rich and can be evaluated very quickly. In our experiment, we used the same three kinds of Haar wavelet features as in [1] which are modeled from adjacent rectangles with the same size and shape. The feature value is defined as the difference of the sum of the pixels within rectangles (see Figure 5).

By using integral image definition [1], these feature rectangle values can be computed very quickly. The integral image at location (x, y) is defined as $ii(x, y) = \sum_{x' <= x, y' <= y} i(x', y')$ where $ii(x, y)$ is the integral image and $i(x, y)$ is the original image. In practice, $ii(x, y)$ can be computed simply by using the following recurrent function: $ii(x, y) = ii(x, y - 1) + ii(x - 1, y) + i(x, y) - ii(x - 1, y - 1)$ and sum of the pixels within a rectangle can be computed from four integral image values of its vertices, for example, $Sum(D) = 1 + 4 - (2 + 3)$.

Because the Haar wavelet feature set defined above is over-complete (close to 200,000 features), to use it with SVM [13], first, the maximum 200 features are selected by AdaBoost [14,1]. Then, from the same feature set, the first-50 features are selected in the order they are added in the training process, and another first-50 features are selected by using our method. The performances of the SVM classifiers trained on these two subsets are shown in Figure 6. This figure indicates that, in terms of performance, using our feature-selection method is slightly better than not using it. In terms of speed, the SVM classifier trained on the feature subset selected by our method has 3,405 support vectors and runs at a speed of 538 PPS, while that trained on the first-50-feature subset has 4,017 support vectors and runs at a speed of 469 PPS (approximately 1.15 times slower). In Figure 7, some face-detection results are shown.

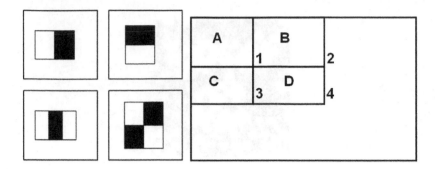

Fig. 5. Haar wavelet features can be evaluated very fast by using the integral image

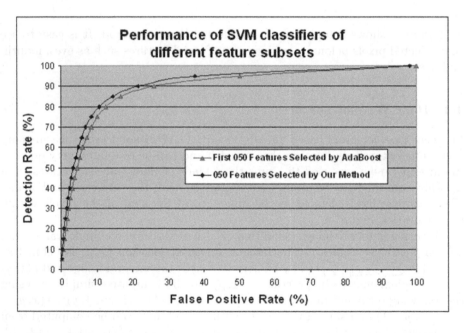

Fig. 6. Performances of two 50-feature subsets selected by different methods

5   Conclusion

We have developed a simple yet efficient method for selecting a good feature subset for building object-detection systems. The method investigates at variance of input data and selects features which are closest to principle components computed by PCA. With this method, by reducing dimensionality of feature vectors, the final classifier runs faster while maintaining high prediction accuracy. In experiments on different kinds of features used for face detection, the method demonstrated promising results.

Fig. 7. Some face detection results

Acknowledgments

The authors thank the anonymous reviewers for their helpful comments on improving this paper.

References

1. Viola, P., Jones, M.: Rapid object detection using a boosted cascade of simple features. In: Proc. Intl. Conf. on Computer Vision and Pattern Recognition (CVPR). Volume 1. (2001) 511–518
2. Bins, J., Draper, B.A.: Feature selection from huge feature sets. In: Proc. Intl. Conf. on Computer Vision (ICCV). Volume 2. (2001) 159–165
3. Sun, Z., Bebis, G., Miller, R.: Object detection using feature subset selection. Pattern Recognition **37** (2004) 2165–2176
4. Li, S., Zhang, Z.: Floatboost learning and statistical face detection. IEEE Transactions on Pattern Analysis and Machine Intelligence **26**(9) (2004) 23–38
5. Turk, M., Pentland, A.: Face recognition using eigenfaces. In: Proc. Intl. Conf. on Computer Vision and Pattern Recognition (CVPR). (1991)
6. Martinez, A., Kak, A.: Pca versus lda. IEEE Transactions on Pattern Analysis and Machine Intelligence **23**(2) (2001) 228–233
7. Bartlett, M., Movellan, J., Sejnowski, T.: Face recognition by independent component analysis. IEEE Transactions on Neural Networks **13**(6) (2002) 1450–1464
8. Jain, A., Zongker, D.: Feature selection: Evaluation, application, and small sample performance. IEEE Transactions on Pattern Analysis and Machine Intelligence **19**(2) (1997) 153–158
9. Pudil, P., Novovicova, J., Kittler, J.: Floating search methods in feature selection. Pattern Recognition Letter **15**(11) (1994) 1119–1125
10. Chang, C.C., Lin, C.J.: LIBSVM: a library for support vector machines. (2001) Software available at http://www.csie.ntu.edu.tw/~cjlin/libsvm.
11. Liu, C., Shum, H.: Kullback-leibler boosting. In: Proc. Intl. Conf. on Computer Vision and Pattern Recognition (CVPR). Volume 1. (2003) 587–594
12. Lin, Y.Y., Liu, T., Fuh, C.S.: Fast object detection with occlusions. In: Proc. Intl. European Conference on Computer Vision (ECCV). Volume 3021. (2004) 402–413
13. Le, D.D., Satoh, S.: Fusion of local and global features for efficient object detection. In: Applications of Neural Networks and Machine Learning in Image Processing IX, IS&T/SPIE Symposium on Electronic Imaging. (2005)
14. Freund, Y., Schapire, R.E.: A short introduction to boosting. Journal of Japanese Society for Artificial Intelligence **14**(5) (1999) 771–780

Multi-SOMs: A New Approach to Self Organised Classification

Nils Goerke, Florian Kintzler, and Rolf Eckmiller

Div. of Neural Computation, Dept. of Computer Science,
University of Bonn, D-53117 Bonn, Germany
{goerke, kintzler, eckmiller}@nero.uni-bonn.de
http://www.nero.uni-bonn.de

Abstract. We propose a method to use self organizing neural networks to extract information out of nonlinear dynamic systems for control. Nonlinear strange attractors are educed by these systems or the attractors can be reconstructed. These attractors are partitioned by a newly developed self organizing neural network. Thus the stream of system states is transformed into a stream of symbols, which can now serve as basis for further investigation or control. We are convinced, that controlling and understanding such nonlinear or chaotic systems is easier, when using the information within the stream of extracted symbols.

1 Introduction

Nonlinear systems can be found in a wide variety of applications. Nonlinearity seems to be a favorite of "Mother Nature"; therefore it is desirable to get control over nonlinear dynamical systems, although modeling isn't available or appropriate. Controlling and describing these systems mostly requires profound knowledge about the governing dynamics [1,2,6]. Since nonlinear dynamic systems tend to educe chaotic attractors that determine the systems development most linear methods fail directly while attempting to gain control over the system. As presented in this paper, these attractors are the key-components in a new control-model, which utilizes almost the entire complexity of the system and is more convenient than linear methods.

The complete dynamics of the attractor is represented within the temporal evolution of the state variables [1,7,8]. A time series of these system variables contains the characteristics of the nonlinear, sometimes even chaotic system. [1,8]. It has been shown, that the reconstruction of an attractor by temporal delayed sampling of the state variables is possible [7]. Such attractors are a reconstruction of the original attractor, with the same dynamical properties than the original one. Thus it is an alternative representation of dynamics of the system, suitable for classification and subsequent control. To take further advantage of the information carried within the attractor, respectively the time series, we transform this information into a stream of symbols. All further classification and control tasks will then use the stream of classified symbols as basis.

S. Singh et al. (Eds.): ICAPR 2005, LNCS 3686, pp. 469–477, 2005.

2   Method

Our purpose is to give an alternative access to investigation of nonlinear dynamical systems, by partitioning the nonlinear attractor into regions. We propose to express and to represent the dynamics of the chaotic system by a stream of symbols which preserves the dynamical properties. This stream of symbols realiseds the basis for subsequent information processing stages like, pattern recognition, control, clustering, time series comparision, ...

In traditional control theory often a priori knowledge about the application or the process is used, e.g. a threshold value. We want our model to be self organised and solely data dependent. So we decided to use self organizing neural networks to do the job of knowledge extraction.

We implemented a Self Organizing Map (SOM) as proposed by T. Kohonen [4], a Neural Gas [5] and a Growing SOM (G-SOM) as proposed by B. Fritzke [3] and investigated the results with respect to partitionning a chaotic, nonlinear dynamical system. In some cases, clustering the Roessler or the Lorenz attractor, the SOM *twisted* and the results weren't usable. It has been often observed that twisting depends on the SOM size [4], and small SOMs ($\leq 5x5$) normally don't twist at all. The Neural Gas model on the other hand, has the disadvantage that it does not realize a topology preserving mapping into a low-dimensional space [3], which is desirable for easy handling of a system of higher dimensionality. To overcome these described disadvantages, we developed an extention of the SOM, a Multiple Self Organizing Map (M-SOM).

2.1   Architecture

The Multi-SOM (M-SOM) is a set of multiple partner-SOMs. It is not necessary that all partner-SOMs within a Multi-SOM are identical, they can be different in topology, size and dimension. Each partner-SOM has the complete functionality of a classical SOM. The novelty of the presented approach arises from the possibility to treat each partner-SOM individually.

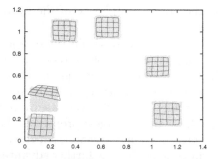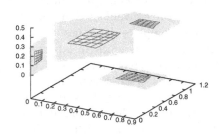

Fig. 1. left: M-SOM containing six 5×5 partner-SOMs, depicted while learning a segmented distribution in 2D. right: M-SOM containing four 2-dimensional 6×6 partner-SOMs, depicted while learning a segmented data distribution in 3D.

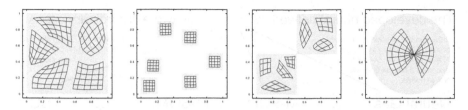

Fig. 2. Several 2-dimensional M-SOMs with different topology and different areas to represent

To avoid the unwanted effect of twisting and to increase adaptation speed, each of these partner-SOMs should be kept small in size, which has been reported several times througout the neural networl community. Extensive simulations yielded a reliable bound for the size of the SOMs to be below $\approx 8 \times 8 \times 8$. A regular topology M-SOM consists of M partner-SOMs with cartesian, rectangular or cuboid dimensions. An M-SOM with M cuboid partner SOMs, organised in K_1 rows with K_2 columns and K_3 layers will be denoted as :

$$M - K_1 \times K_2 \times K_3 \times \ldots$$

and have $(N = M \cdot K_1 \cdot K_2 \cdot K_3 \cdot \ldots)$ neurons in total.

The M-SOM topology has two extraordinary cases:

- If the number of SOMs shrinks to $M = 1$, the M-SOM equals a classical SOM.
- On the other hand, if the number of neurons in every partner-SOM shrinks to 1, with $N = M \cdot 1$, the M-SOM becomes a Neural Gas with $N = M$ nodes.

2.2 Learning

Learning with a Multi-SOM is done using the same methods as for classical SOMs, to benefit from the complete variety of published modifications and enhancements to the basic SOM algorithm.

Two basic M-SOM learning principles arise directly from the Multi-SOM architecture:

I Pure un-supervised M-SOM learning
II Self-organised-supervised M-SOM learning

I) Unsupervised learning is the classical way of adapting SOMs. For pure un-supervised M-SOM adaptation, only the very partner SOM that contains the winning neuron (winning SOM) is changed, following the approved methods of SOM-adaptation. Thus only the reference vectors of the winning SOM are modified to form a topographic mapping of the input space in the direct vicinity of the actual input vector. The other partner SOMs within the M-SOM remain unchanged, and can thus "concentrate" their topological mapping to other regions of the input space. Since only a small fraction of the M-SOM has to be modified

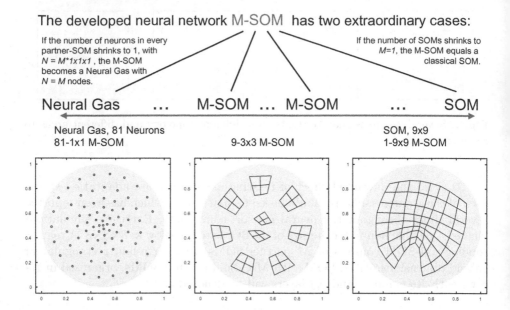

Fig. 3. Depending on the chosen topology the M-SOM can be anything between a classical SOM and a neural gas

during each training step, adaptation is fast and the the elapsed time is low, even if a large number of iterations might be necessary.

II) Supervised training methods require somehow the access to a teacher, or at least a critic to provide a teaching information (e.g. LVQ, LOVQ). M-SOMs give the appealing feature to devise teacher information in an unsupervised way. Thus self-organisation of the individual partner-SOMs can be established using teacher-like information that has been gained through a process of unsupervised classification.

In this case we speak about self-organised-supervised M-SOM learning. Lets suppose that each individual partner-SOM is a representative of an individual class of data, then all other partner-SOMs stand for other classes. This fact can now be used to devise the teaching information. Once the winning SOM has been determined and adapted, the other partner-SOMs are trained following the idea of Learning Vector Quantization. LVQ is designed to be a supervised method for classification which makes benefit of the class-membership of a given input value. Within the self-organised-supervised M-SOM learning the self organised information of being a member of the winning SOM or not is used instead. Of course the complete variety of LVQ methods and enhancements can be applied to the M-SOM.

2.3 Classification

In our task of classifying regions of chaos in the selected attractors, we regarded every partner net of the M-SOM as one class of it's own. Once the M-SOM has

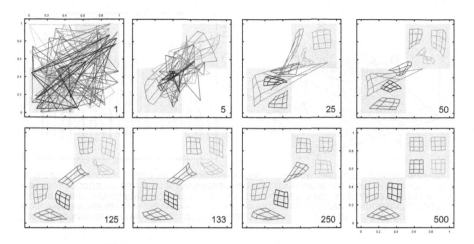

Fig. 4. Unsupervised learning to represent an area that is divided into two parts with an 7-4×4 M-SOM. Depicted are selected snapshots at time $1, 5, 25, 50$ (upper row from left ro right) and $125, 133, 250, 500$ (lower row from left to right). The seven partner-SOMs are arranged to cover the teaching area.

adapted it's shape to the input data, the classification of the stream of derived values is reduced to just notice which partner-net contains the winner-neuron. This leads to a fast computation and good classification-results, as demonstrated below.

Subsequent processes working on the derived stream of symbols can lead to further information extraction. Principles from signal processing can be used to characterize the development within the stream of symbols and thus within the dynamical system. Since the underlying dynamics is deterministic and concentrated to reside on the attractor, it is likely to find a grammar within the sequence of symbols, worth to be further investigated.

3   Simulation

To evaluate and demonstrate the capabilities of our approach, we have decided to use two well investigated nonlinear chaotic systems, the Roessler-System and the Lorenz-System [8] (see figure 5 and 6). These attractors of both systems do have the advantage of being embedded into the 3-dimensional space, so that visualization is still possible. The developed method of information-extraction is not limited to three dimensions, and applies to higher dimesnions as well.

We have calculated data from the chaotic attractors, which - treated as a time-series - represent the evolution of state variables of a fictive nonlinear dynamical system. Starting the calculation with arbitrary parameters thus equals starting this fictive system with an arbitrary starting-configuration. The resulting values were used to train the self organizing neural networks, for generating a

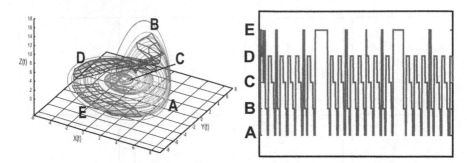

...BBBBBBBBBAADDDDDDDDDDDDDDCCCCCCCCCBBAAAAAAADDDDDDDDDDDDCCCCCCCCCBBBBBBBBAAAA
DDDDDDDDDDDDDCCCCCCCCCBBAAAAAEEEEEEBBBBBBBBBAAADDDDDDDDDDDDDDCCCCCCCCCBBBBA
AAAAEEEEEEEEEEEEEEEEEEEEEEEEEEEEEEEEEEEEEEEEEEEEEEEEEEEBBBBBBBAAAADDDDDDDDDDD
CCCCCCCCCBBBBBBBAAAAAADDDDDDDDDDDDDDCCCCCCCCCBBAAAAAEEEEEEEEBBBBBBBBBAAAADDDD
DDDDDDDCCCCCCCCCBBBBBAAAAAADDDDDDDDDDDDDDCCCCCCCCCBBBBAAAAAAAEEDDDDDDDDDDCCCC...

Fig. 5. Left: the Roessler-attractor classified by a 5-5×10 M-SOM. The five partner-SOMs cover the area of the attractor. Right: stream of five symbols (A,B,C,D,E) derived by classifying subregions of the Roessler-attractor into five classes. Bottom: visualisation of the stream of 5 symbols derived with the 5-5×10 M-SOM.

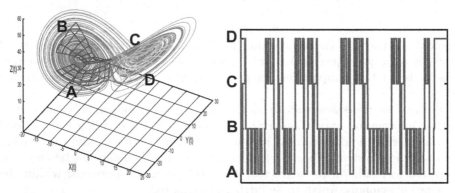

...CCCCCCCCCCCCCCCCCCCCCCCCCCCCCCCCCCCCDDDDDDDDDDDBBBBBBBBBBBBBBBBBBBBBBBBBBB
BBBBBBBBBBBBBBBAAAAAAAAAAAAAAAAAAAAAAAAAAAAAAAAAAAAAAAAAAAAAABBBBBBBBBBBBBBBBBB
BBBBDDDDAAAAAAAAAAAAAAAAAAACCCAAACCCCCCCCCCCCCCCCCCCCCCCCCCCCCCCCCCCCCCCCCC
CCCCDDDDDDDDDDDDDBBBBBBBBBBBBBBBBBBBBBBBBBAAAAAAAAAAAAAAABBBBBBBBBBBBBBBB
BBBBBBBBBBBBBBBBBBBBBBBBBBBBBBBAAAAAAAAAAAAAAAAAAAABBBBBBBBBBBBBBBBBBBBBBB...

Fig. 6. Left: the Lorenz-attractor clustered by a 4-4×8 M-SOM. Each of the two "wings" of the attractor is well represented by two partner-SOMs. Right: stream of four symbols (A,B,C,D) derived from the M-SOM partitionning of the Lorenz-attractor. Bottom: visualisation of the stream of 4 symbols derived with the 4-4×8 M-SOM.

partitioning scheme. In the case of the M-SOM (described above), we have tested various sizes and quantities of regular (2-dimensional) M-SOMs, and found them to be adequate for this task.

The time series, obtained by integration of the differential equation system was thereby translated into a stream of symbols, using the proposed partitioning scheme.

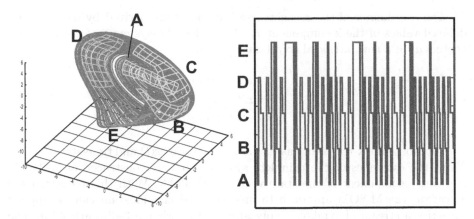

Fig. 7. Left: Reconstructed Roessler-attractor clustered by a 5-5×10 M-SOM Right: stream of symbols derived by clustering the reconstructed Roessler-attractor. The stream of symbols reflect the dynamics of the attractor, that is similiar to the dynamics of the original attractor.

Further work on the M-SOM model will lead to more sophisticated results and finer shaped self organizing neural representations.

In Addidition to thet we have applied the developed method to a three dimensional attractor that has been reconstructed from a one-dimensional time series by time-delayed sampling. Therefore we reconstructed the Roessler attractor out of a time series ($\approx$ 10000 values), which contained the dynamics of one single component of the system's state, by time delayed sampling [8].

4   Results

The chosen nonlinear systems (Roessler-System 5 and Lorenz-System 6) have been partitionned by the Multi-SOMs into 5 and 4 classes respectiveley. The sequence of winning partner-SOMs is the stream of symbols derived from the presented way of M-SOM classification of the data. Sorting the state of the dynamical system $(x, y, z)(t)$ into one of the M-SOM classes, is a variant of vector quantization. The sequence of symbols can be visualised in different ways; see right part, and bottom part of fig. 5 and 6.

Regular structures in both variants of visualizations are obvious, and easy to determine. This validates our assumption, that transforming the information into a stream of symbols via a self organizing neural network still represent the underlying dynamical behavior of the nonlinear and even chaotic system.

For further validation of our approach we used the x-component of the Roessler-system and reconstructed the attractor in 3-D using the method of time delayed sampling [8],[7]. Figure 7 shows the result of the reconstruction together with the trained 5-5×10 M-SOM.

The coordinates of the reconstructed attractor are derived by using time τ delayed values of the x-component x_t of the Roessler-system to form the basis vectors a_i of a new, artificial coordinate-system:

$$a_1 = x(t - 0 \cdot \tau)$$
$$a_2 = x(t - 1 \cdot \tau)$$
$$a_3 = x(t - 2 \cdot \tau)$$

We choose the x-trajectory as the base time-series $x(t)$ and time-steps $\tau = 10$. Thus, the dynamics of the complete system can be reconstruced even if only one single component of the state vector is available. With this reconstructed attractor, the M-SOM approach to unsupervised classification can be applied to derive a stream of symbols, typivcally for the original 3-dim attractor. The visualization of the derived stream in figure 7 is very similar to the one of the original attractor (figure 5), it's dynamics is comparable to the one of the original system.

5   Conclusion

Within this paper we have presented a method of partitioning chaotic attractors of nonlinear dynamical systems into a set of classes. Each class is thereby represented by an individual symbol, accessible for further information processing structures. The dynamic evolution of the system along the attractor is reflected within the stream of symbols generated by the partitioning network. The symbols can now be used for further analysis and for control of the dynamical system.

The process of classifying a nonlinear attractor was performed using different neural self organizing paradigms (SOMs, Neural-Gas, Growing SOMs) showing moderate results. Applying Multi-SOMs for assigning classes to different regions of the attractor, showed to be a powerful and efficient method. The advantage to pre-define the number of classes is obvious.

Several nonlinear dynamical attractors that are known to be chaotic (Lorenz- and Roessler attractor) have been partitioned and transformed into streams of symbols. It has been demonstrated, that the derived stream of extracted symbols is suitable for further tasks of analysis and control, like:

- classification of different forms or quality of chaos,
- qualitative characterization of the attractor into different types,
- basis for performing symbolic dynamic with the stream of symbols,
- analysing the stream of symbols to find, and or create a grammar that governs the nonlinear attractor,
- quantitative calculation of dynamic parameters (fractal dimension, ljapunov exponents, ...),
- input basis for a controller to obtain control over the dynamical system (e.g. switching controller, fuzzy inference system, ...).

The different partitions (classes) the attractor has been subdivided into, are solely formed by the probability distribution of the un-disturbed attractor itself. It is unlikely that such type of segmentation will be optimal suited to control for everytype of dynamical system, but it is the opinion of the authors, that the presented approach is generating a powerfull solution.

At the moment we are working towards including additional constraints directly into the the self organization mechanism of the M-SOM. In addition we are investigating several approaches of incorporating the effectiveness of applied control into the adaptation process of the neural self organizing maps.

The presented method of Multi-SOMs for unsupervised partitionning of a data into a stream of symbols is a novel, easy to implement mathod of classification, that can be applied in a wide variety of applications.

References

1. Bergé, P.; Pomeau, Y.; Vidal, Ch.: Order within Chaos, Towards a deterministic approach to turbulence, John Wiley & Sons, New York 1984.
2. Chen, G.; Dong, X.: From Chaos to Order, Methodologies, Perspectives and Applications, World Scientific Series on Nonlinear Science, Series A, Vol.24., World Scientific Publ., 1998.
3. Fritzke, Bernd: Wachsende Zellstrukturen - ein selbstorganisierendes neuronales Netzwerkmodell, PhD Thesis, Erlangen (1992).
4. Kohonen, T.: Self organizing maps, Springer, Berlin Heidelberg 1995.
5. Martinez, Thomas M.; Berkovich, Stanislav G.; Schulten, Klaus J.: "Neural Gas" Network for Vector Quantization and its Application to Time-Series Prediction, IEEE 1993.
6. Nicolis, G.; Prigogine, I.: Die Erforschung des Komplexen, Pieper, München 1987.
7. Takens, F.: Detecting strange attractors in turbulence. In: Rand,D.; Young, L.S. (Eds.): Neural Information Processing Systems 7, volume 898 of Lecture Notes in Mathematics, pp. 366-381. Springer Verlag, Berlin, 1981.
8. Thomson, J.; Stewart, H.: Nonlinear Dynamics and Chaos, Geometrical Methods for Engeneers and Scientists. John Wiley & Sons, Chinchester, 1986.

Selection of Classifiers Using Information-Theoretic Criteria*

Hee-Joong Kang

Department of Computer Engineering, Hansung University,
389 Samsun-dong 3-ga, Sungbuk-gu, Seoul, Korea
hjkang@hansung.ac.kr

Abstract. Combining multiple classifiers have focused mainly on combination methods, but a few studies have investigated on how to select component classifiers from a classifier pool. Performance by the information fusion varies with the component classifiers as well as the combination method. Previous studies focus on diverse classifiers which accurate and make different errors using the over-produce and choose strategy or the measures of diversity. In this paper, methods based on information theory are proposed for selecting component classifiers by considering the relationship among classifiers. These methods are applied to the classifier pool and examine the possible classifier sets. A classifier set is selected as a candidate and evaluated together with the other classifier sets on the recognition of public unconstrained handwritten numerals.

1 Introduction

Combining multiple classifiers has been studied for more than a decade and has reported improved performance over single classifier approaches [1,2,3,4]. Performance by the information fusion varies with the component classifiers as well as the combination method. A few studies have investigated on how to select the component classifiers from a classifier pool [5,6,7]. Thus, the selection of component classifiers, how to select them, or how many to select remain important research issues. Woods et al. [4] suggested that a strategy should be devised when selecting the mix of classifiers, because they observed that in some cases, fewer classifiers provided superior results. More recently, Kang and Lee reported some strategies for selecting the multiple classifiers [5]. Giacinto and Roli [6] focus to select diverse classifiers which accurate and make different errors using the overproduce and choose strategy [8]. Kuncheva et al. [7] proposed several measures of diversity for diverse ensembles which have a better potential for improvement on the accuracy than non-diverse ensembles.

In this paper, two simple selection criteria and three information-theoretic methods are proposed and reviewed for constructing multiple classifier systems. In order to simplify the selection problem of classifiers, it is assumed that the number of selected component classifiers is fixed in advance. A simple selection approach is to select the component classifiers according to the ranking order of their forced recognition rate or reliability rate up to the fixed number of classifiers. Information-theoretic criteria are

* This research was financially supported by Hansung University in the year of 2005.

S. Singh et al. (Eds.): ICAPR 2005, LNCS 3686, pp. 478–487, 2005.

based on the measure of closeness in [1,9] or the conditional entropy in [2,10] which considers the relationship among classifiers or the minimization of mutual information (mMI) between classifiers. The mMI criterion is proposed to select the component classifiers as complementary to each other as possible.

Three information-theoretic criteria for selecting the component classifiers are evaluated together with two selection criteria for recognition of unconstrained handwritten numerals from the Concordia University [11] and University of California, Irvine (UCI) [12] repositories. The selection criteria are applied to the classifier pool and we examine the possible classifier sets, and select one of the classifier sets as the candidate for building a multiple classifier system (MCS).

The MCS candidates are evaluated by using the combination methods in [1,2] together with the other classifier sets in the experiments. From the experimental results, it was found that the MCS candidates selected by the CE criterion were superior to the other classifier sets selected by the other criteria, with a few exceptions. Thus, the CE criterion is regarded as a promising clue for the selection of classifiers.

The remainder of this paper is organized as follows. Section 2 explains the selection criteria. Experimental results for evaluating the selection criteria are provided in Section 3 and a discussion is given in Section 4.

2   Selection Criteria

Two simple selection criteria are first introduced. One is the forced recognition rate (FRR) criterion and the other is the reliability rate (RR) criterion. The FRR criterion evaluates the classifier forcing a decision for every input, and not allowing rejections. The RR criterion considers the accuracy of all non-rejected decisions. Three information-theoretic criteria are explained by considering the first- and second-order dependencies among classifiers. These dependencies enable us to optimally approximate the high order probability distributions with the product of low distributions for Bayesian decision combination methods as in [1,2].

2.1   Measure of Closeness (MC) Criterion

The measure of closeness (MC) can be used for obtaining the optimal approximations by minimizing the difference between a real distribution $P(C)$ and an approximate distribution $P_a(C)$ where a vector variable C represents both a label class and K classifiers' decisions where K is the number of classifiers. The measure of closeness, $I(P(C), P_a(C))$, is defined in the following expression:

$$I(P(C), P_a(C)) = \sum_c P(c) \log \frac{P(c)}{P_a(c)}. \qquad (1)$$

When the dth-order dependency in the $(K+1)$st-order probability distribution of C is considered for the application of the measure of closeness, an approximate formula is defined by the following expression:

$$P_a(C_1, \cdots, C_{K+1}) = \prod_{j=1}^{K+1} P(C_{n_j}|C_{n_{id(j)}}, \cdots, C_{n_{i1(j)}}), \tag{2}$$

$$(0 \leq id(j), \cdots, i1(j) < j),$$

such that C_{n_j} is conditioned on all d terms from $C_{n_{i1(j)}}$ to $C_{n_{id(j)}}$, and where $(n_1, \cdots, n_K, n_{K+1})$ is an unknown permutation of integers $(1, \cdots, K, K+1)$ and where $P(C_{n_j}|C_0, C_{n_{i \cdot (j)}})$ is defined as $P(C_{n_j}, C_{n_{i \cdot (j)}})$.

Given the order of dependency d and K classifiers, the optimal product approximation for each classifier set is found by the application of the approximate formula P_a of Eq. (2) to Eq. (1), as in the following expressions by dropping the subscript n of C_{n_j}:

$$I(P(C), P_a(C)) = \sum_c P(c) \log \frac{P(c)}{P_a(c)}$$

$$= \sum_c P(c) \log P(c) - \sum_{j=1}^{K+1} \sum_c P(c) \log P(C_j|C_{id(j)}, \cdots, C_{i1(j)})$$

$$= -\sum_{j=1}^{K+1} M(C_j; C_{id(j)}, \cdots, C_{i1(j)}) + \sum_{j=1}^{K+1} H(C_j) - H(C) \tag{3}$$

$$H(C) = -\sum_c P(c) \log P(c)$$

$$M(C_j; C_{id(j)}, \cdots, C_{i1(j)}) = \sum_c P(c) \log \frac{P(C_j|C_{id(j)}, \cdots, C_{i1(j)})}{P(C_j)} \tag{4}$$

From Eq. (3), minimizing $I(P(C), P_a(C))$ is equivalent to maximizing $\sum_{j=1}^{K+1} M(C_j; C_{id(j)}, \cdots, C_{i1(j)})$ which is the total sum of dth-order mutual information, since remaining terms are constant. It is assumed that the larger the total sum of the dth-order mutual information is, the better its associated classifier set. Thus, the MC criterion finds an optimal product approximation relevant to each classifier set by maximizing the total sum of mutual information and then selects as a MCS candidate one classifier set having the largest total sum of mutual information.

2.2   Conditional Entropy (CE) Criterion

The conditional entropy (CE) relevant to the Bayes error rate can be also applied for obtaining the optimal approximations by minimizing the conditional entropy $H(M|E)$ composed of a label class M and a vector variable E of K classifiers' decisions. The Bayes error rate P_e is defined in the following expression by introducing the C-D(Class-Decisions) mutual information $U(M; E)$ as in [2]:

$$P_e \leq \frac{1}{2} H(M|E) = \frac{1}{2}(H(M) - U(M; E)) \tag{5}$$

$$U(M; E) = \sum_m \sum_e P(m, e) \log \frac{P(m, e)}{P(m)P(e)}. \tag{6}$$

When dth-order dependency in the probability distribution of M and E is considered for the application of the minimization of conditional entropy, two approximate formulae are defined by the following expressions, as we consider dependencies among classifiers:

$$P_{\mathrm{a}}(E_1, \cdots, E_K, M) = \prod_{j=1}^{K} P(E_{n_j} | E_{n_{id(j)}}, \cdots, E_{n_{i1(j)}}, M), \tag{7}$$

$$P_{\mathrm{a}}(E_1, \cdots, E_K) = \prod_{j=1}^{K} P(E_{n_j} | E_{n_{id(j)}}, \cdots, E_{n_{i1(j)}}), \tag{8}$$

$$(0 \leq id(j), \cdots, i1(j) < j),$$

such that E_{n_j} is conditioned on all d terms from $E_{n_{i1(j)}}$ to $E_{n_{id(j)}}$, and where $(n_1, \cdots, n_K)$ is an unknown permutation of integers $(1, \cdots, K)$. $P(E_{n_j} | E_0, E_0, M)$ is $P(E_{n_j}, M)$, $P(E_{n_j} | E_0, E_{n_{i \cdot (j)}}, M)$ is $P(E_{n_j} | E_{n_{i \cdot (j)}}, M)$, and $P(E_{n_j} | E_0, E_{n_{i \cdot (j)}})$ is $P(E_{n_j}, E_{n_{i \cdot (j)}})$, by definition.

Given the order of dependency d and K classifiers, the optimal product approximation for each classifier set is found by the application of the approximate formulae P_{a} of Eqs. (7) and (8) to the C-D mutual information, as in the following expressions by dropping the subscript n of E_{n_j}:

$$U(M; E) = \sum_{e} \sum_{m} P(e, m) \log \frac{P(e|m)}{P(e)}$$

$$= \sum_{e,m} P(e, m) \log[\frac{1}{P(m)} \prod_{j=1}^{K} P(E_j | E_{id(j)}, \cdots, E_{i1(j)}, m)]$$

$$- \sum_{e} P(e) \log \prod_{j=1}^{K} P(E_j | E_{id(j)}, \cdots, E_{i1(j)})$$

$$= H(M) + \sum_{j=1}^{K} [D(E_j; E_{id(j)}, \cdots, E_{i1(j)}, m) - D(E_j; E_{id(j)}, \cdots, E_{i1(j)})] \tag{9}$$

$$D(E_j; E_{id(j)}, \cdots, E_{i1(j)}, m) = \sum_{e,m} P(e, m) \log \frac{P(E_j | E_{id(j)}, \cdots, m)}{P(E_j)}$$

$$D(E_j; E_{id(j)}, \cdots, E_{i1(j)}) = \sum_{e} P(e) \log \frac{P(E_j | E_{id(j)}, \cdots, E_{i1(j)})}{P(E_j)}$$

$$\Delta D(E_j; E_{id(j)}, \cdots, E_{i1(j)}) =$$
$$D(E_j; E_{id(j)}, \cdots, E_{i1(j)}, m) - D(E_j; E_{id(j)}, \cdots, E_{i1(j)}) \tag{10}$$

From Eq. (9), maximizing $U(M; E)$ is equivalent to maximizing $\sum_{j=1}^{K} \Delta D(E_j; E_{id(j)}, \cdots, E_{i1(j)})$ which is the total sum of Δ dth-order mutual information, since the remaining term is constant. It is assumed that the larger the total sum of Δ dth-order mutual information is, the better its associated classifier set. Thus, the CE criterion finds an optimal product approximation relevant to each classifier set by maximizing the total sum of Δ mutual information and then selects as a MCS candidate one classifier set having the largest total sum of Δ mutual information.

2.3   Minimization of Mutual Information (mMI)Criterion

The minimization of mutual information (mMI) criterion is proposed to select the component classifiers in a pool as complementary to each other as possible. The mutual information is used to measure the relative complementarity of classifiers. It is assumed that the higher the mutual information is, the lower the complementarity. The mMI criterion selects classifiers in the pool and puts them into the classifier set of multiple classifier system up to the number of classifiers. Initially, a classifier set S is empty, and the mutual information between every classifier and a label class set, and the mutual information between classifiers are computed respectively. A procedure to find the classifier set as a MCS candidate is as follows:

1. For computed mutual information, find a classifier having the maximum mutual information in a pool as to the label class set and then put the classifier into the classifier set S.
2. In order to find a classifier in a pool as complementary to classifiers in the classifier set S as possible, and find a classifier having minimum mutual information in a pool as to the classifiers in the classifier set, and then put the classifier into the classifier set.
3. Until the number of classifier in the classifier set S meet the fixed number of classifiers, repeat the step 2 and then a final classifier set will be found.

3   Experimental Results

A number of MCSs built from the pool of six classifiers, $E1$, $E2$, $E3$, $E4$, $E5$, $E6$, will be evaluated in this section. These classifiers are developed using the features and structural knowledge of numerals such as bounding box, centroid, and the width of horizontal runs, from KAIST and Chonbuk National Universities. Some are singular or modular back-propagation neural networks and the others are modular rule-based classifiers, as shown in Table 1.

The handwritten numeral database is a fairly representative collection of digits. The UCI data sets in [12] are used for optical recognition of handwritten digits and consist of three training data sets *tra*, *cv*, *wdep* and one test data set *windep*. The Concordia data sets consist of two training data sets A, B and one test data set T.

The performance of individual classifiers on test data sets is shown in terms of recognition and reliability rates in Figure 1. We note that classifiers $E4$ and $E5$ were trained using the structural knowledge obtained from the numerals of the Concordia University source, they are not as good on the numerals from UCI. The *reject* results of a classifier were used in the MC criterion.

In our experiments, each neural network based classifier was trained with the training data sets A and *tra*. The optimal product sets were found by using the two data sets A, B and the three data sets *tra*, *cv*, *wdep*. The selection criteria were applied to the possible classifier sets and then we selected the most successful classifier set among them for a fixed number of classifiers. To denote the information-theoretic criteria according to the order of dependency, we use the abbreviations as follows: MC1 stands for a MC criterion by first-order dependency, CMC1 for a MC criterion by conditional

Table 1. Introduction of individual classifiers

	architecture	classifier	distance function	reference
E1	singular	neural network	pixel distance function	[13]
E2	modular	neural network	directional distance distribution	[13]
E3	singular	neural network	mesh feature	[13]
E4	modular	rule-based	modified structural knowledge	[14]
E5	modular	rule-based	structural knowledge	[14]
E6	singular	neural network	contour feature	[15]

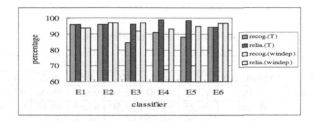

Fig. 1. Results of individual classifiers on test data sets: *T, windep*

first-order dependency, MC2 for a MC criterion by second-order dependency, CE1 for a CE criterion by first-order dependency, and CE2 for a CE criterion by second-order dependency. All the MCSs were evaluated by the following combination methods on the test data sets: voting, Borda count, Bayesian combination methods abbreviated as in Table 2. The Bayesian methods are described in [1,2,3].

Table 2. Bayesian combination methods

method	meaning
CIAB	Conditional Independence Assumption based Bayesian
ODB1	first-Order Dependency based Bayesian
CODB1	Conditional first-Order Dependency based Bayesian
ODB2	second-Order Dependency based Bayesian
DODB1	Δ first-Order Dependency based Bayesian
DODB2	Δ second-Order Dependency based Bayesian

From the possible 20 MCSs consisting of three classifiers for each data set, the classifier sets by the selection criteria are shown in Table 3. Table 4 shows the results of the selected classifier sets and the *best* classifier set which can be dynamically selected by an oracle among the possible MCSs as to the given combination method, in terms of recognition rates. In case of the numerals of Concordia, the CE criterion was slightly better than the other criteria in most combinations, However, in case of the numerals of UCI, the mMI criterion outperforms the other criteria in most combinations and it shows very similar quality of performance to the *best* classifier sets.

Table 3. MCS of three classifiers

data set	selection criterion	classifiers
Concordia	*FRR*	*E1,E2,E6*
	RR	*E3,E4,E6*
	MC1,CMC1,MC2	*E1,E4,E6*
	CE1,CE2	*E2,E4,E6*
	mMI	*E1,E3,E5*
UCI	*FRR,RR,MC1,CMC1,MC2*	*E2,E3,E6*
	CE1,CE2	*E2,E4,E6*
	mMI	*E1,E2,E4*

Table 4. Results of three classifier MCS

data set	combination	selection criterion					
		FRR	*RR*	*MC*	*CE*	*mMI*	*best*
Concordia	*voting*	97.30	95.60	96.50	96.90	96.00	97.30
(*T*)	*Borda*	96.25	97.60	97.50	97.65	96.10	97.75
	CIAB	96.20	96.80	96.55	96.85	96.65	97.50
	ODB1	96.20	96.80	96.55	96.85	96.65	97.10
	CODB1	97.10	97.25	97.45	97.30	97.55	97.70
	ODB2	95.80	97.25	97.45	95.85	97.55	97.70
	DODB1	97.10	97.30	97.55	97.45	97.35	97.65
	DODB2	97.30	97.30	97.50	97.65	97.35	97.65
UCI	*voting*	96.77	96.77	96.77	96.99	97.38	97.38
(*windep*)	*Borda*	96.44	96.44	96.44	94.82	98.33	98.33
	CIAB	96.88	96.88	96.88	96.83	97.77	97.77
	ODB1	96.88	96.88	96.88	96.83	97.83	97.83
	CODB1	96.99	96.99	96.99	97.61	97.77	97.77
	ODB2	97.05	97.05	97.05	97.05	97.77	97.77
	DODB1	97.16	97.16	97.16	97.38	97.66	97.94
	DODB2	97.33	97.33	97.33	97.77	97.33	97.89

For four classifiers, 15 MCSs for each data set were examined, and the selected classifier sets were evaluated in terms of recognition rates, as shown in Tables 5 and 6. While the MCSs by the FRR and CE criteria showed slightly better than those of other criteria in case of the numerals of Concordia, the MCSs by the CE1 and mMI criteria in case of the numerals of UCI showed better results than those of other criteria in most combinations.

From the 6 possible MCSs consisting of five classifiers, the classifier sets were evaluated and their results are shown in Tables 7 and 8. The MCSs by the FRR and CE1 and MC criteria showed better results than those of other criteria in case of the numerals of Concordia, but the MCSs by the CE2 criterion showed better results than those of other criteria in case of the numerals of UCI.

From the results, the CE1 criterion was useful in selecting the most promising classifier sets from the pool of classifiers for constructing a MCS in case of the numerals of Concordia, although the MCS candidates by the CE1 criterion did not necessarily

Table 5. MCS of four classifiers

data set	selection criterion	classifiers
Concordia	RR	E1,E3,E4,E5
	MC1,CMC1,MC2	E1,E4,E5,E6
	FRR,CE1,CE2	E1,E2,E4,E6
	mMI	E1,E2,E3,E5
UCI	FRR,RR,MC1,CMC1,MC2	E1,E2,E3,E6
	CE1	E2,E3,E4,E6
	CE2	E3,E4,E5,E6
	mMI	E1,E2,E3,E4

Table 6. Results of four classifier MCS

data set	combination	selection criterion						
		FRR	RR	MC	CE1	CE2	mMI	best
Concordia	voting	97.80	97.15	97.05	97.80	97.80	96.85	97.80
(T)	Borda	97.80	97.15	97.75	97.80	97.80	97.70	98.00
	CIAB	97.05	97.45	97.00	97.05	97.05	96.70	97.45
	ODB1	95.85	97.00	96.55	95.85	95.85	96.70	97.15
	CODB1	96.30	98.00	97.80	96.30	96.30	97.50	98.00
	ODB2	96.90	98.00	97.45	96.90	96.90	97.50	98.00
	DODB1	98.00	97.60	97.90	98.00	98.00	97.55	98.00
	DODB2	97.85	97.65	97.60	97.85	97.85	97.60	97.95
UCI	voting	97.27	97.27	97.27	97.66	95.83	97.22	97.77
(windep)	Borda	96.66	96.66	96.66	97.44	96.83	97.77	98.05
	CIAB	96.83	96.83	96.83	97.05	96.94	96.77	97.66
	ODB1	96.83	96.83	96.83	97.05	97.38	96.77	97.77
	CODB1	97.11	97.11	97.11	97.16	97.33	97.38	98.16
	ODB2	97.38	97.38	97.38	97.55	97.33	97.38	98.00
	DODB1	97.44	97.44	97.44	97.83	96.88	97.72	98.05
	DODB2	97.38	97.38	97.38	97.50	97.44	98.05	98.05

Table 7. MCS of five classifiers

data set	selection criterion	classifiers
Concordia	RR	E1,E2,E3,E4,E5
	FRR,CE1,MC1,CMC1,MC2	E1,E2,E4,E5,E6
	CE2	E2,E3,E4,E5,E6
	mMI	E1,E2,E3,E5,E6
UCI	FRR,RR,MC1,CMC1,MC2	E1,E2,E3,E5,E6
	CE1,mMI	E1,E2,E3,E4,E6
	CE2	E2,E3,E4,E5,E6

coincide with the *best* classifier set. The mMI criterion proposed for the complementarity was superior to the other criteria in building a MCS of three or four classifiers in case of the numerals of UCI. And the CE1 and CE2 criteria were respectively good for building a MCS of four or five classifiers. Particularly, the mMI criterion for build-

Table 8. Results of five classifier MCS

data set	combination	selection criterion						
		FRR	RR	MC	CE1	CE2	mMI	best
Concordia	voting	97.90	97.40	97.90	97.90	97.45	97.45	97.90
(T)	Borda	97.90	98.15	97.90	97.90	97.85	97.80	98.15
	CIAB	97.45	97.40	97.45	97.45	97.60	96.80	97.60
	ODB1	97.10	96.95	97.10	97.10	96.85	96.80	97.20
	CODB1	97.85	97.70	97.85	97.85	97.60	97.80	98.25
	ODB2	97.35	97.65	97.35	97.35	95.85	97.50	97.95
	DODB1	98.15	97.90	98.15	98.15	98.10	97.65	98.25
	DODB2	97.95	97.80	97.95	97.95	97.95	97.80	97.95
UCI	voting	97.94	97.94	97.94	97.72	98.16	97.72	98.16
(windep)	Borda	97.55	97.55	97.55	97.66	97.22	97.66	97.83
	CIAB	97.22	97.22	97.22	97.05	97.33	97.05	97.33
	ODB1	97.22	97.22	97.22	97.05	97.44	97.05	97.44
	CODB1	97.77	97.77	97.77	97.61	97.38	97.61	98.00
	ODB2	98.11	98.11	98.11	97.77	98.11	97.77	98.11
	DODB1	98.44	98.44	98.44	97.89	98.27	97.89	98.44
	DODB2	98.11	98.11	98.11	98.05	98.22	98.05	98.33

ing a MCS of three classifiers exactly coincided with the *best* classifier set in 6 out of 8 combinations and the CE2 criterion for building a MCS of five classifiers coincided with the *best* classifier set in 4 out of 8 combinations. The selection criteria based on information theory would be one of the promising clues when Bayesian combination methods are considered.

4   Discussion

Although the selection criteria based on information theory showed positive evidence and their utility was supported through the recognition experiments, further studies are needed because the MCS candidates always do not guarantee the best recognition and their performances vary with the source of data, and the limit lies with the fixed number of classifiers except the mMI criterion. As for the mMI criterion, there is a still room to deal with higher order dependency among classifiers, because current version uses only the first-order mutual information between classifiers. Furthermore, three classifiers were sometimes better than four or five classifiers according to the combination methods. It will be useful to deal with the limitation of our approaches as a future work.

References

1. Kang, H.J., Lee, S.W.: A Dependency-based Framework of Combining Multiple Experts for the Recognition of Unconstrained Handwritten Numerals. In: Proc. of 1999 IEEE Comp. Soc. Conf. on CVPR. Volume 2. (1999) 124–129
2. Kang, H.J., Lee, S.W.: Combining Classifiers based on Minimization of a Bayes Error Rate. In: Proc. of the 5th ICDAR. (1999) 398–401

3. Kittler, J., Hatef, M., Duin, R.P.W., Matas, J.: On Combining Classifiers. IEEE TPAMI **20** (1998) 226–239
4. Woods, K., Kegelmeyer Jr., W.P., Bowyer, K.: Combinition of Multiple Classifiers Using Local Accuracy Estimates. IEEE TPAMI **19** (1997) 405–410
5. Kang, H.J., Lee, S.W.: Experimental Results on the Construction of Multiple Classifiers Recognizing Handwritten Numerals. In: Proc. of the 6th ICDAR. (2001) 1026–1030
6. Giacinto, G., Roli, F.: An pproach to the automtic design of multiple classifier systems. Pattern Recognition Letters **22** (2001) 25–33
7. Kuncheva, L.I., Skurichina, M., Duin, R.P.W.: An experimental study on diversity for bagging and boosting with linear classifiers. Information Fusion **3** (2002) 245–258
8. Patridge, D., Yates, W.B.: Engineering multiversion neural-net systems. Neural Computation **8** (1996) 299–314
9. Lewis, P.M.: Approximating Probability Distributions to Reduce Storage Requirement. Information and Control **2** (1959) 214–225
10. Wang, D.C.C., Wong, A.K.C.: Classification of Discrete Data with Feature Space Transform. IEEE TAC **AC-24** (1979) 434–437
11. Suen, C.Y., Nadal, C., Legault, R., Mai, T.A., Lam, L.: Computer Recognition of Unconstrained Handwritten Numerals. Proc. of IEEE (1992) 1162–1180
12. Blake, C., Merz, C.: UCI repository of machine learning databases [http://www.ics.uci.edu/~mlearn/mlrepository.html]. Irvine, CA, Dept. of Infor. and Comp. Sciences (1998)
13. Oh, I.S., Suen, C.Y.: Distance features for neural network-based recognition of handwritten characters. IJDAR **1** (1998) 73–88
14. Oh, I.S., Lee, J.S., Hong, K.C., Choi, S.M.: Class-expert approach to unconstrained handwritten numeral recognition. In: Proc. of the 5th IWFHR. (1996) 35–40
15. Matsui, T., Tsutsumida, T., Srihari, S.N.: Combination of Stroke/Background Structure and Contour-direction Features in Handprinted Alphanumeric Recognition. In: Proc. of the 4th IWFHR. (1994) 87–96

ICA and GA Feature Extraction and Selection for Cloud Classification

Miguel Macías-Macías, Carlos J. García-Orellana, Horacio González-Velasco,
and Ramón Gallardo-Caballero

Departamento de Electrónica e Ingeniería Electromecánica - Universidad de Extremadura,
Centro Universitario de Mérida - Sta. Teresa de Jornet, 38 - 06800 Mérida, (Badajoz) - Spain
{miguel, carlos, horacio, ramon}@nernet.unex.es

Abstract. In this work we tackle a particular case of image segmentation, the automatic detection of the amount and type of clouds over the Iberian Peninsula using satellite images. To segment the images we classify each pixel of the image into one of the classes defined using a neural network and a set of features representative of the pixel. We emphasized in the preprocessing stage, extracting and selecting a suitable set of features from the images to carry out an optimal classification. To carry out the feature extraction we use the independent component analysis (ICA) algorithm. The features extracted with this algorithm are very dependent on the dimension of the patches, so we extract several sets of features, one for each value of the dimension of the patch. All of these sets of features are joined together to form an initial characteristic vector of the pixels of the images. Finally, we reduce the dimensionality of this initial characteristic vector by means of Genetic Algorithms (GA), choosing the best subset of features that offer the best classification results.

1 Introduction

Accurate cloud information is very important to modelling the radiation balance in the climatic system. Clouds play an important role reflecting the solar radiation and absorbing thermal radiation emitted by the land and the atmosphere, therefore reinforcing the greenhouse effect. The contribution of the clouds to the Earth albedo is very high, controlling the energy entering the climatic system. It has been estimated that an increase in the average albedo of the Earth-atmosphere system in only 10 percent could produce a decrease in the surface temperature to levels of the last ice age. Therefore, global change in surface temperature is highly sensitive to cloud amount and type.

For these reasons, numerous works about this topic have been published in the last years, many of them dealing with the search of a suitable classifier, neural networks [1-6] or linear discriminations techniques [7, 8].

Other works are related to the search of an initial feature set that allow obtaining reliable classification results. In the first works, simple spectral features were used, as albedo and temperature. Later studies included textural features [9]. In [8] Welch et al. used statistical measures based on grey level co occurrence matrix

S. Singh et al. (Eds.): ICAPR 2005, LNCS 3686, pp. 488–496, 2005.
© Springer-Verlag Berlin Heidelberg 2005

(GLCM) proposed by Haralick et al. in [10]. In [6] several image transformation schemes as singular value decomposition (SVD) and wavelet packets (WP's) were exploited. In [11] Gabor filters and Fourier features are recommended for cloud classification and in [6] authors showed that SVD, WP´s and GLCM textural features achieved almost similar results.

In [12] the authors showed that the initial feature set extracted with the ICA algorithm applied directly to the images was, from the meteorologist point of view, very good. They used the 10 first independent components extracted from the 5x5 patches defined over the infrared and visible images.

On the other hand there are some works dealing with the reduction of the dimensionality of the original characteristic vector. In that sense, in [13], Doak identifies three different categories of search algorithms: exponential, sequential and randomised. In [14] Aha et al. use the most common sequential search algorithms for feature selection applied to the clouds classification: the forward sequential selection (FSS) and the backward sequential selection (BSS). In [15-17] a genetic algorithm (GA) representative of the randomised category is used for feature selection. They use GA because it is less sensitive than other algorithms to the order of the features that have been selected. On the other hand, in [18, 19] the authors of this work use feature selection algorithms not dependent on the labelling of the prototypes, as principal component analysis (PCA) and independent component analysis (ICA). Also, they compare the classification results with the ones obtained using genetic algorithms.

In this work we have applied the best of these techniques to the cloud cover classification problem. We have used the ICA algorithm in the feature extraction stage, but the features extracted with this algorithm are very dependent on some parameters like the number of extracted components or the dimension of the patches. Therefore, we have applied ICA in the feature extraction stage, obtaining several characteristic vectors, one for each set of ICA parameters. Joining together these characteristic vectors we obtained a unique characteristic vector to carry out the classification. This characteristic vector had a very large dimensionality, so we used GA in the feature selection stage to reduce it. In section 2 we show the methodology followed in all the process, namely, the neural networks usage and the pre-processing stage that includes the feature selection and the feature extraction stages. In section 3 the classification results are given and, finally, the conclusions and comments are presented in section 5.

The ideas showed in this paper can be applied to other similar applications of remote sensing in general, hyper spectral analysis, etc.

2 Methodology

Our main objective is to provide the meteorologists with an automatic classification system to estimate the cloud cover in each image received by the geostationary satellite meteosat. The proposed system has been developed using a feature extractor and a feature selector for pre-processing, and a neural network as classifier. This satellite gives multi-spectral data in three wavelength channels. In this work two of them, the visible and infrared channels, are used. All the process will be described in this section, including the neural network usage and the pre-processing stage.

2.1 Neural Network

In order to train our neural network to be used as a classifier for the pixels of the images we need a large set of prototypes. The set of prototypes is also used to test and compare the classification systems and to build systems with generalization capabilities. The subjective interpretation of these images by Meteorology experts suggested us to consider the following classes: sea (S), land (L), fog (F), low clouds (C_L), middle clouds (C_M), high clouds (C_H) and clouds with vertical growth (C_V).

To implement the mapping imposed by the prototypes set and to carry out the classification of the pixels of the images we used neural networks, in particular one-hidden-layer perceptrons trained with the Resilient Backpropagation RProp algorithm described in [20]. Basically, this algorithm is a local adaptive learning scheme which performs supervised batch learning in multi-layer perceptrons. It differs from other algorithms since it considers only the sign of the summed gradient information over all patterns of the training set to indicate the direction of the weight update.

In order to test the classification results obtained, and to select the best feature set extracted by the GA algorithm, the set of prototypes was divided into a training set, a validation set and a test set.

Optimisation of the neural network is made as follows. First we wish to select the best network, i.e. the network with the best generalization, for each feature set. To obtain that network we use the validation and training subsets of prototypes. For each feature set we start with very few neurons in its hidden layer. Then the network is trained with the learning set, and the sum of the squared error over the validation (SSE_v) set is calculated in each iteration. When the value of SSE_v reaches a minimum the learning process is stopped and the network is saved. To avoid local minimums the process is repeated several times maintaining the topology of the network but randomizing the initial weights. The network with the lowest SSE_v is selected. Then the whole process is repeated by increasing the number of the neurons of the hidden layer, saving the network with the lowest SSE_v. Finally, the test set is used to compare the classification results obtained with the networks representatives of each feature set and, therefore, to select the optimal feature set.

2.2 Preprocessing Stage

The preprocessing stage is the most important step in a problem of data classification and its design is one of the most significant factors in determining the performance of the final system. In our case the preprocessing stage includes a calibration step, because satellite data must be corrected to obtain physical magnitudes which are characteristic of clouds and independent of the measurement process. Afterwards a feature extraction step and a feature selection step have been made.

2.2.1 Calibration

Our final aim is the design of a system to segment images corresponding to different hours of the day and different days of the year. Therefore, satellite data must be corrected in the pre-processing stage in order to obtain physical magnitudes which could be said to be characteristic of the clouds and then, independent of the measuring process.

From the infrared channel, we obtained brightness temperature corrected of the aging effects and the radiometer transfer function. From the visible channel we obtained albedo after correcting it of the radiometer aging effects and considering the viewing and illumination geometry. For each pixel, this correction deals with the sun-earth distance, the solar zenith angle at the image acquisition date and time, and the longitude and latitude of the pixel. In [6] no data correction is made and an adaptive PNN network was proposed to resolve this issue.

2.2.2 Feature Extraction

At this point we have to define the characteristic vector representative of the pixel of the images from the albedo and brightness temperature data. To carry out this task we used the Independent Component Analysis (ICA) in the sense proposed in [21].

Recently, blind source separation by Independent Component Analysis (ICA) has received attention because of its potential applications in signal processing such as in speech recognition systems, telecommunications and medical signal processing. The goal of ICA [22-23] is to recover independent sources given only sensor observations that are unknown linear mixtures of the unobserved independent source signals. In contrast to correlation-based transformations such as Principal Component Analysis (PCA), ICA not only decorrelates the signals (2nd-order statistics) but also reduces higher-order statistical dependencies, attempting to make the signals as independent as possible. This technique can be used in feature extraction essentially finding the building blocks of any given data [21].

The seminal work on blind source separation was by Herault and Jutten [24] where they introduced an adaptive algorithm in a simple feedback architecture that was able to separate several unknown independent sources. Comon [22] elaborated the concept of independent component analysis and proposed cost functions related to the approximate minimization of mutual information between the sensors.

One way of formulating the ICA problem consist in considering the data matrix $\mathbf{X}$ to be a linear combination of non-Gaussian (independent) components i.e. $\mathbf{X} = \mathbf{S} \cdot \mathbf{A}$ where columns of $\mathbf{S}$ contain the independent components and $\mathbf{A}$ is a linear mixing matrix. In short, ICA attempts to 'un-mix' the data by estimating an un-mixing matrix $\mathbf{W}$ where $\mathbf{X} \cdot \mathbf{W} = \mathbf{S}$

Different algorithms for ICA have been proposed [25]. In our case, we used the FastICA algorithm [26] and simulations were performed using the FastICA package [27] for R (A Programming Environment for Data Analysis and Graphics) [28].

To perform ICA each pixel is represented by the values of the pixels in the NxN patch (window centred at the pixel at issue), each pixel covering an area of 7 Km^2 of the Iberian Peninsula. We can define one patch in the albedo data and another in the temperature data so each pixel is defined by a $2 \cdot N^2$ dimensional characteristic vector.

In this study, 80000 pixels were randomly extracted from a set of 40 images chosen to be representative of all types of clouds, land and sea. So, the matrix $\mathbf{X}$ has 80000 rows and $2 \cdot N^2$ columns. This matrix is passed to the fastICA algorithm to extract n independent components. These n independent components are extracted simultaneously and the function used in the approximation to neg-entropy is "log-cosh" with a constant value of 1. The ICA algorithm then estimates an un-mixing matrix $\mathbf{W}$ and a pre-whitening matrix $\mathbf{K}$ s.t. $\mathbf{X} \cdot \mathbf{K} \cdot \mathbf{W} = \mathbf{S}$.

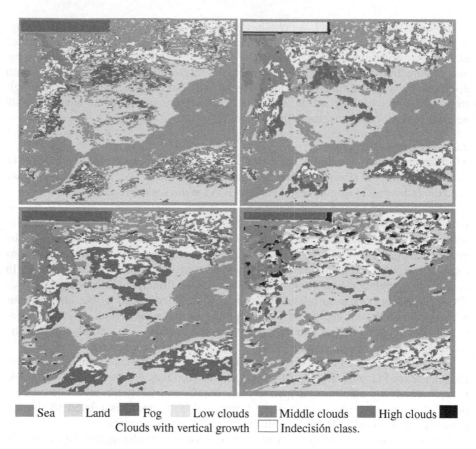

Sea   Land   Fog   Low clouds   Middle clouds   High clouds
Clouds with vertical growth   Indecisión class.

Fig. 1. Segmentation of an Iberian Peninsula Meteosat Image using four different sets of ICA parameters. Top-left: $N=3$, $n=4$, Top-right: $N=5$, $n=6$, Bottom-left: $N=7$, $n=8$ and Bottom-right $N=9$, $n=12$.

Once ICA is performed, we can define a new n dimensional characteristic vector representative of each pixel of the image by multiplying the old $2 \cdot N^2$ characteristic vector of the pixel by the $\mathbf{K} \cdot \mathbf{W}$ matrix estimated by the ICA algorithm.

The independent components are very dependent on the dimension of the patches N and on the number of extracted components n. In [12] the authors showed that using the ICA algorithm in the feature extraction stage, the best results were obtained with $n=6$ and $N=5$. In figure 1, we can observe an example of image segmentation with different patches dimension. We can observe that the classification results are different in each case. For these reasons in this work we considered the independent components extracted using N values of 3, 5, 7 and 9, and we used a Genetic Algorithm to select the best characteristics to make the classification.

2.2.3 Feature Selection

The large dimensionality of the vector obtained in the feature extraction stage and the limited quantity of prototypes available lead us to the case where the sparse data provide a very poor representation of the mapping. This phenomenon has been termed as "the curse of dimensionality" [29]. Thus, in many problems, reducing the number of input variables can lead to improved performances for a given data set, even though some information is being discarded. Therefore, this process constitutes one of the fundamental steps of the preprocessing stage and also one of the most significant factors in determining the performance of the final system.

To reduce the dimensionality we applied GA. We used this algorithm to select a subset of features such that the neural network presents the best generalisation by using the prototypes selected and labelled by the experts in Meteorology. That is, the network which, trained with the prototypes of the learning set, achieves the minimum number of misclassifications over the validation set.

For each subset of features the algorithm uses one hidden layer perceptron where the number of the neurons of the hidden layer changes from 20 to 40. For each topology the training process is repeated 20 times randomizing the weights each time. As fitness we used the sum of squared error (SSE) over the validation set.

The GA was configured using a crossover probability of 0.6, a mutation probability of 0.1, population of 350 individuals, tournament selection method and steady-state population replacement system, with a 30% of replacement.

The simulations were carried out in a Beowulf style cluster with Clustermatic as OS (a patched RedHat 7.2 Linux OS, with *bproc* for cluster management). The cluster is built using a double Pentium III @ 800 MHz with 1 Gbyte of RAM memory on master node, and 25 nodes, with AMD Athlon @ 900 MHz with 512 Mbytes of memory each. For GA simulations we used the PGAPack [30] simulator with MPI enabled.

3 Results

In order to implement the processes described above, the experts in Meteorology selected and labeled 4599 prototypes, 2781 for the training set, 918 for the validation set and 900 for the test set. These subsets are randomly selected from the set of prototypes trying to maintain the equal number of prototypes for all the classes in each subset. On the other hand, the number of prototypes in each subset is chosen according to the 20%, 20% and 60% rule. The prototypes selection was made from the Iberian Peninsula Meteosat images corresponding to the years 1995-1998.

In the feature selection process we used the ICA algorithm to extract four different characteristic vectors of dimensions 4, 6, 8 and 12, for the values of the patch dimension "N" of 3, 5, 7 and 9 respectively. Joining together these vectors we obtained a

Table 1. Parameters selected by the GA algorithm

N=3, n=4				N=5, n=6						N=7, n=8							
1	2	3	4	5	6	7	8	9	10	11	12	13	14	15	16	17	18
1	2	3	4	1	2	3	4	5	6	1	2	3	4	5	6	7	8

N=9, n=12											
19	20	21	22	23	24	25	26	27	28	29	30
1	2	3	4	5	6	7	8	9	10	11	12

Table 2. Classification results over the learning and the validation sets with the N=5 and n=6 ICA parameters

SET	F	C_L	C_M	C_H	C_V	L	S	SSE
Learning	94.4	91.4	98.3	98.9	96.5	100	100	166
Validation	65.9	85.2	96.2	90.5	93.3	92.4	100	167

Table 3. Classification results over the learning and the validation sets with the GA parameters

SET	F	C_L	C_M	C_H	C_V	L	S	SSE
Learning	96.2	91.2	96.5	100	98.1	100	100	133
Validation	93.4	89.9	89.3	96.8	97.8	99.3	100	87

30-th dimensional characteristic vector. From this characteristic vector the GA algorithm extracted seven parameters. These parameters can be observed shadowed in table 1 and they are the first independent component extracted with N =3, the first and fourth components with N=5, the third independent component extracted with N=7 and the third, fifth and seventh components extracted with N=9.

In [12] the authors showed that, using the ICA algorithm in the feature extraction stage, the best results were obtained with $n=6$ and $N=5$. Now we want to compare these classification results with the ones obtained in this work. In table 2 we can observe the classification results over the learning and the validation sets of prototypes presented in [12]. In table 3 the classification results obtained with the parameters selected by the GA algorithm (parameters shadowed in table one) over the learning and the validation sets or prototypes are presented.

4 Conclusions

In tables 2 and 3 we can observe that the classification results obtained with the GA parameters are better than those obtained with the ICA ($N=5$ and $n=6$) parameters. These results are measured over the learning and the validation sets, though the set that must be used to decide which the best classification system is should be the test set. In table 4 we can observe the classification results over this set of prototypes and we can notice that the classification results are a little worse for the GA parameters. To improve the set of parameters selected by the GA we propose, for a future work, to use the SSE over the test set as fitness for the Genetic Algorithm.

Table 4. Classification results over the learning and the validation sets with the GA parameters

Test set	F	C_L	C_M	C_H	C_V	L	S	SSE
ICA, N=5 n=6	68.3	67.3	95.9	95.9	91.1	99.2	100	180
GA	80.5	60	94.1	100	82.1	99.2	100	195

Acknowledgements

This study was supported by the Extremadura University under the Project "Caracterización de cubierta nubosa mediante técnicas de redes neuronales".

References

1. Lee J., Weger R. C., Sengupta S. K. And Welch R.M.: A Neural Network Approach to Cloud Classification. IEEE Transactions on Geoscience and Remote Sensing, vol. 28, no. 5, pp. 846-855, Sept. 1990.
2. M. Macías, F.J. López, A. Serrano and A. Astillero: "A Comparative Study of two Neural Models for Cloud Screening of Iberian Peninsula Meteosat Images", Lecture Notes in Computer Science 2085, Bio-inspired applications of connectionism, pp. 184-191, 2001.
3. A. Astillero, A Serrano, M. Núñez, J.A. García, M. Macías and H.M. Gónzalez: "A Study of the evolution of the cloud cover over Cáceres (Spain) along 1997, estimated from Meteosat images", Proceedings of the 2001 EUMETSAT Meteorological Satellite Data Users' Conference, pp. 353-359, 2001
4. Bankert, R. L et al.,: Cloud Classification of AVHRR Imagery in Maritime Regions Using a Probabilistic Neural Network. Journal of Applied. Meteorology, 33, (1994) 909-918.
5. B. Tian, M. A. Shaikh, M R. Azimi, T. H. Vonder Haar, and D. Reinke, "An study of neural network-based cloud classification using textural and spectral features," IEE trans. Neural Networks, vol. 10, pp. 138-151, 1999.
6. B. Tian, M. R. Azimi, T. H. Vonder Haar, and D. Reinke, "Temporal Updating Scheme for Probabilistic Neural Network with Application to Satellite Cloud Classification," IEEE trans. Neural Networks, Vol. 11, no. 4, pp. 903-918, Jul. 2000.
7. Welch, R.M., Kuo K. S., Sengupta S. K., and Chen D. W.: Cloud field classification based upon high spatial resolution textural feature (I): Gray-level cooccurrence matrix approach. J. Geophys. Res., vol. 93, (oct. 1988) 12633-81.
8. R. M. Welch et al., "Polar cloud and surface classification using AVHRR imagery: An intercomparison of methods," J. Appl. Meteorol., vol. 31, pp. 405-420, May 1992.
9. N. Lamei et al., "Cloud-type discrimitation via multispectral textural analysis," Opt. Eng., vol. 33, pp. 1303-1313, Apr. 1994.
10. R. M. Haralick et al., "Textural features for image classification", IEEE trans. Syst., Man, Cybern., vol. SMC-3, pp. 610-621, Mar. 1973.
11. M. F. Augusteijn, "Performance evaluation of texture measures for ground cover identification in satellite images by means of a neural.network classifier," IEEE trans. Geosc. Remote Sensing, vol. 33, pp. 616-625, May 1995.
12. Independent component analysis for cloud screening of Meteosat images, M. Macías, C.J. García, H.M. Gónzalez and R. Gallardo, Lectures Notes in Computer Science, Vol. 2687, pp. 551-558, 2003
13. 13 Doak J. An evaluatin of feature selection methods and their application to computer security (Technical Report CSE-92-18). Davis, CA: University of California, Department of Computer Science.
14. Aha, D. W., and Bankert, R. L.: A Comparative Evaluation of Sequential Feature Selection Algorithms. Artificial Intelligence and Statistics. V., D. Fisher and J. H. Lenz, editors. Springer-Verlag, New York, 1996.

15. Eng Hock Tay F. and Li Juan Cao, A comparative study of saliency analysis and genetic algorithm for feature selection in support vector machines", Intelligent Data Analysis, vol. 5, no. 3, pp. 191-209, 2001.

16. A. Tettamanzi, M. Tomassini. Soft Computing. Integrating Evolutionary, Neural and Fuzzy Systems. Springer, 2001.

17. F.Z. Brill, D.E. Brown and W.N. Martin. Fast genetic selection of features for neural network classifiers. IEEE Transactions on Neural Networks, 3(2): 324-328, 1992.

18. M. Macías, C. J. Garcia, H. M. Velasco, R. Gallardo, A. Serrano "A comparison of PCA and GA selected features for cloud field classification", Lectures Notes in Artificial Intelligence, vol., 527, pp., 42-49, 2002.

19. C.J. García, M. Macías, A. Serrano, H.M. González and R. Gallardo, "A comparison of PCA, ICA and GA selected features for cloud field classification", Journal of Intelligent & Fuzzy Systems. Vol. 12 pp. 213-219.

20. M. Riedmiller, M., Braun, L.: A Direct Adaptive Method for Faster Backpropagation Learning: The RPROP Algorithm. In Proceedings of the IEEE International Conference on Neural Networks 1993 (ICNN 93), 1993.

21. A. Hyvärinen, E. Oja, P. Hover and J. Hurri. Image feature extraction by sparse coding and independent component analisys. In Proc. Int. Conf. On Pattern Recognition (ICPR'98), pp.1268-1273, Brisbane, Australia, 1998.

22. P. Comon, "Independent component analysis, a new concept?," Signal Processing, vol. 36, no. 3, pp. 287--314, April 1994.

23. Hyvärinen A., Karhunen J., Oja E. Independent Component Analysis. John Wiley and Sons, 2001.

24. C. Jutten and J. Herault, "Blind separation of sources, part I: an adaptive algorithm based on neuromimetic architecture" Signal Processing, vol.24, no. 1, pp. 1-10, july 1991.

25. A. Hyvärinen and E. Oja. Independent Component Analysis: Algorithms and Applications. Neural Networks, 13(4-5):411-430, 2000.

26. A. Hyvärinen .Fast and robust fixed-point algorithms for independent component analysis. IEEE Trans. On Neural Networks, 10(3): 626-634, 1999.

27. R and C code implementation of the fastICA package. http://www.cis.hut.fi/ projects/ica/fastica/.

28. The R project for statistical computing. http://www.r-project.org/

29. R. Bellman: "Adaptive Control Processes: A Guided Tour". New Jersey, Princeton University Press.

30. D. Levine. Users Guide to the PGAPack Parallel Genetic Algorithm Library. Research Report ANL-95/18. Argonne Na

A Study on Robustness of Large Vocabulary Mandarin Chinese Continuous Speech Recognition System Based on Wavelet Analysis

Long Yan, Gang Liu, and Jun Guo

Beijing University of Posts and Telecommunications, Beijing 100876, P.R. China
dragonyan@263.net, lg@pris.edu.cn, guojun@bupt.edu.cn

Abstract. In this paper wavelet decomposition is used to decompose speech signal into five levels. Low-frequency part of the speech signal was reconstructed. Because different frequencies of the speech signal have different influence on the performance of the system, the acoustic model of each level was trained and tested. The experimental results show that the acoustic model of level 1 is the best for clean speech and the acoustic model of level 2 is the best for noisy speech .It is proved that the frequency band of A1 makes a lot of contribution on the performance of clean speech and the frequency band of A2 makes a lot of contribution on the performance of noisy speech.

1 Introduction

Mel-Frequency Cepstral Coefficients(MFCC) have been the most widely used features for speech recognition. A MFCC based speech recognizer outperforms other feature based (such as Linear Prediction Cepstral Coefficients(LPCC),Linear Prediction Coefficients(LPC),Reflection Coefficients(RC) and so on) speech recognizers.

The wavelet analysis has the best features of narrow band and wide band analysis within one transform without assuming a stationary signal. The wavelet analysis of a speech signal produces fine time resolution at high frequencies and fine frequency resolution at low frequencies[1].

Based on wavelet analysis, there are some feature extraction methods in speech recognition[1][2][3][4]. Tufekci Z. proposed a new feature vector consisting of Mel-Frequency Discrete Wavelet Coefficients(MFDWC). The MFDWC are obtained by applying Discrete Wavelet Transform to the mel-scaled log filter bank energies of a speech frame. The purpose of using the Discrete Wavelet Transform is to benefit from its localization property in the time and frequency domains. A MFDWC based speech recognizer outperforms the feature based MFCC[2] .

Speech signal is a none stationary signal. Although applying wavelet transform on noisy speech will lose some high frequency which may make contributions to recognition, the noise in high frequency can be suppressed in great. Therefore the research on the method which can not only de-noise but also keep the valuable high frequency is very meaningful.

S. Singh et al. (Eds.): ICAPR 2005, LNCS 3686, pp. 497 – 504, 2005.

The method in this paper is different from the traditional methods which focus on feature extraction by using wavelet transform. In this paper, the wavelet decomposition is used to decompose speech signal into five levels. Low-frequency parts of the speech signal were reconstructed and different frequencies of the speech signal have different influence on the performance.

The structure of this paper is as follows. Wavelet theory is described briefly in section 2. In Section 3, the wavelet-based de-noising is described. The continuous Speech Recognition System Based on Wavelet Analysis is mentioned in Section 4. The Experiment and Results are given in Section 5. And at last we come to the conclusion in Section 6.

2   Wavelet Theory

Wavelet theory is based on generating a set of filters by dilation and translation of a generating wavelet. All of the wavelets are scaled versions of the "mother wavelet". This means that only one filter needs to be designed and the others will follow the scaling rules in both the time and frequency domain.

A set of wavelets is generated from the mother wavelet $\Psi(t)$ by:

$$\Psi_{a,b}(t) = \frac{1}{\sqrt{a}} \Psi(\frac{t-b}{a}) \tag{1}$$

The wavelets are contracted (a<1) or dilated (a>1) and are moved over the signal to be analyzed by time step b (which is real valued). Contraction and dilation scale the frequency response to allow the set of wavelets to span the desired frequency range.

The set of wavelets can be considered as a filter bank for speech analysis.

For admissibility as a wavelet the following condition has to be met:

$$\int \left| \hat{\Psi}(w) \right|^2 \frac{dw}{|w|} < \infty \tag{2}$$

This implies that if the wavelet is differentiable then:

$$\hat{\Psi}(0) = 0 \tag{3}$$

The continuous wavelet transform(CWT) is defined as:

$$CWT(b,a) = \frac{1}{\sqrt{a}} \int s(t) \Psi(\frac{t-b}{a}) dt \tag{4}$$

The discrete wavelet transform (DWT) is given by:

$$DWT(a^i, a^i n) = \frac{1}{\sqrt{a^i}} \sum_k \Psi(\frac{k}{a^i} - n) s(k) \tag{5}$$

where i is an integer. The DWT computes data points on a dyadic grid if $a = 2$. (A dyadic grid has half of the number of data points at each successive lower octave). This makes it difficult to use the DWT for input to classical recognizers such as Hidden Markov Models (HMM) because they are designed to accept frame synchronous data. A variation of the DWT is the sampled CWT (SCWT). This produces frame synchronous data (redundant at lower frequencies) but retains the features that are offered by the wavelet transform. The sampled CWT is given by:

$$SCWT(a^i, n) = \frac{1}{\sqrt{a^i}} \sum_k \Psi(\frac{k-n}{a^i}) s(k) \tag{6}$$

By restricting the values of i to be integers, each of the wavelets will be an octave space apart if $a = 2$. Choosing other values for a will change the number of wavelets that are required to cover a certain frequency range. Thus if the initial generating wavelet is defined appropriately then sub-octave resolution can be accommodated. The spacing within each octave will need to be preserved so that the given number of voices appears within each octave. This can be varied by changing the value of a or by choosing i to be real and to be a fraction of the number of voices in each octave.

3   Wavelet-Based De-noising

Wavelet transform has recently emerged as a powerful tool for removing noise from speech signal. Such as de-noising based on wavelet decomposition and reconstruction. Donoho et al. developed a nonlinear wavelet shrinkage de-noising method for statistical applications[5]. And multiple wavelet de-noising[6] is coming out recently. The conventional wavelet de-nosing methods only focus on the vision effect or hearing effect of the waveform.

As far as speech recognition is mentioned, each frequency part maybe has different influence on the performance of speech recognition systems. This paper proposed a method based on wavelet analysis to verify which frequency band makes the main contribution to the speech recognition system.

4   Continuous Speech Recognition System Based on Wavelet Analysis

Figure 1 shows the wavelet decomposition of speech signal S. In this paper, 5 levels of decomposition is used, the original speech, S, would be decomposed into a set of sub signals A5, D5, D4,..., D1, where Di is the i level detail and Ai is the i level approximation of the original signal.

In this paper, the sampling frequency of speech signal is 16kHz, so the frequency bandwidth of the speech signal is 8kHz. Table 1 shows the frequency band of each sub signal.

Figure 2 shows the continuous speech recognition system based on wavelet analysis. The wavelet function "sym4" is used. The speech signal was decomposed into five levels. The signals of the low-frequency part were reconstructed. And for each level, the acoustic model was obtained by carrying out the same training steps, AM 0 is the acoustic model of the baseline system. From level 1 to level 5, the acoustic model are presented by AM 1,...AM 5, then the language model(LM) and recognizer are used to test the performance of the acoustic model.

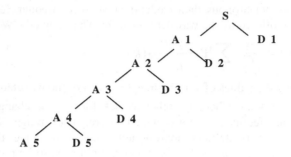

Fig. 1. 5-Level Decomposition Tree

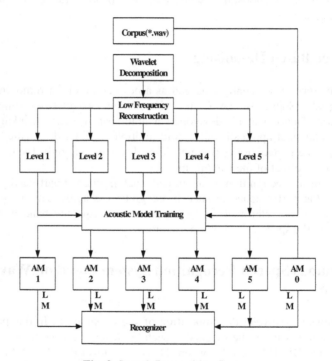

Fig. 2. Speech Recognition System

Table 1. Frequency Band

	Low-Frequency		High-Frequency
A1	0～4000Hz	D1	4000～8000 Hz
A2	0～2000 Hz	D2	2000～4000 Hz
A3	0～1000 Hz	D3	1000～2000 Hz
A4	0～500 Hz	D4	500～1000 Hz
A5	0～250Hz	D5	250～500Hz

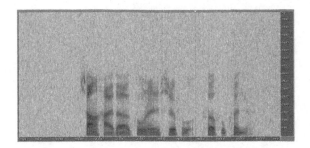

Fig.3a

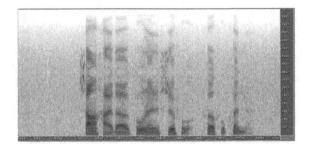

Fig. 3b

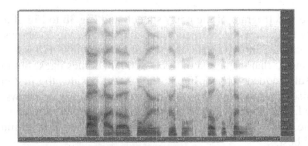

Fig. 3c

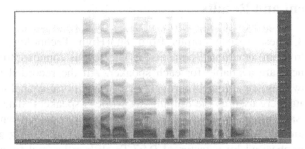

Fig. 3d

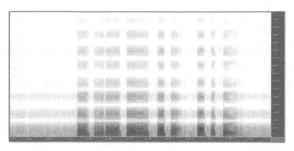

Fig. 3e

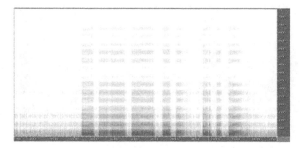

Fig. 3f

Fig.3a is the spectrogram of the noisy speech and Fig.3b to Fig.3f are spectrograms of the five low frequency reconstruction speech signals. Fig.3a shows that the speech signal is concentrated in 0-4000Hz, the signal of 4000-8000Hz was corrupted by the noise. Fig.3b shows that the signal of 4000-8000Hz comes back, but the noise are still strong in the full frequency band. Fig.3c shows that besides the first characteristics of Fig.3b, the noise of high frequency band are suppressed. Fig.3d shows that more noise are suppressed. Fig.3e and Fig.3f show that the signal are also suppressed by suppressing the noise. Therefore for noisy speech, the second level must represent the original signal best.

5   Experiment and Results

A continuous Chinese speech corpus from 863(High Technology Research and Development Program) materials[7] was used. The corpus contains 80 speakers' data and 520 utterances are available for each speaker. All the recorded materials were obtained in a low noise environment through a close-talk noise-canceling microphone. 41 speakers' data were used as the training set while 8 speakers' data were used for testing. At the same time, the clean speech test corpus are added white noise. The Signal noise ratio is 15db.The speech data were sampled at 16KHz and 16bit. The speech is pre-emphasised by a factor of 0.97.Twelfth-order mel-frequency cepstral coefficients (MFCC) and power are computed every 10ms. Temporal difference of the coefficients (□MFCC) and power (□LogPow) are also incorporated. So the feature vector at each frame consists of 26 variables. Each model consists of three states excluding the initial and final states that have no distributions. The state transitions are all left-to-right, and the path from the initial state and that to the final state are limited to one.

Table 2 shows the system performance of clean and noisy speech recognition. The results of clean speech shows that the highest correct rate of Initial/Final is 86.83% which is the performance of the AM 1. The performance of AM 2 is a little bit lower than AM 1. AM 1 and AM 2 are both better than AM 0 whose rate is 82.83%. And the performance of AM 3, AM 4 and AM 5 are too low. The results of noisy speech shows that the highest correct rate of Initial/Final is 56.36% which is the performance of the AM 2. The performance of AM 1 and AM 3 are a little bit lower than AM 2. AM 1 AM 2 and AM 3 are better than AM 0. And the performance of AM 4 and AM 5 are too low.

Table 2. Performance of the System

Acoustic Model	Initial/Final Correct Rate	
	Clean speech(%)	Noisy speech(%)
AM 0	82.83	39.52
AM 1	86.83	44.19
AM 2	83.53	56.36
AM 3	64.97	47.23
AM 4	34.25	27.33
AM 5	23.7	22.75

6 Conclusion

In this paper a wavelet analysis method is proposed for large vocabulary Chinese Mandarin continuous speech recognition. The wavelet decomposition is used to decompose speech signals into five levels. Low-frequency parts of the speech signal were reconstructed. The acoustic model of each level was trained and tested. From the experimental results of clean speech and noisy speech, the acoustic model of level 1 is the best for clean speech and the acoustic model of level 2 is the best for noisy speech. It is proved that the frequency band of level 1 which is between 0 and 4000Hz makes main contribution on the performance of the clean speech recognition system and the frequency band of level 2 which is between 0 and 2000Hz makes main contribution on the performance of the noisy speech recognition system.This paper proved that discarding high frequency part which is corrupted by noise will do some good to speech recognition. This is consistent with the man's hearing effect. For example, when hearing shortwave broadcasting, you can hearing more clearly by discarding high frequency part.

Acknowledgement

The author would like to thank the pattern recognition and intelligent system lab of Beijing University of Posts and Telecommunications. This work was sponsored by Chinese Ministry of Education grants (number: 02029) and Nation Natural Science Fund of China grants (number: 60475007).

References

[1] Favero R.F. and King R.W, "Wavelet Parameterization for Speech Recognition: Variations in Translation and Scale Parameters", *Speech, Image Processing and Neural Networks, 1994.* Proceedings, ISSIPNN '94., 1994 International Symposium on ,13-16 April 1994 Pages:694 - 697 vol.2

[2] Tufekci Z. and Gowdy J.N, "Feature Extraction Using Discrete Wavelet Transform for Speech Recognition", Southeastcon 2000. Proceedings of the IEEE ,7-9 April 2000 Pages:116 – 123

[3] C.J.Long and S.Datta, "Wavelet Based Feature Extraction for Phoneme Recognition", Spoken Language, 1996. ICSLP 96. Proceedings., Fourth International Conference on , Volume: 1 ,3-6 Oct. 1996,Pages:264 - 267

[4] Maya Gupta and Anna Gilbert, "Robust Speech Recognition Using Wavelet Coefficient Features", Automatic Speech Recognition and Understanding, 2001. ASRU '01. IEEE Workshop on ,9-13 Dec. 2001,Pages:445 – 448

[5] Donoho, D.L.; De-noising by soft-thresholding, Information Theory, IEEE Transactions on , Volume: 41 ,Issue: 3 , May 1995 ,Pages:613 - 627

[6] Downie, T.R.; Silverman, B.W.; The discrete multiple wavelet transform and thresholding methods, Signal Processing, IEEE Transactions on [see also Acoustics, Speech, and Signal Processing, IEEE Transactions on] , Volume: 46 , Issue: 9 , Sept. 1998 ,Pages:2558 – 2561

[7] National High Technology Research and Development Program of China(HTRDP) http://www.863data.org.cn

Recognition of Insect Emissions Applying the Discrete Wavelet Transform

Carlos García Puntonet[1], Juan-José González de-la-Rosa[2],
Isidro Lloret Galiana, and Juan Manuel Górriz

[1] University of Granada, Department of Architecture and Computers Technology,
ESII, C/Periodista Daniel Saucedo, 18071, Granada, Spain
carlos@atc.ugr.es
[2] University of Cádiz. Research Group TIC168 - Computational Electronics,
Instrumentation and Physics Engineering, EPSA, Av, Ramón Puyol S/N,
E-11202-Algeciras-Cádiz-Spain
juanjose.delarosa@uca.es

Abstract. The time-domain *fingerprint* of termite alarm signals is enhanced by wavelets and wavelet packets, using multi-resolution analysis. We take advantage of these emission patterns, characterized by four-impulse bursts. Identification has been developed by means of analyzing the impulse response of three sensors undergoing natural excitations. Denoising exhibits good performance up to SNR=-30 dB, in the presence of white Gaussian noise. The test can be extended to similar vibratory or acoustic signals resulting from impulse responses.

1 Introduction

Ultra-sounds signals produced by insects can be detected using ultrasonic sensors [1] which register only the vibratory signals which, in turn, constitute the patterns of the emissions, filtering the audio band of the spectra. When wood fibers are broken by termites (or similar insects) they produce acoustic signals which can be monitored using *ad hoc* resonant AE piezoelectric sensors which include microphones and accelerometers, targeting subterranean infestations by means of spectral and temporal analysis. The drawbacks are the relative high cost and their practical limitations due to subjectiveness [2].

In acoustic emission (AE) signal processing an usual problem is to extract some physical parameters of interest in situations which involve join variations of time and frequency. This situation can be found in almost every nondestructive AE tests for characterization of defects in materials, or detection of spurious transients which reveal machinery faults [3]. The problem of insect detection lies in this set of applications involving non-stationary signals [2].

The prior-art second order methods (spectra and spectrogram) failure in low SNR conditions even with *ad hoc* piezoelectric sensors. Bispectrum have proven to be a useful tool for characterization of termites in relative noisy environments using low-cost sensors [4],[5]. The computational cost could be pointed out as the

S. Singh et al. (Eds.): ICAPR 2005, LNCS 3686, pp. 505–513, 2005.
© Springer-Verlag Berlin Heidelberg 2005

main drawback of the technique. This is the reason whereby diagonal bispectrum have to be used.

Numerous wavelet-theory-based techniques have evolved independently in different signal processing applications, like wavelets series expansions, multiresolution analysis, subband coding, etc. The wavelet transform is a well-suited technique to detect and analyze events occurring to different scales [6]. The idea of decomposing a signal into frequency bands conveys the possibility of extracting subband information which could characterize the physical phenomenon under study [7].

In this paper we show an application of wavelets' de-noising possibilities for the characterization and detection of termite alarm signals in low SNR conditions. Waveforms have been buried in Gaussian white noise. Working with three different vibratory sensors, we find that the estimated signals' spectra matches the spectra of the acoustic emission whereby termite alarms are recognized. The paper is structured as follows: Section 2 summarizes the problem of acoustic detection of termites; Section 3 remembers the theoretical background of wavelets and wavelet packets. Experiments and conclusions are drawn in Section 4.

2   Acoustic Detection of Termites

2.1   Characteristics of the AE Signals

Acoustic Emission(AE) is defined as the class of phenomena whereby transient elastic waves are generated by the rapid (and spontaneous) release of energy from a localized source or sources within a material, or the transient elastic wave(s) so generated (ASTM, F2174-02, E750-04, F914-03 [1]).

Figure 1 shows one impulse in a burst produced by termites and its power spectrum. Significant drumming responses are produced over the range 200 Hz-10 kHz. The carrier (main component) frequency of the drumming signal is around 2600 Hz. The spectrum is not flat as a function of frequency as one would expect for a pulse-like event. This is due to the frequency response of the sensor (its selective characteristics) and also to the frequency-dependent attenuation coefficient of the wood and the air.

2.2   Devices, Ranges of Measurement and HOS Techniques

Acoustic measurement devices have been used primarily for detection of termites (feeding and excavating) in wood, but there is also the need of detecting termites in trees and soil surrounding building perimeters. Soil and wood have a much longer coefficient of sound attenuation than air and the coefficient increases with

---

[1] American Society for Testing and Materials. F2174-02: Standard Practice for Verifying Acoustic Emission Sensor Response. E750-04: Standard Practice for Characterizing Acoustic Emission Instrumentation. F914-03: Standard Test Method for Acoustic Emission for Insulated and Non-Insulated Aerial Personnel Devices Without Supplemental Load Handling Attachments.

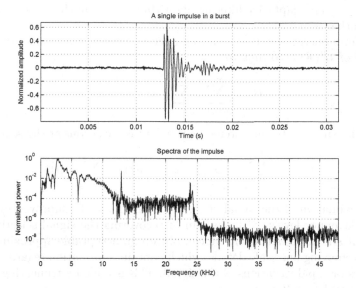

Fig. 1. Normalized power spectrum of a single impulse in a burst

frequency. This attenuation reduces the detection range of acoustic emission to 2-5 cm in soil and 2-3 m in wood, as long as the sensor is in the same piece of material [8]. The range of acoustic detection is much greater at frequencies <10 kHz, and low frequency accelerometers have been used to detect insect larvae over 1-2 m in grain and 10-30 cm in soil [1].

It has been shown that ICA success in separating termite emissions with small energy levels in comparison to the background noise. This is explained away by statistical independence basis of ICA, regardless of the energy associated to each frequency component in the spectra [5]. The same authors have proven that the diagonal bispectrum can be used as a tool for characterization purposes [4]. With the aim of reducing computational complexity wavelets transforms have been used in this paper to de-noise corrupted impulse trains. In section 3 we summarize the theoretical background of wavelet and wavelet packets.

3   Wavelet Packets (WP)

3.1   Wavelet Bases

The WP method is a generalization of wavelet decomposition that offers more possibilities of reconstructing the signal from the decomposition tree. If L is the number of levels in the tree, WP methods yields more than $2^{2^{L-1}}$ ways to encode the signal. The wavelet decomposition tree is a part of the complete binary tree.

When performing a split we have to look at each node of the decomposition tree and quantify the information to be gained as a result of a split. An entropy based criterion is used herein to select the optimal decomposition of a given signal. We use an adaptive filtering algorithm, based on the work by Coifman and Wickerhauser [9].

Any finite energy signal $s(t)$ can be decomposed over a wavelet orthogonal basis [6] [2] of $\mathbf{L}^2(\Re)$ according to:

$$s(t) = \sum_{j=-\infty}^{+\infty} \sum_{k=-\infty}^{+\infty} \langle s, \psi_{j,k} \rangle \psi_{j,k} \tag{1}$$

Each partial sum can be interpreted as the details variations at the scale $a = 2^j$:

$$d_j(t) = \sum_{k=-\infty}^{+\infty} \langle s, \psi_{j,k} \rangle \psi_{j,k} \qquad s(t) = \sum_{j=-\infty}^{+\infty} d_j(t) \tag{2}$$

The approximation of the signal $s(t)$ can be progressively improved by obtaining more layers or levels, with the aim of recovering the signal selectively. For example, if $s(t)$ varies smoothly we can obtain an acceptable approximation by means of removing fine scale details, which contain information regarding higher frequencies or rapid variations of the signal. This is done by truncating the sum in 1 at the scale $a = 2^J$:

$$s_J(t) = \sum_{j=J}^{+\infty} d_j(t) \tag{3}$$

3.2   Multiresolution and Tree Decomposition

We consider the resolution as the time step 2^{-j}, for a scale j, as the inverse of the scale 2^j. The approximation of a function s at a resolution 2^{-j} is defined as an orthogonal projection on a space $\mathbf{V}_j \subset \mathbf{L}^2(\Re)$. $\mathbf{V}_j$ is called the scaling space and contains all possible approximations at the resolution 2^{-j}.

Let us consider a scaling function ϕ. Dilating and translating this function we obtain an orthonormal basis of $\mathbf{V}_j$:

$$\left\{ \phi_{j,k}(t) = \frac{1}{\sqrt{2^j}} \phi \left(\frac{t - 2^j k}{2^j} \right) \right\}_{(j,k) \in \mathbb{Z}^2}. \tag{4}$$

The approximation of a signal s at a resolution 2^{-j} is the orthogonal projection over the scaling subspace $\mathbf{V}_j$, and is obtained with an expansion in the scaling orthogonal basis $\{\phi_{j,k}\}_{k \in \mathbb{Z}}$:

$$P_{\mathbf{V}_j} s = \sum_{k=-\infty}^{+\infty} \langle s, \phi_{j,k} \rangle \phi_{j,k} \tag{5}$$

The inner products

$$a_j[k] = \langle s, \phi_{j,k} \rangle \phi_{j,k} \tag{6}$$

represent a discrete approximation of the signal at level j (scale 2^j). This approximation is low-pass filtering of s sampled at intervals 2^{-j}.

---

[2] $\left\{ \psi_{j,k}(t) = \frac{1}{\sqrt{2^j}} \psi \left(\frac{t - 2^j k}{2^j} \right) \right\}_{(j,k) \in \mathbb{Z}^2}.$

A fast wavelet transform decomposes successively each approximation $P_{\mathbf{V}_{j-1}}s$ into a coarser approximation $P_{\mathbf{V}_j}s$ (local averages) plus the wavelet coefficients carried by $P_{\mathbf{W}_j}s$ (local details). The smooth signal plus the details combine into a multiresolution of the signal. Averages come from the scaling functions and details come from the wavelets.

$\{\phi_{j,k}\}_{k\in\mathbb{Z}}$ and $\{\psi_{j,k}\}_{k\in\mathbb{Z}}$ are orthonormal bases of $\mathbf{V}_j$ and $\mathbf{W}_j$, respectively, and the projections in these spaces are characterized by:

$$a_j[k] = \langle s, \phi_{j,k}\rangle \qquad d_j[k] = \langle s, \psi_{j,k}\rangle \tag{7}$$

A space $\mathbf{V}_{j-1}$ is decomposed in a lower resolution space $\mathbf{V}_j$ plus a detail space $\mathbf{W}_j$, dividing the orthogonal basis of $\mathbf{V}_{j-1}$ into two new orthogonal bases:

$$\{\phi_j(t - 2^j k)\}_{k\in\mathbb{Z}} \qquad and \qquad \{\psi_j(t - 2^j k)\}_{k\in\mathbb{Z}} \tag{8}$$

$\mathbf{W}_j$ is the orthogonal complement of $\mathbf{V}_j$ in $\mathbf{V}_{j-1}$, and $\mathbf{V}_j \subset \mathbf{V}_{j-1}$, thus:

$$\mathbf{V}_{j-1} = \mathbf{V}_j \oplus \mathbf{W}_j. \tag{9}$$

The orthogonal projection of a signal s on $\mathbf{V}_{j-1}$ is decomposed as the sum of orthogonal projections on $\mathbf{V}_j$ and $\mathbf{W}_j$.

$$P_{\mathbf{V}_{j-1}} = P_{\mathbf{V}_j} + P_{\mathbf{W}_j}. \tag{10}$$

The recursive splitting of these vector spaces is represented in the binary tree. This fast wavelet transform is computed with a cascade of filters $\bar{h}$ and $\bar{g}$, followed by a factor 2 subsampling, according with the scheme of figure 2.

Functions that verify additivity-type property are suitable for efficient searching of the tree structures and node splitting. The criteria based on the entropy match these conditions, providing a degree of randomness in an information-theory frame. In this work we used the entropy criteria based on the p-norm:

$$E(s) = \sum_i^N \|s_i\|^p; \tag{11}$$

with p$\leq$1, and where $s = [s_1, s_2, \ldots, s_N]$ in the signal of length N. The results are accompanied by entropy calculations based on Shannon's criterion:

$$E(s) = -\sum_i^N s_i^2 log(s_i^2); \tag{12}$$

with the convention $0 \times log(0) = 0$.

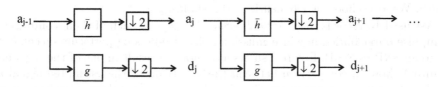

Fig. 2. Cascade of filters and subsampling

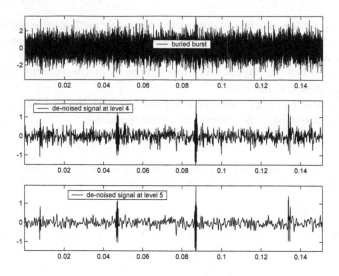

Fig. 3. Limit situation of the de-noising procedure using wavelets (SNR=-30 dB). From top to bottom: a buried 4-impulse burst, estimated signal at level 4, estimated signal at level 5.

Once the mathematica foundations have been established, we described the experienced in 4.

4   Experiments and Conclusions

Two accelerometers (KB12V, seismic accelerometer; KD42V, industrial accelerometer, MMF) and a standard microphone have been used to collect data (alarm signals from termites) in different places (basements and subterranean wood structures and roots) using the sound card of a portable computer and a sampling frequency of 96000 (Hz), which fixes the time resolution. These sensors have different sensibilities and impulse responses. This is the reason whereby we normalize spectra. In fact we are only interested in the frequency pattern of the emissions.

The de-noising procedure was developed using a *sym*8 wavelet, which belongs to the family *Symlets* (order 8), which are compactly supported wavelets with least asymmetry and highest number of vanishing moments for a given support width. We also choose a soft heuristic thresholding.

We used 15 records (from *reticulitermes lucifugus*), each of them comprises a 4-impulse burst buried in white gaussian noise. De-noising performs successfully up to an SNR=-30 dB. Figure 3 shows a de-noising result in one of the registers. Figure 4 shows a comparison between the spectrum of the estimated signal at level 4 and the spectrum of the signal to be de-noised, taking a register as an example.

Significant components in the spectrum of the recovered signal are found to be proper of termite emissions.

The same 15 registers were processed using wavelet packets. Approximation coefficients have been thresholded in order to obtain a more accurate estimation of the starting points for each impulse. Stein's Unbiased Estimate of Risk (SURE) has been assumed as a principle for selecting a threshold to be used for de-noising. A more thorough discussion of choosing the optimal decomposition can be found in [6]. Figure 5 shows one of the 15 de-noised signals using wavelets packets.

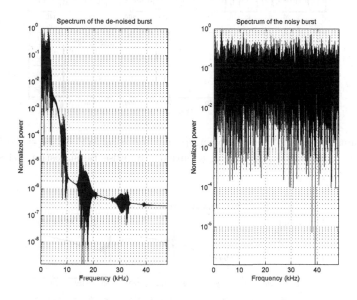

Fig. 4. Spectra of the estimated signal and the buried burst

This result can be see the result of reconstructing progressively each a_j by the filter banks.

To show the importance of the pre-processing high-pass filter, we have included figure 6. We can seen, for the same SNR conditions that the impulses in the burst have not been clearly enhanced, despite the fact that they can be distinguished.

Future effort should be put in the task of simulating with new noise processes. Results obtained with non-Gaussian noise, and with non-symmetrical noise, will be specially welcomed in order to establish the real limits of this application. The objective is to reduce the computational complexity of the algorithms with the goal of implementing the code in a DSP processor. This work has established the basis of the equipment which constitutes the objective of a Spanish project for the transference of technology.

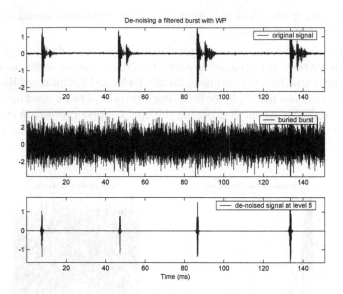

Fig. 5. Limit situation of the de-noising procedure using WP (SNR = -28.5545 dB). From top to bottom: original signal, a buried 4-impulse burst, estimated signal at level 5.

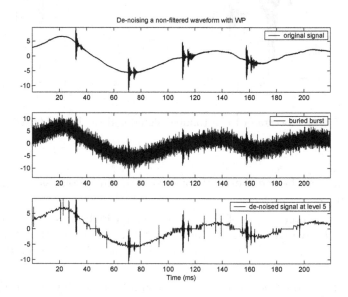

Fig. 6. Limit situation of the de-noising procedure for a non- filterd signal, using WP (SNR = -30 dB). From top to bottom: original signal, a buried 4-impulse burst, estimated signal at level 5.

Acknowledgement

The authors would like to thank the *Spanish Ministry of Education and Science* for supporting the projects DPI2003-00878 and P.E.T.R.I. project PTR1995-0824-OP, and the *Andalusian Autonomous Government Division* for funding the research with *Contraplagas Ambiental S.L.*

References

1. Robbins, W.P., Mueller, R.K., Schaal, T., Ebeling, T.: Characteristics of acoustic emission signals generated by termite activity in wood. In: Proceedings of the IEEE Ultrasonic Symposium. (1991) 1047–1051
2. de la Rosa, J.J.G., Puntonet, C.G., Górriz, J.M., Lloret, I.: An application of ICA to identify vibratory low-level signals generated by termites. Lecture Notes in Computer Science (LNCS) **3195** (2004) 1126–1133 Proceedings of the Fifth International Conference, ICA 2004, Granada, Spain.
3. Lou, X., Loparo, K.A.: Bearing fault diagnosis based on wavelet transform and fuzzy inference. Mechanical Systems and Signal Processing **18** (2004) 1077–1095
4. de la Rosa, J.G., Lloret, I., Puntonet, C.G., Górriz, J.M.: Higher-order statistics to detect and characterise termite emissions. Electronics Letters **40** (2004) 1316–1317 Ultrasosics.
5. de la Rosa, J.J.G., Puntonet, C.G., Lloret, I.: An application of the independent component analysis to monitor acoustic emission signals generated by termite activity in wood. Measurement **37** (2005) 63–76 Available online 12 October 2004.
6. Mallat, S.: A wavelet tour of signal processing. 2 edn. Academic Press (1999)
7. Angrisani, L., Daponte, P., D'Apuzzo, M.: A method for the automatic detection and measurement of transients. part I: the measurement method. Measurement **25** (1999) 19–30
8. Mankin, R.W., Osbrink, W.L., Oi, F.M., Anderson, J.B.: Acoustic detection of termite infestations in urban trees. Journal of Economic Entomology **95** (2002) 981–988
9. Coifman, R.R., Wickerhauser, M.: Entropy-based algorithms for best basis selection. IEEE Trans. on Inf. Theory **38** (1992) 713–718

On the Performance of Hurst-Vectors for Speaker Identification Systems

R. Sant'Ana[1], R. Coelho[1], and A. Alcaim[2]

[1] Instituto Militar de Engenharia (IME),
Electrical Engineering Department
coelho@ime.eb.br
[2] Pontifícia Universidade Católica (PUC-Rio),
Center of Telecommunications Studies (CETUC)
alcaim@cetuc.puc-rio.br

Abstract. The performance of Hurst-Vectors (pH feature) for speaker identification systems is presented and discussed in this paper. The pH feature is a vector of *Hurst* (H) parameters obtained by applying a *wavelet*-based multi-dimensional estimator ($M\_dim\_wavelets$) to the windowed short-time segments of speech. The GMM (*Gaussian Mixture Models*) and the M_dim_fBm (multi-dimensional fractional Brownian motion) classification systems were considered in the performance analysis. The database—recorded from fixed and cellular phone channels— was uttered by 75 different speakers. The results have shown the superior performance of the M_dim_fBm classifier and that the pH feature aggregates new information on the speaker identity.

1   Introduction

In an automatic speaker identification process, a speech utterance has to be identified as to which of the registered speakers it belongs [1]. Important areas of interest are found in law enforcements, such as penitentiary monitoring and forensic applications. Identification systems involve three basic steps: speech acquisition/pre-processing, speech feature extraction and classification.

The most commonly used features employed in speaker recognition are the LPC-derived cepstral parameters and the mel-cepstral coefficients. Generally, physiological features are not robust to the channels acoustic distortion and their extraction from the speech signal requires a high computational load. This is due to the fact that these features model the spectral characteristics of the human vocal mechanism. The statistical (pH) feature proposed in [2][3] consists of a vector of *Hurst* (H) parameters. Unlike the physiological features, the pH feature tends to be robust to channel distortions, since it models the stochastic behavior of the speech signal. The pH feature is not related to the transfer functions of the vocal tract and needs less complex extraction/estimation methods. Additionally, it can be obtained in real-time, i.e., during speakers' activity. The performance of the Hurst-vectors for the GMM (Gaussian Mixture Model) [4] and M_dim_fBm (Multi-dimensional fractional Brownian motion) [2][3] identification systems is

S. Singh et al. (Eds.): ICAPR 2005, LNCS 3686, pp. 514–521, 2005.
© Springer-Verlag Berlin Heidelberg 2005

examined in this paper. The M_dim_fBm models the speech characteristics of a particular speaker using the H parameters along with the statistical means and variances of the input speech matrix features.

2   The Hurst-Vectors or pH Feature

The *Hurst* parameter[1] expresses the time-dependence or scaling degree of a stochastic process. It can also be defined by the decaying rate of the auto-correlation coefficient function $\rho(k)$ $(-1 < \rho(k) < 1)$ as $k \to \infty$. Let the speech signal be represented by a stochastic process $X(t)$, with finite variance and normalized auto-correlation function (ACF) or auto-correlation coefficient $\rho(k) = Cov[X(t), X(t + k)]/Var[X(t)]$, $k = 0, 1, 2, \ldots$ where the $\rho(k)$ belongs to $[-1, 1]$ and $\lim_{k\to\infty} \rho(k) = 0$. The asymptotic behavior of $\rho(k)$ is given by $\rho(k) \sim H(2H - 1)k^{2(H-2)}$. This means that $\rho(k)$ is a slowly decaying function and that when $k \to \infty$, $\rho(k) \sim H(2H - 1)k^{2(H-2)}$ and hence, $\rho(k)/H(2H-1)k^{2(H-2)} \sim 1$. The H parameter is the exponent of the ACF of a stochastic process. Only for fractal or self-similar processes, one can relate the H parameter to a fractal dimension (D_h) through the equation $D_h = 2 - H$. Examples where the fractal dimension are used in pattern recognition studies can be found in [5] and [6]. The fractal dimension has already been used for discriminating fricative sounds and for speaker identification. The studies presented in [7] and [8] assumed the hypothesis that speech is a fractal signal. In the present work, however, although a vector of H parameters is adopted as a speech feature, it is not assumed that the speech signal is a fractal or self-similar signal.

The most known H estimators are the R/S (*ReScaled adjusted range*) statistic [9], the Higuchi [10] and the wavelet-based Abry-Veitch (AV) [11]. The R/S estimator can be used for any type of speech signal distribution. However, the R/S estimation of the H parameter is a time-consuming procedure since it depends on the user visual intervention to define the linear regression region. The Higuchi estimator is only appropriate for fractal stochastic processes and it cannot be proved that speech signals are fractals. For these reasons, the Wavelet-based Multi-dimensional Estimator ($M\_dim\_wavelets$) [3] is based on the AV method. Moreover, it enables the pH feature extraction in real-time and presents a low computational cost when compared to the standard physiological features extraction.

Similar to the H estimator proposed in [12] the *wavelet-based multidimensional* estimator — $M\_dim\_wavelets$ — uses the discrete wavelet transform (DWT) to successively decompose a sequence of samples into the detail and approximation coefficients. From each detail sequence, $d(j, k)$, generated by the filter bank in a given scale j, an H parameter is estimated, H_j. The set of H_j values and the H value obtained for the entire speech signal (H_0) compose the pH feature.

---

[1] The H notation is used for a single *Hurst* parameter. The proposed feature is a vector of H parameters and is denoted by pH.

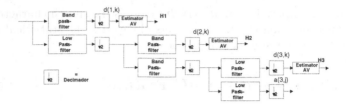

Fig. 1. *M_dim_wavelets* estimation example

The *M_dim_wavelets* estimator (cf. Fig. 1) can be described in two main steps:

1. Wavelet decomposition: the DWT is applied to the speech samples generating the detail sequences $d(j, k)$.
2. pH estimation: application of the AV estimator to the entire speech signal (H_0) and then to each of the J detail sequences obtained in the previous step. The resulting $(J + 1)$ H values will compose the *pH* feature.

The speech signal is split into N frames (with overlapping) and the proposed estimator *M_dim_wavelets* is applied to each speech frame. This means that at each frame n, several H parameter values are estimated. In this study, the *pH* matrix— containing the *pH* vectors along the frames— was obtained from 80ms frames with 50% overlapping. From several experiments, it was found that a good configuration for extraction of the *pH* feature matrix is given by (1) Frame duration: 80ms; (2) *Daubechies* wavelets [13] with 12 coefficients; (3) Number of decomposition scales for the H_0: 6 and (4) Coefficient range from 3 to 5.

3   Classification Schemes

3.1   GMM

The GMM is one of the most widely used classifiers for speaker recognition [14]. A mixture of Gaussian probability densities is a weighted sum of M densities, and is given by $p(\boldsymbol{x}|\lambda) = \sum_{i=1}^{M} p_i b_i(\boldsymbol{x})$ where $\boldsymbol{x}$ is a random vector of dimension D, $b_i(\boldsymbol{x})$, $i = 1, ..., M$, are the density components, and p_i, $i = 1, ..., M$, are the mixture weights. Each component density is a D variate Gaussian function of the form $b_i(\boldsymbol{x}) = \dfrac{e^{\left(-\frac{1}{2}(\boldsymbol{x}-\boldsymbol{\mu})^T K_i^{-1}(\boldsymbol{x}-\boldsymbol{\mu})\right)}}{(2\pi)^{\frac{D}{2}}\sqrt{|K_i|}}$ with mean vector $\boldsymbol{\mu}_i$ and covariance matrix K_i, where T denotes the transpose operation and $|.|$ is the determinant. The Gaussian mixture model, λ, is parametrized by mean vectors, covariance matrices, and mixture weights. These parameters are jointly represented by the following notation: $\lambda = \{p_i, \boldsymbol{\mu}_i, K_i\}$ $i = 1, ..., M$. The model parameters are estimated for a set of training data as the ones that maximize the likelihood of the GMM. In this paper, we obtain the parameter estimates using a special case of the expectation-maximization (EM) algorithm [14]. For a sequence of T

independent training vectors $X = \{x_1, ..., x_T\}$, the normalized log-likelihood of the GMM is given by $\log p(X|\lambda) = \frac{1}{T}\sum_{t=1}^{T} \log p(x_t|\lambda)$. The decision rule for the speaker identification system chooses the speaker model for which this value is maximum.

3.2 M_dim_fBm

The recently proposed M_dim_fBm classifier [2] [3] models each speaker on the basis of the speech features time-dependence or scaling characteristics. The speech signals are not assumed to be fractals. The M_dim_fBm is based on the fBm process. The fBm [15] is a Gaussian stochastic process $(X_H(t))$ indexed in $\Re$ with zero mean and continuous sample path (null at origin). The fBm is known as the unique gaussian H-sssi, i.e., self-similar with self-similarity parameter and stationary increments. The variance of the independent increments is proportional to its time interval accordingly to the expression $Var[X_H(t_2) - X_H(t_1)] \propto |t_2 - t_1|^{2H}$ for all instants t_1 and t_2 and $X_H(t)$ has stationary increments, $X_H(0) = 0$, $E[X_H(t)] = 0$ for any instant t and it presents continuous sample paths. The fBm is considered a self-similar process since its statistical characteristics[2] hold for any time scale. In other words, for any τ and $r > 0$, $[X_H(t + \tau) - X_H(t)]_{\tau \leq 0} \overset{d}{\approx} r^{-H}[X_H(t + r\tau) - X_H(t)]_{\tau \leq 0}$ where $\overset{d}{\approx}$ means equal in distribution and r is the process scaling factor ($r = \tau = |t_2 - t_1|$). Note that $X_H(t)$ is a Gaussian process completely specified by its mean, variance, H parameter and ACF given by [16] $\rho(k) = \frac{1}{2}[(k+1)^{2H} - 2k^{2H} + (k-1)^{2H}]$ for $k \geq 0$ and $\rho(k) = \rho(-k)$ for $k < 0$. Similarly to the GMM classification procedure the M_dim_fBm scheme is based on the input features models. The M_dim_fBm model of a given speaker is generated according to the following steps:

1. *Pre-processing*: the feature matrix formed from the input speech features[3] is split into r regions. This matrix contains c rows, where c is the number of feature coefficients per frame, and N columns, where N is the number of frames.
2. *Decomposition*: for each row of the feature matrix in a certain region the *wavelet* decomposition is applied in order to obtain the *detail* sequences.
3. *Parameters Extraction/Estimation*: from each set of *detail* sequences obtained from each row of step 2, estimate the mean, the variance and the H parameters of the features being used by the identification system. For the H parameter estimation, use the AV *wavelet − based* estimator proposed in [11].
4. *Generation of fBm Processes*: using the *Random Midpoint Displacement* (RMD) algorithm [15] and the three parameters computed in step 3, generate the fBm processes. Therefore, c fBm processes are obtained for a given region.

---

[2] *Statistical characteristics* means marginal distribution and time-dependence degree.
[3] Note that the M_dim_fBm classifier is not constrained to the pH feature. It can be used with any selected set of speech features.

5. *Determining the Histogram and Generating the Models*: compute the histogram of each fBm process of the given region. The set of all histograms defines the speaker c-dimensional model for that region.
6. *Speaker Model*: the process is repeated for all of the r regions. This means that a $r.c$-dimensional fBm process is obtained, which defines the speaker M_dim_fBm model.

In the phase of tests, the histograms of the speaker, obtained from the M_dim_fBm model, are used to compute the probability that a certain c-dimensional feature vector x belongs to that speaker. This is performed to the N feature vectors, resulting in N probability values: $p_1, p_2, ...p_N$. Adding these values, the measure of the likelihood that the set of feature vectors under analysis belongs to that particular speaker is obtained.

4   Experiments Results

Experiments were carried out in order to examined the performance of pH vectors, containing 7 H parameters, vectors of 15 mel-cepstral coefficients and the fusion of the pH and the mel-cepstral coefficients. We have investigated both the GMM and the M_dim_fBm classifiers. It is important to remark that in all experiments the M_dim_fBm classifier was used with $r = 1$ region only. This means that the speaker model is defined by a c-dimensional fBm process, where c is the number of feature coefficients.

The database[4] used in the experiments is composed of 75 speakers (male and female, 2 : 1) from 27 Brazilian regions that read 2 different texts. To record the speech signal the speakers called a free automatic communication center using first a fixed phone and then a cellular phone. Hence, two databases are available in which each speaker recorded four different text files (i.e., two different training and tests speech texts recorded from fixed and cellular phones). The speech average duration for training as well as testing phases, was 170 s. The tests were applied to 20, 10 and 5 seconds speech segments. A separate speech segment of 1 minute duration was used to train a speaker model. The performance results of the identification systems — M_dim_fBm and GMM— are presented in terms of the recognition accuracy. The results using the pH satisfy the low computational cost requirement [17]. On the other hand, the use of the mel-cepstrum and the joint use of the mel-cepstrum and the pH, is useful in applications where the computational cost is not of major concern [18].

4.1   *pH Feature Results*

Tables 1 and 2 show the speaker recognition accuracy of the identification systems based on the pH, for speech signals recorded from a fixed and a cellular telephony channel.

---

[4] This database was developed by the Electrical Engineering Department of the *Instituto Militar de Engenharia (IME)* under a project jointly sponsored by FAPERJ and the Security Department of Rio de Janeiro. The database is available and interested readers can send a request to coelho@ime.eb.br

Table 1. Recognition accuracy (%) of the identification systems based on the pH, for speech signals recorded from a fixed telephony channel

Testing Interval	M_dim_fBm	GMM
20 s	95.48	95.48
10 s	94.22	94.09
5 s	89.98	89.69

Table 2. Recognition accuracy (%) of the identification systems based on the pH, for speech signals recorded from a cellular telephony channel

Testing Interval	M_dim_fBm	GMM
20 s	87.53	86.85
10 s	84.93	84.89
5 s	61.43	61.10

Table 3. Recognition accuracy (%) of the identification systems based on the 15 mel-cepstral coefficients, for speech signals recorded from a fixed telephony channel

Testing Interval	$MdimfBm$	GMM
20 s	98.54	97.95
10 s	97.99	97.99
5 s	97.59	97.46

Table 4. Recognition accuracy (%) of the identification systems based on the fusion use of the pH and the mel-cepstral coefficients, for speech signals recorded from a fixed telephony channel

Testing Interval	$MdimfBm$	GMM
20 s	98.57	98.40
10 s	98.62	98.51
5 s	97.91	97.66

Table 5. Recognition accuracy (%) of the identification systems based on the fusion use of the pH and the mel-cepstral coefficients, for speech signals recorded from a cellular telephony channel

Testing Interval	$MdimfBm$	GMM
20 s	98.19	98.14
10 s	92.56	92.03
5 s	89.96	89.96

As can be seen from these tables, the best results (around 95% recognition) were obtained with the M_dim_fBm and the GMM classifiers for fixed telephone recordings using a testing interval of 20s. The results drop significantly for 5s testing intervals. The performance for cellular telephony recordings is much lower than for fixed telephone speech. This is due to the effects of the cellular channel. Comparing the classifiers performance results, it can be seen that the simpler M_dim_fBm also provided some improvement over GMM (around 0.6% for 20s testing intervals and cellular speech). It is important to remark that the pH feature used only 7 H parameters per speech frame. This implies in a lower complexity of the classifiers as compared to the systems operating on 15 mel-cepstral coefficients per frame. Moreover, it should be reminded that the estimation of the pH feature demands less computational complexity ($O(n)$) than the extraction of the mel-cepstral coefficients (the FFT computational complexity is $O(nlog(n))$).

4.2 pH + mel-cepstral Results

In this second set of experiments, the speaker recognition accuracy of the identification systems was examined for the mel-cepstral coefficients and for the fusion of the mel-cepstral coefficients and the pH. These results are presented in Tables 3 and 4, respectively, for speech signals recorded from a fixed telephony channel. From these tables it can be verified that the best results were achieved by the systems based on the joint use of the mel-cepstral coefficients and the pH. This means that the pH feature aggregates new information on the speaker identity. The recognition accuracies of the identification systems based on the fusion use of the pH and the mel-cepstral coefficients, for speech signals recorded from a cellular telephony channel, are shown in Table 5. Comparing the results of Tables 4 and 5, it can been seen that the M_dim_fBm and GMM systems performances are degraded around 0.3% due to the effects of the cellular telephony channel. Again, the simpler M_dim_fBm yields a small improvement over the GMM.

5   Conclusions

In this paper the performance of Hurst-vectors or the pH feature is presented for M_dim_fBm and GMM identification systems. For applications requiring a low computational cost, the systems employing only the pH feature have shown to be an attractive choice. On the other hand, if computational complexity is not of major concern, the best strategy — due to its highest performance — is the one based on the fusion of the pH feature and the mel-cepstral coefficients. Moreover, it was also shown that the M_dim_fBm classifier yields a better modeling accuracy with a lower computational load. The M_dim_fBm is characterized by only 3 scalar parameters (i.e., mean, variance and H) while the GMM needs 32 gaussians, each one characterized by 1 scalar parameter, 1 mean vector and 1 covariance matrix, to achieve comparable performance results.

   The results presented in this paper show that the M_dim_fBm requires less computational load and provides a more accurate modeling strategy as compared to the GMM. It is also shown that the pH feature adds substantial information to the systems based on the mel-cepstral coefficients.

References

1. O'Shaughnessy, D.: Speech Communication Volume 2. ed. IEEE Press (2000).
2. Sant'Ana, R., Coelho, R., Alcaim, A.: Automatic Speaker Verification Based on Fractional Brownian Motion Process. Electronics Letters 40 (2004) 1232-1233.
3. Sant'Ana, R., Coelho, R., Alcaim, A.: Text-Independent Speaker Recognition Based on the Hurst Parameter and the Multi-Dimensional Fractional Brownian Motion. (To appear in IEEE Transactions on Speech and Audio Processing).
4. Reynolds, D., Quatieri, R., Dunn, R.: Speaker Verification Using Adapted Gaussian Mixture Models. Digital Signal Processing 10 (2000) 19-41.
5. Esteller, R., Vachtsevanos, G., Henry, T.: Fractal Dimensions Characterizes Seizure Onset in Epileptic Patients. IEEE Proceedings, ICASSP99 4 (1999) 2343-2346.
6. Morimoto et al. T.: Pattern Recognition of Fruit Shape Based on the Concept of Chaos and Neural Networks. Computers and Electronics in Agriculture 26 (2000) 171-186.
7. Fernández, S., Feijóo, S., Balsa, R.: Fractal Characterization of Spanish Fricatives. Proceedings of the ICPhS (1999) 2145-2148.
8. Petry, A., Barone, D.: Fractal Dimension Applied to Speaker Identification. Proceedings of the ICASSP (2001).
9. Hurst, E: Long-Term Storage Capacity of Reservoirs. American Society of Civil Engineers Trans. (1951) 770-799.
10. Higuchi, T.: Approach to an Irregular Time Series on the Basis on the Fractal Theory. Physica D. (1988).
11. Veith, D., Abry, P.: A Wavelet-Based Joint Estimator of the Parameters of Long-Range Dependence. IEEE Transactions on Information Theory 45 (1998) 878-897.
12. Roughan, M., Veith, D., Abry, P.: Real-Time Estimation of the Parameters of Long-Range Dependence. IEEE/ACM Transactions on Networking 8 (2000) 467-478.
13. Daubechies, I.: Ten Lectures on Wavelets. SIAM, Philidelphia (1992).
14. Reynolds, D., Rose, R.: Robust Text-Independent Speaker Identification Using Gaussian Mixture Speaker Models. IEEE Transactions on Speech and Audio Processing 3 (1995) 72-83.
15. Barnsley et al, M.: The Science of Fractal Images. Springer-Verlag New York Inc., USA (1988).
16. Beran, J.: Statistics for Long-Memory Processes. Chapman & Hall (1994).
17. Kumagai, J.: Talk to the Machine. IEEE Spectrum 39 (2002) 60-64.
18. Nist: The Nist Year 2001 Speaker Recognition Evaluation Plan. http://www.nist. gov/speech/publications

Transformations of LPC and LSF Parameters to Speech Recognition Features

Vladimir Fabregas Surigué de Alencar and Abraham Alcaim

Pontifícia Universidade Católica do Rio de Janeiro – PUC-RIO,
Centro de Estudos em Telecomunicações – CETUC,
Rua Marquês de São Vicente, 225,
22453-900, Rio de Janeiro / RJ, Brazil
{Vladimir, Alcaim}@cetuc.puc-rio.br

Abstract. In this paper, we describe and present an overall evaluation of several features for distributed speech recognition systems. These systems are based on a client-server architecture. This means that recognizers access only the coded parameters of the speech coder employed in communication networks (e.g., cellular mobile and IP networks). The recognition features considered in this paper are obtained from transformations of codec parameters. In particular, features generated from LPC and LSF parameters, in intervals of 10 ms and 20 ms, are analyzed in a continuous observation HMM-based speaker independent recognizer.

1 Introduction

The growth of the Internet and mobile communication systems has stimulated a great effort to realize speech processing applications in these networks. A particularly important problem is concerned with Automatic Speech Recognition (ASR) in a server system, based on the extracted and quantized acoustic parameters at the user terminal. Such systems, usually known as Distributed Speech Recognition (DSR), are very attractive due to the complexity and large memory requirements of ASR systems.

Speech coding schemes used in mobile communication systems and IP networks operate at low bit rates and utilize, in general, LPC (Linear Predictive Coding) algorithms based on a speech production model. In this model, an excitation signal is applied to an all-pole filter (characterized by the LPC parameters), that represents the spectral envelope information of the speech signal. Usually, the LPC parameters are transformed to LSF (Line Spectral Frequencies), due to attractive properties of the latter to the quantization and interpolation procedures. Speech coders employed in cellular and IP networks use these parameters to caracterize the speech spectral envelope.

In distributed ASR systems, it is preferrable to directly use the codec parameters than to extract them from the decoded signal [1]. Since these parameters are not the most adequate ones for the remote recognition system, it is important to consider and examine different codec parameter transformations, in order to improve the recognition performance. The main contribution of this paper is to provide a global analysis of different speech features reported in the literature, aiming at improving the performance of DSR systems. Moreover, the results are presented at two frame rates: 100 Hz (typical of speech recognizers) and 50 Hz (usually employed by speech codecs).

Features obtained from the LPC and LSF parameters are described in Sections 2 and 3, respectively. Experimental results are presented and analyzed in Section 4. Finally, conclusions are summarized in Section 5.

2 Recognition Features Obtained from Transformations of LPC Parameters

This section deals with the recognition features that can be extracted directly from the LPC parameters, without the need to reconstruct the speech signal. This approach is attractive for DSR due to the speech decoding structures used in mobile and VoIP (Voice over IP) systems. In these structures, LPC parameters are obtained in a stage prior to speech reconstruction. This means that speech features extracted in this stage are computationally more attractive. Moreover, as we have previously mentioned, the use of codec parameters is more efficient for speech recognition than generating features from reconstructed speech.

Recognition features that can be obtained from the LPC parameters are the LPCC (LPC Cepstrum) and the MLPCC (Mel-Frequency LPCC) [2]. The LPCC are computed from the LPC parameters by means of a recursive equation, and the MLPCC are derived from a first-order all-pass filtering operation.

2.1 LPC Cepstrum (LPCC)

The extraction process of the LPCC features from the LPC coefficients is formulated in the z -transform domain, using the complex logarithm of the LPC system transfer function, which is analogous to the cepstrum computation from the discrete Fourier transform of the speech signal [2]. The i-th LPCC parameter is given by the following recursive equation

$$c_i = \begin{cases} \ln(G) & i = 0 \\ a_1 & i = 1 \\ a_i + \sum_{j=1}^{i-1} \dfrac{i-j}{i} c_{i-j} a_j & 1 < i \le p \\ \sum_{j=1}^{p} \dfrac{i-j}{i} c_{i-j} a_j & i > p \end{cases} \tag{1}$$

where a_i is the i-th LPC parameter, p is the LPC system order and G is the gain factor of the system.

2.2 Mel-Frequency LPCC (MLPCC)

The MLPCC feature is obtained by transforming the real frequency axis of the LPCC to the mel frequency scale. This is performed by a bank of n first-order all-pass filters, where n is the number of LPCC features [3]. The filters have their first-order all-pass transfer function $\psi(z)$ [4] given by

$$\psi(z) = \frac{z^{-1} - a^*}{1 - az^{-1}} \tag{2}$$

where a is the all-pass filter coefficient and a^* is the complex conjugate of a. Each LPCC parameter, c_i, is processed by a different filter.

Since the purpose of each filtering operation is to approximate the mel scale frequency, it is important to analyze the relationship of the transfer function given by (2) and the transformation of the frequency axis. In order to simplify the filter implementation, let a be a real number [5]. Now rewrite ψ, as a function of $e^{j\Omega}$, as

$$\psi(e^{j\Omega}) = e^{-j\theta(\Omega)} \tag{3}$$

where Ω is the real frequency. From (2) and (3), we can derive the mel scale frequency as a function of the real frequency Ω:

$$\theta(\Omega) = \arctan\left[\frac{(1 - a^2)\operatorname{sen}\Omega}{(1 + a^2)\cos\Omega - 2a}\right] \tag{4}$$

Changing the value of a it is possible to adjust $\theta(\Omega)$ to the mel scale curve. At an 8kHz sampling frequency, the value of a that best approximates the mel scale curve is 0.3624 [5].

The outputs of the filter bank are the MLPCC features.

3   Recognition Features Obtained from Transformations of LSF Parameters

The Line Spectral Frequencies (LSFs) are often used for speech coding due to their high coding efficiency and their attractive interpolation properties [6].

Extracting recognition features from the LSFs avoids a speech decoding operation, as well as a conversion of LSF to LPC. A distributed speech recognition system that adopts this strategy becomes computationally more efficient than any other one based on speech reconstruction or LPC parameters. The recognition features which can be obtained from LSFs are the PCC (Pseudo-Cepstral Coefficients) [7], MPCC (Mel-Frequency PCC) [7], PCEP (Pseudo-Cepstrum) [1] and the MPCEP (Mel-Frequency PCEP) [1].

It is worth to mention that these features, which are directly obtained from the LSFs, correspond to approximations of the LPCC and MLPCC features. Using these approximations we avoid to recover LPC parameters to obtain the recognition features.

3.1   Pseudo-Cepstral Coefficients (PCC)

The PCC is computed directly from the LSFs. However, its derivation is based on the LPCC. Mathematical manipulations and approximations allow it to be expressed in terms of the LSFs [7]. The n-th PCC is given by the equation

$$\hat{c}_n = \frac{1}{2n}\left(1+(-1)^n\right) + \frac{1}{n}\sum_{i=1}^{p}\cos nw_i \qquad (5)$$

where w_i is the i-th LSF parameter.

3.2 Pseudo-Cepstrum (PCEP)

Using the mathematical expression of the PCC features, it is somewhat trivial to obtain the PCEP [1]. They are derived from the PCC by eliminating the $\frac{1}{2n}\left(1+(-1)^n\right)$ term. Note that this term does not depend on the speech signal, i.e., it does not depend on the LSF parameters. The n-th PCEP expression is given by

$$\hat{d}_n = \frac{1}{n}\sum_{i=1}^{p}\cos nw_i \qquad (6)$$

It is fair to expect a good spectral performance of the PCEP because they provide a spectral envelope very similar to the one provided by the Cepstrum, wich is generated from the original speech signal [1]. The PCEP features have the advantage of presenting a computational load even lower than the PCC.

3.3 Mel-Frequency PCC (MPCC)

To obtain the MPCC features from the PCC, the LSFs w_i are replaced by w_i^m, wich are defined by the transformation

$$w_i^m = w_i + 2\tan^{-1}\left(\frac{0.45\sin w_i}{1-0.45\cos w_i}\right) \qquad (7)$$

This expression transforms the frequency axis of a particular set of parameters to the mel scale frequency axis [8]. The MPCC features are expressed by

$$\hat{c}_n^m = \frac{1}{2n}\left(1+(-1)^n\right) + \frac{1}{n}\sum_{i=1}^{p}\cos nw_i^m \qquad (8)$$

where $\hat{c}_n^m$ is the n-th MPCC.

3.4 Mel-Frequency PCEP (MPCEP)

Following the same procedure described for the MPCC, we can express the MPCEP features by

$$\hat{d}_n^m = \frac{1}{n}\sum_{i=1}^{p}\cos nw_i^m \qquad (9)$$

where $\hat{d}_n^m$ is the n-th MPCEP.

4 Experimental Results

The goal of the experiments carried out in this work is to determine which speech recognition features represent a good trade-off between recognition performance and computational load. Of course, the analysis is performed having in mind that they will be used in distributed speech recognition systems. Figure 1 illustrates the features extractors and systems to be investigated in this section. It should be remarked that the quantization effects are not being taken into account in this work. We focus on a global comparative analysis of the features at two different frame rates.

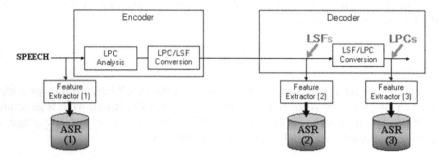

Fig. 1. Features extractors and ASR systems

According to Fig. 1, the following feature extractors will be examined:

- Feature Extractor (1) – provides MFCC (Mel-Frequency Cepstrum Coefficients) features [9]-[10] from the original speech signal in 10 ms and in 20 ms frame intervals
- Feature Extractor (2) – provides the PCC, PCEP, MPCC and MPCEP features from the LSFs in 10 ms and in 20 ms frame intervals
- Feature Extractor (3) – provides the LPCC and MLPCC features from the LPC parameters in 10 ms and in 20 ms frame intervals

It is worth to remark that the MFCC is generated from the original speech signal. It is being considered here, in order to have a performance benchmark for the other features. It is also worth noting that the MFCC is usually employed in speech recognition systems that do not operate in communication networks. Note that this feature cannot be used in communication networks where there is no additional information transmission to the remote ASR system besides the one sent by the encoder.

In all experiments, the feature extractors will generate one set of 10 parameters plus its derivatives (Δ parameters) representing a total of 20 recognition features.

In the simulations carried out in this work, the speech frames have 25 ms duration and the frame rate is either 100 Hz or 50 Hz, depending on the desired rate of the LPC or LSF extractions.

The 100 Hz frame rate was chosen because this is the usual value employed by speech recognizers to provide good performance. The 50 Hz frame rate was chosen because this value is usual in voice coders operating in IP networks and mobile environments.

The ASR system considered in our experiments, is a speaker-independent, isolated word recognizer. The speech database is composed of 50 male speakers and 50 female speakers, each one repeating three times the digits 0,1,2,3,4,5,6,7,8,9 and the word "meia" in Portuguese. This represents a total of 3,300 words. A distribution of 70% and 30% of the speech database was used for training and testing, respectively.

The recognition systems use five-state continuous observation HMMs (Hidden Markov Models) with a mixture of three Gaussians per state. They were implemented with the HTK (HMM Toolkit) software [9].

Table 1 shows the recognition performance results when the features are extracted at each 10 ms and at each 20 ms. This corresponds to 100 Hz and 50 Hz frame rate, respectively. It can be seen that the 20 ms feature generation yields a much lower performance (around 5 %) when compared to the 10 ms feature extraction. We can also verify that the mel scale features (MLPCC, MPCEP and MPCC) always provide better performance than the real frequency features (LPCC, PCEP e PCC). This difference is about 3 %. Moreover, it can be observed that the speech recognition features for distributed environments (MLPCC, MPCEP e MPCC) show fairly good results when compared to the MFCC, obtained from the original speech signal. The difference in recognition rate is around 1 %.

Table 1. Recognition performance

Frame Rate	LPCC	PCC	PCEP	MLPCC	MPCC	MPCEP	MFCC
100 Hz	95.80%	94.60%	95.00%	98.30%	97.50%	98.20%	99.40%
50 Hz	90.80%	90.20%	90.40%	93.80%	93.10%	93.70%	95.00%

It is important to remind that the MPCEP and MPCC features are obtained at the decoder first stage directly from the LSFs. On the other hand, the MLPCC features can only be generated at the second stage of the decoder, i.e., after the LSF/LPC conversion. These characteristics make the MPCEP and the MPCC computationally more efficient than the MLPCC. This is particularly interesting for systems that provide recognition services and do not intend to simultaneously reconstruct the speech signal. It can also be observed from Table 1 that the maximum performance loss of the the MPCC and the MPCEP, compared to MLPCC is 0.7%, at a frame rate of 50 Hz.

An interesting conclusion that can also be drawn from Table 1 is that the MPCEP features always overperform the MPCC, besides being simpler than the MPCC.

Finally, comparing the MPCEP and the MLPCC feature performances of Table 1, it can be seen that the difference in recognition rate is only 0.1% at both frame rates. This particular result is of major concern if we also take into account the computational complexity. Note that the MPCEP is an approximation to the MLPCC and provides a great computational saving over this feature.

5 Conclusions

We have analyzed the impact of various speech features over the performance of speech recognizers. The features were obtained from transformations of the LSF and LPC parameters. The results presented in this paper can be useful to distributed speech recognition systems operating in mobile and IP communication networks. We have concluded that the MPCEP feature, obtained from LSFs, is the one that presents the best trade-off between recognition accuracy and computational load. Comparing the recognition performances for features extracted at 100 Hz and 50 Hz frame rates, we have observed a degradation of approximately 4% of the latter relative to the first. Note that the 50 Hz frame rate is the usual condition in speech codecs. It is clear, therefore, that additional processing techniques, such as parameter interpolation, have to be applied in order to achieve results that might be closer to the ones obtained at 100 Hz frame rate.

References

1. H. S. Choi, H. K. Kim, e H. S. Lee, "Speech Recognition Using Quantized LSP Parameters and their Transformations in Digital Communication", vol. 30, pp. 223-233, Speech Communication, (2000)
2. Y. Ohshima, "Environmental Robustness in Speech Recognition using Physiologically-Motivated Signal Processing," PH. D. Thesis, Carnegie Mellon University, Pittsburgh, Pennsylvanya, December (1993)
3. A. V. Oppenheim e D. H., Johnson, "Discrete Representation of Signals," Proc. IEEE, vol. 60, pp.681- 691, June (1972)
4. S. K. Mitra, Digital Signal Processing: A Computer-Based Approach, McGraw-Hill International Editions, (1998)
5. M. Wölfel, J. McDonough, e A., Waibel, "Minimum Variance Distortionless Response on a Warped Frequency Scale," Eurospeech, Geneva, 2003.
6. W. B. Kleijn e K. K. Paliwal, Speech Coding and Synthesis, Amsterdam, The Netherlands: Elsevier, (1995)
7. H. K. Kim, S. H. Choi e H. S., Lee, "On Approximating Line Spectral Frequencies to LPC Cepstral Coefficients," IEEE Trans. Speech and Audio Processing, vol. 8, pp. 195 – 199, March (2000)
8. F. S. Gurgen, S. Sagayama, e S. Furui, "Line Spectrum Frequency-Based Distance Measures for Speech Recognition," pp.521-524, Proc. ICSLP, Kobe, Japan, November (1990)
9. S. Young, G. Evermann, T. Hain, D. Kershaw, G. Moore, J. Odell, D. Ollason, D. Povey, V. Valtchev and P. Woodland, The HTK Book (for HTK Version 3.2.1), December (2002)
10. S. B. Davies and P. Mermelstein, "Comparasion of Parametric Representations for Mono syllabic Word Recognition in Continuously Spoken Sentences," vol.28, pp.357-366, IEEE Trans. ASSP, August (1980)

Redshift Determination for Quasar Based on Similarity Measure

Fuqing Duan and Fuchao Wu

National Laboratory of Pattern Recognition,
Institute of Automation, Chinese Academy of Sciences,
P.O.Box 2728,Beijing P.R.China,100080
{fqduan, fcwu}@nlpr.ia.ac.cn

Abstract. With the advent of very large redshift surveys, automatic redshift measurement is becoming increasingly important. This paper presents a similarity measure based cross-correlation method for the redshift determination of quasar spectra. Cross-correlation is measured only for the redshift candidates that are determined by the emission line features of the observed spectrum. The similarity measure is defined as the weighted sum of several similarity evidences. Compared with the traditional cross-correlation based methods, our method can be used for higher redshift determination. Compared with the methods based on spectral line matching, our method is less sensitive to the quality of spectral line extraction. Experiment results indicate the high performance of the method.

1 Introduction

The development of fiber-based spectrographs capable of observing hundreds of celestial objects simultaneously has led to many large redshift surveys such as SDSS, 2DF, LAMOST etc. Because of the sheer size of these surveys, it is becoming very important to develop methods of reliable and automated spectral recognition. An astronomical spectrum consists of continuum, spectral lines and noises. The continuum is produced by continuous radiation along wavelength, and it is the low frequency ingredients of a spectrum. Spectral lines include absorption lines and emission lines, which are produced by some atoms in celestial objects for absorbing or radiating energy at fixed wavelengths.An observed spectrum is shown in fig.1, where the horizontal axis denotes wavelength, the vertical axis represents relative flux, the smooth thick curve denotes the continuum, and the spectral lines are indicated. Because of the movement away from the earth of celestial objects, the wavelengths of spectral lines in observed spectra are usually larger than those in the rest frames. That is the redshift phenomenon. Redshift is one of the most important parameters of celestial objects. It can be computed by the formula $Z = (\lambda - \lambda_0)/\lambda_0$, where λ stands for the wavelength of a spectral line in the observed spectrum, λ_0 for its corresponding wavelength in the rest frame, Z for redshift. Quasars (QSO) are the brightest and furthest celestial objects detected up to now. They play a very important role in the search of the

S. Singh et al. (Eds.): ICAPR 2005, LNCS 3686, pp. 529–537, 2005.

universe, and are given more and more attention. Redshift determination of an observed QSO spectrum is a main task in QSO recognition.

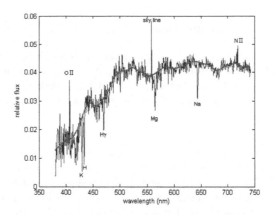

Fig. 1. An observed spectrum

The observed QSO spectra have the following characters: firstly, the spectra are usually contaminated heavily by noise so that the spectra vary drastically from their rest frames; secondly, the observed wavelength range is limited, generally from 370nm to 780nm about, which is a part of overall wavelength of QSO; finally, there are very few spectral lines in most spectra for their large redshifts. These characters make it very difficult to determine the redshift automatically.

Existing automated methods for redshift determination can be divided into two classes. The one is based on cross-correlation [1,2], the other is based on spectral line matching. The most common technique [1] of the first class is the cross-correlation of the observed spectrum with a set of templates, and the redshift is determined by the location of the largest peak in the cross-correlation functions. PCAZ [2] generalizes the cross-correlation approach by replacing the individual templates with a simultaneous linear combination of orthogonal templates gained by using principle component analysis (PCA). Although these methods are regarded as the most successful methods of automatic redshift measurement, unfortunately, they can be used only for those galaxy spectra with small redshifts. The typical methods in the second class include the approaches of Hough transform based [3] and density estimation based [4]. In these methods, spectral line extraction is a key step. However, due to aforementioned characters of the QSO spectra, the result of spectral line extraction usually is not satisfactory so that the result of redshift determination is not reliable. In order to reduce the dependence on the spectral line extraction in redshift determination, this paper presents a novel cross-correlation method by combining the two classes of methods. It determines the redshift candidates by the result of the spectral line extraction firstly, and then measures the cross-correlation for the redshift candidates. The novelty of the proposed method includes the following:

1. Cross-correlation is measured only for the redshift candidates.

2. A new similarity measure is proposed, it can measure the similarity between two spectra more reasonably.

3. The new thresholding in feature extraction can improve the quality of spectral line extraction.

The remainder of this paper is organized as follows. Section 2 introduces the preprocessing of spectra. Section 3 introduces the feature extraction. Section 4 describes the proposed method. Section 5 discusses experimental results. This paper is concluded in Section 6.

2   Preprocessing

The preprocessing of spectra includes sky subtraction, continuum subtraction and de-noising.

Usually the spectra show a number of residual sky features in the regions of strong atmospheric emission and absorption lines. Where these are the strongest features in the spectrum, there is a danger that the correlation between the strong lines in the template and the sky residual will be greater than the correlation between the template and the much weaker spectra. We remove these sky residuals by median filter in 6 nm bands around 557 nm, 630 nm etc.

Continuum subtraction reduces the smoothly varying background to zero and essentially has the same effect as filtering out the long-period Fourier components of the spectra. Without continuum subtraction, the cross-correlation function shows a peak representing the cross-correlation of the two continua, with a small spectral cross-correlation peak superimposed. In addition, the intensities of spectral lines are not shown truly for the existence of continua. The continuum is fitted from the observed spectrum by a median filter with a filter window 60 nm wide. We divide the observed spectrum by the continuum, and subtract one from the spectrum.

The noise in astronomical spectra can be roughly regarded as random white noise, so soft- thresholding [5] is used for noise removal.

The middle part in fig.2 shows the result of the preprocessing of the spectrum shown up in fig.2.

3   Feature Extraction

We call the spectrum after preprocessing line spectrum. Usually there are many pseudo spectral lines on line spectra because of the heavy noise contamination and the rough continuum fitting. The feature used in our method is the feature wavelengths of the emission lines. How to extract all spectral lines from line spectra is a hot potato. Spectral line extraction usually includes two steps: at first, the line spectrum is thresholded; secondly, we search for the locations of local maxima on the line spectrum, which are regarded as the feature wavelengths of spectral line candidates. Because the intensity at peak of every spectral line is different, it's difficult to choose an appropriate universal threshold. If the threshold is too high, it's possible that no spectral line is obtained. Conversely, if the threshold is too low, many pseudo spectral lines will be obtained. Therefore, the conjunction of local thresholding and universal thresholding by the formula (1) is adopted here.

$$s(i) = \begin{cases} s(i), & s(i) > T(i)\,\&\,s(i) > T_0 \\ 0, & else \end{cases} \qquad i = 1, 2, \cdots \qquad (1)$$

Where $s(i)$ is the intensity of the ith point on the spectrum, $T(i)$ is the local threshold at that point, T_0 is the universal threshold, which is the lower limit for intensities at the peak of spectral lines. We set $T_0 = 1.5 * rms$, where $rms = \sqrt{\frac{\sum_{i=1}^{n} s(i)^2}{n}}$ denotes the root mean square (RMS) of the array $\{s(i), i = 1, 2, \cdots, n\}$.

For a point on the spectrum, we take a fixed-width window centered by the wavelength at the point and compute the RMS of the intensities of all the points in the window, and choose $c * RMS$ as the local threshold for that point. By experience, the window width is 100 nm and $c = 2.5$ in this paper. We found in experiments that the result of the feature extraction was less sensitive to the variation of the window width due to the sparse distribution of spectral lines.

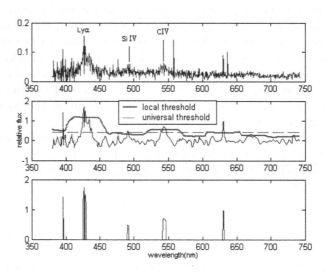

Fig. 2. Preprocessing and feature extraction (up: original spectra; middle: the spectra after preprocessing; low: spectral lines)

The thresholds are shown in the middle part of fig.2 and the emission line candidates are shown low in fig.2.

4   Redshift Determination

4.1   Redshift Candidate and Similarity Measure

Let $L = \{\lambda_i, i = 1, 2, \ldots, N\}$ denote the feature wavelengths obtained from the observed spectrum. Let $L' = \{\lambda'_i, i = 1, 2, \ldots, M\}$ denote the feature wavelengths of all emission lines on QSO template. Set

$$zc = \lambda/\lambda' - 1 \tag{2}$$

where $\lambda \in L, \ \lambda' \in L'$.

Definition 1: *If $zc \geq 0$, we define zc as a redshift candidate of the observed spectrum.*

Let $d-$dimensional vector X and Y denote spectra A and B respectively. We evenly divided the two spectra into K parts. Let $X = [X_1, \cdots, X_K]$ and $Y = [Y_1, \cdots, Y_K]$. Set

$$r_{AB} = \sum_{i=1}^{K} w_i (X_i Y_i^T)/(\|X_i\| \, \|Y_i\|) \tag{3}$$

Where $\sum_{i=1}^{K} w_i = 1, w_i$ denotes the weight.

Definition 2: *Define r_{AB} as the similarity measure between spectra A and B.*

4.2   Redshift Determination

Redshift determination based on similarity measure is composed of the following four steps:

Step1: Perform the continuum subtraction for the QSO template.

Step2: Determine the redshift candidates of the observed spectrum by formula (2).

Step3: Shift the QSO template after continuum subtraction according to every redshift candidate and compute the similarity between the shifted template and the observed spectrum of the sky and continuum subtraction.

Step4: Choose the redshift candidate corresponding to the highest similarity as the redshift of the observed spectrum.

Remark1: The proposed similarity measure becomes the traditional one with $K = 1$. Because the observed spectra always vary drastically from the rest frame, if the similarity is measured traditionally, it's possible that the similarity corresponding to the redshift is lower than others due to the high local correlation caused by noise disturbance or by the two strong lines corresponding to some redshift candidates. Similar to the evidence theory [6], the similarity measure proposed is the weighted sum of several similarity evidences. The principle of setting the weights is that the smaller the angle between X_i and Y_i, the higher the weight w_i.

4.3   Comparison with the Traditional Cross-Correlation Based Methods

This method is different from the traditional cross-correlation based methods (here called CCM) in the following aspects. Firstly, CCM is based on the least squared criterion, so it requires the wavelength range of the templates and the observed spectrum must be identical, while our method has not the limitation because it only needs to shift the template according to the redshift candidates. Due to the limitation, CCM only can be used for low redshift spectra (generally with redshifts lower than 0.3), while our method also can be used for high redshift spectra. Secondly, in fact, the cross-correlation in CCM is measured for the samples produced by uniform sampling in a redshift interval, which can be regarded as the redshift candidates in our method. It's evident that the number of the redshift candidates is smaller in our method than in CCM. Because the value corresponding to the redshift is not necessarily the highest peak of the cross-correlation function for those spectra with high noise contamination, the error risk is lower in our method than in CCM.

5   Experiments

In this section, both simulated spectra and observed spectra are used to verify the effectiveness of the proposed method. The QSO template, as shown in fig.3, is from Vanden Berk et al [7]. Because we can't get the templates used in CCM, in all the subsequent experiments, we don't report the comparison with it. The density estimation method (called DEM) is a typical method based on spectral line matching. It includes the following steps: firstly, the spectral lines are extracted and used to determine the redshift candidates; secondly, the densities of the candidates are estimated; finally, the average of the candidates in a neighborhood of the candidate with maximal density is regarded as the redshift. The comparisons with DEM are reported. In the experiments, the spectra are divided into four segments for the similarity measure, and the weight is set to be $0.1 : 0.2 : 0.3 : 0.4$ under the principle in the *remark 1*.

5.1   Experiment with Simulated Spectra

In this experiment, the simulated spectra are generated by shifting the QSO template with redshift values ranging from $0-2$ with a step of 0.01. All simulated spectra are digitized and linearly interpolated to the wavelength range of $380 - 742nm$ with a step of $0.5nm$.

A Guassian noise with 0 mean and σ standard deviation (noise level) is added to the spectra. The correct rate vs. SNR (SNR=$1/\sigma$) is plotted in fig.4. All the results in fig.4 are the average value of 100 independent trials. It can be seen from the figure that the correct rate is increased with the raising of SNR and the correct rate reaches 97% above at the SNR of 6. This indicates that the proposed technique is robust and effective.

For a comparison with DEM, the corresponding result from DEM is also plotted in fig.4. We clearly see that the two methods are incomparable at every noise level. This is mainly because there are few spectral lines on the QSO spectra when redshifts are larger than 0.5, and it is difficult for a method based on spectral line matching to determine the redshift correctly under this condition.

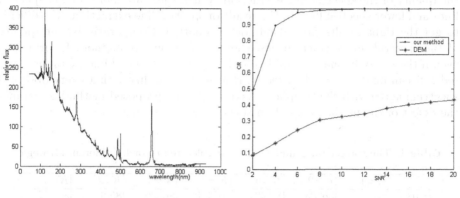

Fig. 3. The quasar template   Fig. 4. Correct rate vs. SNR

5.2   Experiments with Observed Spectra

In the following experiments, the test data includes 3056 observed QSO spectra from SDSS data release2. The redshifts of these test spectra vary from 0 to 5.

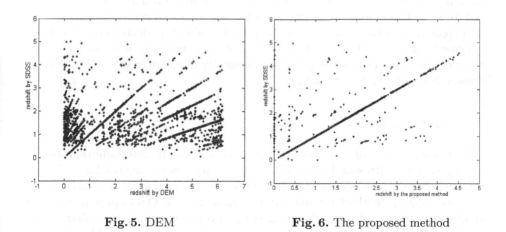

Fig. 5. DEM   Fig. 6. The proposed method

Comparisons with DEM: Fig.5 shows the results from DEM. Fig.6 is from the proposed method. The horizontal axes denote the results obtained by the two methods, and the vertical axes denote the redshifts from SDSS. The points on the line $(y = x)$ denote the obtained redshifts are coincident with the redshifts from SDSS. The correct rates in Figs. 5, 6 are 24.8% and 94.2% respectively. The results show that the proposed method is superior to DEM. From the fig.5, we can see that the error is mainly for those spectra with redshifts larger than 0.5. This indicates DEM is also not fit for high redshift spectra.

All methods based on spectral line matching have a high requirement for the quality of the spectral line extraction. When there are more pseudo spectral lines and fewer spectral lines in the result of spectral line extraction, it's difficult to get the right results for redshift determination. Comparatively, the quality of the spectral line extraction has a lower effect on our method. In principle, even if there is only one spectral line in the result of spectral line extraction, the redshift can be determined correctly. However, it's unlikely that every observed spectrum agree with the template. As a whole, the proposed method overcomes the defect of the methods based on spectral line matching.

Table 1. The correct rates under different tolerated error by different strategies

	$K=1$	$K=2$	$K=3$	$K=4$	$K=5$	$K=6$	$K=7$
$\varepsilon = 0.01$	83.5%	86.1%	87.4%	87.1%	86.7%	85.9%	85.1%
$\varepsilon = 0.02$	89.5%	91.8%	93.1%	93.2%	92.5%	92.1%	91.7%
$\varepsilon = 0.03$	90%	92.4%	94%	94%	93.5%	93.1%	92.5%

Comparisons Among Different Similarity Measure Strategies: In this part, different similarity measure strategies are compared. We evenly divide the spectra into $K = 1 \sim 7$ parts respectively. The weights are chosen as $1/K$ in every similarity measure. Tab.1 shows the correct rates under different tolerated error ε. From the Tab.1, we can see that under every tolerated error, the correct rate is increased with the K varying from 1 to 3, while it is decreased with the K varying from 4 to 7. The result by the traditional similarity measure ($K =1$) is the weakest. This proves the proposed similarity measure is more effective and reasonable than the traditional one.

6   Conclusion

Up to now, nearly all the existing methods of redshift determination are for galaxy spectra with small redshifts, there hasn't been a successful technique for QSO spectra published in the literature. In this paper, we presented a new cross-correlation method for redshift determination of QSO spectra based on similarity measure. The similarity measure proposed, which is similar to the evidence theory, is more reasonable than the traditional ones. The proposed method overcomes the defect that high redshift spectra can't be processed by the traditional cross-correlation based methods. Compared with the methods based on spectral line matching, our method is less sensitive to the quality of spectral line extraction. Experiment results have demonstrated the proposed method is robust and effective. We think that the idea of the similarity measure proposed can also be used in other areas that need similarity measure. Further works will be dedicated to confidence analysis and error estimation.

Acknowledgement

This work is supported by the National 863 High Technology R&D Program under grant No.2003AA133060.

References

1. John.Tonry, Marc.Davis : A Survey of Galaxy Redshifts. I. Data Reduction Techniques. The Astronomical Journal. **58(10)** (1979) 1511–1525
2. Karl.Glazebrook, Alison.R.Offer, Kathryn.DeeleyJ : Determination By Use of Principal Component Analysis. I. Fundamentals. The Astronomical Journal. **492(1)** (1998) 98-109
3. L Y. Huang, F. M. Sun, Z Y. Hu : A new automatic quasar recognition technique based on PCA and the Hough Transform. ICPR2000. **2** (2000) 499-502
4. DUAN Fuqing, WU Fuchao, LUO Ali et al: Density estimation based model matching method for redshift determination Spectroscopy and Spectral Analysis (in Chinese)(to appear)
5. D.L Donoho : De-noising by soft-thresholding. IEEE Trans.on IT. **41(3)** (1995) 613-627
6. J. W. Guan , D. A. Bell. Evidence theory and its applications. Elsevier Science Ltd, New York, USA. (1991)
7. DE Vanden Berk, GT Richards, et al : Composite quasar spectra from the Sloan Digital Sky Survey. The Astronomical Journal. **122(8)** (2001) 549-564

Learning with Segment Boundaries for Hierarchical HMMs

Naoto Gotou, Akira Hayashi, and Nobuo Suematu

Graduate School of Information Sciences, Hiroshima City University, 3-4-1,
Ozukahigashi, Asaminami-Ku, Hiroshima,731-3194, Japan
{goto, akira, suematsu}@robotics.im.hiroshima-cu.ac.jp

Abstract. Hierarchical hidden Markov models (HHMMs) can be used
for time series segmentation. However, it is difficult to obtain a desirable segmentation result, because the form of learning for HHMMs is
unsupervised. In the paper, we present a semisupervised learning algorithm for HHMMs. It is semisupervised in the sense that the supervisor
teaches segmentation boundaries but *not* segment labels. The learning
performance of the proposed algorithm is demonstrated through an experiment using music data.

1 Introduction

The hierarchical hidden Markov model (HHMM) was proposed by Fine et al.
[1] as a generalization of the hidden Markov model (HMM) with a hierarchical
structure. HHMMs can be used for time series *segmentation*. Here, segmentation
refers to the segmentation of a sequence into a number of sub-sequences, and
the sub-sequences are referred to as segments.

HHMMs have a great potential in segmentation, because HHMMs can model
time series with switching dynamics where each segment has its own dynamics. However, it is difficult to obtain a desirable segmentation result in actual
problems using HHMMs, because the meaning of dynamics is subjective, or task
dependent, and the form of learning for HHMMs is unsupervised.

In the paper, we present a semisupervised learning algorithm for HHMMs. It
is semisupervised in the sense that the supervisor teaches segmentation boundaries (i.e. when the dynamics changes), but *not* segment labels (i.e. what dynamics it is). For many time series data, it is much easier for humans to tell when
the dynamics changes than to tell what the current dynamics is.

As a related work, Cohen segmented facial video sequences based on emotion
using the Multilevel-HMM [2]. However, he used segmented data with segment
labels for learning.

2 Hierarchical HMMs

2.1 Review of HHMMs

We briefly review the hierarchical HMM[1,3]. An example of a HHMM is presented
in Figure 1. HHMMs have three kinds of states: internal states, production states

S. Singh et al. (Eds.): ICAPR 2005, LNCS 3686, pp. 538–543, 2005.

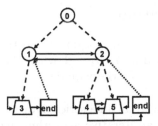

Fig. 1. Example of a HHMM. Circles, trapezoids and squares denote internal, production and end states, respectively, and the solid, dashed and dotted arrows denote horizontal, vertical and forced transitions, respectively.

and end states. State transitions are of three types: vertical, horizontal, and forced transitions. We describe how to generate a sequence from the HHMM as follows:

1. Initialize the time index ($t = 1$), and start at the root state.
2. Vertical transition: perform a transition to a lower state from the current (internal/root) state. If the transited state was internal, repeat vertical transition.
3. Emitting an observation: emit the observation O_t by the production state, and then add 1 to the time index.
4. Horizontal transition: perform a transition within the same level. If the transited state was internal, return to 2. If the transited state was a production state, return to 3 . Otherwise (i.e. transited to end state), go to 5.
5. Forced transition: perform a transition to the parent (i.e. calling) state. Then, go to 4.

We assume for simplicity that all production states are at the bottom of the hierarchy. We also assume that the root state and the first level states do not have transitions to themselves.

2.2 Notations for HHMMs

d denotes the level of hierarchy: the level of the root state is $d = 0$, and the maximum level of hierarchy is $d = D$. q_t^D denotes the production state emitting the observation O_t at t , and q_t^{D-1} denotes its parent state. In order to simplify notations, let $q_{t:t'}^d = \{q_t^d, \ldots, q_{t'}^d\}, O_{t:t'} = \{O_t, \ldots, O_{t'}\}(t < t')$.

In addition, we introduce F_t^d, as defined by Murphy [3]: $F_t^d = 1$ means that the state transition to the end state was performed at level d and time t, and $F_t^d = 0$ means that the state transition was not performed.

The model parameters $b(i, v)$, $A^d(i, j)$, and $\pi^d(k, i)$ denote the observation density of the ith state, which indicates the probability of emitting observation v, the horizontal transition probability into the jth state from the ith state at level d, and the vertical transition probability into the ith state from the kth state at level d, respectively (the forced transition probability is always 1.0).

2.3 Segmentation by HHMMs

We assume that the segmentation occurs when q_t^1 (i.e. level 1 state) makes a horizontal transition. Segmentation is performed by finding the most likely state sequence $\max_{q_{1:T}^{1:D}, F_{1:T}^{1:D}} P(q_{1:T}^{1:D}, F_{1:T}^{1:D} | O_{1:T})$. It can be found by converting the HHMM to an HMM, which has megastates $[q_t^{1:D}, F_t^{1:D}]$, and then using Viterbi algorithm of the HMM.

3   Semisuprevised Learning Algorithm

A segment boundary is defined as the time between one segment and the next. The segment boundary information is described as τ_t: $\tau_t = 1$ when there is a segment boundary between O_t and O_{t+1}, and $\tau_t = 0$ otherwise (i.e. when O_t and O_{t+1} belong to the same segment).

We present a parameter learning algorithm from time series data with segment boundary information. The algorithm is an iterative algorithm that contains two steps in each iteration: an inference step and a parameter update step. In the inference step, the posterior probability of the hidden variables $\{q_t^d, F_t^d\}$ is calculated from evidence variables (i.e. observations and segment boundary information: $O_{1:T}, \tau_{1:T-1}$). In the parameter update step, the parameters are updated from the posterior probability of the hidden variables.

3.1   Converting to an HMM

In order to simplify formulation, we first convert the HHMM to an HMM[3]. The HMM is created using the HHMM stack (vector) state $q_t^{1:D} = [q_t^1, \ldots, q_t^D]'$, to yield the HMM (mega) state. If the megastate is I, then the stack state is $q_t^{1:D} = [i^1, i^2, \ldots, i^D]'$. The HMM parameters, the transition probability $\bar{A}(I, J)$, the observation density $\bar{b}(I, v)$, and the initial state probability $\bar{\pi}(I)$), are given as follows.

$$\bar{A}(I, J) \stackrel{\text{def}}{=} P(q_{t+1}^{1:D} = I | q_t^{1:D} = J) = \sum_{d=0}^{D} A^*(I, J, d) \tag{1}$$

$$A^*(I, J, d) = P(q_{t+1}^{1:D} = J, F_t^{1:d} = 0, F_t^{d+1:D} = 1 | q_t^{1:D} = I)$$

$$= \begin{cases} \prod_{d'=d+1}^{D} [A^{d'}(i^{d'}, end)\pi^{d'}(j^{d'-1}, j^{d'})] A^d(i^d, j^d) & if \ d \neq D \\ A^D(i^D, j^D) & if \ d = D \end{cases}$$

$$\bar{b}(I, v) \stackrel{\text{def}}{=} P(O_t = v | q_t^{1:D} = I) = b(i^D, v) \tag{2}$$

$$\bar{\pi}(I) \stackrel{\text{def}}{=} P(q_1^{1:D} = I) = \pi^0(0, i^1) \prod_{d=1}^{D-1} \pi^d(i^d, i^{d+1}) \tag{3}$$

3.2   Inference Step

In the inference step, we use an algorithm like the forward-backward procedure of the HMM. We define the forward probability α, the backward probability β, and derive the recursion formula as follows:

$$\alpha(I, t) \stackrel{\text{def}}{=} P(q_t^{1:D} = I, O_{1:t}, \tau_{1:t-1})$$

$$= \sum_J P(q_{t-1}^{1:D} = J, O_{1:t-1}, \tau_{1:t-2}) P(q_t^{1:D} = I | q_t^{1:D} = J)$$

$$\times P(O_t | q_t^{1:D} = I) P(\tau_{t-1} | q_{t-1}^{1:D} = J, q_t^{1:D} = I)$$

$$= \sum_J \alpha(J, t - 1) \bar{A}(J, i) \bar{b}(I, O_t) \rho(J, I, t - 1) \tag{4}$$

$$\beta(I, t) \stackrel{\text{def}}{=} P(O_{t+1:T}, \tau_{t:T-1} | q_t^{1:D} = I) = \sum_J \beta(J, t+1) \bar{A}(I, J) \bar{b}(J, O_{t+1}) \rho(I, J, t) \tag{5}$$

where $\rho(I, J, t)$, a Boolean function to prevent a state transition in contradiction to the segment boundary information, is given as

$$\rho(I, J, t) \overset{\text{def}}{=} P(\tau_t | q_t^{1:D} = I, q_{t+1}^{1:D} = J) \tag{6}$$

$$= \begin{cases} P(q_t^1 = q_{t+1}^1 | q_t^{1:D} = I, q_{t+1}^{1:D} = J) \text{ if } \tau_t = 0 \\ P(q_t^1 \neq q_{t+1}^1 | q_t^{1:D} = I, q_{t+1}^{1:D} = J) \text{ if } \tau_t = 1 \end{cases} = \begin{cases} 1 \text{ if } \tau_t = 0, i^1 = j^1 \\ 1 \text{ if } \tau_t = 1, i^1 \neq j^1 \\ 0 \text{ otherwise} \end{cases}$$

The posterior probabilities of $\{q_t^d, F_t^d\}$ are obtained from α and β as follows.

$$\xi^*(I, J, d, t) \overset{\text{def}}{=} P(q_t^{1:D} = I, F_t^{1:d} = 0, F_t^{d+1:D} = 1, q_{t+1}^{1:D} = J | O_{1:T}, \tau_{1:T-1})$$
$$= P(q_t^{1:D} = I, O_{1:t}, \tau_{1:t-1}) P(q_{t+1}^{1:D} = J, F_t^{1:d} = 0, F_t^{1:D} = 1 | q_t^{1:D} = I) P(O_{t+1} | q_{t+1}^{1:D} = J)$$
$$\times P(O_{t+2}, \tau_{t+1:T-1} | q_{t+1}^{1:D} = J) P(\tau_t | q_t^{1:D} = I, q_{t+1}^{1:D} = J)) / P(O_{1:T}, \tau_{1:T-1})$$
$$= [\alpha(I, t) A^*(I, J, d) \bar{b}(J, O_{t+1}) \beta(J, t+1) \rho(I, J, t)] / \sum_{I'} \alpha(I', t) \beta(I', t) \tag{7}$$

$$\gamma^*(I, d, t) \overset{\text{def}}{=} P(q_t^{1:D} = I, F_t^{1:d} = 0, F_t^{d+1:D} = 1 | O_{1:T}, \tau_{1:T-1})$$
$$= \begin{cases} \sum_J \xi^*(I, J, d, t) & \text{if } t = 1, ..., T-1 \\ (1 - A(i^d, end)) \prod_{d'=d+1}^D A(i^{d'}, end) \alpha(I, T) / \sum_{I'} \alpha(I', T) & \text{if } t = T \end{cases} \tag{8}$$

3.3 Parameter Update Step

The model parameters are updated as follows.

$$\gamma(i, d, t) \overset{\text{def}}{=} P(q_t^d = i, F_t^{d+1} = 1 | O_{1:T}, \tau_{1:T-1})$$

$$\xi(i, j, d, t) \overset{\text{def}}{=} P(q_t^d = i, F_t^d = 0, F_t^{d+1} = 1, q_{t+1}^d = j | O_{1:T}, \tau_{1:T-1})$$

$$\xi_e(i, d, t) \overset{\text{def}}{=} P(q_t^d = i, F_t^d = 1, F_t^{d+1} = 1 | O_{1:T}, \tau_{1:T-1})$$

$$\chi(k, i, d, t) \overset{\text{def}}{=} \begin{cases} P(F_{t-1}^d = 1, q_t^{d-1} = k, q_t^d = i, | O_{1:T}, \tau_{1:T-1}) \text{ if } t = 2, 3, \ldots, T \\ P(q_1^{d-1} = k, q_1^d = i | O_{1:T}, \tau_{1:T-1}) & \text{if } t = 1 \end{cases}$$

$$\hat{A}^d(i, j) = \frac{\sum_{t=1}^{T-1} \xi(i, j, d, t)}{\sum_{t=1}^{T-1} \gamma(i, d, t)}, \quad \hat{A}^d(i, end) = \frac{\sum_{t=1}^T \xi_e(i, d, t)}{\sum_{t=1}^T \gamma(i, d, t)}$$

$$\hat{\pi}^d(k, i) = \frac{\sum_{t=1}^T \chi(k, i, d, t)}{\sum_{i'} \sum_{t=1}^T \chi(k, i, d, t)}, \quad \hat{b}(i, v) = \frac{\sum_{1 \leq t \leq T \text{ s.t. } O_t = v} \gamma(i, D, t)}{\sum_{t=1}^T \gamma(i, D, t)}$$

where $\{\gamma, \xi, \xi_e, \chi\}$ is calculated by marginalization of $\{\gamma^*, \xi^*\}$. For example,

$$\xi(i, j, d, t) = \sum_{L \text{ s.t.} l^d = i} \sum_{H \text{ s.t.} h^d = j} \xi^*(L, H, d, t)$$

3.4 Extension to Incomplete Information

In the above, we discussed learning from complete information on segment boundaries. Here, we extend the formulation when the information is incomplete, i.e. when only a (known) part of the time series is segmented. The extension is simple. If there is no information between t and $t+1$ ($\tau_t = null$) , then we simply set $\rho(I, J, t) = 1$ in (6). Note that if $\tau_t = null$ for $1 \leq t \leq T$, then our semisupervised learning algorithm becomes an unsupervised learning algorithm originally developed for HHMMs by Fine and Murphy.

4   Experiment

As a test data, we used a song, 'I want to hold your hand' (The Beatles, 1963). The source CD audio is converted to 2.7562 KHz mono format, and then subdivided into 74.3 ms frames at a rate of 4.4 ms. There were 3289 frames in total. Each frame is Hamming windowed and then parameterized using the Melfrequency cepstral coefficients (MFCC) (with first 5 harmonic numbers) to yield a feature vector [4].

The segment boundaries were given by a supervisor based on phrase. The model structure is illustrated in Figure 2(a). Since submodels at the second level share the observation density (at the third level), the only differences in the

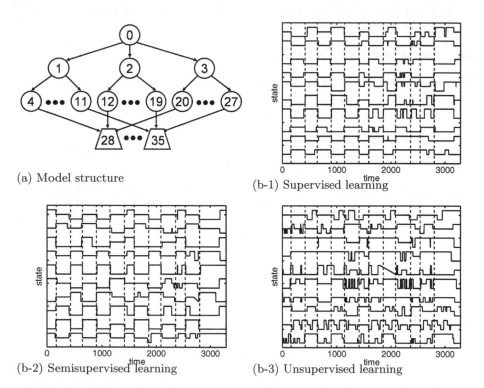

(a) Model structure

(b-1) Supervised learning

(b-2) Semisupervised learning

(b-3) Unsupervised learning

Fig. 2. (a) Model structure presenting only vertical transitions. The horizontal transitions and the end state are omitted. (b) Viterbi segmentation result of all data. Solid curves are segmentation results of 10 trials, vertical dashed lines indicate the segment boundaries based on the phrase given by supervisor.

submodels are their transition probabilities, which determine the segmentation result.

We compared the segmentation results of the three learning algorithms. (1) semisupervised learning with complete information: The training data was $\{O_{1:1649}, \tau_{1:1648}\}$. (2) semisupervised learning with incomplete information: The training data was $\{O_{1:3289}, \tau_{1:1648}\}$. $\tau_{1649:3288} = null$. (3) unsupervised learning: The training data was $\{O_{1:3289}\}$.

In model parameter learning, we used random initial parameters (but the initial transition probability to the end state was limited to less than 0.0001). We then segmented $O_{1:3289}$ using the learned model. We performed 10 trials. The segmentation results of the 10 trials are shown in Figure 2(b), which shows the most likely state sequence for the first level states. We can see that the semisupervised learning with complete/incomplete information gave better result than the unsupervised learning, thanks to the segment boundary information. We can also see that the semisupervised learning with incomplete information is slightly better than that with complete information. We conjecture the reason as follows. Since the data size is small, the observation sequence without segment boundaries $O_{1650:3289}$ helps to learn the better model.

5   Conclusions

We have presented a semisupervised learning algorithm for HHMMs to segment time series data. The learning performance of the proposed algorithm has been demonstrated through an experiment using music data.

References

1. S. Fine, Y. Singer, and N. Tishby, "The hierarchical hidden markov model: analysis and applications", Machine Learning. vol. 32, pp. 41-62, 1998.
2. I. Cohen, A. Garg, and T.S. Huang, "Emotion recognition from facial expressions using multilevel HMM", Neural Information Processing Systems Workshop on Affective Computing, Denver, USA, 2000.
3. K.P. Murphy and M.A. Paskin, "Hierarchical HMMs", Technical reports, 2002.
4. L. Rabiner and B.H. Juang, "Fundamentals of Speech Recognition", Prentice Hall PTR., 1993.

A Bayesian Method for High-Frequency Restoration of Low Sample-Rate Speech

Yunpeng Xu[1], Changshui Zhang[1], and Naijiang Lu[2]

[1] State Key Laboratory of Intelligent Technology and Systems,
Department of Automation, Tsinghua University, 100084, Beijing, P.R. China
xuyp03@mails.tsinghua.edu.cn, zcs@mail.tsinghua.edu.cn
[2] Shanghai Cogent Biometrics Identification Technology Co. Ltd., Shanghai, China
lunj@cbitech.com

Abstract. Compared with high sample-rate speeches, low sample-rate speeches lose all high frequency components that outrange the Nyquist frequency, which might severely impair the speeches' sound effects. To address this problem, this paper proposes a novel High-frequency (HF) restoration method of low sample-rate speech based on Bayesian inference, which turns the restoration problem into a maximizing a posteriori estimation. With this method, the relation between high frequency components and low frequency components is first extracted from the training set. The compatibility between neighboring audio frames is also modelled by a one dimensional Markov Random Field. Then the extracted knowledge is adopted in reconstructing the original high sample-rate signal for the testing low sample-rate audio. Experiments prove the applicability and effectiveness of this method.

1 Introduction

Although the high frequency components of a typical speech audio has little power compared with its lower frequency components, they still preserve rich information and determine the speech audio quality to a large extent. This can be proved by the fact that we always describe low sample-rate audio that has few high frequencies as obscure and blur while associating high sample-rate audio that contains abundant high frequencies with clear and bright.

In many real audio systems, however, high frequencies are quickly attenuated and suppressed due to various reasons. This usually results in deteriorated sound effects. To solve this problem, many EQ-based mechanisms have been introduced to boost and compensate the high frequencies in the audio industry. While these methods alleviate the problem, they can only be adopted to emphasize existing but attenuated high-frequency content of audio files. The increased need for cross-platform working has posed a new set of problems related to low sample rates. For example, we would expect that low sample rate speech from the telephone can have the same quality as the high sample rate audio of CD. However, this cannot be simply achieved by upsampling[6], which carefully remove all spectral components beyond the input signal bandwidth with a low

S. Singh et al. (Eds.): ICAPR 2005, LNCS 3686, pp. 544–552, 2005.

pass filter. To address this problem, in [2], an excitation algorithm is presented in order to extrapolate the high frequencies that outrange the Nyquist frequency from the existing lower frequency content. However, lacking the original high sample rate audio, information concerning the high frequencies is not presented by low sample-rate audio. This means that they just guess the high frequency content in a heuristic way. Therefore, artifacts that are irrelevant to the speaker would be inevitable in this system.

On the other hand, however, as one basic assumption in speech recognition, the spectrum of a specific speaker's voice has a relatively stable pattern of composing, which indicates that the high frequency components and the low frequency components in one's voice are related in a certain way. Therefore, if the relation between high frequency components and low frequency components can be learned from a training set, then the missing high frequency components in a low sample-rate audio can be inferred based on this knowledge. Inspired by the idea, we propose a novel HF restoration method of low sample-rate speech based on Bayesian inference, which turns the restoration problem into a Maximizing a Posteriori (MAP) estimation, and estimate the original high sample-rate speech audio from the training audio.

The rest of this paper is organized as follows. Section 2 describes the principles and the algorithm of the proposed method in details. Experimental results are shown in section 3. Finally, the concluding remarks and future research plans are given in section 4.

2 Bayesian Framework for HF Restoration of Low Sample-Rate Speech

In this section, we first introduce the Bayesian Framework for HF restoration of low sample-rate speech. The two factors, likelihood and prior, which determine the optimization objective are then analyzed, followed by an algorithm to optimize them. Finally, the feature selection of audio signal is discussed.

2.1 The Bayesian Framework

Consider an observed low sample-rate speech audio L composed by n overlapped frames, i.e., $L = \{l_1, l_2, ..., l_n\}$, and denote the corresponding high sample-rate speech audio by $H = \{h_1, h_2, ..., h_n\}$. Then the problem of restoration can be simply described as: infer H with a given L.

For a specific speaker's voice, the low frequency components relate probabilistically to the high frequency components. Such a probabilistic relation can be conveyed by the training audio. A reasonable deduction is that the testing audio of the same person should also preserve this relation. Therefore, the process of inferring H can be comprehended as, given L, to find the optimal series of $\{h_1^*, h_2^*, ..., h_n^*\}$ for H^* as the reconstructed speech audio, such that the probability $P(H|L)$ would be maximized, i.e.,

$$H^* = \arg\max_H (P(H|L)) \tag{1}$$

This is a typical problem of MAP estimation. By Bayesian Theorem, it follows that

$$H^* = \arg\max_{H}(P(L|H)P(H)) \tag{2}$$

where, $P(L|H)$ is called **Likelihood**, i.e., the probability of a given H producing L; $P(H)$ is called **Prior**, i.e., the probability of occurrence for the estimated H.

2.2   Likelihood and Prior

The likelihood function $P(L|H)$ describes the probability of a high sample-rate audio producing the corresponding low sample-rate audio. It can be written as

$$P(L|H) = \prod_{i=1}^{n} p(l_i|h_i) \tag{3}$$

where, $p(l_i|h_i)$ is the probability of h_i producing l_i and depends on the transform between the two. Intuitively, if a high sample-rate audio frame approximates to the observed low sample-rate frame after downsampling, then the likelihood would approximate to 1, per contra to 0. Therefore, it is desirable to model the transform with a Gaussian probabilistic function, i.e., we define $p(l_i|h_i)$ as:

$$p(l_i|h_i) = \frac{1}{Z}\exp\{-||Dh_i - l_i||^2/\sigma^2\} \tag{4}$$

where, D is a downsampling operator, $||\cdot||$ is a certain distance measure to describe the difference between Dh_i and l_i, σ^2 is the variance, and Z is a normalization constant. We denote $\{||Dh_i - l_i||^2/\sigma^2\}$ by $\phi(l_i, h_i)$. Therefore, $P(L|H)$ can be writen as:

$$\begin{aligned}
P(L|H) &= \frac{1}{Z}\exp\{-\Phi(L,H)\} = \frac{1}{Z}\exp\{-\sum_{i=1}^{n}\phi(l_i,h_i)\} \\
&= \frac{1}{Z}\exp\{-\sum_{i=1}^{n}||Dh_i - l_i||^2/\sigma^2\}
\end{aligned} \tag{5}$$

Formula (4) suggests a straightforward nearest neighbor algorithm for this task. For each low sample-rate frame $l_i(i \in 1, 2, ..., n)$, we search in the training set H_{train} for the high sample-rate frame $h_{train,j}(j \in 1, 2, ..., m)$ which can best approximate it after downsampling. $h_{train,j}$ is then used to replace l_i as the restored high sample-rate frame h_i. It should be emphasized that, in practice, to ensure the low frequency component h_i^l of h_i unchanged after upsampling, only the high frequency component $h_{train,j}^h$ of $h_{train,j}$ is used for replacement and is taken as the restored high frequency component h_i^h of h_i. Then h_i^l, which is generated by interpolation from l_i, is added to $h_{train,j}^h$ to generate h_i, i.e., $h_i = h_{train,j}^h + h_i^l$.

However, this simple method cannot preserve the smooth connection at frame joints because it ignores the consistency of neighboring frames. In fact, the local frame information alone is insufficient for HF restoration, which indicates that neighboring primitives must be taken into consideration. Therefore, we propose to model the HF restoration of low sample-rate speech problem with a one-dimensional Markov Random Field (MRF), so as to properly represent the compatibility between frames in a non-parametric way.

The MRF model for HF restoration is shown in Figure 1. Each node in the figure denotes an audio frame. Here, we let the low sample-rate frames be observation nodes, L, while the high sample-rate frames be hidden nodes, H. The lines indicate statistical dependencies between nodes, where function ϕ is the likelihood energy function defined above, and function ψ describes the compatibility of each two neighboring high sample-rate frames. Similarly, two neighboring frames are more compatible if they agree better in the overlap region, then the prior potential takes small value consequently. Therefore, we define the function ψ as:

$$\psi(h_i, h_{i+1}) = ||h_i - h_{i+1}||^2_{h_i \cap h_{i+1}} \tag{6}$$

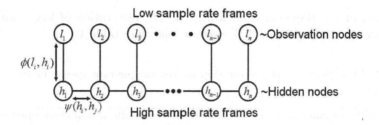

Fig. 1. Illustration of the MRF model for HF restoration of low sample-rate speech

Notice that if each frame is taken as a state and the transition probability between states is defined as a function of ψ, then this model is equivalent to a HMM, which is broadly used in audio modelling. The MRF model for HF restoration of low sample-rate speech also facilitates the computation of the frame prior. By Hammersley-Clifford Theorem, each MRF has a joint probability in Gibbs form[1]. This is so-called Markov-Gibbs Equation. Therefore, the prior $P(H)$ can be expressed as:

$$P(H) = \frac{1}{Z_H} \exp\{-\Psi(H)/T\} \tag{7}$$

where, Z_H is a normalization constant, T is a control parameter. If only one-order neighborhood is considered, then $\Psi(H)$ can be expressed as the following function of the prior potential:

$$\Psi(H) = \sum_{i \in \{1,2,...,n-1\}} \psi(h_i, h_{i+1}) \tag{8}$$

Hence,

$$P(H_{hi}) = \frac{1}{Z_H} \exp\{ \sum_{i \in \{1,2,...,n-1\}} ||h_i - h_{i+1}||^2_{h_i \cap h_{i+1}}/T\} \tag{9}$$

To summarize, with the MRF model for HF restoration and Hammersley-Clifford Theorem, we decompose the complex computation of the joint prior probability into the computation of the local prior potentials $\psi(h_i, h_{i+1}), (i \in 1, 2, ..., n-1)$. Then the constraint of consistency and compatibility between frames is also guaranteed in this way.

2.3   Posteriori and Its Computation

Combine formulae (2)(5)(7), it follows the expression for the posteriori $P(H|L)$, i.e.,

$$P(H|L) = \frac{1}{Z \times Z_H} \exp(-(\Phi(L, H) + \Psi(H)/T))$$

Then the optimization objective is equivalent to

$$H^* = \arg \min_H (\Phi(L, H) + \Psi(H)/T) \tag{10}$$

Therefore, in the Bayesian framework, the HF restoration of low sample-rate speech can be realized through 4 steps, as shown in table 1.

Table 1. Flow chart for HF restoration of low sample-rate speech in the Bayesian framework

1. Divide the training audio H_{train} and the testing audio L into overlapped frames;
2. Separate the high frequency components H_{train}^h from the training audio H_{train};
3. Upsampling the testing audio L by interpolation method to get the low frequency components H^l of the high sample-rate output;
4. Search in H_{train}^h for the optimal combination H^{h*} of $h_{train,j}^h (j = 1, 2, ..., m)$, such that the formula (10) can be minimized by H^*, which is the sum of H^l and H^{h*}. Then H^* is the result of restoration.

The 4th step of the flow compares $\{\Phi(L, H) + \Psi(H)/T\}$ for each possible combination of $h_{train,j}^h$. However, the complexity of this computation increase exponentially as the number of frames grows. To find a good tradeoff between efficiency and effect, we adopt a $one-pass\ algorithm$[4], which only searches for the frame that can best match the previously selected high frequency frame and the current testing frame. We find that the one-pass algorithm is of satisfying quality and utility for this problem, as it can both give good results and be performed in real-time.

2.4   Audio Feature

To measure the difference between two audio frames for matching computation, we need to specify the distance measure in formula (4). Here, Euclidean distance measure is adopted. However, it is inappropriate to measure the Euclidean distance directly using the samples in each frame due to the possible phase shifts and large sample size. Therefore, it is desirable to extract features from each frame and use them to compute the distance instead of using samples directly. In this paper, we adopt the features of MFCCs.

MFCCs features are widely used in the field of speech recognition. They are proved to be very effective in modelling the spectrum magnitude of audio signals. The extraction of these features takes into account the human auditory characteristics by adopting filter banks and transforms that are similar to human auditory systems. A more detailed introduction of MFCCs is presented in [5].

3   Experiments and Results

In this section, experiments on human speech are presented to test the above HF restoration method.

In the experiments, we record the speech of one male speaker in a common meeting room with no other sound source. To improve the speech audio quality, we filter the audio file with a denoiser using the *CoolEdit* software. We also carefully remove all continuous blank frames that exceed 0.5s in length to make the speech audio more compact. This results in a 10 minutes speech audio pool with little backgrounds noise. The format of the original speech audio is 48kHz, 16bits and mono-channel. From the audio pool, we randomly select 4 minutes continuous speech as the training audio, and a distinct 20 seconds continuous speech as the testing audio. Testing audio is then downsampled to 6kHz. Both the training audio and the testing audio are divided into 20ms frames with 5ms hop-size. The choice of parameter T and σ in formula (4) and (7) is empirically dependent. In fact, we can set one parameter to be 1 and estimate another. In the experiments below, we set σ to be 1 and estimate T by a simple heuristic search, which is set to be 0.15 and proves suitable for this problem.

Figure 2(a) shows the restored high sample-rate audio (middle) for the testing set in the first 2 seconds in time domain, in contrast with the original audio (above) and the upsampling result (below). Here, upsampling method is used as a comparison model to produce results that preserve exactly the same sound effects as the input low sample-rate speech. Therefore, comparisons can be made at the same sample-rate. Compared with the original audio, the upsampling result loses many fine details in that some small lumps and spindle structures of the original audio degenerate to lines. Such lumps and spindle structures are related to the most rapidly fluctuant components of the audio wave and therefore correspond to the audio high frequencies. While the HF restored speech by our method recovers these details quite well, which means the matching mechanism is quite effective in retrieving the high frequencies.

In addition, the spectral configurations of these three audios are compared in Figure 2(b). It is obvious that the high frequencies beyond 3kHz rapidly decay in upsampling audio, while the spectra of our restored audio is of approximately the same configuration as that of the original audio. This provides a more convincing proof that the proposed method can well capture the relation between high frequency components and low frequency components of the speaker's voice,

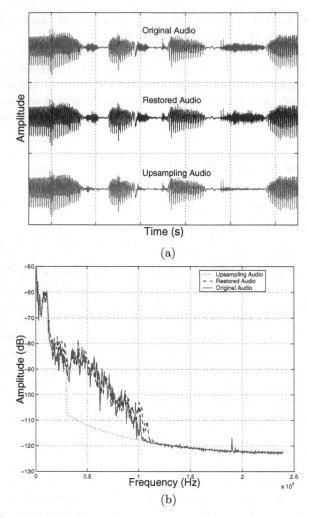

Fig. 2. Comparative Experiments Results. (a). Comparison in time domain. (b). Comparison in frequency domain. increases.

instead of guessing the high frequency content in a heuristic way. Therefore, artifacts that are irrelevant to the speaker can be safely avoided.

To evaluate the quality of the HF restored audio, we adopt the MSE criterion that is frequently used in Speech Enhancement [3]. This criterion computes the mean square error (MSE) between the logarithms of the spectra of the original and estimated signals. Generally, this criterion is believed to be correlated with the quality of the speech signal and more perceptually meaningful than the MSE between the original and estimated signal waveforms. We compute MSE value for each frame and average over all frames. For validation consideration, we also average over 10 runs with randomly selected 4 minute continuous training set and 20 seconds continuous testing set. This produces the MSE values for

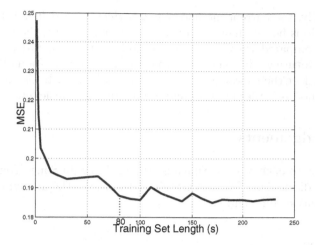

Fig. 3. The MSE value decreases as the length of training set increases

the upsampling audio and our HF restored audio, which are 0.2641 and 0.1862 respectively. This result indicates that the quality of the restored speech does improve. Furthermore, the experience of human auditory test supports the same conclusion, i.e., compared with the upsampling audio, the restored audio by our method sounds more clear and bright with less blur and obscureness that exist in the low sample-rate audio.

Experiments also reveal that, the HF restored results are closely related to the length of the training set. For evaluation, training audio with different sizes are taken to compute the MSE values between output and the original audio. Experimental results are given in Figure 3. It shows that when the length of training set is small, MSE takes large values, while it decreases rapidly as the length increases. When the length exceeds 80 seconds, the curve levels off except some fluctuations within a small range. This also explains the reason why we choose a small training set: 80 seconds would possibly be a sufficient length for this task. However, much more experiments are needed to draw a more confirmative conclusion.

4   Conclusion and Future Work

This paper proposes a novel High-frequency restoration method for low sample-rate speech based on Bayesian inference. With this method, the problem of HF restoration is turned into a MAP estimation, by which the original high sample-rate audio can be estimated as the optimal solution to formula (2). Compared with the upsampling methods, this method can properly reconstruct the high frequency components of the original audio instead of introducing artifacts that are irrelevant to the speaker. The experimental results demonstrate the validity and effectiveness of this method.

Admittedly, the current method is not perfect yet. Although the quality of restored audio is better than that of the upsampling result, there are still much to improve compared with that of the original audio. To further improve the method, our future work would focus on selecting a combination of features that can better reflect the characteristic of the original audio. It also makes sense to test our method on other kinds of audio data such as music, etc.

Acknowledgements

We would like to acknowledge Yungang Zhang, Yangqiu Song, Yonglei Zhou for their helpful discussions. We also gratefully thank the the anonymous reviewers for their valuable comments.

References

1. P. Braud, *Markov chains: Gibbs fields, Monte Carlo simulation and queues*, Springer, NY, 1999.
2. C. I. Cheng, "High-frequency compensation of low sample-rate audio files: A wavelet-based spectral excitation algorithm", *Proc. of International Computer Music Conference*, Sept. 1997.
3. Y. Ephraim, H.L. Ari, W.J.J. Roberts,"A brief survey of speech enhancement", *The Electronic Handbook*, CRC Press, 2003.
4. W. T. Freeman, T. R.Jones, E. C. Pasztor, "Example-based super-resolution", *IEEE Trans. on Computer Graphics and applications*, Vol. 22, Issue:2, pp.56 – 65, March-April, 2002.
5. L. Rabiner, B. H. Juang, *Fundamentals of speech recognition*, Prentice-Hall, 1993.
6. M. Vetterli, "A theory of multirate filter banks", *IEEE Trans. on Acoustics, Speech, and Signal Processing* , Vol.35, No. 3, pp.356 – 372, March 1987.

Probabilistic Tangent Subspace Method for Multiuser Detection

Jing Yang, Yunpeng Xu, and Hongxing Zou

Department of Automation, Tsinghua University,
Beijing, 100084, P.R. China
{yang-jing03, xuyp03}@mails.tsinghua.edu.cn
hongxing_zou@tsinghua.edu.cn

Abstract. The nonlinear Multiuser Detection (MUD) in Direct Sequence Code Division Multiple Access (DS/CDMA) system can be viewed as a two-class classification task. A new classification method called Probabilistic Tangent Subspace (PTS) is introduced to be used as an MUD. Due to the mobility of communicator, wireless communication channels are in fact time variant. The uncertainties of the time-varying channel's coefficients cause the uncertainties of the Multiuser Interference (MUI). On the other hand, the probabilistic tangent subspace method is designed to encode the pattern variations. Therefore, we are motivated to adopt this method to develop a classifier as a multiuser detector for time-varying channels. Simulation results show that this MUD performs better than that based on Support Vector Machine (SVM) for Rayleigh fading channel in DS/CDMA system.

1 Introduction

In Direct Sequence Code Division Multiple Access (DS/CDMA) system [9], all users transmit at the same carrier frequency in an uncoordinated manner. Among these users, only one is the desired user and the others are all interfering users. Therefore, this causes Multiuser Interference (MUI)[10]. Because of MUI, it is difficult to recognize information bits of the desired user from the received sequence. Therefore, to cope with this problem, Multiuser Detector (MUD) [10] has been proposed to detect the users of interest for DS/CDMA systems.

There are two kinds of MUDs: one is linear detector and the other is nonlinear case. The linear detectors include linear minimum mean square error (MMSE) MUD [3] [4] and linear minimum bit error rate (BER) MUD [5] [6]. However, Linear detectors can only work when the signal classes are linearly separable. In fact, in DS/CDMA system, the nonlinear separable cases are common. Then the nonlinear detectors are proposed to solve this problem. The nonlinear multiuser detection essentially can be viewed as a classification task. So, we use pattern recognition methods to design detectors, including neural networks MUD [7] [8] and Support Machine Machine (SVM) MUD [2].

Due to the mobility of communicator, wireless communication channels are in fact time variant rather than time-invariant Additive White Gaussian

S. Singh et al. (Eds.): ICAPR 2005, LNCS 3686, pp. 553–559, 2005.

Noise (AWGN) channels. For example, most digital communication channels are Rayleigh fading channels [11]. The uncertainties in the time-varying channel's coefficients cause the uncertain distortions of the transmitted signals' amplitude and phase. These distortions will increase the complexity and the uncertainties of the MUI, that is why an MUD performs better in AGWN channels than that in Rayleigh fading channels.

In this paper, the nonlinear detector is considered. For time-varying channels, neural network methods often require long convergence time and a large number of neurons, which make the implementation expensive. While SVM method regards the patterns as invariant and tolerates only small variations of input patterns while keeping the class label unchanged. However, it is more desirable to take into account the effects of variations of the channels. This calls for the adoption of invariant classification techniques.

There has been much research addressing to invariant classification problem in machine learning and pattern recognition fields. These methods consider the pattern variations and can give the invariant classification results while the variations are large. we interpret the channel's time variations as the uncertainties of its coefficients, which correspond to the uncertainties of pattern variations. This interpretation motivates us to use these invariant algorithms as an MUD for time-varying channel.

There have been some invariant algorithms for pattern variations, such as Tangent Distance (TD) [13] [14] and Virtual-Support Vector Machine (V-SVM). But these two methods need the prior knowledge of the variations, which is difficult for time-varying channel because of its uncertainties of the channel confidences. To cope with these problems, we introduce Probabilistic Tangent Subspace (PTS) method [1] to the multiuser detection problem in time-varying channel for DS/CDMA system. PTS method is a novel and practical way which can encode the variations and need no prior knowledge of the channel. The simulation results demonstrate that PTS MUD gives better performance than SVM MUD for the time-varying channel.

The rest of the paper is organized as follows. In Section 2, we describe the model of DS/CDMA system and introduce the multiuser detection viewed as pattern recognition problem. Section 3 gives the details about the PTS algorithm. Simulation results will be given in section 4. Finally, we conclude in section 5.

2   MUD for DS/CDMA System

2.1   System Model

For many Code Division Multiple Access (CDMA) systems, it is instructive to examine the structure of the signature waveforms employed, namely, Direct Sequence Spread Spectrum (DS/CDMA) [10]. Spread-spectrum signaling formats feature large duration-bandwidth products. Generally, Barker sequence, M sequence, Gold sequence and Hadamard-Walsh sequence can all be used as the signature code sequence in DS/CDMA system [12].

The Model of discrete-time model of synchronous DS/CDMA downlink is illustrated in Fig. 1.

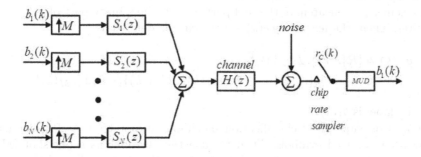

Fig. 1. Discrete-time model of synchronous DS/CDMA downlink

In a Rayleigh channel, by passing through a chip-matched filter, followed by a chip-rat sampler, the discrete-time output of the receiver during one symbol interval can be modelled as

$$r(k) = h(k) \sum_{n=1}^{N} A_n b_n(k) s_n + e(k) \tag{1}$$

where, $e(k)$ is ambient channel noise with $\mathrm{E}\left[e(k)^{\mathrm{H}}e(k)\right] = \sigma^2 I$. N is the number of active users. A_n is the received signal amplitude of the nth user. $h(k)$ is the Raleigh channel's coefficients. $b_n(k) \in \{\pm 1\}$ is information bit from the nth user. M is processing gain, s_n is M-by-1 signature code sequence of user n

$$s_n = [s_{n,1}, \cdots, s_{n,M}]^{\mathrm{T}} \tag{2}$$

and $s_n^{\mathrm{T}} s_n = 1$.

For convenience, we will assume that the user 1 is the user of interest. Consequently, s_1 denotes the signature waveform of desired user 1. Our MUD is required to detect the information symbol of user 1 b_1.

2.2 MUD Viewed as Classification Problem

Through the above analysis, the received sequence r is corrupted by the interfering users and the uncertainties of the time-varying channel's coefficients. In the actual applications, the task of the MUD at a bit period is only to detect the information bit of the desired user (User 1). So the multiuser detection problem can be viewed as a classification problem. Based on this point, the received symbol $r(k)$ is mapped onto a feature space and the input of the classifier is the features. The output of the MUD should match as better as the information bit of user 1 $b_1(k)$.

Due to the information symbol sequence of desired user 1 $b_1 \in \{+1, -1\}$, the multiuser detection can be viewed as a two-class classification tasks. We define the feature vector according to received signal $r(k)$. As the time-varying channel's coefficients are complex, $r(k)$ is M-by-1 complex vector. Therefore, in our scheme, we concatenate the real part and the imaginary part of $r(k)$ as the features. Then, the features $y_\mathcal{F}(k) \in \mathbb{R}^{2M}$ can be defined as

$$
\begin{aligned}
y_\mathcal{F}(k) &= [\mathcal{R}\{r(k)\}, \mathcal{I}\{r(k)\}]^\mathrm{T} \\
&= [\mathcal{R}\{r_1(k)\}, \cdots, \mathcal{R}\{r_M(k)\}, \mathcal{I}\{r_1(k)\}, \cdots, \mathcal{I}\{r_M(k)\}]^\mathrm{T}, \quad (3)
\end{aligned}
$$

and the label is $b_1(k)$.

Based on supervised classification method, we first train a classifier using some prior received symbols. Then we use the classifier as a one-shot MUD. The training process is out-line. However, detection process is on-line, we can adaptively classify the received signal. Therefore, Our scheme is an adaptive MUD.

3 Probabilistic Tangent Subspace Method

Probabilistic Tangent Subspace (PTS) [1] is based on Tangent Distance algorithm. Its basic assumption is that tangent vectors can be approximately represented by the pattern variations. In [1], three subspace models are proposed, including the linear subspace, nonlinear subspace, and manifold subspace models. The features of each sample $y_\mathcal{F}(k) \in \mathbb{R}^{2M}$. If we assume that the feature space $\mathbb{R}^{2M}$ is linear, we can apply the linear subspace method called PTS-I to design the MUD.

3.1 Linear Subspace: PTS-I

Set the training sequence $y_T = \{y_i\}_{i=1}^m$. For each y_i, the features are $y_F \subset \mathbb{R}^{2M}$. First we form the tangent vector set S according to

$$
S = \{z | z = y - y_r, \text{ if } c(y) = c(y_r) \text{ and } y \in \mathcal{N}(y_r)\} \quad (4)
$$

where $c(y)$ denotes the class label of sample y. $\mathcal{N}(y_r)$ indicates the neighbor set of prototype y_r.

If μ is the mean vector of z, and Σ is the covariance matrix of z, then the Mahalanobis distance is

$$
d(z) = (z - \mu)^\mathrm{T} \Sigma^{-1} (z - \mu) = u^\mathrm{T} \Lambda^{-1} u = \sum_{i=1}^{2M} \frac{u_i^2}{\lambda_i} \quad (5)
$$

where $\Lambda = \mathrm{diag}\{\lambda_1, \cdots, \lambda_{2M}\}$ is the eigenvalue matrix of Σ. And, $u = U^\mathrm{T}(z - \mu)$, where U is the eigenvector matrix of Σ.

Table 1. Algorithm for PTS-I MUD

Training:
Training set: T with the label L
Step 1: Obtain Σ by the tangent vector set;
Step 2: Perform eigen-decomposition (PCA) $\Sigma = U \Lambda^{-1} U^{\mathrm{T}}$.
Step 3: Estimate the weight coefficient ρ.
Classification (Detection):
Test sample: $r(k) \in R$
Step4 : Project $z_r = r(k) - y_r$;
Step 5: Compute the error $\varepsilon^2(\cdot)$;
Step 6: Compute the approximate Mahalanobis distance $\hat{d}(z_r)$;
Step 7: Repeat Step 4-Step 6 for each $r(k)$;
Step 8: Return the label of $r(k)$.

PTS-I represents S as a linear space. The principal subspace of S is spanned by the first p components, which is principal component analysis (PCA) on S. Then, the Mahalanobis distance $d(z)$ can be approximated by

$$\hat{d}(z) = \sum_{i=1}^{p} \frac{u_i^2}{\lambda_i} + \sum_{i=p+1}^{2M} \frac{u_i^2}{\lambda_i} = \sum_{i=1}^{p} \frac{u_i^2}{\lambda_i} + \frac{1}{\rho} \sum_{i=p+1}^{2M} u_i^2 = \sum_{i=1}^{p} \frac{u_i^2}{\lambda_i} + \frac{1}{\rho} \varepsilon^2(z) \quad (6)$$

where ρ is a weight coefficient.

$$\rho = \frac{1}{2M - p} \sum_{i=p+1}^{2M} \lambda_i \quad (7)$$

While classifying a sample y_t, we project the linear variation $z_r = y_t - y_r$ into the principal subspace. If the Mahalanobis distance $\hat{d}(z_r)$ is the shortest, the class label of y_t is the same as that of y_r.

3.2   The Algorithm of the PTS-I

Firstly, m received signals are used for training. After training, the following K received signals are detected using PTS-I MUD. $T = \{r(k)\}_{k=1}^{m} \subset \mathbb{R}^{2M}$ denotes the training set and the training label set is $L = \{b_1(k)\}_{k=1}^{m} \in \{+1, -1\}$. $R = \{r(k)\}_{k=m+1}^{K+m}$ denotes the test set. The algorithm of the PTS-I [1] which is used to design the MUD is presented in Table 1.

4   Simulation Results

In this section, we conducted simulations to evaluate the performance of the PTS-I MUD. We simulate the performance PTS-I MUD while the Signal-to-noise ratio (SNR) of the desired user 1 is varying. The increase of the SNR means the decrease of the other interfering users' interference. Also, for comparison, we simulate experiments on SVM MUD, which is proposed in [2].

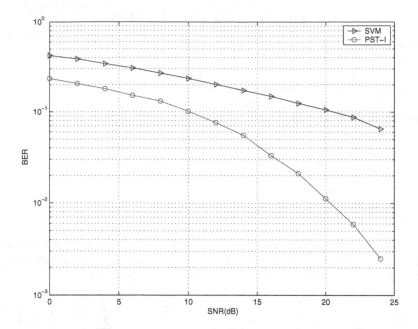

Fig. 2. The comparison of BER for PTS-I MUD and SVM MUD at different SNR level of the desired user 1

We use a Rayleigh fading channel [11] to model the time-varying channel. The normalized Doppler frequency spread is set to be 3.4×10^{-3} Hz. Gold sequences [12] of length $M = 31$ are used as signature code sequences of all users. There are ten active users in the DS/CDMA system. User 1 is desired user. Among other nine interfering users, five users have an SNR of 30 dB each, three users have SNR of 40 dB each, and another user has an SNR of 50 dB.

In the following, training set size is 500 signals, and test size is 5000 signals. We increase the SNR of the desired user 1 and the simulations at each SNR level of desired user 1 are averaged over 100 independent runs. Fig. 2 shows the results of Bit Error Rates (BERs) at each SNR level of desired user 1. BER denotes the total error bit rate of the two classes. At high SNR, the MUI is weak, so the BER is low. By comparing the BERs for the two MUD, it is obvious that the PTS-I MUD outperforms the SVM MUD at each SNR level of the desired user 1.

5   Conclusions

In this paper, we apply two-class PTS-I classifier as an adaptive MUD for time-varying channel in DS/CDMA system. The uncertainties of the channel's coefficients are interpreted as the uncertainties of pattern variations. Therefore, in this scheme, the time variations of the channel can be partly encoded by the x

PTS-I method. Simulation results demonstrate that this MUD provides satisfactory performance and is superior to the SVM MUD.

References

1. Jianguo Lee, Jingdong Wang, Changshui Zhang, and Zhaoqi Bian: Probabilistic Tangent Subspace: A Unified View. In: Proc. 21st Intl. Confs. on Machine Learing (ICML 2004), Banff, Alberta, Canada (2004)
2. Chen, S., Samingan, A.K., and Hanzo, L.: Support vector machine multiuser receiver for DS-CDMA signals in multipath channels. IEEE Trans. Neural Networks, Vol. 12 (2001) 604 – 611
3. S. L. Miller: An adaptive direct-sequence code-division multiple-access receiver for multiuser interference rejection. IEEE Trans. Commun., vol. 43 (1995) 1556 – 1565
4. U. Madhow and M. Honig: MMSE interference suppression for direct sequence spread-spectrum CDMA. IEEE Trans. Commun., Vol. 42 (1994) 3178 – 3188
5. N. B. Mandayam and B. Aazhang: Gradient estimation for sensitivity analysis and adaptive multiuser interference rejection in code-division multi-access systems. IEEE Trans. Commun., Vol. 45, no. 7 (1997) 848 – 858
6. I. N. Psaromiligkos, S. N. Batalama, and D. A. Pados: On adaptive minimum probability of error linear filter receivers for DS-CDMA channels. IEEE Trans. Commun., Vol. 47, no. 7 (1999) 1092 – 1102
7. B. Aazhang, B. P. Paris, and G. C. Orsak: Neural networks for multiuser detection in code-division multiple-access communications. IEEE Trans. Commun., Vol. 40 (1992) 1212 – 1222
8. U. Mitra and H. V. Poor: Neural network techniques for adaptive multiuser demodulation. IEEE J. Select. Areas Commun., Vol. 12 (1994) 1460 – 1470
9. R. Prasad: CDMA for Wireless Personal Communications. Norwood, MA: Artech House (1996)
10. Sergio Verd: Multiuser detection. Cambridge, New York : Cambridge University Press (1998)
11. John G. Proakis: Digital Comunications. New York : McGraw-Hill, 3rd ed (1995)
12. Andrew J. Viterbi: CDMA : principles of spread spectrum communication. Reading, Mass. : Addison-Wesley Pub. Co. (1995)
13. T. Hastie, P. Simard, and E. Saeckinger: Learning prototype models for tangent distance. Advances in Neural Information Processing Systems 7
14. P. Simard, Y. LeCun, J. Denker, and B. Victorri: Transformation invariance in pattern recognition - tangent distance and tangent propagation. Inernational Journal of Imaging System and Technology, Vol. 11 (2001) 181 – 194

Feature Extraction for Handwritten Chinese Character by Weighted Dynamic Mesh Based on Nonlinear Normalization

Guang Chen, Hong-Gang Zhang, and Jun Guo

Beijing University of Posts and Telecommunications, 186 mailbox, Beijing 100876
Chg_pris@126.com

Abstract. This paper describes a new feature extraction method contributing to improvement of the performance of a handwritten Chinese character recognition system. By using enhanced weighted dynamic meshes based on nonlinear normalization, this method not only avoids the zigzags and other undesirable side effects introduced in the original Yamada et al.'s nonlinear normalization method but also avoids additional feature normalization process in the original Lian-Wen Jin et al.'s and WU Tian-lei et al.'s dynamic mesh method. Experiment on HCL2000, a handwritten Chinese character database, shows that our method achieves superior performance.

1 Introduction

During last few years, considerable progress has been made in research on handwritten Chinese character recognition ([1] and [2]), and pattern matching methods has become the main topic of research in this field. In pattern matching methods, statistical features are extracted from handwritten character, and the discriminative capability of these features is important to the performance of an overall character recognition system.

Chinese characters have large number of categories, very complex shape structures and many similar characters as compared with Roman alphabet. For off-line handwritten Chinese characters, there are many kinds of shape variations involving variation of size and density, partial displacement, stroke translation, stroke inclination, stroke length variation, stroke width variation, broken and connected strokes, and so on. These characteristics, especially the shape variations, make recognition of off-line handwritten Chinese character one of the most difficult problems in the area of character recognition. To perform reliably in the presence of all these sources of variability, an off-line handwritten Chinese character recognition system must extract stable features that preserve enough useful information needed to distinguish characters correctly. Many methods, such as nonlinear normalization ([3], [4] and [5]) as a preprocessing step and dynamic mesh method ([6] and [7]) at feature extraction stage, were proposed.

S. Singh et al. (Eds.): ICAPR 2005, LNCS 3686, pp. 560–568, 2005.

Nonlinear normalization is to make feature projection histogram by projecting a certain density feature at each point onto horizontal- or vertical- axis and equalize feature densities by re-sampling the feature projection histogram. By correcting nonlinear shape variations and homogenizing the two-dimensional line density, this preprocessing method improves the stability and reliability of features extracted in following steps and improves the performance of pattern matching classifier. But in the meantime, it introduces some undesired deformations, like change in stroke directions, zigzags at stroke edge, change in relative location of strokes, increase of strokes' cracks and change in thickness of strokes, etc. (see Fig. 1). These deformations may have a considerable impact on feature's ability to represent the original character image, and further affects the recognition accuracy of the total system.

Dynamic mesh method at feature extraction stage is to divide original image into several nonuniform meshes according to the density image of input image. By combining density equalization and nonuniform mesh division, this approach not only homogenizes the two-dimensional line density but also avoids the deformation introduced by nonlinear normalization. Because of nonuniform mesh division, the number of pixels in each mesh is not equal, so feature normalization is necessary after feature extraction. In this normalization process, each pixel is treated equally, so some useful information such as local area stroke density is lost and furthermore affects the accuracy of the total system.

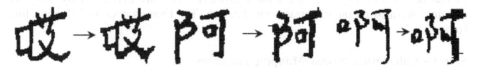

Fig. 1. Some undesired distortion introduced by nonlinear shape normalization – some stroke distortion, some zigzags at stroke edge, etc. Left one of each set is input image and right one is normalized image.

In this paper, we propose a new feature extraction method using weighted dynamic mesh based on nonlinear normalization. Firstly the inverse-mapping function is calculated based on the density image according to our given density definition. The next step is to divide normalized image into uniform meshes, equivalent to dividing the original image into dynamic nonuniform meshes. At last, the features extracted in each mesh are accumulated according to the inverse-mapping function. In this way, each pixel in dynamic meshes achieves respective weight factor. So no more additional normalization is needed. This method keeps the regional weighting information, avoids the deformation introduced by nonlinear normalization, so it can represent the original pattern more effectively, and performs better in classification process. Another advantage of this method is that it can be conveniently applied to all the existing mesh division and feature extraction method without increasing the computation cost.

The flowchart graph for this feature extraction method goes as follows:

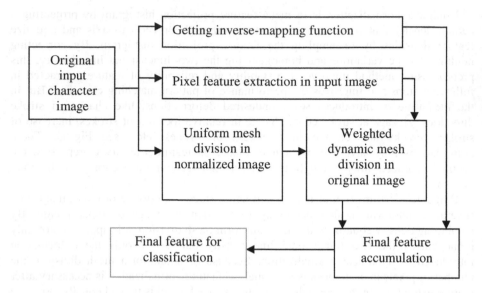

2 Feature Extraction with Weighted Dynamic Mesh

In this section, we introduce the typical process of feature extraction by weighted dynamic mesh based on nonlinear normalization. The Process goes as following four steps.

Step 1 – Calculating Inverse-Mapping Function

In this step, inverse-mapping function will be calculated based upon nonlinear normalization. There are many density definitions ([4] and [5]) available for nonlinear normalization. Lee and Park[5] compared the performance and computational complexity of existing nonlinear normalization methods and experimental results indicated that Yamada et al.'s method based on line density[4] performed better while

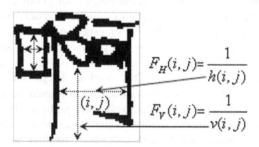

Fig. 2. Definition of improved line density and character density features

having a higher computational cost and side effects of zigzags. For more suitably representing the structural characteristic of Chinese character and reducing the computational cost, we use improved line density by line interval.

To describe the proposed method, we introduce the notation. Let $B(i,j)$ be an input binary character image, whose size is $I \times J$, $i=1,2......I$ and $j=1,2......J$. Let $h(i,j)$ and $v(i,j)$ be horizontal and vertical line density of pixel (i,j), respectively. In our method, $h(i,j)$ and $v(i,j)$ is horizontal and vertical distance between neighboring strokes, respectively. (See Fig. 2)

Let $F_H(i,j)$ and $F_V(i,j)$ be two characteristic density features of pixel (i,j). For pixels in background area, whose binary value is 0, characteristic density features are given as

$$F_H(i,j) = \frac{1}{h(i,j)}$$

$$F_V(i,j) = \frac{1}{v(i,j)}$$

(1)

For pixels in pattern area, whose binary value is 1, characteristic density features are given by the average value of all the characteristic density features of pixels on the same row or the same column, and they reflect stroke densities at different area of character image effectively. The feature projection functions on x-axes and y-axes are defined as

$$H(i) = \sum_{j=1}^{J} F_H(i,j) + \alpha_H, i = 1,2......I$$

$$V(j) = \sum_{i=1}^{I} F_V(i,j) + \alpha_V, j = 1,2......J$$

(2)

Where a_H and a_V are constants.

Let (i',j') be the corresponding pixel in normalized image of pixel (i,j) in input image, and it can be calculated as (forward-mapping function)

$$i' = f_H(i) = \left\lfloor \sum_{m=1}^{i} H(m) \times \frac{I}{\sum_{m=1}^{I} H(m)} + 0.5 \right\rfloor,$$

(3)

$$j' = f_V(j) = \left\lfloor \sum_{n=1}^{j} V(n) \times \frac{J}{\sum_{n=1}^{J} V(n)} + 0.5 \right\rfloor.$$

Where $\lfloor A \rfloor$ is the floor of A. The inverse-mapping function can be define by

$$i = g_H(i') = \min\left\{ k \mid \sum_{m=1}^{k} H(m) \ge i' \cdot \frac{\sum_{m=1}^{I} H(m)}{I} \right\},$$

$$\quad (4)$$

$$j = g_V(j') = \min\left\{ l \mid \sum_{n=1}^{l} V(n) \ge j' \cdot \frac{\sum_{n=1}^{J} V(n)}{J} \right\}.$$

In this way the inverse-mapping function, which mapping pixel *(i',j')* in normalized image to pixel *(i,j)* in input image, is presented. We can see from Eq.(3) and Eq.(4) that each pixel in normalized image has one original pixel in input image, but due to the discrete nature of input image and normalized image, more than one pixel in normalized image may be mapped to the same one pixel in input image. In the following steps we will make use of this characteristic of inverse-mapping function.

Step 2 – DEF Feature Extraction

The directional element feature[9] is one of the most effective features in off-line handwritten Chinese character recognition. Here we extract directional element feature for every pixel on stroke edge in input image.

Chinese characters consist of four kinds of elementary strokes: horizontal strokes, vertical strokes, left diagonal strokes and right diagonal strokes. The directional element feature tests for the presence of these strokes in binary image. Operation for this process includes the following two steps.

1) Contour extraction
    Only pixels on contour will be used to extract feature vector.
2) Dot Orientation extraction
    In this process, four types of line elements - vertical, horizontal and two oblique lines slanted at ±45°, are assigned to each pattern pixel on the contour. For a center pattern pixel in a *3×3* mask, twelve cases are considered (See Fig.3). Here, eight neighbors are used to determine the direction of a pattern pixel. If the other two pattern pixel and the center pixel are on a line, one type of element is assigned; otherwise, two types of elements are assigned simultaneously.

When this step is completed, four directional element features f_0, f_1, f_2, f_3 are generated for every stroke edge pixel *(i,j)* in input image.

Step 3 – Uniform Mesh Division in Normalized Image

At the rough classification stage and the fine classification stage, normalized image is divided into *5×5* and *8×8+7×7* uniform meshes, respectively. (See Fig.4)

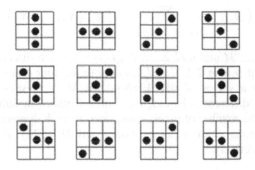

Fig. 3. 12 connection types of stroke edge pixels corresponding to the four elementary strokes

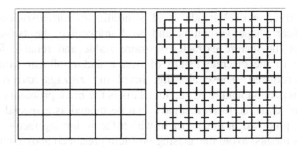

Fig. 4. 5×5 and 7×7+8×8 uniformly divided meshes on normalized image

Let *mesh(m,n)* be one of the *M×N* meshes divided above in normalized image, the grid line for *mesh(m,n)* can be define as

$$x_1(m) = \frac{m-1}{M} \times I, x_2(m) = \frac{m}{M} \times I$$

$$y_1(n) = \frac{n-1}{N} \times J, y_2(n) = \frac{n}{N} \times J$$

(5)

Where *m=1,2......M* and *n=1,2......N*

Then *mesh(m,n)* can be defined as:

$$mesh(m,n) = \left\{ (i', j') \mid x_1(m) \le i' \le x_2(m), y_1(n) \le j' \le y_2(n) \right\}$$

(6)

Where *i'=1,2......I* and *j'=1,2......J*

Step 4 - Final Feature Accumulation

Using the inverse-mapping function calculated in 2.1, we accumulate features for points extracted from input image in corresponding meshes of normalized image to compose the final feature vector *F*. Feature accumulation for the element $F_d(m,n)$ of vector *F* goes as

$$F_d(m,n) = \sum_{(x,y)\in mesh(m,n)} f_d\left(g_H(x), g_V(y)\right)$$

(7)

Where $m=1,2......M$ and $n=1,2......N$, $f_d(i,j)$ is the dth directional element feature extracted for point (i,j) in 2.2, $d=0,1,2,3$ denote the four element directions. So the final feature F is a vector of size $M\times N\times 4$ *for* $M\times N$ meshes division. For stroke density varies in different subareas, the times features in subarea is accumulated varies; therefore the number of pixels considered in each dynamic mesh are equal and further feature normalization is not necessary and the ability of representing original pattern is kept effectively.

3 Discussion

The main motivation of Yamada et al.'s nonlinear normalization method[4] is to correct nonlinear shape variations and to homogenize the two-dimensional line density so that the feature extracted is more stable and reliable for classification. However, due to the discrete nature of image and nonlinear sampling, undesired aliasing is introduced. Those aliasing, involving zigzags and so on, may have considerable impact on feature extraction. This is the main problem of this method.

Conventional dynamic mesh method[6] can be treated as a special case of weighted dynamic mesh proposed in this paper in which the weighting factors for all the pixels are equal. This method avoid the aliasing problem presented above, but feature for each pixel is accumulated only once in a dynamic mesh, and as the size of dynamic mesh varies the number of pixels considered in it varies, so additional feature normalization is necessary and some useful discriminative information is lost in this process.

Weighted dynamic mesh method proposed here combines the advantages of above two methods and avoids main problem of them. By using inverse-mapping function, we effectively homogenizing the two-dimensional line density without introducing any aliasing, and we make full use of original stroke shape characteristic; furthermore, information of local stroke density is well preserved and used and further feature normalization is avoided. It is more suitable for off-line handwritten Chinese character recognition.

4 Experiment

In this section, we apply the proposed feature extraction method to recognition system and compare it with Yamada et al.'s nonlinear normalization method[4] and Lian-Wen Jin et al.'s conventional dynamic mesh method[6] by experiment. HCL2000[8] handwritten Chinese character library collected by PRIS Lab, Dept. of Information Engineering, Beijing University of Posts and Telecommunications, Beijing, P.R.China. There are 1,000 sets of handwritten Chinese character samples written by 1,000 Chinese people respectively in the library. Each set contain all the 3,755 daily used Chinese characters and some information about the writer (sex, age, job etc.). All samples are binary images of size 64 by 64. Some samples of off-line handwritten Chinese character used in our experiment are shown in Fig.5. In our experiment, the

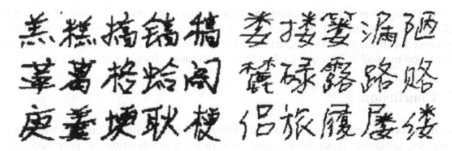

Fig. 5. Some handwritten Chinese character samples in HCL2000 database

first 700 sets (marked as xx001~xx700) out of the 1,000 sets of HCL2000 were used for training, and the other 300 sets (marked as hh001~hh300) were used for testing. The 3,755 daily used Chinese character categories were used in our experiment.

There are 3,755 categories of daily used Chinese character, so we need a pre-classifier to reduce computational cost and speed up the classification process. Here we use 5×5 uniformly divided meshes to extract feature vector of size 100, and apply Euclidean discriminant function[10] to choose 30 candidates from 3,755 categories for fine classification later. 8×8+7×7 uniformly divided meshes are used to extract feature vector of size 452 for fine classifier, and improved Euclidean discriminant function is used to make the final decision from the 30 candidates. Let $x=(x_1,x_2......x_n)$ be an n-dimensional input feature vector, and $u=(u_1,u_2......u_n)$ be the standard feature vector of a category. The improved Euclidean distance is given by

$$D(x,u) = \sqrt{\sum_{i=1}^{n} \frac{(x_i - u_i)^2}{\sigma_i + \theta}} \qquad (8)$$

Where σ_i denotes the standard deviation of jth element, and θ is a constant. By taking account variations of handwritten characters in the Euclidean distance, our improved Euclidean distance of Eq.(8) can detect small changes in character shape and improve the accuracy of overall recognition system with little increase in computational cost. Here we compare the performance of three feature extraction methods: nonlinear normalization method, conventional dynamic mesh method and weighted dynamic mesh method proposed in this paper. The result of this comparison is shown in Table 1.

The recognition rate of system using Weighted Dynamic Mesh method proposed in this paper is 92.72%, 1.14% higher than the recognition rate of system using conventional dynamic mesh method, and 3.36 higher than the recognition rate of

Table 1. Performance comparison of three methods

Feature Extraction Method	Recognition rates (%)
Nonlinear Normalization method	89.36
conventional Dynamic Mesh method	91.58
Weighted Dynamic Mesh method proposed in this paper	92.72

system using nonlinear normalization method. From the point of view of error rate, ER of the new method proposed here reduced 13.5% than that of conventional dynamic mesh method. The experimental result shows that the proposed new feature extraction method is effective for handwritten Chinese character recognition.

5 Conclusion

In this paper, a new feature extraction method for handwritten Chinese character by weighted dynamic mesh based upon nonlinear normalization is proposed. This method has the advantage of avoiding undesired stroke distortion in the peripheral region introduced by nonlinear normalization and making full use of local density information of input image. Experiments on the HCL2000 handwritten Chinese character library have shown that the method outperforms existing nonlinear normalization and dynamic mesh methods. In future study, we will investigate the application of this feature extraction approach to other mesh division methods and other feature types.

References

1. Kimura, Wakabayashi, Tsuruoka and Miyake. "Improvement of Handwritten Japanese Character Recognition Using Weighted Direction Code Histogram." Pattern Recognition 30, no. 8 (1997): 1329-37.
2. Kato, N.; Suzuki, M.; Omachi, S.; Aso, H.; Nemoto, Y. A handwritten character recognition system using directional element feature and asymmetric Mahalanobis distance. Pattern Analysis and Machine Intelligence, IEEE Transactions on Volume: 21 , Issue: 3 , March 1999 258 – 262
3. Tsukumo and Tanaka. "Classification of Handprinted Chinese Characters Using Nonlinear Normalization and Correlation Methods." Pattern Recognition, 1988., 9th International Conference on, no. SN - (1988): 168-71 vol.1.
4. Yamada, Yamamoto and Saito. "A Nonlinear Normalization Method for Handprinted Kanji Character Recognition--line Density Equalization." Pattern Recognition 23, no. 9 (1990): 1023-29.
5. SEONG-WANG LEE, E. P. Nonlinear Shape Normalization Methods for the Recognition of Large-set Handwritten Characters. Pattern Recognition 1994, 27, 895-902
6. Lian-Wen Jin; Bing-Zheng Xu, Directional Cellular Feature Extraction with Elastic Meshing for Handwritten Chinese Character Recognition. JOURNAL OF CIRCUITS AND SYSTEMS 1997,2,7-12
7. WU Tian-lei, MA Shao-ping. "Feature Extraction for Handwritten Chinese Character By Overlapped Dynamic Meshing and Fuzzy Membership." Acta Electronica Sinica 32, no. 2 (2004): 186-90.
8. Jun Guo; Zhi-Qing Lin; Hong-Gang Zhang, A New Database Model of Off-line Handwritten Chinese Characters and Its Applications. Acta Electronica Sinica Vol.28 No.5 2000
9. N. Sun , M. Abe and Y. Nemoto. "A Handwritten Character Recognition System By Using Improved Directional Element Feature and Subspace Method." IEICE J78-D-II, no. 6 (1995): 922-630.
10. K Fukunaga. Introduction to Statistical Pattern Recognition (2nd Edition) [M] .Boston: Academic Press, 1990.

Post Processing of Handwritten Phonetic Pitman's Shorthand Using a Bayesian Network Built on Geometric Attributes

Swe Myo Htwe[1], Colin Higgins [1], Graham Leedham[2], and Ma Yang[2]

[1] The University of Nottingham, School of Computer Science and IT, Wallaton Road,
Nottingham, NG8 1BB, UK
{smh,cah}@cs.nott.ac.uk
[2] Nanyang Technology University, School of Computer Engineering,
N4-#2C-77 Nanyang Avenue, Singapore 639798
{asgleedham, pg01874827}@ntu.edu.sg

Abstract. In this paper, we introduce a new approach to the computer transcription of handwritten Pitman shorthand as a rapid means of text entry (up to 100 words per minute) into today's handheld devices, almost at the rate of speech. It is different from previous applications of the same framework from two aspects: - firstly, a novel idea of using geometric attributes other than phonetic attributes in the abstraction of a phonetic Pitman's shorthand lexicon is proposed. Secondly, a Bayesian network representation for the organisation of shorthand-outline models is introduced, in which natural variability of Pitman shorthand is defined via different nodes and links. Using a probabilistic Bayesian network, the system shows a noticeable robustness not only in transcribing a variety of genuine handwriting, but also in estimating missing vowel components that may have been omitted in speed writing. The accuracy of the new approach (92.86%) is a considerable improvement over previous applications.

1 Introduction

To allow handheld devices to replace desktop computers for running tasks, a means of rapid text entry is necessary, preferably at a comparable or superior rate to typing on a keyboard. The transformation of a standard "QWERTY" keyboard into miniature ones in handheld devices make text input very slow (less than 10 words per minute (wpm)) [1]. Over the past 15 years, solutions that provide slow data input into handheld devices have been carried out by means of handwritten recognition systems: - *Unistroke, Qucikwriting*[2], and predictive text input methods such *as Tegic's T9 and POBox*[1]. Whilst these techniques speed up to an average rate of 25 to 40 wpm, the bottleneck of handheld computing still remains the same. As Handheld devices, like PDAs, become increasingly popular as business appliances, the higher the demand becomes for fast data input, particularly for mobile note takers such as journalists. Today's stenographers cannot take advantage of handheld computing, especially in speed writing, as none of the existing systems enable transcription of shorthand.

S. Singh et al. (Eds.): ICAPR 2005, LNCS 3686, pp. 569–579, 2005.
© Springer-Verlag Berlin Heidelberg 2005

In fact, extensive research [3][4][5] on the computer transcription of handwritten Pitman's shorthand has been carried out for over two decades. However, the recogniser system being unable to detect smooth junctions and a lexicon post processor being unable to detect missing vowel components within an outline, limit the usefulness of existing systems. Recently, Yang et al [6][7] introduced a new rule to detect a smooth junction of a shorthand outline and the new rule improved the classification accuracy by 55%. Another recent work by Nagabhushan et al [8][9] concentrated on the linguistic post processing of shorthand outlines into orthographic English words with the use of modified dictionaries and concluded that further work is required in the homophones (outlines which are written similarly but have different representations) resolution area. In addition, recent research highlights a critical need to cover the loss of data due to phoneme conversion in the linguistic post processing and to establish an efficient probabilistic framework in which uncertainty and dependency between components are well defined.

In this paper, a novel approach to the post processing of handwritten Pitman shorthand by the use of probabilistic Bayesian networks is discussed. Firstly, an overview of the whole system of online recognition of handwritten Pitman shorthand is given and the need for primitive attributes, rather than phoneme attributes, in the linguistic post processor is explained along with examples. Then a closer view of Bayesian network implementation is discussed and experimental results and further work are stated at the end of the paper.

2 Overview of the Whole System

The system consists of two main components: - a recognition and a transcription engine. The recognition engine operates at a low level in which segmentation and classification of pattern primitives are carried out and the transcription engine operates at higher level in which segmented primitives are transliterated into related English words.

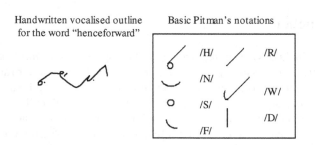

Fig. 1. A sample Pitman shorthand outline for the word "henceforward"

In order to give the reader a quick understanding on the phonetic construction of Pitman shorthand, basic shorthand notations and a script written by a stenographer is shown in *Figure 1*. Pitman outlines can be categorised into two groups: - vocalized outlines and short-forms. A sample script depicted for the word "henceforward" in

Figure 1 is a vocalised outline as its construction adheres to phonetic rules, whereas short-forms are single pen strokes especially defined for the 100 most frequently used English words without following any phonetic rules.

An input outline is firstly distinguished between a vocalised outline (if there is any vowel notation) and a short-form (if there is no vowel notation). The major role of the recognition engine is to detect the dominant points of a vocalised outline and segment it into the most relevant fragments. Then, the segmented primitives are processed via a neural network classifier and a ranked list of pattern primitives, along with related pattern categories, is produced. The role of the transcription engine is to take these classified pattern primitives and estimate the most likely candidate words for a written outline via outline models of the Bayesian network. The candidate words are then put through a sentence level interpreter and the best interpretation for a given outline is finally output using contextual information.

With short-forms, a template-matching approach is used to recognise a limited number of symbols and a ranked list of words for a written short-form is produced at the end of the matching process. In order to choose the correct representation of a written short-form, the candidate words are also forwarded to the sentence level interpreter and the final selection is done by the use of contextual knowledge.

In further sections, a closer view of the implementation of the outline model based on a Bayesian network is discussed. Firstly, reasons why a primitive attribute is preferred over a phonetic attribute to ensure a correct interpretation are discussed.

3 Primitive Attributes vs. Phonetic Attributes

To clarify the difference between "primitive attributes" and "phonetic attributes", refer to an example given in *Figure 2*.

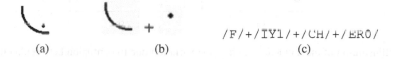

        (a)   (b)   /F/+/IY1/+/CH/+/ER0/
                                                 (c)

Fig. 2. (a) A sample shorthand outline for the word "feature" (b) Two primitives included in an outline for the word "feature" (c) Four phonemes included in an outline for the word "feature"

A primitive attribute relates to a segmented geometrical feature of an outline such as a loop, a circle, a curve or a stroke. *Figure 2 (a)* shows a sample shorthand outline for the word "feature" and the outline includes two primitive attributes i.e., a downward long curve ⌣ and a dot ˙ as shown in *Figure 2 (b)*). A phonetic attribute relates to a phonetic representation of an outline and a sample outline in *Figure 2 (a)* represents four phonemes i.e., a consonant /F/, a vowel /IY1/, a consonant /CH/ and a vowel /ER0/ as shown in *Figure 2 (c)*.

A Bayesian network in our system is based on primitive attributes. The reason why the Bayesian network operates on primitive attributes rather than phoneme attributes is due to the large number of production rules of Pitman shorthand invented for speed improvement purposes. These allow multiple ways of pronouncing a single outline if there is a minor difference between geometric features such as size, length, thickness or the inclination of a stroke. Accurate expression of size, length, or inclination is practical for printed script; however it is less practical in human handwriting, especially if the script is written at speed. In order to clarify the increase in ambiguity due to the conversion of primitives into phoneme values, the following examples are shown.

Example 1: Appearance variation
As shown in *Figure 3(a)*, typical notations for a combination of phonemes /S R/, /S T R/ and /W/ are a combination of an upward stroke ╱ with a preceding circle ○, loop *𝒪* and hook *𝒰* respectively. Assume that a user writes an outline of /S T R / with no clear distinction between a circle, loop and hook as shown in *Figure 3 (b)*, and assume that the circular primitive is classified as a circle ○ primitive instead of a loop *𝒪* primitive by the recogniser, the /T/ phoneme is then lost in the output, resulting in an outline written with three phonemes /S T R/ being wrongly interpreted as an output with two phonemes /S R/. Similarly, if a loop *𝒪* primitive of a written outline *(Figure 3(b))* is classified as a hook *𝒰* by the recogniser, an outline representing the three phonemes of /S T R/ is interpreted as a single phoneme of /W/. According to experimental results, more than 50% of small hooks, loops and circles are thus confused and the direct conversion of primitives, which are prone to minor classification errors, into phonetic values leads to a wrong interpretation.

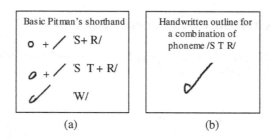

(a)   (b)

Fig. 3. Illustration of the incidence of phoneme variation due to confusion between a circle and a loop

Example 2: Length variation
As shown in *Figure 4(a)*, different lengths of a curve represent different phonemes in Pitman shorthand according to a "double-stroke" rule, saying that if a curve is immediately followed by syllables /TER/, /DER/, /THER/ or /TURE/ the curve should be doubled in length and notations for /TER/, /DER/, /THER/, /TURE/ should be omitted. While writing at speed, length is not always clearly shown in some outlines. In the phonetic based approach, the sample outline shown in *Figure 4 (b)* can be interpreted wrongly as /F IY1/ (e.g., the word "fee") instead of /F IY1 CH ER0/ (e.g., the word "feature") if the curve /F/ is not recognised as a long curve.

In the primitive based approach, a normal-length curve and a double-length curve are simply remarked as a curve and therefore the word "feature" and the word "fee" contain the same types of primitives (i.e., a curve ╰ and a dot ˙). Therefore, a candidate list for the sample outline in *Figure 4 (b)* includes the word "feature" as well as the word "fee" and a correct word can be extracted with the use of contextual knowledge in the sentence level transcription.

In brief, conversion of phoneme attributes from inaccurate handwritten primitives allows the wrong candidates to appear at an early stage which subsequently further affects the transcription processes. Therefore, our new approach works on a primitive level in which the system does not understand any phonetics, but is efficient at interpreting phonetically written Pitman shorthand. The system does support generalization of large vocabularies via conceptual building blocks as in the phonetic representation, but the primitive attributes are more appropriate to cope with the unique features of Pitman shorthand.

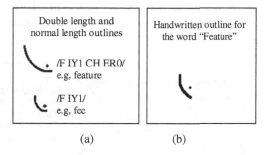

(a)   (b)

Fig. 4. Illustration of the incidence of phoneme variation due to length confusion

4 Bayesian Network

Bayesian network [10] is a directed acyclic graph where each node represents a mutually exclusive and collectively exhaustive set of random variables where the links signify probabilistic dependency between the linked variables. It has been a remarkable tool in the domain of pattern matching for its outstanding ability in defining natural variability. Implementations of Bayesian networks in the domain of computer vision such as signature verification [11], and handwritten character recognition [12] have been addressed and the positive results of the work supports our motive for applying the same tool in resolving similar problems (*natural variability*) in the recognition of handwritten Pitman's shorthand.

Implementation of the network in our system is described in the following three categories: -

1. An outline model: a network clearly showing the relationship and distribution of pattern primitives in the formation of a shorthand outline. In general, an outline model corresponds to one or more words and a lexicon is a collection of more than one outline model in the system.
2. Inference: an algorithm updating the likelihood of individual primitives contained in an outline model with some evidence and a priori probability.
3. Learning and selection: strategies to estimate an optimal likelihood of a primitive appearing in a particular outline model based on the training data and to select the best outline model that maximises the conditional probability of a given set of input primitives.

4.1 Outline Model

An Outline Model is constructed by concatenating segmented primitives of a holistic outline according to their writing order with specified dependency. In every outline model, vowel primitives are located around a tail section of the model, as they are written only after the construction of a whole consonant kernel. Therefore, primitives contained in an outline model do not adhere to a linguistic order. The network is constructed in a hierarchical structure such that the root node corresponds to an outline o and the leaf nodes P_i (i = 1, 2,..n) correspond to primitives contained in the formation of the outline o. Our network architecture is similar to a structure defined in signature verification work by Xiao [11]. The network is constructed with three types of nodes, which are related to three types of dependency between a primitive and a root node. These three types of nodes are: - "Unique" node, "Virtual" node and "Hidden" node.

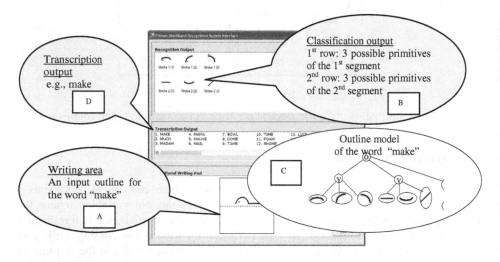

Fig. 5. Illustration of the creation of a virtual node and a hidden node of an outline model with the aid of a screenshot of a demo system

To clarify the creation of a virtual node and a hidden node, consider the example illustrated on a screenshot of our demo system in *Figure 5*. The processing order of the system in *Figure 5* is *Box-A (i.e., ink collection and pre-processing), Box-B (i.e., feature extraction and classification), Box-C (i.e., Outline-model processing)*, and Box-*D (i.e., candidate list creation)* respectively. Detail explanation of the pre-processing, feature extraction and classification can be found in our recent publication [6] and the creation of an outline model is explained in this section. In fact, Bayesian network based outline models in *Box-C* are either updated during the training process or looked up for the N-best candidate words during the interpretation process. The current example in *Box-C* illustrates a sample outline model for the word "make" and the model includes two virtual nodes (denoted as "V") and one hidden node (denoted as "H") as shown.

The creation of "Virtual" nodes relates to an assumption that "If a particular primitive (e.g., the first primitive ⌒﹨ of the sample outline model in *Box C*) is dependent on another primitive (e.g., the second primitive *⌐* of the sample outline model in *Box C*) and there is an optional relationship between them (i.e., either at most one or none of them can be true at a time), there is a mechanism that controls the values of ⌒﹨ and *⌐* , resulting in a virtual node V." In general, a virtual node relates to consonant primitives in our system.

The creation of "Hidden" nodes relates to an assumption that "If a particular primitive (e.g., a vowel primitive • of the outline model in *Box-C*) appears or disappears from time to time and the variation does not adhere to any rule (e.g., omission of vowel components in an outline according to writers' experience and the omitted locations are unpredictable), we can assume that there is a hidden mechanism that controls the value of • primitive resulting in a hidden node H". In general, a hidden node is related to vowel primitives and the /R/ primitive in our system.

The creation of "Unique" nodes relates to an assumption that "If a particular primitive appears in every sample of an outline, the primitive can be considered as independent of other primitives and be linked directly to the root node."

4.2 Inference Algorithm

The inference process of a Bayesian network involves updating the belief of nodes (e.g., primitives in our case) given some evidence and priori probabilities [11]. It is called the belief of a primitive denoted as BEL(P). Among a variety of belief updating algorithms that support the Bayesian network, our work applies the message passing algorithm developed by Pearl[10] in which the belief of every node in the network is taken as the product of π and λ messages, where π is a message received from each of its parents (if any) and λ is a message received from each of its children (if any).

In our system, before the arrival of any evidence, each node is initialised with π and λ messages. The initialisation of π and λ message depends on the type of node in the network. If it is a leaf node, π is set to π, the message received from its parents and λ is set to the confidence score of the node obtained from the training data. None of the π or λ messages are initialised for virtual nodes or hidden nodes as they represent judgemental evidence. The root node is the topmost one in the network and does not have any parent; therefore a π message is set to (0.5,0.5) assuming there is an equal chance of taking a TRUE or FALSE value for this node.

Upon the arrival of evidence, belief updating is done by the formula (1) derived by Xiao[11]: -

$$BEL(x) = \beta\lambda(x)\,\pi(x) \tag{1}$$

where β is a normalization factor, $\lambda(x)$ is a combined message received from all the children of node X and $\pi(x)$ is a combined message received from all the parents of node X.

Depending on the type of a node, $\lambda(x)$ is calculated differently. If it is a Root *Node*, $\lambda(x)$ can be defined by the formula derived by Pearl [10] : -

$$\lambda(x) = \prod_j (\lambda_{Yj}(x)) \tag{2}$$

where $\lambda_{Yj}(x)$ is a message that a node X received from its child node Y_j.

If a node is a hidden node, $\lambda(x)$ is defined as : -

$$\begin{cases} \lambda(x) = & \lambda_{Yj}(x) \text{ if a child } Y_j \text{ of the hidden node X is true} \\ & 0.1 \text{ otherwise} \end{cases} \tag{3}$$

If it is a virtual node, $\lambda(x)$ is defined as : -

$$\lambda(x) = \begin{cases} \lambda_{Yj}(x) \text{ if a child } Y_j \text{ of the virtual node X is true} \\ 0.001 \text{ otherwise} \end{cases} \tag{4}$$

4.3 Learning and Selection

The Learning process includes finding the optimal maximum likelihood estimate of training parameters, and using these estimates to construct new outline models that are not included in the training data. Among a variety of learning algorithms that support a Bayesian network, the use of an appropriate algorithm depends on the structure of the network (*whether it is known or unknown*) and the evidence of nodes (*whether they are fully or partially observable*). In our system, an outline model is firstly constructed based on the knowledge of a primitive lexicon, and therefore the structure of the network is known. Consonant nodes are always observable in the training data and the likelihood of a consonant primitive P to be confused with another consonant primitive Q can be formulated as: -

$$P(P|Q) = N(P=TRUE, Q=q)/N(P) \tag{5}$$

where, $N(P) = N(P=TRUE, Q=TRUE) + N(P=TRUE, Q=FALSE)$. This method is generally denoted as Maximum Likelihood Estimate (MLE).

Vowel primitives are not fully observable in the training data and the estimate of a vowel primitive is not use to train a new outline model in our system. The likelihood of vowel primitives included in the new outline model is set according to the confidence score of a lexicon.

The Searching process includes finding an outline model that produces the highest posterior probability given the input primitives. If the ith outline model is denoted as O_i and the input primitives as $P_1, P_2,.., P_n$, the posterior probability of a written outline can be formulated as:

$$P(O_i| P_1, P_2,.., P_n) = P(O_i)P(P_1, P_2,.., P_n|O_i) \tag{6}$$

Outline models are searched by two level of filters: - firstly models with high joint probability of the first and last primitives are selected as potential candidates, which can be denoted as

$$\text{argmax}_i P(BN_i|F_i, L_i) = \text{argmax}_i P(BN_i)P(F_i| BN_i)P(L_i| BN_i) \tag{7}$$

where BN_i is the ith outline model, and F_i and L_i are the first and last primitives of the ith outline model. This is a unigram approach investigated in our previous work [13] that takes the knowledge of the first and last primitives as a selection factor.

For the second level filter, a model with the highest posterior probability is chosen as a correct representation for a set of input primitives and can be denoted as

$$\text{argmax}_i \ P(BN_i|P_1, P_2,...,P_n) = \text{argmax}_i \ P(BN_i)P(P_1,P_2,..P_n|BN_i) \qquad (8)$$

where BN_i is the ith outline model and $P_1, P_2,..P_n$ are a set of input primitives given by the recognition engine.

5 Experimental Results

We carried out a small experiment to test the improvement made by the Bayesian network and to compare the results with our previous "Approximate Pattern Matching" approach [14]. The test is evaluated from the aspect of genuine handwriting and the experimental results are shown in Table 1 by comparing the two approaches.

In order to test the transcription performance depending on different writers, we collected 423 Pitman shorthand outlines written by three shorthand writers. A WACOM ARTZII digitizing tablet (active area 12 inches x 12 inches; coordinate range 30480 x 30480; resolution 1mm; sampling rate set at 200 points per second) was used. The outlines include the whole range of Pitman primitives and are composed of nine similar sentences with 47 words. 60% of the data is used for training and 40% is used for testing purposes.

The transcription performance of the two approaches is shown in Table 1 and the accuracy rate is based on number of correctly recognised words. For example, if the word "fee" is recognised as the word "feature", then the recognition accuracy is 0%.

As shown in Table 1, the best rate achieved by the Bayesian network approach is 92.86% with the 7.14% error rate mainly due to inconsistent writing i.e. outlines which are legible to human readers, but are not consistent with the writing rules of Pitman shorthand. In order to test the transcription of an outline with missing vowel components, we randomly omitted vowel notations from the test data and evaluated the output. The test showed that a candidate list produced an intended word, however the ambiguity of the list rises by 15% on average due to the omission of vowel components and it is a remit of our collocation analyser to choose the most relevant word in a sentence level transcription.

Table 1. Experimental results of Bayesian network approach and Pattern Matching approach

Average Transcription Accuracy of an intended word in a candidate list	Bayesian Network based approach with the use of primitive attributes (New approach)	Pattern Matching based approach with the use of phonetic attributes (Our Previous approach)
Overall	92.86%	84%
In the presence of vowel omission or confusion	100%	0%
In the presence of inconsistent writing	0%	0%
In the presence of classification error	100%	100%

6 Conclusion

A novel approach on Bayesian network representation for the transcription of handwritten Pitman shorthand is described. The use of primitive attributes prevents unnecessary ambiguity of an early transcription stage and experimental results show that the approach is highly efficient in recognizing genuine handwritten Pitman shorthand. The current approach is tested with a small amount of outline models (47). A large amount of training data is required for the real time transcription of handwritten Pitman shorthand with an average vocabulary of 20k words. The approach is promising, and improvements on the training algorithm can prevent exponential time complexity. Our further work is focused on the implementation of a training algorithm to represent a large lexicon of any domain with the use of a small training data set. This includes further data collection and extensive testing to analyse the performance of the system in any domain.

References

1. Toshiyuki M., 'POBox: An efficient text input method for handheld and ubiquitous computers', Proc. of the ACM Conference on Human Factors in Computing System (CHI'98), Los Angeles, USA, pp. 328-335, April 1998,
2. Perlin K., 'Quick writing: Continuous stylus-based text entry', Proc. Of the ACM Symposium on User Interface Software and Technology (UIST'98), Santafe, NM, pp. 215-216. November 1998
3. Leedham C.G., Downton A.C., 'Automatic recognition of Transcription of Pitman's Handwritten shorthand', In Plamondon R. and Leedham C.G. (Eds), Computer Processing of Handwriting, pp.235-269, World Scientific, 1990
4. Y. Qiao and C.G. Leedham, 'Segmentation and recognition of handwritten Pitman shorthand outlines using an interactive heuristic search', Pattern Recognition, 1993, vol.26, No.3, pp.433-441
5. M. Zhu, Z. Chi, and X.P. Wang, 'Segmentation and recognition of on-line Pitman shorthand outlines using neural network', Proc. International Conference on Neural Information Processing, Singapore, 18-22 Nov. 2002, vol. 5, pp.2454-2458.
6. Ma Yang, Graham Leedham, Colin Higgins & Swe Myo Htwe, 'Segmentation and recognition of phonetic features in handwritten Pitman shorthand', submitted to Pattern Recognition, August 2004, accepted for publication December 2004.
7. Ma Yang, Graham Leedham, Colin Higgins, & Swe Myo Htwe, 'Segmentation and recognition of vocalized outlines in Pitman shorthand', Proceedings of the 17th International Conference on Pattern Recognition, Vol. I, ISBN 0-7695-2128-2, pp. 441-444, Cambridge, UK, 23-26 August 2004.
8. P.Nagabhushan and Basavaraj.Anami, 'A knowledge-based approach for recognition of handwritten Pitman shorthand language strokes', Sadhana, Journal of Indian Academy of Sciences, Vol. 27, Part 5, pp. 685-698, December 2002
9. P.Nagabhushan and Basavaraj.Anami, 'Dictionary Supported Generation of English Text from Pitman Shorthand Scripted Phonetic Text', Language engineering conference, Hyderabad, India, pp.33 December 13-15, 2002
10. J. Pearl, Probabilistic Reasoning in Intelligent Systems: Networks of Plausible Inference, Morgan Kaufmann Publishers, Inc., San Mateo, CA, 1988.

11. X. Xiao, G Leedham, 'Signature verification using a modified Bayesian Network', Pattern recognition, 2002, vol. 35, pp. 983-995
12. S J Cho, J H Kim, 'Bayesian network modelling of strokes and their relatioinships for on-line handwriting recognition', Pattern Recognition, 2004, vol. 37, pp 253-264
13. Swe Myo Htwe, Colin Higgins, Graham Leedham & Ma Yang, Post Processing of Handwriting Pitman's Shorthand using Unigram and Heuristic Approaches, Published in Lecture Notes in Computer Science: Document Analysis Systems VI, 3163, Springer-Verlag, pp. 332-336
14. Swe Myo Htwe, Colin Higgins, Graham Leedham & Ma Yang, Evaluation of Feature Sets in the Post Processing of Handwritten Pitman's Shorthand, Proceedings of the 9th International Workshop on Frontiers in Handwriting Recognition, ISBN 0-7695-2187-8, pp. 359-364, Kokubunji, Tokyo, Japan, 26-29 October 2004.

Ancient Printed Documents Indexation: A New Approach

Nicholas Journet[1], Rémy Mullot[1], Jean-Yves Ramel[2], and Veronique Eglin[3]

[1] L3I, 17042 La Rochelle Cedex 1 - France
{njournet,rmullot}@univ-lr.fr
[2] LI, 64 Avenue Jean Portalis 37200 TOURS - France
Jean-Yves.Ramel@univ-tours.fr
[3] LIRIS INSA de Lyon, Villeurbanne Cedex - France
eglin@rfv.insa-lyon.fr

Abstract. Based on the study of the specificity of historical printed books and on the main error sources of classical methods of page layout analysis, this paper presents a new way to achieve an indexation of ancient printed documents. We have developed an approach based on the extraction and the quantification of the various orientations that are present in printed document images. The documents are initially splitted into homogenous areas in which we analyze significant orientations with a directional rose. Each kind of information (textual or graphical) is typically identified and labelled according to its orientation distribution. This choice of characterization allows us to separate textual regions from graphical ones by minimizing the a priori knowledge. The evaluation of our proposition lies on a document image retrieval using layout extraction criteria and can also be used to precisely localize graphical parts in various types of documents. The system has been tested with success over several ancient printed books of the Renaissance.

1 Introduction

In this paper, we present a work corresponding to a collaboration between three research laboratories dealing with document image analysis and the "Centre d'Etude Superieur de la Renaissance" of Tours. The CESR is a training and research centre which receives students and researchers wanting to work on all the different domains of the Renaissance using a rich library of historical books. The CESR wants to create a Humanistic Virtual Library but, until now, only bitmap versions of historical books that have been scanned or photographed are accessible. So, since a few years, the center is trying to build a more powerful system to index and diffuse their collections through the web.

In this context, we present first a study of historical book specificities in order to infer some invariant characteristics used during the automatic analysis of their layout. Then, we describe the classical extraction methods that are usually applied on such documents by focussing on their drawbacks.

In a second part, we present a new approach of ancient printed documents layout characterization that is robust to noise (in the background but also in the printed text or the graphical areas) and completely independent of the employed typography, the

S. Singh et al. (Eds.): ICAPR 2005, LNCS 3686, pp. 580–589, 2005.
© Springer-Verlag Berlin Heidelberg 2005

characters size, the presence of graphical parts and of any particular editorial chief. It means that interest regions can be localized everywhere in the page with a total typography free consideration. Finally, we show how it is possible to separate text from graphics without any a priori knowledge on the nature of the books and of the typographical tools. We only need some strong hypotheses of text alignment and regularity. In the last part of the paper, we present a synthetic document image retrieval that illustrates the relevance of our layout analysis method.

2 Context of the Study

2.1 Characteristics of the Study

The historians have accumulated many European ancient printed document collections. Those books come from different countries (France, Germany, Italy, Holland,...) and different centuries (from the middle of the 15th to the end of the 17th). Books from the 16th century are characterized by a wide variety of forms and contents, even if it was the beginning of the rationalization and codification of texts, of typography that provided an improved reading comfort.

At the beginning of printing, the fonts and the layouts of the pages were very close to the handwritten books [1]. Later, these printed documents were handcrafted and the technical constraints of the past reduced the regularity of book production (variations in spacing and margins, random alignment, etc.). These documents contain many defects due to the manufacturing process and the conditions in which these books were conserved. The variability of page layouts is due to either technical inaccuracies or liberties taken by the printer. There are no exact rules, but most of the time a body text part covers the majority of the page area with generally some notes in the margins. The page can also contain graphical parts of various sizes and some ornament patterns. In the text, we can find known structures like the titles and the subtitles, the paragraphs, the page numbers, and other more particular structures like the Catchwords. The styles used can alternate, with normal style, justified or aligned on the left. Another characteristic of old printed books comes from weak separations between blocks of text (notes in the margins and body text for example). Lastly, we can notice that on some documents layout rules are always not respected. For example, an illustration can overflow into the margins. To adjust all the lines on a page, the printer could vary the line spacing and the margins. This lack of regularity makes automatic layout analysis difficult. On the other hand, such century books are printed using only strokes and line-art graphics, which can be more easily segmented. Moreover, from one country (or century) to another the printed techniques and layout edition rules were quite different. That is why our corpus presents great variability in pages layout that does not exist in contemporary books.

2.2 Evaluation of the Existing Methods on Ancient Books

The quality of new segmentation methods has significantly increases those last ten years. Consequently, the software that are devoted to contemporain documents recognition are often unsuitable to the processing of books of the Renaissance period even if we corrected the skew and the curvature due to the book binding using Book-restorer software for example.

The methods of structure extraction employed by software dealing with contemporary books can be classified in 3 main categories: bottom-up, top-down and mixed methods [2].

Typical segmentation and information retrieval approaches lie on usual features analysis that are generally based on a connected components analysis or morphological and directional filtering as it is presented in other existing works [3, 4]. These traditional bottom-up methods are not adapted to the historical books with all their specificities presented in section 2.1 (especially non constant spaces between shapes). Thus, the great difficulty of their use lies on the introduction of many parameters that often lead to prohibitive processing times.

In the same way, the top-down and mixed methods [5,6,7] (horizontal and vertical projection, multi-resolution analysis) endures the same weakness: they need to much a priori knowledge about the documents to be effective (number of columns, width of margins, kind and place of ornamental letters…). So, whatever the method is, our tests have shown that these methods are not robust to the variability of our corpus.

That's why, we have decided to use texture-based approaches. Gabor filters, autocorrelation function, fractal or wavelet analysis are interesting methods because they allow a text/graphic separation without using any kind of structural information. In that context, we have finally chosen to use the autocorrelation function that allows us to characterize large area of text and graphics despite digitalization defaults, text skewness and other kinds of ancient document noises (ink dots, background spots…).

2.3   Overview of Our Approach

The aim of our system is to realize a robust indexation system that is adapted to large heterogeneous collections of ancient books. It must be adapted to an end-user, non-specialist in document or image analysis. Thus, no threshold, document model and explicit structure have to be taken into account in the user interface. Due to this high level of constraints, we reduce the indexation process to help the user to build its own indexation. This one is finally based on both his own expertise and on the results of image analysis process. This process has been divided into three parts:

- Unsupervised automatic extraction of homogenous areas in the images using only orientation features (by labelling pixels).
- Classification of the different page layouts for one book using the position and the labels of the different extracted areas. This classification should also underscore different book layout styles.
- Interactive process allowing precise segmentation and semantic labelling of historical book elements.

This paper describes mainly the first stage of the automatic process. This first stage is a labelling process that organizes pages into different large classes of areas: background, text and non-text. This stage does not need any a priori information (except that text is horizontal).

After being resized, a simple separation of foreground and background based on homogeneous high grey level pixels estimation is applied. It can not be considered as a document binarization but it must be seen as a page separation into two main classes. Then, the image is crossed by a window that extracts specific orientations information and marks each foreground pixel in two main classes: the text and the graphics classes.

3 Page Layout Characterization

Our approach lies on the estimation of relevant directions of significant parts of the image: the initial image is then splitted into homogenous regions in which we analyze significant orientations with a directional rose that has been initially proposed by Bres[8].

3.1 Directional Rose Computation

The directional rose computation lies on the use of the autocorrelation function, which correlates the image with itself, highlights periodicities and orientations of texture. This function has been widely used in a context of texture characterization, [9]. Its definition for a bi-dimensional signal is the following.

$$C_{xx}(k,l) = \sum_{k'=-\infty}^{+\infty} \sum_{l'=-\infty}^{+\infty} x(k',l').x(k'+k,l'+l) \ . \tag{1}$$

The autocorrelation function $C_{xx}(i,j)$, applied to an image I, combines this image I with itself after a translation of vector (i,j). The different translations that are considered by the function give information on the different privileged directions in the image. With this principle, it is possible to detect orientations of the texture in the different parts of the image. For example, the translation of a line in the same direction leads to a great correspondence and is expressed by a great value of autocorrelation in the line direction. Inversely, in the orthogonal direction of this line the resulting value will be low. The autocorrelation underlines the objects' overlapping that is obtained by translation. This principle can be generalized to a set of objects having a common direction: in our work, we use it to show that text lines can be characterized by a horizontal privileged direction and can also be considered with a possible skew variation. The determination of relevant orientation is based on the application of successive gliding masks in the original image in which we compute an autocorrelation function that reveals periodicities and orientations in image. The principle of the autocorrelation computation lies on the frequencies decomposition of the analyzed image with a Fourier Transform (FFT) which avoids the highly complex development of the correlation. The Plancherel Theorem [10] is at the basis of this simplification.

The autocorrelation result can be analyzed by the construction of a directional rose that reveals significant directions in the analyzed block image. The rose computation is based on the mean value that is computed from the autocorrelation result. Let's consider I' the block of the image and $\{(x,y)\}$ the set of coordinates in this image. We also consider θ as privileged direction in the area. The mean value E_θ is then defined by the following formula:

$$E_\theta = \left\{ I'(x, y).I'(x + a, y + b) \right\} \text{ with } Arctg(b/a) = \theta . \tag{2}$$

$$R(\theta_i) = \sum_{D_i} C_{xx}(a,b) . \tag{3}$$

The directional rose represents the sum $R(\theta_i)$ of different values $C_{xx}(i, j)$ (defined in formula 1) in a given θ_i direction. So, the directional rose corresponds to the polar diagram where each direction θ_i that is supported by the D_i line, is represented by the sum $R(\theta_i)$. For all points (a,b) of the D_i line we have the following relation:
From this set of values, we only keep relative variations of all contributions of each direction. So, the relative sum $R'(\theta_i)$ is the following :

$$R'(\theta_i) = \frac{R(\theta_i) - R_{min}}{R_{max} - R_{min}} . \tag{4}$$

Figure 1 shows some examples of directional roses corresponding to 3 different initial images.

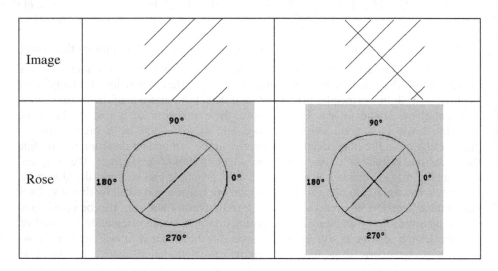

Fig. 1. Directional roses

3.2 Directional Rose Analysis

Due to its mathematical definition, the rose has interesting properties. In a situation of homogenous directions' repartition of the pixels in an image, the rose is a perfect bowl (fig 2 part a). If we add a horizontal line (fig 2 part b), the fact that we only keep relatives variations (formula 4) implies that the rose has two invariant characteristics: one "peak" for the horizontal orientations (0° and 180°) and the same intensity for all the other directions (it gives an impression of "bowl").

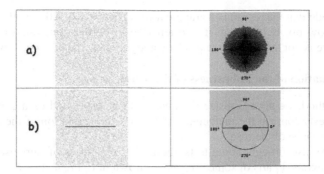

Fig. 2. Text characterisation

A homogenous text area has the same property: all directions are perceptible with a great regularity (that is systematically quantified by a variance measure) and the most important are the horizontal direction that is characterized by a significant characteristic peak with an invariable maximal amplitude (figure 3a. If the text is not homogenous: different sizes of text, non constant spaces....) in the studied area, the formula 4 implies that the size of the "bowl" can decrease or even disappear (figure 3b).

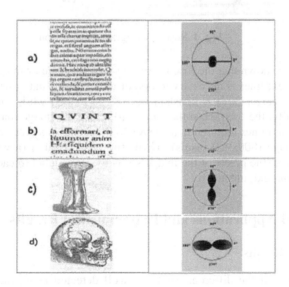

Fig. 3. Graphics characterization

The detection of graphical parts lies on this two following observations: if there are no horizontal significant directions, then we assume that the region contained in the analysed window is a graphical part with a great confidence rate (figure 3c). In the same way, if the main direction is horizontal without a characteristic "peak", then we can conclude that the region is a graphical part (figure 3d). All the other kinds of roses do not allow pixels labelling: it is often due to border zone analysis or to an area where there is not text enough to produce a significant "peak" or a "bowl".

3.3 Evaluation of the Robustness of the Rose

In our method, we always resize the images to be processed on a constant sized image. Our tests confirm the mathematical theory: the resolution of the image does not change the shape of the rose.

We have also brought some tests about the robustness of our rose analysis algorithm on degraded parts of some images as shown in figure 4.

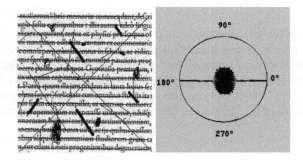

Fig. 4. Rose of a degraded text region

Our tests have shown that the rose is not sensitive to noise or image distortion but a minimum of 4 text lines is necessary to detect a text area. That is why in our method, the size of the analysis area is constant (128X128 pixels) while image sizes can be variable. Actually, the users have the choice to resize the image manually in order to have the minimum lines required in a 128X128 pixel analysis area or to use a constant size for the image (800x900 pixels). Then, the analysis area (a window) is moved iteratively all over the pixels of the image so as to give a label to each of them (background, text, non-text, unknown).

4 Targetted Applications and Experimental Results

4.1 Labelling Results

In most images, all the different parts are well detected using the directional rose analysis. The main problem comes from non constant transitions that exist between blocks and from very huge characters sizes in some titles. Except these two special cases, our method gives quite good results and allows us to realize our purpose: a relevant separation of the pages into pre-labelled regions that also highlights the document visual content with a robust extractor (with same results whatever are the typography, the fonts, the background noise and the resolution).

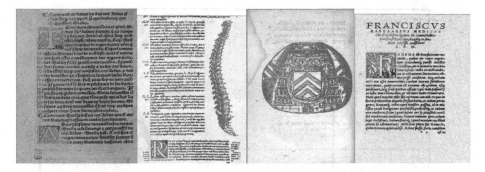

Fig. 5. Examples of results

In the following results, the colors of the pixels are organized as follow: for text and graphics parts we choose respectively blue and red colors. The figure 5 shows a short panel of typical results.

A more consequent test has been realized on over 100 pages from five different books of the Renaissance. The results provided by our algorithm have been compared with a perfect labelling according to a human (our ground truth). So, for text parts, the number of letters which were correctly, wrongly or miss detected has been manually enumerated. For graphical parts, the number of graphical pixels which are correctly, wrongly or not detected has also been manually enumerated. Detection results are summarized in table 1.

Table 1. Detection results (% of correct/wrong/missed detection parts)

Text parts	Graphical parts
95/3/2	88/10/2

4.2 Page Comparison

We think that the complex problem of ancient document images indexation can be simplified by a global analysis of all pages of a book before taking any kind of physical or logical conclusion. As Shin and Maderlechner in [11,12] we want to compute a page classification.

With this page layout classification, we hope to extract editorial rules, special text (with specific typography) or graphical features to adapt the following processing.

By using only the percentages of text and non text pixels obtained with our algorithm, we compute a similarity measure between images. Then, it is possible to find pages with similar layout by providing a request image (as in content based image retrieval). A simple distance is computed by the comparaison of the labelling of two images pixel by pixel using a Hamming distance (figure 6).

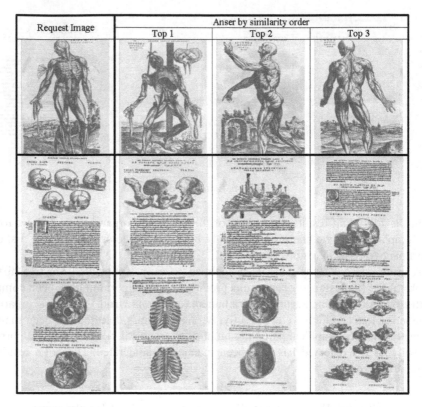

Fig. 6. Pages comparison using only percentages of text and non text pixels

4.3 Precise Page Layout Analysis

The other possible use of the labelling deals with the location of illustrations that must be as precise as possible (sharpness of contours and in the frontiers of graphical parts). Our approach provides valuable information about the possible position of the graphical parts as shown in figure 7.

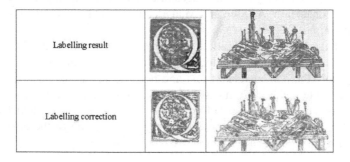

Fig .7. Graphical parts segmentation using our labeling

While being focused successively near all the red areas our experiments show that it is possible to merge this information with an algorithm of extraction of connected components. Thus, by analyzing the labels of pixels contained in all the selected connected components, it is possible to extract in a very fine way the graphical entities without any parameters about the size of the graphical elements in historical books.

5 Conclusion

In the first part of the article, we have highlighted the sources of errors of the traditional methods of page decomposition using a characterization of page layout in the historical books. Then we present an efficient method for pixel labelling adapted to ancient printed documents. The originality of our approach lies on the development of a new extraction and analysis tool which separates textual and graphical areas from different kinds of ancient document images without any knowledge like thresholds, models, or structure information. The resulting document labelling is employed as a new feature for a relevant pages comparison. Current results are very promising. The content based classification of an entire book is a direct perspective of this contribution.

Acknowledgements. To CESR-CNRS (France) for providing the images www.cesr.univ-tours.fr.

References

[1] HJ Martin, La naissance du livre moderne, Editions du Cercle de la Librairie, 2000.

[2] A. Belaid, Computer aided design of models of page for their use in recognition of documents, Workshop one Electronic Page Models, LAMPE' 97. 1997.

[3] L O'Gorman, The Document Spectrum for Page Analysis Layout, Trans IEEE One PAMI.15(11), 1993 P1162-1173.

[4] F Lebourgeois, H Emptoz, E trinh, Compression and accessibility with the images of digitized documents – Application to the Debora project, Numerical Document, Flight 7n°3-4, 2003 p103-127.

[5] Jie Xi, J Hu, L Wu, Page segmentation of chinese newspaper Pattern recognition 2002 2695-2704

[6] D. Malerba, F Esposito Oronzo, Adaptive Layout Analysis of document. Università degli Studi di via Bari, ismis 2002.

[7] P. Duygulu, V. Atalay A Hierarchical Representation of Form Documents for Identification and Retrieval, International Journal on Document Analysis and Recognition IJDAR 5 2002 1, 17-27.

[8] S Bres, Contributions à la quantification des critères de transparence et d'anisotropie par une approche globale. PhD Thesis, 1994.

[9] W.K. PRATT, Digital Image Processing, 2nd edition New-York : Wiley, 1991, p230.

[10] mathworld.wolfram.com/PlancherelsTheorem.html

[11] C. Shin, D. Doermann, Classification of document page images based on visual similarity of layout structures, Language and Media Processing Laboratory Center for Automation Research University of Maryland. 2000

[12] G. Maderlechner, P. Suda, T. Bruckner, classification of documents by form and content, Siemens AG, Corporate Research and Delelopment, Otto-Hahn-Ring 6, D-81730 Munchen, Germany

Applying Software Analysis Technology to Lightweight Semantic Markup of Document Text

Nadzeya Kiyavitskaya[1], Nicola Zeni[1], James R. Cordy[2], and Luisa Mich[1]
and John Mylopoulos[3]

[1] Dept. of Information and Communication Technology, University of Trento, Italy
{nadzeya.kiyavitskaya, nicola.zeni, luisa.mich}@dit.unitn.it
[2] ITC-IRST, Trento, Italy, and School of Computing, Queens University, Canada
cordy@cs.queensu.ca
[3] Dept. of Information and Communication Technology, University of Trento, Italy,
and Dept. of Computer Science, University of Toronto, Canada
jm@cs.toronto.edu

Abstract. Software analysis techniques, and in particular software "design recovery", have been highly successful at both technical and business-level semantic markup of large scale software systems written in a wide variety of programming languages, and in particular have proven efficient and scalable in assisting the resolution of the "year 2000" problem for billions of lines of legacy source code. In this work we describe a first experiment in applying the same technical solutions and tools that have proven so successful in software markup to the more general problem of semantic markup of text documents. In this early report we describe our adaptation of the software analysis techniques, propose a general domain-independent architecture for semantic markup using them, and demonstrate its feasibility in a limited but realistic domain of application by comparison with both raw and tool-assisted human semantic markers.

1   Introduction

Semantic markup [1] is the annotation of world-wide web or other natural language documents to assign explicit real-world semantics to portions of the document in order to allow for rapid identification of documents and parts of documents relevant to a particular question or purpose. Semantic markup represents the essential difference in the vision of the "semantic web" [2].

Given the number and scope of documents on the world-wide web, transition to the semantic web vision cannot be achieved without large-scale efficient automation of semantic markup [3]. It seems clear that full natural language understanding systems will not be ready for this task for some time, and thus lightweight, approximate methods may be our best hope for this immediate and pressing need.

S. Singh et al. (Eds.): ICAPR 2005, LNCS 3686, pp. 590–600, 2005.

2   Software Analysis and Design Recovery

Another domain in which such an immediate and pressing need for large scale analysis of source texts has been faced is legacy software source analysis, which successfully faced the "year 2000" problem only a few years ago. Some of the most successful techniques for automating solutions to that problem utilized "design recovery" [4], the analysis and markup of source code according to a semantic design theory, to assist in this problem [5].

Design recovery from source code poses many problems in common with natural language text processing: the need for robust parsing techniques, because real documents do not always match the grammars of the languages they are written in; the need to understand semantics of the source text according to a semantic theory or ontology; semantic clues drawn from a vocabulary for the domain; contextual clues drawn from the syntactic structure of the source text; and inferred semantics from exploring relationships between semantic entities and their properties, contexts and related entities.

Formal processes for software design recovery utilize a range of tools and techniques designed and proven to address these challenges for many billions of lines of legacy software source code [6]. One of these is the generalized parsing and structural transformation system TXL [7], the basis of the automated year 2000 system LS/2000 [5].

In this work we propose to leverage the highly efficient methods and tools already proven in the software analysis and markup domain as the basis of a new lightweight method for semantic analysis and markup of natural language texts, in the hope that we can attain similar performance and scalability while yielding good quality approximate results.

3   A New Architecture

The architecture of our solution (Fig. 1) is based on the LS/2000 software analysis architecture [5], generalized to allow for easy parameterization by a range of semantic domains. The architecture explicitly factors out reusable domain-independent knowledge such as the structure of basic entities (email and web addresses, monetary formats, date and time formats, and so on) and language structures (object, document, paragraph, sentence and phrase structure), shown on the left hand side, while allowing for easy change of semantic domain, characterized by vocabulary (category word and phrase lists and contra-lists) and ontology (entity-relationship schema and interpretation), shown on the right.

The process uses three phases. In the first stage, an approximate ambiguous context-free grammar is used to efficiently obtain an approximate phrase structure parse of the source text using the TXL parsing engine. Using robust parsing techniques borrowed from compiler technology [8], this stage results in a deterministic maximal parse even for badly malformed text. As part of this first stage, basic entities such as email addresses, web addresses, monetary amounts, dates, times and other word-equivalent objects are recognized grammatically as

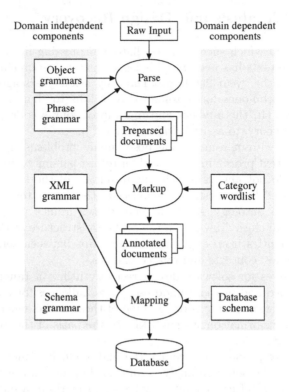

Fig. 1. Architecture of our semantic markup process

would be done in a programming language parser (Fig. 2). The parse is linear in the length of the input and runs at compiler speeds. In our first experiments this parse is relatively coarse-grained, ignoring language structure below the sentence and verb-clause level.

In the second stage, initial semantic markup of the document is derived using a wordlist file specifying both positive and negative indicators for semantic categories (Fig. 3). Indicators can be both literal words and phrases (e.g., "air conditioning") and names of parsed entities (e.g., "email"). Phrases are marked up once for each category they match - thus at this stage a sentence may end up with many different (even conflicting) semantic markups. Vocabulary word and entity lists are derived from the ontology for the target semantic domain. At present this is done by hand, but work is underway on automating vocabulary and schema construction from a formal ontology. This stage uses the structural pattern matching and source transformation capabilities of the TXL transformation engine in much the same way as it is used for software markup [9] to yield a preliminary marked-up text in XML form (Fig. 4).

The third stage uses the XML marked-up text to populate an entity-relationship database according to an ER schema approximating the ontology for the domain. The schema is provided as an XML template (Fig. 5) derived (at present

```
% International phone number grammar
tokens
      longnumber   "\d\d\d\d\d\d\d\d*"
      zeronumber   "0\d\d*"
end tokens

define phone
          '+ [anynumber][opt zerocode][opt phone_separator]
            [repeat number_separator+][anynumber]
      |   '+ [anynumber][opt zerocode][opt phone_separator][repeat longnumber+]
      |   '( [opt space][opt '+][anynumber][opt space]')[space_number]
            [repeat separator_number]
      |   '( [opt space][opt '+][anynumber][opt space]')[opt space]
            [repeat longnumber+]
      |   [anynumber] [opt phone_separator] [repeat longnumber+]
      |   [zeronumber][opt phone_separator][longnumber]
      |   [zeronumber] [separator_number] [repeat separator_number+]
      |   [repeat number_separator][longnumber]
      |   [number_separator][number_separator][repeat number_separator][anynumber]
      |   [opt '+][longnumber] [repeat space_number]
end define
```

Fig. 2. Part of TXL grammar for phone number objects

```
term : date
        [rented by] minimum maximum month months short long term terms
        holidays holiday days lets let period periods
    |   { money price }
```

Fig. 3. Prototype category wordlist description for the "term" concept. This wordlist specifies that a phrase or sentence may relate to the "term" concept if it contains a "date" object and/or one or more of the words and phrases listed, and does not contain any objects of the "money" category or the "price" concept.

```
3348 <type><location> Very elegant apartment located in Piazza Lante,
just a walk from Fosse Ardeatine and 10 minutes to Colosseum by bus
(Bus stop in the square) </location></type>. <facility> 75 smq in a
charming, and full furnished environment </facility>. <type><facility>
The apartment has a large and well-lit living room with sofa bed a dining
area, a large living kitchen with everything you need, a bathroom with tub,
a large double bedroom </facility></type>. <facility> TV, hi-fi and a
washing machine </facility>. <facility><price> 1.200 euro a month,
utilities not included </price></facility>. <contact> Write to
pseudonym@somewhere.it or phone to 347.7894321 </contact>
```

Fig. 4. Example result XML-marked up accommodation advertisement. Low-level objects such as email and phone numbers, while recognized and marked-up internally, are intentionally not part of the result since they are not in the target schema.

by hand) from the ontology for the target domain. Sentences and phrases with multiple markups are "cloned" using TXL source transformation to appear as multiple copies, one for each different markup, before populating the database. In this way we do not prejudice one interpretation as being preferred; rather

```
<ad>
    <location></location>
    <price></price>
    <contact></contact>
    <facility></facility>
    <term></term>
    <type></type>
</ad>
```

Fig. 5. Database template schema for accommodation advertisements

we assume that a single sentence or phrase may in fact be a reasonable answer for all of the semantic categories it is marked as. The result of this stage is an entity-relationship database in XML format suitable for importing into standard database tools such as MySQL or MS Access.

Both the XML marked-up text and the database are products of our process, the former yielding an approximate semantic markup of the original document text and the latter serving as a query answering service yielding relevant sentences and phrases from the document text in response to database queries.

4   A First Experiment

As a first proof-of-concept experiment in the application of our new method, we have been working in the domain of travel documents, and in particular with published advertisements for accommodation drawn from online newspapers. This domain is typical of the travel domain in general and poses many problems commonly found in other text markup problems; partial and malformed sentences, short-forms, location-dependent vocabulary, monetary, date and time conventions, and so on.

In order to make a realistic test of the generality of the method, we restricted ourselves to some constraints: no proper nouns or location-dependent phrases in our vocabulary, raw uncorrected text, and no formatting or structural cues. The human markers against whom we were testing could take full advantage of all of this knowledge in their results, but the tool could not.

In the first instance we used a set of several hundred advertisements for accommodation in Rome drawn from an online newspaper. The task was to identify and mark up several categories of semantic information in the advertisements according to a given accommodation ontology, which was reduced by hand to an entity-relationship schema in XML format for input to our system (Fig. 5). The desired result was a database with one instance of the schema for each advertisement in the input, and the marked-up original advertisements (Fig. 4).

5   Early Results

The evaluation of semantic markup presents a particularly difficult problem. Because human opinions on the "correct" markup can vary widely, ideally we should compare our automated results against a wide range of high quality human opinions. However, in practice the cost of the human work involved is prohibitive for all but the largest companies and projects.

In order to evaluate our initial experimental results, we designed a modest but cost-effective three stage validation. At each stage, we were interested in measuring the precision, recall, fallout, accuracy and error (using the definitions of Yang [10]) for the tool's automated markup compared to human opinions. The tool was allowed 38 randomly chosen advertisements as a "training set", although no real training took place - rather, after encoding the target ontology into the tool's vocabulary and schema tables, the tables were allowed to be tuned to do well on this first set by hand.

In the first stage, the tool and each of two human markers were asked to mark up a sample set of ten advertisements different from the training set used to tune the tool for the domain. The tool was then compared against each of the human markers for this set separately (Fig. 6a), and then calibrated against each of the two as definitive (Fig. 6b,c). By comparison with these (widely differing) two human annotators, the system exhibited a high level of recall (about 92% compared to either human, higher than either human compared to the other), but a lower level of precision (about 75% compared to either human, whereas they each exhibit about 89% compared to the other). However, the system was able to show a 92% accuracy rating compared to either human, extremely high for such a simple system.

Measure	System vs. Human 2	System vs. Human 1
Recall	92.42%	92.19%
Precision	76.25%	73.75%
Fallout	7.54%	8.27%
Accuracy	92.45%	91.82%
Error	7.55%	8.18%
F-Measure	83.56%	81.94%

(a) System vs. Humans

Measure	Human 2 vs. Human 1	System vs. Human 1
Recall	90.63%	92.19%
Precision	87.88%	73.75%
Fallout	3.15%	8.27%
Accuracy	95.60%	91.82%
Error	4.40%	8.18%
F-Measure	89.23%	81.94%

Measure	Human 1 vs. Human 2	System vs. Human 2
Recall	87.88%	92.42%
Precision	90.63%	76.25%
Fallout	2.38%	7.54%
Accuracy	95.60%	92.45%
Error	4.40%	7.55%
F-Measure	89.23%	83.56%

(b) Calibrated against Human 1   (c) Calibrated against Human 2

Fig. 6. First stage experiment - system vs. unassisted human markup

Measure	Rome (10 ads) Training set	Rome (10 ads) Test set-1	Rome (100 ads) Test set-1	Venice (10 ads) Test set-2
Recall	98.73%	94.20%	92.31%	86.08%
Precision	97.50%	97.01%	96.93%	95.77%
Fallout	0.84%	0.87%	0.93%	1.29%
Accuracy	99.06%	98.00%	97.43%	95.51%
Error	0.94%	2.00%	2.57%	4.49%
F-Measure	98.11%	95.59%	94.56%	90.67%

Fig. 7. Third stage experiment - system vs. assisted human opinions

In the second stage evaluation, we were interested in measuring the effect of the initial automated markup of the tool on human markup efficiency. The time taken by an unassisted human marker to semantically annotate a new sample of 100 advertisements was measured, and compared to the time taken by the same human marker when asked to correct the automated markup created by the tool. In this first evaluation he human marker was observed to use 78% less time to mark up text with assistance than without, a significant saving. Because the system was shown in the first evaluation to be more aggressive than humans in markup, the majority of the correction work was removing markup inserted by the tool. With an appropriate interface for doing this easily, the time savings could be even greater than we observed.

In the third stage, we gave the human annotators the advantage of correcting automatically marked up text from the tool to create their markups, and compared the final human markup to the original opinion of the tool. For this test, three sets of documents were used in addition to the original training set, one new set of 10 advertisements from the same online newspaper, another set of 100 from Rome, and a new set of 10 from Venice. The summary of results so far is shown in Figure 7. Accuracy for all of the Rome sets is about 98%, and in the new set from Venice, a completely different location, the accuracy was measured as over 95% with similar precision. A drop in recall to 80% is indicative of locality effects from the original training set - a wider set will be needed to make a general tool for advertisements.

Obviously these tests on this one small domain are insufficient to make any meaningful statements about the generality or applicability of our new architecture and method, and we are presently moving on to large scale tests in both this and other domains, including travel websites, news stories and academic publications in the coming months.

At best what we can say so far is that the results of our small study do validate that there is a real potential for a fast lightweight method based on the software design recovery model. Even without local knowledge and using a very small vocabulary, we have been able to demonstrate accuracy comparable to the best heavyweight methods, albeit thus far on a very limited domain. Performance of our as yet untuned experimental tool is also already very fast, handling for example 100 advertisements in about 1 second on a 1 GHz PC. Performance has also been validated as linear on sets ranging from 38 to 7,600 advertisements (about 2,500 to 500,000 words), at a rate of about 53 kb/sec on a 1 GHz PC.

6   Related Work

Many systems have been shown to do well for various kinds of assisted or semi-automated semantic tagging of large corpora. SHOE [11] and Ontobroker [12] for example are pioneering tools providing machine assistance for manual semantic markup, and AeroDAML [13] automatically generates DAML annotation suggestions for Web pages given an ontology.

Recent work includes fully automated techniques more directly comparable to ours. SemTag [3] for example has been able to process enormous amounts of data, reporting accuracy measures of about 79% in identifying instances of a given set of known entities in web pages. Using compiler-style tokenization of source text followed by a search for entities of the very large TAP taxonomy, SemTag expends much of its effort in disambiguating multiple tags using local context. By contrast our system aims primarily at higher level markup, and tries to minimize ambiguity using a combination of structural information from the parse and contra-indicators in the vocabulary.

The KIM (Knowledge and Information Management) platform [14] is designed for the purpose of implementing the full semantic web vision. Compiler-style tokenization begins the process, followed by a split into sentences and part-of-speech tagging. A gazetteer and ontology-augmented pattern-matching grammars encode rules for markup of a large set of entities of general in interest. In our system phrase structure is identified by the parse, and rules are driven by simple word occurrences.

S-CREAM (Semi-automatic Creation of Metadata) [15] uses a framework that includes a learnable information extraction component. Users hand-annotate a corpus for training the learner, which infers markup rules for a subset of a given ontology. S-CREAM utlilizes the same front end and basic set of steps as KIM, its distinguishing feature being its automated inference of annotation rules from the training set.

Our work differs from all of these approaches in three fundamental ways - first, it uses an extremely lightweight but robust context-free parse in place of tokenization, regular expressions and part-of-speech recognition. Second, it does not use a gazetteer or knowledge base of known proper entities, rather it infers their existence from their structural and vocabulary context, in the style of software analyzers. And third, it has already been shown to handle higher-level semantic markup for concepts above and depending on entities rather than just the entities themselves.

Information extraction [16] is a closely related problem to semantic markup with an even larger base of published work. Rather than markup of the documents themselves, the goal of information extraction is the population of a template with slots for information to be extracted from semi-structured documents such as web pages. This corresponds to the third step in our process (population of the database schema after semantic markup).

In general, techniques used for information extraction use patterns based on syntactic and semantic constraints in some ways similar to our initial phrase and object parsing stage. While much of the work in the information extraction community is aimed at "rule learning", automating the creation of extraction patterns from previously tagged or semi-structured documents [17] and unsupervised extraction [18], issues our work does not address, the actual application of the patterns to documents is in many ways similar to our method.

In particular, ontology-based methods such as Embley et al's [19] are in some ways quite similar, using a relational schema as the target structure where we use an entity-relationship schema, and using keyword lists and constraints quite similar to our own vocabularies. The major differences lie in the implementation - whereas Embley's method relies primarily on regular expressions, ours combines high-speed context-free robust parsing combined with simple word search.

Wrapper induction methods such as Stalker [20] and BWI [21] which try to infer patterns for marking the start and end points of fields to extract, also relate well to our work. When the learning stage is over and these methods are applied, their effect is quite similar to our results, identifying complete phrases related to the target concepts. However, our results are achieved in a fundamentally different way - by predicting start and end points using phrase parsing in advance rather than phrase induction afterwards.

7   Conclusions and Future Work

Obviously this work is only beginning. At most we have thus far demonstrated that applying software design recovery techniques to semantic markup of documents is feasible and has potential. It is also clear that these techniques can retain their efficiency in this new domain, exhibiting very fast linear performance even without tuning, and it seems likely that they could provide high levels of markup accuracy.

However, the work set out for us now is clear - testing and validation of our method on large corpora and richer conceptual spaces so that a more meaningful comparison with the state of the art can be done. While our method has done well for our small but realistic first domain of application, it is by no means clear that it will retain such high levels of accuracy as we scale to larger and richer domains.

There are still a number of techniques used in software analysis that we have not taken advantage of - alias resolution, unique naming, architecture patterns, markup refinement and so on. In future we hope to explore these other techniques to improve our semantic annotation architecture as well.

References

1. Daconta, L., Orbst, L., Smith, K.: The Semantic Web: A guide to the future of XML, web services and knowledge management (2003)
2. Berners-Lee, T., Hendler, J., Lassila, O.: The Semantic Web. Scientific American **284** (2001) 34–43
3. Dill, S., Eiron, N., Gibson, D., Gruhl, D., Guha, R., Jhingran, A., Kanungo, T., McCurley, K., Rajagopalan, S., Tomkins, A., Tomlin, J., Zien, J.: A case for automated large-scale semantic annotation. J. Web Semantics **1** (2003) 115–132
4. Biggerstaff, T.: Design recovery for maintenance and reuse. IEEE Computer **22** (1989) 36–49
5. Dean, T., Cordy, J., Schneider, K., Malton, A.: Experience using design recovery techniques to transform legacy systems. In: Proc. 17th Int. Conference on Software Maintenance. (2001) 622–631
6. Cordy, J., Dean, T., Malton, A., Schneider, K.: Source transformation in software engineering using the TXL transformation system. J. Information and Software Technology **44** (2002) 827–837
7. Cordy, J.: TXL – a language for programming language tools and applications. Proc. 4th Int. Workshop on Language Descriptions, Tools and Applications, Electronic Notes in Theoretical Computer Science **110** (2004) 3–31
8. Dean, T., Cordy, J., Malton, A., Schneider, K.: Agile parsing in TXL. J. Automated Software Engineering **10** (2003) 311–336
9. Cordy, J., Schneider, K., Dean, T., Malton, A.: HSML: Design-directed source code hotspots. In: Proc. 9th International Workshop on Program Comprehension. (2001) 145–154
10. Yang, Y.: An evaluation of statistical approaches to text categorization. J. Information Retrieval **1** (1999) 67–88
11. Sean, L., Lee, S., Rager, D., Handler, J.: Ontology-based web agents. In: Proc. 1st International Conference on Autonomous Agents. (1997) 59–68
12. Decker, S., Erdmann, M., Fensel, D., Studer, R.: Ontobroker: Ontology-based access to distributed and semi-structured information. In: Proc. 8th Working Conference on Database Semantics. (1999) 351–369
13. Kogut, P., Holmes, W.: AeroDAML: Applying information extraction to generate DAML annotations from web pages. In: Proc. KCAP-2001 Workshop on Knowledge Markup and Semantic Annotation. (2001)
14. Popov, B., Kiryakov, A., Ognyanoff, D., Manov, D., Kirilov, A.: KIM: a semantic platform for information extaction and retrieval. J. Web Semantics **10** (2004) 375–392
15. Handschuh, S., Staab, S., Ciravegna, F.: S-CREAM: Semi-automatic CREAtion of Metadata. In: Proc. 13th Int. Conference on Knowledge Engineering and Management. (2002) 358–372
16. Muslea, I.: Extraction patterns for information extraction tasks: A survey. In: Proc. AAAI-99 Workshop on Machine Learning for Information Extraction. (1999) 1–6
17. Nobata, C., Sekine, S.: Towards automatic acquisition of patterns for information extraction. In: Proc. International Conference on Computer Processing of Oriental Languages. (1999)
18. Etzioni, O., Cafarella, M.J., Downey, D., Popescu, A.M., Shaked, T., Soderland, S., Weld, D.S., Yates, A.: Unsupervised named-entity extraction from the web: An experimental study. Artificial Intelligence **165** (2005) 91–134

19. Wessman, A., Liddle, S.W., Embley, D.W.: A generalized framework for an ontology-based data-extraction system. In: Proc. 4th Int. Conference on Information Systems Technology and its Applications. (2005) 239–253
20. Muslea, I., Minton, S., Knoblock, C.A.: Active learning with strong and weak views: A case study on wrapper induction. In: Proc. 18th Int. Joint Conference on Artificial Intelligence. (2003) 415–420
21. Freitag, D., Kushmerick, N.: Boosted wrapper induction. In: Proc. 17th National Conference on Artificial Intelligence. (2000) 577–583

Noisy Digit Classification with Multiple Specialist

Andoni Cortes, Fernando Boto, and Clemente Rodriguez

Computer Architecture and Technology Department, Computer Science Faculty,
UPV/EHU, Aptdo. 649, 20080 San Sebastian, Spain
acbbosaf@si.ehu.es

Abstract. A multi-classifier formed by specialised classifiers for noise produced by an image is shown in this work. A study has been carried out in the
case of structure noisy images. Classifiers based on neighbourhood criteria are
used in this work, the zoning global feature and the Euclidean distance too. The
experiments have been carried out with images of typewritten digits, taken from
forms of the Bank of Spain. Trying to obtain a strong database to support the
experiments, we have added noise to the images of the digits. The recognition
rate improves from 64.58% to 96.18%.

1 Introduction

Human intervention in full scale digitization of documents is tedious because of the
large amount of documents to be processed. Nowadays there are some recognition
systems in the market (OCR) for typewritten texts but still they create many problems. The document digitization process is usually made starting with the isolated
characters and sometimes this isolation process produces some image disturbances.
The noise characters can be obtained through a bad quality of digitization or as a
result of a bad segmentation [1] [2] [3].

Many authors use systems based on the combination of classifiers. These systems
have different aims [4]: Efficiency [5] [6], improved performance [7] [8] [9], generalisation [10]. Besides, [4] [11] makes a survey of some of the possibilities to combine classifiers and the rules to combine them. Another classifier combination strategies are given in [12] [13]. In our case, we need a multi-classifier in order to combine
classifiers with a different purpose. Each classifier will specialise in a type of problem
or distortion and together, by means of a decision rule, will provide a result by common consent.

Another point discussed in this item makes reference to the disturbances produced
by digitization or segmentation defects, which will result in noisy or blurred numbers,
with thickness defects or cuts with loss of structure. Each one of these disturbances
has been solved independently, that is with sub-systems which provide a good rate of
success with some disturbances but with different results before other types of disturbances.

The basic aim of the proposed model of this work is to obtain a multi-classifier
formed by specialised classifiers for each type of noise produced by an image. Each
classifier gives a decision, depending on the applied process adequate for the type of
noise it is treating and which, supposedly, has the image. The decisions adopted by

S. Singh et al. (Eds.): ICAPR 2005, LNCS 3686, pp. 601–608, 2005.
© Springer-Verlag Berlin Heidelberg 2005

each classifier will be treated jointly in a global discriminating function. Should an individual classifier treat images of patterns outside its specialisation, normally the decision of that classifier will not be very reliable and the global discriminating function will disregard the classifier for these patterns.

The paper has the following structure. Section 2 describes the model, section 3 shows types of disturbances and section 4 a study of the model proposed with noisy digits.

2 Specification of the Model

This is a general description of a Multi-classifier (Multi-stage and Multi-specialist) model [10]:

- $MC(s, \{S_0 \ldots\ldots S_S\})$

    where s is the number of stages of the multi-classifier and $\{S_0 \ldots S_s\}$ the stages for the multi-classifier

    A stage is defined as:

- $S_i(nc_i, \{C_k \ldots\ldots, C_l\} f_i) \ 0 \le i \le s$

    where nc_i is the number of channels, $\{C_k \ldots\ldots C_l\}$ is the set of channels of that stage and f_i the decision function of stage i.

    A channel is defined as:

- $C_k(ip_k, fr_k, N_k)$

    Where ip_k is the inverse transformation applied to that channel, fr_k is the type of feature, and N_k the learned set (net) used in channel k. One or several channel can be repeated in different stages. If channel C_k is performed in S_i and S_j for $j > i$, the classification result for C_k is the same in both stages.

Each classifier or channel k returns a confidence value and its output classes ($Class_{k1}$ and $Class_{k2}$) of the two nearest patterns (2-NN classifier [14] [15]), with its corresponding distances, D_{k1} and D_{k2}, given that $Class_{k1} \neq Class_{k2}$. The distance between to patterns is defined as: given a learned pattern belonging to class $Class_k$ (R_1, R_2, $\ldots\ldots R_D$) within a dimensional space D and an entry or test pattern also belonging to class $Class_k$ ($P_1, P_2, \ldots\ldots P_D$), the Euclidean distance is defined as $\sqrt{\sum_{i=0}^{D}(R_i - P_i)^2}$.

The confidence value of each classifier is defined as $V_k = 1 - \dfrac{D_{k1}}{D_{k2}}$.

In all the experiments described in section 4 only multi-specialist systems are used (figure 1), so the channels in the multi-classifiers perform in parallel (the number of stages is one, s = 1). The feature used (fr_k) is the *zoning* global one [5] with 8x5 dimensionality. A decision function determines which classifier provides the classification result as a combination rule. The decision function (F) is the same for all the experiments in this work: decided class $Class_{k1}$ of k channel, with the best confidence value (V_k).

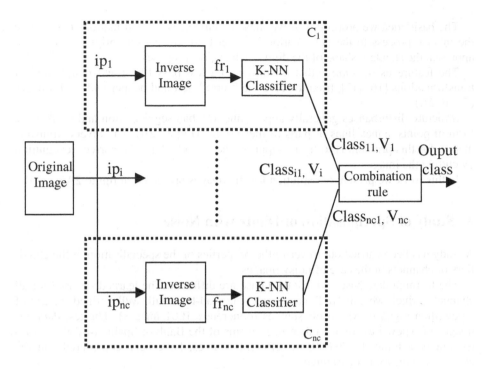

Fig. 1. Multi-specialist classifier description

3   Types of Disturbances

In digitization systems, due to different reasons, we often find that the obtained digit has some structural deformation, added information or some type of disturbance which makes recognition difficult and even annuls it (figure 2).

Fig. 2. Some examples of image disturbances: Blurred, salt and pepper, thick and annex lines

The usual recognition systems have high levels of success with well conformed digits, but the problem arises when recognising badly segmented digits or digits with a bad digitization, even weakened by defective writing instruments (matrix printers generate sources difficult to be recognised).

All these disturbances can appear, either alone or combined with the other, which increases the complexity of the problem because to obtain an adequate inverse disturbance to make the used feature more reliable and discriminating becomes really difficult.

The basic idea we propose to solve these disturbances or deformations is to create the inverse process to the deformation in order to return it to its original state or to approach the standard shape of the digit, such as [10] proposes.

The feature used, *zoning* (that is similar to a low-pass filter) and morphological transformations [16] [17], tried to solve the blurred, salt and pepper and thick images (figure 2A).

Structure disturbances generally appear due to a bad segmentation and it takes the form of points, annex lines or parts of these digits (Figure 2B). The inverse transformation for this problem is to try to suppress the part blocking the correct recognition by cutting the image.

In this work we present our solution for the images presented in figure 2A and 2B.

4   Study of Specialisation of Digits with Noise

A study has been carried out to verify the properties of the specialisation of the classifiers or channels in the case of noisy images.

The learning data base and test data base are different. The learned set used for all channels, which we call NET, has been created ad-hoc, with well defined images of typewritten digits of Microsoft sources (concretely 1001 images). The test data has images of typewritten digits, taken from forms of the Bank of Spain: 14.750 test digits, out of a total of 100.000, of different sources, have been considered and the classes are uniformly distributed.

We have added noise to images of digits, trying to obtain a strong database to support the experiments. Annex lines have been added to 14.740 images, both in their upper and lower parts, simulating typical segmentation errors, where images have noise coming from the cell including the digit. In images with annex lines, both in their upper or lower parts, three groups will be found: little noise (10%), moderate (20%) noise, and excessive noise (30%), as the amount of noise is variable in real cases, depending on the segmentation error.

The disturbed images considerably decrease the reliability of the system because the characteristics of the noisy images confuse the classifier.

The system proposed for the recognition of this type of disturbance contemplates the need to obtain the original image or noise free image, cutting the same in its upper or lower part in a percentage depending on the noise level.

The table 1 shows the classifiers or channels used in this study. The difference among them is the applied process previous to the recognition (ip_k). For example, in C_5 it is supposed that the image has lower noise in a moderate amount, then the image is cut 20% in order to take off the noise of the image.

The simulator of the multi-classifier used has enabled us to specify the parameters of each stage as well as their combination parameters.

The study contemplate several possibilities. In the first place, if the system had knowledge of where it is and how much noise the image has, it would know where to cut to eliminate the noise and how much it should cut, consequently, the recognition would be made easier. The following figure (figure 3) shows results with this premise for the case of upper noise, the behaviour in case of lower noise would be the same.

Table 1. Chanels used in the experiments

C_k	fr_k	N_k	ip_k
C1	Zoning 8x5	NET	∅
C2	Zoning 8x5	NET	cut 10% upper part
C3	Zoning 8x5	NET	cut 10% lower part
C4	Zoning 8x5	NET	cut 20% upper part
C5	Zoning 8x5	NET	cut 20% lower part
C6	Zoning 8x5	NET	cut 30% upper part
C7	Zoning 8x5	NET	cut 30% lower part

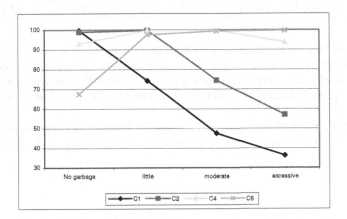

Fig. 3. Results for the case of upper noise. Axis X represents the level of noise of the images and axis Y the percentage of success of the four channels individually, cutting the images 0%, 10%, 20% and 30% respectively.

We can see that the classifier that tries to take only the original image (C_1) reduces the reliability while the disturbance of the image increases, and, on the contrary, channel C_6 increases the rate of success when the noise is higher. It is amazing that channel C_4 provides the best average recognition of the four, with 96.60% of success in all the images. A good result has also been attained and the fact is that, if all the data regarding the noise included in the image would be known, a certitude of 99.76% would be reached, facing 64.58% obtained with channel C_1.

Naturally, the reality is different and we are not going to know always where, and much less how much noise there is.

In the first place, we have created two multi-classifiers. They are defined as follows: MC_1 (4, $\{S_1\}$) and MC_2 (4, $\{S_2\}$), being S_1 (4, $\{C_1, C_2, C_4, C_6\}$, F) and S_2 (4, $\{C_1 C_3, C_5, C_7\}$, F). This means that only the channels of upper or lower noise are competing in an independent way, together with channel C_1. A function determines whether the input image has noise in its upper or lower part so some images are the input to MC_1 and some others to MC_2. This is a theoretical study because it is no easy to determine where is the noise in the image but some segmentation systems provide this information.

The following table shows the results of these two systems. The columns indicate the amount of noise of the digit presented to the system. The two multi-classifiers together attain an average recognition of 97.14%.

Table 2. Results for MC_1 and MC_2

	NO noise	Little	Moderate	Excessive	Average
MC_1	99,46	99,31	98,18	93,99	97,74
MC_2	99,46	99,45	97,42	89,87	96,55
Average	99,46	99,38	97,8	91,93	97,14

Now, we study a multi-classifier where there is no knowledge of the situation or the amount of noise in the image. Seven classifiers will compete in parallel for the recognition of each image, images noise free, with little, moderate and excessive noise. The proposed multi-classifier is defined as MC_3 (1, $\{S_3\}$) being S_3 (7, $\{C_1, C_2, C_3, C_4, C_5, C_6, C_7\}$, F).

The table 3 presents the results for this multi-classifier, and the attained results for each type of noise can be seen, with an average recognition of 96.18% for all the images.

Table 3. Results for MC_3

	No noise	Little	Moderate	Excessive	Average
Upper	99,47	99,12	97,75	92,64	97,25
Lower	99,47	99,17	95,55	86,26	95,11
Average	99,47	99,16	96,65	89,45	96,18

The average percentage of success attained is much higher, comparing it with the results attained by the system with only a classifier, which treats the images as they come (C_1). This mono-classifier system reached an average success of 64.59% for images with upper, and lower noise and noise free, but now this recognition rate soars to 96.18% thanks to the multi-classifier parallel system.

Figures 4 and 5 show the behaviour of the multi-classifier system (MC_3) with two different input images. The inverse transformations applied (ip_k), the output class ($Class_{k1}$) and the confidence value (V_k) are shown for each channel. In the figure 4 the best channel is the C_1 and the channel C_6 has the best confidence value in the figure 5. This is an example that illustrates the goodness of the system when the image has noise and when it have not: the noisy images have a not very good confidence value in the classification, it can see in figure 5. Furthermore, the channels with inverse perturbations provide a low confidence in good images (figure 4).

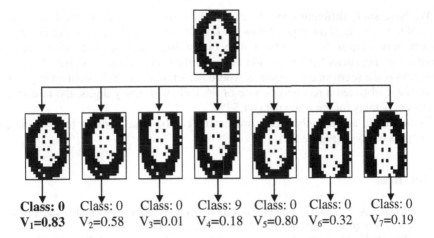

Fig. 4. System behaviour in the case of images with salt and pepper. The output class and the confidence value are shown for each classifier.

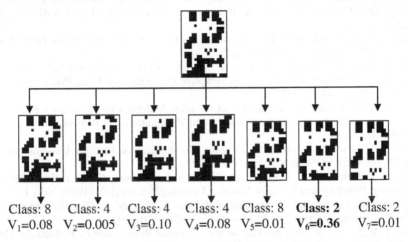

Fig. 5. System behaviour with different types of noise. The output class and the confidence value are shown for each classifier.

5   Conclusions

The noisy patterns has been one of the problems in the pattern recognition. While the human recognition don't have too much problems to recognize this kind of patterns, OCR systems have poor recognition in this field.

For us the specialization in the recognition, for all kind of noise is the base. That is because we treat each case of noise separately in parallel and then a criterion decide. So the system provide two results, the expected class of the pattern and what kind of noise has the image, depending on which specialist has responded.

We have study different specializations, the difference of each one is the knowledge of the context. If the type of noise is known the recognition is easier because the system know what is the most efficient specialist. But when the context information is smaller the specialists have to decide between them in a multi-classifier. This is important in some segmentation systems where the knowledge of the context is possible.

We have obtained a recognition rate of 96.18%, with noisy digits, the non specialist system obtains for the same data 64.58%.

In the future, additionally image processing can be done in order to recognize another types of noise (Figure 2C).

References

1. Whichello, A. P.,Yan, H.: Linking broken character borders with variable sized masks to improve recognition. Pattern Recognition Vol. 29 (8) (1996) 1429-1435
2. Rodriguez, C., Muguerza, M., Navarro, M., Zárate, A., Martín, J. I.,Pérez, J. M.: A Two-Stage Classifer for Broken and Blurred Digits in Forms. ICPR 98 Brisbane, Australia Vol. 2 (1998) 1101-1105
3. Omachi, S., Sun, F.,H., A.: A Noise-Adaptive Discriminant Function and its Application to Blurred Machine-Printed Kanji Recognition. IEEE Transactions PAMI Vol. 22 (3) (2000) 314-319
4. Kittler, J., Hated, M., Duin, R. P. W.,Matas, J.: On Combining Classifiers. IEEE Transactions PAMI Vol. 20 (3) (1998) 226-239
5. Rodriguez, C., Soraluze, I., Muguerza, J., Martín, J. I.,Álvarez, G.: Hierarchical Classifiers based on neighbourhood criteria with Adaptive Computational Cost. Pattern Recognition Vol. 35 (12) (2002) 2761-2769
6. Alpaydin, E., Kaynak, C.,Alimoglu, F.: Cascading Multiple Classifiers and representations for Optical and Pen-Based Handwritten Digit Recognition. 7th IWFHR Amsterdam Vol. (2000) 453-462
7. Ho, T. K., Hull, J. J.,Srihari, S.: Decision Combination in Multiple Classifier Systems. IEEE Transactions PAMI Vol. 16 (1) (1994) 66-75
8. Bauer, E.,Kohavi, R.: An Empirical Comparison of Voting Classification Algorithms: Bagging, Boosting, and Variants. Machine Learning Vol. 36 (1-2) (1999) 105 - 139
9. Aksela, M., Girdziusas, R., Laaksonen, J., Oja, E.,Kangas, J.: Class-Confidence Critic Combining. 8th IWFHR Ontario, Canada Vol. (2002) 201-206
10. Ha, T. M.,Bunke, H.: Off-line, Handwritten Numeral Recognition by Pertubation Method. IEEE Transactions PAMI Vol. 19 (5) (1997) 535-539
11. Erp, M. v., Vuurpijil, L.,Shomaker, L.: An overview and comparison of voting methods for pattern recognition. 8th IWFHR Ontario, Canada Vol. (2002) 195-200
12. Cappelli, R., Maio, D.,Maltoni, D.: A Multi-Classifier Approach to Fingerprint Classification. Pattern Analysis & Applications Vol. 5 (2) (2002) 136-144
13. Grim, J. Ã., Kittler, J., Pudil, P.,Somol, P.: Multiple Classifier Fusion in Probabilistic Neural Networks. Pattern Analysis & Applications Vol. 5 (2) (2002) 221-233
14. Dasarathy, B. V.: Nearest Neighbor (NN) Norms: NN Pattern Classification Techniques. I. C. S. Press (1991)
15. Devroye, L., Györfi, L.,Lugosi, G.: A Probabilistic Theory of Pattern Recognition. N. Y. Springer-Verlag (1996)
16. Haralick, R. M., Stenberg, S. R.,Zhuang, X.: Image analysis using mathematical morphology. IEEE Transactions PAMI Vol. 9 (4) (1987) 532-550
17. Serra, J.: Image Analisys and Mathematical Morphology. L. Academic Press (1982)

Automatic Table Detection in Document Images

Basilios Gatos, Dimitrios Danatsas, Ioannis Pratikakis, and Stavros J. Perantonis

Computational Intelligence Laboratory, Institute of Informatics and Telecommunications,
National Center for Scientific Research "Demokritos",
GR 15310 Athens, Greece
{bgat, dan, ipratika, sper}@iit.demokritos.gr

Abstract. In this paper, we propose a novel technique for automatic table detection in document images. Lines and tables are among the most frequent graphic, non-textual entities in documents and their detection is directly related to the OCR performance as well as to the document layout description. We propose a workflow for table detection that comprises three distinct steps: (i) image preprocessing; (ii) horizontal and vertical line detection and (iii) table detection. The efficiency of the proposed method is demonstrated by using a performance evaluation scheme which considers a great variety of documents such as forms, newspapers/magazines, scientific journals, tickets/bank cheques, certificates and handwritten documents.

1 Introduction

Nowadays, we experience a proliferation of documents which leads to an increasing demand for automation in document image analysis and processing. Automatic detection of subsequent page components like tables gives a great support to fulfill the demand for automation. More specifically, in the case of a table recovery, a great support to compression, editing and information retrieval purposes can be given.

Tables have physical and logical structure [1]. The physical structure concerns the location in an image of all the constituent parts of a table. The logical structure defines the type of the constituent parts and how they form a table. Therefore, all parts in a table have both physical and logical structure.

In this paper, we focus on the detection of all lines, both vertical and horizontal, along with their intersection, which will aid not only to detect a table which consequently can be extracted out of a whole document but also to describe both the physical and logical structure, thus, inferring a table recognition process.

In the literature, other researches have worked to accomplish the goals mentioned above. Zheng et al. [2] proposed a frame line detection algorithm based on the Directional Single-Connected Chain (DSCC). Each extracted DSCC represents a line segment and multiple non-overlapped DSCCs are merged to compose a line based on rules. During our experiments, we have compared this approach with our proposed approach for horizontal/vertical table line detection. Neves and Facon [3] have presented a method for automatic extraction of the contents of passive and/or active cells in forms. This approach is based on the analysis and recognition of the types of intersection of the lines that make up the cells. In the particular domain of business letters,

S. Singh et al. (Eds.): ICAPR 2005, LNCS 3686, pp. 609–618, 2005.
© Springer-Verlag Berlin Heidelberg 2005

Kieninger and Dengel [4] propose the so-called T-Recs Table location that consists of block segmentation and table locator. The table locator is based on simple heuristics that concern the extracted blocks. Finally, Cesarini et al. [5] describe an approach for table location in document images where the presence of a table is hypothesized by searching parallel lines in the modified X-Y tree of the page. Furthermore, located tables can be merged on the basis of proximity and similarity criteria.

In this paper, we propose a novel technique for automatic table detection in document images that neither requires any training phase nor uses domain-specific heuristics, thus, resulting to an approach applied to a variety of document types. Experimental results support the robustness of the method. The proposed approach builds upon several consequent stages that can be mainly identified to the following: (i) image preprocessing; (ii) horizontal and vertical line detection and (iii) table detection. In the following sections, we present our methodology for table detection in document images, as well as our experimental results that demonstrate the efficiency of the proposed method.

2 Methodology

2.1 Pre-processing

Pre-processing of the document image is essential before proceeding to the line and table detection stages. It mainly involves image binarization and enhancement, orientation and skew correction as well as noisy border removal. Binarization is the starting step of most document image analysis systems and refers to the conversion of the gray-scale image to a binary image. The proposed scheme for image binarization and enhancement is described in [6]. It is an adaptive approach suitable for documents with degradations which occur due to shadows, non-uniform illumination, low contrast, large signal-dependent noise, smear and strain. Text orientation is determined by applying an horizontal/vertical smoothing, followed by a calculation procedure of vertical/horizontal black and white transitions [7]. The proposed scheme for skew correction is described in [8] and uses a fast Hough transform approach based on the description of binary images using rectangular blocks. In the pre-processing stage of our approach, the process of noisy borders removal is based on [9] and employs a "flood-fill" based algorithm that starts expanding from the outside noisy surrounding border towards the text region. Fig. 1 illustrates the proposed pre-processing step.

In the proposed methodology, we use a particular parameterization that depends on the average character height of the document image. Therefore, we proceed with an average character size estimation step that is more specifically required for adjusting all line detection algorithm parameters in order to achieve invariance to the scanning resolution or the character font size. Our main intentions are to exclude all short line segments that belong to character strokes and to approximate the maximum expected line thickness. We propose a method to automatically estimate the average character height based on calculating the surrounding rectangles height of the image connected components. We take the following steps:

STEP 1: We pick a random pixel (x,y) that has at least one background pixel in its 4 connected neighborhood.

<u>STEP 2:</u> Starting from pixel (x,y), we follow the contour of the connected component that pixel (x,y) belongs to.

<u>STEP 3:</u> We repeat steps 1, 2 for all existing connected components until we have a maximum number of samples *(MaxSamples)*. During this process we calculate the histogram H_h of the surrounding rectangles height h at the corresponding connected components.

<u>STEP 4:</u> We compute the maximum value of the histogram H_h which expresses the average character height AH. An example of the estimated average character size is illustrated in Figure 2.

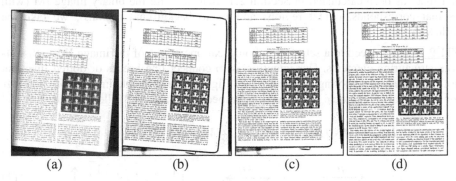

(a)   (b)   (c)   (d)

Fig. 1. Document image pre-processing example. (a) Original gray scale image; (b) resulting image after binarization and image enhancement; (c) resulting image after skew correction; (d) resulting image after noisy border removal.

K	% of print	rank	% of print
19.8	1	16.5	
19.8	8	18.1	
19.9	9	18.4	
19.9	4	17.8	
19.9	4	18.4	
19.9	3	18.0	
19.9	2	17.7	

In all four images, our unsupervi rithm produced clearly better results t algorithm [2].

2) Values of Updated Parameters: parameters (using the training data) b of the updated parameter values afte shown in Tables II and III. The initial was 0.5 for both print and background the contextual classifiers used the sam age and window. As seen from Tabl probability varied more than the mean of each class. This is reasonable sin exclusively of background, while othe

$AH = 16$

Fig. 2. The estimated average character height of a document image

2.2 Line Detection

A novel technique for horizontal and vertical line detection in document images is proposed. The technique is mainly based on horizontal and vertical black runs processing as well as on image/text areas estimation in order to exclude line segments that belong to these areas. Initially, a set of morphological operations with suitable structuring elements is performed in order to connect possible line breaks and to enhance line segments. The distinct steps of the proposed line detection technique are the following: (i) horizontal and vertical line estimation and (ii) line estimation improvement by using image/text areas removal.

Horizontal and Vertical Lines Estimation. At this step, we make a first estimation of horizontal and vertical line segments. The final estimation of lines will be accomplished after a refinement of this result by removing line segments that belong to image/text areas. The proposed line detection algorithm is based on horizontal and vertical black runs processing as well as on a set of morphological operations with suitable structuring elements in order to connect possible line breaks and to enhance line segments. All parameters used in this step depend on the average character height AH that has been calculated in Section 2.1. Starting with the binary image IM (with 1s that corresponds to text regions and 0s to background regions), we take the following steps:

STEP 1: We proceed to a set of morphological operations of the image IM with suitable structuring elements. Our intention is to connect line breaks or dotted lines but not to connect neighboring characters (see Fig. 3). We calculate images IM_H and IM_V for horizontal and vertical line detection, respectively, as in the following Eq. 1, 2:

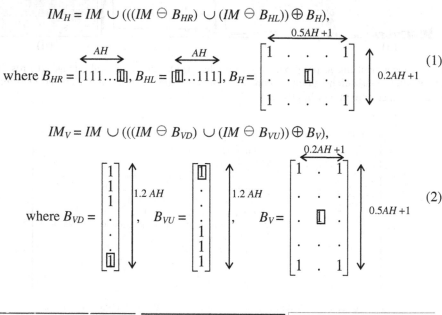

$$IM_H = IM \cup (((IM \ominus B_{HR}) \cup (IM \ominus B_{HL})) \oplus B_H),$$

where $B_{HR} = [111...\boxed{0}]$, $B_{HL} = [\boxed{0}...111]$, $B_H = \begin{bmatrix} 1 & . & . & . & 1 \\ . & . & \boxed{0} & . & . \\ 1 & . & . & . & 1 \end{bmatrix}$ (1)

with AH horizontal, $0.5AH + 1$ wide and $0.2AH + 1$ high.

$$IM_V = IM \cup (((IM \ominus B_{VD}) \cup (IM \ominus B_{VU})) \oplus B_V),$$

where $B_{VD} = \begin{bmatrix} 1 \\ 1 \\ 1 \\ 1 \\ . \\ . \\ \boxed{0} \end{bmatrix}$, $B_{VU} = \begin{bmatrix} \boxed{0} \\ . \\ . \\ 1 \\ 1 \\ 1 \end{bmatrix}$, $B_V = \begin{bmatrix} 1 & . & 1 \\ . & . & . \\ . & \boxed{0} & . \\ . & . & . \\ 1 & . & 1 \end{bmatrix}$ (2)

with $1.2\,AH$ tall, $0.2AH + 1$ wide and $0.5AH + 1$ high.

	MAP1			MAP1			MAP1
μ_p	101.2		μ_p	101.2		μ_p	101.2
(a)			(b)			(c)	

Fig. 3. Morphological processing in order is to connect line breaks or dotted lines. (a) Initial image *IM*; (b) Resulting image IM_H; (c) Resulting image IM_V.

STEP2: All 1s of images IM_H and IM_V that belong to line segments of great length and small width are turned to a label values **L**. In the case of horizontal lines, all 1s of IM_H that belong to horizontal black runs of length greater than AH and to vertical black runs of length less than AH are turned to **L**. In the case of vertical lines, all 1s of IM_V that belong to vertical black runs of length greater than AH and to horizontal black runs of length less than AH are turned to **L**.

STEP3: Images IM_H and IM_V are smoothed in horizontal and vertical directions correspondingly in order to set to **L** all short runs that have a value different than **L**. In the case of horizontal lines, horizontal runs of IM_H pixels with values not equal to **L** and length less than AH are set to **L**. In the case of vertical lines, vertical runs of IM_V pixels with values not equal to **L** and length less than AH are set to **L**.

STEP 4: Horizontal and vertical lines in images IM_H and IM_V, respectively, are defined from all connected components with **L**-valued pixels having length greater than $2\,AH$.

Line Estimation Improvement by Using Image/Text Areas Removal. Image/text areas estimation is accomplished by performing an horizontal and vertical smoothing of image IM_n that has 1's for pixels that do not belong to the detected horizontal or vertical lines. After this smoothing, all connected components of great height ($> 3AH$) belong to graphics, images or text. In this phase, tables will not appear as individual connected components in the final smoothed image since vertical and horizontal lines are excluded. More specifically, we take the following steps:

STEP 1: We proceed to an horizontal smoothing of image IM_V by setting all horizontal runs with 0's that have length less than $1.2AH$ to **L**.

STEP 2: We proceed to a vertical smoothing by setting all vertical runs with 0's that have length less than $1.2AH$ to **L**.

STEP 3: Image/text areas IT are defined from **L**-valued connected components in the resulting IM_V image having surrounding rectangle height greater that $3\,AH$.

From all horizontal lines HL and vertical lines VL we exclude those that lie inside non line entities IT that have been estimated in the previous step. Fig. 4 illustrates the line detection step.

2.3 Table Detection

After horizontal and vertical line detection we proceed to table detection. Our table detection technique involves two distinct steps: (i) Detection of line intersections and (ii) table detection and reconstruction.

Detection of Line Intersections. All possible line intersections (see Table 1) are detected progressively according to the following algorithm. First, we detect all intersections with IDs 1-4. In this case, an end point of an horizontal line and another end point of a vertical line define a line intersection of this type if they have the minimum distance among others around a neighborhood. Thereafter, we trace for intersections with IDs 5-8. In this case, an end point of either an horizontal line or a vertical line is tested against another line point which is not an end point and corresponds to a vertical line or an horizontal line, respectively. A line intersection of this type is defined

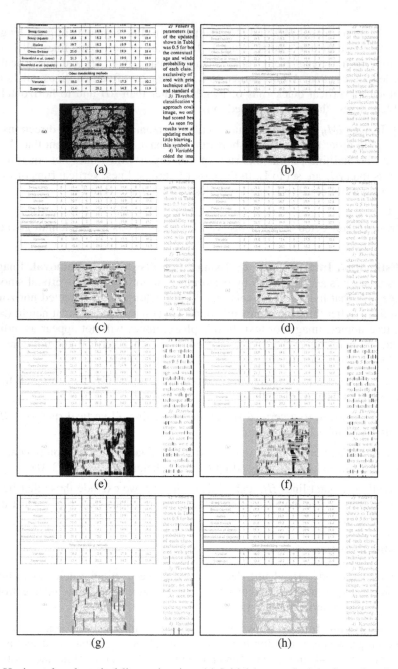

(a)

(b)

(c)

(d)

(e)

(f)

(g)

(h)

Fig. 4. Horizontal and vertical line estimation: (a) Initial image; (b,e) horizontal and vertical line segments of great length; (c,f) line segments of great length and small width; (d,g) detected horizontal and vertical lines after smoothing; (h) line estimation improvement using image/text areas removal.

for such points that have the minimum distance among others around a neighborhood. Finally, we detect intersections with ID 9 that correspond to horizontal and vertical line crossing points.

Table 1. Line intersections

ID	1	2	3	4	5	6	7	8	9
Line inter-sections									

Table Detection and Reconstruction. Table detection and reconstruction involves the following steps: First, all pixels that belong to the detected lines are removed (see Fig. 5(c)). Then, all detected line intersections are grouped first horizontally and then vertically. Each group is further aligned according to the mean value of the vertical or horizontal positions for horizontal and vertical groupings, respectively. Finally, we achieve a table reconstruction by drawing the corresponding horizontal and vertical lines that connect all line intersection pairs. Table detection and reconstruction is illustrated in Fig. 5.

3 Experimental Results

The corpus for the evaluation of the proposed methodology was prepared by selecting 102 images with a total of 2813 ground-truthed horizontal and vertical lines. It consists of scanned forms, newspaper - magazines, scientific papers, tickets – bank checks, certificates and handwritten documents. Most of the images have severe problems such as poor quality, broken lines or overlapping text and line areas. Representative results of the proposed methodology for line and table detection are illustrated in Fig. 6. In order to extract some quantitative results for the efficiency of the proposed methodology, we calculated the recognition rate and the recognition accuracy for horizontal and vertical line detection and compared the results with those of the DSCC algorithm [2] which is a state-of-the-art algorithm for unsupervised horizontal/vertical table line detection and the corresponding source code is available at [11]. The performance evaluation method used is based on counting the number of matches between the detected horizontal/vertical lines and the corresponding horizontal/vertical lines appearing in the ground truth [10]. We use a "MatchScore" table for horizontal and vertical lines whose values are calculated according to the intersection of the resulting line pixels and the ground truth. A global performance metric can be detected if we combine the detection rate and the recognition accuracy results according to the following formula:

$$GlobalPerformanceMetric = \frac{2 DetectionRate * \mathrm{Re}\,cognitionAccuracy}{DetectionRate + \mathrm{Re}\,cognitionAccuracy} \qquad (3)$$

sults of all the thinning algorithms.

TABLE II
STATISTICS ON THINNING ALGORITHMS
(FOR 1,000 CHARACTERS IN DATABASES 1 AND 2)

Algorithm	Database 1				Database 2			
	Time (sec)	Extra Pixels	End Points	Print-tives	Time (sec)	Extra Pixels	End Points	Print-tives
Chen & Hsu [3]	39.0	373	2109	2295	184	431	2284	2440
Chen & Tsai [5]	54.3	19072	2189	2323	179	36169	2541	2510
Chin et al. [6]	15.0	2342	2145	2238	65	3578	2494	2454
Guo & Hall [8]	9.5	1004	2186	2315	41	968	2351	2468
Hilditch [12]	12.9	0	2221	2337	60	0	2384	2486
Holt et al. [13]	12.2	7387	2119	2294	49	13828	2281	2442
Suzuki & Abe [23]	3.9	0	2250	2330	14	0	2447	2511
Wang & Zhang [25]	9.3	1111	2085	2287	41	1118	2229	2438
Wu & Tsai [26]	12.4	433	2190	2305	53	489	2360	2466
Zhang & Suen [28]	10.0	8538	2084	2286	42	15359	2232	2437

C. Number of End Points

(a)   (b)

(c)   (d)

Fig. 5. Table detection and reconstruction: (a) Initial image; (b) detected line intersections; (c) image without horizontal and vertical lines; (d) table reconstruction.

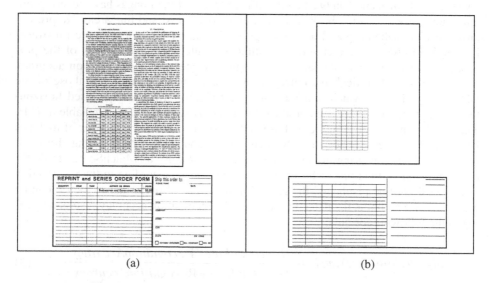

(a)   (b)

Fig. 6. Line detection results: (a).Original image; (b) Detected lines

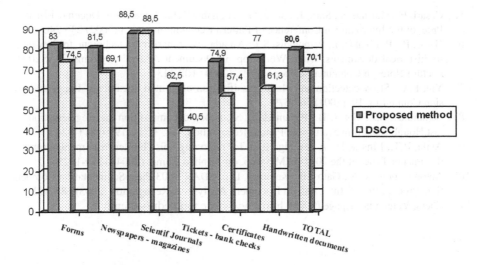

Fig. 7. Evaluation graphs for horizontal and line detection

As shown at Fig.7, for all types of the examined scanned documents, we get higher recognition rates compared to the DSCC algorithm. The global performance metric for all images is 80.6%, while DSCC algorithm achieves 70.1%.

4 Conclusions

This paper strives toward a novel methodology for automatic table detection in document images. The proposed methodology neither requires any training phase nor uses domain-specific heuristics, thus, resulting to an approach applied to a variety of document types. It builds upon several consequent stages that can be mainly identified to the following: (i) image pre-processing; (ii) horizontal and vertical line detection and (iii) table detection. Experimental results demonstrate the efficiency of the proposed method.

References

1. Zanibbi, R., Blostein, D., Cordy, J.: A survey of table recognition. International Journal of Document Analysis and Recogntion (IJDAR), vol. 7 (2004) 1-16
2. Zheng, Y., Liu, C., Ding, X., Pan, S.: Form Frame Line Detection with Directional Single-Connected Chain. Proc. of the 6th Int. Conf. on Doc. Anal. & Recognition (2001) 699-703
3. Neves, L. , Facon, J.: Methodology of Automatic extraction of Table-Form Cells. IEEE Proc. of the XIII Brazilian Symposium on Computer Graphics and Image Processing (SIBGRAPI'00) (2000) 15-21
4. Kieninger, T., Dengel, A.: Applying the T-Recs Table Recognition System to the Business Letter Domain. Proc. of the 6th International Conference on Document Analysis & Recognition, Seattle, (2001) 518-522

5. Cesari, F., Marinai, S., Sarti, L., Soda, G.: Trainable Table Location in Document Images. Proc. of the International Conference of Pattern Recognition, vol. 3 (2002) 236-240

6. Gatos, B., Pratikakis, I., Perantonis, S.J.: An adaptive binarisation technique for low quality historical documents. IARP Workshop on Document Analysis Systems (DAS2004), Lecture Notes in Computer Science (3163), (2004) 102-113

7. Yin, P.Y.: Skew detection and block classification of printed documents. Image and Vision Computing 19, (2001) 567-579

8. Perantonis, S.J., Gatos, B., Papamarkos, N.: Block decomposition and segmentation for fast Hough transform evaluation. Pattern Recognition, vol. 32(5) (1999) 811-824

9. Avila, B.T., Lins, R.D.: A new algorithm for removing noisy border from monochromatic documents. Proc. of the 2004 ACM Symp. on Applied Comp. (2004) 1219-1225

10. Antonacopoulos, A., Gatos, B., Karatzas, D.: ICDAR 2003 Page Segmentation Competition. Proc. of the 7th Int. Conf. on Document Analysis & Recognition (2003) 688-692

11. Zheng Yefeng homepage (2005): http://www.ece.umd.edu/~zhengyf/

High Performance Classifiers Combination for Handwritten Digit Recognition

Hubert Cecotti, Szilárd Vajda, and Abdel Belaïd

READ Group, LORIA/CNRS, Campus Scientifique BP 239,
54506 Vandoeuvre-les-Nancy cedex France

Abstract. This paper presents a multi-classifier system using classifiers based on two different approaches. A stochastic model using Markov Random Field is combined with different kind of neural networks by several fusing rules. It has been proved that the combination of different classifiers can lead to improve the global recognition rate. We propose to compare different fusing rules in a framework composed of classifiers with high accuracies. We show that even there still remains a complementarity between classifiers, even from the same approach, that improves the global recognition rate. The combinations have been tested on handwritten digits. The overall recognition rate has reached 99.03% without using any rejection criteria.

1 Introduction

Multi-classifier systems have shown their high efficiency in different applications [2,5,11]. The difference between classifiers can be explained by their specific algorithms and particular perception of the training data. The advantages of multi-classifier systems can largely be put ahead when several classifiers can complement each one easily. The improvement given by a multi-classifier system can also hide the lack of performance of each used classifier. When one classifier achieves high quality results, how can other classifiers improve it without disabling it? The problem is when the classifiers are already efficient and where the complementarity is not huge and not easy to express. We present 7 different high performance classifiers based on two different approaches. The first one used is a stochastic model based on Markov Random Field. The second one is based on neural networks. The purpose of this work is to show the relationship between these two different techniques in a multi-classifier scheme. In the first part, the multi classifier design will be explained while in the second part each classifier will be described. In the third part the different combination schemes will be shown. Finally, the gain obtained by the different techniques will be discussed.

2 Multi-classifier Design

The effectiveness of a multi-classifier system relies principally on combining complementary classifiers. Several approaches have been proposed to construct different sets made up of complementarity classifiers. Among these methods, the

S. Singh et al. (Eds.): ICAPR 2005, LNCS 3686, pp. 619–626, 2005.

knowledge of the problem and the classifiers behaviors can influence the choice of the classifiers. One of the other methods consists to manipulate the training data to train a classifier with different training sets like the Adaboost algorithm [4]. In this work, we use two accurate approaches in character recognition: a stochastic model and a neural model. The purpose is to highlight the different relationship between the methods. Four different classifiers have been created with the first method and three with the second one.

3 Classifiers Description

Let a database DB of N mutually exclusive sets, $DB = C_1 \cup C_2 \cup ... \cup C_N$, where each of $C_i, \forall i \in \{1, .., N\}$ represents a set of patterns called "Class". Each classifier or expert, denoted as e, assigns to a pattern $x \in DB$ an index $j \in \{1, .., N+1\}$, which represents x as belonging to the class C_j, if $j \neq N+1$. If the classifier does not recognize the class corresponding to x, then x is rejected by e and $j = N + 1$. We note the classifier decision by $e(x) = j$. A classifier e is defined by a triplet (τ_r, τ_s, τ_q) where τ_r, τ_s and τ_q are the recognition rate, the error rate and the rejection rate respectively.

3.1 NSHP-HMM

The NSHP-HMM (Non Symmetric Half Plane Hidden Markov Model) is a powerful stochastic tool, originally designed for handwritten word recognition but used also with the same success in handwritten digit recognition [14]. The originality of this method resides in coupling a context based local vision performed by a NSHP with a HMM giving horizontal elasticity to the model. It operates on pixel level, analyzing pixel columns, which are viewed as the random field realizations.

Let I be the image having m rows and n columns observed by the NSHP. The joint field mass probability $P(I)$ of the image I can be computed following the chain decomposition rule of conditional probabilities:

$$P(I) = \prod_{j=1}^{n} \prod_{i=1}^{m} P(X_{ij} \mid X_{\Theta_{ij}}) \tag{1}$$

As the equation 1 can be written as above, it is possible for the NSHP-HMM to observe the image via its columns considering this like an observation entity. As in our model the Θ_{ij} was chosen as being a 3rd order neighborhood, the NSHP-HMM is sub-divided in 4 NSHP-HMM. Such a division of the model is necessary due the non-symmetric sampling of the image pattern. Each NSHP-HMM can be considered as a separate classifier representing the different reading senses (right to left, left to right, bottom to top and top to bottom) of the model.

3.2 Neural Networks

Neural networks are widely used for classification and have proved their high efficiencies in many applications. The considered neural networks are based on a

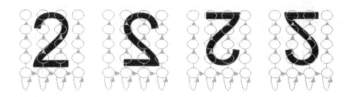

Fig. 1. NSHP-HMM models

multi-layer perceptron (MLP) with the classical back-propagation for the learning. Three neural networks have been created. The first neural network was designed with only one hidden layer containing 500 neurons. This network is fully connected. Such network has been applied with success for postal pin code recognition [13]. The parameters of these neural classifiers have been fixed based on different trial runs.

Convolutional Neural Network. This neural network is designed with a different topology. The goal of the topology based on convolutional neural network is to classify the image given as input by analyzing it through different "receptive fields". Each layer is composed of several maps, each one corresponding to an image transformation. These transformations extract features like edges, strokes, etc [8,16]. The neural network is composed of 5 layers. The first one corresponds to the input image, normalized by its center and reduced to a size of 29*29 pixels. The next two layers correspond to the information extraction, performed by convolutions. The second and third layers are composed of 10 and 50 maps respectively. Each map describes a convolution and a sub-sampling. In these maps, neurons share the input weights represented by a pivot neuron in each map. The last two layers are fully connected and finally the last layer is the output: 10 neurons, one output for each digit.

With this topology two neural networks have been created. This first network has been trained with the original data patterns whereas the second one has been trained with both the original patterns and the patterns with inversed colors. This network was indeed trained to recognize black images on white background and white images and black background. The purpose of this model was to extract features linked to the contrast variation. Instead of letting the model

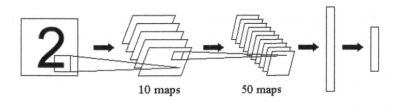

Fig. 2. Convolutional Neural Network topology

learn its own discriminative feature set, the training set has been suited to learn specific characteristics. The outputs of each classifier have to be normalized to obtain values of the same range. While the first model is probabilistic and its outputs correspond naturally to probabilities, the neural network outputs are normalized by the *Softmax* function in order to obtain probabilities.

4   Combination Rules

In the case of several classifiers, the combination of D different classifiers denoted $e_k, k \in \{1, .., D\}$ is defined as E. Each classifier assigns to a pattern x a decision j_k denoted by $e_k(x) = j_k$. The final solution j for the sample x is given by E. Let $v_i^{(k)}(x)$ be the real value computed by the classifier number k for the sample x and the class C_i. This value can represent a probability, a confidence value. It means the degree of membership to one class. In this work we will only discuss about a particularly architecture: the horizontal combination scheme. It corresponds to a topology where classifiers are performed in parallel. The classifiers work independently and concurrently and a fusion module combines their results.

4.1   Results Combination

The outputs of each classifier can be combined by simple rules. These rules merge the outputs value of all the classifier for one class.

- Selection of the maximum result: $\forall i \in \{1, .., N\}, v_i'(x) = max_{k=1,..,D} v_i^k(x)$
- Sum of the results: $\forall i \in \{1, .., N\}, v_i'(x) = \sum_{k=1}^{k=D} v_i^k(x)$
- Median of the results: $\forall i \in \{1, .., N\}, v_i'(x) = median_{(k=1,..,D)} v_i^k(x)$

$$E(x) = \begin{cases} i & \text{if } v_i(x) = max_{k=1,..,D} v_i^{',k}(x) \text{ and } v_i'(x) \geq \alpha \\ N+1 & otherwise \end{cases}$$

Where $\alpha \in [0; 1]$ is a threshold value. These methods allow to merge the results of each classifier but none of them extract knowledge concerning each classifier strength.

4.2   Majority Voting

The majority voting is an easy method to implement and it has shown good results in the literature [1,7,10]. For a multi-classifier system E, the majority voting can be expressed as follows:

$$(E(x) = i) \Leftrightarrow (|\{k \in \{1..D\}, e_k(x) = i\}| \geq ((D/2) + d)), 1 \leq d \leq (D/2)$$

If $d = D + 2$ then the voting corresponds to a consensus: all the classifiers agree to the same solution.

4.3   Behavior Knowledge Space

A behavior knowledge space is a D-dimensional space, each dimension corresponding to the decision of one classifier [6]. Each classifier has as decision values the total number of classes N. Let $x \in C_i$ be the character to be recognized belonging to the class C_i. Let $s_k = j_k, k = 1..D$ be the k^{st} classifier among D and j_k its answer for the current character x. The probability that $x \in C_i$ is defined by the following formula:

$$Belief(C_i) = \frac{P(s_1(x) = j_1, .., s_D(x) = j_D, x \in C_i)}{P(s_1(x) = j_1, .., s_D(x) = j_D)}$$

A cell of the BKS corresponds to the intersection of the individual classifiers decisions. Each point of the BKS is noted by $BKS(j_1, .., j_D), j_i = 1..N$; and contains a vector of size N: $bks(j_1, .., j_D)(i), i = 1..N$.

Let $bks(j_1, ..j_D)(i)$ be the total number of characters x such that $s_1(x) = j_1, .., s_D(x) = j_D$ and $x \in C_i, i = 1..N$. Let $T(j_1, .., j_D)(i)$ be the total number of characters x such that $s_1(x) = j_1, .., s_D(x) = j_D$. The best representative class of $BKS(j_1, .., j_D)$: R is defined by:

$$R = argmax(bks(j_1, .., j_D)(i)), i = 1..N$$

If one cell of the BKS is empty then the pattern is naturally rejected. A small database could be a problem to obtain a good generalization. Many empty cells may occur if the database is not representative. As the BKS size increases exponentially with the number of classifiers, the data sets has to increase in the same way [12]. For BKS cells where the most representative class is defined by a low probability, meaning ambiguous cases, characters are rejected. R is rejected if $Belief(C_i) \leq \alpha$ where α is a threshold representing the desired recognition quality.

5   Experiments

The system has been tested on the MNIST database. This well-known database contains separated handwritten digit images of $28 * 28$ in gray level. The learning set contains 60000 images and the test set contains 10000 images. In the learning set, 50000 images are used for real learning; 10000 images are used to find the best parameters. The first objective is to show the behavior of the different combination methods for these classifiers. Let NSHP1, NSHP2, NSHP3, NSHP4 be the 4 flip of the NSHP-HMM. Let NN1 be the neural network with the fully connected topology, NN2 and NN3 convolutional neural networks. The NN3 neural networks has been trained with both the initial MNIST database and the MNIST database with inverted colors.

The Table 1 shows the results obtained for each classifier for the test database. The different parts of the NSHP-HMM model obtain the lowest recognition rates whereas the different neural networks give the best results. The best

Table 1. Recognition rate for each classifier

	NSHP1	NSHP2	NSHP3	NSHP4	NN1	NN2	NN3
Train	93.69	95.00	94.42	95.22	99.72	99.71	99.57
Test	93.44	94.91	94.00	95.25	98.54	98.73	98.41

classifier in the system is the convolutional neural network. While many classifiers process the images by extracting time-costly features, in our work the considered classifiers are based on pixel level (i.e raw images). The recognition rate of all these classifiers is still low compared to the actual best results reported in the literature [3,8,9]. However, some of the top recognition percentages on the MNIST database have been achieved by using different expansion of the initial MNIST training database [15], or by using SVM, which still suffers of memory space and computational speed issues for classification [9]. In our tests, only the initial training database has been used for all of the 7 classifiers presented.

Table 2. Strength of each classifier

	NSHP1	NSHP2	NSHP3	NSHP4	NN1	NN2	NN3
NSHP1	0	411	304	404	576	587	575
NSHP2	264	0	274	221	436	440	424
NSHP3	248	365	0	378	516	533	520
NSHP4	223	187	253	0	393	414	400
NN1	66	73	62	64	0	98	85
NN2	58	58	60	66	79	0	63
NN3	78	74	79	84	98	95	0

The strength of each classifier is exposed in the Table 2. A cell (i, j) of the table corresponds to the number of pattern recognized by the classifier j and not recognized by the classifier i. It exhibits the strength and the weakness of each classifier versus the others in the test database. Firstly, there is a strong complementarity between the different flips of the NSHP-HMM. Each flip of the NSHP-HMM can contribute with about more than 200 patterns to the other flips. In this case, we have clearly a proof that results must be combined. Moreover, the 4 classifiers extracted from the NSHP-HMM method come from the same method. A little difference between those classifiers, even coming from the same algorithm, leads to obtain a high complementarity. Secondly, in spite of the strength of the different neural networks, all the classifiers can complete them. The contribution is not as significant as between the NSHP-HMM flips but they can be combined as they all give different results. These results display that any classifier makes the same mistake as the others. The results can be combined in order to extract their local strengths.

Without searching the forces and different relationships between classifiers, their results can be fused as described in 4.1. Classifiers have been clustered

Table 3. Combination Results

	Test (all classifiers)	Test (NSHP-HPP 4 flips)	Test (3 NN)
Consensus	87.21/12.75/0.04	87.89/11.37/0.74	97.09/2.65/0.26
Majority Voting	97.93/0.77/1.30	93.97/4.15/1.88	98.91/0.13/0.96
Oracle	99.89/0.00/0.11	98.61/0.00/1.39	99.68/0.00/0.32
Maximum rule	98.54/0.00/1.46	95.66/0.00/4.34	99.03/0.00/0.97
Sum rule	96.76/0.00/3.24	96.44/0.00/3.56	99.03/0.00/0.97
Median rule	95.66/0.00/4.34	96.09/0.00/3.91	98.96/0.00/1.04
BKS	97.94/1.34/0.72	96.11/0.31/3.58	98.42/0.59/0.99

Table 4. Best improvement for each classifier, without rejection

Classifiers used	NSHP1	NSHP2	NSHP3	NSHP4	NN1	NN2	NN3
All	+5.10	+3.63	+4.54	+3.29	0	-0.19	+0.13
4 NSHP	+3.00	+1.53	+2.44	+1.19			
3 NN					+0.49	+0.30	+0.62

in two groups. The first group contains the 4 NSHP-HMM classifiers and the second group is composed of the 3 neural networks. The different fusing methods presented have been tested. The triplet (τ_r, τ_s, τ_q) of each voting method is shown in the Table 3. Each rows gives for each classifiers cluster the triplet (τ_r, τ_s, τ_q) for one fusing techniques. The oracle method simulates the results that could be obtained with an optimal vote: if one of the classifier finds the good class then this class is selected. It allows estimating limits for the voting methods. In the BKS case with just the 4 NSHP-HMM flips and with just the 3 neural networks, the recognition rate did not increase but the error has decreased. It has though improved the relevance of the global results. The best improvement achieved for each classifier is presented in the Table 4. The lowest error rate is obtained by a voting method with the combination of all the classifiers: 0.04%. The best recognition score, without rejection, is obtained by the maximum rule with the combination of the 3 neural networks: 99.03%.

6 Conclusion

We have presented the combination of different kinds of classifier for handwritten digits recognition. These classifiers were from two different approaches: a stochastic model NSHP-HMM and neural network models. They have been combined using different rules. Their strength and weakness have been highlighted. Thanks to the combination, we have obtained good results considering the experimental conditions by combining neural networks. Multi-classifier systems can always improve a recognition system even in a case where the complementarity between classifiers is low. When the ensembles of classifiers may not always directly improve the recognition rate, they can improve the reliability of the results by qualifying the rejection.

References

1. Alpaydin, E.: Improved classification accuracy by training multiple models and taking a vote. In: 6th Italian Workshop. Neural Nets Wirn Vietri-93. (1994) 180–185
2. Bahler, D., Navarro, L.: Methods for Combining Heterogeneous Sets of Classifiers. 17th Natl. Conf. on Artificial Intelligence (AAAI 2000), Workshop on New Research Problems for Machine Learning, (2000)
3. Bortolozzi, F., de Souza Britto Jr., A., Oliveira, L.S., Morita, M.: Recent Advances in Handwritting Recognition. International Workshop on Document Analysis'05, (2005) 1–30
4. Freund, Y., Iyer, R., Schapire, R.E., Singer, Y.: An efficient boosting algorithm for combining preferences. Journal of Machine Learning Research, vol. 4, (2003) 933–969
5. Gunes, V., Ménard, M., Loonis, P., Petit-Renaud, S.:Systems of classifiers: state of the art and trends. International Journal of Pattern Recognition and Artificial Intelligence, vol. 17, no. 8, World-Scientific, (2004)
6. Huang, Y.S., Suen, C.Y.: A method of combining multiple experts for the recognition of unconstrained handwritten numerals. IEEE Trans Pattern Anal Mach Intell vol. 17, no. 1, (1995) 90–94
7. Lam, L., Suen, C.Y.: Application of majority voting to pattern recognition: an analysis of its behavior and performance. IEEE Trans Pattern Anal Mach Intell. vol. 27, no. 5, (1997) 553–568
8. LeCun, Y., Bottou, L., Bengio, Y., Haffner, P.: Gradient-Based Learning Applied to Document Recognition. Proceedings of the IEEE, vol. 86, no. 11, (1998) 2278–2324
9. Liu, C-L., Nakashima, K., Sako, H., Fujisawa, H.: Handwritten digit recognition: benchmarking of state-of-the-art techniques. Pattern Recognition, vol. 36, (2003) 2271–2285
10. Ng, C.S., Singh, H.: Democracy in pattern classifications: combinations of votes from various pattern classifiers. AIE, vol. 12, no. 3, (1998) 189–204
11. Rahman, A.F.R., Fairhurst, M.C: Multiple classifier decision combination strategies for character recognition: A review. International Journal on Document Analysis and Recognition. vol. 5, (2003) 166–194
12. Raudys, S., Roli, F.: The Behavior Knowledge Space Fusion Method: Analysis of Generalization Error and Strategies for Performance Improvement. Multiple Classifier Systems 4. (2003) 55–64
13. Roy, K., Vajda, S., Pal, U., Chaudhuri, B. B.: A System towards Indian Postal Automation. 9th International Workshop on Frontiers in Handwriting Recognition, Tokyo, Japan, October, (2004)
14. Saon, G., Belaïd, A.: High Performance Unconstrained Word Recognition System Combining HMMs and Markov Random Fields. International Journal of Pattern Recognition and Artificial Intelligence, vol. 11, no. 5, (1997) 771–788
15. Simard, P.Y., Steinkraus, D., Platt, J.C.: Best Practices for Convolutional Neural Networks Applied to Visual Document Analysis. 7th International Conference on Document Analysis and Recognition. (2003) 958–962
16. Teow, L., Loe, K-F.: Robust vision-based features and classification schemes for off-line handwritten digit recognition. Pattern Recognition, vol. 35, no. 11, (2002) 2355–2364

A Novel Approach for Text Detection in Images Using Structural Features

H. Tran[1,2], A. Lux[1], H.L. Nguyen T[2], and A. Boucher[3]

[1] Institut National Polytechnique de Grenoble,
Laboratory GRAVIR, INRIA Rhone-Alpes, France
[2] Hanoi University of Technology,
International Research Center MICA, Hanoi VietNam
[3] Institut de la Francophonie pour l'Informatique
thi-thanh-hai.tran@inrialpes.fr

Abstract. We propose a novel approach for finding text in images by using ridges at several scales. A text string is modelled by a ridge at a coarse scale representing its center line and numerous short ridges at a smaller scale representing the skeletons of characters. Skeleton ridges have to satisfy geometrical and spatial constraints such as the perpendicularity or non-parallelism to the central ridge. In this way, we obtain a hierarchical description of text strings, which can provide direct input to an OCR or a text analysis system. The proposed method does not depend on a particular alphabet, it works with a wide variety in size of characters and does not depend on orientation of text string. The experimental results show a good detection.

1 Introduction

The rapid growth of video data creates a need for efficient content-based browsing and retrieving systems. Text in various forms is frequently embedded into images to provide important information about the scene like names of people, titles, locations or date of an event in news video sequences, etc. Therefore, text should be detected for semantic understanding and image indexation. In the literature, text detection, localisation, and extraction are often used interchangeably. This paper is about the problem of detection and localisation. Text detection refers to the determination of the presence of text in a given image and text localisation is the process of determining the location of text in the image and generating bounding boxes around the text.

For text detection we need to define what text is. A text is an "alignment of characters", characters being letters or symbols from a set of signs which we do not specify in advance. In images, text can be characterised by a region of elongated shape band containing a large number of small strokes. The style and the size of characters can vary greatly from one text to another. In images of written documents background as well as text color are nearly uniform, the detection of text can easily be performed by thresholding the grayscale image. However, the task of automatic text detection in natural images or video frames

S. Singh et al. (Eds.): ICAPR 2005, LNCS 3686, pp. 627–635, 2005.

is more difficult due to the variety in size, orientation, color, and background complexity. A generic system for text extraction has to cope with these problems.

2   Methods of Text Detection in Images

Approaches for detecting and localizing text in images in the literature can be classified into three categories: (1) bottom-up methods [6,8], (2) top-down methods [12,14] and (3) machine learning based top-down methods [5,7]. The first category extracts regions in image and then groups character regions into words by using geometrical constraints such as the size of the region, height and width ratio. These methods avoid explicit text detection but they are very sensitive to character size, noise and background complexity. In the second category, characters can be detected by exploiting the characteristics of vertical edge, texture, edge orientation and spatial properties. These methods are fast but give false alarms in case of complex background. The third category has been developed recently and receives much attention from researchers. The evaluation of machine learning based methods showed the best performance in comparison with other approaches [1]. The principle is to extract some characteristics like wavelets [5], statistical measures [2,3] or derivatives [7] from fixed-size blocks of pixels and classify the feature vectors into text or non-text using artificial neural networks. As usual with this kind of learning method, the quality of results depends on the quality of the training data and on the features which are fed into the learning machine.

3   Proposed Approach

The objective of our work presented in this paper is to construct an automatic text detector which is independent with respect to the size, the orientation and the color of characters and which is robust to noise and aliasing artifacts. We propose a new method of text detection in images that is based on a structural model of text and gives more reliable results than methods using purely local features like color and texture.

The structural features used here are ridges detected at several scales in the image. A ridge represents shape at a certain scale. Analyzing ridges in scale space permits to capture information about details as well as global shape. A line of text is considered as a structured object. At small scales we can clearly see the strokes. At lower resolution, the characters disappear and the text string forms an elongated cloud. This situation can be characterized by ridges at small scales representing skeletons of characters and at coarser scale representing the center line of the text string (figure 1). These properties are generic for many kinds of text (scene text, artificial text or targeted scene text), do not depend on the alphabet (e.g. latin characters, ideograms), and also apply for hand written text (figure 2).

The ridge detection operator is iso-symmetric so it can detect a text string as straight line or curve at any orientation (figure 2d). The multi-resolution computation detects a wide variety in text size. Unlike the multi-resolution approach proposed in [11,12], where candidate text is detected at each scale separately and requires an additional scale fusion stage, our work directly exploits the topological change of text over scale. In addition, analyzing the relation between scales and ridge lengths can predict the number of characters in a text line, and character dimensions.

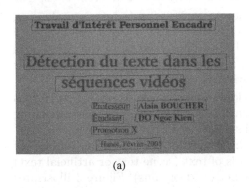

(a)   (b)

Fig. 1. (a) Image of a slide; detected text regions are bounded by red rectangles. (b) Ridges detected at two levels $\sigma_1 = 2\sqrt{2}$ (blue) and $\sigma_2 = 16$ (red): red lines represent the center lines of text strings, blue lines represent skeletons of characters.

The rest of this paper is organized as follows: In section 4, we present briefly the definition of ridge and explain the representation of text line based on ridges. We then analyze in detail the constraints that a text region must satisfy to be discriminated from a non-text region. Some experimental results and conclusions will be shown in sections 5 and 6 respectively.

4   Text Detection Based on Ridges

This section explains the method for finding text regions in images based on ridges. It consists of 2 stages: (1) computing ridges in scale space and (2) classifying regions corresponding to ridges into 2 classes: text or non-text.

4.1   Computing Ridges at Multiple Scales

This section briefly explains ridge detection. For more technical details, see [9]. Given an image $\mathcal{I}(x,y)$ and its laplacian $\mathcal{L}(x,y)$ a point (x_r, y_r) is a *ridge point* if the value of its laplacian $\mathcal{L}(x_r, y_r)$ is a local maximum in the direction of the highest curvature; it is a *valley point* if the value of its laplacian $\mathcal{L}(x_r, y_r)$ is a local minimum. In the sequel, we use the term "ridge" to indicate these two types of points. Ridge points are invariant to image rotation and translation.

To detect ridge points, we compute the main curvatures and associated directions at each pixel using the eigenvalues and eigenvectors of the Hessian matrix [4]. We then link ridge points to form ridge lines by connected components analysis.

Scale space adds a third dimension σ to the image such that $I_\sigma(x, y)$ is the original image I smoothed by a Gaussian kernel with standard deviation σ. In our system, we use a discrete sampling of scale space, explicitly computing $I_\sigma(x, y)$ for a small number of values $\sigma = \sigma_0 \ldots \sigma_{k-1}$; we then compute ridges for each of these smoothed images to capture structures of different sizes. The values of σ we use are: $\sigma_i = \sqrt{2}^{\,i}$ with level $i = 0, \ldots \log_2(\min(w,h))$ where w, h are image width and height. These computations are carried out in a very efficient way using recursive filters [10]. In practice, if we know the dimensions of characters and text strings, values of i can be limited to a small range. For example in our database, scales 2 to 8 are sufficient.

Figure 2 shows several images and ridges detected at two scales on regions extracted from the image. We can see that for each text, one ridge corresponding to the center line of the text and several small ridges corresponding to the skeletons of characters have been detected. The structure "one center line and lots of small skeletons " is present for many kinds of text (scene text or artificial text) with different character sets (latin alphabet or ideograms). Figure 3 illustrates the independence on orientation of the ridge based text representation.

Fig. 2. First line: Images with rectangle showing the text region. Second line: Zoom on text regions. Third line: ridges detected at two scales (red in high level, blue in small level) in the text region that represent local structures of text lines whatever the type of text (handwritten text or machine text, scene text or artificial text, latin alphabet or ideograms).

(a)   (b)

Fig. 3. (a) Image of a plate. (b) Ridges detected at scale $\sigma = 2$ (blue lines) and $\sigma = 8\sqrt{2}$ (red lines). This figure shows the independence on orientation of the ridge-based text representation.

4.2 Classification of Candidate Text Blocks

The output of the previous step is k images containing ridge lines detected at k scale levels. Now, for each ridge at level i, $i = 0 \ldots k - 1$, we classify the region corresponding to the ridge as text region or non-text region. The region corresponding to a ridge detected at scale σ is defined as the set of points such that the distance from each point to the ridge is smaller than σ. We call the ridge to be considered the *central ridge*, the region corresponding to the ridge the *ridge region* and all ridges at smaller scale in the ridge region which best fit character skeletons the *skeleton ridges*. The scale of the skeleton ridges is half the width of their strokes. It is not necessary that the skeletons and the center line be of the same type ("ridge"[1], "valley"[2]). We propose the following criteria to classify a region corresponding to a central ridge as text. Note that all detected ridges may be considered as central ridge starting from the largest scale σ_{k-1}.

- **Ridge Length Constraint**: Generally, the length of skeleton ridges representing the skeleton of the characters is approximately equal to the height of characters, which is 2 times the scale σ of the *central ridge*. For round characters like O, U, the length can reach up to 4 times σ. So the *skeleton ridge* length must be inside the interval $[\sigma, 4\sigma]$.
  Concerning the *central ridge*, supposing that $nbCharacters$ is the minimal number of characters in each text string, $minlength_{wc}$ is the minimal width of a character. Thus the length of the *central ridge* has to be longer than $nbCharacters * minlength_{wc}$.
- **Spatial Constraint**: With printed latin characters, *skeleton ridges* often are perpendicular to the *central ridge* at their center points. A text detector

---

[1] Local maximum of Laplacian.
[2] Local minimum of Laplacian.

should take into account this property. However, this is not true for some fonts (e.g. italic), and for other character sets (e.g. chinese or japanese). To construct a generic text detection system, we weaken the perpendicularity constraint by applying a non-parallel constraint. Thus, a text ridge region must contain an *important number of skeleton ridges* which are not parallel to the central ridge. Above, we supposed that there is at least $nbCharacters$ in the text string, as each character contributes at least one skeleton ridge, so the *number of skeleton ridges* inside the central ridge region has to be bigger than $\max\{nbCharacters, length_{centralridge}/minlength_{wc}\}$.

5   Experimental Results

5.1   Databases for Experiments

The databases for the testing algorithm contain single images and video frames. The first database (DB1) contains 10 images of a slide presentation. These images are taken by a camera with a resolution of 640x480 with various lighting conditions. The second database (DB2) consists of 45 images from news video[3], some of them having very complex background. The third database (DB3) contains 20 images extracted from formula 1 racing video[4] with a resolution of 352x288. Text in these images have different orientations (not limited to horizontal and vertical orientation) and undergo affine distortions. The fourth (DB4) contains 20 frames of film titles [5]. In this database, images contain text of different kinds (scene text, artificial text, and targeted scene text), sizes and styles. Table1 summarize these databases.

Table 1. Text detection result

	#images	#words	#detected words	#False alarms	Recall(%)	Precision(%)
DB1	10	172	172	7	100	96.09
DB2	26	103	99	48	96.11	66.67
DB3	45	217	169	114	77.88	59.71
DB4	20	199	177	18	88.9447	90.7692

5.2   Evaluation

In our experiments, text size (the height of characters in the text in pixel) varies in the interval [4, 73], the text detection algorithm is computed only at $\lceil 2\log_2(73/2) \rceil = 11$ levels (while the maximal level is $N = \log_2(640x480) = 18$ with image of resolution of 640x480). The reason is that at scales coarser than

---

[3] $http://www.cs.cityu.edu.hk/\ liuwy/PE\_VTDetect/$ used in [13] for evaluation of text detection

[4] $http://www.detect-tv.com$

[5] $http://www.informatik.uni-mannheim.de/pi4/lib/projects/MoCA$

Fig. 4. Sample results of text detection. (a,b,c,d) When the background is homogeneous, detection is correct and does not give false alarms. (e) A table with text, without clear line structure (f) irregular background : there are false positives, and pieces of text are missed.

11, detected ridges represent structures of width larger than $2 * \sqrt{2}^{11} = 90$ pixels which are not textual structures, so ridges have no sense in the context of text detection. In fact, the number of levels to be considered can be determined based on prior information about the maximal size of text in image. In case where any information is provided, we use $N = \log_2(\text{wxh})$, with w, h the width and the height of the image.

The minimum number of characters $nbCharacter$ in each text string used is equal to 2 which appears reasonable because that we attempt to detect text lines, not isolated character. Moreover, we did not take into account points having the normalised Laplacian magnitude smaller than a threshold (here we used 5.0) in order to avoid false detections due to noise or aliasing artifact. As we have no information about the width of character stroke, we do not know exactly what is the scale of skeletons ridges. Thus for each central ridge detected at level k, the skeleton ridges at one among 3 levels $k-3$, $k-2$ and $k-4$ are taken as input of text constraints verifier. The choice of these 3 levels is based on hypothesis that the ratio of height and width of character stroke is in the interval $[2, 4]$.

For evaluation, we use recall and precision measures. Table 1 shows the result of text detection from images in the 4 databases listed above. In the case of slides, we obtain the best recall as well as the best precision (figure 4a). All text regions in slide images are detected and localized correctly. The reason is that the background of slide image is well uniform and characters are distinctive from background. The detection was easily performed. With scene text having an orientation like those in images from the second database (Formula 1 car racing), the proposed algorithm had no difficulty (figure 4b). It is also well robust to noise and aliasing artifacts and it performs the detection of scene text as well as embedded text. In figure 4b, the score was not considered as a text because it appears too opaque in the scene. The performance of detection diminishes when the background is complex (images in the news video frame database) where there are cases of missed pieces of text and false alarms (figure 4e,f). The principal reason of false alarms is that the criterion "one center line and numerous small skeletons" also is satisfied by regions with regular grids. We either have to restrict our model, or these false responses have to be eliminated by an OCR system.

To compare with texture based and contour based methods, we implemented the texture based segmentation algorithm proposed in [12]. We found that with images in our databases, the clustering did not help to focus interest regions to be considered in a later stage. The contour based method fails in case where text is too blurred and scattered. In comparison with [12] where regions must be fused between scales because of "scale-redundant" regions, our approach verifies regions at the largest scale first; if it is a text region, this region will be no more considered later on. Without scale integration, the computation time is reduced significantly.

6   Conclusion

In this paper, we have proposed a novel approach for text analysis and text detection. Unlike traditional approaches based mainly on edge detection and texture, we use ridges as characteristics representing the structure of text lines at different scales. The experimental results show good recall and precision of the method using ridges (average of 90.7% and 78.3% respectively). The strengths of the method lie in its invariance to the size and the orientation of characters, its invariance to the form and the orientation of the lines, and that it works without any change in parameters for different writing systems (alphabets, ideograms). In addition, based on the scales at which we detect the central ridges and the skeleton ridges, the height, the number of characters in the text lines are measured. The current method still gives some false alarms, that can be eliminated by adding constraints on color and length between characters in text string or by using an OCR system.

References

1. D. Chen and J.M. Odobez J.P. Thiran. A localization/verification scheme for finding text in images and video frames based on contrast independent features and machine learning methods. *Signal Processing: Image Communication*, (19), 205-217 2004.
2. P. Clark and M. Mirmehdi. Combining statistical measures to find image text regions. In *Proceedings of the 15th International Conference on Pattern Recognition*, pages 450–453. IEEE Computer Society, September 2000.
3. P. Clark and M. Mirmehdi. Finding text regions using localised measures. In *Proceedings of the 11th British Machine Vision Conference*, pages 675–684. BMVA Press, September 2000.
4. David Eberly. *Ridges in Image and Data Analysis*. Kluwer Academic Publicshers, 1996.
5. H. Li and D. Doermann. A video text detection system based on automated training. In *Proceedings of the International Conference on Pattern Recognition ICPR'00*, 2000.
6. R. Lienhart. Automatic text recognition in digital videos. In *SPIE, Image and Video Processing IV*, pages 2666 2675, 1996.
7. A. Wernicke R. Lienhart Localizing and segmentating text in images and videos. *IEEE Trans. Pattern Anal. Mach. Intell*, 18(8):256–268, 2002.
8. K. Sobottka and H. Bunke. Identification of text on colored book and journal covers. In *International Conference on Document Analysis and Recognition*, pages 57–62, Bangalore, India, September 1999.
9. H. Tran and A. Lux. A method for ridge extraction. In *Proceedings of the 6th Asean conference on Computer Vision, ACCV'04*, pages 960–966, Jeju, Korea, Feb 2004.
10. L.J. van Vliet, I.T. Young, and P.W. Verbeek. Recursive gaussian derivative filters. In *ICPR*, pages 509–514, August 1998.
11. V. Wu, R. Manmatha, and E.M.Riseman. Finding text in images. In *Proceedings of the ACM International Conference on Digital Libraries*, pages 23–26, 1997.
12. V. Wu, R. Manmatha, and E.M. Riseman. Textfind: An automatic system to detect and recognize text in image. *IEEE Transaction on Pattern Analysis and Machine Intelligence, PAMI*, Vol. 21(No. 11):1224–1229, November 1999.
13. L. Wenyin X.S. Hua and H.J. Zhang. Automatic performance evaluation for video text detection. In *International Conference on Document Analysis and Recognition (ICDAR 2001)*, pages 545–550, Seattle, Washington, USA, September 2001.
14. A. K. Jain Y. Zhong, K. Karu. Locating text in complex color image. *Pattern Recognition*, pages 1523–1536, 1995.

Optical Flow-Based Segmentation of Containers for Automatic Code Recognition*

Vicente Atienza, Ángel Rodas, Gabriela Andreu, and Alberto Pérez

Department of Computer Engineering, Polytechnic University of Valencia,
Camino de Vera s/n, 46071 Valencia, Spain
{vatienza, arodas, gandreu, aperez}@disca.upv.es

Abstract. This paper presents a method for accurately segmenting moving container trucks in image sequences. This task allows to increase the performance of a recognition system that must identify the container code in order to check the entrance of containers through a port gate. To achieve good tolerance to non uniform backgrounds and the presence of multiple moving containers, an optical flow-based strategy is proposed. The algorithm introduces a voting strategy to detect the largest planar surface that shows a uniform motion of advance. Then, the top and rear limits of this surface are detected by a fast and effective method that searches for the limit that maximizes some object / non-object ratios. The method has been tested offline with a set of pre-recorded sequences, achieving satisfactory results.

1 Introduction

Currently in most trading ports, the entering and leaving of container trucks are controlled by human inspection. Using techniques of computer vision and pattern recognition, it is possible to build systems that, placed at the gates of the port, automatically monitorize this container activity [12][13]. To achieve that, this kind of systems must be able to recognize the character code that identifies each container, usually located near the top-rear corner. The process can be quite complex: the system has to deal with outdoor scenes involving unstable lighting conditions (changes in climatology, day/night cycle) as well as with dirty and damaged container codes (See Fig. 1). It is also necessary to consider that the truck is moving when the images are acquired. These unfavourable conditions make the recognition process prone to errors.

To increase recognition ratio and performance of the system it is very useful the introduction of a previous container segmentation process. This kind of visual process makes unnecessary the installation of sensors of presence, like the light-barrier sensor used to detect the rear part of a container used in [12][13]. On the other hand, by having an adequate estimation of the top-rear corner of the container in the image it is possible to limit the code recognition process to a restricted area, instead of processing the whole image. This speeds up the recognition process and reduces the

---

* This work has been partially supported by grant CICYT DPI2003-09173-C02-01.

S. Singh et al. (Eds.): ICAPR 2005, LNCS 3686, pp. 636–645, 2005.

apparition of errors, because of the presence of less character-like forms that could be confused with true characters. Finally, by applying the container localization process to every acquired frame it is possible to track the position of the container along the sequence. This makes feasible the matching of multiple recognition results corresponding to different frames. In practice it has been observed that a given character can be correctly recognized in a frame and not recognized in other, due to the presence of moving shadows or reflections, not uniform surface structure, etc. The analysis and integration of the whole sequence can potentially obtain much better recognition results than if performed to individual, unconnected frames.

Fig. 1. Examples of container codes

Different segmentation strategies have been tested, all of them aimed to find the top and rear limits of the moving container. Line detection by Hough transform [8] obtains relatively good results for the top limit of the container, but finds difficulties in detecting a reliable vertical line that corresponds to its rear end. This is due to the frequent presence of vertical lines in the background (buildings, other containers) and the repetitive pattern of vertical lines that usually present the surface of the containers (Fig. 2.a). To overcome the problems involved by the presence of a non-uniform background, motion detection techniques have been considered. Fast and simple techniques like image subtraction [9] offered promising results, but some problems were found due to the presence of multiple moving objects and the operation of the auto-iris lens, which induced fast brightness changes detected as false motion (Fig. 2.b). A more complex but also much more reliable motion detection-based segmentation strategy is described in the next section. It is based on optical flow computation by a block-matching procedure (see similar approaches in [4] or [5]) and the use of a voting strategy to determine dominant motion. Voting processes to estimate motion parameters or perform multiple motion segmentation are widely used in different approaches [3] [7] [10] [11]. Experimental results for the proposed optical

(a)   (b)

Fig. 2. Difficulties found by other segmentation strategies. (a) Line detection by Hough transform. It is difficult to discriminate which vertical line corresponds to the end of the container. (b) Motion detection by image subtraction. Brilliant zones correspond to high difference values. False motion appears due to the auto iris operation (static parts like the sky should be black). A second moving truck is also detected.

flow-based method are also presented, which confirm the suitability of this technique for the application of interest.

2   Optical Flow-Based Segmentation

In this section we propose a segmentation strategy that achieve reliable results, overcoming the problems found by the aforementioned methods. This strategy is based on the calculation of the optical flow [1][2] derived from the comparison of two consecutive frames (f_{k-1}, f_k). By considering the optical flow information and adequate object and image formation models it is possible to obtain velocity vectors for the scene points. The segmentation strategy consists on isolate the largest planar surface of the scene that presents a coherent motion of advance. Consequently, we consider as a natural assumption that the container of interest (that nearest to the camera) corresponds to that which occupies the largest portion of the image.

To estimate the dominant motion we propose the use of a parametrization strategy that allows to exploit the available *a priori* knowledge over the physical motion (leftward direction, maximum velocity) and structure of the target (vertical planar surface, admissible range of orientations). We use two parameters directly related to the magnitudes whose variability more strongly affects the characteristics of the observed motion field: maximum allowed advance of the truck in a frame time (Δl) and angle of orientation of the container surface with respect to the camera axis (α). These real-world measures are intuitive and adequate constraints for them can be easily obtained from the observance of truck orientations and speeds at the port gate. They can remain as adjustable parameters for the system operator, who does not need to be aware of the image-formation model and the camera calibration process.

The estimation of these parameters will be done by means of a voting strategy, detailed in Section 2.3. This voting procedure is intended to obtain a robust estimation of the dominant motion, in spite of the existence of image areas corresponding to static background and the presence of other (small) moving surfaces in the image.

2.1 Optical Flow Calculation

Optical flow vectors are obtained by measuring the displacement experimented by image blocks from image f_{k-1} to image f_k. An optical flow vector is calculated for every 8×8 block in an 8-pixel-wide grid. To obtain the flow vector corresponding to a given block, we search for the displacement that minimizes a similarity measure. Correlation coefficient [6] has been chosen as similarity measure to achieve tolerance to brightness and contrast changes like that occurred due to the auto-iris effect. As we are only interested in containers that moves to the left, the location of the matching area in f_k for every block is established in the x-direction from its position in f_{k-1} to a maximum leftward displacement. The height of this searching area is limited to a few pixels as we are only interested in motion parallel to the ground. Moreover, the y-component of the optical flow can not be reliably determined because the surfaces of containers frequently exhibit lack of texture variation in the y-direction.

Only blocks containing enough grey-level variation in the x-direction are considered to avoid false motion estimations. For that, we select the set $P(k-1)$ of central points of 8×8 blocks that fulfil the following condition:

$$P(k-1) = \{ p(x, y) \mid \sum_{i=y-4}^{y+3} \sum_{j=x-3}^{x+3} |f_{k-1}(i, j) - f_{k-1}(i, j-1)| > \tau \} \tag{1}$$

Were τ is a fixed threshold value. The value for this parameter has been selected experimentally to obtain a high number of points corresponding to the container area, while avoiding the selection of low quality points in low-textured areas like the sky zone. Good results have been obtained by setting this parameter to $1.0 \times block\ size$. An example of the motion field obtained by this method is shown in Fig. 3.

2.2 Object and Image Formation Models

Fig. 4 represents the object model and the image formation model (only X and Z components are considered). T is the surface of the container, modelled as a planar surface normal to the ground plane, which forms an α angle with respect to the optical axis of the camera Z. Z_0 is the distance from the image plane I to the container surface T along the optical axis Z. Vector l represents the position of a point over the container surface with respect to the container point that corresponds to the centre of the image. Vector u is the correspondent measure (x-component) observed over the image plane. f is the focal length of the lens for the pinhole model.

By the Law of Sines, it can be derived the equation

$$l = \frac{(Z_0 + f) \sin(\beta)}{\sin(\gamma)} \tag{2}$$

Fig. 3. Optical flow obtained for 8×8 blocks. To enhance clarity of figure, only vectors in a 16×16 grid are shown (25% of total flow vectors).

where $\beta = \arctan(u/f)$ and $\gamma = \pi - \alpha - \beta$. For simplicity we denote this relationship by the expression $l = L(\alpha, u, Z_0, f)$ or, by dropping the known model parameters: $l = L(\alpha, u)$. We shall use this latter expression to indicate the conversion from image plane x-coordinate (u) to feature position over the container surface (l) in the next section.

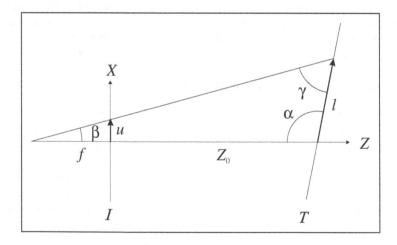

Fig. 4. Image formation model. I: image plane, T: container surface, Z: optical axis.

2.3 Estimating the Motion Parameters of the Largest Container Surface

The structure and motion parameters of the container surface $(\alpha, \Delta l)$ are estimated by a voting strategy. For that, we define a voting matrix $V[\alpha, \Delta l]$, with α and Δl integers in the ranges

$$\alpha \in [90 - \alpha_{max}, 90 + \alpha_{max}], \ \Delta l \in [-\Delta l_{max}, -1]$$

where α_{max} defines the maximum allowed angle deviation (in degrees) for the container surface with respect to the orientation normal to the optical axis Z, and Δl_{max} represents the maximum expected displacement (in centimetres) between consecutive frames. Then, the algorithm works as follows:

Algorithm 1
Initialize voting matrix $V[\alpha, \Delta l]$ to zeroes

$\forall p \in P(k-1)$ **do** /*process all points for which a flow vector is available */

   $\forall \alpha \in [90 - \alpha_{max}, 90 + \alpha_{max}]$ **do** /* try all feasible values for α */

   Compute $\Delta l(p, \alpha) = L[\alpha, u(p + \phi(p))] - L(\alpha, u(p))$

   where:

   $u(\bullet)$ represents the x-coordinate of a point in the image plane, according to the centred coordinate system of the image formation model

   $\phi(p)$ is the optical flow vector obtained for point p

   if $0 < \Delta l(p, \alpha) \le \Delta l_{max}$ **then** $V[\alpha, \Delta l(p, \alpha)] \leftarrow V[\alpha, \Delta l(p, \alpha)] + 1$ /* Vote for $(\alpha, \Delta l(p, \alpha))$ */

Select $(\hat{\alpha}, \Delta \hat{l}) \mid V[\hat{\alpha}, \Delta \hat{l}] = \max(V)$ /*select the most voted pair */

Let $M = \{p \in P(k-1) \mid \Delta l(p, \hat{\alpha}) = \Delta \hat{l}\}$ /*select the points whose corresponding flow vectors match with $(\hat{\alpha}, \Delta \hat{l})$ parameters */

   This strategy assumes that the larger a container surface is, the greater number of optical flow vectors will correspond to it, for a given pair of $(\hat{\alpha}, \Delta \hat{l})$ parameters. As result of this algorithm we obtain the set M of image points whose flow vectors correspond to the largest moving surface (see example of Fig. 5).

2.4 Determining the Limits of the Container

As result of the previous step, we get a set of optical flow vectors that correspond to the container surface. Next, we need to determine the top and rear (right) limits of this surface. We propose an algorithm that will try to find the position of a vertical line that mark the right limit of the container and an horizontal image line that marks the top limit of this area. In other words, these lines will mark the separation between object/non-object image areas in horizontal and vertical directions.

   The algorithm processes the optical flow images (8-times less resolution than the original grey-level images). This optical flow images are binarized to obtain image B in this way: a pixel $B(x, y)$ is set to *object* (container surface) if the point $p(x, y)$

belongs to M , and is set to *background* in other case. Then, the algorithm searches for the rear limit of the container, by maximizing a measure of *object quality* to the left of the tentative limit and *background quality* to its right:

Fig. 5. Optical flow vectors that correspond to the most voted pair of surface parameters. Their base points form set M.

Algorithm 2

$max\_quality = 0$

for $x\_limit = image\_width\text{-}1, image\_width\text{-}2,\ldots,1$

$\quad obj\_quality = obj\_pixels\_to\_the\_left / bckgnd\_pixels\_to\_the\_left$

$\quad bckgnd\_quality = bckgnd\_pixels\_to\_the\_right / obj\_pixels\_to\_the\_right$

$\quad limit\_quality = obj\_quality * bckgnd\_quality$

$\quad$**if** $limit\_quality > max\_quality$ **then**

$\quad\quad max\_quality = limit\_quality;\ best\_limit = x\_limit$

The algorithm that searches for the top limit works in the same way. Fig. 6 shows the limits found by means of this strategy for a frame of example.

3   Experiments and Discussion

This segmentation strategy has been applied to a set of pre-recorded video sequences. A digital video camera was installed in a truck gate at the container terminal of the Valencia port (Spain). This camera was equipped with an auto-iris lens and a monochrome, non-interlaced ½" CCD sensor. The camera captured lateral views of the moving trucks in the selected lane from an approximate distance of 2 meters.

Fig. 6. Top and rear limits found for the container area

Images were digitized at 768×572 pixel resolution by a computer equipped with an acquisition card and stored to hard disk. These images were corrected for the slight barrel distortion effect introduced by the 8 mm lens to assure undistorted straight lines for container limits of the target trucks. The effect of this correction can be noticed by observing the pin cushion effect induced in distant lines like that corresponding to the street lamp to the right end of the image in Fig. 6.

To obtain numerical results, a total of 200 frames corresponding to 18 different sequences were manually inspected and the position of the top-rear corner of the container in every one of them was annotated. This working sequences were acquired under day-light illumination and included situations of multiple moving trucks, cluttered backgrounds and fast global illumination changes. These changes are due to the auto iris operation to compensate for the light increase caused by the motion of a truck that progressively uncovers part of the sky zone. Table 1 shows mean values and standard deviation of errors obtained in the automatic estimation of the corner points for these sequences (errors are computed with respect to the manually estimated position of the points).

Table 1. Statistics for errors committed in the estimation of corner position (pixels)

	mean error	std. deviation
x-direction	-2.2	8.4
y-direction	2.8	7.5

Fig. 7 shows results obtained for two of the processed sequences. The figure depicts segmentation results corresponding to cases for which other segmentation methods found difficulties due to the presence of vertical lines in the background and other moving containers. Flow vectors corresponding to the segmented surface are also shown.

This experiments show satisfactory segmentation results. Low mean error values denote a non-biased estimation (these non-zero mean errors are probably due to the small protrusion that invariably appears in the corners of containers, which affects the manually estimated location of the top-rear corner, as well as to their slightly slanting position). The magnitude of standard deviations are in accordance with the 8-pixel resolution established in the flow estimation process. The flow-based method performs well independently of the static texture present in the background. This is an advantage with respect to static segmentation strategies as those based on edge detection techniques. On the other hand, the systems demonstrates good tolerance to rapid brightness changes as those induced by the operation of the auto iris lens, differently from the inadequate behaviour presented by the image subtraction strategies.

However, optical flow computation by means of block matching techniques had proved to be a time consuming task. The current off-line implementation of the algorithm works at 3 fps on a 2.4 GHz Pentium-4 computer. The current implementation of the recognition module takes a mean of 0.7 seconds to process a whole image (1.4 fps). It is expected that the inclusion of the presented segmentation process will allow the reduction of the size of the area processed for recognition purposes by a factor of 6, obtaining a similar reduction in recognition time. Some optimization efforts would have to be done in the implementation of the segmentation step if higher frame ratios were required.

Fig. 7. Example frames from 2 segmentation sequences

References

[1] Anandan, P., A Computational Framework and an Algorithm for the Measurement of Visual Motion, Int. J. on Comp. Vision , Vol.2 (1989) 283–310

[2] Barron, J.L., Fleet, D.J., Beauchemin, S.S., Performance of optical flow techniques, International Journal of Computer Vision, Vol. 12, n. 1, (1994) 43–77

[3] Bober, M., Kittler, J., Estimation of Complex Multimodal Motion: An Approach Based on Robust Statistics and Hough transform, Image and Vision Computing, Vol. 12 (1994) 661–668

[4] Coifman, B., Beymer, D., McLauchlan, P., Malik, J., A Real-Time Computer Vision System for Vehicle Tracking and Traffic Surveillance, Transportation Research: Part C, Vol 6, no 4 (1998), 271–288

[5] Di Stefano, L., Viarani, E., Vehicle Detection and Tracking Using the Block Matching Algorithm, Proc. of "3rd IMACS/IEEE Int'l Multiconference on Circuits, Systems, Communications and Computer, Vol. 1 (1999) 4491–4496

[6] Gonzalez, R. C., Woods, R. E., Digital Image Processing, Addison-Wesley (1993)

[7] Hill, L., Vlachos, T., Optimal Search in Hough Parameter Hyperspace For Estimation of Complex Motion in Image Sequences, IEE Proc.-Vis. Image Signal Process., Vol. 149, n. 2 (2002) 63–71

[8] Illingworth, J., Kittler, J., A Survey of the Hough Transform. Computer Vision, Graphics, Image Processing, Vol. 44 (1988) 87–116

[9] Jain, R. C., Difference and Accumulative Difference Pictures in Dynamic Scene Analysis, Image and Vision Computing, Vol. 12, n. 2 (1984) 99–108

[10] Kang, E.-Y., Cohen, I., Medioni, G., Non-Iterative Approach to Multiple 2D Motion Estimation, Int. Conf. on Pattern Recognition (ICPR'04), Vol. 4 (2004) 791–794

[11] Nicolescu, M., Medioni, G., A Voting-Based Computational Framework for Visual Motion Analysis and Interpretation, IEEE Transactions on Pattern Analysis and Machine Intelligence, Vol. 27, n. 5 (2005) 739–752

[12] Salvador, I., Andreu, G., Pérez, A., Detection of identifier code in containers, IX Spanish Symposium on Pattern Recognition and Image Analysis, Vol. 2 (2001) 119–124

[13] Salvador, I., Andreu, G., Pérez, A., Preprocessing and Recognition of Characters in Containers Codes, Proceeding of the International Conference on Pattern Recognition (ICPR-2002) (2002) 101–105

Hybrid OCR Combination for Ancient Documents

Hubert Cecotti and Abdel Belaïd

READ Group, LORIA/CNRS, Campus Scientifique BP 239,
54506 Vandoeuvre-les-Nancy cedex France

Abstract. Commercial Optical Character Recognition (OCR) have at lot improved in the last few years. Their outstanding ability to process different kinds of documents is their main quality. However, their generality can also be an issue, as they cannot recognize perfectly documents far from the average present-day documents. We propose in this paper a system combining several OCRs and a specialized ICR (Intelligent Character Recognition) based on a convolutional neural network to complement them. Instead of just performing several OCRs in parallel and applying a fusing rule on the results, a specialized neural network with an adaptive topology is added to complement the OCRs, in function of the OCRs errors. This system has been tested on ancient documents containing old characters and old fonts not used in contemporary documents. The OCRs combination increases the recognition of about 3% whereas the ICR improves the recognition of rejected characters of more than 5%.

1   Introduction

Combining multiple classifiers has been recently a topic of great interest in pattern recognition and character recognition. It has been shown in the literature that several schemes can outperform individual classifiers in order to increase the performance [1,5,10,15]. To obtain optimal classification system, several classifiers can be combined in a first step. However the global performance will depend on how they are complemented. Several problems occur in the process of the obtaining the best result. In the beginning, the needed complement classifiers have to be chosen. After, the multi-classifier architecture has to be created. With a finite number of classifiers, the best combination method has to be found. This choice depends on the particular number of classifiers, their behavior, and the size of the training data available for the combination... We propose a hybrid model where the first stage of the system composed of the classifiers connected in parallel whereas the second part is a special neural network specialized in rejected characters of the first stage. This system has been tested on ancient documents. The classifiers of the first stage are commercial OCRs and the specialized classifier is a convolutional neural network. In a first part we will describe the system and then the strategy used to extract and analyze errors. The third

S. Singh et al. (Eds.): ICAPR 2005, LNCS 3686, pp. 646–653, 2005.

part will describe the relationship between the OCRs combination and the classifier specialized in rejection. Final part shows the recognition improvement given by our approach.

2   System Overview

As was mentioned before the paper describes a character recognition system tailored specifically to the ancient documents. Commercial OCRs are usually trained to recognize all kind of documents and are not specialized for one of them particularly. As a consequence, these generic OCR characteristics ensure a good performance on the majority of the characters. However they unavoidably lead to a low proportion of the documents, to bad performances, as they are less frequent or do not correspond to the trained character models. The use of OCRs for old printed documents is always impeded by the presence of some characters, which cannot be not well recognized because of their unknown patterns or their deformations. OCRs act in different ways depending on the quality of the document. Thus their combination should give a better performance on the classical characters. In order to continue to take advantage of OCRs performance on well-written characters, the first stage is completed by an additional ICR (Intelligent Character Recognition) capable of adapting its topology on the confusion errors. The error analysis is a very important task in qualifying the error type that should be processed further. It highlights rejected patterns during the test phase. The specific ICR is able to correct the error by specifying its own topology. The OCRs combination has to enhance the OCR performances on common characters written in known font styles and to qualify the rejection errors.

3   Combination Strategy

3.1   Combination Topology

Multi-classifiers systems can be divided into three main categories depending on their topology. The first category represents vertical combination schemes: serial combinations. Each classifier is performed sequentially. For example, each classifier is specialized to process the rejected patterns of the previous classifier. In this case, each classifier is tailored in order that participates to a multi-classifier system. The second category represents horizontal combination schemes: parallel combination. The classifiers work independently and concurrently. A fusion module combines their results. In this solution, there are any direct relationships between the classifiers. Some classifiers may behave in the same way whereas others may well complement themselves to achieve great improvement. The choice of this solution can also be driven by practical case. Indeed, only the classifier results are used for the combination. It allows the use of already existing classifiers. The last category, the hybrid combination schemes, corresponds to the use of the two previous schemes. Each combination strategy has its drawbacks;

the first strategy assumes complementary classifiers whereas the second assumes competitive classifiers. In the proposed system, as the recognition algorithm and the training data used for each OCR are not be available; the choice of the combination methods is limited. OCR outputs also give the confidence value of only the first best choice. The system is a hybrid combination. A horizontal topology scheme is used for the OCRs combination as the first step of the system. The relationship between the OCRs combination and the ICR follows a vertical scheme where the ICR processes rejected characters.

3.2   Fusing Rules

Several combination methods are described in the literature for fusing results: voting methods, Bayesian combination, Dempster-Shafer, behavior-knowledge space, neural networks, decision trees, etc. Among them, several methods have been tested for the OCR combination: majority voting, behavior-knowledge space and neural networks allowing a double resolution of the label and the error type. The two first methods work on abstract level and the last one works on measurement level. The majority voting method is an easy method to implement and it has shown good results in the literature [2,7]. The voting method has as consequence to avoid correcting the common errors and leads to reject the maximum of errors instead of correcting them. Considering only two OCRs, the phenomena are accentuated, as there are less correction possibilities. Votes can be weighted with OCR knowledge. Although this method with only two OCRs may not always improve the recognition rate, it can improve the reliability of the results. The classifiers must act independently in order to combine them by the method based on conditional probability as formulated in the Bayes rule. Indeed, this condition is not easy to verify. That is why we preferred to use the behavior-knowledge space (BKS) [6] method that makes no assumption about the classifier dependence. A behavior knowledge space is a D-dimensional space, each dimension corresponding to the decision of one classifier. Each classifier has as decision values the total number of classes N. Let $x \in C_i$ be the character to be recognized belonging to the class C_i. Let $s_k = j_k, k \in \{1..D\}$ be the kth classifier among D and j_k its answer for the current character x. The probability that $x \in C_i$ is defined by the following formulae:

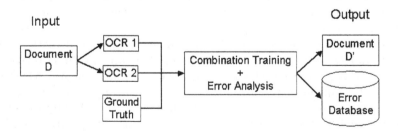

Fig. 1. Errors extraction and combination process

$$Belief(C_i) = \frac{P(s_1(x) = j_1, .., s_D(x) = j_D, x \in C_i)}{P(s_1(x) = j_1, .., s_D(x) = j_D)}$$

A small database may be a problem in obtaining a good generalization and many empty cells may occur if the database is not representative. As the BKS size increases exponentially with the number of classifiers, the data sets has to increase in the same way [12]. For BKS cells where the most representative class is represented by a low probability or where the cell is empty, the class is ambiguous and characters are rejected.

Neural networks can also achieve such combination [14]. Our approach with the neural network consists in extracting simultaneously the label and the error type. The neural network in our experiments takes as input two vectors V_α and V_{error}. V_α is the confusion vector for each character, whereas V_{error} is the error status of the character according to the error types: confusion, addition, deletion, segmentation, etc. The network is able to perform both a simultaneous error and label analysis. Moreover, the neural network can solve some generalization problem of the BKS empty cells. Nevertheless, the neural network learning phase as the testing phase is slower than the BKS ones, which is fully statistics. The final result of the combination is, for each character c, the class C_{m0}. For a text, we note LC_{m0}, the character list results (D' is the output document is the figure 1).

4   Errors Extraction and Categorization

After the combination process, errors have to be extracted properly to specify the ICR. The OCRs combination complements and solves the ambiguous cases; the ICR needs the OCRs combination errors to adapt its topology. This analysis is obtained by differentiating the OCRs results and a ground truth of the document. The error analysis of the differentiation of the OCRs combination or the individual OCR results will provide all the information needed to create the specialized ICR, directly complementary to the combination. OCRs commit several types of errors as: confusion, addition, deletion or segmentation. The ICR will employ these errors in order to adapt its topology. Error types are detected by a comparison between two character lists: L_{GT} representing the ground truth and LC_{m0} obtained by OCRs combination. An appropriate dynamic programming algorithm is used to optimize the alignment of the lists.

The five main types of errors are: the confusion, the addition, the deletion, the fusion and the cutting. These errors correspond to rules extracted by using the edit distance [3,13]. The error types mentioned below are easy to determine when the error chains are small. Inversely, the errors meaning are very difficult to locate when the erroneous chains are large. The error covers several contiguous characters making the error type difficult to determine, as the correspondence between characters is not obvious. The alignment problem is transforming in locating the error origin in this long erroneous chain. The errors are located recursively based on the erroneous chain lengths, starting from the small errors to detect the biggest ones. The procedure starts by locating the small erroneous chains in the entire document. If one of the found errors occurs in the largest erroneous chain, this chain is divided in two parts: prefix before the known error, and suffix, after the error, which are both recursively analyzed according to other smaller errors detected in the document. It is obvious that this approach can work only when the erroneous chain length is reasonably large. Errors that cannot be analyzed properly are ignored for the estimation. If a very noisy textual part of the document is recognized as an image, there will be a long erroneous chain corresponding to a deletion. However, this deletion is produced by a global mistake of the OCRs and does not correspond to the real OCRs behavior for each letters of the chain. Once the errors are detected, a probability function defined by: M_r: confusion matrix, M_f: fusion matrix, M_s: cutting matrix, V_a: addition vector, V_d deletion vector is generated. For a character x recognized as belonging the class j, if its recognition rate is greater than a threshold representing the desired document quality, then the character is accepted. Otherwise the characters image and the different confusion classes of j are given as input of the ICR.

5   Error Correction

A specialized classifier: ICR is dedicated to the character recognition for the error correction, acting directly on the image pixels of the rejected characters. This ICR is a modified multi-layer Perceptron with convolutional layers [8,16]. The neural network is composed of 5 layers. The first one corresponds to the input image, normalized by its centering. The next two layers correspond to the information extraction, performed by convolutions layers using weight sharing. The fourth layer is composed of neurons pool. Each pool is specialized for a class. The links between the fourth and the last layers are function of the error previously detected. The last one corresponds to the output with a number of neurons equal to the total number of classes. The confusion is the error that can be performed by the ICR. However, as we do not know if the image corresponds to a character image or to a character portion occurred as a result of a segmentation

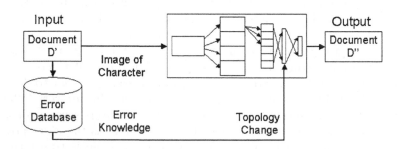

Fig. 2. Rejection processing

problem, the confusion matrix M_r is weighted by the addition and deletion vectors. The remaining errors are not integrated because they do not directly step at the character level (i.e. image level) but they allow better to qualify the confusion measures. The erroneous character image is taken into account by the ICR if and only if the maximum of the confusion rate is lower than a fixed threshold S (i.e. $max_i(M_r(i, C_{m0})) < \alpha$). Considering the image and its associated result j. In the fourth and last layers, for each neurons pool k, they are weighted by $M_r(j, k), k \in \{1..N\}$. The ICR does not work alone for the rejected patterns but uses the OCR behavior knowledge.

6   Experiments

The system has been tested on ancient documents [11]. These documents are extracted from a French dictionary of the XVIII century: "dictionnaire de Trevoux" containing special characters not used in actual documents any more, so naturally disturbing the OCRs. For example, the letter "s" can be written in two shapes: the standard "s" and the long "s". In this case the OCR has a confusion problem between the character "s" and "f", which looks like the long "s". Then the neural network specializes its topology to differentiate shapes of "s" and "f" to reduce the confusion. In this case, the neural network is a tool used not only to solve ambiguous case, but also to differentiate classes. The database is composed of 8 pages of the dictionary chosen to be the best representatives. The pages are scanned at 300dpi and were binarized. Each selected pages is in two columns and contains about 8000 characters. Half of the documents are used for training and the other half is used for testing. The results obtained just by combining 2 OCRs are shown in table 1. The used commercial OCRs are FineReader7 [4] and

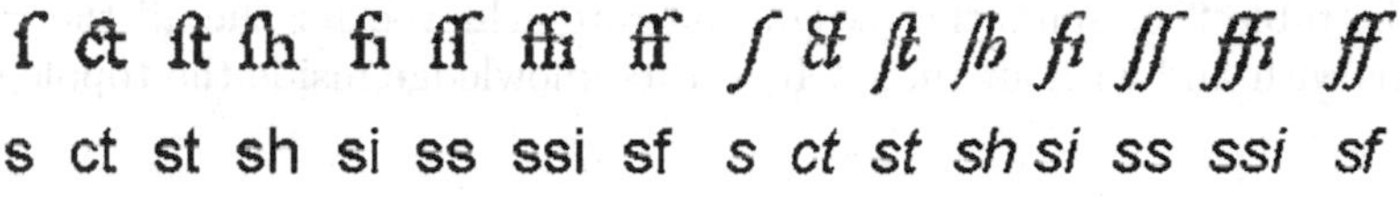

ſ & ﬅ ﬂ ﬁ ﬀ ﬃ ﬀ   ſ & ﬆ ﬂ ﬁ ﬅ ﬃ ﬀ
s ct st sh si ss ssi sf   s ct st sh si ss ssi sf

Fig. 3. Peculiar characters

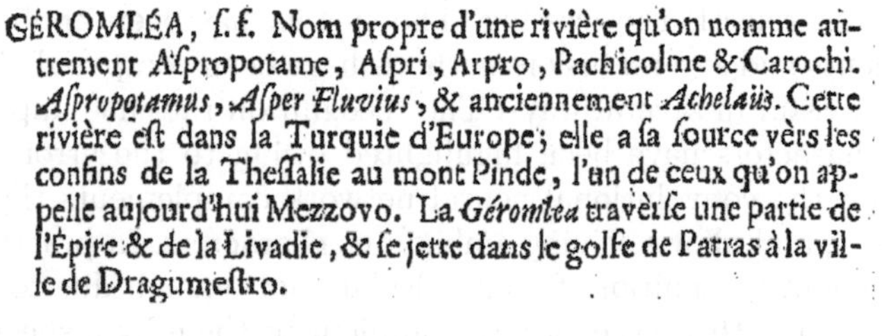

GÉROMLÉA, ſ. f. Nom propre d'une riviére qu'on nomme autrement Aſpropotame, Aſpri, Arpro, Pachicolme & Carochi. *Aſpropotamus, Aſper Fluvius,* & anciennement *Achelaüs.* Cette riviére eſt dans la Turquie d'Europe; elle a ſa ſource vèrs les confins de la Theſſalie au mont Pinde, l'un de ceux qu'on appelle aujourd'hui Mezzovo. La *Géromlea* travèrſe une partie de l'Épire & de la Livadie, & ſe jette dans le golfe de Patras à la ville de Dragumeſtro.

Fig. 4. A document extract

Table 1. OCR and OCRs combination results on the test database

	Recognition	Rejection	Confusion	Addition	Deletion	Cut	Fusion
OCR 1	85,14	0,06	6,71	3,28	2,83	0,92	1,06
OCR 2	87,28	0,04	6,12	3,34	1,61	1,21	0,40
Majority Voting	78,59	11,72	2,08	6,84	2,08	0,10	0,03
Oracle Voting	94,07	5,16	0,00	0,77	0,00	0,00	0,00
BKS	88,48	0,01	6,05	0,73	3,45	0,28	0,99
NN	88.18	0,00	5,88	0,87	3,61	0,41	1,05

Table 2. Recognition rate for ambiguous characters

	Classical Topology	Adaptive Topology
Train	96.63	99.28
Test	93.59	99.27

Omnipage12 [9]. No special dictionary has been used for the OCRs recognition. OCRs were first trained on the documents before the combination. The BKS and the neural network give the best improvements. In the rejected pattern and the confusion classes, the main errors are due to noise, the presence of accents, or the new classes not found by OCR. The more representative error for these documents is the confusion between the characters "f" and "long s", which have almost the same shape. Figure 3 presents some special characters like the long "s" and ligature characters with their corresponding in Arial font. If we consider only the 4 classes: "f,i,l,s" the first OCR has 91.03% and the second has 93.27%. The two "s" are considered as the same character during the OCR combination as they cannot separate the classes. With the OCR combination, the recognition rate is 99.09% for the 4 previous classes. It solves the ambiguity between the 2 "s". The table 2 presents the recognition rate achieved for the all the rejected patterns with and without the OCRs errors knowledge inside the topology.

7   Conclusion

A hybrid multi-classifier model has been presented using a specialized neural network for the rejection processing. The system has been applied on the ancient documents and several fusion methods have been compared. This approach has been successful in several ways. The recognition rate has improved and the ambiguous characters have been highlighted owing to the error analysis. The ICR based on the convolutional neural network complements the OCR owing to its topology, and allows solving ambiguous characters. In practical cases, like ancient document recognition, fusion rules are not enough due to the OCR behaviors. It becomes then necessary to complement them by a specific tool when the system can be tailored for the one kind of documents.

Acknowledgements

We wish to thank the ATILF laboratory (*Analyse et Traitement Informatique de la Langue Franaise*) and especially Prof. J.-M. Pierrel and I. Turcan for providing the input images.

References

1. Bahler, D., Navarro, L.: Methods for Combining Heterogeneous Sets of Classifiers. 17th Natl. Conf. on Artificial Intelligence (AAAI 2000), Workshop on New Research Problems for Machine Learning. (2000)
2. Belaid, A., Anigbogu, J.C.: Use of many classifiers for multifont text recognition. Traitement du signal, vol. 11, no. 1, (1994) 57–75
3. Damereau, F.: A technique for computer detection and correction of spelling errors. Communications of the ACM, vol. 7 (1964) 649–664
4. FineReader 7, ABBYY, http://www.abbyy.com/finereader_ocr/
5. Gunes, V., Menard, M., Loonis, P., Petit-Renaud, S.: Systems of classifiers: state of the art and trends. International Journal of Pattern Recognition and Artificial Intelligence (IJPRAI), 17(8), World-Scientific. (2004)
6. Huang, Y.S., Suen, C.Y.: A method of combining multiple experts for the recognition of unconstrained handwritten numerals. IEEE Trans. On Pattern Analysis and Machine Intelligence, vol. 17, no. 1 (1995) 90–94
7. Lam, L., Suen, C.Y.: Application of majority voting to pattern recognition: an analysis of its behavior and performance", IEEE Trans Pattern Anal Mach Intell, vol. 27, no. 5, (1997) 553–568
8. LeCun, Y., Bottou, L., Bengio, Y., Haffner, P.: Gradient-Based Learning Applied to Document Recognition. Proceedings of the IEEE, vol. 86, no. 11, (1998) 2278–2324
9. Omnipage 12, Scansoft, http://www.scansoft.com/omnipage/
10. Rahman, A.F.R., Fairhurst, M.C.: Multiple classifier decision combination strategies for character recognition: A review. International Journal on Document Analysis and Recognition (IJDAR) vol. 5 (2003) 166–194
11. Ribeiro, C.S., Gil, J.M., Caldas Pinto, J.R., Sousa, J.M.: Ancient document recognition using fuzzy methods. Proc. of the 4th International Workshop on Pattern Recognition in Informations Systems. (2004) 98–107
12. Raudys, S., Roli, F.: The Behavior Knowledge Space Fusion Method: Analysis of Generalization Error and Strategies for Performance Improvement. Multiple Classifier Systems, 4th International Workshop, Multiple Classifier Systems (2003) 55-64
13. Seni, G., Kripasundar, V., Srihari, R.: Generalizing edit distance for handwritten text recognition. In Proceedings of SPIE/IS&T Conference on Document Recognition. San Jose, CA. (1995) 54–65
14. Sharkey, AJC.: Combining artificial neural nets: ensemble and modular multi-net systems. Perspectives in neural computing. Springer, Berlin Heidelberg New York, (1999)
15. Suen, C.Y., Lam, L.: Multiple classifier combination methodologies for different output levels. Springer-Verlag Pub., Lectures Notes in Computer Science, Vol. 1857 (J.Kittler and F.Roli Eds.) (2000) 52–66
16. Teow, L., Loe, K.-F.: Robust vision-based features and classification schemes for off-line handwritten digit recognition. Pattern Recognition vol. 35, no. 11, (2002) 2355–2364

New Holistic Handwritten Word Recognition and Its Application to French Legal Amount

Abderrahmane Namane[1,2], Abderrezak Guessoum[1], and Patrick Meyrueis[2]

[1] Université de Saâd Dahleb de Blida, Faculté des Sciences de l'Ingénieur,
département d'Electronique, Laboratoire de Traitement du Signal et de l'Image,
route de Soumaâ, BP. 270, Blida, Algeria
Tel : (213) 25433850; fax : (213) 25431164 / 433850
namane_a@yahoo.fr
[2] Université Louis Pasteur, Ecole Nationale Supérieure de Physique,
Laboratoire des Systèmes Photoniques, bld. S. Brant,
67400 Illkirch, Strasbourg, France
Tel : (33) 390244618, fax : (33) 390244619

Abstract. This paper presents a holistic recognition of handwritten word based on prototype recognition. Its main objective is to arrive at a reduced number of candidates corresponding to a given prototype class and to determine from them the handwritten class to be recognized. The proposed work involves only an accurate extraction and representation of three zones namely; lower, upper and central zones from the off-line cursive word to obtain a descriptor which provides a coarse characterization of word shape. The recognition system is based primarily on the sequential combination of Hopfield model and MLP based classifier for prototype recognition yielding the handwritten recognition. The handwritten words representing the 27 amount classes are clustered in 16 prototypes or models. These prototypes are used as fundamental memories by the Hopfield network that is subsequently fed to MLP for classification. Experimental results carried out on real images of isolated wholly lower case legal amount bank checks written in mixed cursive and discrete style are presented showing an achievement of 86.5 and 80.75 % rate for prototype and handwritten word recognition respectively. They confirm that the proposed approach shows promising performance results and can be successfully used in processing of poor quality bank checks.

1 Introduction

The automatic reading of handwritten writing is of considerable interest in the achievement of the tiresome tasks such as those which one meets in certain fields: reading of the postal checks, bank checks, reading of command... etc. The reading of the bank checks is one of the most significant applications of the writing recognition. Each day, a bank sorts thousands of checks, which makes the operation of treatment fairly expensive. The recognition of the bank checks presents a big challenge of research in the field of recognition and document analysis. A reasonably high rejection rate could be allowed for the system of treatment of a bank check, but the error rate in the recognition must be as small as possible. Thus, the system of

S. Singh et al. (Eds.): ICAPR 2005, LNCS 3686, pp. 654–663, 2005.
© Springer-Verlag Berlin Heidelberg 2005

treatment must be able to effectively treat the various styles of the written data. Moreover, it should carry out the exact signature checking by using only a small number of authentic reference specimens. Therefore algorithms of high exactitude are employed for the recognition of writing and signature checking [1]. The courtesy amount, the legal amount, and the signature zone are treated independently. The results obtained by the courtesy amount recognition and the legal amount recognition module are then evaluated by a validation module. The handwritten techniques of the writing recognition can be classified in two principal categories: analytical and global [2]. In the analytical approach, the word is first segmented in characters or in parts of characters (pseudo-character), and then the various characters (or pseudo-characters) are identified with specified models [3]. Since many combinations of characters are not readable, contextual post processing is carried out to detect errors and to correct them using a dictionary [4]. The advantage of this approach is that only a little number of models or references is necessary for all the words, and the principal disadvantage is that the approach is likely to lead to segmentation errors. In order to reduce segmentation errors, some methods employ implicit segmentation techniques. They carry out recognition and segmentation at the same time [5]. However, they cannot completely avoid segmental errors. The individual character models ignore the relationship among neighboring characters in a cursive word. Figure 1 shows the influence on the characters "u" by its preceding characters.

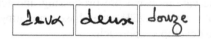

Fig. 1. Influence on the character "u" by its preceding character

Another type of method is the global or holistic solution [6][7][8], which identifies a word as a simple entity. The global approach recognizes a word as only one entity by the use of its characteristics in entirety without consideration of the characters. The word is represented by a vector or a list of primitives independent of the identity of the characters present. The global solution can avoid the segmental errors, but it needs at least a prototype or model for each word. Because this approach does not treat characters or pseudo-characters and does not employ the relationship among neighboring characters, they are usually regarded as tolerant with the dramatic deformations which affect the cursive unconstrained writings [9]. A principal disadvantage of the global methods is that the lexicon can only be updated by addition of word samples. On the other hand, it is considered to be tolerant with the deformations which relate to cursive scripts.

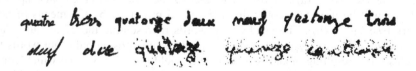

Fig. 2. Poor quality handwritten French legal amounts

The analytical approaches are sensitive to the style and the quality of writing (see Fig. 2), as they are strongly dependent on the effectiveness of the segmentation procedure. Whereas the global solutions are generally employed in the fields with a lexicon of reduced size. Although many algorithms were developed by using the two approaches [10][11], the handwritten word recognition still represents a challenge for the scientific community.

D. Guillevic and C.Y. Suen [10], used a combination of a global feature scheme with a hidden Markov model (HMM) module. The global features consist of the encoding of the relative position of the ascenders, descenders and loops within a word. The HMM uses one feature set based on the uncertain contour points as well as their distance to the baselines. The developed system was also applied to a balanced French database of approximately 2000 checks with specified amounts. Paquet and al. [11], investigate three different approaches for the global modeling and recognition of words used to write the legal amount on French bank checks. A lexicon of 27 amounts was used, written in mixed cursive and discrete styles. The first model is a global one since it does not require any explicit letter level and the two others are based on an analytical approach. The three approaches have been tested on real images of bank checks scanned for the French Postal Technical Research Service (SRTP). De-Almendra-Freitas-CO and al. [12], presented a system for the recognition of the handwritten legal amount in Brazilian bank checks. Their recognizer, based on hidden markov models, does a global word analysis. The word image is transformed into a sequence of observations using pre-processing and feature extraction stages. Their experimental results, when tested on database simulating Brazilian bank checks, show the viability of the developed approach.

The paper is organized as follows: section 2 presents briefly the proposed method. Section 3 presents the preprocessing of handwritten word. Section 4 describes briefly the prototypes creation. Finally, section 5 presents experimental results.

2 Proposed Method

In this paper we present a new holistic recognition of handwritten words applied to French legal amount of bank checks containing wholly lower case characters. The proposed system is addressed to poor and medium quality handwritten word recognition where generally the analytical approach fails through the segmental errors (see Fig. 2). The proposed system is based primarily on two main phases; preprocessing phase yielding a prototype pattern and a recognition phase which classifies the incoming prototype pattern (preprocessing output) as illustrated by Fig. 3. The preprocessing applied consists in filtering, binarizing and correcting the handwritten word in order to enhance the three zones namely; upper, lower and middle zone (see Fig. 3 (a)). Starting from these zones we end to a global shape of the word that will be used to recover the memorized prototype by the means of neural network (see Fig. 3 (b)). The handwritten words representing the 27 amount classes are clustered in 16 prototypes or models created by superposing only four words from each class. The resulting model is processed to yield a final prototype which consists of one memory of Hopfield model of the sequential method [13] as described by Fig. 3. Due to some degradation in the handwritten word we have avoided the use of handwritten contour and the structural components as it was usually used. The main objective of the present recognition system is to arrive at a reduced number of candidates (β_k) and to determine from them the handwritten class to be recognized. Each prototype or model corresponds to a finite number of candidates. As an example, the second prototype corresponds to the handwritten words ($\beta_2=3$); "deux", "trois" and "dix" that have the same global shape. Once the prototype is classified using the prototype recognition system, the candidate choice is based on the presence of dot in the upper zone and the transition number of character-background of the first word character.

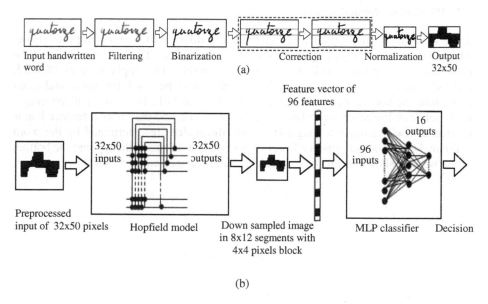

Fig. 3. Proposed method illustration example (a) Handwritten word preprocessing (b) sequential combination method for prototype recognition

3 Handwritten Word Preprocessing

3.1 Word Horizontal Slant

In handwritten word, one of the major variation in writing style is caused by slant, which is defined as the slope of the general writing trend with respect to the vertical line. This operation is performed after skew correction and central zone localization (SC-CZL) [14]. The horizontal slant correction operation consists in correcting the tilted character in order to return it right. The slant correction method [15], which is based primarily on the lower and upper centroid (SC-LUC) is used. The application of the slant correction on the vertical segment separately of the image shown in Fig. 4 (d) yields a slant estimation angle of 50.0 degrees. In our case the horizontal slant correction consists in rotating rather shearing horizontally this segment using as origin the upper reference line. A comparison result of the application of this method and contour method is shown in Fig. 4 (e) and (f) respectively. We note that in Fig. 4 (e), the ascender is well pronounced to the image top than in Fig. 4 (f).

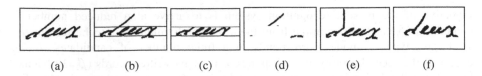

(a)   (b)   (c)   (d)   (e)   (f)

Fig. 4. Slant correction (a) Original image (b-d) result of application of SC-CZL (e) Result of application of SC-LUC on (d). (f) Slant correction based on contour slant estimation

3.2 Word Size Normalization

Since word sizes differ significantly, this makes it difficult to feed the Hopfield model which requires a standard pattern size for all entries. Thus, images should be normalized to a given standard size of 32x50 pixels. The application of SC-CZL method, yields the three zones of interest that will be used for horizontal slant correction and normalization as shown in Fig. 5 (c) and (d). In the normalized image, the baseline or lower reference line is located at y_{lo}=12, and the upper reference line is located at y_{hi}=21 from top image. However, the scales are determined by the word width and the distances between lower-upper reference lines to the top and the bottom of the word as shown in Fig. 5. The scales, S_x and S_y are calculated as follow:

$$S_x = \frac{w_{norm}}{W} \tag{1}$$

$$s_y = \begin{cases} \dfrac{h_{hi}}{yt-yh} & if \;\; pixels \in [\, yh \,,\, yt\,] \\[2mm] \dfrac{h_{cent}}{yh-yl} & if \;\; pixels \in [\, yl \,,\, yh\,] \\[2mm] \dfrac{h_{lo}}{yl-yb} & if \;\; pixels \in [\, yb \,,\, yl\,] \end{cases} \tag{2}$$

where $W=xr-xl$, and w_{norm}=50 pixels is the normalized width, and h_{hi}=11 pixels, h_{cent}=10 pixels and h_{lo}=11 pixels are, upper, central and lower zones height of the normalized word respectively.

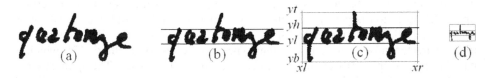

(a)   (b)   (c)   (d)

Fig. 5. Preprocessing (a) binarized image (b) Application of SC-CZL method (c) Horizontal slant correction (d) Normalized version of (c) in the frame 32-by-50 pixels

4 Prototype Creation

The realization of a prototype is carried out by the superposition of four normalized word images within the frame of 32-by-50 pixels of various writings of a word class written in wholly lower case letters; the resulting form is then dilated (see Fig. 6). Figure 6 (e) shows four samples from each word class in the training set used for prototype creation. The vocabulary of our application is composed of 27 words. During the realization of the 27 prototypes corresponding to these word classes, it can be noticed that certain prototypes have similar shapes. Therefore, it is interesting to group prototypes that have similar global shapes corresponding to different word classes by a unique prototype. This clustering operation yields 16 prototypes (see Fig. 7) representing the 27 classes (N_H) of French legal amounts with β_k, $k=1,2,.....,P$ ($P=16$), the number of handwritten word class per prototype class.

$$N_H = \sum_{k=1}^{P} \beta_k \quad , \quad k = 1,2,...,P \tag{3}$$

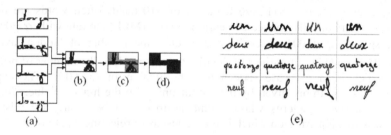

Fig. 6. Prototype creation (a) Four different normalized words from handwritten word class "douze" (b) Superposition operation (c) Dilation and prototype processing (d) Prototype result.(e) Samples of handwritten words used for prototype creation

1	2	3	4	5	6	7	8
un six euro	deux trois dix	quatre	cinq	sept vingt	huit trente	neuf	onze seize
$\beta_1=3$	$\beta_2=3$	$\beta_3=1$	$\beta_4=1$	$\beta_5=2$	$\beta_6=2$	$\beta_7=1$	$\beta_8=2$

9	10	11	12	13	14	15	16
douze treize	quatorze	quinze	quarante	cinquante	cent soixante	mille et	million centime
$\beta_9=2$	$\beta_{10}=1$	$\beta_{11}=1$	$\beta_{12}=1$	$\beta_{13}=1$	$\beta_{14}=2$	$\beta_{15}=2$	$\beta_{16}=2$

Fig. 7. Prototypes used for holistic recognition and their corresponding handwritten classes and β_k (below each prototype image)

5 Experimental Results

5.1 Training Phase

The prototypes shown in Fig. 7 were created using section 4, then fed to Hopfield model and memorized for $P=16$ (number of memorized patterns), and $N=1600$ neurons (32x50 pixels), then after the output from Hopfield model is down sampled in 4x4 pixel block. The MLP was trained only with 810 handwritten words, which corresponds to 30 (810/27) words by each handwritten class. This training set was not chosen to be large due to the strong similarity of global shapes existing among the handwritten words belonging to the same prototype class. However, the global shape of words of the same prototype does not change significantly, particularly when the word image is down sampled in 8x12 yielding feature vectors of 96 features (see Fig. 3 (b)). The down sampling operation is acting as a second noise filter which makes the patterns of the same prototype class to be very similar in the input of the MLP-based classifier. Thus, we found experimentally that when increasing the number of words per prototype beyond 30=810/27 (for a prototype having one handwritten class), the error-reject plot of the proposed method do not change. Finally, the resulting feature vectors (8x12=96 features) of 810 handwritten words are used to train a two layers of fully connected neural network (MLP) module with $\alpha=0.9$ (The training speed), $\xi=0.001$ (viscosity coefficient), the number of hidden layer was fixed to 40, the number of outputs of the MLP corresponds to the number of prototype classes which is in our case sixteen and the total error to 0.001. The expected prototype class is simply given by the output unit with the highest value. The MLP classifier can reject patterns whose membership cannot be clearly established. A typical classification criterion which is used consists of rejecting a pattern if:

$$\bar{y} = \max_{i=1,2,\ldots,P} \{ y_i \} < R_M \tag{4}$$

where P is the number of prototype classes, $y_i \in (0,1)$ is the ith output of the network, and R_M is a proper threshold. An unknown pattern is accepted if only at least one output is greater or equal than R_M.

The associative matrix $W^{(o)}$ is directly calculated and stored, whereas the connection weights v_{ij} and w_{ij} corresponding to the two fully connected MLP are obtained after convergence and also stored for recognition process. We precise, that the Hopfield model memorises the 16 created prototypes as classes. The handwritten images were scanned under 300 dpi, which is appropriate for processing. Two word characteristics were used to distinguish word classes belonging to the same prototype class namely; the dot and the starting loop in case of words starting with "d" (see Fig. 8) as follows:

1- The existing dot is detected using an empty mask of dimension 20-by-20 pixels, fixed experimentally in which the detected dot lies inside.
2- The starting "d" is detected using the intersection of a vertical line with the left part of the word in the lower case zone.

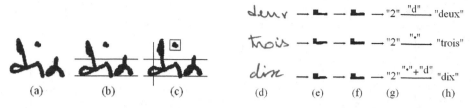

Fig. 8. Starting "d" and dot detection (a) Original image word (b) Application of SC-CZL method (c) Starting "d" and dot detection (d) Original images (e) Dilated version of the normalized image (32-by-50). (f) Hopfield model output (g) Result of prototype recognition by Hopf-MLP classifier. (h) Results of holistic word recognition by combining result of (g) and word characteristics ("d" and dot)

These two characteristics are combined with the prototype recognition results to make a final decision of the unknown handwritten word. Figure 8 (d-h) illustrates the combination of these two characteristics and the results of the combined neural network for prototype recognition. All the 27 legal amounts are distinguished using prototype recognition results associated with the "d" and dot characteristics. Another characteristic was introduced to discriminate classes belonging to the prototype "1" namely, "un", "six" and "euro". It consists of intersecting a horizontal line at the half lower case zone with word parts.

5.2 Recognition Phase

The system has been tested using our proper database of 27 basic words written by 200 writers. The test database consists of 5400 (27x200) legal amounts collected and extracted manually from 20 A4 format papers that were scanned at 300 dpi. The experiments are conducted in two parts, the first part consists in testing the performance of the prototype recognition and the second part presents results of the handwritten word recognition. Figure 9 (a) summarizes the results obtained on the same database of handwritten words of medium and low quality by the sequential combination method. The handwritten words in the test set are uniformly distributed (200 words/class). The plots for the prototype and handwritten methods are obtained for different values of the rejection threshold R_M ($0.3 \leq R_M \leq 0.9$). When using the sequential combination method, the preprocessed handwritten word character (global shape) is processed with Hopfield model for T=1 (Hopfield iteration number), and then the resultant output is down sampled before it is fed to the MLP. According to the results obtained, it can be noticed that for the same rejection rate, the handwritten error rate is always greater than those obtained by prototype recognition. Thus the handwritten recognition is strongly dependant on the results of the prototype recognition used as global shape classifier. When the rejection rate of the prototype recognition system is very low (lower R_M), this ensures higher errors for handwritten classification when comparing to prototype errors. Whereas for higher rejection rate (higher R_M), the proposed system ensures lower errors for handwritten classification which are similar to those obtained with the prototype recognition. This could be explained by the fact that the proposed system rejects (for higher R_M) not only the poor quality words but also a great part of those of low-medium quality and accepts

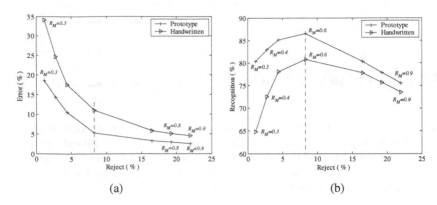

Fig. 9. Recognition results. (a) Error-reject plots (b) Recognition-reject plots.

high-medium quality words, and gives less errors in prototypes and consequently in handwritten words (dots and starting "d" clearly detected).

The error-reject plots of Fig. 9 (a) shows that the recognition system gives best results for prototype recognition when comparing to handwritten recognition. This difference in recognition rate is due to starting "d" and dot detection to determine the handwritten class from the recognized prototype. The handwritten recognition rate converges to the prototype recognition rate when the above mentioned characteristics are correctly detected. Hence the handwritten word recognition can only be as good a solution as the prototype recognition, or it can be worse. Figure 9 (b) presents recognition-reject plots showing best results for both prototype and handwritten recognition for a threshold reject $R_M=0.6$ (in dashed line) corresponding to the same global reject of 8.25%, and global errors of 5.25 % and 11.00 % for both prototype and handwritten recognition respectively. A recognition rate of 86.50 and 80.75 % for prototype and handwritten recognition respectively is achieved. Figure 10 shows results of correctly recognized handwritten words.

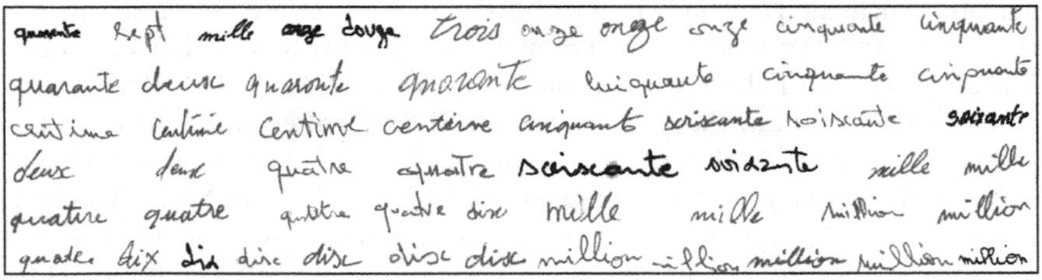

Fig. 10. Handwritten word samples correctly recognized

6   Conclusion

In this paper holistic handwritten word recognition based primarily on accurate zones detection namely; upper, lower and central zones is presented. Experimental results conducted on prototypes and handwritten word show the robustness of the proposed system. It has been shown that the holistic handwritten word recognition with combined neural network involves only few training words. Experiments were conducted on data set of handwritten words of medium and low qualities have shown promising results.

References

1. G. Dimauro, S. Impedovo, G. Pirlo, A. Salzo, "Automatic bank check processing a new engineered system," *World scientific publishing company*, vol 11, N° 4 (1997)
2. Sargur N. Srihari, "Recognition of handwritten and machine- printed text for postal address interpretation ," *Pattern Recognition Letters* 14 (1993) 291-302
3. R. M. Bozinovic and S. N. Srihari, "Off-line cursive script word recognition," *IEEE Trans. Patt. Anal. Mach. Intell.* 11 (1989) 68-83
4. R. Buse, Z. Q. Liu and T. Caelli, "A structural and relational approach to handwritten word recognition," *IEEE Trans. Syst. Man, Cyber.* Part B, 27 (1997) 847-861
5. M. Mohamed and P. Gader, "Handwritten word recognition using segmentation-free hidden Markov modeling and segmentation-based dynamic programming techniques," *IEEE Trans. Patt. Anal. Mach. Intell.* 18 (1996) 548-554
6. S. Madhavanath and V. Govindaraju, "Holistic Lexicon Reduction," *Proc. Third. Int. Workshop Forntiers in Handwriting Recognition,* Buffalo, N.Y. (1993) 71-81
7. S. Madhavanath and V. Govindaraju, "Contour-Based Image Processing for Holistic Handwritten Word Recognition," *Proc. Fourth. Int. Conf. Document Analysis and Recognition (ICDAR 97),* Ulm, Germany (1997)
8. S. Madhvanath, V. Krpasundar and V. Govindaraju, " Syntactic methodology of pruning large lexicons in cursive script recognition," *Pattern-Recognition.* vol.34, N°1 (2001)
9. E. Lecolinet and O. Baret, "Cursive word recognition: methods and strategies," Fundamentals in Handwriting Recognition,, ed. S. Impedovo, NATO ASI Series F, vol. 24, Springer-Verlag (1994) 235-263
10. D. Guillevic and C. Y. Suen, "Recognition of legal amounts on bank checks," *Pattern-Analysis-and-Applications,* vol.1, N°1 (1998) 28-41
11. T. Paquet, M. Avila and C. Olivier, " Word modelling for handwritten word recognition," *Proceedings Vision Interface '99. Canadian Image Process. & Pattern Recogniton Soc,* Toronto, Ont., Canada (1999) 49-56
12. Co. de-Almendra-Freitas, A. El-Yacoubi, F. Bortolozzi and R. Sabourin, " Brazilian bank check handwritten legal amount recognition," *Proceedings 13th Brazilian Symposium on Computer Graphics and Image Processing,* (2000) 97-104
13. A. Namane, M. Arezki, A. Guessoum, E.H. Soubari, P. Meyrueis and M. Bruynooghe, "Sequential neural network combination for degraded machine-printed character recognition," *Document Recognition and Retrieval XII, Proc.* SPIE 5676 (12) (2005) 101-110
14. A. Namane, A. Guessoum and P. Meyrueis, "New skew correction and central zone localization for handwritten word and its application to French legal amounts," *International Conference on Multimedia, Image Processing and Computer Vision,* Madrid, Spain, April (2005) (to be appeared)
15. A. Namane, M. Arezki, A. Guessoum, E.H. Soubari, P. Meyrueis, and M. Bruynooghe, "Off-line unconstrained handwritten numeral character recognition with multiple Hidden Markov models," *Proceeding of the 4th IASTED International Conference on Visualization, Imaging, and Image Processing, VIIP '04* (2004) 269-276

Handwriting Documents Denoising and Indexing Using Hermite Transform

Stéphane Bres, Véronique Eglin, and Carlos Rivero

LIRIS, INSA de Lyon, Batiment Jules VERNE,
69621 VILLEURBANNE CEDEX, France
{stephane.bres, veronique.eglin, carlos.rivero}@liris.cnrs.fr

Abstract. This paper presents a new system for handwriting documents denoising and indexing. This work is based on the Hermite Transform, which is a polynomial transform and a good model of the human visual system (HVS). We use this transformation to decompose handwriting documents into local frequencies and using this decomposition, we analyze the visual aspect of handwritings to compute similarity measures. A direct application is the management of document databases, allowing to find documents coming from the same author or to classify documents containing handwritings that have similar visual aspect. Moreover, ancient documents can contain degradations from different origins. It is often necessary to clean the backgrounds of those degraded documents before analysing them. The current results are very promising and show that it is possible to characterize handwritten drawings without any a priori graphemes segmentation.

1 Introduction

There are many different kinds of databases around the world, and all of them have to deal with the same problem, what ever information they hold: how to organize this information cleverly and how to retrieve visually similar information. It is a great challenge and it can not be resolved with a unique generic solution but it must be adapted to each kind of information. In this paper, we are working on handwriting documents corpus. Our purpose here is to characterize precisely handwritings whatever their authors are and to classify them into visual writers' families. Our approach considers handwritings as special drawings that create a specific texture we want to analyse by considering orientations at different scales. Orientations are considered as sufficiently relevant perceptual features to characterize the special texture of handwritten drawings. These orientations information are extracted by using the Hermite transform which is a particular polynomial transform and a good model of the receptive field profiles of the human visual system. This model leads to the development of an original method of handwriting classification by the computation of handwritings signature and similarity measures that reveal their "visual textural aspects".

1.1 Specificities of Patrimonial Handwritings Documents

The databases we want to treat contain historical handwritings documents and the characteristics of these documents had a direct influence on the approach we choose

S. Singh et al. (Eds.): ICAPR 2005, LNCS 3686, pp. 664–673, 2005.

for our orientations extraction. Many digital images of documents and more generally ancient manuscripts are degraded by the presence of strong artefacts in the background (see figure 1). This can either affect the readability of the text and, in our case, it compromises a relevant handwriting characterization. Consequently, most of the time, it is very difficult to directly extract the handwritings in those images. It becomes necessary to pre-process the images with a cleaning and denoising first step.

Most generally, we consider the documents as a mixed signal composed by a textured background with a superimposed high frequency handwriting signal. The use of thresholding techniques is often not effective since the intensities of background can often be close to those of the foreground text. Some approaches for text and background separation have been proposed in [5] where multistage thresholding techniques have been investigated to segment parts. Other techniques based on adaptative filtering have been tested on forensic documents to separate homogeneous textured background from handwriting marks, [2]. Some approaches consider a physical model of degradation to propose a mathematical model for text enhancement and background cleansing, [9]. In [11], the authors propose a decomposition of the signal into two blind sources where the overlapping texts and the supports (paper) texture are the unknown sources to be recovered with the consideration of different spectral bands of the documents (bands of colours).

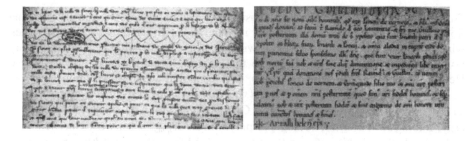

Fig. 1. Examples of ancient manuscripts degraded by strong artefacts [3]

1.2 Texture Feature Extraction and Human Visual System

Texture features extraction is usually performed by linear transformation or image filtering, [1,7], followed by some energy measures or non-linear operator application (e.g. rectification). In this paper, we focus on the multi-channel filtering (MCF) approach. It is inspired by the MCF theory for processing visual information in the early stages of the human visual system, [1,7], where receptive field profiles (RFPs) of the visual cortex can be modelled as a set of independent channels. Each of these channels is tuned on a specific orientation and frequency. The use of these filters leads to the decomposition of an input image into multiple features images. Each of these images captures textural features occurring in a narrow band of spatial frequency and orientation. Among the MCF models having the above properties, Gabor filters have been widely used in texture feature extraction, [4], image indexing and retrieval, [12]. Another model corresponds to Hermite filters of the Hermite transform [6] that agrees with the Gaussian derivative model of the HVS. It has also been shown analytically that Hermite and Gabor filters are equivalent models of receptive field profiles (RFPs), [6],[8]. However, Hermite filters have some advantages over Gabor ones, like being an orthogonal basis leading to information decorrelation and perfect image reconstruction after decomposition. This is the main reason why we are interested in this transform. Moreover, a discrete representation of Hermite filters exits (the Krawtchouk polynomials) with the property of separability for an efficient implementation.

2 Hermite Transform

In this paper we present a method for image document cleaning and indexing based on the Hermite transform. It is exploited here to decompose the initial signal into different parts depending on their frequencies characteristics (high or low). Most of the time, the noise or degradations that appear on ancient documents have low frequencies characteristics, while the writing by itself is composed of high frequencies. It is of a great interest to separate them. This is exactly what we want to achieve with the Hermite transform. In the following paragraph, we present the definition of the Hermite transform.

2.1 Cartesian Hermite Filters and Krawtchouk Filters

Polynomial transforms are the decomposition of a signal $l(x,y)$ into a linear combination of polynomials. The original signal is locally treated, window by window. These windows are positioned on the signal with a constant translation step, and the polynomials are orthogonal with respect to this specific window shape. In the case of Hermite transform, the window $v(x,y)$ is a Gaussian window. Hermite filters $d_{n-m,m}(x,y)$ decompose the original signal $l(x,y)$ by computing a localized signal $l_v(x-p,y-q) = v^2(x-p,y-q) \, l(x,y)$ where $v(x,y)$ is a Gaussian window with spread σ and unit energy, into a set of Hermite orthogonal polynomials $H_{n-m,m}(x/\sigma, \, y/\sigma)$. Coefficients $l_{n-m,m}(p,q)$ at lattice positions $(p,q) \in P$ are then derived from the signal $l(x,y)$ by convolving with the Hermite filters. These filters are equal to Gaussian derivatives where $n-m$ and m are respectively the derivative orders in x- and y-directions, for $n=0,\ldots,D$ and $m=0,\ldots,n$. Thus, the two parameters of Hermite filters are the maximum derivative order D (or polynomial degree) and the scale σ. Hermite filters are separable both in spatial and polar coordinates, so they can be implemented very efficiently. Thus, $d_{n-m,m}(x,y) = d_{n-m}(x) \, d_m(y)$, where each 1-D filter is:

$$d_n(x) = \left((-1)^n / (\sqrt{2^n \cdot n!} \sqrt{\pi} \sigma) \right) H_n(x/\sigma) e^{-x^2/\sigma^2} \tag{1}$$

where Hermite polynomials $H_n(x)$ are orthogonal with respect to the weighting function $exp(-x^2)$, and are defined by Rodrigues' formula in [6] by:

$$H_n(x) = (-1)^n e^{x^2} \frac{d^n}{dx^n} e^{-x^2} \tag{2}$$

In the frequency domain, these filters are Gaussian-like band-pass filters with extreme value for $(\omega\sigma)^2 = 2n$, [8], and hence filters of increasing order analyze successively higher frequencies in the signal.

Krawtchouk filters are the discrete equivalent of Hermite filters. They are equal to Krawtchouk polynomials multiplied by a binomial window $v^2(x) = C_N^x / 2^N$, which is the discrete counterpart of a Gaussian window. These polynomials are orthonormal with respect to this window and they are defined by :

$$K_n(x) = \frac{1}{\sqrt{C_N^n}} \sum_{\tau=0}^{n} (-1)^{n-\tau} C_{N-x}^{n-\tau} C_x^{\tau} \tag{3}$$

for $x=0,\ldots,N$ and $n=0,\ldots,D$ with $D \le N$. It can be shown that the Krawtchouk filters of length N approximates the Hermite filters of spread $\sigma = \sqrt{N/2}$. In order to achieve fast computations, we present a normalized recurrence relation to compute these filters, see [8] :

$$K_{n+1}(x) = \frac{1}{\sqrt{(N-n)(n+1)}} \left[(2x-N)K_n(x) - \sqrt{n(N-n+1)}K_{n-1}(x) \right] , \ n \ge 1 \tag{4}$$

with initial conditions $K_0(x) = 1$, $K_1(x) = \frac{2}{\sqrt{N}} \left(x - \frac{N}{2} \right)$.

2.2 Steered Hermite Filters and Gabor-Like Hermite Filters

In order to have a multi-channel filtering (MCF) approach based on Hermite filters, they must be adapted to orientation selectivity and multi-scale selection. For that purpose, we apply their property of steerability, [6,8]. The resulting filters may be interpreted as directional derivatives of a Gaussian (i.e. the low-pass kernel).

Since all Hermite filters are polynomials times a radially symmetric window function (i.e. a Gaussian), it can be proved that the $n+1$ Hermite filters of order n form a steerable basis for every individual filter of order n. More specifically, rotated versions of a filter of order n can be constructed by taking linear combinations of the filter of order n. The Fourier transform of Hermite filters $d_{n-m,m}(x,y)$ can be expressed in polar coordinates $\omega_x = \omega \cos\theta$ and $\omega_y = \omega \sin\theta$ as $\hat{d}_{n-m,m}(\omega_x, \omega_y) = \hat{d}_n(\omega)\alpha_{n-m,m}(\theta)$

where $\hat{d}_n(\omega)$, which expresses radial frequency selectivity, is the 1-D Fourier transform of the nth Gaussian derivative in (1) but with radial coordinate r instead of x. The cartesian angular functions of order n for $m=0,\ldots,n$, are given as

$$\alpha_{n-m,m}(\theta) = \sqrt{C_n^m} \cos^{n-m} \theta \cdot \sin^m \theta \tag{5}$$

which express the directional selectivity of the filter.

Steered coefficients $l_n(\theta)$ resulting of filtering the signal $l(x,y)$ with these steered filters can be directly obtained by steering the cartesian Hermite coefficients $l_{n-m,m}$ as:

$$l_n(\theta) = \sum_{m=0}^{n} l_{n-m,m} \cdot \alpha_{n-m,m}(\theta) \tag{6}$$

scale representation that fulfils the desired constraints in the frequency domain, which are mainly the number of scales S (radial frequencies ω_0) and the number of orientations R in the filter bank. Since previous works have been done essentially with Gabor filters, we have then adopted a similar multi-channel design. Moreover, both Hermite and Gabor filters are similar models of the RFPs of the HVS [8]. For these reasons, we have named the resulting filters as Gabor-like Hermite filters.

In summary, construction of a Gabor-like Hermite filter bank requires the following procedure. First of all, set the number of desired scales S and orientations R and for each of the scales $s=0,\dots,S-1$ compute:

- the radial central frequency ω_0 and the spatial spread σ_x of respective filters.
- Krawtchouk parameters such as window length N and filter order D.
- Krawtchouk filters: get the corresponding Krawtchouk polynomials through (4) and multiply them by a binomial window of length N.
- Input image convolutions with Krawtchouk filters to obtain cartesian coefficients.
- Steering coefficients to desired orientations through (6) and (5) to obtain the equivalent multi-channel outputs.

3   Patrimonial Documents Denoising Results

Our proposition uses the Cartesian Hermite transform (computed throw the Krawtchouk filters) that extract the local frequencies of a signal. Figure 2 presents the Hermite decomposition of a document at a given scale N=16 and up to degree 2. The most top left image is equivalent to a Gaussian low pass filtered image. Using the higher degrees in both directions allows extracting high frequencies of the original image. As we explain earlier, low frequencies contain information on the background and high frequencies contain information on the writings we want to keep, if their levels are sufficient (above a certain threshold).

Fig. 2. 2D - Hermite transform using Krawtchouk filters for N=16 and up to degree n=2 for the rows and the columns

The first step of our denoising process is then to decompose the original image using the Hermite decomposition. In the second step, we reconstruct the high pass image I_H using degrees higher than N/4 in both directions. Too small values are filtered at this step (values less than 10% of the maximum). We obtain an image with a cleaner background: low frequencies and small high frequencies variations have disappeared (see Figure 3.1). A more detailed view is presented on figure 3.4. This image I_H is used as a mask that localizes the writings we want to keep. Pixels belonging to the background have low values in a low values neighbourhood. Pixels belonging to writings have a highly contrasted neighbourhood. An example of document denoising is shown on figure 3.2. The original image is presented figure 2. Details of these images are presented on figure 3.3 and 3.5. Using denoised images allows focusing on the handwritings by itself.

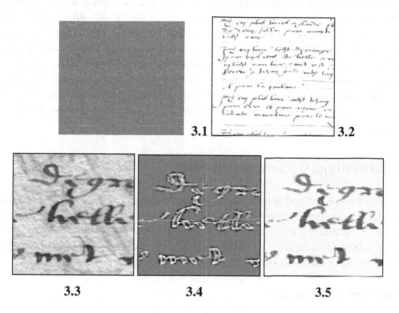

Fig. 3. Example of document denoising. High pass image based on Hermite decomposition (3.1) - Denoised document (3.2) - Detail of original document (3.3) - Image (3.1) detail (3.4) - Image (3.2) detail (3.5).

The main difference between this Hermite based approach and a classical adaptative thresholding comes from the local frequency decomposition we make here. Most of the time, adaptative thresholding methods classify pixels as background pixels or foreground pixels (handwriting lines) depending on local statistical values. Our method uses local frequency decomposition. On their principle, these methods are different, because Hermite based approach allows filtering that take into account the local frequency to decide between as background pixels or foreground pixels. This could be especially interesting in case of wide degradation areas containing black low frequencies. Such areas will not respond to high frequencies Hermite filters but adaptative thresholding can still keep some pixels as foreground pixels.

4   Handwritings Document Signature

Handwriting characterization will be done using orientations extraction by Gabor-like Hermite filters. The two parameters (number of scales S and the number of orientations R) needed for Gabor-like Hermite filters are fixed to S=4 and R=6. This will lead to 24 oriented filters. For a given pixel, each of these 24 filters will give responses that characterize a given orientation at a given scale. We only keep responses on pixels identified as handwritings lines pixels in the denoising step. Then, we have a 24 values vector for each handwritings lines pixel. All these vectors can be represented as a cloud, in a 24 dimensions space, which is a good characterization of the analyzed handwriting. Unfortunately, these signatures are far too big to used directly, and have to be reduce to something as small as possible with a minimal information loss. We choose to keep geometrical information of the clouds, like their gravity center (mean values on each of the 24 coordinates) and main axis (eigenvectors and eigenvalues) after an PCA-like step. Our signature for a given handwriting document is then the 24 means values of the results coming out of the filters bank, 24 normalized eigenvectors and the 24 corresponding eigenvalues of the covariance matrix computed from the centered cloud of orientations vectors. Moreover, experiences show that we do not need to keep all the eigenvectors and associated eigenvalues : only the 3 or 4 greater values need to be stored.

5 Handwritings Document Indexing

Now that we have defined a possible signature for every document in a database, we need to define a distance between these signatures to introduce the similarity notion in the database. Similarity leads to indexing which is the goal we want to reach. With a similarity measure, it is easy to build an indexing motor that can classify the documents and retrieve the most similar documents to a requested one.

5.1 Similarity Computation

In practice, our signature for image number i is made of 24 mean values $M_i(n)$, 4 eigenvalues and the 4 normalized eigenvectors V_i corresponding to the 4 greater eigenvalues L_i. L_{io} quantifies the importance of the vector V_{io} in the shape of the cloud. The distance D we choose to define uses both information of mean values M_i and the couples vectors V_i and values L_i. This distance D is the combination of the distance D_M between the mean values M_i and a multiplicative normalized coefficient $\overline{D_E}$ coming from the eigenvectors and eigenvalues. The D_M (Hi, Hj) distance between handwriting i and handwriting j is defined by :

$$D_M(H_i, H_j) = \sum_{n=1}^{24} |M_i(n) - M_j(n)| \tag{7}$$

The multiplicative normalized coefficient $\overline{D_E}$ coming from the eigenvectors and eigenvalues is based on the non normalized distance D_E between weighted eigenvectors. The weights we use here are their corresponding eigenvalues :

$$D_E(H_i, H_j) = \sum_{n=1}^{4} |L_i(n).V_i(n) - L_j(n).V_j(n)| \tag{8}$$

We obtain $\overline{D_E}$ after a normalization step. For that purpose, we divide D_E by its maximum value to have a value between 0 and 1. Thus :

$$\overline{D_E}(H_i, H_j) = D_E(H_i, H_j) \ / \ \sum_{n=1}^{4} \sqrt{L_i^2(n) + L_j^2(n)} \tag{9}$$

Finally, the distance $D(H_i, H_j)$ between handwriting H_i and handwriting H_j is can be expressed as :

$$D(H_i, H_j) = D_M(H_i, H_j) \cdot \overline{D_E}(H_i, H_j) \tag{10}$$

This distance is symmetrical, which is a good property to assure coherent results during multiple comparisons of databases documents. A small distance means high similarity.

Fig. 4. Examples of images coming from the same authors (one author per column)

5.2 Practical Results and Evaluation

We have tested the whole system on our personal database composed of documents coming from different authors but mainly patrimonial handwritings documents. Most of the time, we have full pages of the same author and for evaluation purpose, these pages are divided into smaller images, 9 per page. Then, most pages give us 9 images from the same author, containing what we can suppose to be similar handwritings. This is how we build our "ground truth": images coming from the same original page image should look the same and have similar handwritings. It is difficult to complete this ground truth with similarities between different author's handwritings because of the subjective judgment involved in such estimation. Figure 4 gives some examples of images coming from the same original page. Our database contains 1438 images coming from 189 different authors, in different languages and alphabets.

To illustrate the discrimination possibilities of our signatures, we present, on figure 5, a 2D representation of the 24D signature space. This 2D representation is obtained by PCA on the mean values M_i of the text images of figure 4. The 9 points in the ellipse #1 correspond to the 9 text images of the author presented in the row #1 of figure 4. The 9 points in the ellipse #2 correspond to the 9 text images of the author presented in the row #2 of figure 4 and so on …

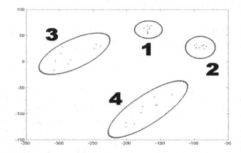

Fig. 5. 2D projection obtained by PCA of the 24D signature space. Each point is the mean value of a text image. They are grouped by author presented on figure 4: ellipse #1 corresponds to the author of row #1, ellipse #2 corresponds to the author of row #2, and so on.

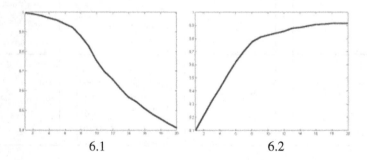

6.1   6.2

Fig. 6. 6.1. Precision curve. 6.2. Recall curve computed on the entire database containing more than 1400 handwriting documents.

The global results we obtain are really promising because, according to our ground truth, a given request has in the ten first better answers (documents with the higher similarity or equivalently the smaller distance) in average more than 83% of correct responses, see recall curve on figure 6. This is an average value computed on the documents that have 9 similar images in the database. These precision and recall curves are a common way to show the efficiency of an indexing system. They have been computed using the 20 first responses. Let's remember that we only have 9 images for each handwriting. That is the reason why the precision decreases strongly after the 9th response.

6 Conclusion

This work is a response to scientific problems of historical handwritten corpus digitalization. It deals with the handwriting denoising and indexation and is applied here to a multi-language and multi-alphabet corpus. We propose here a biological inspired approach for images denoising (by a background *cleaning*) and handwriting characterization for corpus indexing. The developed perception based model lies on the Hermite frequencial decomposition for image denoising and indexing. Our motivation is directly linked to the difficulty to perform efficient image processing on degraded handwriting historical documents without a priori knowledge on the image content. In that way, we have chosen a segmentation free approach that is global and generic. The current results of handwriting denoising and classification with orientation Hermite based features are very promising. We are currently working on an enlarged database in connection with recent digitalization European project.

References

1. Bovik, A.C., Clark, M., and Geisler, W.S., Multichannel texture analysis using localized spatial filters, *IEEE Trans. Pattern Analysis Mach. Intell.*, vol. 12, pp. 55-73, 1990.
2. Franke, K., Koppen, M., A computer-based system to support forensic studies on handwritten documents, Int. Jour. Doc. Anal. Reco., 3:218-231, 2001.
3. GALLICA : digital library of the BNF : *http://gallica.bnf.fr*
4. Grigorescu, S.E., Petkov, N., and Kruizinga, P., Comparison of texture features based on Gabor filters, *IEEE Trans. Image Processing*, vol. 11, pp. 1160-1167, 2002.
5. Leedham G., Varmas,S., Patnkar, A., Govindaraju,V., Separating text and background in degraded document images- a comparison of global thresholding techniques for multistage thresholding. In Proceedings of the 8th inter. Workshop on frontiers in handwriting recognition, 2002, pp. 244-249.
6. Martens J.-B., The Hermite transform – Theory, IEEE Trans. Acoust., Speech, Signal Processing, vol. 38, no. 9, pp. 1595-1606, 1990.
7. Randen, T. and Husøy, J.H., Filtering for texture classification: A comparative study, *IEEE Trans. Pattern Analysis Mach. Intell.*, vol. 21, pp. 291-310, 1999.
8. Rivero-Moreno C.J., Bres S., Conditions of similarity between Hermite and Gabor filters as models of the human visual system, In Petkov, N.& Westenberg, M.A. (eds.): Computer Anal. of Images & Patterns, Lectures Notes Computer Sci., vol. 2756, pp.762-769, 2003.
9. Sharma, S. Show-through cancellation in scans of duplex printed documents. IEEE Trans. Image Process, 10(5):736-754.
10. Strouthopoulos, C. and N. Papamarkos, *Image and Vision Computing,* Text identification for document image analysis using a neural network,pp879-896, vol.16, 1998.
11. Tonazzini A., Bedini L., Salerno E., Independant component analysis for document restoration, Inter. Jour. on Doc. Anal. and Reco., vol.7, N°1, 2004, pp. 17-27.
12. Wu, P., Manjunath, B.S., Newsam, S., and Shin, H.D., A texture descriptor for browsing and similarity retrieval. *Signal Proc.: Image Comm.*, vol. 16, no. 1,2, pp. 33-43, 2000.

Evaluation of Commercial OCR: A New Goal Directed Methodology for Video Documents

Rémi Landais[1,2], Laurent Vinet[1], and Jean-Michel Jolion[2]

[1] Institut National de l'Audiovisuel,
Direction de la recherche et de l'expérimentation, 4, Av. de l'Europe,
94366 Bry-sur-Marne cedex, France
{rlandais, lvinet}@ina.fr
http://www.ina.fr
[2] Lyon Research Center for Images and Intelligent Information Systems (LIRIS),
INSA de Lyon, Bât. J. Verne,
20, rue Albert Einstein, F-69621 Villeurbanne cedex, France
jolion@rfv.insa-lyon.fr
http://rfv.insa-lyon.fr/~jolion

Abstract. Texts embedded in video streams convey crucial information for documentation. Many text detection and recognition systems have been designed to automatically extract such documentary data from video streams. Most of the research teams involved argue that commercial OCR[1] do not work properly on images extracted from a video stream. They thus concieve their own detection systems. Nevertheless, commercial OCR have never been evaluated on such corpora. This article details a new methodology to evaluate a commercial OCR on a video document. This methodology is goal directed: the system is penalized proportionally to TFIDF (Term Frequency Inverse Document Frequency) scores of texts [1]. We experiment our methodology on Abbyy FineReader 6.0[2].

1 Introduction

At the National Institute of Audiovisual (INA), video documentation is completely manual and relies on describing forms containing crucial information about documents (title, summary, ...). The task of writing these describing forms can be ease by the automatic recognition of embedded texts which often convey information about names, places, etc. Commercial OCR can be adapted to video streams and could thus help to fill describing forms. Nevertheless, most research teams working on automatic text detection and recognition [2,3,4] explain that commercial OCR systems perform poorly on images extracted from video streams because of the poor resolution of such images. They thus propose systems which perform text detection and resolution improvement and finally send extracted zones to a commercial OCR which performs recognition. Such considerations earned attention

---

[1] Optical Character Recognition.
[2] www.abbyy.com

S. Singh et al. (Eds.): ICAPR 2005, LNCS 3686, pp. 674–683, 2005.

in the last decade but video formats have changed and OCR systems have evolved in such a way that it is relevant to evaluate them on videos.

This article details a new methodology to evaluate a commercial OCR. This methodology is goal directed: we penalize the OCR proportionally to TFIDF (Term Frequency Inverse Document Frequency) scores of texts. The first section of the article will focus on former methods of ground truth building and evaluation. We will then expose our methodology and apply it to Abbyy Finereader 6.0.

2   State of the Art

Evaluation of text detection and recognition systems requires a ground truth (expected outputs) and a measure to compare results obtained by the system with this ground truth. Several data models used to define ground truths and several evaluation measures will be next discussed.

2.1   Building Ground Truth

Basically, texts are represented in a ground truth by the position of their bounding boxes and by their transcription [5,6,7]. In [8], some other features are computed in order to process finer evaluation: character height variation, skew angle, color and texture, background complexity, string density, contrast and a recognizability index which reflects the difficulty for a text to be read by a human. The VIPER system [9] allows the user to design its own temporal ground truth model for any kind of video object. Concerning layout analysis, ground truth models must contain relationships between zones. In [10], documents are described in accordance with the RDIFF format (*Region Description Information File Format*). Different partial orderings of the regions can be defined to reduce the potential ambiguity of a global ordering. In [11], the data model is hierarchical: a document is composed of pages, a page is divided into several zones that contain many lines... Each of these entities has a unique ID which allows the user to store an ordering. As in the ViPER system, the user can define its own ground truth data model. DAFS format [12] (used for instance in [13]) qualifies each zone according to its position, its content and its relationships with other zones.

2.2   Detecting and Recognizing Texts in Images or Documents: The Evaluation Issue

Layout Analysis. Usually, criteria of evaluation are based on mutual overlap rates between the zones detected by the system ($\{D_i\}_i$) and the zones in the ground truth ($\{G_j\}_j$) [5,14,6]. These criteria are developped in the equation 1.

$$Condition\ \ 1:\quad \frac{A(G_j \cap D_i)}{A(D_i)} > Th \ and \ \frac{A(G_j \cap D_i)}{A(G_j)} > Th$$

if $\exists$ i \ condition 1 is satisfied for $\{D_i, G_j\}$ then G_j is correctly detected   (1)

if $\forall$ i condition 1 is not satisfied for $\{D_i, G_j\}$ then G_j is missed

if $\forall$ j condition 1 is not satisfied for $\{D_i, G_j\}$ then D_i is a false alarm.

The results are then exploited to compute *precision* and *recall* measures. In [14], two levels of analysis are considered: the pixel based level and the text box one. Pixel based information are then combined with text box based information to get high level interpretations. Cases of multiple matching are quite difficult to be treated. The proposition exposed in [6] is based on Liang proposal [13]. Two matrices σ *and* τ are defined as follows:

$$\sigma_{ij} = \frac{A(G_i \cup D_j)}{A(G_i)} \text{ and } \tau_{ij} = \frac{A(G_i \cup D_j)}{A(D_j)}$$

Precision and recall are then computed from these matrices assigning costs depending on the overlap case: one (many) ground truth text box(es) match(es) against one (many) result text box(es). Multiple matching is also treated in [8,15]. Besides this issue, two indexes are proposed in [8] to reflect text difficulty to be detected and how readable the text is (these indexes are fixed empirically). Concerning layout analysis, performance can be defined according to the number of lines which have been uncorrectly detected (the most disadvantageous cases are considered: horizontally merged, horizontally or vertically split or missed) [16]. In [10], a segmentation evaluation protocol is exposed: each pixel is labelled depending on the type of the zone it belongs to (split or merged zone...). Each kind of error is assigned a cost chosen by the user depending on his target application. [17] details a measure to evaluate the performance of the layout analysis stage of an OCR based on the results of its recognition stage. The efficiency of the layout analysis module is established comparing recognition results obtained in automatically segmented zones and recognition results obtained in the corresponding ground truth zone: the smaller the difference between these results is, the more effective the layout analysis module is.

Text Recognition. Output strings and ground truth strings are usually compared with the Levenstein distance [18]. The principle is to compute the minimal cost of transformations needed to transform the result string s_{res} into the ground truth string $s_{gt} : \delta(s_{gt}, s_{res})$. The tranformations are composed of a sequence of three elementary operations: substitution of a character, insertion of a character and deletion of a character. These operations are associated with the following costs:

$$Substitution \ (a \to b) : cost \ \gamma(a, b)$$
$$Insertion \ (\lambda \to b) : \ cost \ \gamma(\lambda, b)$$
$$Deletion \ (a \to \lambda) : \ cost \ \gamma(a, \lambda)$$
$$with \ \delta(a, b) = \gamma(a, b)$$

$\delta(s_{gt}, s_{res})$ is defined regarding the sequence of elementary operations needed. To compare results of recognition between strings of different length, $\delta(s_{gt}, s_{res})$ must be normalized regarding the number of characters in s_{gt}.

Precision and *recall*, based on the number of correctly recognized characters can also be computed [14,2]. The word recognition rate can also be distinguished from the character recognition rate [2].

Most of the methods exposed in this section do not focus on application issues. Furthermore, no considerations are given about repairing issues. In the case of the simplest applications as video indexing (*when does text appear on screen?*), evaluation can be based on criteria detailed in equation 1 since the semantic content of the texts is useless. Nevertheless, if this content need to be involved in the evaluation process, no evaluation protocol mentionned in this section would be sufficient. We expose in the next section our methodology which is goal oriented: the system is penalized regarding the degree of importance of the texts. Errors are also assigned to the detection, the localization or the recognition stage of the system in a reparing perspective. Finally, two indices are defined to handle both the technical evaluation and the application-driven evaluation. As in [10], the user is asked for fixing the strictness degree of the evaluation.

3   A Commercial OCR Evaluation Methodology

3.1   Building Ground Truth

Only texts which convey describing information about the document are retained in the ground truth. In this way, the system is not penalized for having failed in recognizing irrelevant texts. Documents are divided into temporal segments which are homogeneous in a semantic content point of view. The describing forms of these segments underlie our sorting: only texts appearing both in the describing form of the concerned segment and on screen during the same segment are retained. This selection is performed over the *word level*.

Our data model contains basic information (bounding boxes positions, times of appearing and disparition, transcription) about each text. If text is moving, key positions and their temporal references are stored. As our evaluation is goal directed, some other semantic information are added. The figure 1. shows the XML schema we applied to store our ground truth. **The TypologyIndex** denotes if the text belongs to one of the next classes: *artificial* (superimposed on the signal) or *scene text* (effectively contained in the scene which has been filmed). **The semantic segment** designates the segment in which the text appears. **TextType** denotes if the text is a block, a line or a word. As no assumption should be made on the way the OCR performs the layout analysis, words, lines and blocks are stored. This hierarchy is inspired from the relationships used in the building of layout analysis ground truths [11].

We compute for each word his TFIDF score, reflecting if the word may be efficient to discriminate the document with other documents if the final application is content-based retrieval. Scores are computed according to the Okapi formula [1]:

$$TFIDF = qtf_t \underbrace{\frac{(K+1)*tf_{t,d}}{K*(1-b+b*L_d)+tf_{t,d}}}_{TF} \underbrace{log(\frac{N}{N_t})}_{IDF} \qquad (2)$$

where qtf_t (set to 1) is the number of occurrences of term t in the query, $tf_{t,d}$ is the number of occurrences of term t in document d, N_t is the number of

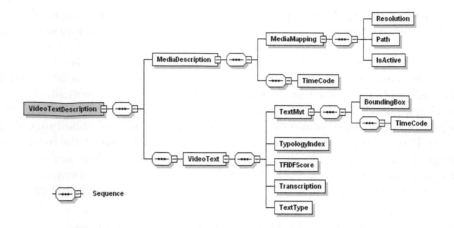

Fig. 1. Ground Truth Annotation XML-Schema

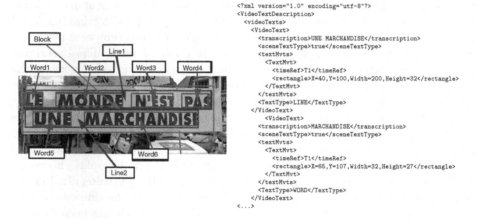

Fig. 2. A complete text *family* (block, lines and words) and the associated XML code of the ground truth

documents in the considered collection containing term t at least once, N is the total number of documents in the collection and L_d is the length of document d divided by the average length of the documents of the collection. b and K parameters were fixed empirically to 0,86 and 1,2.

We implemented our own tool *GTEditor* to build ground truths. Figure 2 shows an example of the XML code which is generated in the ground truth.

3.2   Evaluation

We consider that OCR systems are composed of a sequence of three modules: detection (coarse location of text zones), localization (creation of precise bounding boxes) and recognition. The evaluation of Abbyy FineReader 6.0 is based on these modules. First of all, the distinction between perfectly recognized texts and undetected texts (forgotten texts) is performed. Remaining texts are further qualified according to whether the localization stage or the recognition stage failed. It exists two types of false alarms: either recognition is applied on detected zones which do not contain text (a post-processing would be sufficient to filter these outputs) or recognition is applied on texts which have not been retained in the ground truth. False alarms will not be considered since they are not an obstacle to the final application (content based retrieval).

In this section, a text denotes either the word, its line or its block. Regarding temporal redundancy of texts, an occurence denotes a precise spatiotemporal zone.

Correctly Recognized Texts. A text appearing many times in a different fashion during a same semantic segment is correctly recognized if at least one occurence of this text is correctly recognized (a null Levenstein distance). To handle the case when the OCR recognizes lines, we also browse the results to find the word as it appears in its line (it can be followed by a comma for instance).

Forgotten Texts and Wrongly Recognized Texts

Forgotten Texts. Forgotten texts have been improperly localized. A text is correctly localized if one of its occurences spatially matches with a zone detected by the OCR. The matching criterion relies on barycenters: the detected zone must contain the barycenter of the ground-truth zone and conversely.

Wrongly Recognized Texts. The error is imputed to the recognition stage if the obtained recognition rate would not be higher for any other localization. The behaviour of the OCR is assumed to be homogeneous over all occurences of a same text. Then we can restrict ourselves to explore the neighborhood of one single localization result: the localization which is the closest to the ground truth (OCR_{best}). The ground truth zone which matches with OCR_{best} is called GT_{best}. This zone can correspond to a word, a line or a block. We use the distance d_{recouv} as a criterion to obtain OCR_{best} and GT_{best}:

$$d_{recouv}(GT, OCR) = \frac{A((GT \cup OCR) \setminus (GT \cap OCR))}{A(GT \cap OCR)}$$

Various recognition strings are obtained applying recognition on some localizations randomly chosen in the neighborhood of OCR_{best}. These strings are compared to the transcription of the text contained in GT_{best} (st_{GT}). The minimum Levenstein distance $LD_{min}^{neighborhood}$ is compared to the distance LD computed while comparing the string recognized in OCR_{best} to st_{GT}. If $LD_{min}^{neighborhood}$ is higher than LD, then the error is imputed to the recognition stage.

Goal Directed Evaluation Index. A complete evaluation index must answer two questions: *Have the texts been correctly detected and recognized?*, *Are the* *results useful for our final application ?* Formula 3 details the components of the global success index I_{OCR} which is defined regarding both algorithmic and semantic aspects.

$$\mathbf{I_{OCR}} = \mathbf{s_{TFIDF}} * \mathbf{I_{OCR}^{semantic}} + (1 - \mathbf{s_{TFIDF}}) * \mathbf{I_{OCR}^{algorithmic}} \ where$$

$$I_{OCR}^{algorithmic} = \frac{N - (w_{RI}N_{RI} + w_{LI}N_{LI} + w_F N_F)}{N}$$

$$I_{OCR}^{semantic} = \frac{2 * precision * recall}{precision + recall} \ where \qquad (3)$$

$$precision = \frac{N_R^{high}}{N_R} \ and \ recall = \frac{N_R^{high}}{N_{W+F}^{high} + N_R^{high}}$$

N_R, N_{RI}, N_{LI} and N_F refer respectively to the number of texts belonging to the following classes: *Recognized, Recognition Issue* (error imputed to the recognition stage), *Localization Issue* (error imputed to the localization stage) and *Forgotten*. Special cases are considered: $I_{OCR} = 1$ if $N_R = N$ and $I_{OCR} = 0$ if $N_F = N$. Coefficients w_{RI}, w_{LI} and w_F ($w_{RI} + w_{LI} + w_F = 1$) are chosen according to the final application. $I_{OCR}^{semantic}$ is derived from the harmonic mean used in information retrieval [19]. It quantifies the degree of success of the OCR in a goal directed point of view. This index varies from 0 to 1 and answers the questions: *Has the OCR recognized interesting texts?*, *Has the OCR missed interesting texts?* N_R^{high} (N_{W+F}^{high}) refers to recognized (forgotten or wrongly recognized) texts with a high TFIDF score. These texts are supposed to be interesting in a documentary point of view. s_{TFIDF} denotes the normalized standard deviation of TFIDF scores. This parameter allows to balance the influence of the semantic index $I_{OCR}^{semantic}$ as it can be overestimated when TFIDF scores do not much vary over the corpus.

4   Experimentations

A French news program is selected. This document is segmented into the 26 reports which compose this document, each one being documented in a describing form. Many resolutions were available. Resolution 720*576 was chosen to be close to broadcasting resolution. The VideoTextDescription corresponding to the ground truth contains 477 VideoTexts: 286 words, 118 lines and 73 blocks. 66 words belong to the *scene text* category. 58093 describing forms are stored in a XML database to compute TFIDF scores of the texts retained in the ground truth.

Abbyy FineReader 6.0 was applied (both layout analysis and recognition stages) every ten frames to reduce the processing time (a news program contains about 60 000 frames). Given that texts must stay on screen longer than ten frames so that they could be read, we obtain a representative sampling of possible OCR outputs. The resulting VideoTextDescription contains 4183 VideoTexts. Because of the temporal redundancy of the texts, a same text may be detected and recognized several times. A manual tracking could have been done, but we restrict our evaluation to available outputs.

4.1   Classification of Texts

Correctly Recognized Texts.   167 (N_R) words are correctly recognized (58,40% of the words in the ground truth). 13 of these words belong to the *scene text* class. Restricting to artificial text, the system reaches 70% of correctly recognized words.

Forgotten Texts and Wrongly Recognized Texts

- Forgotten texts: 63 (N_F) words have been forgotten (22% of the words in the ground truth). 42 words belong to the *scene text* class. 63% of *scene text* words (68 *scene text* words in the ground truth) are forgotten. This reinforces the presumption that *scene text* is more difficult to detect than *artificial text*.
- Wrongly recognized texts: 56 texts must be further qualified according to the error typology exposed in 3.2. 23 (N_{LI}) errors are imputed to the localization stage (6 *scene texts*). 33 errors (N_{RI}) are imputed to the recognition stage (5 *scene texts*).

4.2   Computation of the Global Success Index

Our aimed final application is to make use of recognized texts in a content-based retrieval perspective. Hence, the recognition stage is considered to play the most crucial part in the OCR system and errors regarding this stage must be penalized in a more stricly fashion than the others. The following constraint is applied to coefficients w_{RI}, w_{LI} and w_F in order to express this point of view: $w_{RI} = 3 * w_{LI}$ and $w_F = 5 * w_{RI}$. The factors 3 and 5 are empirically chosen. A value of **0.82** is computed for $I_{OCR}^{algorithmic}$.

Concerning the *semantic* component $I_{OCR}^{semantic}$, s_{TFIDF} is computed over the set of terms conserved in the ground truth. A value of 0.33 is obtained. The most crucial stage is then to fix Th_{TFIDF} which denotes the threshold used to determine which texts are *interesant*. Rather than choosing this value empirically, the range of TFIDF values is sampled into 5 segments and several levels of evaluation are proposed, Th_{TFIDF} being set to the mean value of each segment. The user can then choose to evaluate the system in a more or less strict fashion. Each segment is assigned an index called *SemanticTolerance* which

Table 1. Evolution of $I_{OCR}^{semantic}$ and I_{OCR} according to SemanticTolerance

SemanticTolerance	5	4	3	2	1
TFIDF Segment Limits	0.00058|2.88	2.88|5.76	5.76|8.64	8.64|11.52	11.52|14.40
Th_{TFIDF}	1.44	4.32	7.20	10.08	12.96
$I_{OCR}^{Semantic}$	0.69	0.64	0.43	0.17	0.04
I_{OCR}	0.77	0.76	0.69	0.60	0.56

varies from 1 to 5. The descriptive importance of texts decreases proportionally to this index. Variations of I_{OCR} according to $I_{OCR}^{semantic}$ values taken for each level of evaluation are exposed in table 1. These results show that $I_{OCR}^{semantic}$ allows to balance $I_{OCR}^{algorithmic}$. The global success index I_{OCR} decreases from 0.77 in the case of the less strict evaluation fashion (no restriction on the nature of texts which must be recognized and $I_{OCR}^{semantic}=0.69$) to 0.56 if only the most representative texts are considered ($I_{OCR}^{semantic} = 0.04$).

5   Conclusion and Future Work

This article presents a goal directed evaluation methodology applied to Abbyy Finereader 6.0. Two indices are designed: $I_{OCR}^{algorithmic}$ refers to the capacity of the system to detect and recognize properly texts stored in the ground truth, $I_{OCR}^{semantic}$ is computed at different levels of strictness and gives indices about the utility of the results obtained by the system. Depending on the target application, the global success index I_{OCR} can be tuned to reflect the capacity of the system to reach the aimed objectives.

Our methodology requires outputs of every module of the evaluated system to be easily accessed. Thus, any OCR system can be evaluated if an API is available. Then, OCR systems performances could be compared, regarding a precise application . We intend to practice such comparisons in the future. This study will lead to several potentially different rankings of OCR systems, each ranking being dedicated to a particular target application. Furthermore, future work will be dedicated to investigate the influence of post-processing methods ([20]...) on commercial OCR results.

References

1. Jones, K.S., Walker, S., Robertson, S.: A probabilistic model of information retrieval: development and status. Technical Report Technical Report 446, University of Cambridge Computer Laboratory (1998)
2. Wu, V., Manmatha, R., Riseman, E.: Textfinder: An automatic system to detect and recognize text in images. IEEE Transactions on Pattern Analysis and Machine Intelligence **21** (1999) 1224–1229
3. Wolf, C., J-M.Jolion: Extraction and recognition of artificial text in multimedia documents. Pattern Analysis and Applications **6** (2003) 309–326
4. Li, H., Doermann, D.: Automatic text detection and tracking in digital video. IEEE Transactions on Image Processing **9** (2000) 147–156
5. Chen, D., Odobez, J.M., Boulard, H.: Text detection and recognition in images and video frames. Pattern Recognition **37** (2004) 595–608
6. Wolf, C.: Détection de textes dans des images issues d'un flux vidéo pour l'indexation sémantique. PhD thesis, Institut National de Sciences Appliquées de Lyon, France (2003)
7. Li, H.: Automatic processing and analysis of text in digital video. PhD thesis, University of Maryland, College Park (2000)

8. Hua, X.S., Wenyin, L., Zhang, H.J.: An automatic performance evaluation protocol for video text detection algorithms. IEEE Trans. on Circuits and Systems for Video Technology **14** (2004) 498–507
9. Doermann, D., Mihalcik, D.: Tools and techniques for video performance evaluation. In: Proceedings of the ICPR 2000, IEEE Computer Society. Volume 4. (2000) 4167–4170
10. Yanikoglu, B., Vincent, L.: Pink panther: A complete environment for ground-truthing and benchmarking document page segmentation. Pattern Recognition **31** (1998) 1191–1204
11. Lee, C.H., Kanungo, T.: The architecture of trueviz : a groundtruth/metadata editing and visualizing toolkit. Pattern Recognition **36** (2003) 811–825
12. Fruchterman, T.: Dafs: A standard for document and image understanding. In: Proceedings of the Symposium on Document Image Understanding Technology. (1995) 94–100
13. Liang, J., Philips, I., Haralick, R.: Performance evaluation of document layout analysis algorithms on the uw data set. In: In Document Recognition IV, Proceedings of the SPIE. (1996) 149–160
14. Lienhart, R., Wernike, A.: Localizing and segmenting text in images, videos and web pages. IEEE Transactions on Circuits and Systems for Video Technology **12** (2002) 256–268
15. Mariano, V.Y., Min, J., Park, J.H., Kasturi, R., Mihalcik, D., Li, H., Doermann, D.: Performance evaluation of object detection algorithms. In: International Conference on Pattern Recognition. (2002)
16. Mao, S., Kanungo, T.: Empirical performance evaluation of page segmentation algorithms. In: In Proceedingsof SPIE Conference on Document Recognition,San Jose CA. (2000)
17. Kanai, J., Rice, S., Natker, T., Nagy, G.: Automated evaluation of ocr zoning. IEEE Transactions on Pattern Analysis and Machine Intelligence **17** (1995) 86–90
18. Wagner, R., M.J.Fisher: The string to string correction problem. Journal of Assoc. Comp. Mach. **21** (1974) 168–173
19. Van Rijsbergen, C.J.: Information Retrieval, 2nd edition. Dept. of Computer Science, University of Glasgow (1979)
20. Jolion, J.: The deviation of a set strings. Pattern Analysis And Application **6** (2004) 224–231

Author Index

Lecture Notes in Computer Science

For information about Vols. 1–3558

please contact your bookseller or Springer